Principles of Electron Optics

Volume Three: Fundamental Wave Optics

Principles of Electron Optics

Volume Three: Fundamental Wave Optics

Second Edition

Peter Hawkes
CEMES–CNRS, Toulouse, France

Erwin Kasper
Formerly Institute of Applied Physics, University of Tübingen, Germany

ELSEVIER

ACADEMIC PRESS
An imprint of Elsevier

Academic Press is an imprint of Elsevier
125 London Wall, London EC2Y 5AS, United Kingdom
525 B Street, Suite 1650, San Diego, CA 92101, United States
50 Hampshire Street, 5th Floor, Cambridge, MA 02139, United States
The Boulevard, Langford Lane, Kidlington, Oxford OX5 1GB, United Kingdom

British Library Cataloguing-in-Publication Data
A catalogue record for this book is available from the British Library

Library of Congress Cataloging-in-Publication Data
A catalog record for this book is available from the Library of Congress

ISBN: 978-0-12-818979-5

Publisher: Matthew Deans
Acquisitions Editor: Kayla Dos Santos
Editorial Project Manager: Emily Thomson
Production Project Manager: Vijayaraj Purushothaman
Cover Designer: Christian J. Bilbow

Typeset by MPS Limited, Chennai, India

Dedication

In memory of my son Sebastian Hawkes and his mother Françoise.

Contents

Preface to the Second Edition

The revisions of these volumes on Wave Optics are different in nature from those of Volumes 1 and 2 on Geometrical Optics. For the latter, revision consisted mainly in extending the material in the first edition: in the chapter on the aberrations of round lenses, for example, the integrals for the fifth-order geometrical aberration coefficients were added to those for the primary geometrical aberrations. Moreover, although there were of course revisions throughout, the arrival of aberration correctors cast a beneficent shadow over much of the books. For wave optics, there is no obvious single influence but if a candidate had to be found, there would be a good case for the direct detector. In the proceedings of the 2020 Microscopy and Microanalysis Congress of the Microscopy Society of America, "4D-STEM" appears 64 times in the document index and the theme of the 2020 EMAG conference was 'Microscopy Enabled by Direct Electron Detection', which attracted 36 abstracts.

In Wave Optics, new subjects have been added and the coverage of certain topics that were barely mentioned has been greatly expanded. In Chapter 56, some account of the systematic study by Ramaswamy Jagannathan and Sameen Ahmed Khan of electron optics based on the Dirac equation has been included. The chapters on interferometry and holography (Chapters 62 and 63) have been considerably enriched. A long section on phase plates is now present in Chapter 66; the very notion of phase plates has been widened by the realization that lens aberrations can be thought of as a form of phase plate since they too change the shape of the electron wavefront. A section is devoted to contrast transfer in aberration-corrected instruments: where does the contrast come from when $C_s = 0$? The chapter on image formation in the STEM (Chapter 67) now includes a full study of the quadratic term in the theory of image formation, of which we wrote "The quadratic term cannot be simplified in any useful way" in the first edition. A long section on ptychography is also included in this chapter as is an account of the physics and technology of the direct detectors that are revolutionizing scanning transmission electron microscopy. Chapter 68 is new: statistical parameter estimation theory had not yet been applied to electron image interpretation when the first edition was written. This chapter is inevitably only an interim account of the subject, which is in very rapid development (De Backer et al., 2021).

The chapters on beam–specimen interactions (Chapters 68 and 69 in the first edition) are now combined into Chapter 69 and have been substantially revised. The section on the multislice method, for example, has been much improved, and the method is shown to be capable of establishing the optics of electron lenses. Recent work on crystal structure reconstruction in the presence of multiple scattering has been accorded a section here and reappears in the later chapter on three-dimensional reconstruction (Chapter 75).

The changes to the first chapters of Part X on image processing are only minor but the chapter on nonlinear restoration and the phase problem (Chapter 74) has been extended to cover the transfer-of-intensity equation in detail. The notion of phase is considered critically and the elegant concept of aberration space, in which the aberration coefficients are the axes, in introduced. The section on exit-wave reconstruction has been thoroughly revised. There is new material in Chapter 75 on three-dimensional reconstruction, where compressed sensing is proving useful to counter the artefacts arising from the 'missing wedge' or 'cone'. The projection requirement breaks down when multiple scattering occurs in the specimen and artificial neural networks offer a way of combating the problem, already mentioned above for crystal specimens. In the early years, three-dimensional reconstruction was applied almost exclusively to biological specimens but is now widely practised on inorganic material. A new section describes the original aspects of the subject that this entails. The account of microscope parameter measurement and instrument control in Chapter 77 has been updated. The chapter on coherence (Chapter 78) now includes more discussion of the Wigner function.

The last two chapter are new. Chapter 79 on Wigner optics, largely inspired by the work of Alexander Lubk and Falk Röder, shows how image formation can be usefully described in terms of the Wigner function. Entanglement, which arises when the wave function of the electron beam can no longer be treated separately from that of the specimen, raises no particular difficulty and the family connections between ptychography, tomography and image formation emerge naturally.

The book ends with an account of a very new topic, electron orbital angular momentum and vortex beams. The first demonstration of 'twisted electrons' is little more than a decade old (Uchida and Tonomura, 2010) and the subject has since generated a substantial literature. The interactions between vortex beams and materials, especially magnetic specimens, are proving very fruitful. I have also chosen to place an account of the plans for a quantum electron microscope in this chapter, another subject that did not exist when the first edition was published.

It is a pleasure to acknowledge the many forms of help I have received when preparing this new edition. My greatest debt is to Angus Kirkland who contributed to the section on the physics and technology of the new detectors. He has made many improvements to the short chapter on microscope parameter measurement and instrument control and to the section on exit-wave reconstruction as well as numerous constructive suggestions *passim*. But for his many other responsibilities, he would have joined Kasper and me as a coauthor. I must also single out the late John Spence who, in a long review of the first edition, pointed out tactfully that we were sometimes floundering out of our depth in Chapter 69 and suggested many improvements to that chapter and others. He has drawn my attention to many relevant publications that might easily have escaped my notice, and it is thanks to him that the multislice theory of lens optics, barely mentioned in the first edition, is now fully described. Many other colleagues have provided invaluable information and original figures for which I am extremely grateful. At the risk of forgetting some names, it is a pleasure to thank R.W. Glaeser and his colleagues Jeremy Axelrod and Holger Müller, Vincenzo Grillo, Sarah Haigh, Maurice Krielaart, Michael Mousley, Dagmar Gerthsen, Giulio Pozzi and Andrea Parisini, and Ondrej Krivanek. Isabelle Labau (CEMES–CNRS) has procured many elusive documents and the Technische Informationsbibliothek in Hannover is a mine of information, which its staff are happy to share, I am truly grateful for their helpfulness. Ms. Kimi Matsuyama (Hitachi) has supplied information about elusive Japanese material. The late Ugo Valdrè commented constructively on the account of electron interferometry, and the present description of the beginnings of the subject is now closer to the truth. I am greatly indebted to Ms. Mélanie Ravoisier, who re-drew many of the line-diagrams at short notice.

No list of acronyms is provided as I have used very few; only acronyms that have become part of the vocabulary of electron optics, such as STEM, SEM and (C)TEM, will be encountered. Nevertheless, for terms for which an acronym is often employed, such as TIE (transport-of-intensity equation), I have always mentioned the acronym, which thus appears in the index. In their very long list, Dave Williams and Barry Carter include TMBA[1] (Williams and Carter, 2009) and I wholeheartedly share this sentiment.

Finally a note on my co-author Erwin Kasper. He contributed fully to the first edition and is hence also co-author of much of this edition even though he was not free to participate in the revision.

Peter Hawkes

Toulouse, November 2021

[1] Too many bloody acronyms.

Preface to the First Edition

The last attempt to cover systematically the whole of electron optics was made by the late Walter Glaser, whose *Grundlagen der Elektronenoptik* appeared in 1952; although a revised abridgement was published in the *Handbuch der Physik* four years later, we cannot but recognize that those volumes are closer to the birth of the subject, if we place this around 1930, than to the present day.

The difference between Glaser's work and our own is much greater in the present volume than in the two volumes on geometrical optics, for whole branches of the subject have come into being since 1956. The representation of the image-forming process by transfer functions has yielded a much deeper understanding of the notion of resolution. The development of highly coherent light and electron sources has made holography possible and the invention of the electron biprism has rendered it practical. The widespread availability of large fast computers and the gradual introduction of microscope–computer links, as well as the peculiarities of the electron image-forming process, have generated considerable interest in digital image processing. Lastly, a theoretical advance in optical coherence theory clarifies some obscure points in the related electron optical theory.

Only a very small fraction of the present text finds a counterpart in Glaser's work, therefore, and Parts XII–XVI are entirely new. Part XI, on the basic wave-mechanical formalism, follows the original work of Walter Glaser and Peter Schiske quite closely, however, for there is little new to report there apart from the investigations based on the Dirac equation.

Like Volumes 1 and 2, this work is intended to be both a textbook and a source-book. The fundamentals of the topics covered are presented in detail but the reader who wishes to go more deeply into a particular subject will need to examine the original articles, review articles and more specialized textbooks. This is particularly true of themes that are closer to electron microscopy than to electron optics and of image processing. For the former, we have included two chapters on beam–specimen interactions but these are intended merely to initiate the reader into the beginnings of this vast subject. For the latter, and despite the length of Part XV, we have had to be selective and refer frequently to other texts in this field. Furthermore, image processing is a field in rapid growth and the reader will need to complement our account of the fundamentals with the contents of conference proceedings and the journals that specialize in these themes.

The earlier work on wave optics is all brought together in Glaser's two great texts of 1952 and 1956. Of the other earlier works on electron optics, only those of Picht (1939, 1957, 1963) and de Broglie (1950) devote much space to the subject. The subjects of Parts XII–XVI are all more recent and no single volume attempts to cover them all. Specialized texts do of course deal with their particular topics in much more detail than we can and we refer to—and lean on—these in the relevant chapters. Only Part XIV has generated a classic textbook, *Electron Microscopy of Thin Crystals* by Hirsch *et al.* (1965), the record of courses given at a summer school in 1963. Once again, therefore, we can say that, although 'standing on the shoulders of giants', the present volume differs from other books on the themes discussed here in that developments of the past 30 years are set out in detail and in a uniform presentation.

For whom is this work intended? Knowledge of physics and mathematics to the first-degree level is assumed, though many reminders and brief recapitulations are included. It would be a suitable background text for a postgraduate or final year course in electron optics or electron image formation or electron image processing and some of the material has indeed been taught for some years in the Universities of Tübingen and Toulouse. Its real purpose, however, is to provide a self-contained, detailed and above all modern account of wave-mechanical electron optics, with image formation, interference, holography and coherence as the principal examples of application. This is complemented by a survey of the ways in which digital image processing can be made to participate in image interpretation and microscope control. The basic equations are given, the applications are discussed at length and ample guidance to the related literature is provided.

Composition of a volume such as this puts us in debt to a host of colleagues: many have permitted us to reproduce their work and have often provided illustrations; the librarians of our institutes and the Librarian and Staff of the Cambridge Scientific Periodicals Library have been unflagging in their pursuit of recondite and elusive early papers; Mrs Ströer has uncomplainingly word-processed hundreds of pages of mathematical and technical prose, aided by Mrs Lannes and Mrs Davoust; most of the artwork has also been produced by Mrs Ströer, using computer software rather than the drawing pen; the remainder has been prepared by Miss Quessette and Mr Caminade; the references have been typed by Mrs Bret and her assistants; Mr Aussoleil and Mrs Altibelli have provided computer expertise; Academic Press has generously supported the production costs. We are extremely grateful to all of these. We also thank the many authors and publishers who have been good enough to allow us to reproduce published drawings.

CHAPTER 54

Introduction

54.1 Organization of the Subject

The behaviour of beams of free electrons, released from a source and propagating through a vacuum region in some device, is of interest in many diverse fields of instrumentation and technology. The study of such beams forms the subject of electron optics, which divides naturally into geometrical optics, where effects due to wavelength are neglected, and wave optics, where these effects are primordial. Volumes 1 and 2 were devoted to geometrical optics. Volumes 3 and 4 are concerned with wave optics. A knowledge of this branch of the subject is essential in microscopy, to understand the propagation of electrons from the source to the specimen, through the latter and from it to the image plane of the instrument. It is also needed to explain all interference phenomena, notably holography and ptychography, and electron image formation is of course an interference phenomenon. It is also central to the formal theory of coherence and the study of orbital angular momentum.

The various branches of the subject have reached different degrees of sophistication. The laws that govern wave propagation are closely analogous to those already familiar in light microscopy, provided that electron spin is neglected, and can be regarded as well established. Many of the applications are, conversely, in rapid evolution. We have therefore concentrated on the principles, which should remain largely unaffected by the passage of time. We have not, of course, neglected their practical exploitation, in holography for example, and in image formation and processing.

Until very recently, texts on electron optics followed the traditional pattern: geometrical optics based on classical mechanics and Hamilton's theory, wave optics setting out from Schrödinger's equation, with an occasional aside on Dirac's theory. Among the classic authors, only Walter Glaser began working on the new subject of quantum mechanics in the 1920s before turning to electron optics, another new subject, in the next decade. With the appearance of publications on vortex studies by Peter Schattschneider and Johan Verbeeck for example, and on the design of a quantum microscope by Pieter Kruit, a fresh approach is emerging. Their work accords quantum mechanics the leading role with optics as a derived field. The chief exponents of this school are Alexander Lubk and his colleagues and we are fortunate that his *Habilitationsschrift*, containing a full account of this new approach, is available (Lubk, 2018). Although we cannot do full justice to these highly

Principles of Electron Optics.
DOI: https://doi.org/10.1016/B978-0-12-818979-5.00054-1

original studies (Lubk's account fills more than 300 pages), we do draw attention to it on many occasions. In particular, we explain why the Wigner function is omnipresent; it reveals that holography, ptychography and the transfer-of-intensity equation can be thought of as cousins, or even siblings, for example.

We shall not repeat here the general remarks on the classification of electron optical studies to be found in the Preface to Volumes 1 and 2. Our theme in these last volumes is the study of electron propagation through static electric or magnetic fields, including those inside specimens, based largely on Schrödinger's equation. A part covers digital image processing which, if not in the main line of electron optics, is an inevitable preoccupation of anyone concerned with electron microscope imaging. It is here that the phase problem is analysed and tomography is studied with sections on compressed sensing and the role of artificial neural networks.

The wave theory of electron optics is founded on the Dirac equation but, in practice, it is almost always permissible to replace this by the relativistic form of Schrödinger's equation, for spin is negligible except in a few very specialized situation. Volume 3 therefore opens (Part XI) with an account of the relevant material from *quantum mechanics*. A paraxial form of Schrödinger's equation is derived in Chapter 58, which enables us to understand elementary image formation in terms of the wavefunction. Nevertheless, there are branches of electron optics where the Dirac equation is required and we therefore provide the necessary background information and show how paraxial electron optics and aberration theory can be derived directly from the Dirac equation. In the last two chapters of this part (Chapter 59 and Chapter 60) the laws of diffraction and interference are studied.

In the remaining parts, the laws of propagation established in Part XI are applied to a variety of different situations, directly in the case of Parts XII, XIII, XIV and XVI, indirectly in the case of Part XV on image processing. *Interference effects* are the subject of Part XII, which is divided into *interferometry* and *holography*. The distinction between the two is not sharp but, in interferometry, we are not concerned with fine diffraction effects in the specimen, whereas in holography it is precisely these effects that render the technique valuable. Holography has developed enormously since the first edition of this volume appeared and the length of Chapter 61, which has been considerably extended, reflects this. The role of the Wigner function is explained and, although ptychography is treated in depth in Section 67.4, it also appears here as a form of generalized holography. A whole chapter is later devoted to Wigner optics (Chapter 79).

Part XIII, which fills a substantial fraction of the book, is devoted to *image formation*, in the fixed-beam or conventional transmission electron microscope (TEM or CTEM) and in the scanning transmission electron microscope (STEM). Here, the relation between the intensity at the image and the wavefunction at the specimen is explored in great detail and the linear theory that is applicable to a certain class of specimen is presented at length.

The effects of source size and energy spread are examined, as are less conventional imaging modes, using tilted or hollow-cone illumination in particular. Phase plates are now in regular use, especially in the life sciences, and a long section describes the different types of plate. A section is also devoted to contrast formation in aberration-corrected instruments. The chapter on the STEM concentrates on the differences between this instrument and the conventional microscope, notably, the possibility of controlling the detector response, either by configuring the detector surface or by recording the two-dimensional signal generated by each specimen element as though it were an image and combining the intensity values of this image in any way that seems helpful. This has come to be known as 4D-STEM, occasionally treated as a subject in its own right. Lupini et al. (2016) have introduced the word *ptychogram* to describe this four-dimensional dataset. The development of fast direct detectors with extremely wide dynamic range has transformed this activity and a section is devoted to their physics and technology. This leads naturally to a discussion of the many imaging modes – bright field, annular bright field, annular dark field, high-angle annular dark field and confocal. A long section contains an account of ptychography. We also draw attention to the information that can be obtained about crystalline specimens when the area illuminated coherently by the probe is appreciably smaller than the unit cell. Part XIII ends with an account of a family of statistical methods that are of rapidly growing importance in electron microscopy. Here, probability theory makes it possible to obtain information about the sample that is beyond the formal limit of resolution and may require a much smaller electron dose than when a traditional imaging mode is employed.

Part XIV is a brief reminder of the ways in which the propagation of the electron wavefunction through the *specimen* is analysed. Superficial though this presentation inevitably is, for the subject is not central to the theme of the book, we felt that some account of this material was indispensable, for without it certain notions introduced elsewhere, the specimen transparency for example, would remain mysterious. The theory is presented separately for *amorphous* and *crystalline* specimens, for the collective effects in the latter require us to analyse them in terms of concepts totally inappropriate for amorphous materials. We do insist, however, that this part can do no more than bridge the gap between our detailed presentation of imagery and other specialized texts on the microscopy of specimens of a particular kind.

In Part XV, we turn to *digital image processing*. We cover, even if unevenly, the whole field of image processing and make no apology for including this in a book on electron optics for much of the material presented has been a major preoccupation of microscopists over the years: the phase problem and three-dimensional reconstruction are obvious examples and the current studies that aim to use all the information from every object element in the STEM, so that the image becomes four-dimensional, provide an even more persuasive justification. At a humbler level, image enhancement has been practised in scanning electron microscopy since the earliest days of the instrument.

Image processing divides naturally into four large sections: *acquisition, sampling, quantization and coding; enhancement; restoration* and *image analysis*. We have adopted these divisions, adding to them a chapter on *instrument control* and on the measurement of microscope *operating parameters*, during image acquisition in particular. We have also included a short introduction to *image algebra* for, although this subject has not had much impact on electron image processing as yet, we anticipate that some familiarity with it will be required to read the image processing literature of the future. In this part, we describe many of the procedures that are used to improve images in some way or render them more informative; in particular, we devote considerable space to the ideas of mathematical morphology, which are already important in scanning electron microscopy, to the work on the phase problem and to three-dimensional reconstruction.

The book continues with a part of a rather theoretical nature devoted to *coherence* and, in particular, to the relation between coherence and *radiometry* (Chapter 78). A short section is also devoted to instrumental aspects of coherence, notably the effect of partial coherence on image formation in terms of the transmission cross-coefficient. The discussion of the various brightness functions in Chapter 78 is inspired by the groundbreaking work of Emil Wolf and his school, who were concerned with light sources. The translation to electron sources is, however, immediate, since the latter are quasi-monochromatic and only the spatial partial coherence raises problems: the contributions from different wavelengths can safely be added 'incoherently'. Nevertheless, some questions still remain without a fully satisfactory answer; in particular, we have preferred to describe the work of Agarwal et al. (1987) out of context in Section 78.10.1, for although it provides a transparent gateway between light and electron optics, some further elucidation is required before we dare pass through.

A chapter on the role of the Wigner function in electron optics is also included in this part. We show how image formation is expressed in terms of this function and how holography can be studied in terms of the density matrix.

In the final part (Chapter 80), we describe work on electron beams possessing orbital angular momentum. Many features of vortex beams are examined here in considerable detail since the subject is still relatively unknown in electron optics. We have also chosen to describe projects for a quantum electron microscope here.

Most aspects of wave electron optics have thus been covered, some more thoroughly than others. The emphasis throughout is on physical principles and on their theoretical formulation while technical details of microscopes or ancillary equipment are kept to a minimum. Even more than in Volumes 1 and 2, the inclusion of such details would not only have rendered the books even larger but would also have shortened their useful lifetime, for such instrumentation is in rapid development, especially in the fields of holography, high-resolution imagery, STEM imagery and image processing, which already fill so many pages.

Again as in the earlier volumes, we have adopted a compromise towards the voluminous literature of the subject, though the decisions were hard to take for no single attempt to treat all the subjects of these books has hitherto been made. The coverage of the literature is therefore quite full for all parts except that on interactions in thin specimens, where we list only the seminal papers and refer to the specialized textbooks for more information; there are many excellent titles available here. There are a few deliberate omissions in other parts too. In the sections in which the Aharonov–Bohm effect is examined, we have preferred to refer to the book by Peshkin and Tonomura (1989) rather than list all the contributors to the lively but ultimately bootless controversy that raged around this phenomenon. Even in the chapter on holography, the listing of references is far from exhaustive for several very complete review articles have been devoted to this subject.

We conclude this introductory chapter with a brief historical account of the various themes of these volumes. Even more than in Volumes 1 and 2, one is struck by the way in which the human mind invents tools that permit or encourage it to conceive original ideas: the deficiencies of the microscope objective lens, which Scherzer showed by theory to be intrinsic, inspired Gabor to invent holography, long before it could work. The field-emission gun and the laser rendered experimental tests possible but physical limitations made digital rather than optical reconstruction essential – if the computing power needed for digital image processing had not been available, progress in holography would have been far slower. We could trace a similar evolution, original thinking alternating with experimental progress, in many fields: in the history of the STEM from the use of a field-emission gun to the exploitation of Rutherford scattering and of the far-field diffraction pattern generated by each object element for example, or in image simulation or again in the phase problem. Electron optics is an art as well as a science.

54.2 History

The most important step in wave electron optics was taken before geometrical optics came into being, for the notions of electron frequency and subsequently wavelength were introduced by Louis de Broglie in a series of notes to the Académie des Sciences de Paris in 1923, which preceded the full account in his thesis of 1924, published in 1925. When Ernst Ruska learnt of de Broglie's ideas several years later, in 1932, his first reaction was dismay: "I have a lively memory even today [1979] of the first discussion between Knoll and myself about this new kind of wave, for I was, at the time, extremely disappointed that once more a wave phenomenon would limit the resolution". This feeling was fortunately short-lived: "I was immediately heartened, though, directly I had satisfied myself with the aid of the de Broglie equation, that the waves must be around five orders of magnitude shorter in wavelength than light waves" (Ruska, 1979, 1980).

The pre-war years are characterized by a flurry of activity in electron diffraction but only isolated publications appeared on electron image formation in which the wavelength was not neglected. The main stages in the understanding of electron diffraction are well known, notably the classic papers of Clinton J. Davisson and Lester H. Germer (Davisson and Germer 1927), George Paget Thomson and Alexander Reid (Thomson and Reid 1927; Thomson, 1927, 1928) and Seichi Kikuchi and Shoji Nishikawa (Kikuchi 1928; Nishikawa and Kikuchi 1928); their ideas were taken up in many laboratories and this period can be pleasurably explored with the aid of the many historical essays and reminiscences collected by Goodman (1981). The books of Pinsker (1949, 1953) and Vainshtein (1956, 1964) are useful for tracing the early Russian studies. A paper by Elsasser, published as early as 1925, foreshadowed later developments and although it has been thought that "no notice seems to have been taken of the Elsasser letter, and that in particular it had no influence on the work of either Davisson or G.P. Thomson" (Blackman, 1978), it is included in the historical section of the long account of particle diffraction by Frisch and Stern (1933). The papers by Walter Kossel and Gottfried Möllenstedt (Kossel and Möllenstedt, 1938, 1939, 1942; Möllenstedt, 1941; Kossel, 1941) on convergent-beam electron diffraction, which inspired the work of Carolina MacGillavry (MacGillavry, 1940a,b) on the underlying theory, are particularly relevant for modern electron microscopy (Bristol, 1984; Tanaka and Terauchi, 1985; Tanaka et al., 1988). Möllenstedt's recollections of that period may be read in his paper of 1989.

In electron image formation the high points are the realization that electron image contrast is due not to absorption but to scattering, the recognition that the haloes around specimens are Fresnel fringes and the appreciation that a diffraction pattern is formed in the back focal plane of the microscope (for a parallel beam incident on the specimen).

There is some uncertainty in the literature as to who first became aware that contrast is due to scattering. Ladislaus Marton believed that "the first clear statement ascribing the contrast in the image to differences in the scattering properties of the object was contained in [his] paper of 8 May 1934 . . ." (Marton, 1968) and claimed that a statement in a slightly earlier paper of Ruska (1934) showed that the latter had not appreciated this point. Ruska pointed out, however, in a private communication to Charles Süsskind (cited by the latter, 1985) that he had in fact given an explicit description of what we now call scattering contrast:

> Wegen der an sich kleinen Absorption sind Absorptionsunterschiede praktisch von geringem Einfluss auf die Bildkontraste. Dagegen ist bei relativ zur Strahlspannung sehr dünnen Objekten trotz praktisch konstanter durchsetzender Strahlstromdichte die Bildhelligkeit deswegen stark verschieden, weil die einzelnen Objektstellen mit entsprechend ihrer Massendicke verschiedener Apertur (Intensitätsverteilung auf die Streuwinkel) strahlen, so daß bei genügend kleinen Spulenöffnungen verschieden große Ströme von den Objektpunkten auszugehen scheinen bzw. in das Bild gelangen. Man könnte im Grenzfall von Absorptionsund Diffusionsbildern sprechen, doch überwiegt in ihrer Wirkung meist die Diffusionserscheinung. Wegen der großen Bedeutung dieser

> Erscheinung für das Zustandekommen der Bildkontraste hängt der Kontrastreichtum der Bilder so wesentlich von der Apertur der einfallenden Objektstrahlung (Brennweite der Kondensorspule) ab.

Moreover, a careful reading of Marton's paper of 1934 (Marton, 1934a) and in particular of his slightly later paper (1934b) does not clearly support his claim. In the earlier paper the calculation of scattering angle as a function of thickness is made in order to establish the limiting thickness of the metal support film, the purpose of which was to cool the biological specimen to protect it from damage. The notion of scattering contrast is not mentioned, even indirectly. The later paper states unequivocally: "Nous pouvons donc former les images électroniques de chaque objet qui émet des électrons ou faire traverser des objets par des électrons et ainsi rendre visible les objets par absorption". Two years later, however (Marton, 1936), scattering contrast *is* described: "En traversant l'objet, les électrons sont dispersés dans toutes les directions. A la formation de l'image ne contribueront que ceux qui sont dispersés dans l'angle solide délimité par le diaphragme de l'objectif ... Les détails de l'image seront perceptibles, si les quantités d'électrons dispersés dans le diaphragme et provenant de deux points voisins de l'objet sont différentes". Nevertheless, it is agreed that Marton's role in understanding the origin of microscope contrast was a central one: "The main source of contrast was *scattering*, and the first to recognize this fact clearly was Bill Marton", wrote Gabor in 1968 and "Ruska does acknowledge that immediately following his brief reference to these questions [quoted above], Marton pursued them more fully" (Süsskind, 1985). See in particular Marton and Schiff (1941).

Another very relevant event in 1936 was the publication by Hans Boersch of two major papers in which the formation of a diffraction pattern in the back focal plane of the objective was demonstrated and a form of selected-area diffraction was adumbrated. From then on, thanks to the work of Ruska, Marton and Boersch, there could be no doubt that electron image contrast is the result of scattering within the specimen and interception of some of the scattered electrons by the objective aperture. Energy was of course deposited in the specimen by virtue of inelastic scattering, especially in those early years before the development of techniques and equipment for preparing very thin specimens.

The haloes seen around the edges of specimens were interpreted as Fresnel fringes a few years later by Boersch (1940, 1943a,b) and by James Hillier (Hillier, 1940).

For our present purposes, the 1940s are the decade in which Walter Glaser, first with Peter Schiske and later with Günther Braun, developed the theory of image formation on the basis of the Schrödinger equation. The result of this work was a detailed account of paraxial electron optics in wave-optical terms and the beginnings of a study of the effect of aberrations and of the source brightness function. Although Glaser failed to take the vital step, the Fourier transformation of his equation representing the convolutional relation between object and image wavefunctions, his wave-optical analysis occupies a central position in the contrast-transfer theory of image formation. This work may be traced through the work of Glaser (1943, 1949, 1950a,b) and of Glaser and Schiske (1953) and Glaser and Braun (1954, 1955).

The problem of microscope resolution was studied at several levels. Aberration correction and improvements in specimen preparation both played an important part, of course, and the former has been examined at length in Volume 2. The possibility of high, even atomic, resolution in the electron microscope was studied by Hillier (1941) and Leonard I. Schiff (Schiff, 1942a,b) and in the immediate post-war years, Hans Boersch published a series of studies on the possibility of imaging atoms (Boersch, 1946a,b, 1947a,b, 1948). In the paper of 1947b, Boersch introduced the Zernike phase plate into electron microscopy; not only did he discuss a physical plate, in the form of a thin film with a central hole, but he also pointed out that the electrostatic field of a very small einzel lens enclosing the undeflected beam could act as an electrostatic phase plate. The latter are now known as Boersch plates. In 1949 Otto Scherzer analysed "the theoretical resolution limit of the electron microscope". This important paper foreshadows the contrast-transfer theory that was introduced 20 years later for the role of the wave aberration is clearly recognized – we meet the formula $\gamma = s^2\Theta^4 - \tau s\Theta^2$, identical apart from notation with our equation (65.30b), for the first time in the electron optical literature though it was of course known in light optics. The optimum value of defocus that we now call Scherzer focus appears there, in the form $2.5(\lambda C_A/2\pi)^{1/2}$, where $C_A = C_s$ and we note that $2.5/(2\pi)^{1/2} \approx 1$. The resolution of individual atoms is again discussed.[1]

One way of correcting the unfortunate effects of spherical aberration was not examined in Chapter 41 however: holography. The in-line form of this was introduced in 1948 by Dennis Gabor in an attempt to remedy the undesirable consequences of this aberration. The early attempt of Michael E. Haine and Tom Mulvey (Haine and Mulvey, 1952) and later, Tadatoshi Hibi (Hibi, 1956), recollected in Hibi (1985), to put this idea into practice fell victim to the relatively large emissive area of the thermionic sources and their equally large energy spread, which impaired the quality of the holograms recorded. But even if highly coherent field-emission sources had been available in 1950, the fact that the laser still lay in the future would have vitiated the reconstruction step. Although holography had to await technological developments and the invention of the maser and its successor, the laser, simpler kinds of electron interference did not. In 1952 Marton had attempted to use a crystal as a beam splitter, capable of producing two beams with a fixed phase difference from a single source (Marton, 1952; Marton et al., 1953); the scattering process in the

[1] Although most authors refer to $(C_s\lambda)^{1/2}$ as Scherzer (de)focus, other definitions are to be found in the literature, based on different interpretations of the optimum. Even the otherwise authoritative text of Williams and Carter (2009) prefers a different definition.

crystal rendered this way of creating two coherent beams inefficient but, a few years later, Gottfried Möllenstedt and Heinrich Düker split an electron beam with an 'electron biprism', a fine thread held at a positive potential relative to its surroundings (Möllenstedt and Düker, 1955, 1956, see recollections by Möllenstedt, 1991 and Lichte, 1998). Thereupon, electron interferometry became a subject in its own right. Many references to this work and to the interferometric studies of Charles Fert, Jean Faget and Monique Fagot are to be found in Part XII.

During this same period, a calculation of the elastic and inelastic scattering parameters based on a simple but adequate model of the atomic potential was made by Friedrich Lenz (Lenz, 1953, 1954); his formulae were heavily used at a time when computing power was modest and even today they are regularly employed, when extreme accuracy is not required, to study the relative magnitudes of different cross-sections for example.

A series of experimental studies of image formation with crystalline specimens stimulated theoretical work that is the basis of our understanding of these materials. Among the experimental papers, we single out those of Robert D. Heidenreich (Heidenreich, 1949, 1951); James Menter (Menter, 1956) who obtained the first micrographs of edge dislocations in his work on platinum phthalocyanine using a newly acquired Siemens Elmiskop I; Hatsujiro Hashimoto (Hashimoto, 1954) and Peter Hirsch and colleagues (Hirsch et al., 1956), in which dislocation glide was reported (see Hirsch, 1980, 1986). Soon after, all the basic theory of the kinematical theory of diffraction contrast of dislocations in crystalline specimens was developed (Hirsch et al., 1960) and the dynamical theory was published immediately after by Archie Howie and Michael Whelan (Howie and Whelan, 1960a,b, 1961, 1962). Related studies appeared in the same years (Hashimoto et al., 1960, 1961a,b, 1962).

Meanwhile, an observation that was to have a long-lasting effect on image simulation was made by Alex Moodie in Australia: a chance sighting of a series of images of wire-mesh covering a window, illuminated by a distant street lamp, led him and John Cowley to develop a theory of "Fourier images", which subsequently proved to have been noticed by H. Fox Talbot (Talbot, 1836) and discussed long after by Lord Rayleigh (Rayleigh, 1881), Weisel (1910) and Wolfke (1913).[2] The work on Fourier images may be traced through the papers of Cowley and Moodie (1957a,b, 1958, 1959, 1960); see also Sanders and Goodman (1981) and Cowley (1981) as well as Cowley's *Diffraction Physics* (Cowley,1975, 1981). From these ideas, the multislice method of image simulation emerged (Cowley and Moodie, 1957a,b); it was successfully put into practice when the necessary computing power became available (Lynch and O'Keefe, 1972; O'Keefe, 1973) and improved by Ishizuka and Uyeda (1977)

[2] Of these papers, only that of Wolfke is listed by Czapski and Eppenstein (1924), in the context of Abbe's theory of microscope resolution. The paper by Wolfke is the last of a series of papers on the imaging of gratings.

who exploited the mapping of convolution products into direct products by the Fourier transform.

Further attempts to speed up the calculation were made by Dirk Van Dyck, who introduced a real-space procedure in 1980, subsequently perfected by him in collaboration with Wim Coene (Van Dyck and Coene, 1984; Coene and Van Dyck, 1984a,b).

The next major development concerns Part XIII on image formation but first we must return to the 1940s when Fourier optics was born. In 1940 Pierre-Michel Duffieux published two papers on the harmonic analysis of optical images, and a third followed in 1942 (Duffieux, 1940a,b, 1942). These appear to have passed unnoticed and in France, Duffieux's own presentation of his ideas was found incomprehensible (Duffieux, 1970). He was urged to write them out clearly and the result was *L'Intégrale de Fourier et ses Applications à l'Optique* (1946), produced privately for Duffieux by a Rennes printer. For some years, his work was little known but in 1959 it was fully described in Born and Wolf's *Principles of Optics* and in 1960 in *Diffraction, Structure des Images* by Maréchal and Françon. Soon after, Karl-Joseph Hanszen and colleagues introduced the idea of characterizing the transfer of information to the image from the object when the latter scatters weakly by means of a linear transfer theory (Hanszen and Morgenstern, 1965; Hanszen, 1966). The experiments of Friedrich Thon (Thon, 1966a,b) confirmed the correctness of the theory and drew attention to it vividly, after which it was extended to many other practical situations, tilted and hollow-cone illumination for example. The effects of source size and energy spread were likewise explored in great detail, notably by Hanszen and Ludwig Trepte and by Joachim Frank, later in collaboration with Richard Wade (Hanszen and Trepte, 1971; Frank, 1973; Wade and Frank, 1977). The arrival of the STEM (Crewe et al., 1968) was soon followed by the full study of its image-forming mechanism by Elmar Zeitler and Michael Thomson and in the language of transfer theory by Harald Rose (Zeitler and Thomson, 1970; Rose, 1974). The reciprocity between the image-forming mechanisms in the TEM and the STEM was pointed out by Cowley (1969). The possibility of using the detector response as a free parameter was recognized by Dekkers and de Lang (1974) and the repercussions of this observation are with us still.

The idea that the microscope is characterized by a transfer function gave rise to several suggestions for altering the transfer characteristics of the instrument. An early example is the zone plate suggested by Walter Hoppe in 1961. All these ideas may be traced back to the work of André Maréchal and Paul Croce (Maréchal and Croce, 1953, see Croce, 1956) on filtering in the light microscope.

The years around 1970 witnessed developments of the highest importance in image processing: the first attempts to exploit the sequential nature of the image-forming process in the scanning electron microscope with a view to improving the image in various ways;

the first three-dimensional reconstruction from transmission images and the iterative solution of the phase problem.

The scanning microscope was first made available commercially in 1965, when the Cambridge Instrument Co. launched the Stereoscan. The first attempts to alter the image contrast electronically were made soon after (MacDonald, 1968, 1969; White et al., 1968) and although specially designed circuits were used for the purpose at that time, these efforts heralded the digital image processing of today.

It was in 1968 that the first three-dimensional reconstructions were performed by David de Rosier and Aaron Klug. Although this first reconstruction leaned heavily on the known symmetry of the specimen (tail of the bacteriophage T4), the general reconstruction procedure was set out in full and even the possibility of reconstruction from a "field of particles" was described. We quote from their paper (de Rosier and Klug, 1968) for comparison with the Hoppe quotation below:

> The electron microscope image represents a projection of the three dimensional density distribution in the object at all levels perpendicular to the direction of view. According to a theorem familiar to crystallographers, the Fourier coefficients calculated from a projection of a three dimensional density distribution form a section through the three dimensional set of Fourier coefficients corresponding to that distribution. By collecting many different projections of a structure in the form of electron microscope images, it should therefore be possible to collect, section by section, the full set of Fourier coefficients required to describe that structure ... The number of projections needed to fill Fourier space roughly uniformly depends on the size of the particle and on the resolution ... If more than one projection of an object is needed to resolve its structure, such projections can be obtained in two ways. The most obvious one is to systematically tilt and photograph a single particle in the electron microscope. If all the necessary projections are collected regardless of particle symmetry, no assumptions are needed to calculate the three dimensional structure ... Alternatively, different images from a field of particles are, in principle, projections of the same structure ... The exact orientation of each particle in relation to the direction of view must be determined in order to relate correctly the Fourier space section obtained from it to those obtained from other particles.

Their method was soon extended and improved (Crowther, 1971; Crowther et al., 1970a,b). A paper by Hoppe et al. that also appeared in 1968 likewise contained a clear description of the principle of the method:

> Die Dichteverteilung in einem Gitter lässt sich mathematisch in ihre Fourier-Komponenten zerlegen. Diese Daten können wieder in einem dreidimensionalen "Gitter" (dem reziproken Gitter) geometrisch übersichtlich geordnet werden. Der Fourier-Zerlegung jeder Projektion (also jedes elektronenmikroskopischen Bildes eines Kristallgitters in entsprechender Orientierung) entspricht eine durch den Ursprung gehende Gitterebene in diesem reziproken Gitter. Andererseits lässt sich jedes Gitter aus Bündeln von Gitterebenen aufbauen. Die für das dreidimensionale Bild erforderlichen

> Daten kann man also erhalten, wenn man elektronenmikroskopische Aufnahmen von Kristallgittern in verschiedenster Orientierung herstellt, diese mathematisch zerlegt, die Fourier-Komponenten zum dreidimensionalen reziproken Gitter ordnet und schließlich die dreidimensionale Dichteverteilung durch eine Fourier-Synthese berechnet.

We quote it at length here for the late Walter Hoppe always cited it together with the paper of de Rosier and Klug and felt that other authors did not always do justice to it.[3] It does not, however, contain any actual reconstruction.

Before leaving these first three-dimensional reconstructions, we must also mention the isolated attempt by Roger G. Hart to combine several views of a specimen taken at different angles (Hart, 1968). His original aim was not to obtain information about the three-dimensional structure of the object (a dilute suspension of tobacco mosaic virus and colloidal gold particles, which served as fiduciary marks, sprayed onto a support film and air-dried) but to enhance the contrast. Nevertheless, he did realise that three-dimensional reconstruction should be possible:

> . . . if the tilt angle were increased from its present 20 deg to 45 deg, the depth discrimination, for the finest details observable, would be comparable to the lateral resolution and would be limited by only the quality and number of the original micrographs.

> Thus the polytropic montage seems to offer a means of determining the three-dimensional structures of low-contrast biological specimens at a resolution of 3 Å. . . . I have not yet reached this point but preliminary efforts have produced images of tobacco mosaic virus comparable in fineness of detail to those obtained by shadowing . . . Still to be determined is the extent to which the fine details appearing in the montage represent real structures of the virus rather than residual noise that may have survived this attempt at its elimination.

In the following years, three-dimensional reconstruction progressed from being a difficult exercise to which only a very small number of laboratories could aspire to the major activity that it has become today. The structure of the Covid-19 virus, which achieved pandemic status in 2020, was established within weeks of its appearance (Walls et al., 2020; Wrapp et al., 2020; Subramanian, 2020). The subject gradually separated into two parts, electron crystallography and electron tomography. An important step in the latter was the introduction by Marin van Heel and Joachim Frank of a statistical technique known as correspondence analysis (van Heel and Frank, 1980, 1981; Frank and van Heel, 1980).

[3] In a circular letter dated October 1986, which accompanied his retrospective account of 1983, Hoppe wrote: "Similarly disconcerting is also the citation of our first works on 3-dimensional electron microscopy. Almost nowhere is mentioned that (Hoppe et al., 1968) is not only a parallel paper to (de Rosier and Klug, 1968), but theoretically and experimentally showed the way to 3-dimensional analysis of native structures . . . Apparently somewhat better known is that we carried out the first true 3-dimensional reconstruction." (The last remark refers to Hoppe et al., 1974, 1976).

This enabled them to classify images of poor visual quality into groups, each corresponding to a particular view through the particle being studied, an important part of the pre-processing stage.

The "phase problem" was not new when the first successful iterative solution was proposed in 1972 by Ralph Gerchberg and W. Owen Saxton (Gerchberg and Saxton, 1972, 1973). The difficulty is easily stated: how can we obtain the modulus and phase of a complex signal when only the intensity (i.e. the square of the modulus) can be recorded? It had arisen in coherence theory, in X-ray crystallography and in optics, to cite only closely related fields, but little progress had been made in the search for a solution, though ways of circumventing it had been devised by the crystallographers. The original feature of the problem in electron microscope imagery, where we should like to know the complex wavefunction emerging from the specimen, is that both diffraction patterns and images can be recorded. After a first attempt to obtain a direct, noniterative solution, which ran into the same difficulties as earlier efforts (some of which are listed in Chapter 74), Gerchberg and Saxton (1972, 1973) devised an iterative solution, constrained by the measured moduli in the image and diffraction pattern of the same specimen area. This generated a vast activity that goes well beyond the electron microscope community, from which other algorithms (associated in particular with James Fienup) and detailed studies of the uniqueness of the solutions in one and more dimensions emerged. Other ways of solving the phase problem have been found, notably ptychography and holography, all of which illustrate the role of redundancy: if much of the information available is redundant it should be possible to calculate phase and amplitude.

In this connection, we should also mention here the extension by Peter Schiske of the Wiener filter to complex, weakly scattering objects, for which there is a linear relation between complex object transparency and (real) image contrast. His first publication (1968) neglected noise but, in a later paper (1973), noise is included and the analogy with the simple Wiener filter is exact.

While digital electron image processing was coming into being, an old proposal for remedying some of the defects of the electron microscope image was at last successfully tested: holography. With the development of first pointed-filament sources and later field-emission guns and the introduction of the electron biprism, the conditions necessary for electron hologram formation could be met. The laser was by this time in widespread use and so the conditions for reconstruction could likewise be satisfied. Finally, Emmett Leith and Juris Upatnieks (Leith and Upatnieks, 1962, 1963) had devised an "off-axis" method of separating the two images that are superimposed in the original in-line procedure of Gabor and the time was therefore ripe to resume attempts to put the technique into practice. Before mentioning the early landmarks, however, we must just draw attention to a semantic difficulty: in its primitive form, the hologram is formed by interference between the part of the beam that traverses the specimen without being scattered by the atoms that make up the object – this is the reference beam – and those electrons that have been scattered. The hologram is thus essentially the same as the bright-field image. There is hence a large body of literature, especially dating from the 1970s, in which no distinction is made between holography and bright-field imagery and inverse filtering of the image of weakly scattering specimens is described as holography. The interplay between these different points of view is examined with great care by Hanszen in his various surveys (Hanszen, 1971, 1973, 1982).

The earliest example of Fraunhofer in-line electron holography was published by Akira Tonomura and Hiroshi Watanabe (Tonomura and Watanabe, 1968) and their remarkable reconstruction is reproduced here (Fig. 54.1); see also Tonomura et al. (1968) and Watanabe and Tonomura (1969). Several further attempts to use this approach were made, notably by Gallion et al. (1975), Troyon et al. (1976), Bonnet et al. (1978) and Bonhomme and Beorchia (1980) and by Munch (1975), who used a field-emission gun.

The first off-axis electron hologram was made with the aid of a biprism and reconstructed using a He-Ne laser by Möllenstedt and Herbert Wahl (Möllenstedt and Wahl, 1968), closely followed by Tonomura (1969) using a crystal to split the beam into two coherent beams. The Möllenstedt and Wahl experiment only partially fulfilled the conditions for holography as the two virtual sources were lines rather than points. The first off-axis hologram using point sources was made by Hiroshi Tomita and colleagues (Tomita et al., 1970a,b, 1972)

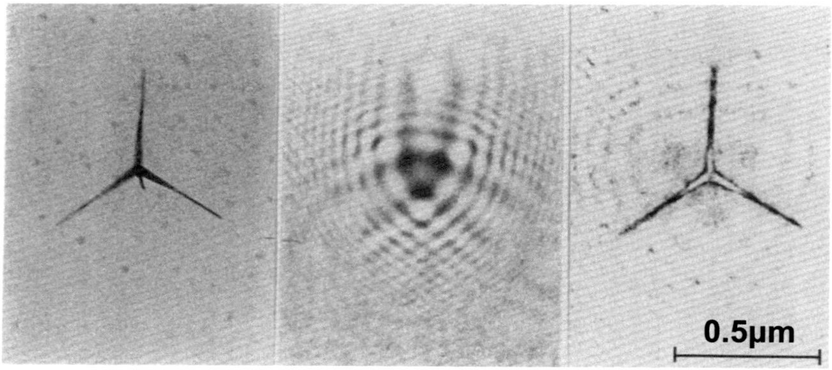

Figure 54.1
In-line Fraunhofer hologram (centre) and optical reconstruction (right), obtained by Tonomura and Watanabe (1968). In fact, the Fraunhofer condition is not satisfied for the whole specimen, a zinc oxide crystal (left), but is satisfied for the needles since their diameter is small. *After Tonomura and Watanabe (1968), Courtesy Physical Society of Japan, https://doi.org/10.11316/butsuri1946.23.683.*

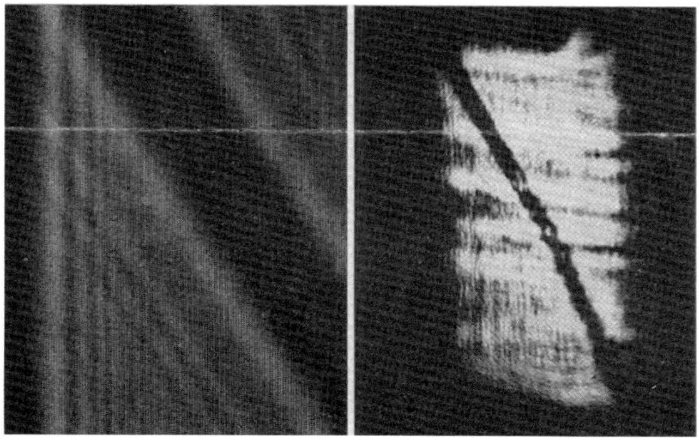

Figure 54.2
Off-axis Fresnel hologram of a metallized quartz fibre (left) and optical reconstruction (right), obtained by Möllenstedt and Wahl (1968). The virtual sources are lines, not points, and the fibre is oblique to these lines. *After Möllenstedt and Wahl (1968), Courtesy Springer, https://doi.org/10.1007/BF00600454.*

and we reproduce in Fig. 54.2 the reconstructions of Möllenstedt and Wahl and Tomita et al. Wahl then went on to study off-axis image plane holography much more fully (Wahl, 1974, 1975). In the following years, holography was explored in depth by Akira Tonomura and colleagues in Japan, by Hannes Lichte and colleagues in Tübingen and by Karl-Joseph Hanszen, Georg Ade and Rolf Lauer in Braunschweig, with more isolated work elsewhere.

One side effect of the experimental developments that contributed to the progress of holography was the incontrovertible demonstration by Tonomura et al. (1986) that the Aharonov–Bohm effect really exists. This interference phenomenon, in fact noticed by Werner Ehrenberg and Raymond Siday in 1949 and rediscovered by Yakir Aharonov and David Bohm in 1959, who gave credit to Ehrenberg and Siday as soon as they became aware of the earlier work of their colleagues (Ehrenberg and Siday, 1949; Aharonov and Bohm, 1959, 1961), has a considerable literature. The effect is a shift of the interference fringes formed when two beams that have passed on either side of a local magnetic field are made to overlap; even if the magnetic field is negligibly small in the regions traversed by the electrons, the fringe structure changes if the magnetic field is altered. This is due to the fact that the vector potential $\boldsymbol{A}$ is different along the two paths even if $\boldsymbol{B} = \mathrm{curl}\,\boldsymbol{A}$ effectively vanishes. A fierce controversy over the reality of the effect raged for many years, the Italian group around P. Bocchieri in Pavia and A. Loinger in Milan manifesting the most obdurate incredulity; the experimental results of Tonomura et al. that put the reality of the phenomenon beyond further doubt are reproduced as Fig. 62.11.

The search for solutions to the phase problem has given rise to a new technique, ptychography. This was originally proposed (and named) by Walter Hoppe and Reiner Hegerl (Hoppe, 1969a,b; Hegerl and Hoppe, 1970) but lay fallow until Richard Bates and John Rodenburg put it on a clear theoretical basis (Bates and Rodenburg, 1989; Rodenburg and Bates, 1992). It is now in widespread use in X-ray microscopy and is gaining acceptance among electron microscopists, essentially thanks to the development of large, fast direct electron detectors (Rodenburg and Maiden, 2019). Its only drawback is that diffraction patterns have to be recorded from a large number of positions of the probe in STEM. This implies that the specimen is subjected to a high dose, too high for beam-sensitive material. This can be mitigated to some extent by defocusing the probe.

An important addition to the electron microscopy toolbox is the family of methods based on statistical parameter estimation theory; these offer alternative ways of extracting information from electron micrographs with high precision, often with lower electron doses than regular imaging. They were introduced by Dirk Van Dyck, Sandra Van Aert and colleagues in the University of Antwerp (Bettens et al, 1999; van den Bos and den Dekker, 2001; Van Aert et al., 2002a,b,c; Van Dyck et al., 2003); their spread is illustrated by De Backer et al. (2021).

The relation between source coherence and the radiometric quantities has been explored in depth for light sources but has been neglected for electron emitters. The account here therefore leans heavily on the literature of light optics. The fact that traditional radiometry is essentially applicable only to incoherent sources and that some new definition of brightness was needed was first pointed out by Adriaan Walther (1968) but the principal developments are due to Emil Wolf and a series of colleagues, E.W. Marchand and W.H. Carter in particular. We cite especially the papers by Marchand and Wolf (1972a,b, 1974) upon which most subsequent developments repose and the recognition of the importance of the quasihomogeneous source by Carter and Wolf (1977).

Many of the themes of this volume can be united with the aid of the Wigner function as we have mentioned earlier. We conclude this very selective historical section with a reminder that electron beams with orbital angular momentum were first demonstrated by Masaya Uchida and Akira Tonomura (2010) and that the possibility of creating a quantum electron microscope, in which the electron beam would not see the specimen, was suggested by Pieter Kruit (Kruit et al., 2016).

A future polyvalent electron microscope is imagined by Hattar et al. (2021).

PART XI

Wave Mechanics

CHAPTER 55

The Schrödinger Equation

55.1 Introduction

Electrons, like all other kinds of elementary particles, have a dual character: they behave like *corpuscles* or like *waves*, depending on the particular experimental conditions. This double nature of electrons can be completely understood only within the frame of the general quantum theory and a rigorous treatment would therefore require the whole subject of electron optics to be cast into quantum theoretical form. Such a treatment is extremely complicated and unnecessarily detailed for almost all practical problems but a full account is now available (Jagannathan and Khan, 2019). We thus seek reasonable simplifications.

Wave mechanics is a part of the general quantum theory, in which the *wave nature* of the radiation is of paramount interest. Concepts typical of wave physics, such as frequency, wavelength, diffraction and interference, and experimental situations in which these concepts are useful are studied. Like geometrical electron optics, which is based on classical mechanics, this is again an incomplete description. The full theory, in which attempts are made to unify the particle and wave physics, is practically never needed in electron optical practice.

Electrons have spin and hence satisfy not Schrödinger's but Dirac's equation. In electron optics, effects caused by spin are usually negligible with a few exceptions, such as very low-voltage scanning electron microscopy and vortex studies (Chapter 80). A critical account of the transition from the Dirac formalism to the Schrödinger equation is to be found in Chapter 56.

55.2 Formulation of Schrödinger's Equation

The material presented in the remainder of this chapter is dealt with extensively in all the numerous textbooks on quantum mechanics and is quite familiar; the following account is therefore very concise.

We start from the *Hamilton equation* for the motion of a single electron in a stationary electromagnetic field:

$$H(\boldsymbol{r},\boldsymbol{p}) = \frac{1}{2m_0}\{\boldsymbol{p}+e\boldsymbol{A}(\boldsymbol{r})\}^2 - e\Phi(\boldsymbol{r}) = E = \text{const} \tag{55.1}$$

Principles of Electron Optics.
DOI: https://doi.org/10.1016/B978-0-12-818979-5.00055-3

in which we have retained the notation used in Volume 1: m_0 denotes the rest mass of the electron, e the absolute elementary charge, Φ the electrostatic potential and $\boldsymbol{A}$ the vector potential, while $\boldsymbol{p}$ is the *canonical* momentum.

In wave mechanics, the momentum $\boldsymbol{p}$ and the total energy E are replaced by *operators*, which act on a *wavefunction* $\Psi(\boldsymbol{r}, t)$. These operators are chosen as

$$\boldsymbol{p} \rightarrow -\mathrm{i}\hbar\nabla \equiv -\mathrm{i}\hbar \,\mathrm{grad} \tag{55.2a}$$

$$E \rightarrow \mathrm{i}\hbar\frac{\partial}{\partial t} \tag{55.2b}$$

$\hbar = h/2\pi = 1.05 \times 10^{-34}$ J s being Dirac's constant. Replacing $\boldsymbol{p}$ and E in (55.1) by these operators and applying the resulting operator equation to a wavefunction Ψ, we arrive at Schrödinger's equation

$$\frac{1}{2m_0}\left\{-\mathrm{i}\hbar\nabla + e\boldsymbol{A}(\boldsymbol{r})\right\}^2\Psi(\boldsymbol{r},t) - e\Phi(\boldsymbol{r})\Psi(\boldsymbol{r},t) = \mathrm{i}\hbar\frac{\partial\Psi(\boldsymbol{r},t)}{\partial t} \tag{55.3}$$

This is a *complex* linear partial differential equation of second order in the space coordinates and of first order in time. The complex nature of (55.3) implies that the solution Ψ cannot have an immediate physical meaning; the calculation of observable quantities is the topic of the next section.

In very many practical applications, it is sufficient to consider *time-independent* solutions. These are obtained by seeking a solution in the separated form

$$\Psi(\boldsymbol{r},t) = \psi(\boldsymbol{r})\exp(-\mathrm{i}\omega t) \tag{55.4}$$

On relating the oscillation frequency $\omega = 2\pi\nu$ to the total energy E by Einstein's relation

$$E = \hbar\omega \tag{55.5}$$

and cancelling out a common factor $\mathrm{e}^{-\mathrm{i}\omega t}$, we arrive at the *time-independent* Schrödinger equation

$$\hat{H}\psi(\boldsymbol{r}) := \left\{\frac{1}{2m_0}(-\mathrm{i}\hbar\nabla + e\boldsymbol{A})^2 - e\Phi\right\}\psi(\boldsymbol{r}) = E\psi(\boldsymbol{r}) \tag{55.6}$$

The operator $\hat{H}$, defined to be the expression in braces, is called the *Hamilton operator*. This partial differential equation and its relativistic generalization, derived in the next chapter, are the starting point of most subsequent calculations.

A still simpler form is obtained for the motion of electrons in purely electrostatic fields. We may then assume that $\boldsymbol{A}(\boldsymbol{r}) = 0$ and can hence cast (55.6) into the form

$$\nabla^2\psi(\boldsymbol{r}) + \frac{2m_0}{\hbar^2}\left\{E + e\Phi(\boldsymbol{r})\right\}\psi(\boldsymbol{r}) = 0 \tag{55.7}$$

This has the form of a *Helmholtz equation*

$$\nabla^2\psi(\boldsymbol{r}) + k^2(\boldsymbol{r})\psi(\boldsymbol{r}) = 0 \tag{55.8}$$

with a wavenumber depending on position:

$$k(\boldsymbol{r}) := \frac{\sqrt{2m_0\{E + e\Phi(\boldsymbol{r})\}}}{\hbar} \tag{55.9}$$

In nonrelativistic classical mechanics the conservation of energy takes the form

$$\frac{1}{2m_0}g^2 - e\Phi(\boldsymbol{r}) = E$$

and the *kinetic* momentum (2.14) is thus given by

$$g = \sqrt{2m_0\{E + e\Phi(\boldsymbol{r})\}} \tag{55.10}$$

Comparing this with (55.9), we immediately obtain de Broglie's relation

$$g = \hbar k \tag{55.11}$$

The validity of this relation will be studied later.

55.3 The Continuity Equation

An analogue of the well-known continuity equation of classical electrodynamics can be derived in wave mechanics. To obtain this, we multiply both sides of (55.3) by the complex conjugate function Ψ^*, giving

$$\Psi^*\frac{1}{2m_0}(-\mathrm{i}\hbar\nabla + e\boldsymbol{A})^2\Psi - e\Phi\Psi^*\Psi = \mathrm{i}\hbar\Psi^*\frac{\partial\Psi}{\partial t}$$

The complex conjugate of the whole equation is

$$\Psi\frac{1}{2m_0}(+\mathrm{i}\hbar\nabla + e\boldsymbol{A})^2\Psi^* - e\Phi\Psi^*\Psi = -\mathrm{i}\hbar\Psi\frac{\partial\Psi^*}{\partial t}$$

Subtraction of the second equation from the first yields an equation from which the terms in Φ have cancelled out and the right-hand side is the time-derivative of a product:

$$\frac{1}{2m_0}\left\{\Psi^*(-\mathrm{i}\hbar\nabla + e\boldsymbol{A})^2\Psi - \Psi(\mathrm{i}\hbar\nabla + e\boldsymbol{A})^2\Psi^*\right\} = \mathrm{i}\hbar\frac{\partial}{\partial t}\left(\Psi^*\Psi\right) \tag{55.12}$$

The evaluation of the left-hand side proceeds as follows: expansion of the first quadratic term gives

$$\Psi^*(-\mathrm{i}\hbar\nabla + e\boldsymbol{A})^2\Psi = \Psi^*\left(-\hbar^2\nabla^2\Psi + e^2\boldsymbol{A}^2\Psi - 2\mathrm{i}e\hbar\boldsymbol{A}\cdot\mathrm{grad}\Psi - \mathrm{i}e\hbar\Psi\mathrm{div}\boldsymbol{A}\right)$$

with an analogous result for its complex conjugate*. The terms in e^2A^2 cancel and we can rewrite (55.12) in the form

$$\frac{\partial}{\partial t}|\Psi|^2 + \frac{\hbar}{2m_0\mathrm{i}}\left(\Psi^*\nabla^2\Psi - \Psi\nabla^2\Psi^*\right) + \frac{e}{m_0}\left\{\boldsymbol{A}\cdot\left(\Psi^*\nabla\Psi + \Psi\nabla\Psi^*\right) + |\Psi|^2\mathrm{div}\boldsymbol{A}\right\} = 0$$

The terms involving spatial derivatives can be expressed as the divergence of a vector and we thus arrive at a continuity equation (cf. 47.6)

$$\frac{\partial}{\partial t}\rho(\boldsymbol{r},t) + \mathrm{div}\,\boldsymbol{j}(\boldsymbol{r},t) = 0 \tag{55.13}$$

with the scalar density function

$$\rho(\boldsymbol{r},t) := |\Psi(\boldsymbol{r},t)|^2 \tag{55.14}$$

and the current density vector

$$\boldsymbol{j}(\boldsymbol{r},t) := \frac{\hbar}{2m_0\mathrm{i}}\left\{\Psi^*\mathrm{grad}\,\Psi - \Psi\mathrm{grad}\,\Psi^*\right\} + \frac{e\rho}{m_0}\boldsymbol{A} \tag{55.15}$$

Both density functions are always real and hence have a physical significance. The function $\rho(\boldsymbol{r}, t)$ is the *particle density*, which here means the probability of finding a single electron at the given point in space; $\boldsymbol{j}(\boldsymbol{r}, t)$ is then the corresponding particle current density. If the wavefunction can be normalized to unity

$$\int_{\mathbb{R}^3}\rho(\boldsymbol{r},t)d^3r \equiv \int_{\mathbb{R}^3}|\Psi|^2d^3r = 1 \tag{55.16}$$

the continuity Eq. (55.13) ensures the conservation of each individual particle in space. Creation and annihilation processes cannot occur.

Some other density functions of practical interest are the electric space charge density

$$\rho_{el}(\boldsymbol{r},t) = -e\rho(\boldsymbol{r},t) \tag{55.17}$$

and the electric current density vector

$$\boldsymbol{j}_{el}(r,t) = -e\boldsymbol{j}(\boldsymbol{r},t) \tag{55.18}$$

We recall that in the physics of electron guns (Part IX of Volume 2), the opposite sign convention for $\boldsymbol{j}_{el}$ is more usual.

An important difference between electron optics and the physics of atoms and molecules is that *unbounded* solutions of (55.3) are of greatest interest in charged particle optics, which implies that the normalization (55.16) is often formally not possible; physically, wavefunctions are always truncated, by apertures for example, and are hence square integrable.

55.4 The Gauge Transformation

The wave-optical concepts introduced in the context of Schrödinger's equation have *no* physical meaning in the sense that they are not observable quantities. Only the density functions ρ and $\boldsymbol{j}$ (and consequently ρ_{el} and $\boldsymbol{j}_{el}$) can be considered as observable in principle. To show this, we subject all the quantities appearing in the theory to a gauge transformation.

Since (55.3) is a *homogeneous* linear partial differential equation, it is immediately clear that its solution may contain an arbitrary complex factor. To keep the density functions ρ and $\boldsymbol{j}$ invariant, this factor must have unit norm.

This is, however, not the only possible degree of freedom; within the frame of time-independent potentials, the following gauge transformations are allowed:

$$\begin{aligned}
\boldsymbol{A}(\boldsymbol{r}) &= \boldsymbol{A}'(\boldsymbol{r}) + \operatorname{grad} F(\boldsymbol{r}) && \text{(a)}\\
\Phi(\boldsymbol{r}) &= \Phi'(\boldsymbol{r}) + \Phi_0 && \text{(b)}\\
\boldsymbol{E} &= \boldsymbol{E}' - e\Phi_0 && \text{(c)}\\
\omega &= \omega' - e\Phi_0/\hbar && \text{(d)}\\
\Psi(\boldsymbol{r},t) &= \Psi'(\boldsymbol{r},t)\exp(\mathrm{i}\varphi(\boldsymbol{r},t)) && \text{(e)}\\
\varphi(\boldsymbol{r},t) &= \alpha + e\{\Phi_0 t - F(\boldsymbol{r})\}/\hbar && \text{(f)}
\end{aligned} \tag{55.19}$$

The real constants α and Φ_0 and the scalar function $F(\boldsymbol{r})$ can be chosen at will. The relation between E and Φ must be such that the classical kinetic energy

$$T = E + e\Phi(\boldsymbol{r}) = E' + e\Phi'(\boldsymbol{r}) \tag{55.20}$$

remains invariant; the *kinetic* momentum g (55.10) is hence invariant but not the canonical momentum $\boldsymbol{p}$.

From (55.19a) and (55.19f), a minor calculation shows that

$$(-\mathrm{i}\hbar\nabla + e\boldsymbol{A})\Psi = \mathrm{e}^{\mathrm{i}\phi}(-\mathrm{i}\hbar\nabla + e\boldsymbol{A}')\Psi' \tag{55.21a}$$

and consequently

$$(-\mathrm{i}\hbar\nabla + eA)^2\Psi = \mathrm{e}^{\mathrm{i}\phi}(-\mathrm{i}\hbar\nabla + eA')^2\Psi' \tag{55.21b}$$

The time-dependent term in (55.19f) is chosen so that

$$e\Phi\Psi + \mathrm{i}\hbar\frac{\partial\Psi}{\partial t} = \left\{e\Phi'\Psi' + \mathrm{i}\hbar\frac{\partial\Psi'}{\partial t}\right\}\mathrm{e}^{\mathrm{i}\phi} \tag{55.22}$$

On introducing all this into (55.3), we notice that a common factor $e^{i\varphi}$ appears throughout and can therefore be cancelled. This means that the Schrödinger equation takes the same form for Ψ' as for Ψ: it is *form invariant*.

It is obvious that the phase factor $e^{i\varphi}$ cancels out from (55.14), leaving ρ invariant: $\rho' = \rho$. After a minor calculation, we find that $\boldsymbol{j}' = \boldsymbol{j}$ is also an invariant. For *unbounded* states, these (and ρ_{el} and $\boldsymbol{j}_{el}$) are the essential observables.

55.5 Wave–Particle Duality

The coexistence of waves and particles has been a source of numerous theories since the beginnings of quantum mechanics. We draw attention to a recent attempt to resolve the enigma since its authors are close to the domain of electron optics.

The propagation of electrons or light is described in terms of a two-point correlation function, similar in appearance to that encountered in coherence theory (Eq. 78.21a,b). However, the new theory is causal and deterministic and excludes the stumbling blocks of the familiar theory, wave–particle duality and self-interference in particular. Space does not permit us to present it here. It is very clearly described in the long account by Castañeda and Matteucci (2019). The progress of the theory is charted in the following publications: Castañeda (2016, 2017), Castañeda et al. (2016a,b, 2021) and Castañeda and Matteucci (2017). It is used to demystify the Aharonov–Bohm effect by Castañeda et al. (2022).

CHAPTER 56

The Relativistic Wave Equation

The Schrödinger equation, dealt with in the previous chapter, is valid only for *nonrelativistic* electron motion. In practice, this means a restriction to kinetic energies not exceeding about 100 keV. In many cases the electron energy reaches the relativistic domain and in high-voltage electron microscopy (HVEM), it may even exceed the threshold $2m_0c^2 \approx 1$ MeV. It is therefore obviously necessary to consider *relativistic* wave equations.

56.1 The Dirac Equation

The relativistically correct form of the wave equation is Dirac's equation. This can be written in various ways, which show its Lorentz covariance. A fairly simple form is

$$c\boldsymbol{\alpha}\cdot(-\mathrm{i}\hbar\nabla + e\boldsymbol{A})\hat{\Psi} + \alpha_4 m_0 c^2\hat{\Psi} - e\Phi\hat{\Psi} = \mathrm{i}\hbar\frac{\partial\hat{\Psi}}{\partial t} \tag{56.1}$$

where $\hat{\Psi}(\boldsymbol{r}, t)$ is a four-element *spinor*, that is a column vector the four elements of which are complex wavefunctions $\Psi_j(\boldsymbol{r}, t)$, $j = 1 \ldots 4$. The symbol $\boldsymbol{\alpha} = (\alpha_1, \alpha_2, \alpha_3)$ denotes a vector operator; its components α_j ($j = 1, 2, 3$) and also α_4 are Hermitian 4×4 matrices satisfying

$$\alpha_j\alpha_k + \alpha_k\alpha_j = 2\delta_{jk} \quad (j, k = 1\ldots4) \tag{56.2}$$

A simple explicit representation of these Clifford matrices is given by

$$\alpha_1 = \begin{pmatrix} 0 & 0 & 0 & 1 \\ 0 & 0 & 1 & 0 \\ 0 & 1 & 0 & 0 \\ 1 & 0 & 0 & 0 \end{pmatrix} \qquad \alpha_2 = \begin{pmatrix} 0 & 0 & 0 & -\mathrm{i} \\ 0 & 0 & \mathrm{i} & 0 \\ 0 & -\mathrm{i} & 0 & 0 \\ \mathrm{i} & 0 & 0 & 0 \end{pmatrix}$$

$$\alpha_3 = \begin{pmatrix} 0 & 0 & 1 & 0 \\ 0 & 0 & 0 & -1 \\ 1 & 0 & 0 & 0 \\ 0 & -1 & 0 & 0 \end{pmatrix} \qquad \alpha_4 = \begin{pmatrix} 1 & 0 & 0 & 0 \\ 0 & 1 & 0 & 0 \\ 0 & 0 & -1 & 0 \\ 0 & 0 & 0 & -1 \end{pmatrix}$$

They are simultaneously Hermitian and unitary and have unit determinant and vanishing trace.

Principles of Electron Optics.
DOI: https://doi.org/10.1016/B978-0-12-818979-5.00056-5

The Dirac equation satisfies all the requirements of relativistic wave mechanics: it is Lorentz-covariant, gauge-covariant and describes correctly all spin interactions. Moreover, it provides correct expressions for the particle and current densities.

Below the threshold $2m_0c^2$, creation and annihilation processes cannot take place and even in HVEM they are quite unimportant. Hence, Dirac's equation describes the propagation of electrons in practically all electron optical systems to a very high degree of accuracy. Nevertheless, it is little used in electron optics because the calculations become exceedingly and unwarrantably complicated.

56.2 The Scalar Wave Equation

It is standard practice in electron optical calculation to start with Schrödinger's equation and then apply some relativistic corrections to the results obtained. This is certainly justified if the corrections are chosen in such a way that good agreement with experimental results is achieved. From the theoretical standpoint, however, such a procedure is unsatisfactory and can be avoided, as we shall soon see.

Simplification of Dirac's equation that reduces it to a scalar wave equation is only possible if we are willing to sacrifice some essential aspects of the relativistic theory.

1. A wavefunction having only *one* component $\Psi(\boldsymbol{r}, t)$ is acceptable only if we ignore completely the electron spin. In fact, spin effects are unimportant in electron optics except in some very specialized situations (see Reimer and Kohl, 2008, Section 7.3.5 and Samarin et al., 2018) so that this simplification is usually justified. Orbital angular momentum and related topics are considered in Chapter 80.
2. There is no need for a Lorentz-covariant formulation. In practically all situations, the description in the *laboratory frame* is perfectly adequate. We may hence sacrifice the possibility of performing Lorentz transformations.
3. As in geometrical electron optics, it is convenient to choose a fixed gauge for the electrostatic potential: the cathode surface is assigned the potential $\Phi = 0$. The quantity $e\Phi(\boldsymbol{r})$ is then a direct measure of the energy acquired during acceleration in the electrostatic field.

With these assumptions, the formulation of a scalar relativistic wave equation is possible. The latter can be directly derived from Dirac's equation (Kasper, 1973; Lubk, 2018; Jagannathan and Khan, 2019) but the following argument, originally published in the first edition of this book, is much simpler.

We start from the familiar classical Hamiltonian

$$H(\boldsymbol{r},\boldsymbol{p}) = c\sqrt{(\boldsymbol{p}+e\boldsymbol{A})^2 + m_0^2c^2} - e\Phi(\boldsymbol{r}) - m_0c^2 \tag{56.3}$$

This is obtained from Eq. (4.28) of Volume 1 by setting $Q = -e$. A consequence of the assumption (3), which requires (2), is that the numerical value of H is very small. At the cathode surface $\Phi = 0$, the value of H is the *kinetic emission energy*, which is usually less then 1 eV. Since H represents the total energy, which is conserved, it must *always* remain very small, and so $|H| \ll m_0c^2$.

In order to obtain a form that can be replaced by an operator, the square-root term must be eliminated. Solving (56.3) for this term and then squaring the whole result, we obtain, still exactly,

$$2H\left(m_0 + \frac{e}{c^2}\Phi\right) + \frac{H^2}{c^2} = (\boldsymbol{p}+e\boldsymbol{A})^2 - 2em_0\Phi - \frac{e^2\Phi^2}{c^2}$$

It is now advantageous to introduce a *mass function*

$$m(\boldsymbol{r}) := m_0 + \frac{e}{c^2}\Phi(\boldsymbol{r}) \tag{56.4}$$

which is the classical value of the relativistic electron mass (cf. Eqs. 2.3 and 2.10 of Volume 1). The second term represents the mass equivalent of the acceleration energy according to Einstein's energy–mass relation. Using (56.4) and ignoring the extremely small term H^2/c^2, we can cast the Hamiltonian into the form

$$H(\boldsymbol{r},\boldsymbol{p}) = \frac{1}{2m(\boldsymbol{r})}(\boldsymbol{p}+e\boldsymbol{A})^2 - \frac{1}{2m}\left(2m_0e\Phi + \frac{e^2\Phi^2}{c^2}\right)$$

By introducing the relativistic acceleration potential (Eq. 2.18)

$$\hat{\Phi}(\boldsymbol{r}) := \Phi(\boldsymbol{r})\left\{1 + \frac{e}{2m_0c^2}\Phi(\boldsymbol{r})\right\} \tag{56.5}$$

(see also Section 2.3), we obtain the convenient result

$$H(\boldsymbol{r},\boldsymbol{p}) = \frac{1}{2m(\boldsymbol{r})}(\boldsymbol{p}+e\boldsymbol{A})^2 - \frac{m_0}{m}e\hat{\Phi}(\boldsymbol{r})$$

This is still not quite suitable for replacement by an operator, since m appears unsymmetrically. This minor deficiency is removed by writing

$$H(\boldsymbol{r},\boldsymbol{p}) = \frac{1}{2\sqrt{m}}(\boldsymbol{p}+e\boldsymbol{A})^2\frac{1}{\sqrt{m}} - \frac{m_0e}{m}\hat{\Phi} \tag{56.6}$$

We are now in a position to define the corresponding Hamilton operator $\hat{H}$:

$$\hat{H} = \frac{1}{2\sqrt{m}}(-\mathrm{i}\hbar\nabla+e\boldsymbol{A})^2\frac{1}{\sqrt{m}} - \frac{m_0e\hat{\Phi}}{m} \tag{56.7}$$

and finally obtain the wave equation

$$\hat{H}\chi = \frac{1}{2\sqrt{m}}(-\mathrm{i}\hbar\nabla + e\boldsymbol{A})^2\left(\frac{\chi}{\sqrt{m}}\right) - \frac{m_0 e\hat{\Phi}}{m}\chi = \mathrm{i}\hbar\frac{\partial\chi}{\partial t} \tag{56.8}$$

Obviously this simplifies to Schrödinger's equation when $|e\Phi| \ll m_0c^2$, for which $m \to m_0$. The wavefunction χ $(\boldsymbol{r}, t)$, being a solution of (56.8), leads to the correct particle density $\rho = |\chi|^2$. The factor $m^{-1/2}$, however, is inconvenient in practical calculations. To remove it, we define a new wavefunction,

$$\Psi(\boldsymbol{r}, t) := \sqrt{\frac{m_0}{m(\boldsymbol{r})}}\chi(\boldsymbol{r}, t) \tag{56.9}$$

which satisfies the modified wave equation

$$\frac{1}{2m_0}(-\mathrm{i}\hbar\nabla + e\boldsymbol{A})^2\Psi - e\hat{\Phi}\Psi = \mathrm{i}\hbar\frac{m(\boldsymbol{r})}{m_0}\frac{\partial\Psi}{\partial t} \tag{56.10}$$

This is more convenient in practice.

56.3 Properties of the Relativistic Wave Equation

In order to demonstrate that (56.10) is a reasonable approximation, we now study its consequences. First of all, we introduce the product separation (55.4) with (55.5), giving

$$\frac{1}{2m_0}(-\mathrm{i}\hbar\nabla + e\boldsymbol{A})^2\psi(\boldsymbol{r}) - e\hat{\Phi}(\boldsymbol{r})\psi(\boldsymbol{r}) = \frac{Em(\boldsymbol{r})}{m_0}\psi(\boldsymbol{r}) \tag{56.11}$$

In the absence of vector potentials ($\boldsymbol{A} \equiv 0$), this can be cast into the form of a Helmholtz equation (55.8) with the wavenumber

$$k(\boldsymbol{r}) = \frac{1}{\hbar}\sqrt{2m_0e\hat{\Phi}(\boldsymbol{r}) + 2Em(\boldsymbol{r})} \tag{56.12}$$

where $|E| \ll m_0c^2$ is necessary for reasons of consistency, since $E^2/c^2 = H^2/c^2$ was neglected in the derivation of (56.10). Within this limit, the square-root factor is still the relativistically correct expression for the kinetic momentum $g(\boldsymbol{r})$. According to the laws of relativistic kinematics (Section 2.3), we have exactly

$$\begin{aligned} g^2(\boldsymbol{r}) &= 2m_0(E + e\Phi(\boldsymbol{r}))\left[1 + \frac{1}{2m_0c^2}\{E + e\Phi(\boldsymbol{r})\}\right] \\ &= 2m_0e\Phi\left(1 + \frac{e}{2m_0c^2}\Phi\right) + 2E\left(m_0 + \frac{e}{c^2}\Phi\right) + \frac{E^2}{c^2} \\ &= 2m_0e\hat{\Phi}(\boldsymbol{r}) + 2Em(\boldsymbol{r}) + \frac{E^2}{c^2} \end{aligned}$$

The only error in (56.12) is indeed the omission of the very small and quite unimportant term E^2/c^2. Hence, de Broglie's relation (55.11) is still valid *relativistically*.

A further consequence is the existence of a continuity equation (55.13). This can be derived just as in Section 55.3. The expressions for ρ and $\boldsymbol{j}$ are now slightly different:

$$\rho(\boldsymbol{r},t) = |\chi(\boldsymbol{r},t)|^2 = \frac{m(\boldsymbol{r})}{m_0}|\Psi(\boldsymbol{r},t)|^2 \tag{56.13}$$

$$\begin{aligned} j(\boldsymbol{r},t) &= \frac{\hbar}{2m\mathrm{i}}(\chi^*\nabla\chi - \chi\nabla\chi^*) + \frac{e}{m}A|\chi|^2 \\ &= \frac{\hbar}{2m_0\mathrm{i}}(\Psi^*\nabla\Psi - \Psi\nabla\Psi^*) + \frac{e}{\sqrt{m_0}}\boldsymbol{A}|\Psi|^2 \end{aligned} \tag{56.14}$$

The mass function $m(\boldsymbol{r})$ cannot be eliminated completely, whether we use χ or Ψ.

When we come to consider the effect of a gauge transformation, a modification is necessary. Since we have ascribed the potential $\Phi = 0$ to the cathode surface, which is the natural gauge in electron optics, it makes little sense to depart from this choice. We therefore set $\Phi_0 = 0$ in (55.19), whereupon all the kinematic functions become invariants. The gauge transformation then reduces to

$$\boldsymbol{A}(\boldsymbol{r}) = \boldsymbol{A}'(\boldsymbol{r}) + \mathrm{grad}\, F(\boldsymbol{r}) \tag{56.15a}$$

$$\Psi(\boldsymbol{r},t) = \Psi'(\boldsymbol{r},t)\exp\{\mathrm{i}(\alpha - eF/\hbar)\} \tag{56.15b}$$

It is easily verified that (56.10) is form invariant with respect to this transform and that ρ and $\boldsymbol{j}$ are again invariants.

A special form of the wave equation, which is frequently needed in the physics of electron scattering and electron diffraction, arises from (55.8) taken together with (56.12). The electrostatic potential then consists of a large and constant acceleration term U and a small atomic contribution $V(\boldsymbol{r})$. A constant energy shift can always be incorporated in the constant U and we may hence set $E = 0$ in (56.12) without loss of generality.

In the physics of electron diffraction, it is often convenient to use an alternative definition of the wavenumber, which we write $\bar{k}$:

$$\bar{k} = \frac{k}{2\pi} = \frac{1}{\lambda} = \frac{g}{h} \tag{56.16}$$

λ denoting the wavelength, g the kinetic momentum and h Planck's constant. Eq. (56.16) is again the de Broglie relation.

The atomic potential $V(\boldsymbol{r})$ has a short spatial range and is quite often very weak, which means that $|V(\boldsymbol{r})| \ll U$. In these conditions, it is advantageous to define the quantities in (56.16) for the *asymptotic* domain, where they become constants. Setting thus $E = 0$ and

$$\hat{\phi} = \hat{U} = U\left(1 + eU/2m_0c^2\right) = U(1 + \varepsilon U)$$

in (56.12), we obtain the *asymptotic* wavenumber

$$k_\infty = \frac{2\pi}{\lambda_\infty} = \frac{1}{\hbar}\sqrt{2m_0 e\hat{U}} \tag{56.17}$$

and the *asymptotic* mass

$$m_\infty = m_0 + \frac{e}{c^2}U = \text{const} \tag{56.18}$$

In the wave equation itself, the term in $V(\boldsymbol{r})$ must be retained but, because $|V| \ll U$, a linearization is allowed. With (56.17) and (56.18), we then obtain

$$\nabla^2\psi(\boldsymbol{r}) + \left\{k_\infty^2 + \frac{2em_\infty V(\boldsymbol{r})}{\hbar^2}\right\}\psi(\boldsymbol{r}) = 0 \tag{56.19}$$

Later we shall drop the subscript ∞, whenever this does not cause confusion.

The wavelengths encountered in electron microscopy are several orders of magnitude smaller than those of visible light. From (56.17), we see that

$$\lambda_\infty/\text{nm} \approx \frac{1.2}{\hat{U}^{1/2}/\text{V}^{1/2}} \tag{56.20}$$

so that at 100 kV, $\lambda_\infty = 3.7$ pm and at 500 kV, $\lambda_\infty = 1.4$ pm. (Lists of values are to be found at the end of Grivet, 1965, 1972, Table 3).

56.4 Rigorous Approach

The foregoing reasoning is adequate for almost all practical purposes but its purpose is to justify using the relativistic form of the Schrödinger equation rather than the full Dirac equation. Apart from some preliminary studies by Rubinowicz (1934, 1957, 1963, 1965, 1966), Durand (1953) and Phan-Van-Loc (1953, 1954, 1955, 1958a,b, 1960), no attempts to work with the Dirac equation directly and hence to check the correctness of using the relativistic Schrödinger equation were made until the 1980s, when first Ferwerda et al. (1986a,b) and subsequently Jagannathan (Jagannathan et al., 1989; Jagannathan, 1990) reopened the question and investigated it with great thoroughness (Khan and Jagannathan, 1995, 2020, 2021; Jagannathan and Khan, 1996; Khan, 2006, 2008, 2017, 2018). Ferwerda et al. present a fully relativistic version of the theory of microscope image formation, which essentially vindicates the use of the Klein–Gordon equation; Jagannathan does not even use the latter but derives the focusing theory for electron lenses, and in particular for magnetic and electrostatic round lenses and quadrupole lenses, from the Dirac equation. Ramaswamy Jagannathan and Sameen Ahmed Khan have brought all their work together in a monograph (Jagannathan and Khan, 2019), which provides a very complete self-contained account. We now reproduce some essential steps in their reasoning, first summarizing the argument that leads from the full Dirac equation to the time-independent form and its consequences, after which the special case of the magnetic lens will be examined. The same authors also provide an equally full analysis of bending magnets, not considered here (Jagannathan and Khan, 2019; Khan and Jagannathan, 2020). They use Dirac's bra and ket notation, which we also employ here.

Propagation of electrons in a rotationally symmetric magnetic field with a distribution typical of that in a magnetic lens has been studied by Löffler et al. (2020). They point out the problems that arise when the familiar Fresnel propagator is adopted and explain why it is preferable to use the Laguerre–Gauss basis functions that will be studied in detail in connection with vortex beams (Chapter 80) and had already been evoked by Howie (1983). We cannot retrace their study here and simply reproduce two key findings. The transverse time-dependent Hamiltonian operator $\hat{H}(t)$ may be written

$$\hat{H}(t) = \frac{1}{2m}\left(-\frac{\hbar^2}{w^2}\nabla_\rho^2 + \frac{w^2e^2\{B_z(vt)\}^2}{4}\rho^2\right) + i\hbar\Omega_L(B_z(vt))\partial_\varphi$$

$$\Omega_L := \frac{eB_z}{2m}, \qquad w := \left(\frac{4\hbar}{eB_0}\right)^{1/2}$$

(cf. Eq. 15.16b of Volume 1) or, with $\rho = r/w$,

$$\hat{H}(t) = \hbar\Omega_L(B_0)\left(-\frac{1}{4}\nabla_\rho^2 + g^2(t)\hat{\rho}^2 + [ig(t)\partial_\varphi]\right)$$

$$g(t) := \frac{\Omega_L(B_z(vt))}{\Omega_L(B_0)} \approx \frac{B_z(vt)}{B_0}$$

The term in brackets disappears in the absence of topological charge (see Chapter 80 for an explanation of this quantity). The electron propagator, $\hat{P}(t) = \exp(i\hat{\Phi})$, in which $\hat{\Phi} = (1/\hbar)\int_0^t \hat{H}dt$, thus becomes

$$\hat{\Phi} = \frac{1}{\hbar}\int_0^t \hat{H}(\tau)d\tau = \Omega_L(B_0)t\left\{\hat{\rho}^2 G(t) - \frac{1}{4}\nabla_\rho^2\right\}$$

$$G(t) := \frac{1}{t}\int_0^t g^2(\tau)d\tau$$

For Glaser's bell-shaped field, Eq. (17.55) of Volume 1, Löffler et al. find

$$G(t) = \frac{1}{2}\frac{1}{1 + (\upsilon t/a)^2} + \frac{a}{2\upsilon t}\arctan\left(\frac{\upsilon t}{a}\right)$$

The need for the Dirac equation in electron scattering theory has also been noticed and studied in depth by several authors, notably Rother and Scheerschmidt (2009) and Majert and Kohl (2019). We consider image formation and scattering separately in Sections 56.4.1 and 56.4.2.

Note. We can give no more than the basic ideas of the studies summarized in the following sections. The notation of the original publications has therefore been largely retained and often differs from that employed in later chapters.

56.4.1 Image Formation

We set out from the free-space Dirac equation in the form

$$\begin{aligned} i\hbar\frac{\partial\Psi(t)}{\partial t} &= \hat{H}_D\Psi(t) \\ \hat{H}_D &= c\boldsymbol{\alpha}\cdot(\hat{\boldsymbol{p}} + e\boldsymbol{A}) + \alpha_4 m_0 c^2 \equiv c\boldsymbol{\alpha}\cdot(\hat{\boldsymbol{p}} + e\boldsymbol{A}) + \beta m_0 c^2 \end{aligned} \tag{56.21}$$

in which Ψ is the four-component spinor. Jagannathan and Khan (2019) use slightly different forms of α_i, $i = 1, \ldots, 4$ and we adopt their definition since the reader will need to use their text to complete our short version. They write (Pauli matrices)

$$\begin{gathered} \boldsymbol{\alpha} = \begin{pmatrix} 0 & \boldsymbol{\sigma} \\ \boldsymbol{\sigma} & 0 \end{pmatrix} \\ \beta = \begin{pmatrix} \mathbf{1} & \mathbf{0} \\ \mathbf{0} & -\mathbf{1} \end{pmatrix} \\ \sigma_x = \begin{pmatrix} 0 & 1 \\ 1 & 0 \end{pmatrix}, \quad \sigma_y = \begin{pmatrix} 0 & -i \\ i & 0 \end{pmatrix}, \quad \sigma_z = \begin{pmatrix} 1 & 0 \\ 0 & -1 \end{pmatrix} \\ \mathbf{0} = \begin{pmatrix} 0 & 0 \\ 0 & 0 \end{pmatrix} \quad \text{and} \quad \mathbf{1} = \begin{pmatrix} 1 & 0 \\ 0 & 1 \end{pmatrix} \end{gathered} \tag{56.22}$$

Time is now eliminated by substituting

$$\Psi(\boldsymbol{r}, z, t) = \exp(-iEt/\hbar)\psi(\boldsymbol{r}, z) \tag{56.23}$$

After some rearrangement, the Dirac equation becomes

$$i\hbar\frac{\partial\psi}{\partial z} = \hat{H}\psi(z)$$

$$\hat{H} = -p_0\beta\chi\alpha_z + eA_z\boldsymbol{J} - \left(\frac{e}{c}\right)\phi\alpha_z + \alpha_z\boldsymbol{\alpha}_\perp\cdot\hat{\boldsymbol{\pi}}_\perp + \text{AMM} \tag{56.24}$$

$$\chi := \begin{pmatrix} \xi\mathbf{I} & 0 \\ 0 & -\xi^{-1}\mathbf{1} \end{pmatrix}, \quad \xi := \left(\frac{E+mc^2}{E-mc^2}\right)^{1/2}, \quad \hat{\boldsymbol{\pi}} := \hat{\boldsymbol{p}} + e\boldsymbol{A}$$

$\boldsymbol{J}$ is the 4×4 identity matrix; $\boldsymbol{\alpha}_\perp$ is the two-component vector (α_x, α_y) and likewise for $\hat{\boldsymbol{\pi}}_\perp$. AMM represents terms arising from the anomalous magnetic moment of the electron, which we do not consider here. It proves convenient to introduce a new wavefunction, which Jagannathan and Khan denote ψ' but which we write $\overline{\psi}$ to avoid confusion with derivatives. This is defined in terms of the operator $\hat{M}$,

$$\hat{M} = \frac{1}{\sqrt{2}}(\boldsymbol{J} + \chi\alpha_z), \quad \hat{M}^{-1} = \frac{1}{\sqrt{2}}(\boldsymbol{J} - \chi\alpha_z) \tag{56.25}$$

and we note that

$$\hat{M}\beta\chi\alpha_z\hat{M}^{-1} = \beta \tag{56.26}$$

Then

$$|\overline{\psi}(z)\rangle := \hat{M}|\psi(z)\rangle \tag{56.27}$$

and satisfies

$$i\hbar\frac{\partial\overline{\psi}(z)}{\partial z} = \left(-p_0\beta + \hat{\mathscr{E}} + \hat{\mathscr{O}}\right)\overline{\psi} \tag{56.28}$$

We do not reproduce the new terms in this equation here as we shall reconsider them in the section on the magnetic lens. They can be usefully simplified by making one or more Foldy–Wouthuysen transforms, the first of which is defined by

$$|\overline{\psi}^{(1)}\rangle = \exp\left(-\frac{1}{2p_0}\boldsymbol{\beta}\hat{\mathscr{O}}\right) \tag{56.29}$$

The axially symmetric magnetic lens. In this case the terms $\hat{E}$ and $\hat{O}$ in Eq. (56.28) collapse to

$$\hat{\mathscr{E}} = eA_z\boldsymbol{J}, \quad \hat{\mathscr{O}} = \chi\boldsymbol{\alpha}_\perp\cdot(\hat{\boldsymbol{p}}_\perp + e\boldsymbol{A}_\perp) \tag{56.30}$$

As before, a series of Foldy–Wouthuysen transforms leads to manageable differential equations, from which an explicit expression for the Dirac Hamiltonian up to third-order terms is established. This consists of four terms:

$$\hat{H}_0 = \hat{H}_{0,p} + \hat{H}'_0 + \hat{H}_0^{(h)} + \hat{H}_0^{(h)}$$

$$\hat{H}_{0,p} = \left\{ \frac{1}{2p_0}\hat{p}_\perp^2 + \frac{1}{2}p_0\alpha^2(z)r_\perp^2 - \alpha(z)\hat{L}_z \right\} \boldsymbol{J} \quad \text{(paraxial term)}$$

$$\begin{aligned} \hat{H}'_0 = \Bigg\{ & \frac{1}{8p_0^2}\hat{p}_\perp^4 - \frac{\alpha(z)}{2p_0^2}\hat{p}_\perp^2\hat{L}_z - \frac{\alpha^2(z)}{8p_0}(r_\perp \cdot \hat{p}_\perp + \hat{p}_\perp \cdot r_\perp)^2 + \frac{3\alpha^2(z)}{8p_0}(r_\perp^2\hat{p}_\perp^2 + \hat{p}_\perp^2 r_\perp^2) \\ & + \frac{1}{8}(\alpha''(z) - 4\alpha^3(z))\hat{L}_z r_\perp^2 + \frac{p_0}{8}(\alpha^4(z) - \alpha(z)\alpha''(z))r_\perp^4 \Bigg\} \boldsymbol{J} \quad \text{(third-order geometrical aberrations)} \end{aligned} \tag{56.31}$$

$$\hat{H}_0^{(h)} = \frac{\hbar^2}{32p_0}\left\{ 4(\alpha'^2(z) - 2\alpha(z)\alpha''(z))r_\perp^2 + (\alpha''^2(z) - \alpha'(z)\alpha'''(z))r_\perp^4 \right\} \boldsymbol{J} \quad \text{(spin-related term)}$$

$$\alpha(z) := -\frac{eB(z)}{2p_0} \tag{56.32}$$

The first enables us to calculate the paraxial properties. The second term contains the information needed to calculate the third-order geometric aberrations. The third term is a small perturbation proportional to $\hbar^2$. All three terms are essentially scalar, in the sense that they act separately on the components of the spinor wavefunction and do not introduce any mixing between them. Nevertheless, if the third term is retained, the resulting Klein–Gordon equation will not be exactly the same as the usual equation. When this term and the fourth term can be neglected, the Klein–Gordon equation and hence the 'relativistically corrected' equation can be used in electron optics: the effect of spin is then negligible. We now turn to the fourth term (not reproduced in full here). Unlike the first three, this does cause mixing of the components of the spinor wave-vector and all the aberration coefficients will be affected, differently for each component.

Although we have of necessity omitted most of the steps in the mathematical analysis, it should be clear that a thorough study of the paraxial properties and third-order geometrical aberrations of the more important electron optical systems is now available and ready for use in circumstances in which spin must be taken into account.

56.4.2 Scattering Theory

Elastic scattering. Rother and Scheerschmidt (2009) have pointed out that relativistic effects cannot always be neglected in scattering theory. They therefore present two fully relativistic theories of the elastic scattering process. In one case, the Dirac equation is integrated numerically subject to suitable boundary conditions. In the other, the multislice approach (Section 69.3.8) is employed.

As expected, the starting point is the Dirac equation in the form

$$E\psi(\boldsymbol{r}) = \left\{c\boldsymbol{\alpha}\cdot\hat{\boldsymbol{p}} + mc^2\alpha_4 + V(r)\right\}\psi(\boldsymbol{r}) \tag{56.33a}$$

or

$$\left\{(E-V)^2 - c^2\hat{\boldsymbol{p}}^2 - m^2c^4 - \mathrm{i}e\hbar c\boldsymbol{E}\cdot\boldsymbol{\alpha} - e\hbar\boldsymbol{B}\cdot\boldsymbol{\sigma}\right\}\psi = 0 \tag{56.33b}$$

The differential scattering cross-section is given by the Mott formula

$$\frac{d\sigma}{d\Omega}(\theta) = \frac{Z^2\alpha_0^2\{1-\beta^2\sin^2(\theta/2)\}}{4\beta^2|\hat{\boldsymbol{p}}|^2\sin^4(\theta/2)} \tag{56.34}$$

As usual, $\beta = v/c$ and α_0 is the Sommerfeld fine-structure constant $(\approx 1/137)$; $\hat{\boldsymbol{p}}$ is the momentum operator. When $\beta \ll 1$ or θ is small, Eq. (56.34) collapses to the Rutherford cross-section

$$\frac{d\sigma}{d\Omega}(\theta) = \frac{Z^2\alpha_0^2}{4\beta^2|\boldsymbol{p}|^2\sin^4(\theta/2)} \tag{56.35}$$

Rother and Scheerschmidt examine the relative magnitudes of the terms in the Dirac equation for ferromagnetic specimens and for spin-polarized beams.

For a plane beam incident on a crystal specimen and with the forward scattering assumption, the Dirac equation can be reduced to four coupled first-order ordinary differential equations as we now show. We substitute

$$\begin{aligned} V(\boldsymbol{r}) &= \textstyle\sum_{\boldsymbol{q}} V(\boldsymbol{q},z)\exp(\mathrm{i}\boldsymbol{q}\cdot\boldsymbol{d}) \\ \boldsymbol{A}(\boldsymbol{r}) &= \textstyle\sum_{\boldsymbol{q}} \boldsymbol{A}(\boldsymbol{q},z)\exp(\mathrm{i}\boldsymbol{q}\cdot\boldsymbol{d}) \\ \psi(\boldsymbol{r}) &= \textstyle\sum_{\boldsymbol{q}} \psi(\boldsymbol{q}-\boldsymbol{k}_0,z)\exp\{\mathrm{i}(\boldsymbol{q}-\boldsymbol{k}_0)\cdot\boldsymbol{d}\} \end{aligned} \tag{56.36}$$

in the Dirac equation; $\boldsymbol{d}$ is a vector in the detector plane, $\boldsymbol{q}$ is as usual the reciprocal coordinate and $\boldsymbol{k}_0$ is the incident beam direction (set equal to zero below). After some manipulation, we find

$$\frac{\partial\psi(\boldsymbol{q},z)}{\partial z} = -\mathrm{i}\alpha_z\left[\boldsymbol{\alpha}_d\cdot\left\{\boldsymbol{q} - \frac{1}{\hbar}\boldsymbol{A}_d(\boldsymbol{q},z)\otimes\right\} - \frac{1}{\hbar}\alpha_z\cdot A_z(\boldsymbol{q},z)\otimes + \frac{mc\alpha_4}{\hbar} - \frac{E}{c\hbar} + \frac{1}{c\hbar}V(\boldsymbol{q},z)\otimes\right]\psi(r) \tag{56.37}$$

In the same way, the Klein–Gordon equation can be reduced to a set of four ordinary, coupled, second-order differential equations:

$$\begin{aligned} &\left\{-c^2\hbar^2\frac{\partial^2}{\partial z^2} - 2\mathrm{i}c^2\hbar A_z(\boldsymbol{q},z)\otimes\frac{\partial}{\partial z}\right\}\psi(\boldsymbol{q},z) = K\psi(\boldsymbol{q},z) \\ &K := E^2 - 2EV(\boldsymbol{q},z)\otimes + V^2(\boldsymbol{q},z)\otimes - c^2A_z^2(\boldsymbol{q},z)\otimes - c^2\hbar^2\left(\boldsymbol{q}-\tfrac{1}{\hbar}\boldsymbol{A}_d(\boldsymbol{q},z)\right)^2\otimes - m^2c^4 \end{aligned} \tag{56.38}$$

It can be shown straightforwardly that the relativistically corrected Schrödinger equation is obtained from the Klein–Gordon equation when $V \ll 2E$. This equation, in the form

$$\{(E-V)^2 - c^2\hat{\boldsymbol{p}}^2 - m^2c^4\}\psi = 0 \tag{56.39}$$

reduces to

$$\{E^2 - m^2c^4\}\psi = (2EV + c^2\hat{\boldsymbol{p}}^2)\psi \tag{56.40a}$$

or

$$\left(\frac{E^2 - m^2c^4}{2\gamma mc^2}\right)\psi = \left(\frac{\hat{\boldsymbol{p}}^2}{2\gamma m} + V\right)\psi = \left(-\frac{h^2\nabla^2}{2\gamma m} + V\right)\psi \tag{56.40b}$$

as expected.

For the accelerating voltages of electron microscopes, it seems reasonable to assume that $\partial^2\overline{\psi}/\partial z^2$ and $A_z\partial\overline{\psi}/\partial z$ can be neglected, where we have written

$$\psi(\boldsymbol{r}) = \overline{\psi}(r)\psi_Q\exp(\mathrm{i}k_0z) \tag{56.41}$$

$\psi_Q\exp(\mathrm{i}k_0z)$ is the plane incident wave. A formal solution of the differential equation for $\overline{\psi}(\boldsymbol{r})$ is

$$\begin{aligned}
\overline{\psi}(\boldsymbol{d}, z+\Delta z) &= \mathscr{F}^{-}\left\{\exp\overline{\psi}^{(1)}\mathscr{F}\left(\exp\overline{\psi}^{(2)}\exp\overline{\psi}^{(3)}\overline{\psi}(\boldsymbol{d}, z)\right)\right\}\\
\overline{\psi}^{(1)} &= \frac{\mathrm{i}\Delta z\hbar^2\boldsymbol{q}^2 + 2\hbar\boldsymbol{A}_d\cdot\boldsymbol{q}}{2\hbar^2k_0}\\
\overline{\psi}^{(2)} &= \frac{\mathrm{i}\int\left(EV - V^2 - c^2\boldsymbol{A}^2 - c^2\hbar k_0A_z\right)dz'}{2c^2\hbar^2k_0}\\
\overline{\psi}^{(3)} &= \frac{\mathrm{i}\int\left(e\hbar c\boldsymbol{\alpha}\cdot\boldsymbol{E} + e\hbar c^2\boldsymbol{\sigma}\boldsymbol{B}\right)dz'}{2c^2\hbar^2k_0}
\end{aligned} \tag{56.42}$$

This is the new form of the multislice algorithm. The term $\exp\overline{\psi}^{(1)}$ is known as the *propagator function* and is the same as in the familiar multislice algorithm. The second exponential term is the *transmission function*. The last term is new and arises because the wavefunction is now a spinor; its role is to couple the components of the spinor when electrostatic and magnetic fields are present. A slightly simpler form is obtained when the fields are sufficiently small.

Rother and Scheerschmidt insist that it has been possible to derive these formulae on the assumption that only forward scattering occurs, that the small-angle approximation is valid and that the vector potential $\boldsymbol{A}$ is periodic. In the remainder of their study, they compare the results obtained with the various expressions and conclude that 'The validity of the standard formalism currently used in electron scattering simulations could be verified in case of practically all electromagnetic potentials present in the microscope. Even in the case of strong scatterers like gold atoms or ferromagnetic materials like α-Fe, the influence of the spin and the V^2 term stays small. The error introduced by using the multislice formalism instead of a numerical forward integration is more serious, in particular when regarding higher order reflections. If only small angle scattering is regarded [as] in HRTEM, the multislice formalism yields equivalent results to the forward integration. If, however, large-angle scattered beams are measured, [as] in diffraction analysis under large-angle conditions (HAADF), the multislice formalism might yield inaccurate results in particular when scattering on heavy atoms. Depending on the zone axis orientation and the sample thickness the error can reach several 10% .... a realistic image simulation based on numerical forward integration would require time-consuming frozen lattice summations (Loane et al., 1991), preventing a standard application of the formalism'.

Relativistic effects in energy-filtering transmission electron microscopy. An important study by Majert and Kohl (2019) elucidates the differences between relativistic and nonrelativistic calculations of cross-sections for inner-shell ionization. The fact that neglect of relativistic effects can result in misleading results was first pointed out by Knippelmayer (1996) and Knippelmayer et al. (1997). In these papers, however, multiple elastic scattering was not considered and the dipole approximation was used to calculate the cross-sections. Pokroppa (1999) and Frigge (2011) avoided using the dipole approximation but did not go on to simulate elemental maps. Dwyer (2005a,b) and Dwyer and Barnard (2006) found that dynamical relativistic effects can have some effect on simulations that include inner-shell ionization but they did not consider spin. The full study by Majert and Kohl goes far beyond all these earlier studies. Multiple elastic scattering events are correctly included and the dipole approximation is not used. We have seen that Rother and Scheerschmidt (2009) developed a relativistic multislice algorithm on the basis of the Dirac equation. Majert and Kohl establish a multislice procedure that "considers the generation of magnetic fields, the retardation of the interaction between electron and specimen and the electron spin". For inelastic scattering, a transition formalism is employed. They compare elemental maps simulated with their algorithm and the regular multislice method and can thus "gauge the impact of (nonconventional) relativistic effects".

We cannot reproduce their work here – their account fills more than 100 pages. They conclude that "the relativistic effects not included in the conventional approach to image simulation can have a significant impact on the intensity distribution in an atomic-resolution EFTEM elemental map. In particular, this statement is valid for very thin specimens as well as thick specimens where the consideration of multiple elastic scattering events becomes important". Many other aspects of the new algorithm are recapitulated in their 'Summary and outlook', which interested readers are encouraged to consult.

CHAPTER 57

The Eikonal Approximation

Exact solutions of the Schrödinger equation or the relativistic wave equation are known only in very few exceptional cases, which are far from fitting the experimental conditions. A systematic method of finding an approximate solution in realistic conditions is therefore required.

One such procedure is the eikonal method, closely associated in general quantum mechanics with the Wentzel–Kramers–Brillouin method. Its main purpose is to show that classical mechanics is an approximation to the more correct quantum mechanical formulation. Moreover, it also offers a way of establishing approximate wavefunctions which may then be used in the theory of electron diffraction.

57.1 The Product Separation

The solution of (56.10) is a complex-valued function $\Psi(\boldsymbol{r}, t)$. Any such function can certainly be represented in terms of its modulus or 'amplitude' $a = |\Psi|$ and its phase. Two sign conventions for wave propagation are in use in the literature of optics and electron optics, corresponding to (57.1) (quantum mechanical convention) and its complex conjugate. We are in agreement with Glaser (1952, 1956) and Born and Wolf (1959, 2002) but not with (most of) Spence (2013). For a detailed examination of these conventions, see Saxton et al. (1983), whose table of sign conventions is reproduced with corrections by Spence (2013 p. 126).

We make use of (55.4) with (55.5) and come thus to the product separation

$$\Psi(\boldsymbol{r}, t) = a(\boldsymbol{r})\exp\left[\frac{\mathrm{i}}{\hbar}\left\{S(\boldsymbol{r}) - Et\right\}\right] \tag{57.1}$$

with real functions $a(\boldsymbol{r})$ and $S(\boldsymbol{r})$. After introducing this into Eq. (56.10), carrying out the necessary differentiations and cancelling a common exponential factor, we obtain

$$\begin{aligned}&-\hbar^2\nabla^2 a + a(\nabla S)^2 - \mathrm{i}\hbar\left(a\nabla^2 S + 2\nabla a\cdot\nabla S\right) - 2\mathrm{i}e\boldsymbol{A}\cdot(\hbar\nabla a + \mathrm{i}a\nabla S)\\ &+ a\left(e^2A^2 - \mathrm{i}e\hbar\nabla\cdot\boldsymbol{A} - 2m_0e\hat{\Phi} - 2mE\right) = 0\end{aligned}$$

Principles of Electron Optics.
DOI: https://doi.org/10.1016/B978-0-12-818979-5.00057-7

This complex equation can be satisfied only if its real and imaginary parts vanish separately. This gives

real part:

$$a\left\{(\nabla S)^2 + 2e\boldsymbol{A}\cdot\nabla S + e^2A^2 - 2m_0e\hat{\Psi} - 2mE\right\} - \hbar^2\nabla^2 a = 0 \tag{57.2}$$

imaginary part:

$$\hbar\left\{a\nabla^2 S + 2\nabla a\cdot\nabla S + 2e\boldsymbol{A}\cdot\nabla a + ea\nabla\cdot\boldsymbol{A}\right\} = 0 \tag{57.3}$$

The imaginary part can be cast into a more convenient form by multiplication by $a/\hbar$. Recalling that $\nabla(a^2) = 2a\nabla a$ and using the product differentiation rules of vector analysis, we may write (57.3) as a *divergence equation*:

$$\operatorname{div}\left\{a^2(\operatorname{grad} S + e\boldsymbol{A})\right\} = 0 \tag{57.4}$$

The real part (57.2) can be cast into the more concise form

$$(\operatorname{grad} S + e\boldsymbol{A})^2 = 2m_0e\hat{\Phi} + 2mE + \hbar^2\frac{\nabla^2 a}{a}$$

The first two terms on the right-hand side are just the square of the classical *kinetic* momentum, $g^2(\boldsymbol{r})$ and the real part therefore becomes

$$(\operatorname{grad} S + e\boldsymbol{A})^2 = g^2 + \hbar^2\frac{\nabla^2 a}{a} \tag{57.5}$$

These two differential equations (57.4) and (57.5) hold *exactly*, but they are not simpler to solve than (56.10), as they are coupled and not even linear. Some simplifications are therefore necessary to come to a practical solution.

57.2 The Essential Approximation

The coupling between (57.4) and (57.5) is essentially due to the last term in (57.5). This is the only term containing $\hbar$. An important simplification will be achieved if this term can be neglected. This requires that $|a^{-1}\nabla^2 a| \ll g^2/\hbar^2 = k^2$ (56.16). Since $4\pi^2 \approx 40 \gg 1$, a sufficient limit is

$$\left|\frac{\nabla^2 a}{a}\right| < \left(\frac{k}{2\pi}\right)^2 = \frac{1}{\lambda^2} \tag{57.6}$$

This is the eikonal approximation in its proper sense. When (57.6) is satisfied, (57.5) becomes independent of the amplitude function $a(\boldsymbol{r})$ and takes the form

$$\left\{\operatorname{grad} S(\boldsymbol{r}) + e\boldsymbol{A}(\boldsymbol{r})\right\}^2 = g^2(\boldsymbol{r}) \tag{57.7}$$

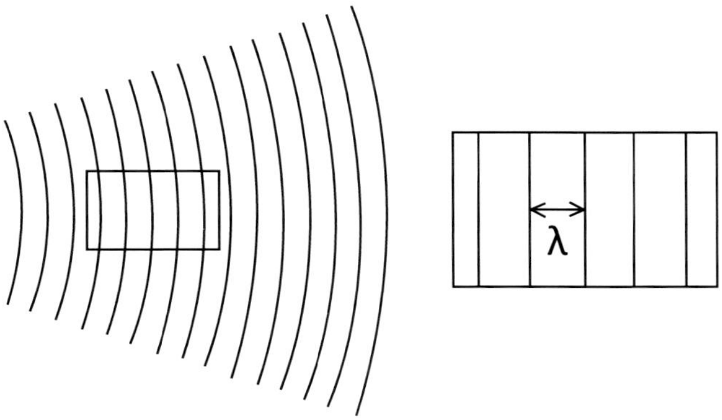

Figure 57.1
Illustration of the eikonal approximation. Macroscopically, the wavefronts are curved and the amplitude varies in space. The radius of curvature, however, is much larger than the local wavelength. In every sufficiently small domain, shown in the enlarged rectangle, the wavefronts are nearly planar and the amplitude is practically constant.

This is the reduced *Hamilton–Jacobi equation* derived in Chapter 5. Eq. (57.7) is identical with Eqs (5.10) and (5.11) of Volume 1 for $Q = -e$ and the nominal value $E = 0$, if S is identified[1] with $\overline{S}$. The solution of the partial differential Eq. (57.7) gives the *classical electron trajectories* in a time-independent representation. The surfaces $S = \text{const}$, considered in Hamiltonian optics, are now seen to be *surfaces of constant wave phase* and thus take on a wave-optical meaning. We have thus justified some fundamentals of geometrical electron optics in the sense that they emerge from the more rigorous wave mechanics.

In the general case, it is difficult to specify concrete limits for the validity of (57.6). It is trivially obvious that (57.6) is satisfied for every plane or spherical wave. Moreover, (57.6) is justified if the amplitude $a(\boldsymbol{r})$ varies very little in domains that are a few wavelengths in geometrical extent. The true wave can then always be approximated locally by a plane wave

$$\psi(\boldsymbol{r}) = a_0 \exp\{\mathrm{i}\boldsymbol{k}_0 \cdot (\boldsymbol{r} - \boldsymbol{r}_0)\}$$

the constants to be chosen appropriately. This approximation is sketched in Fig. 57.1.

The eikonal approximation breaks down in the following cases:

1. In the vicinity of interfaces: here the incident wave splits abruptly into a reflected and a refracted wave.
2. In the vicinity of sharp edges: these cause diffraction.

[1] In Chapter 5 of Volume 1, we used the notation $\overline{S}$ so as to be able to write $S := \overline{S}/(2m_0 e)^{1/2}$ in the remainder of the book. This is not convenient here and S will henceforward denote the function defined in Eqs. (4.32) and (4.34) of Volume 1.

3. In the vicinity of foci of every kind: here the amplitude becomes very large in a very small domain.
4. In all atomic fields: these vary considerably over distances of the order of an ångström (100 pm) and cause scattering. (We recall that 100 pm is about 30 wavelengths at 100 kV.)

These situations are sketched in Fig. 57.2A–D. In view of all these exceptional cases, it is obvious that the eikonal approximation is valid only for motion in *macroscopic* fields, far from their singularities. Even then it may break down, since a focus with singular amplitude can be produced even in field-free space.

All this does not mean that the eikonal approximation is useless. Reflection, refraction and diffraction can be treated, not in terms of a *single* amplitude–phase solution, but by *superposition* of several solutions. The eikonal method then supplies the necessary tools for calculating the elementary waves.

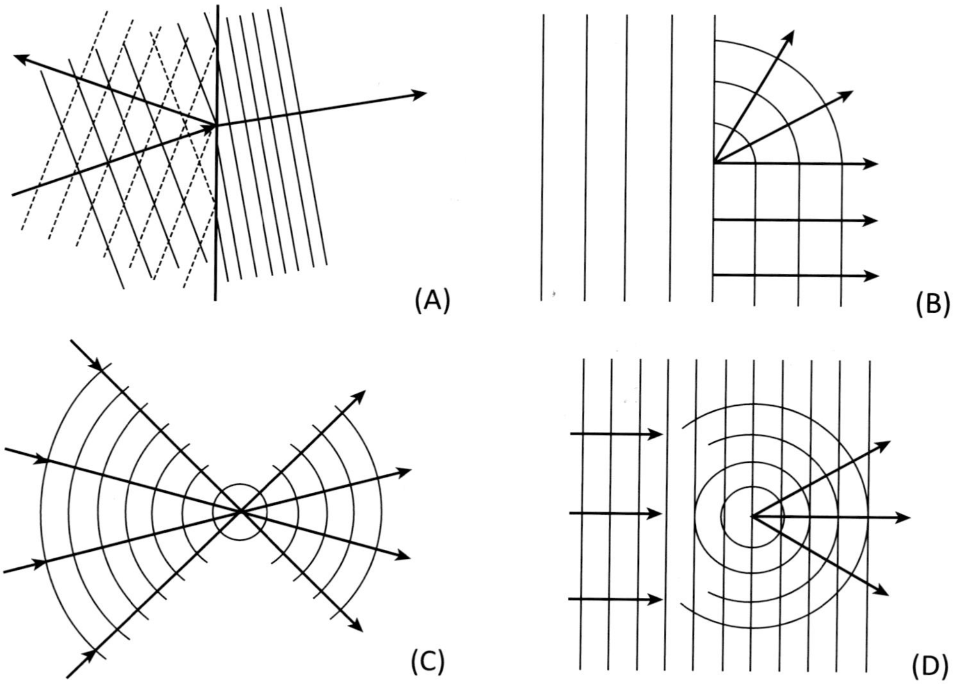

Figure 57.2
Cases in which the eikonal approximation breaks down: (A) reflection and refraction at surfaces, (B) diffraction at sharp edges, (C) high intensity in foci and (D) scattering in atomic fields.

57.3 The Variational Principle

In Part I, we regarded Hamilton's variational principle as fundamental and derived from it all the classical equations of motion including the Hamilton–Jacobi equation; now, however, it is only an approximation. Moreover, the Hamilton principle is *not* a direct consequence of the wave equation. In this situation, the next natural step is to establish a variational principle from the Hamilton–Jacobi equation, which will prove to be the electron optical form of the Euler–Maupertuis principle of least action.

In the following derivations we assume that a finite domain can be found in which the solution $S(\boldsymbol{r})$ of (57.7) is unique. Eq. (57.7) prescribes only the *length* of the vector (grad $S + e\boldsymbol{A}$). We can convert (57.7) to a vector equation by writing

$$\text{grad}\, S(\boldsymbol{r}) + e\boldsymbol{A}(\boldsymbol{r}) = \boldsymbol{g}(\boldsymbol{r}) = :g(\boldsymbol{r})\boldsymbol{t}(\boldsymbol{r}) \tag{57.8}$$

with a unit vector $\boldsymbol{t}(\boldsymbol{r})$. This is initially not known but must exist and can be determined *after* the solution of (57.7) has been found. Later this vector $\boldsymbol{t}$ will be identified with the tangent vector of the classical electron trajectory. In any case, there must exist lines (A) along which the local tangent $\boldsymbol{t}_A(s)$ always coincides with the vector $\boldsymbol{t}$ of (57.8). We now study the situation sketched in Fig. 57.3. Curve (A) has the above property and runs from a fixed starting point P_1 to a fixed terminal point P_2. Line (B) is an arbitrary smooth curve joining the same points.

We now evaluate the eikonal function $S(P_1, P_2)$ between these points and satisfying $S(P_1, P_1) = 0$. The assumed uniqueness means that

$$S(P_1, P_2) = \int_{P_1}^{P_2(A)} \text{grad}\, S \cdot d\boldsymbol{r} = \int_{P_1}^{P_2(B)} \text{grad}\, S \cdot d\boldsymbol{r}$$

Substituting for grad S from (57.8) and using $d\boldsymbol{r} = \boldsymbol{t}_A\, ds$ on (A) and $d\boldsymbol{r} = \boldsymbol{t}_B\, ds$ on (B), we obtain

$$S = \int_{P_1}^{P_2(A)} (g\boldsymbol{t}\cdot\boldsymbol{t}_A ds - e\boldsymbol{A}\cdot d\boldsymbol{r}) = \int_{P_1}^{P_2(B)} (g\boldsymbol{t}\cdot\boldsymbol{t}_B ds - e\boldsymbol{A}\cdot d\boldsymbol{r})$$

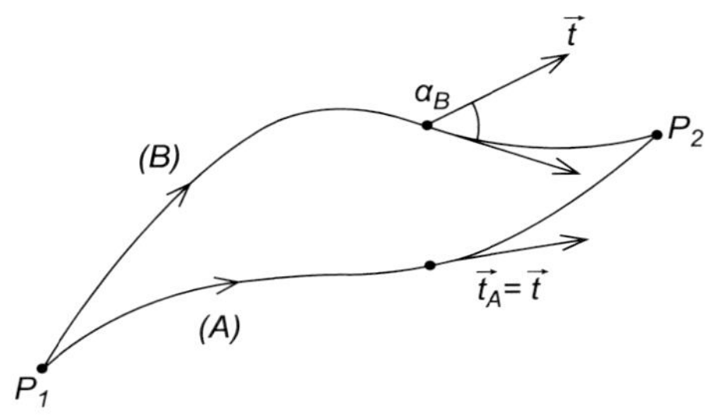

Figure 57.3
Two paths of integration (A) and (B) joining the points P_1 and P_2. Line (A) is an extremal curve.

From the initial assumption, we have $\boldsymbol{t}=\boldsymbol{t}_A$, $\boldsymbol{t}\cdot\boldsymbol{t}_A=1$ on line (A), whereas $\boldsymbol{t}\cdot\boldsymbol{t}_B=\cos\alpha_B\leq 1$ on line (B). We may therefore conclude that

$$S=\int_{P_1}^{P_2(A)}(g\,ds-e\boldsymbol{A}\cdot d\boldsymbol{r})=\int_{P_1}^{P_2(B)}(g\cos\alpha_B ds-e\boldsymbol{A}\cdot d\boldsymbol{r})$$
$$\leq\int_{P_1}^{P_2(B)}(g\,ds-e\boldsymbol{A}\cdot d\boldsymbol{r})$$

Since this is true for any test curve (B) joining the same endpoints P_1 and P_2, we find:

$$S(P_1,P_2)=S=\int_{P_1}^{P_2(A)}(g\,ds-e\boldsymbol{A}\cdot d\boldsymbol{r})=\text{extremum} \qquad (57.9)$$

which is a form of the Euler–Maupertuis principle of least action. The curve (A) along which the vector $\boldsymbol{t}(s)$ satisfies (57.8) is an extremal curve. Introducing the vector $\boldsymbol{g}=g\boldsymbol{t}$ from (57.8) and noting that $\boldsymbol{p}=\boldsymbol{g}-e\boldsymbol{A}$ is the canonical momentum, we find

$$S(P_1,P_2)=S=\int_{P_1}^{P_2}\boldsymbol{p}\cdot d\boldsymbol{r}=\text{extr.} \qquad (57.10)$$

in agreement with (4.33). Eq. (57.10) is an expression of Fermat's principle.

It is easy to derive the time-independent ray equation directly from (57.8). We rewrite this as

$$\text{grad}\,S=g\boldsymbol{t}-e\boldsymbol{A}$$

and form the curl of both sides, giving

$$\text{curl}(g\boldsymbol{t})-e\,\text{curl}\,\boldsymbol{A}=\text{curl}(g\boldsymbol{t})-e\boldsymbol{B}=0$$

$\boldsymbol{B}$ being the magnetic field strength. Forming the vector product with $\boldsymbol{t}$, we find

$$\boldsymbol{t}\times\text{curl}(g\boldsymbol{t})+e\boldsymbol{B}\times\boldsymbol{t}=0$$

Expansion of the curl of a product gives

$$\text{curl}(g\boldsymbol{t})=-\boldsymbol{t}\times\text{grad}\,g+g\,\text{curl}\,\boldsymbol{t}$$

and so

$$\boldsymbol{t}\times\text{curl}(g\boldsymbol{t})=\text{grad}\,g-(\boldsymbol{t}\cdot\text{grad}\,g)\boldsymbol{t}+g\boldsymbol{t}\times\text{curl}\,\boldsymbol{t}$$

The rules of vector analysis tell us that

$$\boldsymbol{t}\times\text{curl}\,\boldsymbol{t}=\frac{1}{2}\text{grad}(\boldsymbol{t}\cdot\boldsymbol{t})-\boldsymbol{t}\cdot\nabla\boldsymbol{t}=-\boldsymbol{t}\cdot\nabla\boldsymbol{t}$$

since $t^2 = 1$. Putting all this together, we obtain

$$\text{grad } g - (\boldsymbol{t} \cdot \text{grad } g)\boldsymbol{t} - g\boldsymbol{t} \cdot \nabla\boldsymbol{t} = \text{grad } g - \boldsymbol{t} \cdot \nabla(g\boldsymbol{t}) = -e\boldsymbol{B} \times \boldsymbol{t}$$

Next we observe that $\boldsymbol{t} = d\boldsymbol{r}/ds$ for the tangent to the curve and that $\boldsymbol{t} \cdot \nabla = d/ds$ is the operator of differentiation with respect to the arc length s. Thus we finally arrive at

$$\frac{d}{ds}\left(g(\boldsymbol{r})\frac{d\boldsymbol{r}}{ds}\right) = \text{grad } g(\boldsymbol{r}) + e\boldsymbol{B}(\boldsymbol{r}) \times \frac{d\boldsymbol{r}}{ds} \tag{57.11}$$

which agrees with (3.1) for $Q = -e$.

The derivation of the fundamental laws of geometrical electron optics as an *approximate* consequence of wave mechanics is now complete. All this is contingent on the approximation (57.6), which allowed us to replace (57.5) by (57.7).

57.4 The Calculation of Eikonal Functions

The eikonal functions play an important role as elementary waves in the theory of diffraction, as we have already mentioned. We now enquire how they can be determined in practice.

In general, this is a very difficult task. In order to determine the point eikonal $S(P_1, P_2)$ between two points P_1 and P_2, we have first to establish the classical ray joining these two points, which requires numerical solution of (57.11). Since we are given two endpoints rather than a starting point and an initial direction, this is mathematically a two-point *boundary-value problem* for trajectories. As this may have more than one solution, the point eikonal $S(P_1, P_2)$ can have several values.

Once the trajectory (or the set of trajectories) had been determined, the eikonal itself would be obtained by numerical evaluation of the integral (57.9). Since electron wavelengths are so short, this would require very high precision of about 14–15 significant digits. With modern computers this is feasible – undeniably very hard but not unrealistic. Since the entire procedure cannot be carried out for *all* point-pairs (P_1, P_2) in practice, it would have to be performed for a suitable pattern of discrete points. The values for intermediate points would then have to be determined by means of suitable interpolation techniques.

It is not always necessary to start from a fixed point P_1. We may equally well prescribe a surface $S(\boldsymbol{r}) = S_0 = \text{const}$ and start the ray trace from this. For simplicity we shall assume here that $\boldsymbol{A}(\boldsymbol{r}) \equiv 0$, at least in the domain of interest. The rays are then orthogonal to surfaces of constant S, as shown in Fig. 57.4. Such a procedure is advantageous whenever the propagation of wavefronts is to be studied only over a short distance. Certain

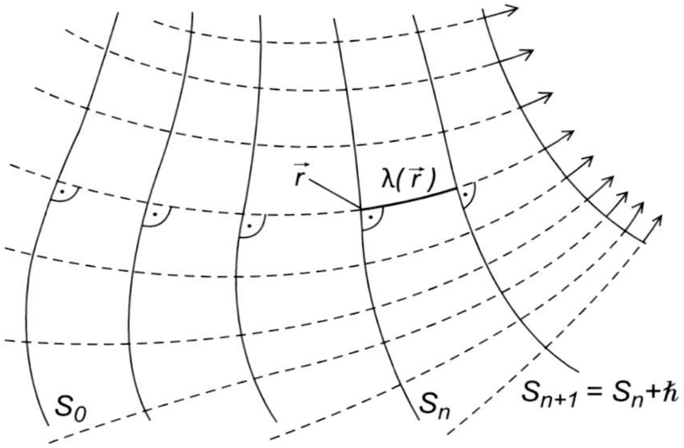

Figure 57.4
Propagation of wavefronts starting from a surface $S(r) = S_0 =$ const. The local distance between neighbouring surfaces is approximately the wavelength λ ($\boldsymbol{r}$), if the eikonal difference is Planck's constant $\hbar$. The broken lines indicate the classical orthogonal trajectories. The sketched orthogonality holds only for $\boldsymbol{A}(\boldsymbol{r})\equiv 0$.

simplifications that considerably facilitate the calculation are then sometimes permissible. This will become more clear when we deal with applications in the physics of scattering.

A third way of determining eikonal functions involves series expansions. These will be dealt with in the context of electron diffraction and aberration theory.

We conclude this section with a brief remark on the gauge transformation. The eikonal S is *not* gauge invariant. If we introduce (56.15b) into (57.1), we find

$$S(\boldsymbol{r}) = S'(\boldsymbol{r}) - eF(\boldsymbol{r}) + \hbar\alpha \tag{57.12}$$

which means that not even eikonal *differences* between two points, from which α cancels out, are gauge invariant! However, the *kinetic* momentum *is* gauge invariant thanks to (56.15a):

$$\boldsymbol{g}(\boldsymbol{r}) = \text{grad}\, S(\boldsymbol{r}) + e\boldsymbol{A}(\boldsymbol{r}) = \text{grad}\, S'(\boldsymbol{r}) + e\boldsymbol{A}'(\boldsymbol{r}) \tag{57.13}$$

as it must be.

57.5 The Calculation of Wave Amplitudes

We shall now discuss the interpretation and solution of the divergence relation (57.4). From (57.8) we can write this in the form

$$\text{div}\left\{a^2(\boldsymbol{r})\boldsymbol{g}(\boldsymbol{r})\right\} = 0 \tag{57.14}$$

and (57.13) shows that this is a gauge-invariant relation. The physical meaning of (57.14) is that particle current is conserved. To see this, we differentiate (57.1) and multiply by Ψ^*, giving

$$\Psi^* \nabla \Psi = a\left(\nabla a + \frac{\mathrm{i}a}{\hbar}\nabla S\right)$$

Introducing this and its complex conjugate into the second form of (57.14), we see that ∇a cancels out and find

$$\boldsymbol{j}(\boldsymbol{r}) = \frac{1}{m_0}a^2(\nabla S + e\boldsymbol{A}) = \frac{a^2}{m_0}\boldsymbol{g} \tag{57.15}$$

Hence (57.14) can be rewritten as $\operatorname{div}\boldsymbol{j} = 0$ and is evidently the stationary case of the continuity equation (55.13).

Introducing (57.1) into (56.13), we find

$$\rho(\boldsymbol{r}) = \frac{m(\boldsymbol{r})}{m_0}a^2(\boldsymbol{r}) \tag{57.16}$$

Since $\boldsymbol{v}(\boldsymbol{r}) = \boldsymbol{g}(\boldsymbol{r})/m(\boldsymbol{r})$ is the classical particle velocity, we can cast (57.15) into the familiar classical form

$$\boldsymbol{j}(\boldsymbol{r}) = \rho(\boldsymbol{r})\upsilon(\boldsymbol{r}) \tag{57.17}$$

We come now to the solution of (57.14). This can be obtained by formally integrating and then using Gauss's integral theorem

$$\oint a^2(\boldsymbol{r})\boldsymbol{g}(\boldsymbol{r})d\boldsymbol{\sigma} = 0 \tag{57.18}$$

in which $d\sigma$ denotes the surface-element vector. We choose now a section cut from a current tube, as shown in Fig. 57.5A. The mantle surfaces are formed by rays and do not contribute to the integral (57.18), since $\boldsymbol{g} \perp d\boldsymbol{\sigma}$ on them. The two cross-sections $\Delta\sigma_1$ and $\Delta\sigma_2$ are orthogonal to the mantle and so small that the midpoint rule of integration may be applied. We find then that the quantity

$$a_1^2 g_1 \Delta\sigma_1 = a_2^2 g_2 \Delta\sigma_2 = a^2(s)g(s)\Delta\sigma(s) \tag{57.19}$$

is constant along the tube, the latter expression giving this quantity at an arbitrary cross-section specified in position by the arc length s along the tube.

The amplitude becomes singular if the rays start from a common point P_1 as shown in Fig. 57.5B. We then choose two concentric spheres of radii R_1 and R_2. If these are so small

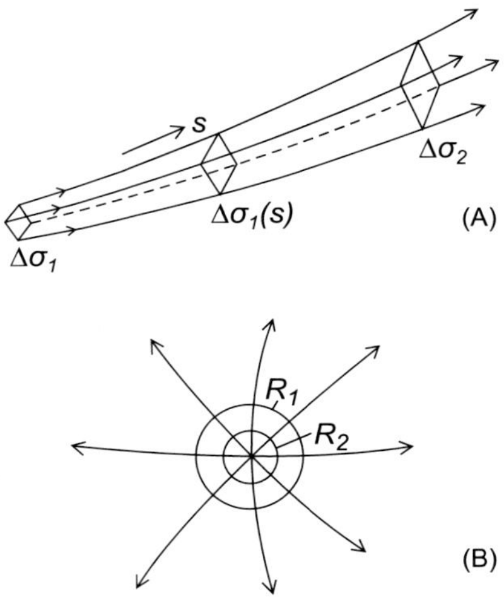

Figure 57.5
Domains of integration for the continuity law: (A) section of a current tube with endfaces $\Delta\sigma_1$, $\Delta\sigma_2$ and a cross-section $\Delta\sigma\ (s)$ at an arbitrary position; (B) spherical shell with radii R_1 and R_2 round a point source.

that the curvature of the rays can be ignored and the absolute momentum g is a constant, we find

$$4\pi R_1^2 a_1^2 = 4\pi R_2^2 a_2^2 = 4\pi R^2 a^2 = \text{const}$$

hence,

$$a(R) = \text{const}/R \quad \text{for} \quad R \to 0 \tag{57.20}$$

Fortunately, the eikonal approximation (57.6) does not break down close to such a singularity, since for $R > 0$ the expression in (57.20) is a solution of the Laplace equation in spherical coordinates. This is the basis for the calculation of a *Green's function*, which will be needed in the next chapter.

If the rays do not intersect exactly in a common point, the reasoning based on Laplace's equation breaks down. This occurs in all real sources or foci, which must therefore be treated carefully in a more rigorous way with a more accurate solution of (56.10).

Another consequence of the continuity equation is that the amplitude–phase solutions cannot in general be normalized. This is immediately obvious for the purely spherical wave, for which $a = \text{const}/R$ holds for all $R > 0$. Amplitude–phase solutions with a *continuous energy spectrum* must therefore be superposed to build up a wave-packet. It is therefore necessary to use the time-dependent form (56.10) of the wave equation with an energy constant E that does not vanish identically.

CHAPTER 58

Paraxial Wave Optics

58.1 The Paraxial Schrödinger Equation

The relativistically corrected Schrödinger equation (56.10) and its approximate solution in the form of an amplitude-phase product (57.1) are valid whenever (57.6) is satisfied. So far as image formation in electron microscopes is concerned, this constraint is still unnecessarily weak. In order to keep the electron optical aberrations sufficiently small, the beam must be confined to a very narrow domain in the vicinity of the optic axis, and this limitation should be incorporated in the search for an approximate solution of (56.10).

The wave-optical calculations pass essentially through the same stages as the geometrical calculations. In the first step the spatial domain considered is so close to the optic axis that lens aberrations can be ignored; this is the *paraxial domain*. The corresponding solution describes *perfect* image formation in wave-optical terms. When the corresponding wavefunctions have been established, the second step is to use them in a perturbation calculus and thereby include the effect of the lens aberrations. This corresponds to the geometrical perturbation theory.

There are two ways of obtaining an approximate solution of Schrödinger's equation. The first, which will be followed extensively in the later chapters, involves finding a general solution and then confining it to the appropriate domain. The alternative is to *simplify* the wave equation before attempting to solve it. The fact that the solutions are applicable only in the paraxial region is then self-evident. This approach was followed by Glaser and Schiske (1953) and is presented in detail by Glaser (1952, 1956). We now give a brief account of this approach, including relativistic correction terms.

We set out from the wave equation (56.10), which it is convenient to reorganize in the form

$$\nabla^2\Psi + \frac{2ie}{\hbar}\boldsymbol{A}\cdot\nabla\Psi + \frac{1}{\hbar^2}\left(2m_0e\hat{\Phi} - e^2\boldsymbol{A}^2\right)\Psi = -\frac{2im}{\hbar}\frac{\partial\Psi}{\partial t} \tag{58.1}$$

The potentials $\boldsymbol{\Phi}(\boldsymbol{r})$ and $\boldsymbol{A}(\boldsymbol{r})$ are now assumed to be rotationally symmetric and A_z to vanish (see Section 7.3 of Volume 1). On the optic axis, we then have

$$\frac{\partial^2\Psi_0(z,t)}{\partial z^2} + \frac{2m_0e}{\hbar^2}\hat{\phi}(z)\Psi_0 = \frac{2m(z)}{i\hbar}\frac{\partial\Psi_0}{\partial t} \tag{58.2}$$

Principles of Electron Optics.
DOI: https://doi.org/10.1016/B978-0-12-818979-5.00058-9

where $\phi(z)$ is as usual the axial potential and we recall that

$$\hat{\phi}(z) := \phi(1 + \varepsilon\phi) \quad , \quad m(z) = m_0(1 + 2\varepsilon\phi) = \gamma m_0$$
$$\varepsilon = \frac{e}{2m_0c^2} \quad , \quad \gamma = 1 + 2\varepsilon\phi$$

See Eqs. (2.16) and (2.21) of Volume 1.

For monoenergetic electrons the time dependence can be represented by the familiar exponential factor, $\exp(-\mathrm{i}Et/\hbar)$. In accordance with the conventions of Section 56.3, the very small energy E can be set to zero for electrons having negligible energy at the cathode surface; $E = e\Delta\phi$ is then the chromatic energy shift if emission spectra are taken into account.

After substituting the harmonic time dependence in (58.2), we find

$$\frac{\partial^2 \Psi_0}{\partial z^2} + \frac{g^2(z)}{\hbar^2}\Psi_0 = 0 \tag{58.3}$$

with the axial momentum

$$g(z) = \left\{2m_0e\hat{\phi}(z) + 2me\Delta\phi\right\}^{1/2} \tag{58.4}$$

Apart from quadratic terms in $\Delta\phi$, which can be ignored, this is in agreement with

$$g^2(z) = 2m_0e(\phi + \Delta\phi)\{1 + \varepsilon(\phi + \Delta\phi)\}$$

which again confirms the consistency of the approximations made in the previous chapter. A highly accurate solution of (58.3) is

$$\Psi_0(z,t) = \frac{C}{\sqrt{g(z)}}\exp\left(\frac{\mathrm{i}}{\hbar}\int g\,dz - \frac{\mathrm{i}e\Delta\phi}{\hbar}t\right) \tag{58.5}$$

The necessary approximation

$$\left|\frac{3g'^2}{4g^2} - \frac{g''}{2g}\right| \ll \frac{g^2}{\hbar^2} = \frac{4\pi^2}{\lambda^2}$$

is entirely justified in practice: the left-hand side is of the order of 1 mm^{-2}, while the right-hand side approaches 1 $\mathrm{fm}^{-2} = 10^{24}\ \mathrm{mm}^{-2}$. Even the less stringent condition

$$\left|\frac{g'}{g}\right| \ll \frac{2\pi}{\lambda} = \frac{g}{\hbar} \tag{58.6}$$

is perfectly justified.

We now seek a solution of (58.1) of the form

$$\Psi(\boldsymbol{r}, t) = \Psi_0(z, t)\psi_p(\boldsymbol{r}) \tag{58.7}$$

The suffix p for 'paraxial' is a reminder that – together with the time dependence – the rapidly varying axial phase factor has been removed. We shall later drop this suffix, whenever this does not cause confusion.

Differentiation of (58.7) using (58.5) gives

$$\begin{aligned}
\frac{\partial \Psi}{\partial t} &= -\frac{\mathrm{i}e}{\hbar}\Delta\phi\psi_p\Psi_0 \\
\nabla\Psi &= \left(\nabla\psi_p + \frac{\mathrm{i}}{\hbar}g\psi_p\boldsymbol{e}\right)\Psi_0(z, t) \\
\nabla^2\Psi &= \left(\nabla^2\psi_p + \frac{2\mathrm{i}}{\hbar}g\frac{\partial\psi_p}{\partial z} - \frac{g^2}{\hbar^2}\psi_p\right)\Psi_0(z, t)
\end{aligned}$$

$\boldsymbol{e}$ denoting the unit vector in the axial direction. Introducing this into (58.1) and taking (58.2)–(58.5) into consideration, a common factor $\Psi_0(z, t)$ cancels, leaving

$$\nabla^2\psi_p + \frac{2\mathrm{i}}{\hbar}g\frac{\partial\psi_p}{\partial z} + \frac{2\mathrm{i}e}{\hbar}\boldsymbol{A}\cdot\nabla\psi_p + \frac{1}{\hbar^2}\left\{2m_0e\hat{\Phi}(\boldsymbol{r}) + 2me\Delta\phi - p^2 - e^2A^2\right\}\psi_p = 0 \tag{58.8}$$

The quantity $\hat{\Phi}(\boldsymbol{r})$ can be replaced by its radial series expansion. In the paraxial approximation this is truncated after the term in r^2:

$$\Phi(r, z) = \phi(z) - \frac{r^2}{4}\phi''(z) + O\left(r^4\right)$$

and consequently

$$\hat{\Phi} = \Phi(1 + \varepsilon\,\Phi) = \hat{\phi}(z) - \gamma\frac{r^2}{4}\phi'' + O\left(r^4\right)$$

From (58.4) and $m = \gamma m_0$, we find

$$2m_0e\hat{\Phi} + 2m_0e\Delta\phi - g^2 = -\frac{1}{2}m(z)er^2\phi''(z) \tag{58.9}$$

which vanishes everywhere on the optic axis. In the paraxial approximation, we may also write

$$A^2(r, z) = \frac{1}{4}r^2B^2(z) + O\left(r^4\right)$$

$B(z)$ being as usual the axial magnetic field strength. Introducing all this into (58.8), we find:

$$\nabla^2\psi_p + \frac{2\mathrm{i}}{\hbar}\left\{g\frac{\partial\psi_p}{\partial z} + e\boldsymbol{A}(r)\cdot\nabla\psi_p(r)\right\} - \frac{r^2}{\hbar^2}\left\{\frac{em}{2}\phi''(z) + \frac{e^2B^2}{4}\right\}\psi_p(r) = 0 \tag{58.10}$$

The term involving the vector potential $\boldsymbol{A}(\boldsymbol{r})$ is inconvenient in practical calculations but can be removed by transforming the wave equation from the laboratory frame to the *rotating frame* (see Eq. 15.7 of Volume 1), as we did systematically in the geometrical treatment of magnetic lenses.

We again adopt the notation introduced in Chapter 15 of Volume 1: the Cartesian coordinates in the laboratory frame are X, Y, and z and the coordinates of the rotating system x, y, and z. The latter are non-Cartesian and defined by

$$x + \mathrm{i}y := (X + \mathrm{i}Y)\exp\{-\mathrm{i}\theta(z)\} \tag{58.11}$$

A consequence of this definition is that we have to consider the derivatives

$$x'(z) = \theta'(z)y(z), \quad y'(z) = -\theta'(z)x(z) \tag{58.12}$$

obtained from (58.11) if we hold X and Y constant. This implies that we have to distinguish carefully between partial and total derivatives with respect to z, as will shortly become obvious.

We now write the paraxial wavefunction as

$$\psi_p(X, Y, z) =: \psi(x, y, z) \tag{58.13}$$

It is easy to verify that

$$\frac{\partial^2\psi_p}{\partial X^2} + \frac{\partial^2\psi_p}{\partial Y^2} = \frac{\partial^2\psi_p}{\partial x^2} + \frac{\partial^2\psi_p}{\partial y^2} \tag{58.14a}$$

$$X\frac{\partial\psi_p}{\partial Y} - Y\frac{\partial\psi_p}{\partial X} = x\frac{\partial\psi_p}{\partial y} - y\frac{\partial\psi_p}{\partial x} \tag{58.14b}$$

are scalar invariants. The differentiation is straightforward, since z is kept constant.

In the paraxial approximation the vector potential is given by Eqs. (7.43) and (7.44) of Volume 1

$$A_X = -\frac{1}{2}Y\,B(z), \quad A_Y = \frac{1}{2}X\,B(z), \quad A_z \equiv 0$$

and the corresponding term in Eq. (58.10) hence becomes

$$\begin{aligned}\boldsymbol{A}\cdot\nabla\psi_p &\equiv A_X\psi_{p|X} + A_Y\psi_{p|Y} \\ &= \frac{1}{2}B\left(X\psi_{p|Y} - Y\psi_{p|X}\right) = \frac{1}{2}B(z)\left(x\psi_{|y} - y\psi_{|x}\right)\end{aligned} \tag{58.15}$$

in which (58.14b) has been used. (The subscripts behind the vertical bars denote partial derivatives with respect to the corresponding coordinates.)

In partial differentiations with respect to z, we have to remember that

$$\psi_p(X, Y, z) = \psi(x(z), y(z), z)$$

and that X and Y are to be kept constant. With (58.12), we find

$$\psi_{p|z} = \psi_{|z} + \psi_{|x}x' + \psi_{|y}y' = \psi_{|z} - \theta'(z)\left(x\psi_{|y} - y\psi_{|x}\right) \tag{58.16}$$

Introducing this and (58.15) into (58.10), we see that the mixed term can be eliminated by choosing

$$\theta'(z) = \frac{eB(z)}{2g(z)} \tag{58.17}$$

This is seen to be in agreement with Eq. (15.9) of Volume 1 if we recall (58.4); the term in $\Delta\phi$ describes the effect on $g(z)$ of chromatic aberrations, which can easily be included here.

The derivative $\psi_{p|zz}$ transforms to a more complicated expression if we repeat the operations of (58.16); it is, however, not really necessary to perform this calculation. The whole expression can be ignored since it is much smaller than the derivatives $\psi_{|xx}$ and $\psi_{|yy}$. This can be easily understood if we consider the special case of field-free motion ($\phi = \text{const}$, $B \equiv 0$). We then have *eigen equations*

$$\psi_{|xx} = -\hbar^2 g_x^2 \psi \ , \quad \psi_{|yy} = -\hbar^2 g_y^2 \psi$$

but as a consequence of (58.7):

$$\psi_{|zz} = -\hbar^2 (g_z - g)^2 \psi$$

From $g^2 = g_x^2 + g_y^2 + g_z^2$ and $g_x^2 + g_y^2 = \alpha^2 g^2$ with a typically very small slope ($\alpha \approx 10^{-2}$), we can conclude that

$$(g - g_z)^2 \approx \frac{1}{4} g^2 \alpha^4 \ll g_x^2 + g_y^2$$

can indeed be ignored in the paraxial approximation. This together with (58.13)–(58.17) reduces (58.10) to the simpler wave equation

$$\frac{\partial^2 \psi}{\partial x^2} + \frac{\partial^2 \psi}{\partial y^2} + \frac{2i}{\hbar} g(z)\psi - \frac{r^2}{\hbar^2}\left\{\frac{em(z)}{2}\phi''(z) + \frac{e^2 B^2(z)}{4}\right\}\psi = 0$$

This, in turn, can be cast into a more convenient form by the introduction of the lens function [Eq. (15.13) of Volume 1]

$$\begin{aligned} F(z) &:= \frac{1}{g^2(z)}\left\{\frac{e}{2}m(z)\phi''(z) + \frac{e^2}{4}B^2(z)\right\} \\ &= \frac{2(1+2\varepsilon\phi)\phi'' + eB^2/m_0}{8\phi(1+\varepsilon\phi+2\varepsilon\Delta\phi)} = \frac{\gamma\phi'' + \eta^2 B^2}{4\hat{\phi}_c} \end{aligned} \tag{58.18}$$

with $\hat{\phi}_c = \phi(1+\varepsilon\phi+2\varepsilon\Delta\phi)$. Finally, we obtain the paraxial Schrödinger equation

$$\frac{\partial^2\psi}{\partial x^2} + \frac{\partial^2\psi}{\partial y^2} + \frac{2\mathrm{i}}{\hbar}g(z)\frac{\partial\psi}{\partial z} - \frac{g^2 r^2}{\hbar^2}F(z)\psi(x,y,z) = 0 \tag{58.19}$$

This differs from Glaser's expression (Glaser, 1952, eq. 159.31 or 1956, eq. 45.10) in the following respects:

1. Relativistic correction terms are included.
2. Chromatic effects are included.
3. A term involving the derivative $g'(z)$ has been removed by the inclusion of the square-root factor (58.5): $\psi_{\mathrm{Glaser}} \equiv \psi g^{-1/2}$.

The physical meaning of the different terms in (58.19) becomes more clear when we rewrite it thus:

$$-\frac{\hbar^2}{2m}\left(\frac{\partial^2\psi}{\partial x^2} + \frac{\partial^2\psi}{\partial y^2}\right) + \frac{g^2 F}{2m}\left(x^2+y^2\right)\psi = \frac{\mathrm{i}\hbar g}{m}\psi_{|z} \equiv \mathrm{i}\hbar v\psi_{|z} \tag{58.20}$$

The first term represents the kinetic energy of the *transverse* motion, the second a purely quadratic *focusing potential*. The right-hand side of the wave equation shows that the role of the time t has been taken over by the axial coordinate z: the transformation $dz = v\,dt$ with the velocity $v = g/m$ is self-explanatory.

A continuity equation can be derived by a calculation quite analogous to that of Section 55.3. We multiply (58.19) by $\hbar\psi^*/2\mathrm{i}g$ and add the result to its complex conjugate. The terms containing the factor F cancel out and we obtain

$$\frac{\hbar}{2\mathrm{i}g}\left(\psi^*\psi_{|xx} - \psi\psi^*_{|xx} + \psi^*\psi_{|yy} - \psi\psi^*_{|yy}\right) + \psi^*\psi_{|z} + \psi\psi^*_{|z} = 0$$

This can be cast into the form of a two-dimensional continuity equation:

$$\frac{\partial}{\partial x}J_x + \frac{\partial}{\partial y}J_y + \frac{\partial}{\partial z}\left(\psi^*\psi\right) = 0 \tag{58.21}$$

with a flux density vector given by

$$J_x = \frac{\hbar}{g(z)}\Im\left(\psi^* \psi_{|x}\right), \quad J_y = \frac{\hbar}{g(z)}\Im\left(\psi^* \psi_{|y}\right) \tag{58.22}$$

From (58.21) the conservation law of intensity

$$I = \iint |\psi(x, y, z)|^2 dx\, dy = \text{const} \tag{58.23}$$

can be derived, provided that ψ is square-integrable in a natural way, that is, without imposing some cut-off that would violate (58.19). Generally, this is possible only if $\psi(x, y, z)$, or more correctly $\Psi(\boldsymbol{r}, t)$, represents a *wave packet* with a finite transverse spectrum.

58.2 Particular Solution of the Paraxial Schrödinger Equation

The solution is obtained in much the same way as in Section 57.1; the only novel aspect is the reduction of dimensions, as the Laplacian in (58.19) is only two-dimensional.

We seek a solution of the form

$$\psi(x, y, z) = C(z)\exp\left\{\frac{\mathrm{i}}{\hbar}\overline{S}(x, y, z)\right\} \tag{58.24}$$

and shall see that this is possible in closed form and with no further approximations. Substitution of (58.24) into (58.19) results in a complex differential equation, which is equivalent to two coupled real ones. Separation into real and imaginary parts gives.

$$\overline{S}^2_{|x} + \overline{S}^2_{|y} + 2g\overline{S}_{|z} + g^2 r^2 F = 0 \tag{58.25a}$$

$$\left(\overline{S}_{|xx} + \overline{S}_{|yy}\right)C + 2gC' = 0 \tag{58.25b}$$

We notice that $C(z)$ does not appear in (58.25a), which means that this equation is, in fact, *uncoupled* and that we should try to solve it first.

In the paraxial approximation, the eikonal $\overline{S}(x, y, z)$ must be a quadratic form in x and y and, to be consistent with the assumption of rotational symmetry, the terms in x^2 and y^2 can appear only in the form $x^2 + y^2$; hence

$$\overline{S}(x, y, z) = \frac{1}{2}Q(z)\left(x^2 + y^2\right) + \alpha(z)x + \beta(z)y + \gamma(z) \tag{58.26}$$

with coefficient functions to be determined from (58.25a). Since this must be valid for *all* values of x and y, we obtain the four conditions

$$\begin{aligned} &gQ' + Q^2 + g^2F = 0 && \text{(a)} \\ &g\alpha' + Q\alpha = 0, \quad g\beta' + Q\beta = 0 && \text{(b)} \\ &g\gamma' + \left(\alpha^2 + \beta^2\right)/2 = 0 && \text{(c)} \end{aligned} \tag{58.27}$$

Eq. (58.27a) is a Riccati equation and can be transformed into a homogeneous, *linear* differential equation; indeed, if we write

$$Q(z) = :g(z)h_p'(z)/h_p(z)$$

in (58.27a) and multiply the result by h_p/g^2, we arrive at

$$\frac{1}{g(z)}\frac{d}{dz}\left\{g(z)h_p'(z)\right\} + F(z)h_p(z) = 0 \tag{58.28}$$

This is the familiar *paraxial ray equation*,[1] as can be seen by comparison with Eq. (15.13) of Volume 1; we have to bear in mind here that $g \propto \hat{\phi}^{1/2}$. The general solution of the paraxial ray equation requires a second, linearly independent partial solution, which we shall denote here by $g_p(z)$. The Wronskian

$$w = g\left(g_p h_p' - g_p' h_p\right)$$

is then a constant and must not vanish.

If we eliminate $Q(z)$ from (58.27b), we obtain the double equation

$$\frac{\alpha'}{\alpha} = \frac{\beta'}{\beta} = -\frac{h_p'}{h_p}$$

which is readily integrated to give

$$\alpha = -\frac{a}{h_p(z)}, \qquad \beta = -\frac{b}{h_p(z)}$$

with arbitrary constants of integration, a and b. Eq. (58.27c) now becomes

$$\gamma' = -\frac{\left(a^2 + b^2\right)}{2gh_p^2}$$

[1] We use $g_p(z)$ and $h_p(z)$ for the solutions of the paraxial equation as the boundary conditions associated with these solutions (15.56) will prove convenient here. The suffix p is added to distinguish these paraxial solutions from the kinetic momentum and Planck's constant.

and this too can be integrated in closed form, as follows: we divide the Wronskian by gh_p^2, giving

$$\frac{w}{gh_p^2} = \frac{g_p h_p'}{h_p^2} - \frac{g_p'}{h_p} = -\frac{d}{dz}\left(\frac{g_p}{h_p}\right)$$

and so

$$\gamma' = -\frac{a^2+b^2}{2w}\frac{d}{dz}\left(\frac{g_p}{h_p}\right), \quad \gamma = \frac{a^2+b^2}{2w}\frac{g_p(z)}{h_p(z)} + S_0$$

in which S_0 is a constant.

Substituting all these expressions into (58.26), we find

$$\overline{S} = \frac{1}{h_p(z)}\left\{\frac{g(z)h_p'(z)}{2}\left(x^2+y^2\right) - ax - by + \frac{a^2+b^2}{2w}g_p(z)\right\} + S_0 \qquad (58.29)$$

In order to apply this formula to the theory of image formation, it is advantageous to choose the functions $h_p(z)$ and $g_p(z)$ and the constants of integration in such a way that $\overline{S}(x, y, z)$ vanishes if the reference point (x, y, z) approaches an object point (x_o, y_o, z_o). This is easy with the standard initial conditions (Fig. 58.1)

$$g_p(z_o) = 1, \quad g_p'(z_o) = 0, \quad h_p(z_o) = 0, \quad h_p'(z_o) = 1 \qquad (58.30a)$$

We then find

$$w \equiv g(z)\left\{g_p(z)h_p'(z) - h_p(z)g_p'(z)\right\} = g(z_o) = :g_o \qquad (58.30b)$$

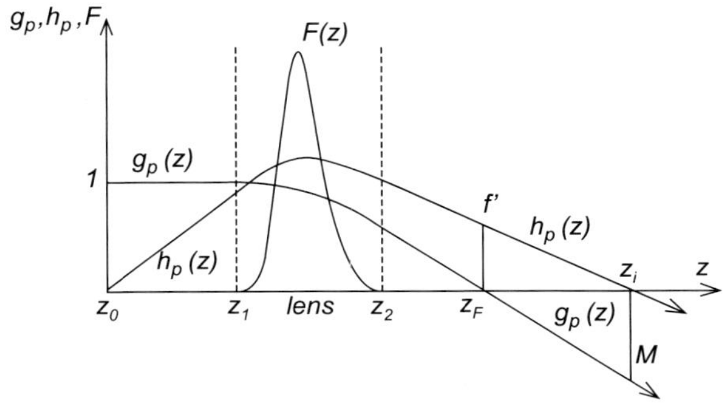

Figure 58.1
The pair of basic solutions $g_p(z)$ and $h_p(z)$ of the paraxial ray equation with standard initial conditions adapted to the object plane $z = z_o$. The image plane is defined by $h_p(z_i) = 0$, and $g_p(z_i) = M < 0$ is then the magnification. The back focal plane is $z = z_F$ with $g_p(z_F) = 0$. In the most frequent case $g'_p(z_o) = 0$, the value $f' := h_p(z_F)$ is simply the focal length $f = -1/g'_p(z_F)$. More generally we have $f = f' g_i / g_o$.

and

$$a = g_o x_o, \quad b = g_o y_o, \quad S_o = 0$$

Hence

$$\overline{S} = \frac{g_o}{2h_p(z)}\left\{\frac{g(z)}{g_o}h_p'(z)\left(x^2 + y^2\right) - 2(x_o x + y_o y) + \left(x_o^2 + y_o^2\right)g_p(z)\right\} \tag{58.31}$$

This formula is *exactly* valid within the paraxial approximation.

For very small values of $|z - z_o|$, we can set $g(z) \to g_o$, $g_p(z) \to 1$, $h_p(z) \to z - z_o$, giving

$$\overline{S} \approx \frac{g_o}{2(z - z_o)}\left\{(x - x_o)^2 + (y - y_o)^2\right\} \tag{58.32}$$

This formula demonstrates an important difference between the wave-optical theory and the geometrical approximation. In the latter, we have to choose $x(z)$ and $y(z)$ as rays passing through the point (x_o, y_o, z_o) and hence $x - x_o = x_o'(z - z_o)$ $y - y_o = y_o'(z - z_o)$. We then obtain the well-behaved result

$$\overline{S} = \frac{1}{2}g_o(z - z_o)\left(x_o'^2 + y_o'^2\right)$$

In the wave-optical theory, however, the coordinate pairs (x, y) and (x_o, y_o) are independent and $\overline{S}$ may hence become *singular* as $z - z_o$ vanishes. This singularity does not really appear in the final results because integrations remove it, as will become clear in the later chapters.

It still remains to determine the amplitude $C(z)$ from (58.25b). The Laplacian of $\overline{S}$ from (58.29) is found to be

$$\frac{\partial^2 \overline{S}}{\partial x^2} + \frac{\partial^2 \overline{S}}{\partial y^2} = \frac{2gh_p'}{h_p}$$

On introducing this into (58.25b), we see that the factor $2g$ cancels out and so

$$C'/C = -h_p'/h_p$$

This is readily integrated to give

$$C(z) = C_0/h_P(z) \tag{58.33}$$

with an arbitrary constant of integration C_0. This amplitude also becomes singular for $z \to z_o$, but this singularity too will be removed by integration. Introducing (58.33), (58.24) and (58.31) into (58.22), we obtain the density functions

$$\overline{\rho}(z) := |\psi|^2 = \frac{|C_0|^2}{h_p^2(z)} \tag{58.34a}$$

$$J_x = \overline{\rho}\frac{h'_p x - g_o x_o/g}{h_p} \tag{58.34b}$$

$$J_y = \overline{\rho}\frac{h'_p y - g_o y_o/g}{h_p} \tag{58.34c}$$

It is easy to verify explicitly that the continuity equation (58.21) is satisfied. Here, we have an obvious example of the fact that (58.21) does not always permit us to derive (58.23): the wavefunction is not square-integrable since $\overline{\rho}$ is independent of x and y. The density functions take a slightly different form if we include the factor Ψ_0 of (58.5) and (58.7) and are then in agreement with Glaser's formulae.

58.3 Paraxial Image Formation

From (58.31) and (58.32), it is obvious that the partial solution (58.33) becomes singular as $z \to z_o$ and is thus not at all satisfactory. This difficulty can be easily circumvented by forming a suitable *wave packet*. We now regard the object coordinates x_0 and y_o as free parameters over which we may integrate. The wave packet is then constructed by forming a weighted linear superposition of solutions of (58.19) of the form

$$\psi(x, y, z) = \iint C(x_o, y_o)\exp\left\{\frac{\mathrm{i}}{\hbar}\overline{S}(x, y, z; x_o, y_o, z_o)\right\}\frac{dx_o dy_o}{h_p(z)} \tag{58.35}$$

The amplitude function $C(x_o, y_o)$ can be chosen at will provided that it is sufficiently well-behaved for the integration. Glaser and Schiske (1953) were able to show that the function $C(x_o, y_o)$ is proportional to the wavefunction ψ itself in the object plane and exploited this to explain the process of image formation within the frame of wave mechanics. We briefly outline this theory and then discuss its inevitable shortcomings.

With the aid of (58.30b), we can cast the eikonal $\overline{S}$ (58.31) into the more convenient form

$$\overline{S} = \frac{g g'_p}{2g_p}\left(x^2 + y^2\right) + \frac{g_o g_p}{2h_p}\left\{\left(x_o - x/g_p\right)^2 + \left(y_o - y/g_p\right)^2\right\} \tag{58.36}$$

The terms that do not depend on x_o and y_o are taken outside the integral, giving

$$\psi(x, y, z) = \frac{1}{h_p}\exp\left\{\frac{\mathrm{i} g g'_p}{2\hbar g_p}\left(x^2 + y^2\right)\right\}\iint C(x_o, y_o)\exp\left[\frac{\mathrm{i} g_o g_p \kappa}{2\hbar h_p}\left\{\left(x_o - x/g_p\right)^2 + \left(y_o - y/g_p\right)^2\right\}\right]dx_o dy_o \tag{58.37}$$

We have to assume here that $g_p \neq 0$. The factor κ that we have included will be useful later; here it is equal to unity.

If we now examine the behaviour of this integral as $z \to z_o$, we see that the factor $h_p = z - z_o$ in the denominator causes huge but always imaginary exponents, which means that the whole exponential function is an extremely rapidly oscillating factor. The contributions to the integral cancel, unless $x = x_o/g_p$ and $y = y_o/g_p$. The approximation

$$C(x_o, y_o) \to C(x/g_p, y/g_p)$$

is therefore justified and this factor can then be taken outside the integral. The remaining integral is easily evaluated by introducing polar coordinates

$$x_o = x/g_p + r_o\cos\varphi, \quad y_o = y/g_p + r_o\sin\varphi$$

and then writing $r_o^2/2 = :s, r_o\, dr_o = ds$. The integrand is independent of the azimuth φ and we find then, with a truncation radius R_0,

$$\begin{aligned} &\iint \exp\left[\mathrm{i}\frac{\Lambda}{2}\left\{ (x_o - x/g_p)^2 + (y_o - y/g_p)^2 \right\} \right] dx_o dy_o \\ &= \int_0^{R_0}\int_0^{2\pi} \exp\left(\mathrm{i}\Lambda r_o^2/2\right) r_o dr_o d\varphi \\ &= 2\pi \int_0^{R_0^2} \exp(\mathrm{i}\Lambda s) ds = \frac{2\pi\mathrm{i}}{\Lambda}\left\{1 - \exp\left(\mathrm{i}\Lambda R_0^2/2\right)\right\} \\ \Lambda: &= \frac{g_o g_p \kappa}{\hbar h_p} \end{aligned}$$

The surviving exponential term depends very sensitively on R_0 if Λ is real. This difficulty can be removed by the following trick: we allow the constant κ to become complex, $\kappa = 1 + \mathrm{i}\delta$, where δ is a very small but positive quantity. We now can proceed to the limit $R_0 \to \infty$, whereupon the exponential term vanishes. The remainder $2\pi/\Lambda$ does not depend on R_0 and we can now set $\delta \to 0$.

This procedure is called a *limitation*; it is mathematically correct and physically justified, since we know that for $R_0 \to \infty$ the paraxial approximation has no meaning; the oscillations that have been eliminated are thus unphysical artefacts.

The quantity $h_p(z)$, which vanishes for $z = z_o$, now cancels in (58.37) and we obtain the well-behaved result

$$\psi(x, y, z) = \frac{2\pi\mathrm{i}\hbar}{g_o g_p} C\left(\frac{x}{g_p}, \frac{y}{g_p}\right) \exp\left\{\frac{\mathrm{i}gg_p'}{2\hbar g_p}\left(x^2 + y^2\right)\right\} \tag{58.38}$$

This formula can be used in various ways. First, we evaluate it in the *object plane*, $g_p = 1, \; g_p' = 0, \; g = g_o$ and obtain immediately

$$C(x_o, y_o) = \psi(x_o, y_o, z_o)\frac{g_o}{2\pi\mathrm{i}\hbar} = -\frac{\mathrm{i}}{\lambda_o}\psi(x_o, y_o, z_o) \tag{58.39}$$

in which $\lambda_o = 2\pi\hbar/g_o$ is the de Broglie wavelength.

There is, however, a second case of interest: the only condition for the validity of (58.38) was $h_p(z_o) \to 0$ and this is true also for the *image plane* $z = z_i$. There we have $g'_p = g'_{pi}$, $g = g_i$ and the magnification $g_{pi} = M$. In the rotating frame the relations between the lateral coordinates take the familiar form $x_i = Mx_o$, $y_i = My_o$. From (58.38), we now obtain the simple formula

$$\begin{aligned}\psi(x_i, y_i, z_i) &= \frac{2\pi i\hbar}{Mg_o} C(x_o, y_o) \exp\left\{\frac{ig_i g'_{pi}}{2\hbar M}\left(x_i^2 + y_i^2\right)\right\} \\ &= \frac{i\lambda_0}{M} C(x_o, y_o) \exp\left\{\frac{i\pi g'_{pi}}{\lambda_i M}\left(x_i^2 + y_i^2\right)\right\}\end{aligned}$$

The quadratic phase factor is irrelevant, since only intensities can be measured. In fact, it represents the spherical wave surface centred on the image point. This phase term has been observed by Ishizuka et al. (2018) by recording the image of the selected-area aperture in focus and slightly out of focus. The transport-of-intensity equation (Section 74.3.8) was then solved iteratively and the resulting phase shifts correspond to a spherical wave. Experimental evidence for this curvature is reproduced in Fig. 58.2, where the diameter of

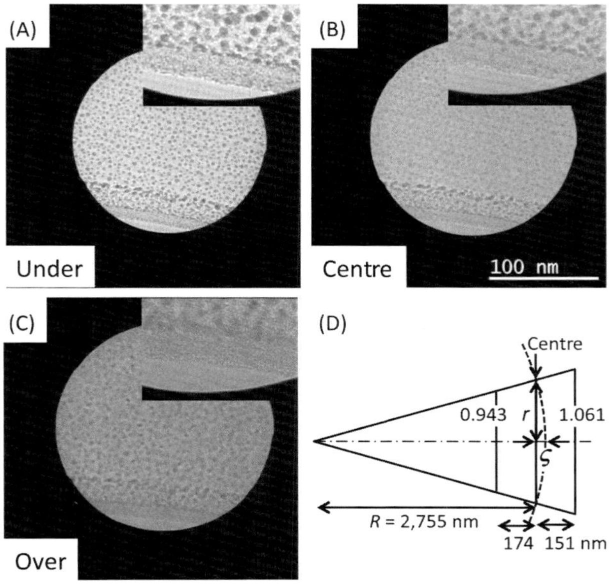

Figure 58.2

(A–C) Experimental images of a 10-μm selected-area aperture at underfocus, in focus and at overfocus. (D) Geometrical construction showing how the size varies as a result of the curvature of the wavefront. The radius of the wave is found to be 2.76 μm. *After Ishizuka et al. (2018), Courtesy Elsevier.*

the image of the selected-area aperture is seen to change as the defocus is altered, as it would for a spherical wave.

After elimination of $C(x_o, y_o)$ by means of (58.39), we obtain the very simple relation:

$$|\psi(x_i, y_i, z_i)|^2 = \frac{|\psi(x_i/M, y_i/M, z_o)|^2}{M^2} \tag{58.40}$$

This expression demonstrates clearly that the intensity distribution at the object plane is reproduced exactly at the image plane with magnification M.
The factor $1/M^2$ guarantees that the total intensity is conserved in the sense that (58.23) is satisfied. We can hence conclude that the paraxial approximation is *self-consistent* in the sense that, after the truncation of the radial series expansions for the potentials beyond the quadratic terms and the omission of $\psi_{|zz}$, no further simplifications are necessary.

Returning to (58.35), we can use (58.39) to obtain an expression for the wavefunction in an arbitrary plane in terms of that in the object plane:

$$\begin{aligned}
\psi(x, y, z) &= \frac{g_o}{2\pi i\hbar h_p(z)} \iint \psi(x_o, y_o, z_o) \exp\left\{\frac{i}{\hbar}\bar{S}(x, y, z; x_o, y_o, z_o)\right\} dx_o dy_o \\
&= \frac{1}{i\lambda_o h_p(z)} \iint \psi(x_o, y_o, z_o) \\
&\quad \times \exp\left[i\pi\left\{\frac{h'_p}{\lambda h_p}(x^2 + y^2) - \frac{2}{\lambda_o h_p}(xx_o + yy_o) + \frac{g_p}{\lambda_o h_p}(x_o^2 + y_o^2)\right\}\right] dx_o dy_o \\
&= \frac{e^{i\zeta}}{i\lambda_o h_p(z)} \iint \psi(x_o, y_o, z_o) \exp\left[i\pi\left\{-\frac{2}{\lambda_o h_p}(xx_o + yy_o) + \frac{g_p}{\lambda_o h_p}(x_o^2 + y_o^2)\right\}\right] dx_o dy_o
\end{aligned} \tag{58.41}$$

with $\zeta := (\pi h'_p/\lambda h_p)(x^2 + y^2)$. This is the general *law of propagation* of the wavefunction in integral form. It enables us in principle to calculate the wavefunction anywhere in space provided that we know it in a particular plane. This is a consequence of the fact that the wave Eq. (58.19) is of first order in z; the omission of $\psi_{|zz}$ is hence a very powerful simplification.

As another example of the usefulness of the law of propagation, we consider the back focal plane $z = z_F$ in which $g_p(z)$ vanishes: $g_p(z_F) = 0$ (Fig. 58.1); we write $g(z_F) = g_f$ and $g_f = 2\pi\hbar/\lambda_f$. For simplicity, we assume that the space between the focal plane and image plane is field-free so that $g_f h_p(z_F)/g_o = \lambda_o h_p(z_F)/\lambda_f$ is equal to the focal length $f = -1/g'_p(z_F)$.

From (58.41), we have

$$\psi(x,y,z_F)=\frac{g_f}{2\pi\mathrm{i}\hbar f}\exp\left\{\frac{\mathrm{i}g_f h_p'(z_F)}{2\hbar f}\left(x^2+y^2\right)\right\}$$

$$\times\iint\psi(x_o,y_o,z_o)\exp\left\{-\frac{\mathrm{i}g_f}{\hbar f}(x_o x+y_o y)\right\}dx_o dy_o$$

$$=\frac{1}{\mathrm{i}\lambda_f f}\exp\left\{\frac{\mathrm{i}\pi h_p'(z_F)}{\lambda_f f}\left(x^2+y^2\right)\right\}$$

$$\times\iint\psi(x_o,y_o,z_o)\exp\left\{-\frac{2\pi\mathrm{i}}{\lambda_f f}(x_o x+y_o y)\right\}dx_o dy_o$$

The quadratic phase factor exp iζ in front of the integral is of no immediate importance since it disappears when we form the intensity distribution. The integral has the form of a two-dimensional Fourier transform and can be cast into a convenient form by introducing the variables

$$q_x:=x/\lambda_f f,\quad q_y:=y/\lambda_f f \tag{58.42}$$

which are the components of the *spatial frequency*. Thus

$$\psi(x,y,z_F)=\frac{e^{\mathrm{i}\zeta}}{\mathrm{i}\lambda_f f}\iint\psi(x_o,y_o,z_o)\exp\left\{-2\pi\mathrm{i}\left(x_o q_z+y_o q_y\right)\right\}dx_o dy_o$$

The inverse Fourier transform of any two-dimensional function $U(x_o, y_o)$ is defined by

$$\tilde{U}\left(q_x,q_y\right):=\mathscr{F}^-(U)\equiv\iint\limits_{-\infty}^{\infty}U(x_o,y_o)\exp\left\{-2\pi\mathrm{i}\left(q_x x_o+q_y y_o\right)\right\}dx_o dy_o \tag{58.43}$$

and so

$$\psi(x,y,z_F)=-\frac{\mathrm{i}e^{\mathrm{i}\zeta}}{\lambda_f f}\mathscr{F}^-(\psi(z=z_o)) \tag{58.44}$$

Apart from unimportant factors, therefore, the wavefunction in the back focal plane is equal to the inverse Fourier transform of the object wave.

We state without proof a minor generalization of this result. The quadratic phase factor is absent from the wavefunction at the object plane, which is equivalent to illumination by a plane wave and hence to an electron source effectively at infinity. Let us now place the source at a finite distance from the object, and replace the g_p-ray by another solution of the paraxial equation, $\bar{g}(z)$, that again intersects the object plane at unit height, $\bar{g}(z_o) = 1$ but intersects the axis in the source plane. A quadratic phase factor now does appear at the object plane and the Fourier transform of the object wave is formed in the plane conjugate to the source plane. Only in the special case of illumination by a plane wave or infinitely distant source does this plane coincide with the back focal plane. We note that Komrska and Lenz (1972) found that the diffraction pattern is always in the back focal plane because only a plane incident wave was considered.

The intensity corresponding to (58.44) is proportional to $\tilde{\psi}\tilde{\psi}^*$, where $\tilde{\psi}$ denotes $\mathscr{F}^-(\psi(z = z_o))$; this is the Fourier transform of the autocorrelation function of the object wavefunction.

58.4 Concluding Remarks

The paraxial approximation of the wave theory has enabled us to derive some important laws, notably (58.40), which states that a sharp image should be obtained on the basis of this theory. Although this is in full agreement with the corresponding classical corpuscular theory, it is certainly *wrong*: we have completely ignored the inevitable *lens aberrations* and all *diffraction* at beam-confining apertures. The latter could be incorporated into the paraxial theory, as Glaser and Schiske (1953) did, but this is still not entirely satisfactory.

An attempt to extend the paraxial Schrödinger equation (58.19) further, to include lens aberrations, reveals a severe weakness of this theory (Glaser, 1953, 1954; Glaser and Braun, 1954, 1955). It is no longer permissible to ignore the term $\psi_{|zz}$, so that we have no simple propagation law like Eq. (58.41). The quantity $\boldsymbol{A} \cdot \nabla\psi_p$ cannot be completely eliminated since we have nonlinear terms in the vector potential. Consequently, mixed second-order derivatives of ψ appear in a complicated manner. The theory then loses all its attraction, and we shall therefore not pursue it further.

In the following chapters, we shall deal with diffraction processes in a fairly general manner, before we again start to specialize.

CHAPTER 59

The General Theory of Electron Diffraction and Interference

In this chapter, we shall develop the fundamentals of electron diffraction and interference. Practical examples are collected in Chapter 60 in order to avoid too many distractions from the general theme by technical details.

Although electron optical diffraction phenomena closely resemble those of light optics, the underlying theory is much more complicated than Kirchhoff's approach. This is a consequence of the fact that an electron diffraction pattern is separated from the object generating it by an electromagnetic field. We shall nevertheless start with the derivation of Kirchhoff's formula, in order to avoid too much abstraction at the beginning.

Electron interference patterns differ from the corresponding light-optical phenomena in two important respects: polarization effects are usually absent and magnetic flux affects the location of the interference fringes. This so-called Aharonov–Bohm effect will be dealt with in Section 59.6.

59.1 Kirchhoff's General Diffraction Formula

Kirchhoff's formula is an integral that provides an *approximate* expression for diffraction phenomena *in field-free space*. The propagation of electrons in field-free domains is described by a solution of the wave equation (55.8)

$$\nabla^2 \psi(\boldsymbol{r}) + k^2 \psi(\boldsymbol{r}) = 0 \tag{59.1}$$

Here, we have again used the conventional wavenumber

$$k = \frac{g}{\hbar} = \frac{2\pi}{\lambda} \tag{59.2}$$

g being the constant kinetic momentum. A simple solution of (59.1) in spherical coordinates (R, ϑ, φ) is evidently

$$\hat{G}(R) = \frac{1}{4\pi R} \mathrm{e}^{\mathrm{i}kR} \tag{59.3}$$

Principles of Electron Optics.
DOI: https://doi.org/10.1016/B978-0-12-818979-5.00059-0

for $R > 0$. This can be generalized to a function of *two* positions $\boldsymbol{r}$ and $\boldsymbol{r}'$:

$$G(\boldsymbol{r}, \boldsymbol{r}') = \frac{1}{4\pi|\boldsymbol{r} - \boldsymbol{r}'|} e^{ik|\boldsymbol{r}-\boldsymbol{r}'|} \qquad (\boldsymbol{r} \neq \boldsymbol{r}') \tag{59.4}$$

which obviously satisfies (59.1) with respect to *both* positions, $\boldsymbol{r}$ and $\boldsymbol{r}'$, since it exhibits the symmetry property

$$G(\boldsymbol{r}, \boldsymbol{r}') = G(\boldsymbol{r}', \boldsymbol{r}) = :\hat{G}(|\boldsymbol{r} - \boldsymbol{r}'|) \tag{59.5}$$

The function G becomes *singular* for $\boldsymbol{r} \to \boldsymbol{r}'$, when $R \to 0$. We shall see that this leads to a partial differential equation for G containing Dirac's δ-function. To show this, we keep the singularity at position $\boldsymbol{r}'$ fixed and consider a sphere S with centre at $\boldsymbol{r}'$ and radius $R = |\boldsymbol{r} - \boldsymbol{r}'|$ in the $\boldsymbol{r}$-space. Integration of the normal derivative $\boldsymbol{n} \cdot \nabla G = d\hat{G}/dR$ over the surface ∂S gives

$$\oint_{\partial S} \boldsymbol{n} \cdot \nabla G \, d^2\boldsymbol{r} = (-1 + ikR)e^{ikR}$$

As $R \to 0$, this tends to a finite limit:

$$\lim_{R \to 0} \oint_{\partial S} \boldsymbol{n} \cdot \nabla G \, d^2\boldsymbol{r} = -1 \tag{59.6}$$

The theory of distributions shows that Gauss's integral theorem can be generalized to include singular functions. It now tells us that

$$\oint_{\partial S} \boldsymbol{n} \cdot \nabla G \, d^2\boldsymbol{r} = \int_S \nabla^2 G \, d^3\boldsymbol{r}$$

and consequently

$$\lim_{R \to 0} \int_S (\nabla^2 G + k^2 G) d^3\boldsymbol{r} = -1 \tag{59.7}$$

The additional term vanishes linearly with R and therefore does not alter the limit. Dirac's δ-function has the property

$$\int f(\boldsymbol{r})\delta(\boldsymbol{r} - \boldsymbol{r}')d^3\boldsymbol{r} = f(\boldsymbol{r}') \tag{59.8}$$

so that if we choose $f = -1$, we see that the integrand of (59.7) is $-\delta(\boldsymbol{r} - \boldsymbol{r}')$. Hence

$$\nabla^2 G(\boldsymbol{r}, \boldsymbol{r}') + k^2 G(\boldsymbol{r}, \boldsymbol{r}') = -\delta(\boldsymbol{r}, \boldsymbol{r}') \tag{59.9a}$$

and, owing to the symmetry (59.5), G also satisfies

$$\nabla'^2 G(\boldsymbol{r}, \boldsymbol{r}') + k^2 G(\boldsymbol{r}, \boldsymbol{r}') = -\delta(\boldsymbol{r}, \boldsymbol{r}') \tag{59.9b}$$

A function satisfying these inhomogeneous partial differential equations is called a *Green's function*; in the present context, it is the 'free' Green's function.

It is now fairly simple to derive an integral equation for the solution of (59.1). We begin by writing $u = \psi(\boldsymbol{r})$ and $v = G(\boldsymbol{r}, \boldsymbol{r}')$ with fixed $\boldsymbol{r}'$ in Green's integral theorem

$$\int_D (v\nabla^2 u - u\nabla^2 v)d^3\boldsymbol{r} = \oint_{\partial D} n\cdot(v\nabla u - u\nabla v)d^2\boldsymbol{r}$$

where D denotes a closed domain (containing the point $\boldsymbol{r}$) and ∂D its surface. On the left-hand side, the terms in k^2, arising from (59.1) and (59.9a) cancel out and only $\psi(\boldsymbol{r})\delta(r - \boldsymbol{r}')$ remains in the integrand. Using (59.8) with $f = \psi$, integration gives $\psi(\boldsymbol{r}')$ and hence

$$\psi(\boldsymbol{r}') = \oint_{\partial D} n\cdot\{G(\boldsymbol{r},\boldsymbol{r}')\nabla\psi(\boldsymbol{r}) - \psi(\boldsymbol{r})\nabla G(\boldsymbol{r},\boldsymbol{r}')\}d^2\boldsymbol{r} \tag{59.10}$$

This is Kirchhoff's formula in its most general form. It is exact and, in principle at least, it allows values of the function $\psi(\boldsymbol{r})$ to be calculated inside the domain from a knowledge of the boundary values of ψ and $\boldsymbol{n}\cdot\nabla\psi$ on ∂D.

Unfortunately, this integral formula is of no practical use, since the boundary values of ψ and $\boldsymbol{n}\cdot\nabla\psi$ cannot be chosen independently and consistent sets of them of any physical interest are unknown (cf. Sommerfeld, 1954, Section 34). Thus major simplifications are necessary to generate a useful formula.

59.2 Necessary Simplifications

We now suppose that the electron wave falls on a large opaque screen with small openings, through which the wave can propagate from the exterior into the domain D of diffraction (Fig. 59.1). Complete opacity means that we can assume that ψ and $\boldsymbol{n}\cdot\nabla\psi$ vanish on the

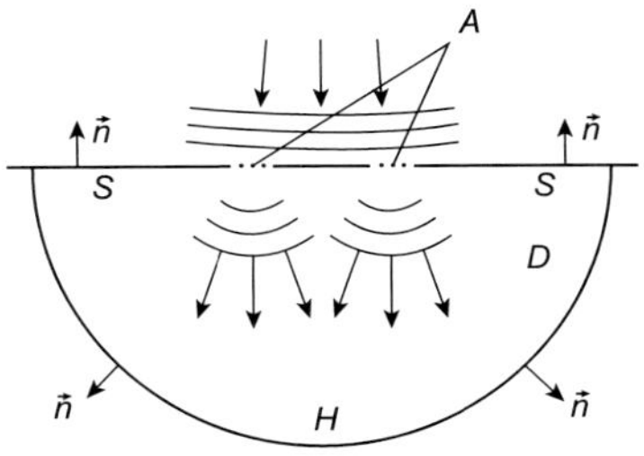

Figure 59.1
A typical diffraction configuration consisting of a screen with openings A and opaque parts S. The domain D is closed by a large hemisphere H. A wave propagates from outside through the openings A into the interior of D.

inner side of the corresponding parts S. The hemisphere H can be chosen so large that it does not contribute to the diffraction surface integral. This is justified, since for $R \to \infty$ the Green's function and the diffracted wavefunction become asymptotically proportional, so that the integrand of (59.10) vanishes at least as R^{-3}.

The openings A are very small in comparison with the distance of the observation point $\boldsymbol{r}'$ from the screen. On the other hand, their diameters are much larger than the wavelength. In the plane of the screen, the incident wave is then very little affected by the edges of the apertures. This influence does not extend beyond about one wavelength. It is hence reasonable to assume that the boundary values in the apertures A are equal to the *unperturbed* incident wave $\psi = \psi_0(\boldsymbol{r})$. The integral Eq. (59.10) then reduces to the well-defined integral

$$\psi(\boldsymbol{r}') = \boldsymbol{i}_z \cdot \iint_A \{ - G(\boldsymbol{r}, \boldsymbol{r}') \nabla \psi_0(\boldsymbol{r}) + \psi_0(\boldsymbol{r}) \nabla G(\boldsymbol{r}, \boldsymbol{r}') \} dx\, dy \tag{59.11}$$

the screen now being located in the coordinate plane $z = 0$ (Fig. 59.2).

Although this integral contains only well-known functions, it is still too complicated for exact evaluation. The assumption that the point of observation is far from the screen implies that $|\boldsymbol{r} - \boldsymbol{r}'| \gg \lambda$. This distance is also much larger than the geometrical size of the apertures, which are located near the origin of the coordinate system; this implies that $|\boldsymbol{r} - \boldsymbol{r}'| \gg \boldsymbol{r}$ in the area of integration. These very strong assumptions are often satisfied in practice.

The following approximations for the Green's function and its normal derivative are then justified:

$$G(\boldsymbol{r}, \boldsymbol{r}') = \frac{1}{4\pi R'} e^{ik|\boldsymbol{r} - \boldsymbol{r}'|} \tag{59.12a}$$

$$\boldsymbol{i}_z \cdot \nabla G = - ik \cos\vartheta' G(\boldsymbol{r}, \boldsymbol{r}') \tag{59.12b}$$

with

$$R' = |\boldsymbol{r}'| = (x'^2 + y'^2 + z'^2)^{1/2} \tag{59.12c}$$

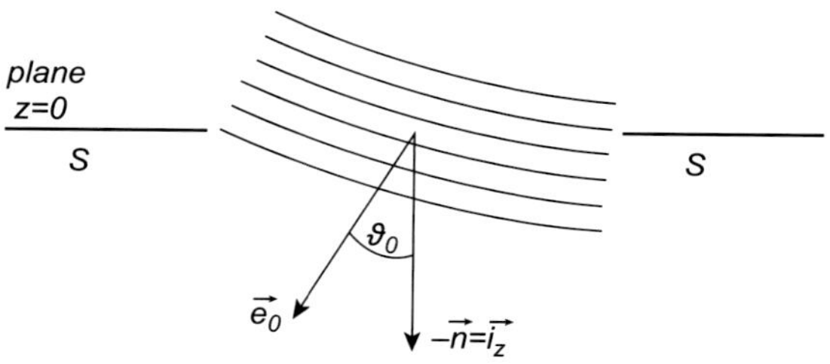

Figure 59.2
The undiffracted wave ψ_0 in the vicinity of an opening and its mean wave-normal e_0.

and

$$\cos\vartheta' = \frac{z'}{R'} \tag{59.12d}$$

Moreover, we make the approximation

$$\begin{aligned} \nabla\psi_0(\boldsymbol{r}) &= \mathrm{i}k\boldsymbol{e}_0\psi_0(\boldsymbol{r}) && \text{(a)} \\ \boldsymbol{i}_z\cdot\nabla\psi_0 &= \mathrm{i}k\cos\vartheta_0\psi_0(\boldsymbol{r}) && \text{(b)} \end{aligned} \tag{59.13}$$

and consider the unit vector $\boldsymbol{e}_0$ (see Fig. 59.2) to be so slowly varying that it may be replaced by its mean value. This implies that the curvature of the wavefronts may be neglected in the slowly varying amplitude factor (but not in the rapidly varying phase factor). Bringing all this together, Kirchhoff's formula (59.11) simplifies further to

$$\psi(\boldsymbol{r}') = \frac{k(\cos\vartheta_0 + \cos\vartheta')}{4\pi \mathrm{i}R'}\iint_A \psi_0(\boldsymbol{r})\mathrm{e}^{\mathrm{i}k|\boldsymbol{r}-\boldsymbol{r}'|}dx\,dy \tag{59.14}$$

It is the rapidly varying exponential factor in (59.14) that creates difficulties. These can be circumvented, as we shall show in the next section.

59.3 Fresnel and Fraunhofer Diffraction

The assumption that we have made concerning the magnitude of $|\boldsymbol{r}-\boldsymbol{r}'|$ suggests that we should expand the exponential term in (59.14) as a power series. Using (59.12c) and (59.12d) and recalling that $z=0$, we have

$$|\boldsymbol{r}-\boldsymbol{r}'| = R' - \frac{1}{R'}(xx'+yy') + \frac{1}{2R'}(x^2+y^2) - \frac{1}{2R'^3}(xx'+yy')^2 + \ldots \tag{59.15}$$

In almost all experimental situations the diffraction angles are small, which means that $|\vartheta_0| \ll 1$, $|\vartheta'| \ll 1$ and hence that $x'^2 + y'^2 \ll z'^2$. Consequently

$$R' = z' + \frac{x'^2+y'^2}{2z'} + O\left(\frac{1}{z'^3}\right) \tag{59.16a}$$

We may then drop the last term in (59.15) and can rewrite this expansion more concisely as

$$|\boldsymbol{r}-\boldsymbol{r}'| = z' + \frac{(x-x')^2+(y-y')^2}{2z'} + O\left(\frac{1}{z'^3}\right) \tag{59.16b}$$

Substituting all this into (59.14) yields *Fresnel's diffraction formula*

$$\psi(\boldsymbol{r}') = \frac{1}{\mathrm{i}\lambda z'}\mathrm{e}^{\mathrm{i}kz'}\iint_A \psi_0(x,y)\mathrm{e}^{\mathrm{i}\varphi}dx\,dy \tag{59.17a}$$

where

$$\varphi = \frac{k}{2z'}\left\{(x-x')^2 + (y-y')^2\right\} = \frac{\pi}{\lambda z'}\left\{(x-x')^2 + (y-y')^2\right\} \tag{59.17b}$$

Important examples of the application of this diffraction formula will be found in Chapter 60 and we shall meet it in several other contexts.

A still further simplification is obtained if the diffracting object is so small that even the second-order terms in x and y can be ignored. It is then not necessary to assume that x' and y' are also small, although this will quite often be the case. The result takes a very convenient form, if we introduce the notion of *spatial frequency* (Eq. 58.43)

$$\begin{aligned} q_x &:= \frac{kx'}{R} = \frac{2\pi}{\lambda}\sin\alpha_x \\ q_y &:= \frac{ky'}{R} = \frac{2\pi}{\lambda}\sin\alpha_y \end{aligned} \tag{59.18}$$

$\alpha_x := x'/R'$ and $\alpha_y := y'/R'$ being the diffraction angles (Fig. 59.3). With $\cos\vartheta \approx \cos\vartheta'$, we obtain the *Fraunhofer formula*

$$\psi(\boldsymbol{r}') = \frac{1}{\mathrm{i}\lambda R'}\mathrm{e}^{\mathrm{i}kR'}\cos\vartheta' \int\!\!\int_A \psi_0(x, y)\exp\{-\mathrm{i}(xq_x + yq_y)\}dx\,dy \tag{59.19}$$

The essential part of this diffraction formula is now a *two-dimensional Fourier integral*. Since the mathematical techniques for the calculation of Fourier integrals are fast and efficient, it is easy to evaluate (59.19). This is one reason why many optical techniques rely on Fraunhofer diffraction, as will become more evident in many later chapters.

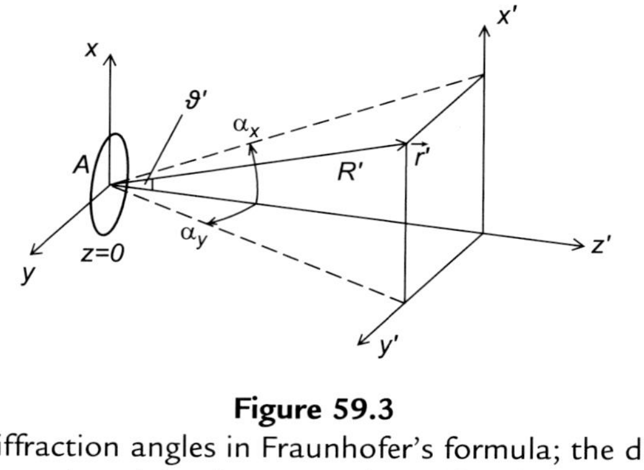

Figure 59.3
Coordinate system and diffraction angles in Fraunhofer's formula; the diffracting aperture lies in the plane $z = 0$ and close to the origin; the vector $\boldsymbol{r}'$ specifies the observation point. The angles satisfy the relation $\sin^2\vartheta' = \sin^2\alpha_x + \sin^2\alpha_y$.

A straightforward generalization of this analysis extends the reasoning to the case of spherical incident waves. This shows that the Fraunhofer diffraction pattern is formed in the plane conjugate to the centre of the spherical incident wavefront, a result of considerable practical importance. In a microscope, for example, this diffraction pattern is formed in the plane conjugate to the crossover of the gun, which may not coincide exactly with the back focal plane of the objective. The presence of the quadratic phase factors in the propagation laws of Chapter 58 is another manifestation of this. For details of this generalization, see for example Section 8.3.3 of Born and Wolf (2002).

59.4 Electron Diffraction in the Presence of Electromagnetic Fields

The theory of diffraction outlined above is mathematically identical with the corresponding theory in light optics and fairly simple. We have recapitulated it to familiarize the reader with the concepts and approximations of the diffraction theory. In reality, this field-free approach is unsatisfactory since the diffracted electron waves are strongly influenced by the fields in electron lenses and other electron optical elements. It is, therefore, necessary to generalize the theory for these cases. The first, fully developed account of this topic appeared in the 1952 treatise of Glaser, who had been pondering over the wave theory of electron optics since 1943 (Glaser, 1943, 1950a, which is based on a course given in Prague in 1942/1943; 1950b, 1951c). The detailed theory was developed in collaboration with Schiske (Glaser and Schiske, 1953), as Glaser makes clear (note [236] of Glaser, 1952; Glaser, 1951c, p. 111). The present account proceeds along similar lines to theirs.

The starting point is the time-independent Schrödinger equation (56.11) and the aim is to establish an approximate solution in the form of a diffraction formula. It is convenient to use the expression (56.12) for the wavenumber k even in the presence of a magnetic vector potential, and we can then rewrite Eq. (56.11) more concisely:

$$\hat{D}\psi := \left[\left\{\nabla + \frac{\mathrm{i}e}{\hbar}\boldsymbol{A}(\boldsymbol{r})\right\}^2 + k^2(\boldsymbol{r})\right]\psi(\boldsymbol{r}) = 0 \tag{59.20}$$

This is evidently a generalization of the Helmholtz equation (59.1) in which the vector potential is present and the wavenumber is position-dependent.

A new aspect is that the operator $\hat{D}$, as defined by (59.20), is *not self-adjoint*. The adjoint operator is identical with the complex conjugate:

$$\hat{D}^{\dagger} = \left\{\nabla - \frac{\mathrm{i}e}{\hbar}\boldsymbol{A}(\boldsymbol{r})\right\}^2 + k^2(\boldsymbol{r}) \tag{59.21}$$

Green's integral theorem is now not directly applicable but must be replaced by a more general integral formula containing the vector potential. In order to find this, we have to recast the alternating bilinear form $v\hat{D}u - u\hat{D}^{\dagger}v$ as a divergence expression, which can then be integrated.

Evaluation of the squares of the operators appearing in $\hat{D}$ and $\hat{D}^{\dagger}$ gives

$$vDu - uD^{\dagger}v = v\nabla^2 u - u\nabla^2 v + \frac{2ie}{\hbar}\{\boldsymbol{A}\cdot(v\nabla u - u\nabla v) + uv\nabla\cdot\boldsymbol{A}\}$$

This can be rewritten as

$$v\hat{D}u - u\hat{D}^{\dagger}v = \mathrm{div}\left(v \;\mathrm{grad}\; u - u \;\mathrm{grad}\; v + \frac{2ie}{\hbar}\boldsymbol{A}uv\right)$$

Gauss's integral theorem then gives

$$\int_D (v\hat{D}u - u\hat{D}^{\dagger}v)d^3\boldsymbol{r} = \oint_{\partial D} \boldsymbol{n}\cdot\left(v\nabla u - u\nabla v + \frac{2ie}{\hbar}\boldsymbol{A}uv\right)d^2\boldsymbol{r} \tag{59.22}$$

We now identify u with the required solution ψ of Eq. (59.20) and v with the corresponding Green's function $G(\boldsymbol{r}, \boldsymbol{r}')$; the latter has to satisfy the *adjoint* partial differential equation:

$$\hat{D}^{\dagger}G(\boldsymbol{r}, \boldsymbol{r}') = -\,\delta(\boldsymbol{r} - \boldsymbol{r}') \tag{59.23}$$

Introducing these choices into Eq. (59.22), we find immediately

$$\begin{aligned}\psi(\boldsymbol{r}') = &\oint_{\partial D} \boldsymbol{n}\cdot\{G(\boldsymbol{r}, \boldsymbol{r}')\nabla\psi(\boldsymbol{r}) - \psi(\boldsymbol{r})\nabla G(\boldsymbol{r}, \boldsymbol{r}')\}d^2\boldsymbol{r}\\ &+ \frac{2ie}{\hbar}\oint G(\boldsymbol{r}, \boldsymbol{r}')\psi(\boldsymbol{r})\boldsymbol{n}(\boldsymbol{r})\cdot\boldsymbol{A}(\boldsymbol{r})d^2\boldsymbol{r}\end{aligned} \tag{59.24}$$

which is evidently a generalization of Eq. (59.10).

The next task is thus to find the appropriate Green's function, which will meet with only partial success. We consider the reference point $\boldsymbol{r}'$ as the singularity of G and look for a solution of the form

$$G(\boldsymbol{r}, \boldsymbol{r}') = a(\boldsymbol{r}, \boldsymbol{r}')e^{iS(\boldsymbol{r}, \boldsymbol{r}')/\hbar} \tag{59.25}$$

$\boldsymbol{r}$ being the variable to which the operator $\hat{D}^{\dagger}$ refers. The function $S(\boldsymbol{r}, \boldsymbol{r}')$ is the point eikonal from the integration point $\boldsymbol{r}$ to the reference position $\boldsymbol{r}'$. A solution of Eq. (59.20) of the form (59.25) would require an eikonal S with $\boldsymbol{r}$ as the terminal position. This is consistent with (59.25), since the eikonal is unaltered if the two points are exchanged and $\boldsymbol{A}(\boldsymbol{r})$ changes its sign, as required by (59.21):

$$S(\boldsymbol{r}, \boldsymbol{r}')_{(+A)} = S(\boldsymbol{r}', \boldsymbol{r})_{(-A)} \tag{59.26}$$

Eq. (59.25) hence has the appropriate form.

The singularity of the Green's function must have the same strength as in the field-free case, since the fields are unimportant in the vicinity of the singularity:

$$\lim_{|\boldsymbol{r}-\boldsymbol{r}'|\to 0}\{4\pi|\boldsymbol{r}-\boldsymbol{r}'|a(\boldsymbol{r}-\boldsymbol{r}')\}=1 \tag{59.27}$$

The amplitude has to satisfy a continuity equation corresponding to (59.12a–d), in which the vector potential term has a negative sign. Since $\boldsymbol{r}$ is now the starting position in $S(\boldsymbol{r}, \boldsymbol{r}')$. ∇S changes sign and the kinematic momentum is hence given by

$$\boldsymbol{g}(\boldsymbol{r})=-\left(\frac{\partial S}{\partial \boldsymbol{r}}+e\boldsymbol{A}(\boldsymbol{r})\right) \tag{59.28}$$

Thus (57.14) is again satisfied. To obtain a closed formula for the amplitude, we apply (57.19) to the configuration shown in Fig. 59.4. The quantities with label 1 refer to the element of area $d\sigma$ in the plane $z=$ const, while the quantities with label 2 refer to the element of solid angle subtended at P'. The surface normal at P is inclined at an angle ϑ to the beam axis, and so

$$a_1^2 g_1 \Delta\sigma_1 \to a^2(\boldsymbol{r},\boldsymbol{r}')g(\boldsymbol{r})\Delta\sigma\cos\vartheta$$

From (59.27), and considering the radius $R=1$:

$$a_2^2 g_2 \Delta\sigma_2 \to \frac{1}{(4\pi)^2}g(\boldsymbol{r}')\Delta\Omega'$$

Equating these two expressions, we find

$$a^2(\boldsymbol{r},\boldsymbol{r}')=\frac{g(\boldsymbol{r}')}{16\pi^2 g(\boldsymbol{r})\cos\vartheta}\frac{\Delta\Omega'}{\Delta\sigma} \tag{59.29}$$

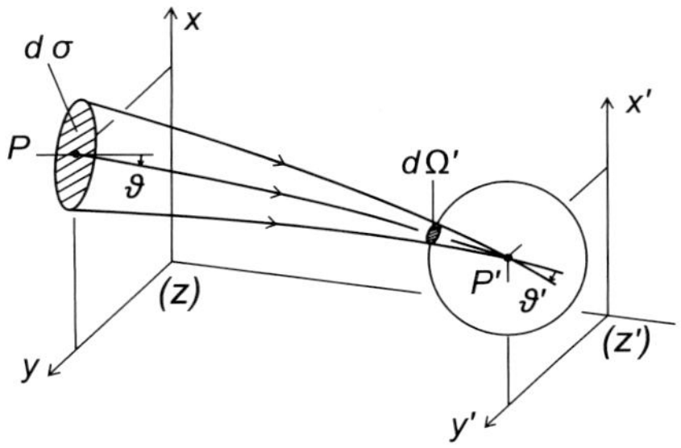

Figure 59.4
Definition of geometrical parameters: The rays starting in the area $d\sigma$ with centre P intersect at point P'; $d\Omega'$ is the element of solid angle subtended by these rays.

In the limit $\Delta\Omega' \to 0$, $\Delta\sigma \to 0$, the expression $\Delta\Omega'/\Delta\sigma$ becomes a Jacobian determinant, which can be expressed in terms of derivatives of S. The ray tangent at $\boldsymbol{r}'$ has the components

$$\sin\alpha_k = \frac{g_k(\boldsymbol{r}')}{g(\boldsymbol{r}')} = -\frac{1}{g(\boldsymbol{r}')}\left\{\frac{\partial S}{\partial x_k'} + eA_k(\boldsymbol{r}')\right\} \quad (k = 1, 2, 3)$$

with $\vartheta' = \pi/2 - \alpha_3$. Evaluation of dg_x and dg_y with respect to $\boldsymbol{r}'$, while retaining x and y as parameters, gives

$$\begin{aligned} d\Omega' &= \frac{dg_x dg_y}{g^2(\boldsymbol{r}')\cos\vartheta'} = \frac{1}{g^2(\boldsymbol{r}')\cos\vartheta'}\frac{\partial(g_x, g_y)}{\partial(x, y)}dx\,dy \\ &= \frac{1}{g^2(\boldsymbol{r}')\cos\vartheta'}\begin{vmatrix} \partial^2 S/\partial x\partial x' & \partial^2 S/\partial y\partial x' \\ \partial^2 S/\partial x\partial y' & \partial^2 S/\partial y\partial y' \end{vmatrix} d\sigma \end{aligned}$$

(The vector potential terms do not depend on x and y.) Introducing this into (59.29), we obtain the final result

$$a(\boldsymbol{r}, \boldsymbol{r}') = \frac{\begin{vmatrix} \partial^2 S/\partial x\partial x' & \partial^2 S/\partial y\partial x' \\ \partial^2 S/\partial x\partial y' & \partial^2 S/\partial y\partial y' \end{vmatrix}^{1/2}}{4\pi\{g(\boldsymbol{r})g(\boldsymbol{r}')\cos\vartheta\cos\vartheta'\}^{1/2}} \tag{59.30}$$

Obviously, this expression is *symmetric* with respect to the two positions:

$$a(\boldsymbol{r}, \boldsymbol{r}') = a(\boldsymbol{r}', \boldsymbol{r}) \tag{59.31}$$

The general theory now passes through essentially the same stages as the field-free case. We again make the various simplifications and arrive at a formula containing the undiffracted wave ψ_0. This differs from Eq. (59.11) in that an additional term involving the vector potential is now present:

$$\begin{aligned} \psi(\boldsymbol{r}') = \boldsymbol{i}_z \iint_A \{-G(\boldsymbol{r}, \boldsymbol{r}')\nabla\psi_0(\boldsymbol{r}) + \psi_0(\boldsymbol{r})\nabla G(\boldsymbol{r}, \boldsymbol{r}')\}d^2\boldsymbol{r} \\ -\frac{2\mathrm{i}e}{\hbar}\iint_A G(\boldsymbol{r}, \boldsymbol{r}')\psi_0(\boldsymbol{r})A_z(\boldsymbol{r})d^2\boldsymbol{r} \end{aligned} \tag{59.32}$$

Simplifying assumptions concerning the local gradients can now be made. For $\nabla\psi_0$ we have

$$\begin{aligned} \{\hbar\nabla + \mathrm{i}e\boldsymbol{A}(\boldsymbol{r})\}\psi_0(\boldsymbol{r}) &= \mathrm{i}g(\boldsymbol{r})\boldsymbol{e}_0(\boldsymbol{r})\psi_0(\boldsymbol{r}) \\ \boldsymbol{i}_z\cdot(\hbar\nabla + \mathrm{i}e\boldsymbol{A})\psi_0 &= \mathrm{i}g(\boldsymbol{r})\cos\vartheta_0\psi_0(\boldsymbol{r}) \end{aligned} \tag{59.33}$$

which is a generalization of (59.13). The angle ϑ_0 now denotes the angle of *incidence*, which may be different from the angle ϑ in Fig. 59.4. The corresponding derivative of the Green's function can be approximated by

$$\boldsymbol{i}_z\cdot(\hbar\nabla - \mathrm{i}e\boldsymbol{A})G(\boldsymbol{r}, \boldsymbol{r}') = -\mathrm{i}g(\boldsymbol{r})\cos\vartheta\, G(\boldsymbol{r}, \boldsymbol{r}') \tag{59.34}$$

since the contribution from ∇a can again be ignored in the far zone. Introducing all this into (59.32), we obtain the still passably accurate diffraction formula

$$\psi(\boldsymbol{r}') = -\frac{\mathrm{i}}{\hbar}\int\int_A (\cos\vartheta_0 + \cos\vartheta)g(\boldsymbol{r})G(\boldsymbol{r}, \boldsymbol{r}')\psi_0(\boldsymbol{r})dx\, dy \tag{59.35}$$

In practice, all direction cosines can be replaced by unity. The diffracting object is so small that the amplitude $a(\boldsymbol{r}, \boldsymbol{r}')$ can be replaced by its *axial* value and the kinetic momentum is then given by

$$g(z) = \{2m_0 e\hat{\phi}(z)\}^{1/2}, \quad g(z') = \{2m_0 e\hat{\phi}(z')\}^{1/2}$$

The Jacobian appearing in (59.30) is replaced by its axial value

$$J(z, z'): = \begin{vmatrix} \partial^2 S/\partial x \partial x' & \partial^2 S/\partial y \partial x' \\ \partial^2 S/\partial x \partial y' & \partial^2 S/\partial y \partial y' \end{vmatrix} \tag{59.36}$$

For a diffracting object in the plane $z = 0$, we finally arrive at the diffraction formula

$$\psi(\boldsymbol{r}') = -\,\mathrm{i}b(z')\int\int_A \psi_0(x, y)\mathrm{e}^{\mathrm{i}S(\boldsymbol{r}, \boldsymbol{r}')/\hbar}dx\, dy \tag{59.37a}$$

with

$$b(z') = \frac{1}{\hbar}\left\{J(0, z')\frac{\hat{\phi}(0)}{\hat{\phi}(z')}\right\}^{1/2} \tag{59.37b}$$

In the case of field-free diffraction, we have

$$S(\boldsymbol{r}, \boldsymbol{r}') = g|\boldsymbol{r} - \boldsymbol{r}'|$$

for which the Jacobian has the exact form

$$J(\boldsymbol{r}, \boldsymbol{r}') = \frac{g^2(z - z')^2}{|\boldsymbol{r} - \boldsymbol{r}'|^4}$$

Making the small-angle approximation and recalling that $g = \hbar k$ (59.2), we recover (59.14) with $R' = |z'|$: Eqs. (59.37a) and (59.37b) essentially contain Eq. (59.14) as a special case.

In Glaser's book (1952) the assumption $G(\boldsymbol{r}, \boldsymbol{r}') = 0$ for $z = 0$ was made to eliminate the term involving $\nabla\psi$ from (59.24). The corresponding Green's function then becomes more complicated by virtue of the mirror operations which then become necessary. We have preferred to circumvent this difficulty by making a reasonable and simple assumption (59.33), which leads practically to the same result.

Although we have succeeded in casting the solution of the diffraction problem into the fairly compact form of Eqs. (59.37a) and (59.37b), we are still not in a position to carry out the necessary integrations. As we have seen in Section 59.3, even Fresnel diffraction in field-free space is fairly complicated. We cannot expect to obtain simple explicit results in closed form. The next steps are, therefore, *series expansions* of S in the exponential term. Since S is the *characteristic function* (apart from a factor $\sqrt{2m_0e}$, omitted in Parts III and IV of Volume 1), we can invoke the vast mass of material on such series expansions derived in geometrical electron optics.

Even this will be of no immediate practical help, since the integrals that arise are still far too complicated to be evaluated in closed form. Efficient numerical integration requires entirely new concepts, which will be developed in Part XII.

59.5 Asymptotic Diffraction Formulae

Quite often, the beam-confining aperture in an electron microscope is located in the weak fringe-field domain behind the magnetic objective lens. Not only are the electron waves irradiating the aperture from the specimen side influenced by the field, but so too are the diffracted waves. The surfaces of equal phase, and by implication the rays, are weakly curved until they become indistinguishable from their exit asymptotes. This situation is sketched in Fig. 59.5.

A correct treatment of this diffraction problem would be extremely complicated and has never, as far as we know, been attempted. Glaser's contribution, which is presented in considerable detail in his textbook (1952), deals only with *paraxial* imaging and therefore leaves the difficult part of the problem untouched, valuable though his analysis is.

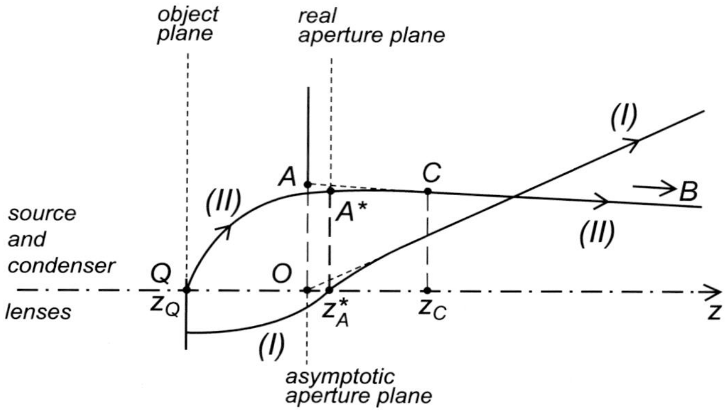

Figure 59.5
Classical paraxial electron trajectories I and II with their exit asymptotes (*broken lines*). The physical aperture must be located in the plane $z = z_A^*$, while the asymptotic aperture lies in the plane through the intersection point O.

Clearly, some intermediate approach is required for it is certainly not permissible to ignore the field beyond the aperture completely, though this is often done for simplicity. The approximate reasoning that follows seems a good compromise between total neglect of the fringing field beyond the aperture and the extreme complication of a rigorous analysis.

We consider a very small specimen close to the optic axis, as is the case in high-resolution electron microscopy. Its axial point is Q in Fig. 59.5. The family of rays (I) pass through the specimen without any interaction; the slope in the plane $z = z_Q$ is determined by the condenser system; the axial intersection at z_A^* specifies the plane of the real (physical) aperture. The asymptotic aperture, considered later, is located in the plane through the intersection point O of the corresponding *exit* asymptote; without loss of generality, we can choose this point as the origin of the coordinate system.

Ray (II) in Fig. 59.5 is scattered at the axial point Q in the specimen; it intersects the real aperture plane in A^*, the corresponding asymptotic intersection point is A. We assume now that for $z \geq z_C$ all real rays are practically identical with their asymptotes. The whole configuration can easily be calculated by numerical ray tracing, which does not raise any problems.

Let us now consider the eikonal S_{II} from the starting point Q to the terminal point B; this is

$$S_{\mathrm{II}} = S_{QA^*} + S_{A^*C} + S_{CB}$$

with

$$S_{CB} = g|\boldsymbol{r}_B - \boldsymbol{r}_C|$$

$g = \left(2m_0 e\hat{U}\right)^{1/2}$ being the asymptotic momentum. This is not yet a favourable representation; we therefore add and subtract again the optical path from A to C, thus obtaining

$$S_{\mathrm{II}} = S_{QA^*} + S_{A^*C} - g|\boldsymbol{r}_C - \boldsymbol{r}_A| + g|\boldsymbol{r}_B - \boldsymbol{r}_A|$$

The first three terms together converge rapidly to a unique function

$$S_a(x_a, y_a) := S_{QA^*} + A_{A^*Q} - g|\boldsymbol{r}_C - \boldsymbol{r}_A|$$

x_a, y_a being the Cartesian coordinates of the point A in the plane $z = 0$; the subscript a indicates the *asymptotic* character of this function. If we form the line integral over the canonical momentum $\boldsymbol{p}$, as in (57.9) and (57.10) with terms of *higher order*, but along the *paraxial* ray, we admit the same errors as in the standard aberration theory (see Part IV of Volume 1); these errors are acceptably small.

Another quite useful representation of the eikonal S_a is obtained in the following way. Here, it is not even necessary for the source Q to be situated on the optic axis. In the case of a *perfect* lens, all rays emerging from the arbitrary point Q in the object plane $z = z_o < 0$ would intersect exactly in the conjugate image point with position vector $\boldsymbol{r}_i^{(0)}$, the eikonal along these rays would have the same value $S_{QI}^{(0)}$ and the asymptotic eikonal in the aperture plane would consequently be $S_a^{(0)} = S_{QI}^{(0)} - g|\boldsymbol{r}_i^{(0)} - \boldsymbol{r}_A|$.

For a real lens with its inevitable aberrations, this is not correct and we have instead:

$$S_a(\boldsymbol{r}_Q, \boldsymbol{r}_A) = S_{QI}^{(0)} - g\left(|\boldsymbol{r}_i^{(0)} - \boldsymbol{r}_A| + W(\boldsymbol{r}_Q, \boldsymbol{r}_A)\right) \tag{59.38}$$

where the deviation term is called the *wave aberration*. This quantity has the dimension and meaning of a length and is the additional optical path caused by aberrations. The function $W(\boldsymbol{r}_Q, \boldsymbol{r}_A)$ can be evaluated by means of power series expansions. Apart from the scale and the notation, these are the same as in Part IV of Volume 1. Further examples will be given in Part XIII.

We are now in a position to derive a fairly accurate integral expression for the wavefunction $\psi_a(x_a, y_a)$. We identify the 'object' in Eq. (59.37a) with the source plane $z = z_Q$ so that $\psi_0(x, y) \rightarrow \psi_Q(x_Q, y_Q)$. The recording plane is now our aperture plane, hence $\boldsymbol{r}' \rightarrow \boldsymbol{r}_A = (x_a, y_a, 0)$. The eikonal S is identified with the S_a of (59.38), in which the additive constant $S_{QI}^{(0)}$ is irrelevant and ignored. We include a complex amplitude factor c_a, and, using $k = g/\hbar$, obtain

$$\psi_a(x_a, y_a) = c_a \iint_Q \psi_Q(x_Q, y_Q) \mathrm{e}^{-iks} dx_Q\, dy_Q \tag{59.39a}$$

with

$$s := |\boldsymbol{r}_i^{(0)} - \boldsymbol{r}_A| + W(\boldsymbol{r}_A, \boldsymbol{r}_Q) \tag{59.39b}$$

This wavefunction is now to be introduced into the integrand of (59.14), where the integration is taken with respect to x_a, y_a. The position $\boldsymbol{r}'$ is not restricted to the image plane, which means that a defocus $z' \neq z_I$ is still allowed. Once again, we collect all constant amplitude factors in a single constant c and assume small-angle diffraction. The result is then

$$\psi(\boldsymbol{r}') = \frac{c}{z'} \iint_A \iint_Q \psi_Q(x_Q, y_Q) \mathrm{e}^{-ikD'} dx_Q\, dy_Q\, dx_a\, dy_a$$

with $D' := -|\boldsymbol{r} - \boldsymbol{r}_A| + |\boldsymbol{r}_i^{(0)} - \boldsymbol{r}_A| + W$.

Apart from W, this expression for D' is the small difference between two large distances. This unfavourable form can be improved in the following way:

$$W - D' = \frac{|\boldsymbol{r}' - \boldsymbol{r}_A|^2 - |\boldsymbol{r}_i^{(0)} - \boldsymbol{r}_A|^2}{|\boldsymbol{r}' - \boldsymbol{r}_A| + |\boldsymbol{r}_i^{(0)} - \boldsymbol{r}_A|} = \frac{(\boldsymbol{r}' - \boldsymbol{r}_i^{(0)})(\boldsymbol{r}' + \boldsymbol{r}_i^{(0)} - 2\boldsymbol{r}_A)}{|\boldsymbol{r}' - \boldsymbol{r}_A| + |\boldsymbol{r}_i^{(0)} - \boldsymbol{r}_A|}$$

which still holds exactly and lends itself to simplification.

A small defocus, $\Delta z := z' - z_I \neq 0$, could be considered by evaluation of (59.39a) and (59.39b) in the corresponding out-of-focus plane. It is, however, usual and certainly more convenient to choose the image plane $z = z_I =: b$ and to compensate the ensuing error by adding a term

$$\Delta W = -\Delta z + \frac{\Delta x^2 + \Delta y^2}{2\Delta z}$$

to the wave aberration W, Δx and Δy being the lateral shifts due to the defocus Δz. The length b, which measures the distance from the aperture to the screen, is always much larger than all other geometrical parameters. We can thus replace the denominator by $2b$. Moreover, we can assume that the lateral extent $|\boldsymbol{r}' - \boldsymbol{r}_i^{(0)}|$ of the diffraction spot is very small, so that terms in $\boldsymbol{r}'^2 - \boldsymbol{r}_i^{(0)2}$ can be ignored. We thus arrive at

$$\begin{aligned} D' &= W + \frac{1}{b}\boldsymbol{r}_A \cdot (\boldsymbol{r}' - \boldsymbol{r}_i^{(0)}) \\ &= W(\boldsymbol{r}_Q, \boldsymbol{r}_a) + \frac{1}{b}\left\{x_a(x' - x_i^{(0)}) + y_a(y' - y_i^{(0)})\right\} \end{aligned}$$

Introducing this into the integral expression for ψ and using the notation $\boldsymbol{r}_i$ instead of $\boldsymbol{r}'$ for position in the image plane, we obtain

$$\psi(\boldsymbol{r}_i) = \frac{c}{b}\iint_A \iint_Q \psi_Q(x_Q, y_Q) e^{-ikD} dx_Q \, dy_Q \, dx_a \, dy_a \tag{59.40a}$$

$$D := W(\boldsymbol{r}_Q, \boldsymbol{r}_a) + \frac{1}{b}\left\{x_a(x_i - x_i^{(0)}) + y_a(y_i - y_i^{(0)})\right\} \tag{59.40b}$$

In practice, this integral is never evaluated in this explicit form, since that would be far too laborious. Some further simplifications are possible for a point source Q located on the optic axis. These already allow a fairly accurate estimate of the resolution achieved in electron optical imaging to be made. This topic is dealt with in Sections 60.5 and 60.6.

A much more important aspect of Eqs. (59.40a) and (59.40b) is that they play a fundamental role in the *transfer calculus* dealt with in Part XIII, especially in Section 65.3. We shall see that this calculus can be justified if W does not depend on $\boldsymbol{r}_Q$. This results in such a far-reaching simplification that the assumption $W = W(\boldsymbol{r}_a)$ is practically always made whenever this is not too misleading. We shall see that this is tantamount to limiting the discussion to isoplanatic systems.

59.6 The Observability of Diffraction and Interference Fringes

Electron interference is mathematically included in the solutions of Eqs. (59.37a) and (59.37b), or (59.40a) and (59.40b), the domain A of integration then consisting of several separate openings $A_1 \ldots A_N$. Thus diffraction and interference appear together as one wave-propagation process. The basic arrangement of an interferometer of biprism type is shown in Fig. 59.6. An electron source Q illuminates two slits A_1, A_2 *coherently*. For the moment, this just means that the incident wave emerges from a point source at Q and has an exact energy E, so that a simple spherical wave can be used. The concept of coherence will be examined more carefully in the later chapters, especially Chapter 78.

The two slits A_1 and A_2 act as apertures for the diffraction process. In the absence of any field the diffraction fringes associated with the edges of the slits will not overlap. If a suitable electrostatic field is created by the electron optical element B (a biprism or a sequence of biprisms), the two initially separated electron waves will be deflected in such a manner that they overlap further downstream. An interference pattern can then be observed on the recording screen R.

This interference pattern can be altered by introducing a specimen S into one of the beams. Macroscopically, its action is due to the electromagnetic fields obtained by averaging over all atomic fields. A magnetic flux Φ_M, concentrated entirely in the nonoverlap domain

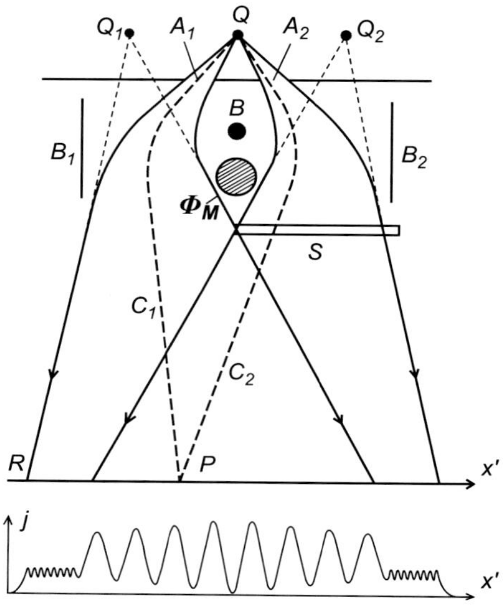

Figure 59.6
Principal features of an electron biprism interferometer with point source Q, virtual sources Q_1, Q_2, effective entrance slits A_1, A_2, biprism wire B, electrostatic plates B_1, B_2, magnetic flux Φ_M, specimen S and recording screen R. C_1 and C_2 denote two classical electron trajectories starting at Q and ending at P on the screen R. The curve $j(x')$ below the ray diagram shows the recorded intensity. For clarity, the scales in the x-direction are drastically exaggerated. Further details are given in Chapter 62.

between the two beams, likewise has an influence on the interference pattern, as will be explained in detail below.

The intensity distribution, recorded on the screen R, is essentially the normal component j_z of the current density $\boldsymbol{j}$. According to (56.14) and with harmonic time dependence, this is given by

$$j_z = \frac{\hbar}{2m_0\mathrm{i}}\left(\psi^*\frac{\partial\psi}{\partial z'} - \psi\frac{\partial\psi^*}{\partial z'}\right) + \frac{e}{m_0}A_z(\boldsymbol{r}')\psi^*\psi \tag{59.41}$$

Differentiation of (59.37a) with respect to z' and multiplication of the result by $\hbar\psi^*/\mathrm{i}$ gives

$$\frac{\hbar}{\mathrm{i}}\psi^*\frac{\partial\psi}{\partial z'} = \frac{\hbar}{\mathrm{i}b}\frac{db}{dz'}|\psi|^2 - \mathrm{i}b\psi^*\iint\psi_0(x,y)\frac{\partial S}{\partial z'}\exp\left(\frac{\mathrm{i}}{\hbar}S\right)dx\,dy$$

We now expand $\partial S/\partial z'$ in terms of x and y. The slopes of the classical trajectories are so small that this series expansion can be truncated after the zero-order term

$$S_{z0} := \left(\frac{\partial S}{\partial z'}\right)_{(0,0,z;x',y',z')}$$

This factor can be taken outside the integral, which then becomes proportional to ψ, hence

$$\frac{\hbar}{\mathrm{i}}\psi^*\frac{\partial\psi}{\partial z'} = \frac{\hbar}{\mathrm{i}b}\frac{db}{dz'}|\psi|^2 + S_{z0}|\psi|^2$$

Introducing this and its complex conjugate into (59.41), we obtain

$$j_z = \frac{1}{m_0}\{S_{z0} + eA_z(\boldsymbol{r}')\}|\psi|^2 = \frac{1}{m_0}g_z(\boldsymbol{r}')|\psi|^2$$

Again thanks to the very small slopes, there is little error in replacing g_z by $|\boldsymbol{g}|$, giving

$$j_z \approx j = \frac{1}{m_0}(2m_0e\hat{\phi})^{1/2}|\psi(\boldsymbol{r}')|^2 \tag{59.42}$$

To establish the *gauge invariance* of this result, it is sufficient to show that (56.15b) is valid for the time dependence of (55.4). We use (56.15a) and set

$$S(\boldsymbol{r},\boldsymbol{r}') = S'(\boldsymbol{r},\boldsymbol{r}') - eF(\boldsymbol{r}') + \hbar\alpha \tag{59.43}$$

as the generalization of (57.12). On introducing this into (59.37a), we see that we can take the exponential factor $\exp\{-\mathrm{i}eF(\boldsymbol{r}')/\hbar + \mathrm{i}\alpha\}$ outside the integral. Moreover, we see that the determinant (59.36) is unaltered, since F does not depend on x and y. Bringing all this together does indeed confirm the transformation (56.15b), from which the invariance of $|\Psi|^2 = |\psi|^2$ is immediately obvious. This shows that all the approximations were made consistently.

We now investigate the influence of a locally confined magnetic flux Φ_M on the interference pattern. Fig. 59.6 shows a cross section through this flux. The B-lines are perpendicular to the plane of the drawing. Such a flux can be produced to a good approximation by a current-carrying solenoid with a sufficiently small diameter and pitch, closed by magnetic yokes outside the electron beam. This arrangement does not produce a $\boldsymbol{B}$-field in the region of wave propagation; the corresponding vector potential $\boldsymbol{A}$ is such that curl $\boldsymbol{A} = 0$ (locally) and

$$\oint \boldsymbol{A}\cdot d\boldsymbol{r} = \Phi_M \tag{59.44}$$

The current density at any point P on the recording screen is determined by the phase *difference* between the two classical trajectories C_1 and C_2 starting from Q and intersecting again at P. We consider first the case $\boldsymbol{A}(r) = 0$, for which Φ_M is absent. The corresponding phase difference is then given by

$$\varphi_0 = \frac{1}{\hbar}(S_1^{(0)} - S_2^{(0)}) = \frac{1}{\hbar}\left(\int_Q^{P(C_1)} g\,ds - \int_Q^{P(C_2)} g\,ds\right) \tag{59.45}$$

A gauge transformation (59.43) with position vectors $\boldsymbol{r}$ for Q and $\boldsymbol{r}'$ for P does not alter this result.

We now suppose that paths C_1 and C_2 enclose magnetic flux, as described above. Since the $\boldsymbol{B}$-field vanishes in the region of the electron beam, the classical trajectories remain unaffected and neither of the integrals in (59.45) is altered. The complete eikonals are, however, given by

$$S_j = \int_Q^{P(C_j)} g\,ds - \int_Q^{P(C_j)} e\boldsymbol{A}\cdot d\boldsymbol{r} \quad (j = 1, 2)$$

in which the second term *is* gauge-dependent. The phase difference is now given by

$$\begin{aligned}\varphi = \frac{1}{\hbar}(S_1 - S_2) &= \varphi_0 - \frac{e}{\hbar}\left(\int_Q^{P(C_1)} \boldsymbol{A}\cdot d\boldsymbol{r} - \int_Q^{P(C_2)} \boldsymbol{A}\cdot d\boldsymbol{r}\right)\\ &= \varphi_0 - \frac{e}{\hbar}\left(\int_Q^{P(C_1)} \boldsymbol{A}\cdot d\boldsymbol{r} + \int_P^{Q(C_2)} \boldsymbol{A}\cdot d\boldsymbol{r}\right)\\ &= \varphi_0 - \frac{e}{\hbar}\oint \boldsymbol{A}\cdot d\boldsymbol{r}\end{aligned}$$

in which the path of integration for the circuit integral is formed by C_1 from Q to P and by C_2 back from P to Q. This may be written as

$$\varphi = \varphi_0 - \frac{e}{\hbar}\int \mathrm{curl}\, \boldsymbol{A}\cdot d\boldsymbol{S} = \varphi_0 - \frac{e}{\hbar}\int \boldsymbol{B}\cdot d\boldsymbol{S}$$

in which the integral is taken over the surface between C_1 and C_2. We thus obtain the simple result

$$\varphi = \varphi_0 - \frac{e}{\hbar}\Phi_M, \quad \Phi_M := \int \boldsymbol{B}\cdot d\boldsymbol{S} \tag{59.46}$$

(cf. Section 5.6 of Volume 1). Since the shift $\Delta\varphi = e\Phi_M/\hbar$, is independent of the position P, this causes all the maxima and minima in the interference pattern to be shifted by $\Delta\varphi/2$, and this is indeed observed.

The formula (59.46) was first derived by Ehrenberg and Siday (1949) and later found again by Aharonov and Bohm (1959, 1961), after whom the effect then became named. It is interesting to note that the reasoning of Ehrenberg and Siday was based almost wholly on classical mechanics, the only wave-optical notion being the elementary interference relation that phase difference $= k\times$ (path difference). It was, however, the paper by Aharonov and Bohm that attracted attention and the first experimental confirmation was published by Chambers (1960), followed shortly after by Fowler et al. (1961) and Boersch et al. (1961, 1962, see also 1981). In 1962 Möllenstedt and Bayh demonstrated the fringe-shift associated with enclosed magnetic flux in a way that was free of the objections to which the earlier experiments had given rise, the interference effect and other classical electromagnetic effects being well separated. A voluminous literature rapidly grew up, enlivened by vigorous controversy. All doubt was finally dispelled by Tonomura et al. (1982, 1984). Discussion of this is deferred to Chapter 62, in which electron interferometry is dealt with in more detail, and to the bibliography of Part XII.

An interesting observation of Lenz (1962) makes this effect easier to understand. He transformed the experiment from the laboratory frame to a frame of reference moving with the electron, so that electric field $\boldsymbol{E}$ in the Maxwell equation $\boldsymbol{E} + \partial\boldsymbol{A}/\partial t = -\,\mathrm{grad}V$ is zero. If we consider two points situated at $\boldsymbol{r}_l$ and $\boldsymbol{r}_r$ on either side of the moving component, we see that

$$\int_{-\infty}^{\infty} V(\boldsymbol{r}_l, t)dt - \int_{-\infty}^{\infty} V(\boldsymbol{r}_r, t)dt = \oint \boldsymbol{A}\cdot d\boldsymbol{r} = \Phi_M$$

The rate of change of the phase of the electron wave with time is given by $eV(\boldsymbol{r}, t)/\hbar$ and we, therefore, arrive at $e\Phi_M/\hbar$ as before.

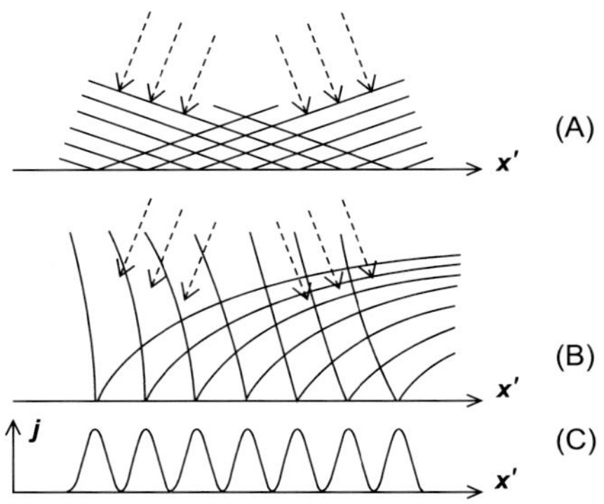

Figure 59.7
Effect of a gauge transformation: (A) plane wavefronts and orthogonal trajectories; (B) transformed wavefronts, the trajectories (*broken lines*) remain unaltered; (C) corresponding current density distribution.

We close this chapter with some remarks about the experimental significance of such wave-optical concepts as wavefronts, wavelength, phase, frequency, and eikonal. We have seen that all these are gauge-dependent. The only invariant quantities are the particle and current density and related densities, and the expectation values of the classical kinematic functions. These wave-optical concepts serve only to illuminate vividly the otherwise very abstract quantum theory but do not correspond to any quantity that could be measured.

The influence of a gauge transformation is illustrated in Fig. 59.7. Case (A) shows two interfering plane waves with corresponding straight orthogonal trajectories. This picture is adequate in field-free space and represents the conventional ideas. Case (B) shows the *same* situation after the introduction of a vector potential $\boldsymbol{A}(\boldsymbol{r}) = \nabla F(\mathbf{r})$. The wavefronts can be *arbitrarily distorted*, as long as the intersection points on the x'-axis (indicating the screen) are unaltered, but the trajectories are no longer orthogonal to the wavefronts (cf. Section 5.4 of Volume 1). The trajectories and the intensity distribution likewise remain unaltered: the wavefronts cannot have any real significance. This fact was very clearly pointed out by Lenz (1962), Rang (1964, 1977), and Boersch (1968). The ambiguity, introduced by gauge transformations, can of course be removed by adopting some *fixed* gauge, for instance by setting div $\boldsymbol{A} = 0$ and $\boldsymbol{A} \to 0$ as $|\boldsymbol{r}| \to \infty$, which seems to be the natural choice. There is, however, no real need to do this, and it is more a philosophical question whether or not one should sacrifice the freedom of gauge transformation.

CHAPTER 60

Elementary Diffraction Patterns

In the previous chapter, we derived general diffraction formulae without offering any practical examples. We now examine some real situations that can be calculated in closed form. This is successful in practice only for diffraction in field-free space and we shall consider mainly the Fresnel and Fraunhofer diffraction treated in Section 58.3, therefore.

60.1 The Object Function

We consider the configuration shown in Fig. 60.1 and adopt the corresponding notation. A point source Q, located at the position $\boldsymbol{r}_s = (x_s, y_s, z_s)$ with $z_s =: -a < 0$, emits a spherical wave

$$\psi_s(\boldsymbol{r}) = \frac{C}{|\boldsymbol{r} - \boldsymbol{r}_s|} e^{ik|r - r_s|} \tag{60.1}$$

In the Fresnel approximation, this simplifies to

$$\psi_s(\boldsymbol{r}) = \frac{C}{|z - z_s|} \exp(ik|z - z_s|) \exp\left[\frac{ik}{2|z - z_s|}\left\{(x - x_s)^2 + (y - y_s)^2\right\}\right] \tag{60.2}$$

In the plane of the diffracting object, $z = 0$, and with object coordinates x_o, y_o, this formula reduces to

$$\hat{\psi}_o(x_o, y_o) = \frac{C}{a} \exp(ika) \exp\left[\frac{ik}{2a}\left\{(x_o - x_s)^2 + (y_o - y_s)^2\right\}\right] \tag{60.3}$$

This is the wavefunction incident on the object plane, on the far side with respect to the observer.

We now assume that the object is effectively plane and so thin that its extent in the z-direction can be ignored. It causes a modulation of any wave passing through it, which we can describe by a complex transmission function[1]

$$O(x_o, y_o) = e^{-\sigma(x_o, y_o) + i\eta(x_o, y_o)} \tag{60.4}$$

[1] Note that two conventions are in use for the sign of the phase term, corresponding to the choice of sign in Eq. (57.1).

Principles of Electron Optics.
DOI: https://doi.org/10.1016/B978-0-12-818979-5.00060-7

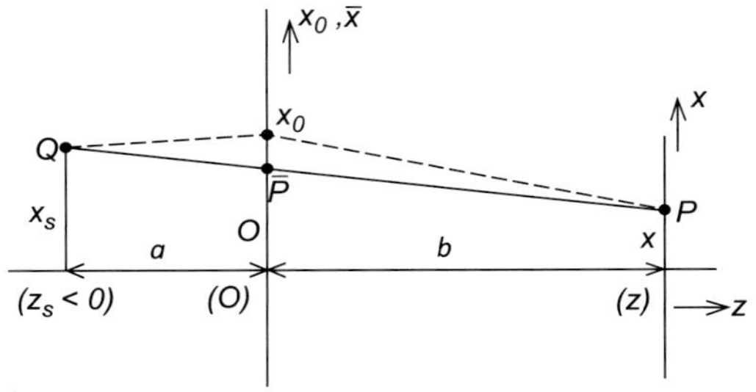

Figure 60.1
Notation and choice of the coordinate system in a Fresnel-diffraction configuration; *Q:* point source, *O:* diffracting object, *P:* observation point.

This will be referred to as the *object function*. The factor $\exp(-\sigma)$ describes an attenuation and $\exp(\mathrm{i}\eta)$ a phase shift, both locally; these are to be understood as macroscopic effects. The following special cases are of particular interest:

$$\left.\begin{array}{lcl} \sigma=0, \eta=0 & : & \text{free opening} \\ \sigma=\infty & : & \text{opaque screen} \\ \sigma=0, \quad \eta\neq 0 & : & \text{pure phase object} \end{array}\right\} \tag{60.5}$$

The wavefunction on the exit side of the object is now given by

$$\psi_o(x_o, y_o) = \hat{\psi}_o(x_o, y_o) O(x_o, y_o) \tag{60.6}$$

This is to be introduced into Fresnel's or Fraunhofer's formula as appropriate. At the observation point with coordinates $\boldsymbol{r}=(x, y, z)$, $z=b>0$, Eqs. (59.17a) and (59.17b) with (60.3) and (60.6) now take the explicit form

$$\psi(\boldsymbol{r}) = \frac{C}{\mathrm{i}\lambda ab} \mathrm{e}^{\mathrm{i}k(a+b)} \iint\limits_{-\infty}^{\infty} O(x_o, y_o) \exp\{\mathrm{i}ks/2\} dx_o dy_o$$

with the length

$$s = \frac{1}{a}\left\{(x_o - x_s)^2 + (y_o - y_s)^2\right\} + \frac{1}{b}\left\{(x_o - x)^2 + (y_o - y)^2\right\}$$

This can be rewritten as

$$s = \frac{1}{f}\left\{(x_o - \bar{x})^2 + (y_o - \bar{y})^2\right\} + \frac{1}{a+b}\left\{(x - x_s)^2 + (y - y_s)^2\right\}$$

Here, we have introduced the abbreviations

$$\frac{1}{f} = \frac{1}{a} + \frac{1}{b} \tag{60.7}$$

$$\bar{x} = \frac{ax + bx_s}{a + b}, \bar{y} = \frac{ay + by_s}{a + b} \tag{60.8}$$

The quantity f will sometimes be identified with a focal length. The coordinates $\bar{x}, \bar{y}$ are those of the point $\bar{P}$ at which the line $\overline{QP}$ intersects the screen (see Fig. 60.1). In the Fresnel approximation the identification

$$D: = a + b + \frac{(x - x_s)^2 + (y - y_s)^2}{2(a + b)} = \overline{QP} = |\boldsymbol{r} - \boldsymbol{r}_s| \tag{60.9}$$

can be made without introducing any additional error; D is thus the distance between the source and the observation point. The diffraction integral now becomes

$$\psi(\boldsymbol{r}) = \frac{C}{\mathrm{i}\lambda ab} \mathrm{e}^{\mathrm{i}kD} \iint\limits_{-\infty}^{\infty} O(x_o, y_o) \exp\left[\frac{\mathrm{i}k}{2f}\left\{(x_o - \bar{x})^2 + (y_o - \bar{y})^2\right\}\right] dx_o dy_o$$

To simplify this integral further, we make the substitution

$$x_o: = \bar{x} + Lu, \quad y_o: = \bar{y} + Lv \tag{60.10a}$$

with the diffraction length

$$L: = \left(\frac{\pi f}{k}\right)^{1/2} = \left(\frac{\lambda f}{2}\right)^{1/2} \tag{60.10b}$$

whereupon the diffraction integral takes its final form

$$\psi(\boldsymbol{r}) = \psi_s(\boldsymbol{r}) M(\boldsymbol{r}) \tag{60.11}$$

with

$$M(\boldsymbol{r}) = \frac{1}{2\mathrm{i}} \iint\limits_{-\infty}^{\infty} O(\bar{x} + Lu, \bar{y} + Lv) \exp\left\{\frac{\mathrm{i}\pi}{2}\left(u^2 + v^2\right)\right\} du\, du \tag{60.12}$$

The modulation factor $M(\boldsymbol{r})$ describes the influence of the object on the incident wave $\psi_s(\boldsymbol{r}) = C\exp(\mathrm{i}k\, D)/(a + b)$ and is therefore of particular interest.

The formula includes the degenerate case of an incident *plane* wave with small transverse components of the wave vector. In this case, we have

$$a \to \infty, \quad b = f, \quad \bar{x} = x - b\alpha_x, \quad \bar{y} = y - b\alpha_y \tag{60.13}$$

α_x, αy being the angles of incidence. The factor $\psi(\boldsymbol{r})$ is then, of course, a *plane* wave.

60.2 Rectangular Structures

These are characterized by the property that the object function separates into two independent factors:

$$O(x_o, y_o) = O_1(x_o) O_2(y_o) \tag{60.14}$$

whereupon the integral (60.12) also separates into two independent factors M_1, M_2 with

$$
\begin{aligned}
M &= M_1 M_2 \\
M_k &= \frac{1-\mathrm{i}}{2} \int_{-\infty}^{\infty} O_k(\bar{x}_k + Lu) \exp\left(\frac{\mathrm{i}\pi}{2} u^2\right) du
\end{aligned} \tag{60.15}
$$

and $\bar{x}_1 = \bar{x}$, $\bar{x}_2 = \bar{y}$. This is a major simplification, since it suffices now to calculate two one-dimensional integrals.

The formulae simplify still further if the object function depends only on one coordinate, x_o say, while $O_2(y_o) \equiv 1$. Using the well-known formula

$$
\int_{-\infty}^{\infty} \exp\left(\frac{\mathrm{i}\pi}{2} u^2\right) du = 1 + \mathrm{i} \tag{60.16}
$$

we see from (60.15) that $M_2 = 1$ and $M(\boldsymbol{r}) = M_1(x, z)$, the dependence on x and z being given by Eqs. (60.7), (60.8), (60.10a) and (60.10b).

If we have no diffracting object at all, $O_1(x_o)$ is also equal to unity everywhere; we can apply (60.16) a second time and find that $M(\boldsymbol{r}) = 1$, as it must be. The Fresnel approximation is thus self-consistent.

We consider now diffraction at the edge of an opaque half-plane. The object function is then

$$
O_1(x_o) = \begin{cases} 0 & \text{for } x_o < 0 \\ 1 & \text{for } x_o \geq 0 \end{cases} \tag{60.17}
$$

We assume that the incident wave is plane and parallel to the object with unit amplitude,

$$
\psi_s(\boldsymbol{r}) = \mathrm{e}^{\mathrm{i}kz} = :\psi_s(z)
$$

the angles α_x, α_y in (60.13) are zero and so $\bar{x} = x$. With these assumptions (60.15) collapses to

$$
M = M_1(x, z) = \frac{1-\mathrm{i}}{2} \int_{-x/L}^{\infty} \exp\left(\frac{\mathrm{i}\pi}{2} u^2\right) du \tag{60.18}
$$

This integral can be expressed in terms of well-known analytic functions known as the *Fresnel integrals*. These are defined by

$$
F(w) \equiv C(w) + \mathrm{i}S(w) := \int_0^w \exp\left(\frac{\mathrm{i}\pi}{2} v^2\right) dv \tag{60.19}
$$

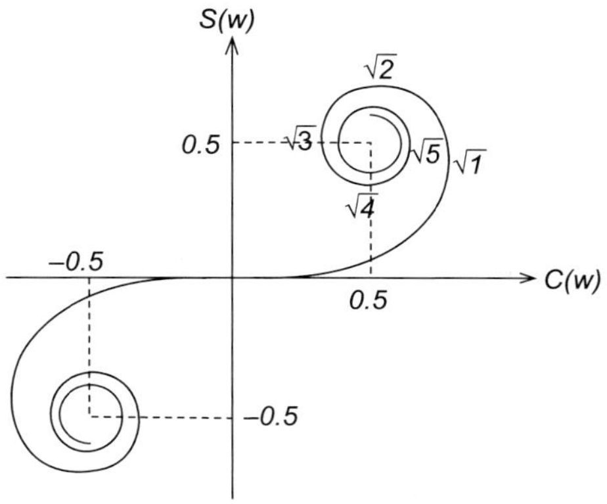

Figure 60.2
Parametric representation of the Fresnel integrals $C(w)$ and $S(w)$.

The behaviour of this function is shown in Fig. 60.2, where the values of $C(w)$ and $S(w)$ are regarded as Cartesian coordinates of points corresponding to all possible values of w. The resulting curve is the Cornu spiral. It is easy to show that

$$F(-w) = -F(w), \quad F(0) = 0 \tag{60.20a}$$

$$F(\pm\infty) = \pm\frac{1+\mathrm{i}}{2} \tag{60.20b}$$

The series expansion for $|w| \gg 1$ is

$$F(w) - F(\pm\infty) = \frac{1}{\mathrm{i}\pi w}\exp\left(\frac{\mathrm{i}\pi}{2}w^2\right)\left\{1 + \frac{1}{\mathrm{i}\pi w^2} + \frac{1\cdot 3}{(\mathrm{i}\pi w^2)^2} + \frac{1\cdot 3\cdot 5}{(\mathrm{i}\pi w^2)^3} + \ldots\right\} \tag{60.21}$$

but this series is asymptotic, which means that the most accurate result is obtained by halting the series after a certain number of terms, which depends on $|w|$. Beyond this, the error increases and the series finally diverges. For $|w| = 2$, the relative error is about 10^{-3} with the terms in (60.21), which is sufficient in practice.

For $|w| \leq 1$ the Taylor series expansion

$$F(w) = \sum_{m=0}^{\infty}\left(\frac{\mathrm{i}\pi}{2}\right)^m \frac{w^{2m+1}}{m!(2m+1)} \tag{60.22}$$

can be evaluated. The accuracy is about 10^{-5} or better if the expansion is truncated after $m = 7$. Other approximations are given by Abramowitz and Stegun (1965) and Olver et al. (2010).

In terms of Fresnel integrals the solution (60.18) can be rewritten as

$$M(\boldsymbol{r}) = \frac{1}{2} + \frac{1-\mathrm{i}}{2} F\left(\frac{x}{\sqrt{\lambda z/2}}\right) \tag{60.23}$$

The normalized intensity distribution is then given by

$$|M(\boldsymbol{r})|^2 = \frac{1}{4}\left|1+(1-\mathrm{i})F\left(\frac{x}{\sqrt{\lambda z/2}}\right)\right|^2 \tag{60.24}$$

This intensity distribution is depicted in Fig. 60.3. For $x < 0$ (in the domain of the geometric shadow), the intensity decreases as $(|x| + L)^{-2}$ towards zero, where $L := \sqrt{\lambda z/2}$. For $x > 0$, an interference pattern is observable in which the amplitude and fringe spacing gradually decrease. If we truncate the asymptotic expansion (60.21) after the first term, we obtain the simple approximation

$$|M(\boldsymbol{r})|^2 =: J(w) = 1 - \frac{\sqrt{8}}{\pi w}\cos\frac{\pi}{2}\left(w^2 + \frac{1}{2}\right) + O\left(w^{-2}\right) \tag{60.25}$$

From this, it can be concluded that the maxima and minima are located at

$$w_n = \sqrt{2n - \frac{1}{2}}, \quad x_n = \sqrt{\lambda z\left(n - \frac{1}{4}\right)}, \quad n = 1, 2, 3, \ldots \tag{60.26}$$

with odd values of the integers n for maxima and even values for minima. These formulae are also valid for spherical incident waves if z is replaced by f of (57.9) with $b = z$ so that $1/f = 1/a + 1/z$.

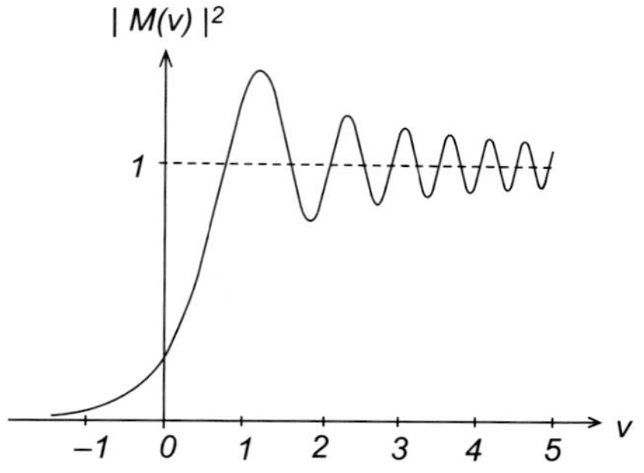

Figure 60.3
Intensity distribution behind a straight edge of an opaque screen, $v = x/\sqrt{\lambda z/2}$.

On the basis of the solution (60.23), more sophisticated situations can be analysed. Often the screen is not completely opaque but is described by an object function of the general form (60.4). Lenz (1961) has investigated the case of a semitransparent half-plane, described by

$$O(x_o, y_o) = \begin{cases} \exp(-\sigma + \mathrm{i}\eta) & x_o < 0 \\ 1 & x_o \geq 0 \end{cases} \tag{60.27}$$

The solution is then given by

$$M(\boldsymbol{r}) = \frac{1}{2}\left(1 + \mathrm{e}^{\mathrm{i}\eta - \sigma}\right) + \frac{1-\mathrm{i}}{2}\left(1 - \mathrm{e}^{\mathrm{i}\eta - \sigma}\right) F\left(\frac{x}{\sqrt{\lambda z/2}}\right) \tag{60.28}$$

where (60.20a) has been used. It is easy to verify that $M(\infty) = 1$ and $M(-\infty) = \exp(\mathrm{i}\eta - \sigma)$ with the aid of (60.20b), as expected. Lenz has cast the relation (60.28) into a form that permits the complex vector representing the local wave excitation to be directly measured from a Cornu spiral. This construction is shown in Fig. 60.4.

Another elementary case is a bar of breadth $2w$ illuminated by a plane wave propagating in the z-direction. The corresponding solution is given by

$$M_b(\boldsymbol{r}) = 1 + \frac{1-\mathrm{i}}{2}\left\{F\left(\frac{x-w}{\sqrt{\lambda z/2}}\right) - F\left(\frac{x+w}{\sqrt{\lambda z/2}}\right)\right\} \tag{60.29}$$

In the complementary case of a slit of breadth $2w$, we have

$$M_s(\boldsymbol{r}) = \frac{1-\mathrm{i}}{2}\left\{F\left(\frac{x+w}{\sqrt{\lambda z/2}}\right) - F\left(\frac{x-w}{\sqrt{\lambda z/2}}\right)\right\} \tag{60.30}$$

These two solutions satisfy $M_b + M_s = 1$ as they must do in agreement with Babinet's principle (Babinet, 1837). In practice, the conditions in which Babinet's principle is valid are rarely satisfied. See Boersch (1951), Hosemann and Joerchel (1954) and Ditchburn (1963) and, for telling examples, Lipson and Walkley (1968). For further discussion of Babinet's principle, see Born and Wolf (2002), Sections 8.3.2 and 11.3, where the boundary conditions of Kirchhoff's theory are examined carefully.

Eq. (60.30) simplifies considerably when the slit is very narrow. After some calculation the approximation

$$M_s(\boldsymbol{r}) = \frac{(1+\mathrm{i})w}{\sqrt{\lambda z/2}} \exp\left(\frac{\mathrm{i}\pi x^2}{\lambda z}\right) \frac{\sin(2\pi x w/\lambda z)}{2\pi x w/\lambda z} = \frac{(1+\mathrm{i})w}{\sqrt{\lambda z/2}} \exp\left(\frac{\mathrm{i}\pi x^2}{\lambda z}\right) \operatorname{sinc}(2xw/\lambda z) \tag{60.31}$$

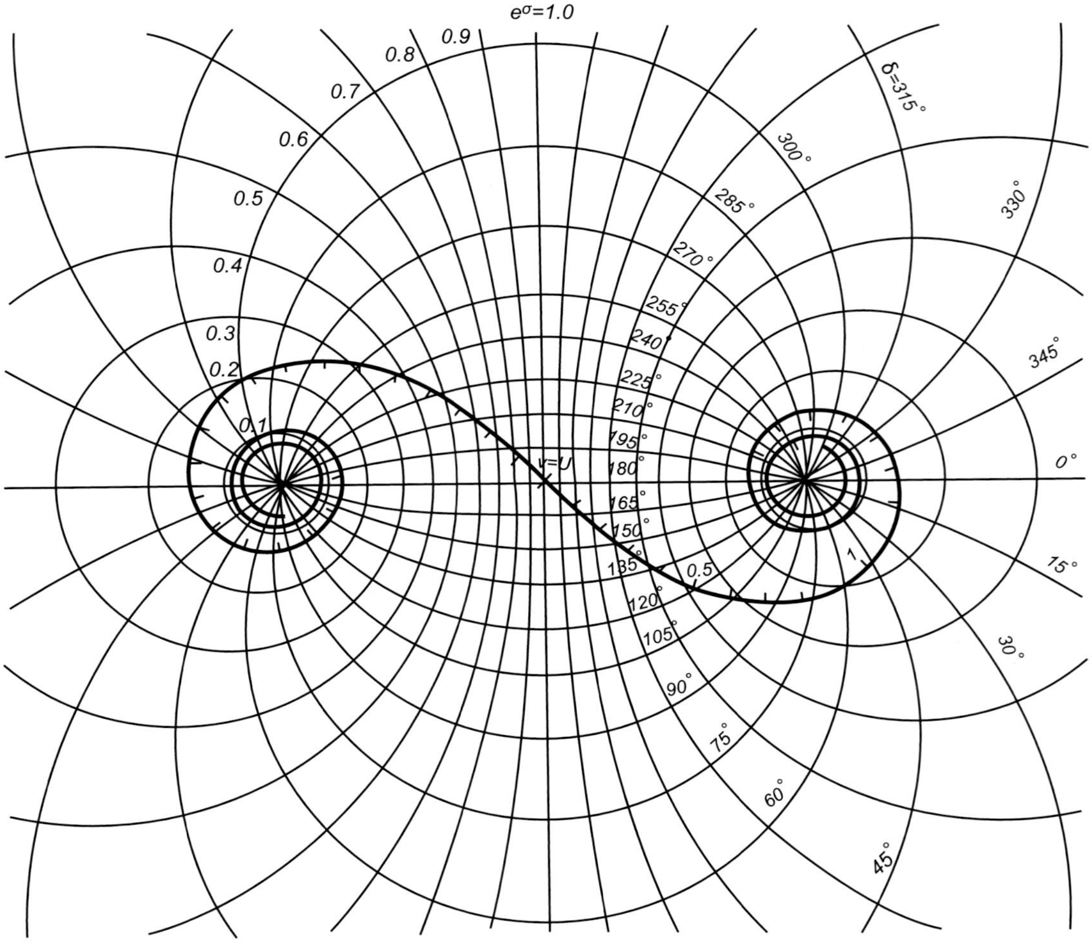

Figure 60.4
Cornu spiral for reading the amplitude and phase of the wave diffracted by a semitransparent phase-shifting half-plane together with the circles $\sigma = \text{const}$ and $\eta = \text{const}$. The length and direction of the vector from the point with coordinates σ and η to the current point with the parameter value w on the Cornu spiral represent the complex amplitude of the wave function in the plane of observation. *After Lenz (1961), Courtesy Springer Verlag, https://doi.org/10.1007/BF01679802.*

is obtained, which represents essentially the Fraunhofer diffraction at the slit. The wave amplitude decreases as $z^{-1/2}$, like that of a cylindrical wave. This is a consequence of the assumption that the slit is infinitely long in the y-direction.

As an obvious generalization, systems of parallel slits or bars can be treated by appropriate superposition of Fresnel integrals. The final solution does of course become more complicated.

60.3 Circular Structures

60.3.1 General Expression for M(r)

Circular structures are described by a rotationally symmetric object function

$$O(x_o, y_o) = O(r_o) = \mathrm{e}^{-\sigma(r_o)+\mathrm{i}\delta(r_o)} \tag{60.32}$$

$$r_o = \sqrt{x_o^2 + y_o^2} \tag{60.33}$$

being the radial coordinate in the object plane. We examine a reasonably simple configuration, which leads to *rotationally symmetric* diffraction patterns, namely a point source on the axis of symmetry, the optic axis; thus $x_s = y_s = 0$. From (60.8), we obtain the simple proportionality relations

$$\overline{x} = \mu x, \quad \overline{y} = \mu y, \quad \mu := \frac{a}{a+b} \tag{60.34}$$

x and y being the transverse coordinates in the plane of observation $z = b$. It is now advantageous to introduce polar coordinates (r_o, φ_o) in the object plane and (r, φ) in the plane of observation. From (60.10a) and (60.34), the relations

$$u = \frac{1}{L}\left(r_o\cos\varphi_o - \mu r\cos\varphi\right)$$

$$v = \frac{1}{L}\left(r_o\sin\varphi_o - \mu r\sin\varphi\right)$$

$$u^2 + v^2 = \frac{1}{L^2}\left\{r_o^2 + \mu^2 r^2 - 2\mu r_o r\cos\left(\varphi_o - \varphi\right)\right\} =: A$$

$$du\, dv = \frac{1}{L^2} r_o dr_o\, d\varphi_o$$

can then be derived. Introducing these into (60.12), we find

$$M = \frac{1}{2\mathrm{i}L^2}\int_0^\infty\int_0^{2\pi} O(r_o)\exp\left(\frac{\mathrm{i}\pi A}{2}\right) d\varphi_o r_o dr_o$$

or explicitly

$$M = \frac{1}{2\mathrm{i}L^2}\exp\left(\frac{\mathrm{i}\pi\mu^2 r^2}{2L^2}\right)\int_0^\infty O(r_o)\exp\left(\frac{\mathrm{i}\pi r_o^2}{2L^2}\right) I(r_o) r_o dr_o$$

with

$$I(r_o) := \int_0^{2\pi} \exp\left\{-\frac{\mathrm{i}\pi\mu r_o r}{L^2}\cos(\varphi_o - \varphi)\right\} d\varphi_o$$

This latter integral may be written as a Bessel function of zero order, since

$$J_0(x) = \frac{1}{2\pi}\int_0^{2\pi} \exp(\mathrm{i}x\cos\alpha)\,d\alpha \tag{60.35}$$

Thus

$$M(r) = \frac{\pi}{\mathrm{i}L^2}\exp\left(\frac{\mathrm{i}\pi\mu^2 r^2}{2L^2}\right)\int_0^{\infty} O(r_o)\exp\left(\frac{\mathrm{i}\pi r_o^2}{2L^2}\right)J_0\left(\frac{\pi\mu r_o r}{L^2}\right)r_o dr_o \tag{60.36a}$$

or explicitly (60.10a, 60.10b, and 60.34)

$$\begin{aligned} M(r) &= \frac{2\pi}{\mathrm{i}\lambda}\left(\frac{1}{a}+\frac{1}{b}\right)\exp\left(\frac{\mathrm{i}\pi a r^2}{\lambda b(a+b)}\right) \\ &\times \int_0^{\infty} O(r_o)\exp\left\{\frac{\mathrm{i}\pi r_o^2}{\lambda}\left(\frac{1}{a}+\frac{1}{b}\right)\right\}J_0\left(\frac{2\pi r_o r}{\lambda b}\right)r_o\,dr_o \end{aligned} \tag{60.36b}$$

The remaining integration can in general not be performed in closed form, but there are two important special cases that do allow further simplification. These are the *axial value* $M(0)$ and the *Fraunhofer approximation.*

60.3.2 Zone Lenses

For $r = 0$, we have $J_0(0) = 1$ and consequently

$$M(0) = \frac{2\pi}{\mathrm{i}\lambda}\left(\frac{1}{a}+\frac{1}{b}\right)\int_0^{\infty} O(r_o)\exp\left\{\frac{\mathrm{i}\pi r_o^2}{\lambda}\left(\frac{1}{a}+\frac{1}{b}\right)\right\}r_o\,dr_o$$

Using again $1/f = 1/a + 1/b$ and the substitution $u = r_o^2$, we obtain

$$M(0) = \frac{\pi}{\mathrm{i}\lambda f}\int_0^{\infty} O(\sqrt{u})\exp\left(\frac{\mathrm{i}\pi u}{\lambda f}\right)du \tag{60.37}$$

This is essentially a *Fourier integral*. As an example, we consider now a system of concentric rings, specified by the object function

$$O(r_o) = \begin{cases} 1 & \text{for } r_{j1} \le r_o \le r_{j2}, \quad j = 1\ldots N \\ 0 & \text{for } r_{j2} < r_o < r_{j+1,1}, \quad j = 1\ldots N-1 \end{cases} \tag{60.38}$$

Evaluation of (60.37) gives

$$M(0) = \sum_{j=1}^{N}\left\{\exp\left(\frac{\mathrm{i}\pi r_{j1}^2}{\lambda f}\right) - \exp\left(\frac{\mathrm{i}\pi r_{j2}^2}{\lambda f}\right)\right\} \tag{60.39}$$

This formula helps to explain the focusing action of a Fresnel zone lens. Clearly, $M(0)$ assumes its maximum value if the phases of all the exponential terms are such that they sum up constructively. The simplest possible choice is

$$r_{j1} = \{(2j-2)\lambda f\}^{1/2}, \quad r_{j2} = \{(2j-1)\lambda f\}^{1/2}, \quad j = 1\ldots N \tag{60.40}$$

so that all the open and opaque rings have the same area $A = \pi\lambda f$; the maximum is then $M(0) = 2N$. Such a ring system is shown in Fig. 60.5. It is also possible to interchange the open and the opaque rings; the result is then $M(0) = -2N$. Moreover, diffraction foci of higher orders can be obtained for

$$\frac{1}{a} + \frac{1}{b} = \frac{1}{f_m} = \frac{(2m+1)\lambda}{r_{1,2}^2} \tag{60.41}$$

m being a positive or negative integer.

Fresnel zone lenses have become of importance in X-ray optics, where they are useful in X-ray telescope and microscope design. In electron optics, there is no practical need for Fresnel lenses but a proposal has been made for reducing the effect of spherical aberration, from which all conventional lenses suffer, by means of a suitable ring-shaped diffraction grid.

This idea was first published by Hoppe (1961, 1963) and pursued by Lenz (1963, 1964, 1965). It is sketched in Fig. 60.6. The spherical aberration causes a deviation of the

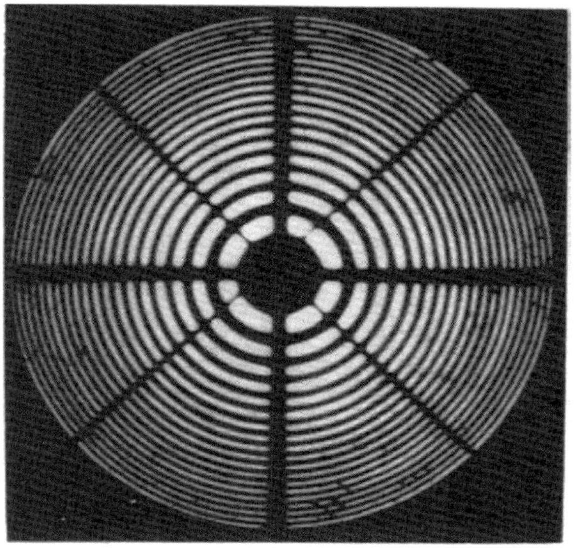

Figure 60.5
A Fresnel zone plate.

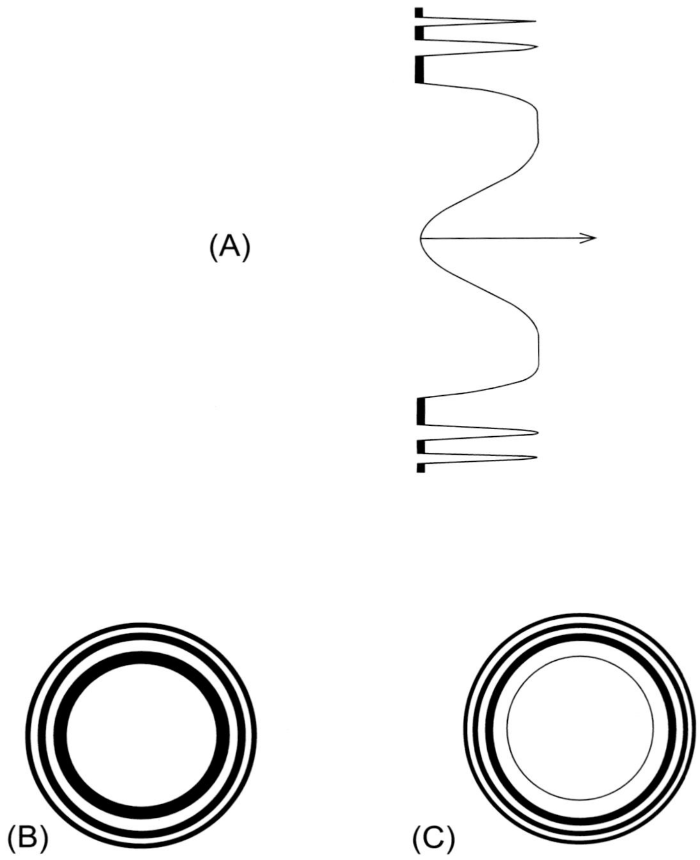

Figure 60.6
Correction plates: (A) construction, (B) Hoppe-plate, and (C) Lenz-plate.

emergent wavefronts from an exactly spherical shape (cf. Fig. 24.5). For a perfect sphere, all the contributions at the image point (the centre of the sphere) would be equidistant from the surface of equal phase and hence all would have the same phase. For a surface that departs progressively farther from a sphere as we move away from the axis, the contributions arrive with different phases. The idea behind the zone-plate corrector is to suppress zones that (on average) would contribute with a phase difference of π, 3π, ... $(2n+1)\pi$, ... while allowing in-phase contributions ($2n\pi$) to pass through and reach the image plane.

Lenz (1963, 1964) has suggested that the complementary ring system, for which the outer open zones interfere destructively with the central zone, might be useful to enhance the image contrast but subsequently withdrew this proposal.

Although all these zone plates should work in principle, they have not been successful in practice. The main reasons are that a grid within the electron beam causes a strong scattering background and is, moreover, soon deformed by strong heating. For attempts to use zone plates in the microscope, see Section 66.6, where many kinds of phase plates are examined.

60.3.3 Fraunhofer Diffraction

The outer radius of the diffracting object is now so small that the exponential term in the integrand of (60.36b) can be replaced by unity to a good approximation. This requires

$$r_o^2 \ll \frac{\lambda ab}{a+b}$$

which shows that the diffracting area must be much smaller than a Fresnel zone. We thus rewrite (60.36b) in the form

$$M(r) = \frac{2\pi}{\mathrm{i}\lambda}\left(\frac{1}{a}+\frac{1}{b}\right)\exp\{\mathrm{i}\delta(r)\}\int_0^\infty O(r_o)J_o\left(\frac{2\pi r_o r}{\lambda b}\right)r_o dr_o \tag{60.42}$$

in which the unimportant exponential factor in front of the integral is denoted by $\exp(\mathrm{i}\delta)$. Eq. (60.42) is essentially a *Fourier–Bessel transform* of the object function $O(r_o)$.

An example for which the result can be expressed in closed form is again the object function specified by (60.38). Using the familiar integral formula

$$\int_{x_1}^{x_2} J_0(x)x\,dx = x_2J_1(x_2) - x_1J_1(x_1) \tag{60.43}$$

J_1 being the first-order Bessel function and identifying x with $2\pi r_o r/\lambda b$, we obtain

$$M(r) = \frac{a+b}{\mathrm{i}ar}\exp(\mathrm{i}\delta)\sum_{j=1}^{N}\left\{r_{j2}J_1\left(\frac{2\pi r r_{j2}}{\lambda b}\right) - r_{j1}J_1\left(\frac{2\pi r r_{j1}}{\lambda b}\right)\right\} \tag{60.44}$$

This formula has a finite value at $r=0$; using $J_1(x) = x/2 + O(x^3)$, we obtain the simple relation

$$M(0) = \frac{A}{\mathrm{i}\lambda}\left(\frac{1}{a}+\frac{1}{b}\right), \quad A := \pi\sum_{j=1}^{N}\left(r_{j2}^2 - r_{j1}^2\right) \tag{60.45}$$

The factor A is the total open area of the ring system; $|M(0)|^2$ usually represents a maximum of the intensity distribution, which obviously increases with the *square* of the open area.

There are two special cases of importance. The first is a simple *circular hole* of radius R and area $A=\pi R^2$. We set $N=1$, $r_{1,1}=0$ $r_{1,2}=R$ and obtain the formula

$$M(r)=M(0)\exp\{i\delta(r)\}\frac{2J_1(x)}{x}, \quad x:=\frac{2\pi Rr}{\lambda b} \tag{60.46}$$

The function $2J_1(x)/x$ and its square, which determines the intensity pattern, the Airy disc and rings, are depicted in Fig. 60.7, curves (a) and (b). The central peak extends to the zero at $x_1^{(1)}=3.83$ and covers 83% of the total intensity. This is of some importance in connection with elementary estimates of the resolution limit of imaging systems.

The other case of interest is a narrow ring with mean radius R and half-width $h \ll R$. We now specialize (60.44) to the case $N=1$, $r_{1,1}=R-h$, $r_{1,2}=R+h$. Using the approximation

$$(x+\Delta x)J_1(x+\Delta x)-(x-\Delta x)J_1(x-\Delta x)=2x\Delta xJ_0(x)+O\big((\Delta x)^3\big)$$

and noting that the area is now $A=2\pi Rh$, we can cast the result into the form

$$M(r)=M(0)e^{i\delta(r)}J_0(x), \quad x=\frac{2\pi Rr}{\lambda b} \tag{60.47}$$

This allows a comparison to be made with (60.46). The central peak is now bounded by the zero at $x_1^{(0)}=2.40$, and is hence narrower than that of the Airy disc, see Fig. 60.7 curve (c), but the outer fringes contain much more intensity. This situation was investigated in detail by Lenz and Wilska (1966/67), who came to the conclusion that for the latter reason an electron microscope with an annular objective aperture will not give better resolution than a conventional instrument.

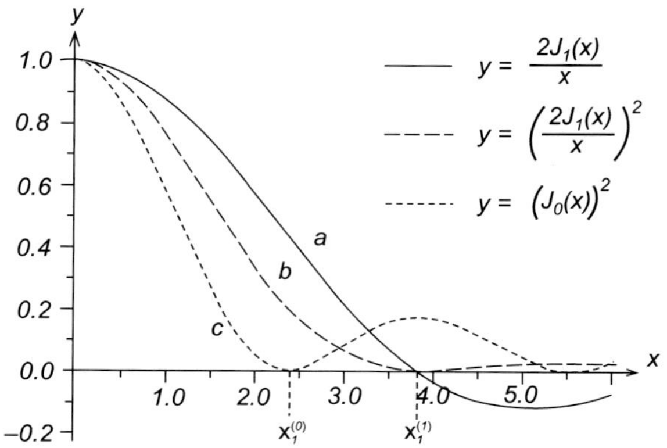

Figure 60.7
Normalized distribution functions for the diffraction at a circular hole (Airy disc, a, b) and for a ring (c) of equal radius.

60.4 Caustic Interferences

We consider now the situation sketched in Fig. 60.8. An electron beam passes through a slit aperture of width $2a$ in the object plane $z_o = -D$ and forms an imperfect line focus in the reference plane $z = 0$. In a purely geometrical description (Chapter 42) the rays form a caustic $x = x_c(z)$. This is a singular situation in which the approximations given so far must break down. The geometrical theory fails, since it predicts infinite electron intensity on the caustic itself. This is a consequence of the fact that the eikonal function then becomes singular. For the same reason the approximation of Chapter 57 also breaks down, even for very wide slits. It becomes obvious that only a wave mechanical treatment as a diffraction problem can be adequate. Since each point of the reference plane in the classically allowed region can be reached by two rays, we can expect that in the wave-optical formulation, two corresponding partial waves will interfere; we may hence anticipate that a characteristic interference phenomenon will be seen.

For conciseness, we assume that the waves propagate in field-free space. In the entrance plane $z_o = -D$, a unique definition of an eikonal is still possible:

$$L(x_o) := \frac{S(x)}{g} = -\frac{x_o^2}{2D} + \frac{1}{6C}\left(\frac{x_o}{D}\right)^3 \tag{60.48}$$

This is independent of y_o and thus represents a cylindrical wavefront; $g = \hbar k$ denotes the constant magnitude of the kinetic momentum.

The slope of the trajectory passing through the point $\boldsymbol{r}_o = (x_o, 0, -D)$ is given by

$$x' = \frac{\partial L}{\partial x_o} = -\frac{x_o}{D} + \frac{x_o^2}{2CD^3}$$

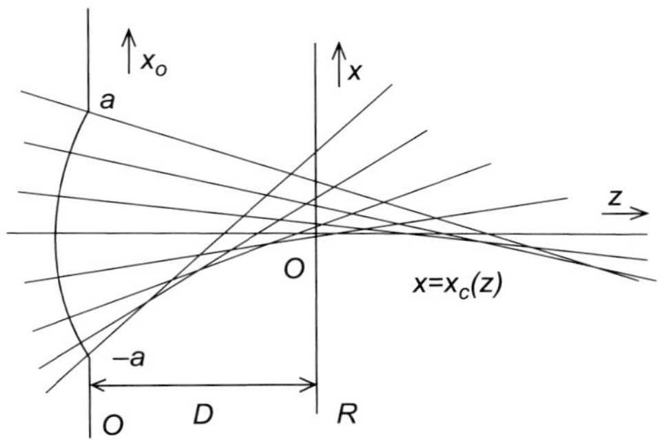

Figure 60.8
Formation of a caustic $x = x_c(z)$ by rays passing through a slit in the object plane O; R denotes the plane chosen for recording the caustic interferences. The inclination of the rays with respect to the z-axis is exaggerated for clarity.

and the Cartesian representation of this particular ray is hence

$$x(z) = x_o + x'(z + D) = -\frac{x_o z}{D} + \frac{x_o^2(z + D)}{2CD^3} \tag{60.49}$$

The whole family of all such rays, sketched in Fig. 60.8, is obtained by regarding x_o as a *parameter*, defined in the interval $[a, -a]$. The caustic is obtained as the singularity of this family from the condition $\partial x/\partial x_o = 0$, giving

$$-\frac{z}{D} + \frac{x_o(z + D)}{CD^3} = 0 \tag{60.50}$$

The Cartesian representation of the caustic is found by elimination of x_o between (60.49) and (60.50), which results in

$$x_c(z) = -\frac{z^2 CD}{2(z + D)} \tag{60.51}$$

This is valid for sufficiently small $|z|$, so that $|x_o| \leq a$; we see from this representation that the coefficient C is the *curvature* of the caustic in the observation plane $z = 0$.

The incident wave is now given by

$$\psi_o(x_o) = e^{ikL(x_o)} \tag{60.52}$$

where we have dropped an unimportant amplitude term. After making the appropriate change of notation for the coordinates, the Fresnel integral (58.17a) and (58.17b) takes the form

$$\psi(x, y, 0) = \frac{\exp(ikD)}{i\lambda D} \int_{x_o=-a}^{a} \int_{y_o=-\infty}^{\infty} \exp(i\Phi) dx_o\, dy_o$$

with

$$\Phi = \frac{k}{2D}\{(x_o - x)^2 + (y_o - y)^2 - x_o^2\} + \frac{k}{6C}\left(\frac{x_o}{D}\right)^3$$

The integration over y_o is again possible in closed form and results in a constant factor $(1 + i)(\pi D/k)^{1/2}$; ψ becomes independent of y, as it must. Combining all factors that do not depend on x_o in a common factor A, we obtain

$$\psi(x) = A \int_{-a}^{a} \exp\left[ik\left\{\frac{1}{6C}\left(\frac{x_o}{D}\right)^3 - x\frac{x_o}{D}\right\}\right] dx_o$$

This integral is not soluble in closed form. It was first evaluated numerically by Lenz and Krimmel (1961, 1963). Following their paper, we cast this integral into a standard form by the introduction of new dimensionless variables:

$$t := \frac{x_o}{D}\left(\frac{\pi}{3C\lambda}\right)^{1/3}, \quad b := \frac{a}{D}\left(\frac{\pi}{3C\lambda}\right)^{1/3}$$

$$X := 2x\left(\frac{3\pi^2 C}{\lambda^2}\right)^{1/3}$$

whereupon the diffraction integral becomes

$$\Psi(X) = \int_0^b \cos\left(t^3 - Xt\right) dt \tag{60.53a}$$

This integral can be evaluated by numerical quadrature, which is, however, very slow. Lenz and Krimmel, therefore, recommend solving the ordinary differential equation satisfied by $\Psi(X)$. Differentiation under the integral gives

$$\Psi'(X) = \int_0^b t\sin\left(t^3 - Xt\right) dt$$

$$\Psi''(X) = -\int_0^b t^2\cos\left(t^3 - Xt\right) dt$$

It is easy to verify that

$$\begin{aligned}\Psi''(X) + \frac{1}{3}X\Psi(X) &= \int_0^b \left(\frac{1}{3}X - t^2\right)\cos\left(t^3 - Xt\right) dt \\ &= \frac{1}{3}\int_0^b \frac{d}{dt}\sin\left(Xt - t^3\right) dt = \frac{1}{3}\sin\left(Xb - b^3\right)\end{aligned} \tag{60.53b}$$

is the required differential equation. This is an inhomogeneous, linear second-order equation and can hence be solved once solutions of the corresponding homogeneous equation have been found. The latter is related to Bessel's equation (Eq. 2.162.11 in Part C of Kamke, 1977) and its solutions can be expressed in terms of the tabulated functions $Z_{1/3}(X)$, $Z_\nu := C_1J_\nu + C_2Y_\nu$ in which C_1, C_2 are arbitrary constants and J_ν, Y_ν are Bessel functions of the first and second kinds. Solutions in this form are tabulated under the name of Airy functions. Lenz and Krimmel solved (60.53b), after determining $\Psi(0)$ and $\Psi'(0)$ by numerical quadrature. The results for the corresponding intensity distribution $I(X, b) = |\Psi(X, b)|^2$ are presented in Fig. 60.9. They show a very clear interference pattern which is most pronounced in the classically accessible interval $0 \leq X \leq 3b^2$ (*dashed lines* in Fig. 60.9). The appearance of weak fringes in the classically forbidden domain is common to all diffraction phenomena. For $b \ll 1$, the solution $\Psi(X, b)$ converges to the Fraunhofer solution for a narrow slit, $\Psi(X, b) = X^{-1}\sin bX = b\,\mathrm{sinc}\,(bX/\pi)$, as it must.

Lenz and Krimmel (1961, 1963) observed such an interference pattern in the line focus of a spectrometer having an aperture defect of second order. This of course required a high degree of monochromatism of the incident beam for otherwise the interference fringes would have been smeared out by chromatic effects.

This kind of diffraction pattern is characteristic of any cross-section through a *regular* part of the caustic, as is sketched in Fig. 60.8. It is unnecessary to assume wave propagation in field-free space – this was only done for reasons of conciseness; the same formulae are

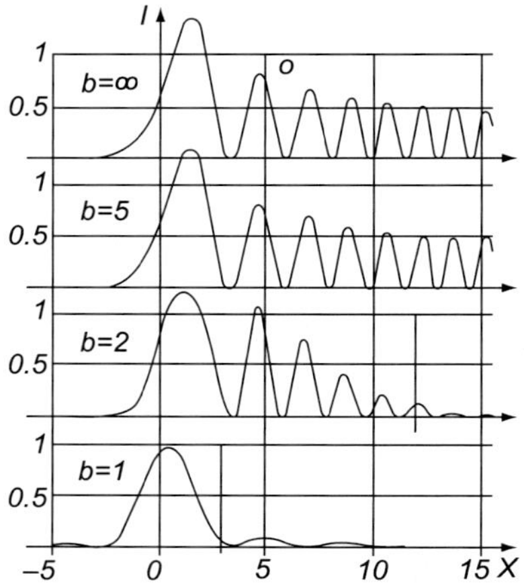

Figure 60.9
Intensity distributions in caustic interferences for various values of the parameter *b*. *After Lenz and Krimmel (1963), Courtesy Springer Verlag, https://doi.org/10.1007/BF01380585.*

obtained for curved trajectories if X is identified with the *difference* between the curvature of the caustic and the rays tangent to it. The diffraction phenomenon naturally becomes more complicated if the caustic consists of several branches, which may even intersect. Such examples were, for instance, studied by Krimmel (1960, 1961). A highly complicated situation arises in the vicinity of caustic cusps formed by spherical aberration and astigmatism. Even the geometrical theory is then extremely complicated. The corresponding wave-optical problem is soluble only by numerical techniques or via crude estimates.

Section 6.4 of Rose (2012) gives an extensive account of caustics with many examples of caustic surfaces

60.5 Diffraction Disc with Lens Aberrations

We now study the situation sketched in Fig. 60.10. A point source is located near the object focal point of a round electron lens, which produces a crossover in a recording plane, usually thought of as the image plane. This crossover is affected by axial lens aberrations and by diffraction, the latter arising at the edge of the aperture confining the electron beam. In what conditions will the diameter of the crossover be smallest for given focal length, magnification and coefficient C_s of spherical aberration?

This is a standard problem in microscopy of all kinds. Its solution is of great importance for the resolution of the electron microscope. The results obtained, however, can furnish only a

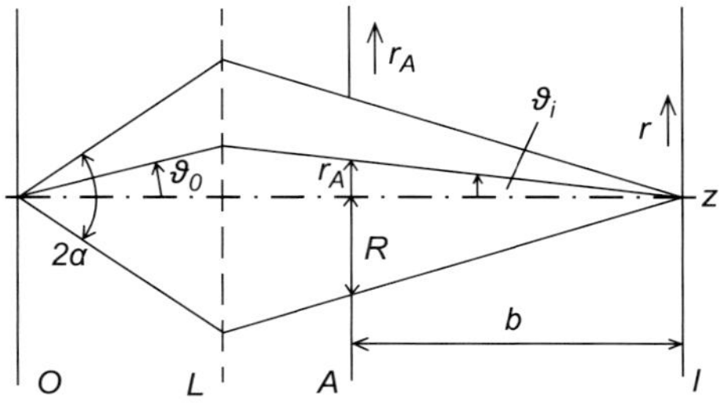

Figure 60.10
Imaging system with notation: *O*: object plane, *L*: lens, *A*: aperture and *I*: image plane. The aperture angles are exaggerated.

preliminary understanding of the resolution limit. The rigorous theory of image formation in the electron microscope is dealt with in Chapters 66 and 67.

In order to confine the topic in a reasonable manner, we consider only the spherical aberration of third order and a small defocus in addition to the diffraction. The aperture is located in the field-free space behind the lens, so that we need to consider only the simpler Kirchhoff formula. The aperture angles ϑ_0 and ϑ_i (Fig. 60.10) are related by the Smith–Helmholtz formula (Eq. 15.55 of Volume 1):

$$\vartheta_i = \frac{r_A}{b} = \frac{g_0 \vartheta_0}{g_i} M \tag{60.54}$$

g_0 and g_i being the corresponding kinetic momenta and M the lateral magnification. If the aperture is located in the lens field, then the corresponding asymptotic values for r_A and b are to be taken.

In the absence of aberrations the wave incident on the aperture is spherical with its centre at the axial image point; the contribution of interest to the eikonal of the real wave, therefore, consists only of *aberration terms*. It is convenient to express these in terms of the aperture angle ϑ_o on the object side. The eikonal $S(\vartheta_o)$, to be introduced into the exponent of the incident wave, is then

$$S(\vartheta_o) = g_o \left(\frac{1}{2} \Delta_o \vartheta_o^2 - \frac{1}{4} C_s \vartheta_o^4 \right) \tag{60.55}$$

Δ_o being the axial defocus; all coefficients and factors refer to the object side. Apart from an unimportant amplitude factor, omitted here, the incident wave is given by

$$\psi_0(\vartheta_o) = \exp\left\{ \frac{\mathrm{i}\pi}{\lambda_o} \left(\Delta_o \vartheta_o^2 - \frac{1}{2} C_s \vartheta_o^4 \right) \right\} \tag{60.56}$$

with the wavelength $\lambda_o = h/g_o$.

This function is now to be identified with the 'object function' $O(r_o)$ in (60.42). Our diffracting object is now the aperture of the system (Fig. 60.10), and consequently, the variable (r_o) is to be identified with r_A in (60.54), which can be expressed in terms of ϑ_o. Apart from factors of no interest, the wavefunction in the image plane is then

$$\psi(r) \propto \int_0^\alpha \exp\left\{\frac{\mathrm{i}\pi}{\lambda_o}\left(\Delta_o\vartheta_o^2 - \frac{1}{2}C_s\vartheta_o^4\right)\right\} J_0\left(\frac{2\pi r\vartheta_o}{\lambda_o|M|}\right)\vartheta_o d\vartheta_o \tag{60.57}$$

where we have made use of $\lambda = \lambda_i = \lambda_o g_o/g_i$ and α is defined in Fig. 60.10.

In order to cast this integral into a more concise form, we introduce dimensionless variables and coefficients as follows [cf. Eq. (65.32)]:

$$\begin{aligned} \Theta: &= \vartheta_o\left(C_s/\lambda_o\right)^{1/4} && (a)\\ \Gamma: &= \alpha\left(C_s/\lambda_o\right)^{1/4} && (b)\\ D: &= \Delta_o/(C_s\lambda_o)^{1/2} && (c)\\ \rho: &= \frac{2\pi r}{\left(C_s\lambda_o^3\right)^{1/4}|M|} && (d) \end{aligned} \tag{60.58}$$

whereupon (60.57) becomes

$$\psi(r) \propto \int_0^\Gamma J_0(\Theta\rho)\exp\left\{\mathrm{i}\pi\left(D\Theta^2 - \frac{\Theta^4}{2}\right)\right\}\Theta d\Theta$$

With the substitution $u := \Theta^2 - D$ and omission of an unimportant constant factor, $\exp(-\mathrm{i}\pi\, D^2/2)$, the complex conjugate wavefunction $\hat{\psi}(\rho) \propto \psi^*(r)$ then takes the final form

$$\hat{\psi}(\rho) = \int_{-D}^{\Gamma^2-D} J_0\left(\rho\sqrt{u+D}\right)\exp\left(\frac{\mathrm{i}\pi}{2}u^2\right)du$$

For $D > 0$ and $\Gamma^2 > D$ this can be written as

$$\hat{\psi}(\rho) = \int_0^{\Gamma^2-D} J_0\left(\rho\sqrt{D+u}\right)\exp\left(\frac{\mathrm{i}\pi}{2}u^2\right)du + \int_0^D J_0\left(\rho\sqrt{D-u}\right)\exp\left(\frac{\mathrm{i}\pi}{2}u^2\right)du \tag{60.59}$$

We shall deal only with this case, since the case $D < 0$ is of little practical interest; the reasons will be obvious later.

Apart from the axial value $\hat{\psi}(0)$, these integrals cannot be evaluated in closed form. The axial value is given in terms of Fresnel integrals (60.19):

$$\hat{\psi}(0) = F\left(\Gamma^2 - D\right) + F(D) \tag{60.60}$$

Another useful result can be derived from the conservation of integrated intensity:

$$\int_0^\infty \rho\left|\hat{\psi}(\rho)\right|^2 d\rho = 2\Gamma^2 \tag{60.61}$$

This simply means that the integrated intensity increases as the area of the aperture.

These two results enable us to deduce optimum working conditions for the lens. It is tempting to make $\left|\hat{\psi}(0)\right|^2$ as large as possible. This gives a unique answer:

$$D_m = 1.2, \quad \Gamma_m = D_m\sqrt{2} = 1.7, \quad \left|\hat{\psi}_m(0)\right|^2 = 3.6 \tag{60.62}$$

and corresponds to the point on the Cornu spiral (Fig. 60.2) that is farthest from the origin. This is, however, *not* necessarily the best choice; perhaps a slightly smaller value of $\left|\hat{\psi}(0)\right|^2$ would give a higher concentration of intensity in the central disc relative to the outer rings. We, therefore, seek the maximum of $\left|\hat{\psi}(0)\right|^2/\Gamma^2$. This quest again has a unique outcome:

$$D_{\text{opt}} = 1, \quad \Gamma_{\text{opt}} = \sqrt{2}, \quad \left|\hat{\psi}(0)\right|^2_{\text{opt}} = 3.2 \tag{60.63}$$

and $|\psi(0)|^2_{\text{opt}}/\Gamma^2_{\text{opt}} = 1.6$ is indeed larger than $\left|\hat{\psi}_m(0)\right|^2/\Gamma^2_m = 1.25$ from (60.62): the choice (60.63) does lead to a better intensity concentration in the central disc. This is hence the required optimum. The quantity $\left|\hat{\psi}(0)\right|^2/\Gamma^2$ is very similar to the Strehl intensity ratio (*Definitionshelligkeit*) used in optical instrument design (Strehl, 1902; Born and Wolf, 2002, Section 9.1.1),

Introducing (60.63) into (60.58b) and (60.58c), we find the aperture angle

$$\alpha_{\text{opt}} = \left(\frac{4\lambda_o}{C_s}\right)^{1/4} \tag{60.64}$$

and the object defocus

$$\Delta_{\text{opt}} = \sqrt{\lambda_o C_s} \tag{60.65}$$

This is the Scherzer focus, which we shall meet again in Section 66.4. Moreover, it is possible to estimate a mean square radius from (60.61). To find this, we replace $\hat{\psi}(\rho)$ by a piecewise constant function:

$$\hat{\psi}(\rho) \approx \begin{cases} \frac{1}{2}\hat{\psi}(0) & \text{for } \rho \le \overline{\rho}, \\ 0 & \text{otherwise} \end{cases}$$

Eq. (60.61) then gives $\left|\hat{\psi}(0)\right|^2 \rho^2/8 = 2\Gamma^2$, and so $\overline{\rho} = 4\Gamma/\left|\hat{\psi}(0)\right| = 3.16 \approx \pi$. Substitution of this into (60.58d) gives the average disc radius

$$\bar{r} = 0.5|M|\left(C_s\lambda_o^3\right)^{1/4} \tag{60.66}$$

which agrees fairly well with experimental findings.

The intensity distribution $\left|\hat{\psi}(\rho)\right|^2$ for the optimum case is plotted in Fig. 60.11. This function has a first minimum at $\rho_m = 2$, which is distinctly smaller than $\bar{\rho}$. For sufficiently small values of ρ this function is represented approximately by

$$\left|\psi(\rho)\right|^2 = \frac{3.2J_0^2(1.2\rho)}{1 + \rho^2/144} \tag{60.67}$$

which satisfies (60.61) acceptably for $\Gamma = 2$.

The optimum defocus $(C_s\lambda)^{1/2}$ was found by Scherzer (1949) and is widely known as *Scherzer focus*. However, some authors, notably Spence (2013), Williams and Carter (2009) and Danev and Nagayama (2001) include the factor 1.2 (60.62) and refer to 1.2 $(C_s\lambda)^{1/2}$ as Scherzer (de)focus.

60.6 The Rayleigh Rule and Criterion

Rayleigh's quarter-wavelength rule is well-established in optics and frequently used to estimate the parameters of complicated aberration patterns. It states that the quality of an image is not seriously affected if the distance between the spherical wave surface and the true wave surface, distorted by aberrations, nowhere exceeds $\lambda/4$ and, hence, if the maximum absolute phase difference relative to the same point nowhere exceeds $\pi/2$.

We shall now reconsider the aberration disc produced by diffraction, spherical aberration and defocus, which we have investigated in the previous section, in the light of this rule.

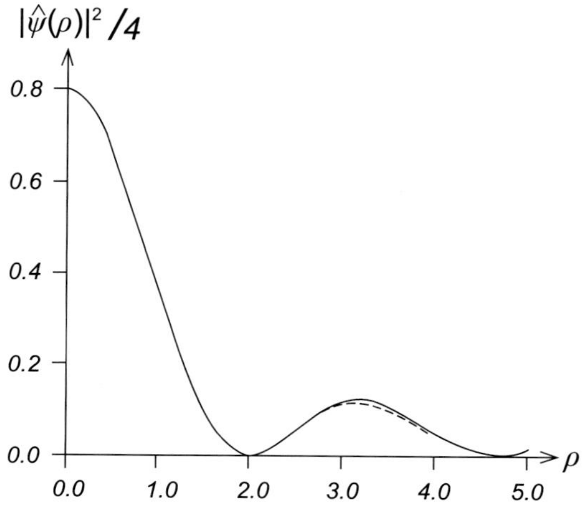

Figure 60.11
Intensity distribution in a focus with optimum defocus. Broken lines: exact numerical solution; full lines: approximation by formula (60.67).

The reference pattern for the comparison is the Airy disc for aberration-free diffraction, given by (60.46).

The Airy disc is produced by a constant object function $O(r_o) = \psi_o(\vartheta_o) = 1$. Hence, (60.56) would lead to essentially the same pattern if, according to the Rayleigh rule,

$$\left|\Delta_o \vartheta_o^2 - C_s \vartheta_o^4/2\right| \leq \lambda_o/2 \tag{60.68}$$

is satisfied for all $\vartheta_o \leq \alpha$. In the reduced variables (60.58) this is equivalent to

$$\left|2D\Theta^2 - \Theta^4\right| \leq 1 \quad \forall \Theta \leq \Gamma \tag{60.69}$$

The task of optimization is now to find those pairs of values Γ and D for which the greatest intensity is concentrated in the central disc subject to the constraint (60.69). The answer can be found from Fig. 60.12, which shows the function $w(\Theta) = 2D\Theta^2 - \Theta^4$ for various values of the parameter D.

The optimal situation corresponds to the curve that just reaches unity at its maximum. For this, $D = \Theta_M^2$, $w_M = \Theta_M^4 = 1$ and hence $\Theta_M = 1$, $D_{\text{opt}} = 1$. The maximum allowed aperture is then determined by the next zero of $w(\Theta)$: $w(\Theta) = 0$ gives $\Gamma_{\text{opt}} = \sqrt{2}$. These are the same as the parameters given in (60.63), which are thus confirmed by a different argument.

The Rayleigh rule also allows us to estimate the radius of the central diffraction spot. The radius of the Airy disc is traditionally chosen as the first zero of the Bessel function in (60.46); setting $x = 3.83$ gives

$$\bar{r} = \frac{3.83\lambda b}{2\pi R} = \frac{0.61\lambda b}{R}$$

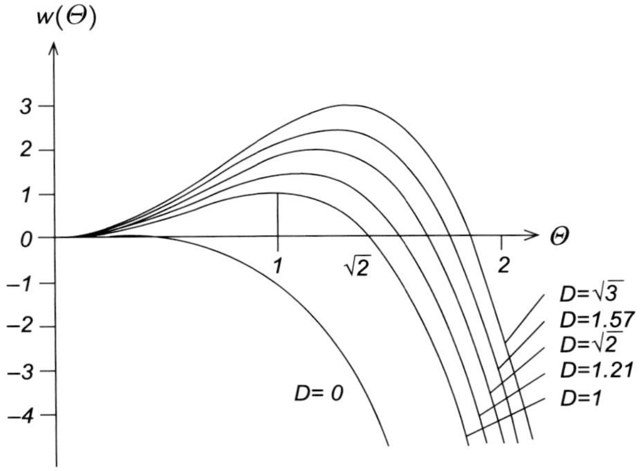

Figure 60.12
Reduced wave aberration $w(\Theta)$ for various values of the defocus parameter D.

If we use $\vartheta_i = R/b$, $\vartheta_o = \alpha$ and $r_A = R$ in (60.54) and recall that $g_o \lambda_o = g_i \lambda$, we obtain

$$\bar{r} = \frac{|M| 0.61 \lambda_o}{\alpha} \tag{60.70}$$

which will be recognized as Abbe's resolution limit for an ideal lens. Inserting now the optimum aperture angle from (60.64), we arrive at

$$\bar{r} = 0.43|M|\left(C_s \lambda_o^3\right)^{1/4} \tag{60.71}$$

which agrees rather well with (60.66). An error of 15% in such an estimate is quite acceptable since the accuracy of measurements is generally not better.

Traditionally, these formulae are used to determine the resolution limit of the electron microscope or any other imaging instrument. This is an everyday procedure in conventional light optics and has been transferred into electron optics by analogy. According to Rayleigh's criterion, two overlapping diffraction patterns are considered to be separated if the intensity at the saddle point is at most 75% of that at the two maxima (see Fig. 60.13). For Airy discs, the distance $\bar{r}$ between the two maxima is then given by (60.70) and (60.71). This is then regarded as the classical resolution limit of the instrument.

All these arguments are based on the tacit assumption that the object can be treated as a collection of individual point sources that radiate incoherently. This means that it is the intensities and not the wavefunctions, produced by the individual sources, that superimpose linearly, as is shown in Fig. 60.13. This assumption is frequently unjustified in electron optics and the reasoning leading to the estimate (60.71) for the resolution limit is extremely over-simplified. Moreover, the incoherent assumption is particularly unlikely to be true in high-resolution conditions. Much of the present volume is devoted to the task of developing a more satisfactory approach to image formation. It is, perhaps, reassuring that (60.64) and

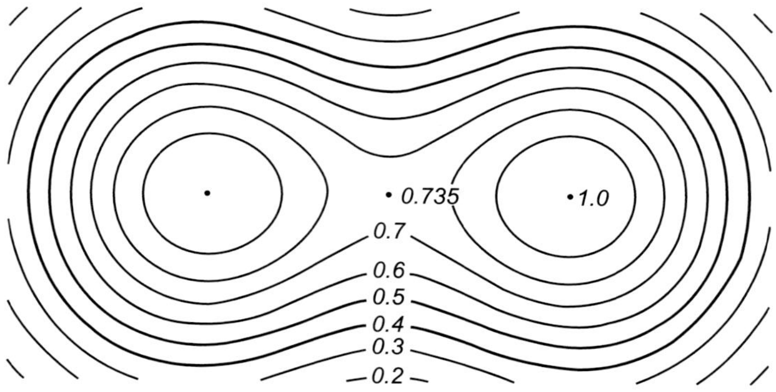

Figure 60.13
Superposition of the intensities of two Airy discs at a mutual distance of $\delta = 0.61\lambda_o/\sin\alpha$.

(60.65) also prove to be optimum conditions when we discuss resolution in terms of the contrast transfer function.

We close this section with an estimate of the maximum allowable astigmatism for which there is no significant image blurring. With axial astigmatism, the wavefunction in the aperture plane is

$$\psi_o(\vartheta_o, \rho_o) = \exp\left[\frac{\mathrm{i}\pi}{\lambda_o}\left\{\Delta_o\vartheta_o^2 - \frac{1}{2}C_s\vartheta_o^4 + A_1\vartheta_o^2\cos2(\varphi_o - \beta)\right\}\right] \tag{60.72}$$

instead of (60.56), $A_1 > 0$ being the astigmatism coefficient referred to the object side (Haider notation, see Table 31.1 of Volume 1). In principle the astigmatism can be compensated but this does not always succeed perfectly and an estimate of its influence on the resolution is hence useful.

The corresponding diffraction integral cannot be evaluated in closed form; even the integration over the azimuth φ_o cannot be carried out analytically. We can, however, again apply Rayleigh's quarter-wave rule. Taking the diffraction pattern without astigmatism as reference for comparison, we see that the astigmatism term in (60.72) must satisfy

$$A_1\vartheta_o^2 \le \frac{\lambda_o}{4}$$

since the phase difference between the value at $\varphi_o = \beta$ and the value at $\varphi_o = \beta + \pi$ must not exceed $\pi/2$. Introducing α_{opt} from (60.64) for ϑ_o, we obtain the estimate

$$A_1 \le \frac{1}{8}\sqrt{\lambda_o C_s} = \frac{\Delta_{\mathrm{opt}}}{8} \tag{60.73}$$

This little example demonstrates clearly that Rayleigh's rule may be quite helpful when seeking preliminary estimates in complicated situations.

PART XII

Electron Interference and Electron Holography

CHAPTER 61

General Introduction

When a coherent or partially coherent electron wavefront is divided into separate parts, which traverse regions of different refractive index and are subsequently recombined, interference fringes will be created. This situation may arise in many ways. In conventional electron microscopy, the electron beam may traverse a wedge-shaped crystal (Fig. 61.1A), in which case fringes will be formed by interference between the part of the beam that has passed through the specimen and the part that has not. Alternatively, the beam may be deliberately divided into two or more parts, by means of an electron biprism (Fig. 61.1B) or a crystal or a structure such as a grating. The phase-contrast bright-field image in a conventional electron microscope may even be regarded as the interference pattern created by the undiffracted beam and the electrons scattered elastically by the specimen.

Interference patterns are of course intensity distributions but their detailed structure is governed by the phase differences between the waves that interfere. A particularly important family of interference phenomena is now known as *holography*. The impossibility of designing a rotationally symmetric electron lens free of third-order spherical aberration (see Section 24.3 of Volume 1, Scherzer's theorem) and the experimental difficulty of introducing correctors such as those proposed by Scherzer (see Chapter 41 of Volume 2) and by himself led Gabor to enquire whether electron images degraded by aberrations could not be reconstituted optically. He argued that the object information in an image obtained with a lens with aberrations was not lost but entangled – wrongly presented to the eye – and suggested that a correct image could be reconstructed by a two-stage holography process (Gabor, 1948, 1949a, 1949b, 1950, 1951a, 1951b). His publications provoked other suggestions and comments (Dyson, 1950; Rogers, 1950a, 1950b, 1952; Haine and Dyson, 1950) and attempts to test the idea experimentally were made by his colleagues at the Associated Electrical Industries' Research Laboratory in Aldermaston (Haine and Mulvey, 1950, 1951, 1952). These could not succeed owing to the lack of coherence[1] of both the electron source used to form the hologram and the light source employed in the reconstruction (the laser had not yet been invented) and the idea lay fallow for many years.

[1] The term 'coherence' is used loosely throughout this part in connection with non-vanishing source size and energy spread. The related effects are discussed in Section 66.2 (contrast-transfer function) and more formally in Part XVI.

Principles of Electron Optics.
DOI: https://doi.org/10.1016/B978-0-12-818979-5.00061-9

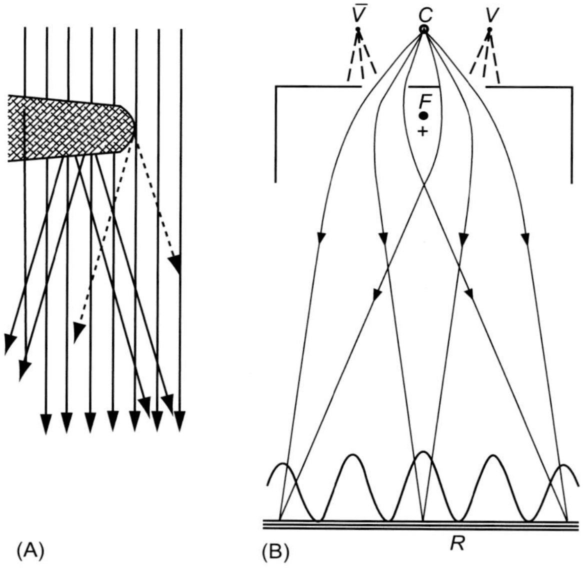

Figure 61.1
Possible ways of producing electron interference: (A) *diffraction* in a crystalline specimen and at its edges; the diffracted waves interfere with the undiffracted ones. The resulting fringes are visible in a defocused image. (B) *Electron biprism* with real source C, virtual sources V, $\overline{V}$ entrance slits, filament F and recording medium R.

The basic idea was as follows. A thin specimen, transparent to electrons, is irradiated and traversed by an electron wave, the *object wave*. The latter is modulated by the electric field inside the specimen and thus contains information about the object structure. This object wave is then brought into coincidence with a second wave, which has not been affected by the specimen and is known as the *reference wave* (Fig. 61.2A). The resulting interference pattern is then recorded on a glass plate, the standard recording medium in Gabor's time. It is this pattern that is known as a *hologram* and it is of course affected by the properties of the device used to create the interference.

In a second quite separate step, the hologram is trans-illuminated by a coherent wave, which need not be an electron wave and will in practice be coherent light from a laser. As we shall see in more detail later, diffraction in the hologram generates *three* waves. One has the structure of the reference wave. A second wave is a reconstruction of the object wave and seems to emanate from a virtual object O' (Fig. 61.2B). The third wave, the *conjugate wave*, converges to a 'real' object O'', which has a different orientation from the specimen O. These three waves must now be separated from one another as completely as possible so that one of the object waves can be isolated.

This stage is known as *reconstruction*. An advantage of this procedure in light optics is that we have complete control over the basic geometrical aberrations, and in particular over the spherical

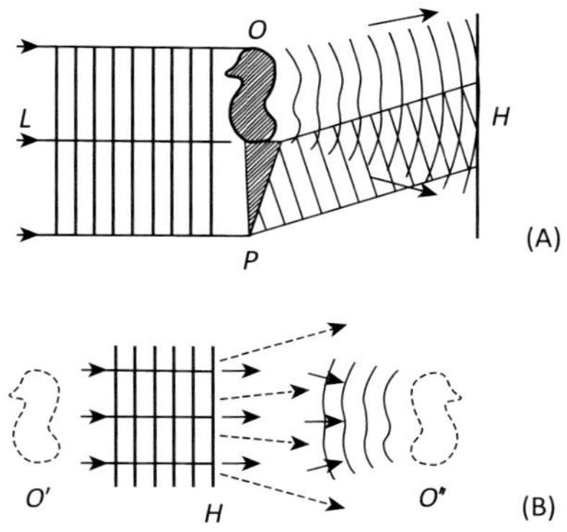

Figure 61.2
Simplified presentation of light-optical holography: (A) formation of the hologram; (B) object reconstruction. *L*: laser-beam; *O*: physical object (specimen); *P*: prism; *H*: hologram and *O*′, *O*″: virtual and real reconstructed object, respectively.

aberration; in principle, therefore, the electron-optical aberrations can be compensated by suitable choice of the glass lenses employed in the reconstruction.

The principle of holography in this simple form is illustrated in Fig. 61.2, where light is used for both the recording of the hologram and for the reconstruction. Technical details (the laser, lenses and diaphragms) are omitted. In the reconstruction step, irradiation of the hologram *H* generates the virtual (*O*′) and real (*O*″) images shown.

Holography is indeed a two-stage process, therefore. In the first stage, a reference wave and an object wave interfere, producing a hologram, and in the second stage, information about the object wave is extracted from this hologram. There are numerous ways of performing each step and in practice it is convenient to distinguish between *in-line* holography (Fig. 61.3A and B) and *off-axis* holography (Fig. 61.3C and D). Each technique has specific advantages and also disadvantages.

In the *in-line* arrangement, the optic axis is a symmetry axis and passes through the points *O*, *O*′ and *O*″ of Fig. 61.3A and B. The aberrations of the lenses (not shown here) have less effect than in the off-axis arrangement but it is not easy to separate the wavefronts *R*, *W*′ and *W*″ sufficiently in the reconstruction stage.

There is no such difficulty in the off-axis arrangement (Fig. 61.3D) but the effect of the lens aberrations in the recording stage is greater, owing to the highly asymmetric disposition of the wavefronts (Fig. 61.3C). The choice between the two arrangements may well be governed by the availability and properties of the appropriate optical elements: sources, beam splitters, apertures and recording media.

The two wavefronts *W*′ and *W*″ of Fig. 61.3B and D are completely equivalent so far as the reconstruction is concerned. It is the subsequent arrangement of lenses and apertures that

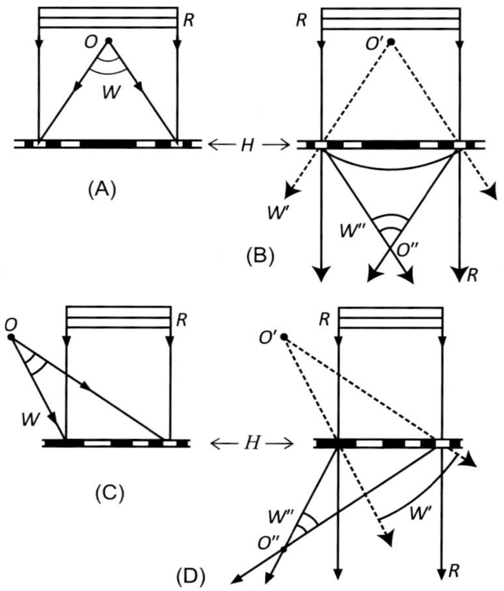

Figure 61.3
Principal types of holography: (A and B) *in-line* set-up; (C and D) *off-axis* set-up; (A and C) recording; (B and D) reconstruction; *R*: reference wave; *H*: hologram; *W*: recorded object wave; *O*: physical specimen; *W′*, *W″*: diverging and converging reconstructed waves and *O′*, *O″*: reconstructed virtual and real objects.

determines which of them is finally recorded. In Chapter 63, we shall see that the three waves *R*, *W′* and *W″* can be considered as three orders, zero and ± 1, in the Fresnel diffraction pattern of the illuminating wave at the hologram film *H*. (The diffraction angles in Fig. 61.3 are highly exaggerated – in reality, they will be extremely small.) In a first very simple approximation, the waves may be regarded as practically plane. The necessary focusing by a lens and beam selection by an aperture are then as shown in Fig. 61.4. The lens is adjusted in such a way that all the beams belonging to the same diffraction order are focused onto the aperture plane. Simultaneously, it must image the hologram with all the beamlets that pass through the aperture onto the image plane. This particular arrangement is known as *image-plane holography*. Other arrangements for producing and reconstructing holograms are described in Chapter 63.

We have followed the traditional development of holography in the foregoing introduction: electron-optical hologram formation, light-optical reconstruction. It has long been apparent, however, that many of the practical difficulties of the reconstruction step could be avoided by digitizing the hologram and performing the reconstruction step computationally. The flexibility would thereby be increased and the only noise present would be that due to the digitization and that of the hologram. Correction of the high spherical aberration of an electron lens by means of a glass lens with equal and opposite aberration is far from easy (see Rogers, 1978, 1980, for

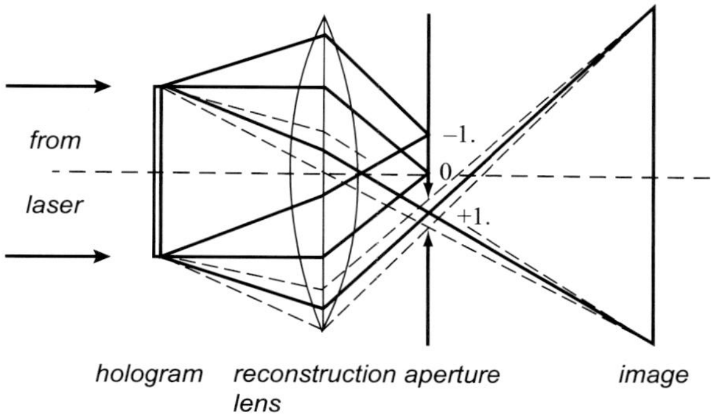

Figure 61.4
Light-optical reconstruction: the foci of the reference wave (0) and of the diffraction order (−1) are excluded by an aperture-containing diaphragm. The rays that pass through the aperture produce a sharp image of the object (image-plane holography).

example, who had a special lens with abnormally high primary aberrations made for this purpose) but is trivial digitally. We shall frequently describe reconstruction in light-optical terms but all of these procedures are in practice performed by digital routines. Moreover, the hologram is digitized immediately by the direct detectors in use today (see Section 67.5). In earlier times, it had to be recorded photographically and digitized by means of a microdensitometer.

It should be clear from this introduction that electron interferometry, electron holography and the phase-contrast bright-field image-forming process are all intimately related (Harada, 2021). We now attempt to disentangle the various strands by considering the main stages in the development of these themes. We stress that only the most important publications are mentioned here. Further references are included in the list for this part. First, however, we examine a question that arose as soon as electron interference became perceptible: if we think of electrons as charged particles, how can they form interference patterns? Many attempts have been made to shed light on this paradox. Biberman et al. (1949) recorded the diffraction pattern of manganese oxide crystallites in conditions in which only one electron was present in the column of the microscope at any one time. An early interference experiment of Marton and colleagues, in which a crystal was used as beam splitter, was inconclusive, but, in 1956, Jean Faget and Charles Fert, working in the Toulouse Laboratory of Electron Optics, published examples of Fresnel diffraction fringes (already well known), the fringe pattern between two holes in an opaque screen (Young's fringes) and interference created by the electron biprism recently introduced by Düker (1955) and Möllenstedt and Düker (1955, 1956). Longer accounts followed (Faget and Fert, 1957a, 1957b; Faget, 1959, 1961) and a book chapter by Fert brought all this work together (Fert, 1961). Two of their experiments are particularly interesting. In one, the fringes between two holes of diameter 1 μm, the centres of which were 1.5 or 2 μm apart, were recorded and an intensity variation can just be discerned although the experimental conditions were

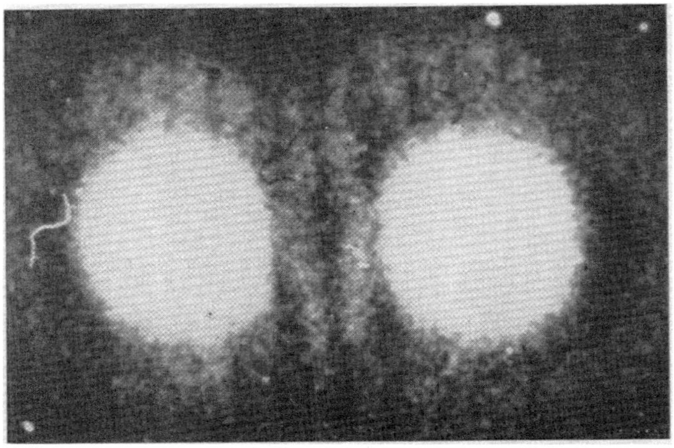

(A)

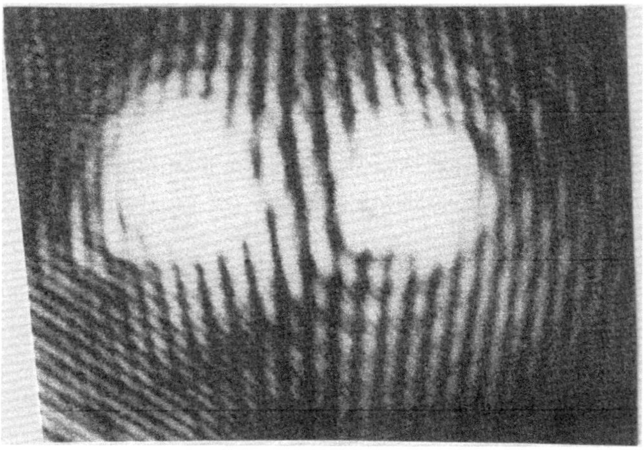

(B)

Figure 61.5

(A) Young's fringes created by two holes, about 1 μm in diameter; their centres are between 1.5 and 2 μm apart. The authors comment that 'La nécessité d'une source ponctuelle, le caractère fortuit de la rencontre de deux trous ayant le diamètre et l'intervalle convenables font de cette expérience fondamentale une expérience difficile. Elle mérite cependant d'être refaite dans des conditions expérimentales meilleures, dès qu'un progrès technique le permettra'.[2] (B) Fringe patterns generated by holes about 1 μm in diameter in an opaque gold film illuminated by an electron beam from a field-emission source (angular spread less than 1 μrad). *(A) After Faget and Fert (1956), Courtesy Académie des Sciences de Paris. (B) After Ohtsuki and Zeitler (1977), Courtesy Elsevier.*

primitive (Fig. 61.5A). A similar experiment was performed many years later by Ohtsuki and Zeitler (1977); their much sharper fringe patterns are shown alongside those of Faget

[2] The need for a point source, the chance of finding two holes with suitable diameter and separation make this fundamental experiment a difficult experiment. It deserves to be repeated in better experimental conditions as soon as technical progress makes this possible.

and Fert in Fig. 61.5B. In the second of Faget and Fert's experiments, the electrons were sufficiently well separated for the authors to write 'Les électrons franchissent un à un l'interféromètre et sont enregistrés un à un par l'émulsion[3]' (Faget and Fert, 1957a).

In 1959 Gottfried Möllenstedt and Claus Jönsson in Tübingen succeeded in producing multiple slits and recorded fringe patterns from one, two, three, four and five slits (Jönsson, 1961). In 1969 Herrmann et al. were able to observe the build-up of fringe patterns with the aid of an image intensifier, only recently available for civilian use (Herrmann et al., 1969, 1971, 1972). A decade later, several papers showing interference fringes were obtained by scientists in the University of Bologna (Donati et al., 1973; Pozzi et al., 1974, 1976; Merli et al., 1976a, 1976b). A film showing the gradual build-up of the interference pattern was recorded and the subject was surveyed by Missiroli et al. (1981) who included many related studies not described here. Another film was produced by Tonomura et al. (1989) in conditions that made it certain that there was at most one electron in the column at any given time. It was this paper and that of Jönsson that were cited in response to an invitation in *Physics World* to choose 'the most beautiful experiment in physics'. A lively correspondence followed (gathered together by Rodgers, 2003), in which the Bologna work was recognized but the pioneering work of Faget and Fert was not mentioned. (Faget and Fert, 1957a had been cited by Tonomura et al. and by Missiroli et al. but not in the other Bolognese publications.) See also Merli et al. (2003). Frabboni, Gabrielli and colleagues have recorded the build-up of the fringes with a very fast detector, which confirms that the electrons do indeed arrive one at a time (Frabboni et al., 2012; Gabrielli et al., 2013, cf. 2011; Matteucci et al., 2013). Simonsohn and Weihreter (1979) performed the double-slit experiment with single-electron detection (see also Frémont et al., 2008).

The first electron interferometer was built by Marton and colleagues at the US National Bureau of Standards (now the National Institute of Science and Technology, NIST); this was an *amplitude-division* design (Fig. 61.6) in which the wavefront was divided and recombined by Bragg reflection within a thin crystalline film (Marton, 1952, 1954; Marton et al., 1953, 1954; Simpson, 1954, 1956). The problems of alignment and lack of intensity of their device were considerable. Such interferometers have been revived, the crystals used by Marton often replaced by fine gratings. Their evolution can be traced through the following papers: Rackham et al. (1978), Buxton et al. (1978), Matteucci et al. (1980, 1981, 1982a, 1982b, 1982c), Ru et al. (1994a, 1994b), Ru (1995), Mertens et al. (1999), Mertens and Kruit (1999), Hansson et al. (2000), Zhou (2001), Gronniger et al. (2006, 2000), Hirayama et al. (2005), Dronyak et al. (2009), Bach et al. (2013), Tavabi et al. (2016, 2017) and Agarwal et al. (2017). Real progress in interferometry was first made when Möllenstedt and Düker in the University of Tübingen introduced the *electron biprism*, in which a fine, charged wire mounted between earthed plates splits the incident beam into two parts by

3 The electrons traverse the interferometer one by one and are recorded by the emulsion one by one.

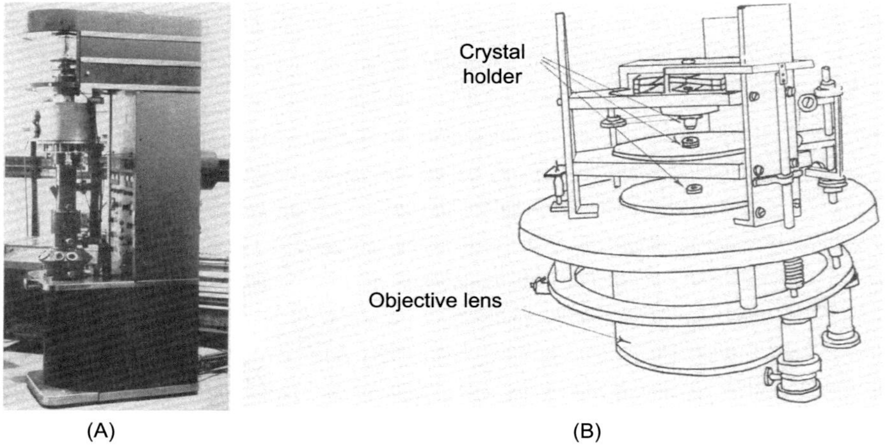

Figure 61.6
Marton's interferometer: (A) electron microscope in which the specimen chamber has been replaced by the interferometer; (B) the interferometer. *After Marton (1954), Courtesy Royal Microscopical Society.*

wavefront-division; these are subsequently reunited by a lens (Düker, 1955; Möllenstedt and Düker, 1955, 1956; Möllenstedt, 1987a, 1987b, 1991). These early biprism experiments were performed with electrostatic lenses; similar work was undertaken with a magnetic column very soon after by Faget and Fert in Toulouse (Faget and Fert, 1956, 1957a, 1957b; Faget et al., 1958; Faget, 1961; Fert, 1961, 1962). Fig. 61.7 shows how the fringes build up as the biprism voltage is increased.

With the introduction of the biprism by Möllenstedt and Düker, interference microscopes were constructed in many laboratories and ingenious combinations of biprisms and other optical elements were tested. Arrangements using more than one biprism were employed by Möllenstedt and Bayh (1961a, 1961b) and Schaal et al. (1966, 1967) and by Lichte who built an electron mirror interference microscope as shown in Fig. 18.2 of Volume 1 (Lichte et al., 1972; Lichte, 1979, 1980a; Lichte and Möllenstedt, 1977, 1979). In Japan, a source using a pointed filament developed by Hibi (1956) enabled Hibi and Takahashi (1963) to obtain many sharp fringes thanks to its brightness and several versions were subsequently built, with high magnification in mind (Tomita et al., 1970a, 1970b, 1972; Yada et al., 1973). The advent of the field-emission gun brought major improvements (Brünger, 1968, 1972; Crewe and Saxon, 1971). Other interference microscopes were developed by Boersch et al. (1960, 1962a), who used anti-parallel magnetic domains in a ferromagnetic foil as a beam splitter and by Buhl (1958, 1959), Keller (1958, 1961a, 1961b), Krimmel (1960), Feltynowski (1963), Drahoš and Delong (1963, 1964),

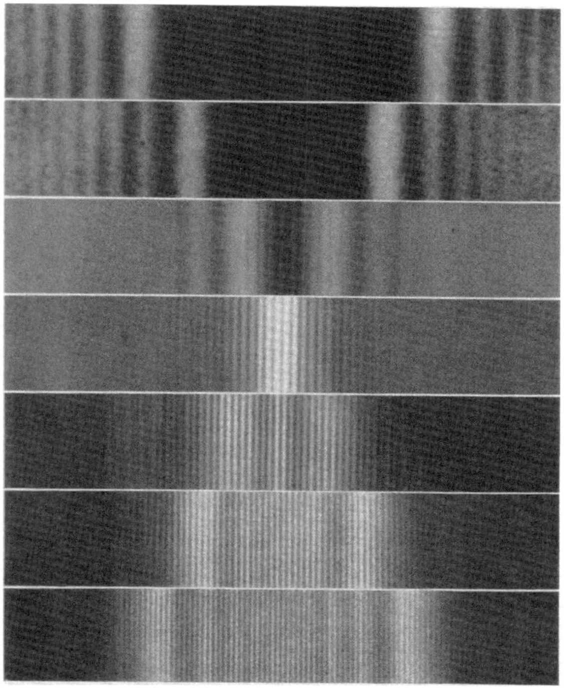

Figure 61.7
The influence of the filament potential of an electron biprism. As the potential is increased from zero (top) to 7 V (bottom), the two sets of Fresnel fringes of the filament approach each another. When they overlap, the Fresnel fringes are superimposed and equidistant interference fringes become visible. In this early experiment of Möllenstedt and Düker (1956), the filament was 2 μm in diameter. *After Möllenstedt and Düker (1956), Courtesy Springer, https://doi.org/10.1007/BF01326780.*

Anaskin et al. (1966), Kerschbaumer (1967), Komrska et al. (1964a, 1964b, 1967), Sonier (1968, 1971) and Merli et al. (1974). The natural extension to multiple-beam interference was made by Möllenstedt and Jönsson (1959), using several slits and by Anaskin and Stoyanova (1967, 1968a, 1968b, 1968c) and Anaskin et al. (1968), using several wires in a different configuration (Fig. 61.8) (see also Jönsson, 1961; Buhl, 1961a, 1961b; Jönsson et al., 1974; Frabboni et al., 2010; Frabboni, 2011). A magnetic biprism in which no part of the incident beam is obscured was examined by Krimmel (1960, 1961). A mixed interferometer incorporating both a crystalline film and a biprism is used by Matteucci and Pozzi (1980). These instruments have been used to measure the degree of coherence of electron sources (see Hibi and Takahashi, 1969; Hibi and Yada, 1976, Section IV.B of Hawkes, 1978, for many early references and also Braun, 1972; Speidel and Kurz, 1977, Möllenstedt and Wohland, 1980; Lenz and Wohland, 1984; Schmid, 1984; Lenz, 1987;

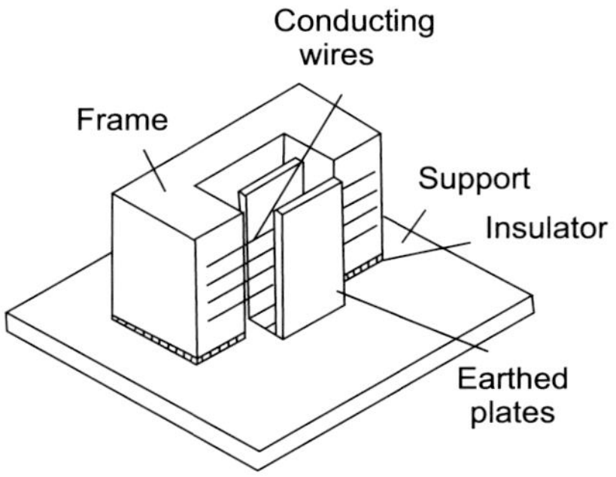

Figure 61.8
Multiple-beam interference using several wires. *After Anaskin et al. (1968).*

Medina and Pozzi, 1990) and the mean inner potential (see Section 69.3.3 and Section 66.6) of thin specimens (Möllenstedt and Buhl, 1957; Möllenstedt and Keller, 1957; Durand et al., 1958; Langbein, 1958; Buhl, 1959; Keller, 1956, 1961a, 1961b, 1962; Hoffmann and Jönsson, 1965; Jönsson et al., 1965; Kerschbaumer, 1967; Tomita and Savelli, 1968; Anaskin and Stoyanova, 1968b; Sonier, 1970, 1971; Yada et al., 1973; Herring et al., 1992; Müller et al., 2005; Wanner et al., 2006). The use of an interferometer for mapping electric field distributions has been considered by Borzjak et al. (1977) and Kulyupin et al. (1978, 1979) and, in a different context, by Frabboni et al. (1987), Matteucci et al. (1987, 1988a, 1989, 1992b, 1997), Missiroli et al. (1991) and by Zhang et al. (1992, 1993a, 1993b, cf. Spence et al., 1993a, 1993b). Contact potential differences have been studied by Brünger (1972), Brünger and Klein (1977) and Krimmel et al. (1964). The fields at $p-n$ junctions are examined by Frabboni et al. (1985, 1986), Matteucci et al. (1988b), Merli and Pozzi (1978), Merli et al. (1976b), Vanzi (1981) and Pozzi and Vanzi (1982). The electron analogue of the Michelson interferometer, in which high temporal coherence is vital, has been studied by Lichte et al. (1972), Lenz (1972) and Lichte (1980a, 1980b) while Ohtsuki and Zeitler (1977) have repeated Young's experiment, using electrons, as mentioned earlier (Fig. 61.5B). Stoyanova and Anaskin (1968) have attempted to estimate the ratio of coherently scattered electrons to those scattered incoherently in a specimen by interferometry and Menu and Evrard (1971) have studied the decline in fringe visibility as the specimen thickness is increased. The book by Stoyanova and Anaskin (1972) has chapters on interference and holography; a full review of the earlier work on interferometry and interference microscopy is given in Missiroli et al. (1981), which includes some applications not considered here. A later review by Matteucci et al. (1984) deals with the extensive work on interferometry with magnetic specimens, which became a major application of holography as we shall see. The survey by Hasselbach (2010) is the most complete. More isolated studies have been made of glass knives for ultramicrotomy (Lauer and Lickfeld, 1988), effects in crystals (Ade, 1986; Kawasaki et al., 1991a, 1991b, 1992b), edges (Subbarao, 1991), interfaces (Weiss et al., 1991) and of possible uses in biology (Lichte and Weierstall, 1988; Matsumoto et al., 1991). The following papers are also relevant: Gabor (1956), Takeda et al. (1982), Ru (1994), Rodgers (2002), Spence (2009), Rosa (2012), Schütz et al. (2014) and the many papers from the Bologna group (Fazzini et al., 2004, 2006; Frabboni et al., 2007, 2015; Matteucci, 1990, 2007, 2011, 2013; Matteucci and Beeli, 1998; Matteucci and Pozzi, 1978, 1985; Matteucci et al., 1982a, 1982b, 1982c, 1982d, 1982e, 1982f, 2009, 2010a, 2010b; Merli et al., 1974, 2003, 1976; Pozzi, 1980a, 1980b, 1974; Pozzi et al., 2004, 2006).

Some other interference effects must also be mentioned: the Aharonov–Bohm effect (Ehrenberg and Siday, 1949; Aharonov and Bohm, 1959, 1961), around which a lively polemic developed, discussed in Section 62.4, and the attempts to detect the Sagnac effect, using electrons. The latter effect is a phase shift caused by rotation of interfering beams and has been demonstrated for electrons by Hasselbach and Nicklaus (1990); for a very detailed account, see Nicklaus (1989). We return to this briefly in Section 62.5, where an electrostatic form of the Aharonov–Bohm effect is also discussed.

Apart from an attempt by Hibi (1956) to repeat the early experiments of Haine and Mulvey with a pointed filament, electron holography attracted no interest until the late 1960s. In 1968 Möllenstedt and Wahl in Tübingen and Tonomura et al. in Tokyo described off-axis and on-axis electron hologram formation (Möllenstedt and Wahl, 1968; Tonomura et al., 1968) and the subject was extensively developed in the following decades, largely in Germany and Japan with some work in France and the United States. At about the same time, Hanszen and colleagues in Braunschweig explored a holographic interpretation of bright-field electron microscope image formation in great detail. Such images may indeed be regarded as fringe patterns formed by interference between the unscattered wave and the wave describing electrons scattered within the specimen; these two waves may be thought of as a reference wave and an object wave so that the bright-field image may legitimately be described as an in-line hologram. Irradiation of this image with laser light and a suitable filter may thus be identified with the reconstruction step, which has led Hanszen (1971, 1982a, 1986a), Stroke et al. (1971a, 1971b, 1973, 1974, 1977) and Stroke and Halioua (1972a, 1972b, 1973a, 1973b) to identify many of the linear reconstruction algorithms described in Part XV with holography. We have to say that this is really a matter of personal preference and is helpful only in so far as one is more familiar with holography than with electron image formation. In the present century, electron holography has become a major activity (for more references, see Dunin-Borkowski et al., 2019).

The reconstruction stage of many holographic processes makes severe demands on the light-optical arrangement employed and it was realized soon after Wahl described his experimental difficulties in detail (1974, 1975) that digital reconstruction would solve many of these (e.g. Hawkes, 1980: Preface); the only drawback would be the size of the matrices corresponding to holograms digitized on a fine enough sampling grid. Digital reconstruction was successfully implemented by Lichte (1988, 1991b) and Fu et al. (1991) and for interference holography by Ohshita et al. (1990) (see also Hanszen, 1982b; Franke et al., 1986, 1987; Ade, 1988; Daberkow et al., 1988; Ade and Lauer, 1992a, 1992b).

Owing to the complexity of the subject, we have divided the material into two chapters. Chapter 62 is devoted to the classical wave mechanical aspects of interferometry, in which the roles played by diffraction in the specimen and by lens aberrations are unimportant. A certain amount of space is devoted to the optics of biprisms, since these are the most important electron-optical elements involved and were not treated in Volumes 1 and 2. The chapter ends with several examples of electron interference phenomena.

In Chapter 63, on holography in the proper sense of the term, diffraction in the object plays a central role. We begin by showing that in-line holography can be understood in terms of Fresnel zone lenses, as shown in Fig. 61.3A and B. We then study the many modes of hologram formation in microscopes equipped with a biprism and digital reconstruction in the computer.

CHAPTER 62

Interferometry

The first stage in electron interferometry requires the separation of an electron beam into two or more parts. This is achieved by *amplitude division*, in which the intact beam is separated into two partial beams by diffraction at a crystal or a synthetic structure that mimics the action of a crystal or by *wavefront division*, using an electron biprism. The latter is by far the more common and we therefore devote the opening sections of this chapter to a full study of the optics of such biprisms. These are followed by a short section on amplitude-division devices, largely confined to recent work. A section is then given to the Aharonov–Bohm, or more correctly, the Ehrenberg–Siday effect. We first concentrate on the work that led to the definitive proof of the existence of the effect; it has, however, generated much discussion of the underlying quantum mechanisms and we include some account of this. The electric and electrostatic counterparts of the magnetic effect are also described. This is followed by a section on the Sagnac effect, in which we follow the arguments of Hasselbach closely – the sceptical comments of Malykin are mentioned briefly. The chapter concludes with short descriptions of some other aspects of electron interference: convergent-beam electron diffraction and interferometry, biprism design, two-filament biprisms, the tandem formed by a biprism and a Wien filter, the electron Hanbury Brown and Twiss experiment and decoherence.

62.1 The Electrostatic Biprism

Together with a sufficiently coherent source and lenses to provide the necessary magnification, the electrostatic biprism is an important component of electron interference instruments of any kind. Its task is to split the coherent wave emitted by the electron source into two partial waves that are still coherent, as shown schematically in Fig. 61.1B. This separation can be made so large that a specimen can be brought into the path of one of the partial beams without loss of the necessary coherence.

The deflection of the electrons in the electrostatic biprism can be treated entirely within the frame of classical mechanics. The reason for this is that the important diffraction and interference events occur up- and downstream from the field of the biprism. This is confirmed by the theoretical studies of Komrska et al. (1964a, 1964b, 1964c, 1967), Komrska and Vlachová (1973) and Komrska and Lenc (1970). The influence of source size on biprism interference fringes was examined by Drahoš and Delong (1964) and Komrska et al. (1964a, 1964b, 1964c) showed that the diameter of the biprism filament

Principles of Electron Optics.
DOI: https://doi.org/10.1016/B978-0-12-818979-5.00062-0

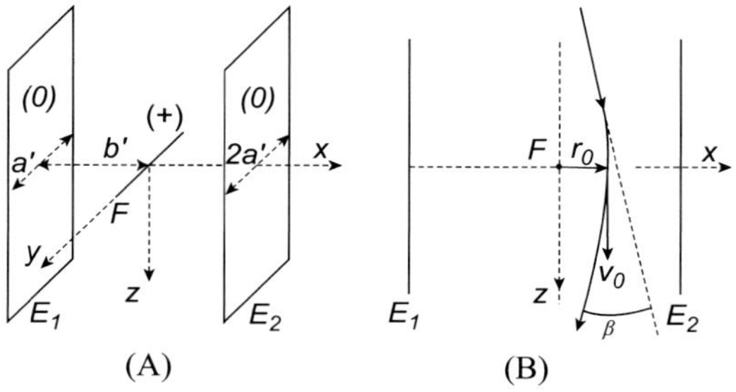

Figure 62.1
Basic shape and physical action of an electrostatic biprism: (A) simplified perspective view. *F*: filament (length $2a'$, charge $2a'q > 0$, potential $V_0 > 0$). E_1, E_2: screening electrodes (charge $-a'q$), each at zero potential; the connection to the voltage supply is omitted; (B) deflection of a trajectory with apex distance r_0, apex velocity v_0 and total deflection angle β.

could be estimated from the form of the fringes. The instrument in which the related experiments were performed is described by Drahoš and Delong (1963).

Electron biprism design had scarcely changed since the device was invented by Möllenstedt and Düker in 1955 until Duchamp et al. (2018) designed a radically new geometry, described in Section 62.6. The basic arrangement is shown schematically in Fig. 62.1A. A straight and very thin metallic filament is situated midway between two plates serving as screening electrodes. These are held at the potential of the instrument column, so that the asymptotic field vanishes. We set this potential equal to zero, and in practice the plates are earthed. This is not the same choice of potential origin as in Volumes 1 and 2, where zero voltage corresponded to zero electron velocity and hence to the potential of the filament of the gun. The biprism filament is *positively* charged when we wish to make the partial beams overlap further downstream; by charging it negatively, the separation of the partial beams can be increased.

62.1.1 Field Model

Accurate calculation of the electrostatic field in this configuration, although complicated, is now included in many program suites for electron optical calculations. In order to display the principal deflecting effect of the field on the electrons reasonably straightforwardly, we employ a simple model, which nevertheless describes the essentials correctly.

The central filament has a finite length $2a'$ and is *uniformly* charged with a total charge $Q > 0$. The charge per unit length, or line-charge density, is hence $q = Q/2a'$. The two electrodes are simulated by two parallel line-charge distributions of equal length $2a'$ and line-charge density $-q/2$. This is necessary to achieve *neutrality* of the total system. Without loss of generality we can choose the coordinate system such that the central filament lies along the y-axis and the two screening lines lie in the $x-y$ plane at positions $x = \pm b'$.

The potential produced by this charge distribution is given by

$$V(x, y, z) = \sum_{n=-1}^{1} \frac{q(1 - 3n^2/2)}{4\pi\varepsilon_0} \int_{-a'}^{a'} \frac{dy'}{\{(y' - y)^2 + (x + nb')^2 + z^2\}^{1/2}}$$

The integral can be evaluated in closed form, giving

$$V(x, y, z) = \sum_{n=-1}^{1} \frac{q(1 - 3n^2/2)}{4\pi\varepsilon_0} \ln \left| \frac{y + a' + \{(y + a')^2 + (x + nb')^2 + z^2\}^{1/2}}{y - a' + \{(y - a')^2 + (x + nb')^2 + z^2\}^{1/2}} \right| \tag{62.1}$$

From this general but still fairly complicated solution, expressions for two important special cases can be derived: the *asymptotic* behaviour and the potential in the vicinity of the origin.

An expansion for large distances

$$R = \sqrt{x^2 + y^2 + z^2} \gg \max(a', b') \tag{62.2a}$$

shows that a *quadrupole field* is the lowest nonvanishing multipole term:

$$V(x, y, z) = -\frac{q}{2\pi\varepsilon_0} \frac{a'b'^2}{R^3} \left(\frac{3}{2} \frac{x^2}{R^2} - \frac{1}{2} \right) \tag{62.2b}$$

The quadrupole, characterizing the electrostatic fringe field, is orientated in the x-direction.

In the immediate vicinity of the central filament, which is the most important region for the deflection effect, we see from (62.1) that the potential, and consequently the field strength, depends on the coordinate y. This is a perturbation which causes electron optical aberrations, since the interference fringes will be straight only if the field does *not* depend on y. In a more realistically calculated field, this perturbation is less pronounced, since the filament surface is then an equipotential. Even so, the effect cannot be completely eliminated, being an inevitable consequence of the finite length of the filament and the surrounding screening plates.

It is hence necessary to confine the domain traversed by the electron beam in the y-direction to a slit, the length of which is considerably smaller than a'. We can then set $y = 0$ in (62.1) and in the corresponding formulae for the field strength to a good approximation.

In the practical realization of biprisms, the radius ρ of the central filament is much smaller than the lengths a' and b'; the electric field strength is therefore high only in the immediate vicinity of this filament. This implies that we can set

$$r := \sqrt{x^2 + y^2} \ll \min(a', b') \tag{62.3}$$

throughout the domain in which the electrons are effectively deflected. In the terms for which $n \neq 0$ in (62.1), we may therefore set $x = z = 0$ and the general formula simplifies to

$$V(x, 0, z) = \frac{q}{2\pi\varepsilon_0}\left\{\ln\left(\frac{2a'}{r}\right) - \ln\left(\frac{a' + \sqrt{a'^2 + b'^2}}{b'}\right)\right\} \tag{62.4a}$$

The *filament potential* V_0 is obtained by setting $r = \rho$, giving

$$V_0 = \frac{q}{2\pi\varepsilon_0}\ln\left[\frac{2b'}{\rho\{1 + \sqrt{1 + (b'/a')^2}\}}\right] \tag{62.4b}$$

Usually this potential is fixed by the experimental conditions; Eq. (62.4b) can then be used to calculate the line-charge density if the above field model is accepted. We can also use the constant V_0 to simplify (62.4a), giving finally

$$V(r) = V_0 - \frac{q}{2\pi\varepsilon_0}\ln\frac{r}{\rho} \tag{62.4c}$$

We shall now use this ideal cylindrical potential, obtained after extreme simplification, to calculate the beam deflection in interferometers. We note that (62.4c) has the same form as that of a cylindrical condenser, in agreement with the observations of Möllenstedt and Düker (1956) who studied the biprism field with the aid of an analogue device commonly used at that time, the electrolytic tank (Francken, 1967). For good agreement, the outer radius of the condenser needed to be slightly smaller than the distance between the plates. Later, Septier (1959) obtained an expression for the field of a wire of negligible radius between earthed plates.

62.1.2 Asymptotic Deflection of Electron Trajectories

In view of the many simplifications already made, there seems little hope of obtaining an accurate solution but, surprisingly, this is not so (Kasper, 1992). The only indispensable approximation is the reduction of the problem to two dimensions, which means that we continue to ignore the confinement of the field in the y-direction.

We start from the ray equations derived in Volume 1 (3.22). Since there is no magnetic field and no motion in the y-direction, these equations reduce to

$$\frac{x''(z)}{1+x'^2} = \frac{1}{2\hat{\phi}}\left(\frac{\partial\hat{\phi}}{\partial x} - x'\frac{\partial\hat{\phi}}{\partial z}\right)$$

On linearizing the accelerating potential, $\hat{\phi} = (U+V)\{1+\varepsilon(U+V)\}$ with respect to V, which is permissible since $|V| \ll U$ and writing $\gamma = 1 + 2\varepsilon U$ as usual, we obtain

$$\frac{x''}{1+x'^2} = \frac{\gamma}{2\hat{U}}\left(\frac{\partial V}{\partial x} - x'\frac{\partial V}{\partial z}\right) \tag{62.5}$$

Since $V(z, x)$ satisfies the two-dimensional Laplace equation, there must be an associated potential $W(z, x)$, related to $V(z, x)$ via the Cauchy–Riemann equations,

$$\frac{\partial V}{\partial x} = \frac{\partial W}{\partial z}, \qquad \frac{\partial V}{\partial z} = -\frac{\partial W}{\partial x}$$

If we use these to replace V by W in (62.5), we find that both sides can be rewritten as total derivatives with respect to z:

$$\frac{x''}{1+x'^2} = \frac{d}{dz}\{\arctan x'(z)\} = \frac{\gamma}{2\hat{U}}\left(\frac{\partial W}{\partial z} + x'\frac{\partial W}{\partial x}\right) = \frac{\gamma}{2\hat{U}}\frac{dW(z, x(z))}{dz}$$

We can integrate this immediately and obtain

$$\alpha(z) := \arctan x'(z) = \frac{\gamma}{2\hat{U}} W(z, x(z)) + \text{const} \tag{62.6}$$

This equation is not limited to the biprism but is true for any two-dimensional electric field; nor is it necessary to assume that the slopes are small.

The potential $W(z, x)$ must satisfy the Neumann condition $\partial W/\partial n = 0$ on all electrode surfaces and has to be made unique by suitable choice of cuts in the z–x plane. Let us now consider the specific case of the biprism. For a trajectory passing on one side of the central filament, we consider the closed loop consisting of the trajectory from the field-free domain on the entrance side to that on the exit side and a path outside the biprism in the positive sense. In this way, we have encircled the line singularity $-q/4\pi\varepsilon_0$ of the shielding plate on the same side of the filament as the trajectory. Stokes' integral theorem then tells us that the line integral of grad W around this loop is equal to $-q/2\varepsilon_0$. This value is thus the potential difference between the two sides of the cut running from the singularity to infinity. We can hence conclude that

$$W(L, x(L)) - W(-L, x(-L)) = -q/2\varepsilon_0, \quad L \to \infty$$

and from (62.6), that

$$\beta := \alpha(L) - \alpha(-L) = -\frac{\gamma q}{4\varepsilon_0 \hat{U}} \quad \text{for } L \to \infty \tag{62.7}$$

The deflection angle β is hence *exactly* proportional to the line-charge density q of the central filament, if we adopt the earlier assumptions that we can neglect the fringe fields in the transverse (y) direction and the nonlinear terms in V. These latter simplifications are, indeed, not severe.

It is important to notice that the deflection angle β depends neither on the distance from the apex, the impact parameter r_0, nor on the asymptotic initial slope. It has the same value for all trajectories passing on the same side of the biprism filament and differs only in sign on the other side. An immediate consequence is that an axial point source above the biprism will be seen as a pair of *virtual* point sources, symmetrically placed about the z-axis, from below the biprism. These are labelled V and $\overline{V}$ in Fig. 62.2. This is the fundamental reason for the success of biprism interferometers.

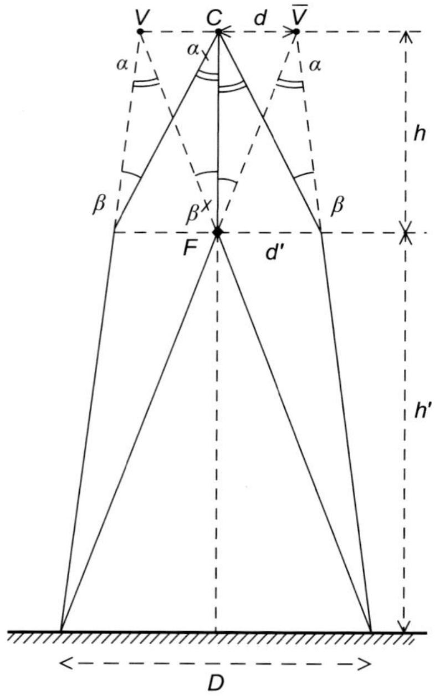

Figure 62.2
Geometrical construction of the two virtual sources V and $\overline{V}$, located symmetrically at a distance d from the real source C. The rays (*full lines*) are schematically refracted in the plane of the filament F. The deflection angle β and the aperture angle a are much exaggerated for clarity. With $|\alpha| \ll 1$, $|\beta| \ll 1$ we have $d = h\beta$, $d' = h\alpha$, $D = (h + h')\,\alpha$.

62.1.3 Applications with Real Interferometers

The simple configuration shown in Fig. 62.2 is not suitable for electron beams. A more realistic, though still schematic, arrangement is shown in Fig. 62.3. A highly coherent electron wave is emitted by an electron source *Q*, which should be as small as possible; in practice, a very bright field-emission gun with a fine cathode tip is used. Nowadays, virtual source radii of the order of ångströms are attainable (see Part IX); for a cold field-emission tip, the true physical radius is of the tip is in the range 10–100 nm.

The real object *S*, that is, the specimen to be investigated, is situated on one side of the axis in front of the objective lens *L*. The latter produces a highly demagnified crossover *C*, to which all the rays from the source are focused. This point is now the real source denoted by *C* in Fig. 62.2. The biprism is placed further downstream (as in Fig. 62.2) and forms two virtual

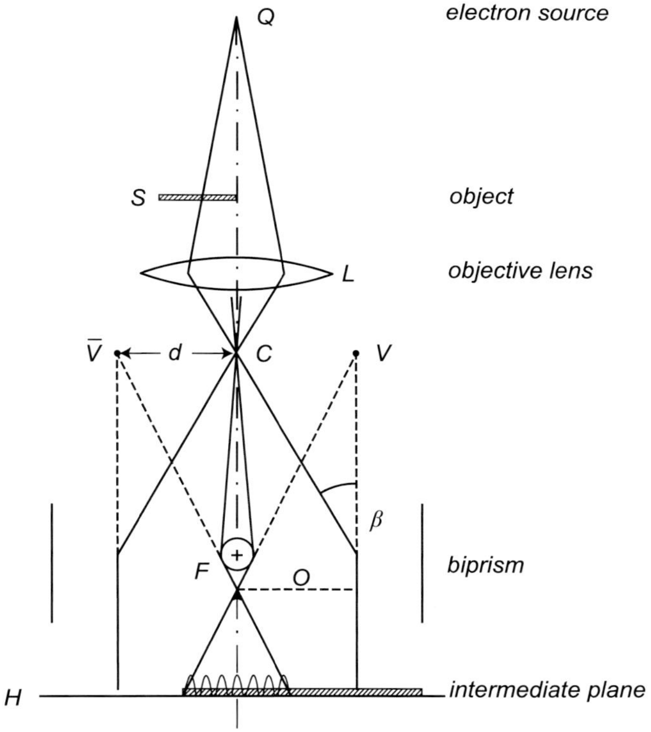

Figure 62.3
Typical arrangement of an electron interferometer, simplified presentation. *Q:* electron source, *S:* real object (specimen), *L:* objective lens, *C*: real crossover of *Q*. *V*, $\overline{V}$: virtual sources (see also Fig. 62.2), *F:* filament of the biprism, β: deflection angle (62.7), *O*: object in the form of the image of *S*, *H*: primary (intermediate) hologram plane. The entrance slits shown in Fig. 61.1B and Fig. 62.4 are omitted here for clarity.

sources V and $\overline{V}$; their positions are found by backward continuation of the deflected rays. To a good approximation, the asymptotes to *all* the deflected rays intersect at these points.

In the absence of deflection, the objective lens would image the specimen as an 'object' O; it will be convenient to regard this below as a 'real object', though it is of course only an intermediate image.

An interference pattern is now formed in the plane H at the bottom of Fig. 62.3 and often called the primary hologram plane. In principle, this intensity distribution could be recorded, but the interference fringes are in fact far too fine to be resolved. A lens system is therefore placed beyond this plane to provide sufficient magnification to resolve the fringes. This lens system and the true recording plate are not shown here. So far as the final intensity distribution is concerned, it is of no importance whether the waves emanated from real or virtual sources or whether they were diffracted in real or virtual objects. The only important requirement is that the wave amplitudes and phases be correctly transferred in the imaging process.

The electron waves that interfere below the biprism are *coherent* if C is a coherent point source. This follows from the symmetry of the device: for instance, along all pairs of mirror-symmetric rays, the electron optical path lengths between point C and the plane H are exactly equal in zero diffraction order.

62.2 Quasi-Homogeneous Interference Fringes

The recorded intensity pattern can become extremely complicated if an arbitrary specimen is brought into the interferometer, so complicated indeed that a lengthy computation is needed to calculate it. We therefore set out from a quite simple configuration and then gradually generalize.

We first consider the two-beam configuration shown in Fig. 62.4. Let us assume that two monochromatic spherical waves $\psi_V(\boldsymbol{r})$ and $\psi_R(\boldsymbol{r})$ emerge from the sources V and $\overline{V}$, respectively and that they are perfectly *coherent*. This simply means that the two sharply defined wavelengths and amplitudes are exactly equal and that there is a *constant* phase difference between these waves. Without loss of generality, we may set this phase difference equal to zero, and we shall also omit a common time-dependent factor exp $(-\mathrm{i}\omega t)$. Adopting the Cartesian (x, y, z) system with origin at C (Fig. 62.4), we then have

$$\psi_V(\boldsymbol{r}) = \frac{A}{r}\exp(\mathrm{i}kr), \quad r = \{(x-d)^2+y^2+z^2\}^{1/2}$$

$$\psi_R(\boldsymbol{r}) = \frac{A}{\overline{r}}\exp(\mathrm{i}k\overline{r}), \quad \overline{r} = \{(x+d)^2+y^2+z^2\}^{1/2}$$

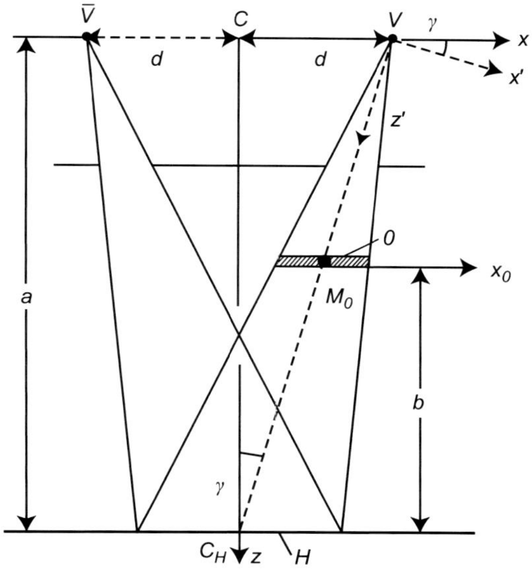

Figure 62.4
Schematic presentation of a two-beam interferometer. V and $\overline{V}$: virtual sources; C: origin of the (x, y, z) coordinate system; z', x': rotated coordinate system; O: object (specimen), H: interferogram; M_0: midpoint of the object.

This is still exactly true in the domains located in front of the two entrance slits of the interferometer. In practice, the angle between the wave normal and the z-axis is always so small that the Fresnel approximation is justified, which is a simplification analogous to the step leading from (60.1) to (60.2). In the present case we obtain

$$\psi_V(\boldsymbol{r}) = \frac{A}{z}\exp\left(\mathrm{i}kz + \frac{\mathrm{i}k}{2z}y^2\right)\exp\left\{\frac{\mathrm{i}k}{2z}(x-d)^2\right\} \tag{62.8a}$$

$$\psi_R(\boldsymbol{r}) = \frac{A}{z}\exp\left(\mathrm{i}kz + \frac{\mathrm{i}k}{2z}y^2\right)\exp\left\{\frac{\mathrm{i}k}{2z}(x+d)^2\right\} \tag{62.8b}$$

62.2.1 No Object Present

Let us now suppose that the object O in Fig. 62.4 is absent. If we also disregard diffraction at the edges of the entrance slits, we achieve the utmost simplification: the intensity distribution in the recording plane $z = a$ is then given by

$$J_H(x, y) = |\psi_V(x, y, a) + \psi_R(x, y, a)|^2 = 4J_R\cos^2(kdx/a) \tag{62.9}$$

with the constant reference value

$$J_R = |\psi_R(x, y, a)|^2 = A^2/a^2$$

This describes a perfect pattern of straight and equidistant fringes, in which the intensity minima fall to zero. The distance between adjacent maxima or minima is simply

$$\Delta x = \lambda a/2d \tag{62.10}$$

In practice, this formula holds only in the *central* part of the fringe system. Near the boundaries, which are approximately determined by the geometrical shadows of the slit edges, the diffraction of waves at these edges must be taken into account. The mathematical tools for this were given in Section 60.2.

Here however the situation has become much more complicated owing to the fact that *two* slit functions for ψ_V and ψ_R are to be superimposed prior to forming $|\psi|^2$. Moreover, in a real interferometer, the two shadow boundaries will not coincide perfectly, as was tacitly assumed in the construction of Fig. 62.4. Thus the intensity calculation becomes exceedingly complicated, even in this simple configuration.

An example of such an intensity distribution is given in Fig. 62.5. In practice, the shadow zones in which Fresnel fringes are formed are disregarded if the recorded intensity patterns are to be evaluated numerically.

62.2.2 Quasi-Homogeneous Object

We now consider the introduction of a specimen S into the interferometer, as in Fig. 62.3. This corresponds to the object O in Fig. 62.4. Its optical properties are described by a transparency or object function.

$$O(x_o, y_o) = \exp\{-\sigma(x_o, y_o) + i\eta(x_o, y_o)\} \tag{62.11}$$

which is the same as that defined by Eq. (60.4). The origin of the (x_o, y_o) system is taken to be the midpoint M_0 of the object, as indicated in Fig. 62.4. This point is projected into the centre C_H of the plane H.

We now make the simplifying assumption that the function $O(x_o, y_o)$ varies so slowly with x_o and y_o that diffraction in the object does *not* need to be taken into account; this is what we mean by the term 'quasi-homogeneous object'. The object function is then transferred to the recording plane by a simple central projection with magnification M:

$$O_H(x, y) = O(x_o, y_o) = O(x/M, y/M) \tag{62.12a}$$

$$M = a/(a - b) \tag{62.12b}$$

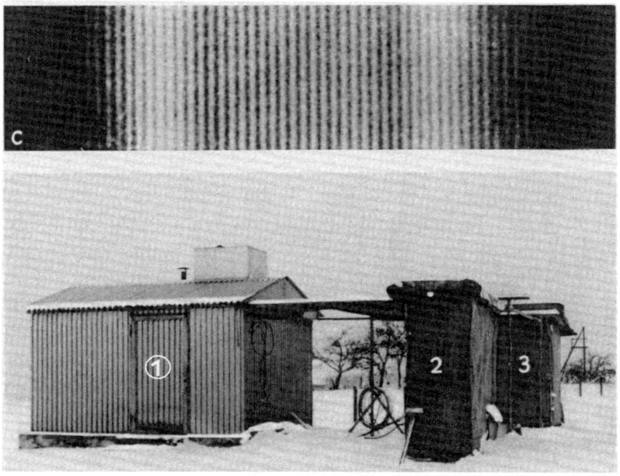

Figure 62.5
(Above) an early example of biprism fringes, obtained in a special laboratory (below), in 'a meadow far from all municipal mains'. Building 1 housed the interferometer; the other huts contained the high-voltage source and the vacuum equipment. *After Möllenstedt (1965), Courtesy Société Française de Microscopie Electronique.*

The *object wave* at the screen H now becomes

$$\psi_o(x, y, a) = O(x/M, y/M)\psi_V(x, y, a) \tag{62.13}$$

with ψ_V given by (62.8a), while the *reference wave* ψ_R remains unaffected. Instead of (62.9), we find

$$J_H(x, y) = |\psi_R + O\psi_v|^2$$

Evaluation of this expression using (62.8) and (62.11) results in

$$J_H(x, y) = J_R\{1 + \mathrm{e}^{-2\sigma} + 2\mathrm{e}^{-\sigma}\cos(2kxd/a - \eta)\} \tag{62.14}$$

with $\sigma = \sigma(x/M, y/M)$ and $\eta = \eta(x/M, y/M)$. It is obvious from this formula that the object phase $\eta(x_o, y_o)$ causes a corresponding shift of the interference fringes in the recording plane:

$$\delta x_\eta = \frac{\lambda a}{4\pi d}\eta\left(\frac{x}{M}, \frac{y}{M}\right) \tag{62.15}$$

For $\eta = 2\pi$ we find $\delta x_\eta = \Delta x$ (Eq. 62.10), as expected.

The attenuation factor $\exp(-\sigma)$ in (62.11) describes two effects at once, a decrease of the mean intensity and a loss of contrast.

The local *mean intensity* $\overline{J}(x, y)$ is obtained by averaging $J_H(x, y)$ over x in intervals that are not only large enough for the cosine term to cancel out but also so small that σ and η do not alter significantly. We then find:

$$\overline{J}(x, y) = J_R[1 + \exp\{-2\sigma(x/M, y/M)\}] \tag{62.16}$$

The local *contrast* is defined by

$$C(x, y) = \frac{J_{\max} - J_{\min}}{J_{\max} + J_{\min}} \tag{62.17}$$

where we again consider σ and η to be slowly varying functions. From (62.14) we find

$$\begin{aligned} J_{\max} &= J_R(1 + e^{-2\sigma} + 2e^{-\sigma}) = J_R(1+e^{-\sigma})^2 \\ J_{\min} &= J_R(1 + e^{-2\sigma} - 2e^{-\sigma}) = J_R(1-e^{-\sigma})^2 \end{aligned}$$

and hence

$$C(x, y) = \frac{2e^{-\sigma}}{1 + e^{-2\sigma}} < 1 \tag{62.18}$$

This contrast formula is still incomplete. There are stronger effects that reduce the contrast, as we shall see in the next sections.

The fringe shift $\delta x_\eta(x, y)$ can be immediately measured from recorded intensity patterns and thus enables us to deduce the phase shift $\eta(x_o, y_o)$. The latter results from changes in the wavenumber $k(\boldsymbol{r})$ associated with local variations of the electrostatic potential in the specimen:

$$\eta(x_o, y_o) = \int_{z_1}^{z_2} \{k(x_o, y_o, z) - k_\upsilon\} dz \tag{62.19}$$

where k_υ denotes the vacuum value of k and z_1, z_2 the local coordinates of the upper and lower specimen surfaces. If the potential is $U = \text{const}$ outside and $U + \Phi(\boldsymbol{r})$ with $|\Phi| \ll U$ inside the specimen, we obtain (with $k_\upsilon = 2\pi/\lambda$)

$$\eta(x_O, y_O) \approx \frac{\gamma\pi}{\lambda\hat{U}} \int_{z_1}^{z_2} \Phi\, dz \tag{62.20}$$

On introducing (62.20) into (62.15), we see that the wavelength λ cancels out and thus

$$\delta x_\eta = \frac{\gamma a}{4d\hat{U}} \int_{z_1}^{z_2} \Phi\, dz$$

In practice, it is commonly assumed that the physical properties of the specimen do not alter significantly with the local depth z; we may then replace the integral by $\Phi\tau$, τ being the local *thickness* of the specimen, and finally arrive at

$$\delta x_\eta = \frac{a\gamma\tau}{4d\hat{U}} \Phi \tag{62.21}$$

Provided that the thickness τ is independently known, the measurement of δx_η from interferograms gives us a method of determining the quantity Φ. This is the '*mean inner potential*' of the solid material (see Section 69.3.3). It is positive and $e\Phi$ ($e > 0$) is a measure of the volume average of the atomic electrostatic potentials.

A typical example of the fringe shift by a mean inner potential is shown in Fig. 62.6. Such measurements and evaluation of Φ were first published by Möllenstedt and Keller (1957); more recent measurements have been reported by Gajdardziska–Josifovska et al. (1993), Wang et al. (1997, 1998), Harscher (1999), Li et al. (1999), Wang (2003), Müller et al. (2005), Chung et al. (2007), Gan et al. (2015) and Ding et al. (2015). Since the wavelength λ cancels out from (62.21), the temporal coherence requirements are less stringent than in modern holography but see Kohn et al. (2020). This was one of the reasons for the success of these early measurements.

The inner potential is not the only source of phase shifts between the two interfering beams. Variations in specimen thickness can also be detected as fringe shifts, and we shall return to this aspect when we consider interference holography (see Section 63.3.2), since this permits very small thickness differences to be detected. A rich field of application of interference microscopy is concerned with magnetic specimens; we refer to Missiroli et al. (1981), Matteucci et al. (1984), Tonomura (1986a, 1992a) and Shindo and Murakami (2008) for detailed accounts. Many more examples are mentioned in Chapter 63.

62.3 Coherence Problems

In the theory presented so far, we have explicitly assumed that the two sources V and $\overline{V}$ are point emitters of coherent waves. This coherence is a consequence of the symmetrical

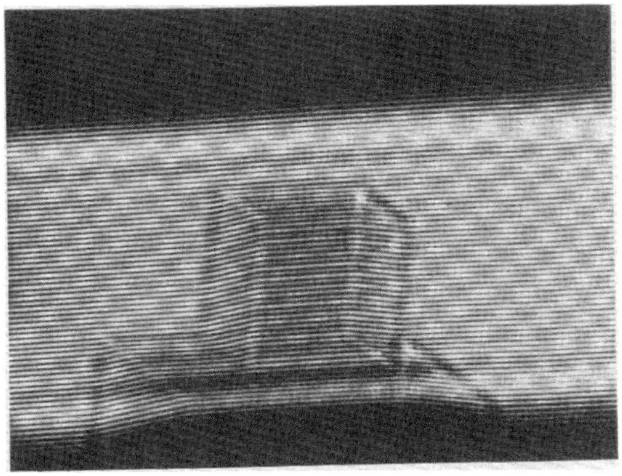

Figure 62.6
Interference fringes; the shift is caused by the 'inner potential' of a solid specimen, here a MgO crystal. *After Völkl (1991), Courtesy Edgar Völkl.*

splitting of *one* spherical wave by the biprism and could not be achieved by placing two real sources at V and $\overline{V}$. But even the assumption of one ideally monochromatic point source, located at the point C (see Fig. 62.3), is unrealistic and will now be abandoned. There are two new features, which lead to a loss of contrast: the *lateral extent* of the real electron source and the width of the *energy spectrum* do not vanish. In reality, both effects occur simultaneously, but for ease of presentation we study them separately. We shall see, in Section 66.2, in a different context, that this is usually legitimate.

62.3.1 Partial Lateral Coherence (Spatial Coherence)

In order to obtain a very small effective electron source, the cathode tip G (Fig. 62.3) is demagnified into the crossover C, but even with the best field electron emission sources, the lateral extent of C, small though it is, cannot be ignored. The consequence of this is shown in Fig. 62.7: lateral displacement of a point source in combination with the biprism causes a *tilt* of the corresponding beam axis.

How will this affect the interference pattern? Different areas of the cathode surface emit electron waves that are completely *uncorrelated* in phase, since the emission of single particles is a stochastic process. We can hence consider the small cross-section in C, from which the electron beam emanates, as a large number of *incoherent* point sources. The tilted axis (dashed line in Fig. 62.7) then refers to one of these point sources.

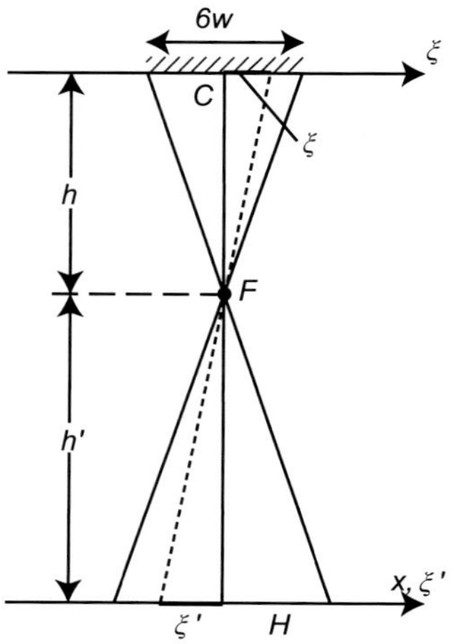

Figure 62.7
Tilt of the local beam axis when the crossover is not vanishingly small. C: Centre of the crossover, F: filament axis, H: interferogram plane, ξ, ξ': displacement coordinates; dashed line: axis corresponding to a coordinate ξ. The breadth $6w$ corresponds to $\exp(-9) \approx 10^{-4}$.

We now introduce new transverse Cartesian coordinates (ξ, ρ) and an intensity distribution $\hat{F}(\xi, \rho)$, such that $\hat{F}(\xi, \rho)\ d\xi\ d\rho$ is the *relative* intensity coming from the element of area $d\xi\ d\rho$ at position (ξ, ρ). This is expressed by the normalization condition

$$\iint\limits_{-\infty}^{\infty} \hat{F}(\xi, \rho) d\xi\ d\rho = 1 \tag{62.22}$$

Since the deflecting field in the essential part of the biprism is cylindrical, this deflection is independent of the direction of incidence: the whole electron bundle starting from a common off-axis point (ξ, ρ) has an axis which is simply tilted round the filament F, as is shown in Fig. 62.7. This causes a corresponding shift

$$\xi' = -\mu\xi, \quad \rho' = 0, \quad \mu := h'/h \tag{62.23}$$

in the recording plane. (The distances h and h' are the same as in Fig. 62.4.)

The total intensity distribution is obtained by superposing all the partial intensities, which gives:

$$\hat{J}(x, y) = \iint J_H(x + \xi', y + \rho')\hat{F}(\xi, \rho) d\xi\ d\rho = \iint J_H(x - \mu\xi, y)\hat{F}(\xi, \rho) d\xi\ d\rho$$

Any shift in the ρ-direction is clearly immaterial and we therefore define a one-dimensional distribution function

$$F(\xi) := \int\limits_{-\infty}^{\infty} \hat{F}(\xi, \rho) d\rho, \qquad \int\limits_{-\infty}^{\infty} F(\xi) d\xi = 1 \tag{62.24}$$

whereupon

$$\hat{J}(x, y) = \int\limits_{-\infty}^{\infty} J_H(x - \mu\xi, y) F(\xi) d\xi \tag{62.25}$$

Once the distribution function $F(\xi)$ is known, the remaining task is reduced to an integration; this has the form of a convolution of the distribution $F(\xi)$ with the 'coherent' intensity J_H. The situation becomes even simpler if the electron source is *symmetric*, which implies that

$$F(-\xi) = F(\xi) \tag{62.26}$$

In practice, perfect mirror symmetry cannot be reached by adjustment of the instrument, but the coordinate system at the final recording screen can always be laterally shifted so that the centre of the fringe system in the absence of any specimen coincides with the origin. This is always done in practice and is equivalent to the assumption in (62.26).

The convolution becomes fairly simple in the case of a quasi-homogeneous object. The object functions σ and η then vary so slowly that they may be considered as constant with respect to the convolution. Using (62.14) for J_H and (62.24) and (62.26) for F, we find

$$\hat{J}(x,y) = J_R\{1 + \mathrm{e}^{-2\sigma} + 2\mathrm{e}^{-\sigma}\cos(2kxd/a - \eta)K_T\} \tag{62.27}$$

which now includes a lateral contrast factor

$$K_T := \int_{-\infty}^{\infty} F(\xi)\cos(2kd\mu\xi/a)d\xi \tag{62.28}$$

The effect of this on fringe contrast is explored further in Section 62.3.3. (Note that although (62.28) appears at first sight to be a cosine Fourier transform, it is in fact a constant since x is not present in the argument of the cosine.)

62.3.2 Partial Longitudinal Coherence (Temporal Coherence)

We now take into account the fact that the electrons in the interfering beams are not exactly monoenergetic, as we have hitherto always assumed, but have a finite – albeit narrow – *energy spectrum*. The wavenumber $k = 2\pi/\lambda$ also has a spectrum, therefore, which we describe by a distribution $G(k)$, normalized so that

$$\int_0^{\infty} G(k)dk = 1 \tag{62.29}$$

We make the simplifying assumption that this function $G(k)$ is the same for all surface elements of the electron source and is thus independent of the coordinates ξ and ϱ introduced above. This assumption is not at all trivial or self-evident but is a reasonably good approximation if the emitting area remains small and if the source is monocrystalline.

The intensity $\hat{J}(x, y)$ obtained after averaging over the lateral source coordinates will now be a function of the wavenumber k and should therefore be denoted by $\hat{J}(x, y, k)$. In order to obtain the total intensity distribution $J(x, y)$ on the screen, we have to integrate over k:

$$J(x,y) = \int_0^{\infty} \hat{J}(x,y,k)G(k)dk \tag{62.30}$$

This formula and all the intensity formulae already discussed imply that the intensity is really *recorded* in the plane H. If this is not the case, some modifications are necessary, which will be examined next.

The integration in (62.30) can be simplified and finally carried out in closed form if we treat the spectral distribution as a very sharp peak in a small interval $|k-\bar{k}| < 3\kappa \ll \bar{k}$ outside which it effectively vanishes. The value $\bar{k}$ refers to the intensity maximum, and κ is a measure of the spectral width. We now substitute $\hat{J}(x, y)$ given by (62.27) into the integral in (62.30). We use the fact that the distribution $F(\xi)$ is very narrow and its Fourier transform $K_T(k)$ is hence a very slowly varying function of k. ($K_T(k)$ is defined by (62.28), in which k is now regarded as a variable instead of a constant.) It is then quite sufficient to replace k by $\bar{k}$ in (62.28), which again becomes a constant and can hence be taken outside the resulting integral. With the normalization condition (62.29) we obtain the intermediate result

$$J(x, y) = J_R \left\{ 1 + e^{-2\sigma} + 2K_T e^{-\sigma} \int_0^\infty \cos\left(\frac{2kxd}{a} - \eta\right) G(k) dk \right\}$$

With the substitution $k = \bar{k} + u$, the remaining integral can be rewritten as

$$\int_0^\infty \cos\left(\frac{2kxd}{a} - \eta\right) G(k) dk = \cos\left(\frac{2\bar{k}xd}{a} - \eta\right) \int_{-\infty}^{\infty} \cos\left(\frac{2uxd}{a}\right) G(\bar{k} + u) du - \sin\left(\frac{2\bar{k}xd}{a} - \eta\right) \int_{-\infty}^{\infty} \sin\left(\frac{2uxd}{a}\right) G(\bar{k} + u) du$$

In practice, the assumption that the spectral function is symmetric is well justified:

$$G(\bar{k} - u) = G(\bar{k} + u) \tag{62.31}$$

The sine terms then vanish in the integration and we are led to define a *longitudinal* contrast factor

$$K_L := \int_{-\infty}^{\infty} G(\bar{k} + u) \cos\left(\frac{2uxd}{a}\right) du \tag{62.32}$$

62.3.3 Superposition

Definition of a *total* contrast factor is straightforward:

$$K(x) = K_T K_L(x) \tag{62.33}$$

The intensity distribution in the fringe system can then be cast into the very compact form

$$J(x,y) = J_R\left\{1 + e^{-2\sigma} + 2Ke^{-\sigma}\cos\left(\frac{2\bar{k}xd}{a} - \eta\right)\right\} \tag{62.34}$$

Note that the familiar assumption of weakly scattering objects, $\sigma \ll 1$ and $|\eta| \ll 1$, has *not* been necessary in the derivation of this formula. A careful investigation shows that the fringe shift δx_η in (62.15) and the mean intensity $\bar{J}$ in (62.16) remain unaltered by the factor K, while the contrast becomes

$$C(x,y) = 2K(x)\frac{e^{-\sigma}}{1 + e^{-2\sigma}} \tag{62.35}$$

The factors K_L and K_T, determining the decrease of contrast, are both Fourier-like transforms of emission intensity distributions. In order to estimate their effect, we consider the case in which the latter are Gaussian:

$$F(\xi) = \frac{1}{w\sqrt{\pi}}\exp\left(-\frac{\xi^2}{w^2}\right) \tag{62.36a}$$

$$G(k) = \frac{1}{\kappa\sqrt{\pi}}\exp\left\{-\frac{(k-\bar{k})^2}{\kappa^2}\right\} \tag{62.36b}$$

w and κ are the widths at which these functions have decreased to $e^{-1} = 0.37$ of their maximum values. Evaluation of the corresponding integrals gives

$$K = K_T K_L(x) = \exp\left\{-\left(\frac{\bar{k}d\mu w}{a}\right)^2\right\}\exp\left\{-\left(\frac{\kappa xd}{a}\right)^2\right\} \tag{62.37}$$

The first exponential factor K_T describes a *uniform* loss of contrast, while the second factor $K_L\,(x)$ describes an attenuation, with full contrast $K_L = 1$ in the centre and a decay to invisibility as $|x| \to \infty$. This is shown in Fig. 62.8, which corresponds to the case in which no specimen is present, $\sigma \equiv 0$.

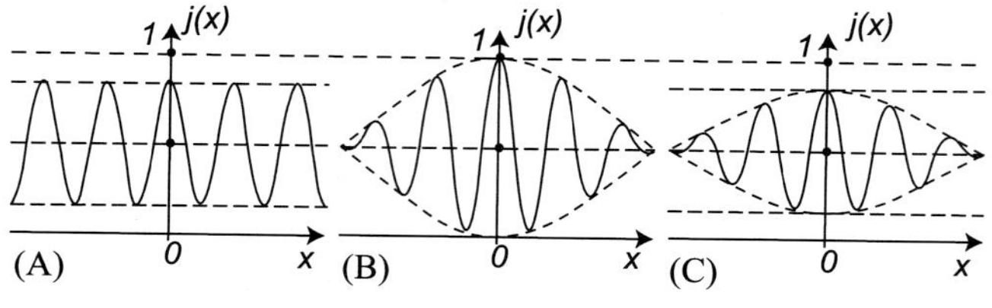

Figure 62.8
Normalized intensity distributions $j(x) = J(x)/2\bar{J}$: (A) extended source only; (B) polyenergetic source only and (C) superposition of both effects.

In practice, this theory contains unacceptable simplifications. The crossover C is produced by a system of electron lenses and is hence affected by their aberrations. In particular, the lateral distribution function $F(\xi)$ is broadened by transverse lens aberrations and by the Boersch effect. Furthermore, the plane of the source C, tacitly assumed to be at $z = 0$, is subject to a defocus and longitudinal chromatic aberrations. For a cold field-emission gun, the distribution is not Gaussian.

Moreover, the intensity is not recorded in the plane H, since the fringes are far too narrow there. This plane is, instead, highly magnified onto the detector. The necessary projector system is not shown in the figures. It is evident that these lenses must cause some further loss of fringe contrast as a consequence of their inevitable aberrations. In particular, the lateral chromatic aberrations, both isotropic and anisotropic cause a loss of contrast for large values of $|x|$ and $|y|$, respectively. These effects have been studied in detail by Lenz and Wohland (1984).

Other sources of deterioration of the fringe contrast are the noise of the electron beam and mechanical vibrations, as Hasselbach (1988) has emphasized. The superposition of all the lateral aberrations gives rise to a decrease of the maximum contrast at $x = 0$, while the combination of longitudinal or chromatic aberrations reduces the number of fringes visible. A rough estimate of this number is given by $N = U/\Delta U$, where ΔU is the effective energy spread. Möllenstedt and Wohland (1980) and Schmid (1984, 1985) used this to determine the spread ΔU. This quantity is not just determined by the energy spectrum of the emission process at the cathode but is also affected by the numerous aberrations, as explained above. Nowadays, values $N > 1000$ are feasible in well-designed interferometers.

62.4 The Ehrenberg–Siday or Aharonov–Bohm Effect

We have already briefly mentioned this effect in Section 5.6 of Volume 1, where we showed that if an electron beam is divided and passes on either side of a local magnetic field, the lengths of the optical paths will be different by an amount that depends on the magnitude of the enclosed field, even if the field is negligibly small in the regions traversed by the electrons. The electrons are affected by the vector potential, $\boldsymbol{A}$, which is different in the two regions. The difference in path length can be rendered visible as an interference effect.

We shall not go into the details of the extensive literature of this subject here but shall simply recall that the effect was first noticed by Ehrenberg and Siday in 1949 but attracted no attention. It was rediscovered by Aharonov and Bohm, who used very different arguments, in 1959 and first demonstrated by Chambers (1960), followed soon after by Fowler et al. (1961), Boersch et al. (1961, 1962b; cf. Lischke, 1969, 1970a, 1970b), Möllenstedt and Bayh (1962a, 1962b) and Bayh (1962). These experiments are discussed in

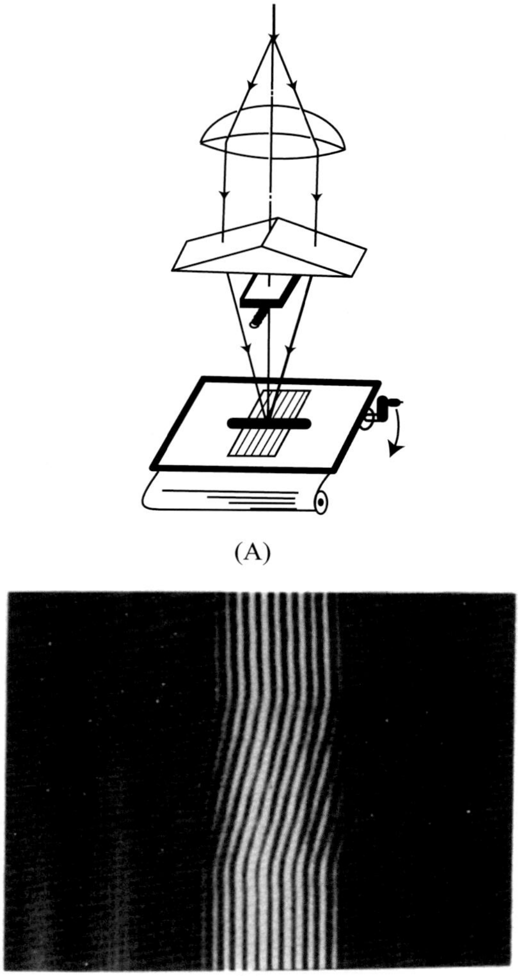

(A)

(B)

Figure 62.9

The Möllenstedt–Bayh experiment. As the current in the solenoid is increased, the film is advanced and the fringes that pass through the slit move laterally: (A) geometry and (B) fringe shift. *After Bayh (1967), Courtesy Springer, https://doi.org/10.1007/BF01377927.*

detail in the monograph of Peshkin and Tonomura (1989) and more recently by Batelaan and Tonomura (2009). In Fig. 62.9, the Möllenstedt and Bayh experiment and the resulting fringe shift are reproduced. A voluminous literature grew up during the 1960s and 1970s, consisting essentially of three groups of papers: by authors who believed that the effect did not exist, theoretical defences and experimental justifications of the effect and finally, abstract discussion of its theoretical implications. The first group is now of largely historic

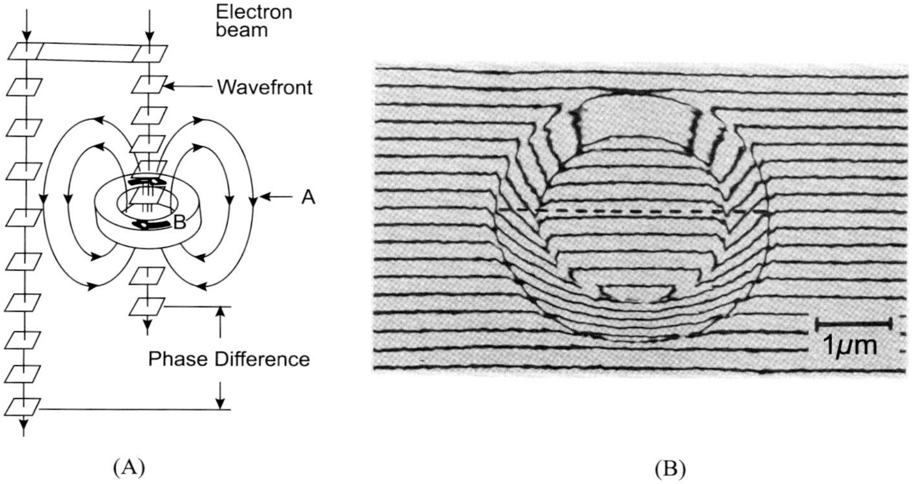

Figure 62.10
(A) Vector potentials near a ferromagnetic particle with a hole (***A***: vector potential; ***B:*** magnetic flux). (B) Interferogram showing the phase distribution of an electron beam passing through a toroidal ferromagnet. *After Tonomura (1984), Courtesy Oxford University Press.*

interest, since Tonomura has shown that the effect persists even when the beams are completely protected from any residual magnetic field (Tonomura et al., 1982a, 1984). The experiment uses the shielding properties of the superconducting state (Kuper, 1980) and is conclusive. First, however, we consider a ferromagnet in the normal state. An electron beam is divided into two partial beams, one of which passes through the central hole in a toroidal ferromagnet while the other passes outside (Fig. 62.10A). The resulting interferogram (Fig. 62.10B) shows that there is a fringe shift, and hence a phase difference, between these two regions. In the improved form of the experiment, described in full detail in Tonomura et al. (1986, 1987, 1990), Osakabe et al. (1986) and Osakabe (2014), the toroid consisted of a permalloy core enclosed in niobium, which becomes superconducting at 9.2 K. The interference fringes obtained with the niobium in the normal and superconducting states (Fig. 62.11) are consistent with the phase differences predicted by the Aharonov–Bohm effect. The magnetic field is far too small in this experiment to cause the fringe shift.

Attempts have been made, so far abortive, to persuade the physics community to attribute this effect to its discoverers, Werner Ehrenberg and Raymond Siday, and not to its rediscoverers. Sturrock and Groves (2010) argue that, even if the work of Aharononov and Bohm has had a wider impact, there is no gainsaying the fact that Ehrenberg and Siday published it 10 years earlier (see Berry, 2010; Peshkin, 2010). The paper by Sturrock and Groves was a reaction to a publication by Batelaan and Tonomura (2009); in a later paper,

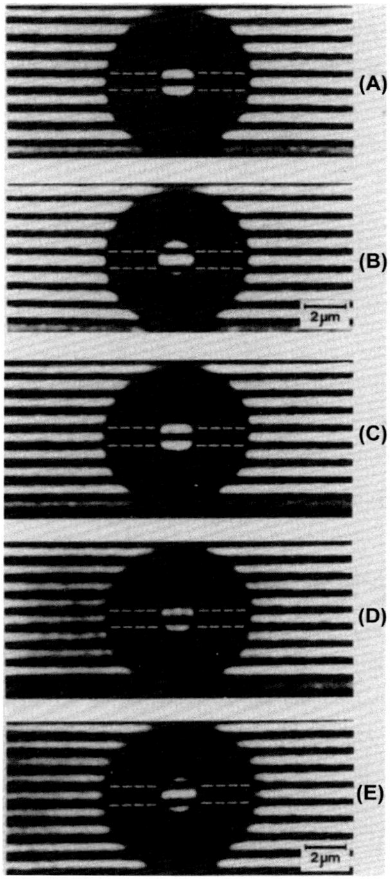

Figure 62.11
Interferograms of toroidal samples. (A) and (B), different samples at 4.5 K. In (A), the phase difference between outside and inside is $\pi(\text{mod } 2\pi)$ corresponding to the quantized flux $nh/2e$ in the superconducting niobium (n odd). In (B) n is even and so the phase difference is $2\pi(\text{mod } 2\pi)$. (C) $T = 4.5$ K. Fringe displacement between inside and outside is half a fringe spacing. (D) $T = 15$ K (above the critical temperature). Fringe displacement can now vary continuously and is here equal to 0.4 of the fringe spacing. (E) Room temperature. Further change in fringe displacement, due to a small decrease in the magnetization of the permalloy core. *After Tonomura et al. (1987), Courtesy Physical Society of Japan.*

these authors mention that a sentence on the work of Ehrenberg and Siday had to be removed to save space (Batelaan and Tonomura, 2010). Carles et al. (2021) refer to 'ESAB' instead of 'AB'.

For extremely full discussion of the earlier literature on this effect, we refer to the monograph of Peshkin and Tonomura (1989). Subsequent developments are recorded in

Selleri (1992) and the proceedings of the International Symposia on the Foundations of Quantum Mechanics (see Kamefuchi et al., 1984; Namiki et al., 1987; Kobayashi et al., 1990; Tsukada et al., 1992; Fujikawa and Ono, 1996; Ono and Fujikawa, 1999, 2002; Ishioka and Fujikawa, 2006, 2009). See also Silverman (1983, 1984a, 1984b, 1985, 1986), O'Raifeartaigh et al. (1991), Herman (1992), Home and Selleri (1992), Kobe et al. (1992), Tonomura (1992c), Tonomura et al. (1995) and Matteucci et al. (2003). Harris and Semon (1980), Blanco (1999), Caprez et al. (2007), Caprez and Batelaan (2009) and Carles et al. (2021) are also of interest in this context.

Several important aspects of the effect are analysed in the work of Semon (Semon, 1982; Semon and Taylor, 1986, 1987a, 1987b, 1988, 1994; Harris and Semon, 1980). In particular, the *modular variable* or *modular momentum*, suggested by Aharonov et al. (1969), (not discussed here) is examined more fully than by Aharonov et al. Their work suggests "that a further study of the 'modular momentum' might provide further insight into the Aharonov–Bohm interaction" between the electrons and the source of the vector potential.

The interpretation of the effect at the quantum level has generated a substantial literature (e.g. Boyer, 1972, 1987, 2000a, 2000b, 2002a, 2002b, 2008, 2015; Walstad, 2010, 2017; Batelaan and Becker, 2015; Wang, 2015). A major preoccupation is the deflection of the electrons and the force involved (Shelankov, 1998; Berry, 1999; Keating and Robbins, 2001; Batelaan and McGregor, 2014), re-examined by Becker et al. (2019); the latter have observed an asymmetry in the recorded fringe patterns that is compatible with the theory of Shelankov and Berry. It therefore seems that some deflection force is present but it cannot be a 'classical' force – instead, Shelankov and Berry speak of a 'quantum force'. We now present these ideas concisely.

Consider an incoming family of plane waves with a Gaussian distribution of wave vectors in the x–y plane. In the paraxial approximation, this leads to

$$\psi_o(\alpha,\theta) = \exp\left(-\frac{1}{2}\theta^2\omega^2\right)\left\{\cos(\pi\alpha) + \sin(\pi\alpha)\operatorname{erfi}\left(\frac{w\theta}{\sqrt{2}}\right)\right\} \tag{62.38}$$

in which $\psi_o(\alpha,\theta)$ is the probability amplitude for electrons to be scattered in the direction θ with respect to the x-axis when the paths are separated by a magnetic flux Φ and $\alpha := -e\Phi/h$. The rms angular spread of the incident beam is $1/w\sqrt{2}$ and

$$\operatorname{erfi}(x) := \operatorname{erf}(\mathrm{i}x) = \frac{2}{\sqrt{\pi}}\int_0^x \exp(\xi^2)\,d\xi \tag{62.39}$$

The function $|\psi_o(\alpha,\theta)|^2$ is shown in Fig. 62.12 for $\alpha = 1$, 1/2 and 1/4, where it is seen to be asymmetric for $\alpha = 1/4$. We now examine the evolution of the wave packet corresponding

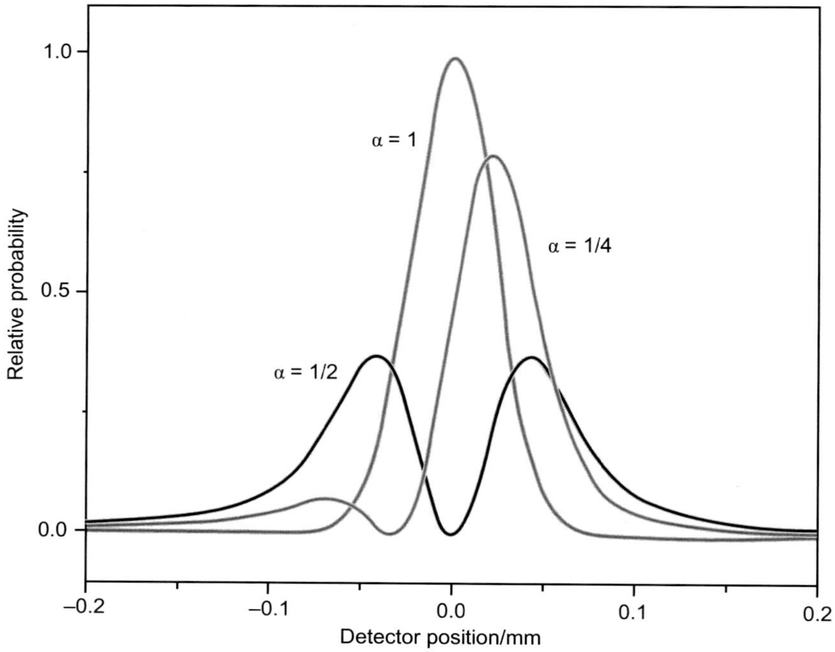

Figure 62.12
The function $|\psi_o(\alpha,\theta)|^2$ for three values of the enclosed flux. The distribution is seen to be asymmetric for $\alpha = 1/4$. *After Becker et al. (2019), Courtesy Nature, https://doi.org/10.1038/s41467-019-09609-9.*

to the set of incident plane waves. In terms of momentum, the wave packet, width $1/a$, is described by

$$\varphi(k,0) = \left(\frac{a^2}{\pi}\right)^{1/4} \exp\left\{-\frac{1}{2}(k-k_0)^2\right\} \exp\{-\mathrm{i}(k-k_0)x_0\} \tag{62.40}$$

the last term characterizing a linear phase ramp. After interaction with the biprism filament, this becomes

$$\varphi_A(k,0) := \varphi(k,0)F(k) \tag{62.41}$$

where $F(k)$ is some complex function. Reverting to the position representation, we find

$$\psi(x_j,t) = \frac{1}{\sqrt{2\pi}} \int \varphi_A(k,0) \exp\{-\mathrm{i}(kx_j - \omega(k)t)\} \tag{62.42}$$

where x_j is a transverse (x_T) or longitudinal (x_L) position coordinate; $\omega(k) = \hbar k^2/2m$. The expectation value of $x_j = \mathrm{i}\partial/\partial k$ for the wavefunction in (62.42) is hence

$$\langle x_j \rangle = x_{j0} + \frac{\hbar\langle k\rangle}{m}t + \frac{a}{\sqrt{\pi}}\int \frac{\partial\kappa}{\partial k}|R|^2 \exp\{-(k-k_0)^2\}dk \tag{62.43}$$

in which we have put $F(k) = R(k)\exp\{\mathrm{i}\kappa(k)\}$. Becker et al. examine the implications of (62.43) for two earlier theories. Zeilinger (1986), supported by Peshkin (1999), assumes that the interaction leads to a pure phase shift, $\kappa(k) = \kappa(k_L), R(k_L) = 1$; the longitudinal momentum has a Gaussian distribution, $\exp\{-(k_L - k_{L0})^2 a^2/2\}$. Hence,

$$\langle x_L \rangle = x_0 + \frac{\hbar k_0}{m}t + \frac{a}{\sqrt{\pi}}\int \frac{\partial\kappa}{\partial k}\exp\{-(k-k_0)^2 a^2\}dk \tag{62.44}$$

This means that if the interaction is nondispersive $(\partial\kappa/\partial k = 0)$, the expectation value of the position of the wave packet will be the same as the classical result: the Aharonov–Bohm effect is force-free.

Shelankov (1998) and Berry (1999) assume that the interaction causes a phase step:

$$F(y) = \exp\{2\pi\mathrm{i}\alpha(H(y) - 1/2)\} \tag{62.45}$$

$H(y)$ is the Heaviside step function. The transverse coordinate x_T is now y. The transverse momentum k_T has a Gaussian distribution proportional to $\exp\{-(k_T - k_{T0})^2 a^2/2\}$. For this case,

$$\varphi_A(k_T, 0) = \left(\frac{\beta^2}{2\pi}\right)^{1/4} \exp\left(-\frac{\beta^2 k_T^2}{4}\right)\{\cos(\alpha\pi) + \sin(\alpha\pi)\ \mathrm{erfi}(\beta k_T/2)\} \tag{62.46a}$$

(with $k_{T0} = 0$); β is a measure of the (transverse) width of the wave packet. More compactly,

$$\varphi_A(k_T, 0) = \left(\frac{\beta^2}{2\pi}\right)^{1/4} \exp\left(-\frac{\beta^2 k_T^2}{4}\right)F(k_T), \quad F(k_T) = R(k_T) \text{ and } \kappa(k_T) = 0 \tag{62.46b}$$

Finally,

$$\begin{aligned}\langle y \rangle &= y_0 + \frac{\hbar\langle k_T\rangle}{m}t \\ &= y_0 + \frac{\hbar}{m\beta}\sqrt{\frac{2}{\pi}}\sin(2\pi\alpha)t\end{aligned} \tag{62.47}$$

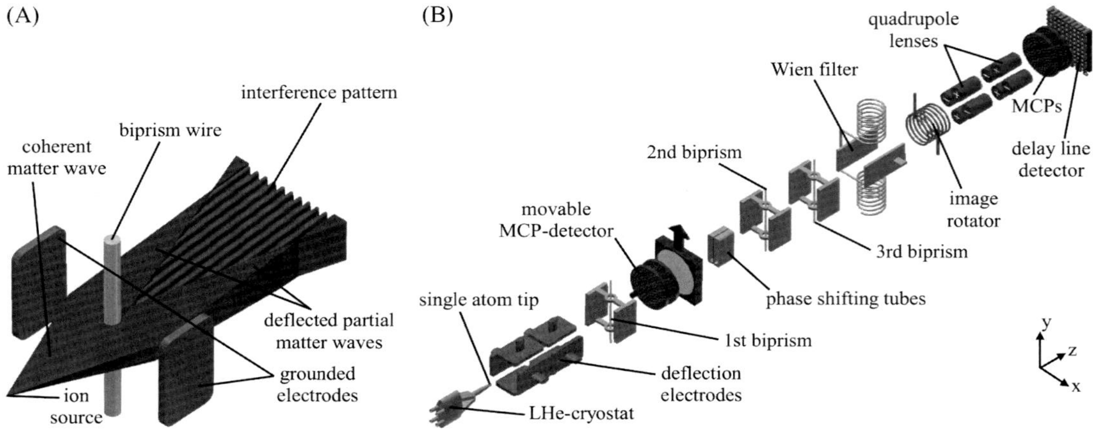

Figure 62.13
Ion interferometer for measurement of the electric Aharonov–Bohm effect: (A) creation of the interference pattern and (B) components of the interferometer. *After Schütz et al. (2015), Courtesy Elsevier.*

In their experiments, Becker et al. did not succeed in observing the deflection, since errors introduced between the recordings with the rod magnetized and neutral were greater than the predicted effect. They were, however, able to record an asymmetric profile consistent with the predicted deflection. This is (indirect) evidence for a 'quantum force'. Caprez and Batelaan (2009) and Carles et al. (2021) are also of interest in this context.

Aharonov and Bohm discussed not only the fringe shift caused by the magnetic vector potential but also an electrostatic effect. Here, a pulsed electron beam is divided as usual by a biprism. The partial beams then pass through metal tubes and while they are inside, different potentials are applied to the tubes. This has proved too difficult to put into practice for electrons, since they spend such a short time inside the tubes and moreover the pulsed beam would have an undesirable energy spread (Groves, personal communication 2021). A feasible experiment has been proposed for ions (Schütz et al., 2015). The optical arrangement is shown in Fig. 62.13. For ions, it is not convenient to pulse the source. Instead, Schültz et al. plan to use a pulse generator to create two correlated signals, one of which is connected to one of the tubes. The other triggers a logic module in a delay-line detector, which therefore accepts only ions that were in the tube while the pulse was applied. Fig. 62.14 shows three possible arrangements and simulations indicate that Fig. 62.14C would be the best. For subsequent work, see Rembold et al. (2014, 2017a, 2017b), Rembold (2017), Günther et al. (2015), Schütz (2018), Pooch (2018) and Pooch et al. (2017, 2018). For the Aharonov–Carmi effect, see Aharonov and Carmi (1973) and Harris and Semon (1980).

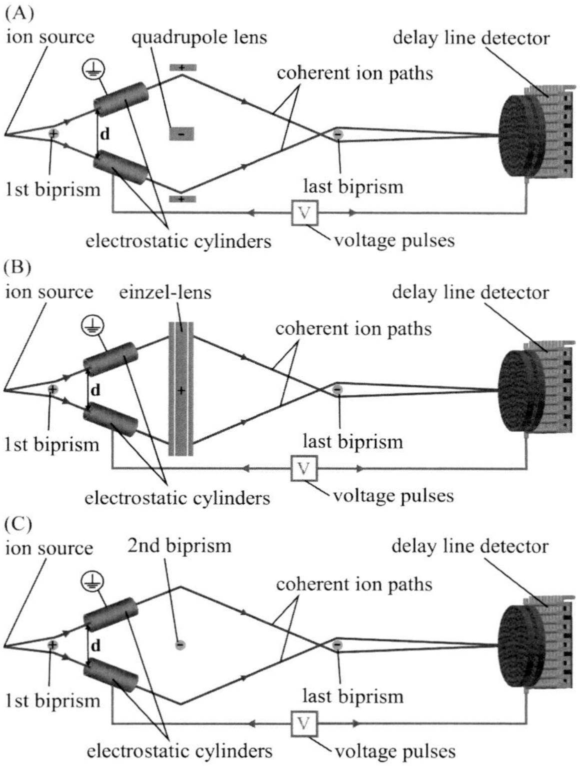

Figure 62.14
Configurations suitable for measurement of the electric Aharonov–Bohm effect. (A) Quadrupole lens. (B) Einzel lens. (C) Second biprism. *After Schütz et al. (2015), Courtesy Elsevier.*

A phase shift has been observed with a static electric field placed between separated partial beams by Matteucci et al. (1982a, 1982b, 1982c, 1982d, 1982e, 1982f) and, in an improved version, by Matteucci and Pozzi (1985). In the first paper (Matteucci et al., 1982a), the partial beams pass a thin metal cylinder, radius a, the two halves of which are made of different metals (Fig. 62.15). The contact potential V_c gives rise to a potential

$$\Phi(x,z) = \frac{V_c}{\pi}\arctan\left(\frac{2ax}{x^2+z^2-a^2}\right) \tag{62.48}$$

We assume that the cylinder is so long that a two-dimensional solution is acceptable. This results in a phase difference between the partial beams given by

$$\Delta\varphi = \frac{4\pi a}{\lambda\Phi_0}V_c \tag{62.49}$$

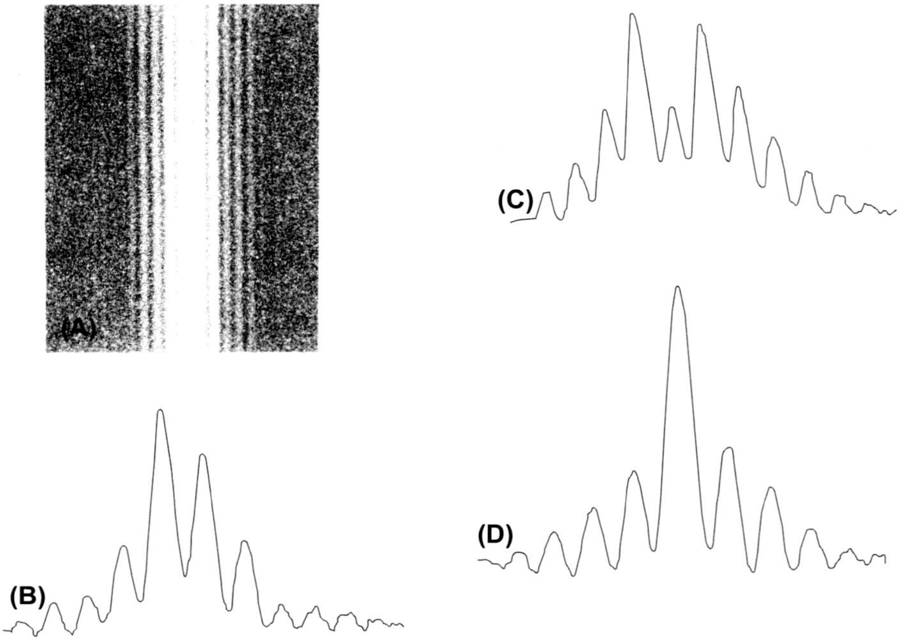

Figure 62.15
Electrostatic phase shift for a platinum filament partially coated with gold. (A) Fringe pattern and (B) densitometer trace from the gold-coated region. Biprism potential −2 V, accelerating voltage, 20 kV. (C) and (D) Densitometer traces from the uncoated region for biprism potentials of −2 V (C) and −1.5 V (D). Note the appearance of a central minimum. *After Matteucci et al. (1982b), Courtesy Elsevier.*

where Φ_0 is the accelerating voltage. For $\Phi_0 = 100$ kV, $V_c = 0.5$ V and $a = 0.5$ μm, we find $\Delta\varphi \approx 3\pi$. The authors tested this using a platinum wire, part of which was coated over half its surface with gold. Interferograms were recorded at $\Phi_0 = 20$ and 40 kV at filament potentials of −2, −1.5 and −2.5 V. Microdensitometer traces of the fringe intensities from the coated and uncoated regions confirmed that a constant phase shift is created. Such a phase shift had been predicted using classical mechanics by Boyer (1973), who discusses the work of Matteucci and Pozzi in a later paper (Boyer, 2002b). For further discussion of this electrostatic effect, see Matteucci et al. (1998, 2003) and Moreau and Ross (1994).

62.5 The Sagnac Effect

If a coherent beam of light or electrons, or indeed other particles, is divided into two coherent partial beams, which later recombine to produce interference fringes, a fringe shift can be caused by rotating the whole structure about an axis perpendicular to the area

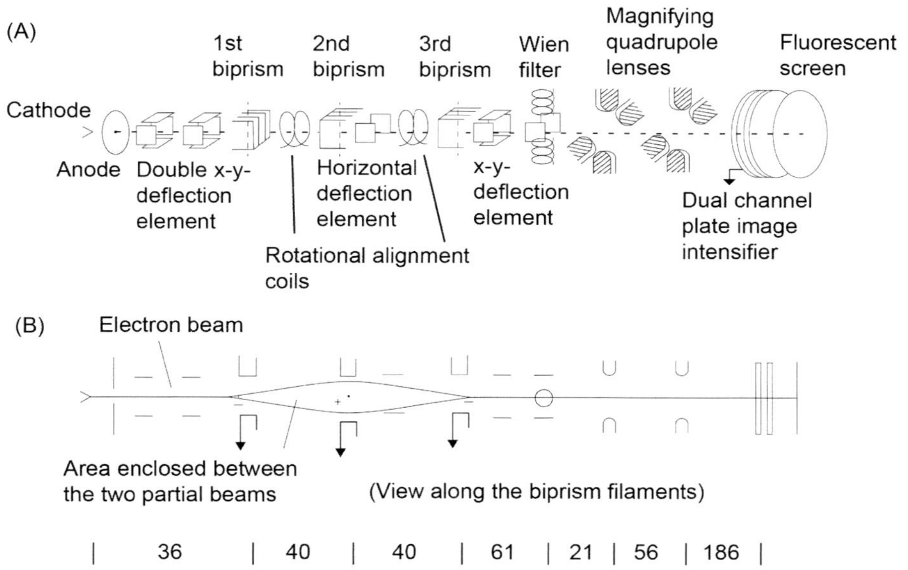

Figure 62.16
Rugged interferometer for detection of the electron Sagnac effect: (A) general view and (B) dimensions in millimetres. *After Hasselbach and Nicklaus (1993), Courtesy American Physical Society, https://doi.org/10.1103/PhysRevA.48.143.*

enclosed by the separate beams. If the area enclosed is A, the angular velocity Ω and the particle energy E, the phase shift $\Delta\varphi$ will be given by

$$\Delta\varphi = \frac{2}{\hbar c^2} EA\Omega \tag{62.50}$$

and for electrons of relativistic mass m,

$$\Delta\varphi = \frac{2m}{\hbar} A\Omega \tag{62.51}$$

(See Nicklaus, 1989 for a full derivation and an account of earlier work and Post, 1967, for a survey.) The technical problems to be overcome are severe (Hasselbach and Nicklaus, 1988), for an entire electron optical bench with field-emission gun, biprisms, Wien filter and detection unit has to be rotated at a frequency of the order of a revolution per second. The Wien filter is included owing to its ability to shift the partial wave packets longitudinally, in the direction of the optic axis, and hence to ensure that they superimpose satisfactorily (Hasselbach and Nicklaus, 1988, 1990; Nicklaus and Hasselbach, 1993, cf. Möllenstedt and Wohland, 1980). An interferometer of rugged design was built for the purpose by Hasselbach (1988). As shown in Fig. 62.16, the electron beam is diverged by a first biprism then reconverged by a second one. The whole system is rotated in the plane normal to the biprism filaments (the plane of the

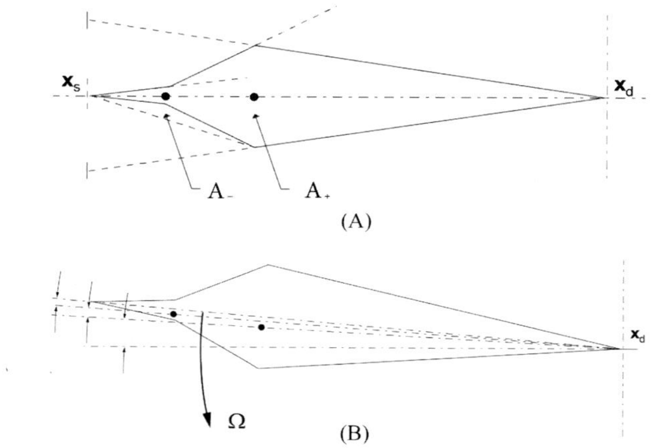

Figure 62.17
Schematic beam paths before and during rotation of the interferometer: (A) interferometer stationary and (B) interferometer rotating. *After Neutze and Hasselbach (1998), Courtesy American Physical Society, https://doi.org/10.1103/PhysRevA.58.557.*

page), with the result that, for one sense of rotation, the lower beam passes closer to the biprisms while the upper beam moves away from them (Fig. 62.17). For the other sense of rotation, the effect is reversed. The phase shift caused by the rotation is thus of electrostatic origin – it can be shown that other potential sources of the shift are negligible. The effect was detected and the magnitude of the phase shift agreed with the theoretical prediction to within the experimental error (Nicklaus, 1989; Hasselbach and Nicklaus, 1990, 1993; Hasselbach, 1992, 1997; Nicklaus and Hasselbach, 1995). The original experiment is described in Hasselbach's review (2010), where the Sagnac effects for light, neutrons and charged particles are described and contrasted. The theory was re-examined and perfected by Neutze and Hasselbach (1998) using a path-integral formalism. They show that rotation does indeed move classical electron paths closer to the biprism or farther from it. The ensuing phase shift is hence definitely electrostatic in origin. Neutze and Hasselbach obtain a corrected expression for the projected area of the interferometer; this has a small effect on the numerical values found in earlier publications. Work on the Sagnac effect has been surveyed and examined critically by Malykin (1997, 2000), who later studied the relativistic Zeno paradox (Malykin, 2002).

62.6 Other Topics in Electron Interference

62.6.1 Convergent-Beam Electron Diffraction Combined with an Electron Biprism

The principle of the interferometric mode is shown in Fig. 62.18. A convergent incident beam is focused on a plane below the specimen. In the plane of the selected-area aperture,

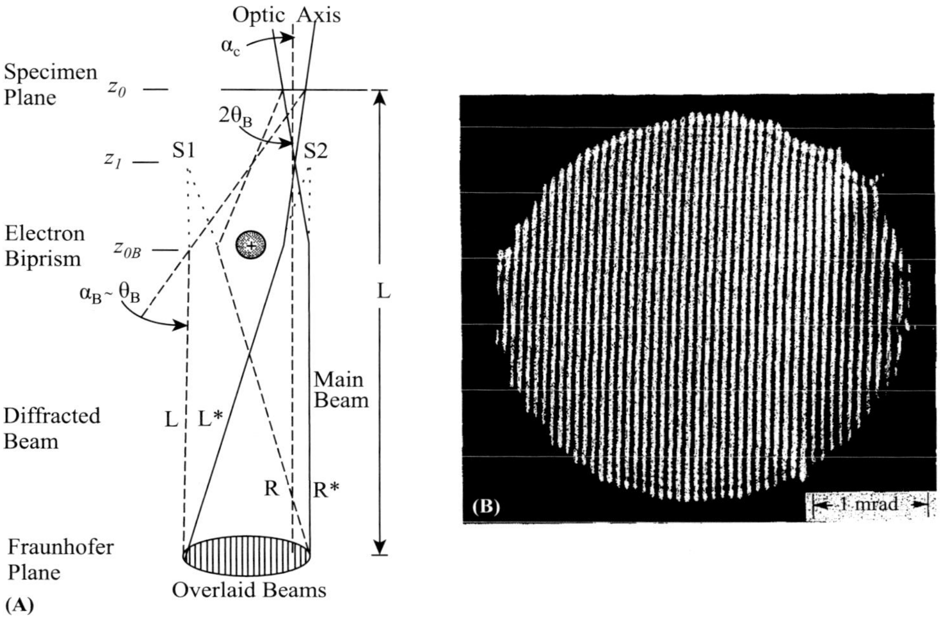

Figure 62.18

Convergent-beam diffraction combined with a biprism. (A) Ray diagram and (B) interferogram in which the main beam interferes with the 111 beam of GaAs. *After Herring et al. (1995), Courtesy Elsevier.*

a biprism is introduced, with the result that the main beam and the beam(s) diffracted in the crystalline specimen interfere in the intermediate image plane. The spacing of the resulting fringe pattern gives information about the phase difference between the convergent-beam diffraction beams. Such a scheme was proposed by Pozzi (1983a, 1983b) and developed by Herring et al. (1993a, 1993b, 1995), who provide extensive discussion and examples.

A different way of using a biprism in conjunction with convergent-beam illumination has been implemented by Houdellier et al. (2015). By inserting a biprism in the space between the gun and the first condenser (Tanigaki et al., 2012), the conical illumination is split into two half-cones. A suitable choice of condenser lens currrents ensures that the latter fall on the specimen with no additional tilt. In the back-focal plane of the objective lens, two semicircular discs will be seen, corresponding to the half-cones. The detailed structure seen in these discs can be used for measurement of strain in the specimen, for example. Ray diagrams for two values of the biprism excitation are shown in Fig. 62.19 together with examples of the resulting patterns. The technique is known as *splitting convergent-beam electron diffraction*.

62.6.2 Biprism Design

The earthed plates on either side of the biprism filament are usually vertical. This is not essential, as Duchamp et al. (2018) have shown. In their design, the plates or *counter-electrodes* are horizontal, lying in the $x-y$ plane, where the z-axis as usual coincides with the optic axis. Duchamp et al. also replace the cylindrical filament by a rectangular electrode, the z-dimension (or height) of which is thick enough to make it opaque to electrons. The complete structure is fabricated from a silicon-on-insulator wafer, consisting of a conducting doped single-crystalline layer on an insulating amorphous silicon oxide film; below this is a silicon substrate. A pattern representing the filament, counter-electrodes and connections is imprinted on the upper silicon layer; the filament is insulated from the counter-electrodes by empty trenches. The whole

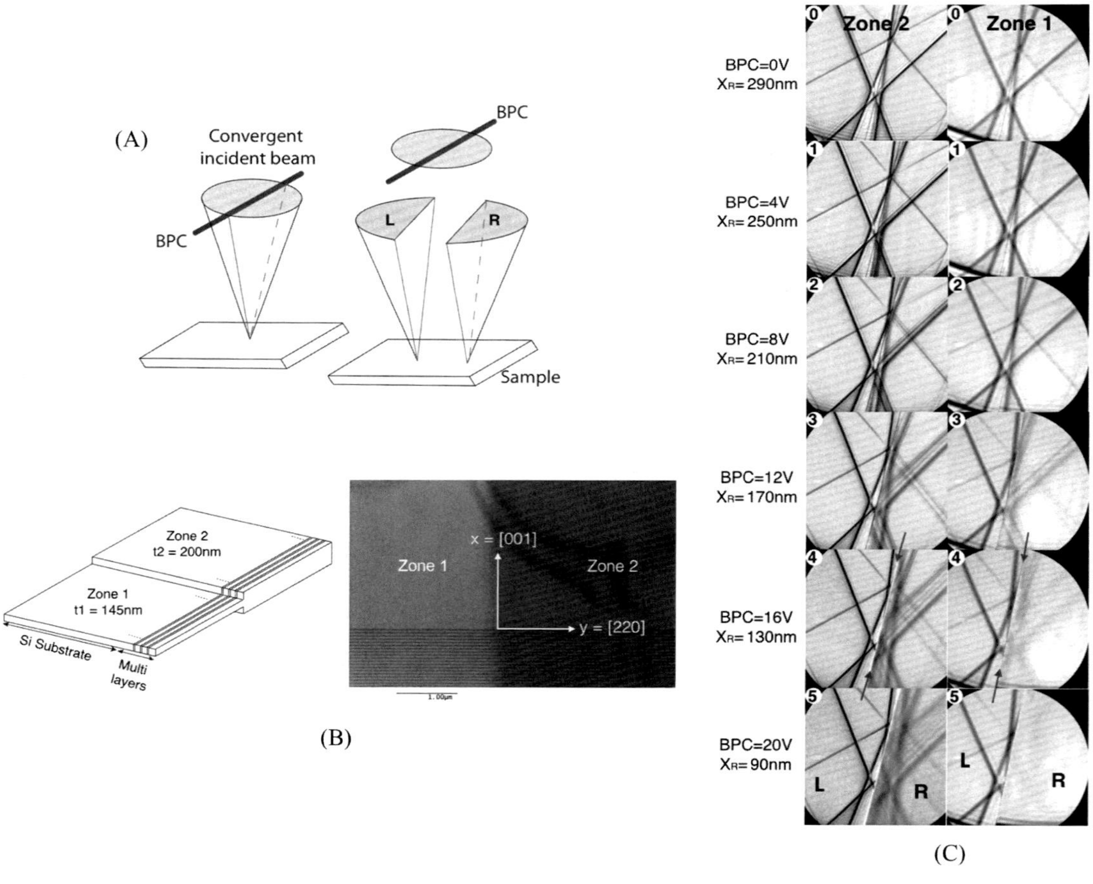

Figure 62.19
Splitting convergent-beam electron diffraction: (A) the effect of the biprism; (B) a multilayer silicon–germanium (Si_{1-x}, Ge_x) test specimen; (C) examples of the resulting patterns. X_R denotes the distance of the right half-probe from the site; BPC = biprism voltage. (D and E) Ray diagrams showing the biprism not excited (D) and excited at 10 V (E). *After Houdellier et al. (2015), Courtesy Elsevier.*

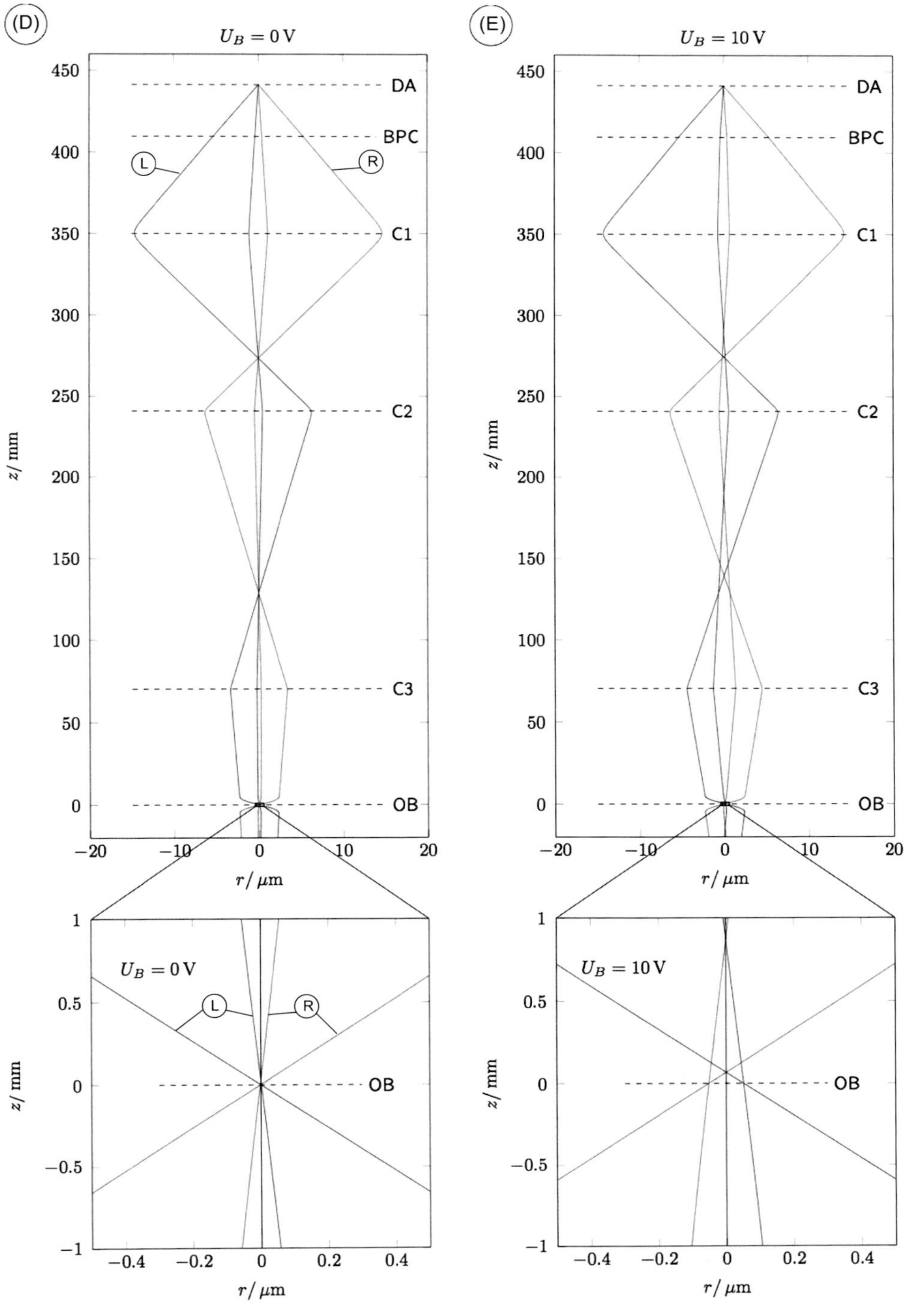

Figure 62.19
(Continued).

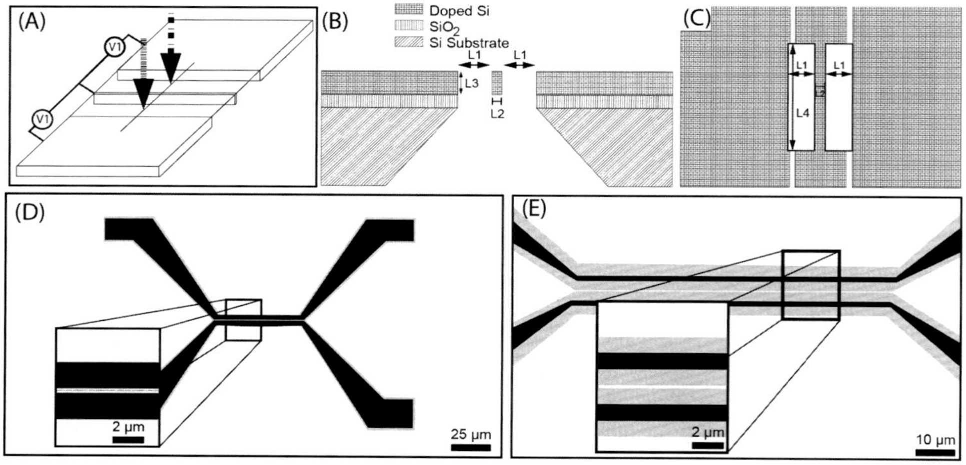

Figure 62.20
Alternative biprism design. Geometry of the device: (A) rectangular biprism between counter-electrodes; (B) cross-section; (C) top view; (D and E) traced scanning electron micrographs of the device [(D), top view and (E) tilted view]. *After Duchamp et al. (2018), Courtesy Elsevier.*

biprism is then extracted by photolithography. Fig. 62.20A–E describes the geometry of the device. The authors have calculated the electrostatic field and the resulting deflection of the electron beam both analytically and using the finite-element method (Volume 1, Chapter 10). The results are similar and we reproduce only the boundary-element values of the deflection as a function of the distance between the filament and the counter-electrodes for four heights and two widths of the rectangular filament (Fig. 62.21). Duchamp et al. have also considered the situation in which one of the counter-electrodes is held at the same potential as the filament. Experiments with this type of biprism installed in a 200 kV instrument with a Schottky field-emission source are described in their paper.

An advantage of this design is the uniformity and reproducibility of the filament. The surface of the traditional cylindrical thread may be rough and its diameter will depend on the mode of preparation. At higher accelerating voltages, it may not be completely opaque. Filament preparation techniques are described by Möllenstedt and Düker (1956), Matteucci (1978), Ohshita et al. (1984), Ogai et al. (1991), Nakamatsu et al. (2008), Matsui et al. (2009) and Schütz et al. (2014).

62.6.3 Two-Filament Biprisms

A device in which two filaments are used to create a trapezoidal prism (Fig. 62.22) has been described by Tanji et al. (1999) and employed by Hirayama (1999) to study electric field distributions. Tanji et al. use their device to produce two holograms. The reference beam passes between the two filaments while the deflected beam passes through first one and then the other

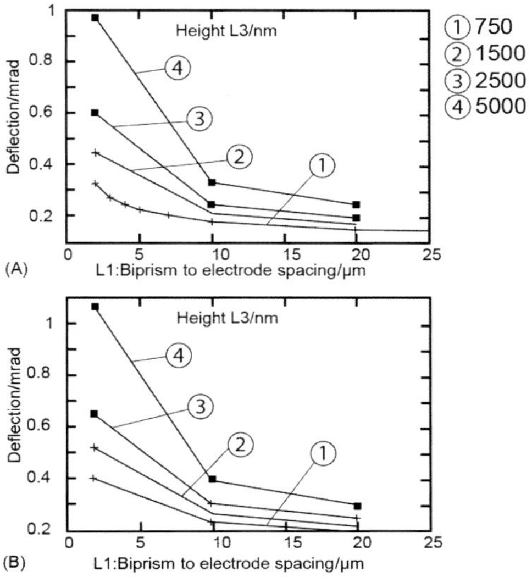

Figure 62.21
Alternative biprism design. Deflection of the electron beam at 200 kV for various geometries: (A) Biprism width $(L_2) = 120$ nm and (B) $L_2 = 250$ nm. *After Duchamp et al. (2018), Courtesy Elsevier.*

of the outer zones. The resulting hologram fringe patterns can then be used to create a differential phase image. Hirayama exploits the three-wave interference pattern illustrated in Fig. 62.23. The three waves ψ_o (object wave) and ψ_1, ψ_{-1} (reference waves) are represented by

$$\begin{aligned}\psi_o &= \psi_Q(x,y)\exp\{(i\eta(x,y)\}\\ \psi_1 &= \psi_Q(x-d,y)\exp(ik_x x)\\ \psi_{-1} &= \psi_Q(x+d,y)\exp(-ik_x x)\end{aligned}$$

in which $\psi_Q(x,y)$ represents the wave incident at the specimen. The current density in the image plane is proportional to

$$\langle|\psi_3(x,y)|^2\rangle = \left\langle\left|\psi_o+\psi_1+\psi_{-1}\right|^2\right\rangle \tag{62.53a}$$

which can be written

$$\left\langle\left|\psi_3\right|^2\right\rangle = 3 + 4Re\{\Gamma(d)\}\cos(k_x x)\cos\eta + 2Re\{\Gamma(d)\}\cos(k_x x) \tag{62.53b}$$

in which $\Gamma(d)$ denotes $\left\langle\psi_Q(x\cdot y)\psi_Q^*(x+d,y)\right\rangle$. The second term shows that lines of equal phase are mapped into intensity modulations of the cosine fringes, damped by the term $\Gamma(d)$. Hirayama illustrates this technique with a three-wave interference pattern created by a latex particle that has been charged by the electron beam. The equipotentials of the resulting electric field are seen as intensity modulations of the fringes (Fig. 62.24).

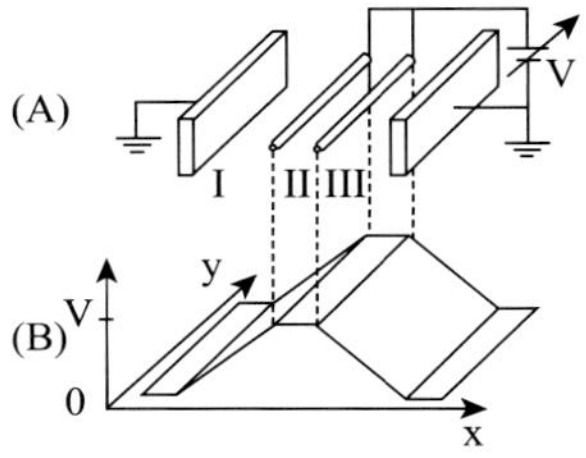

Figure 62.22
Trapezoidal biprism: (A) geometry and (B) ideal potential distribution. *After Tanji et al. (1999), Courtesy Elsevier.*

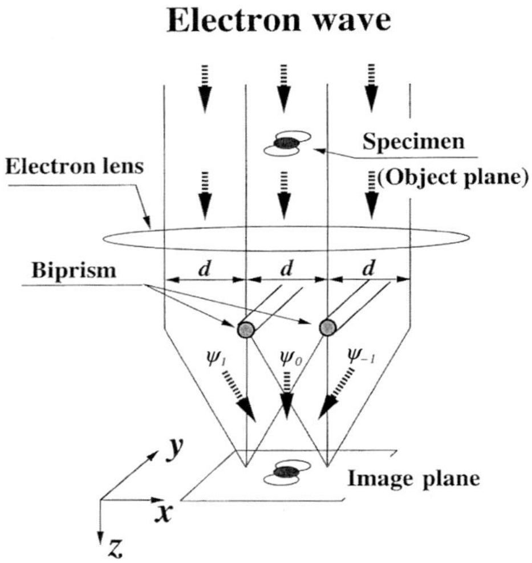

Figure 62.23
Detailed view of a twin-filament biprism. *After Hirayama (1999), Courtesy Elsevier.*

62.6.4 Biprism and Wien Filter

If a beam is separated by a biprism and the two partial beams are directed onto a Wien filter (Volume 2, Chapter 38), the contrast of the interference fringes is found to vary with the excitation of the filter. The reason for this is that the temporal coherence (energy spread) is finite and the wave packets of which the beams are composed are hence of limited length. As the excitation of the Wien filter is increased, the wave packets in the two branches move apart in the axial direction, since their group velocities are different. When they no longer overlap, the fringes vanish.

This observation is interesting for two reasons. First, it offers a sensitive method of measuring the coherence length of the beam (e.g. Möllenstedt and Wohland, 1980;

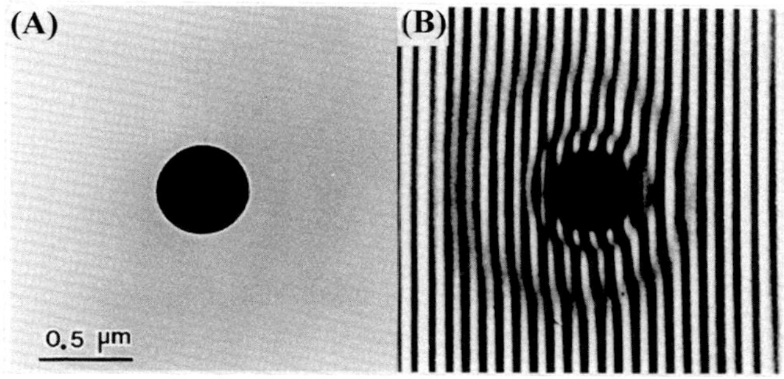

Figure 62.24
Interference pattern created by a latex particle (A) that has been charged by the electron beam. The equipotentials of the electric field are seen as modulations of the fringe intensity (B). *After Hirayama (1999), Courtesy Elsevier.*

Wohland, 1981). A refined arrangement (Fig. 62.25) has demonstrated that the coherence length is improved by introducing a monochromator before the biprism, as we expect (Bauer et al., 1996). Second, the Wien filter can be used when we deliberately wish to manipulate wave packets and may offer a way of transferring information with a high degree of security (Röpke et al., 2020).

Since the length and shape of the wave packet in the direction of travel are defined by the energy spread of the beam, it is possible to deduce the energy spectrum of the beam from measurements of the fringe contrast and spacing as a function of the spacing between the wave packets in the two arms. This has long been known in light optics, where it is known as Fourier spectroscopy (Michelson, 1891a, 1891b; Rubens and Wood, 1911; Vanasse and Sakai, 1967; for the history of the subject, see Loewenstein, 1966). It has been tested for electrons by Hasselbach and Schäfer (1990) and Hasselbach et al. (1995).

62.6.5 Correlation Between Separate Detectors (The Hanbury Brown and Twiss Effect)

In 1956 Robert Hanbury Brown and Richard Twiss demonstrated that correlations can be detected between the signals recorded by two coherently illuminated detectors (Hanbury Brown and Twiss, 1956; Twiss et al., 1957, cf. Hanbury Brown and Twiss, 1954). Attempts have been made to detect such correlations with electrons (Kiesel, 2000; Kiesel et al., 2002). The interpretation of their findings has not been accepted unreservedly and Keramati et al. (2020) suggest that the availability of very short electron pulses (cf. Volume 2, Section 37.5) makes an experiment for which the results would be unambiguous attainable. The work of Silverman (1987a, 1987b, 1987c, 1987d, 1988, 2008, 1987e), Osakabe et al. (1995), Kodama et al. (2011), Baym and Shen (2014), Tizei and Kociak (2017) and Meuret (2020a, 2020b, 2020c) is also relevant here.

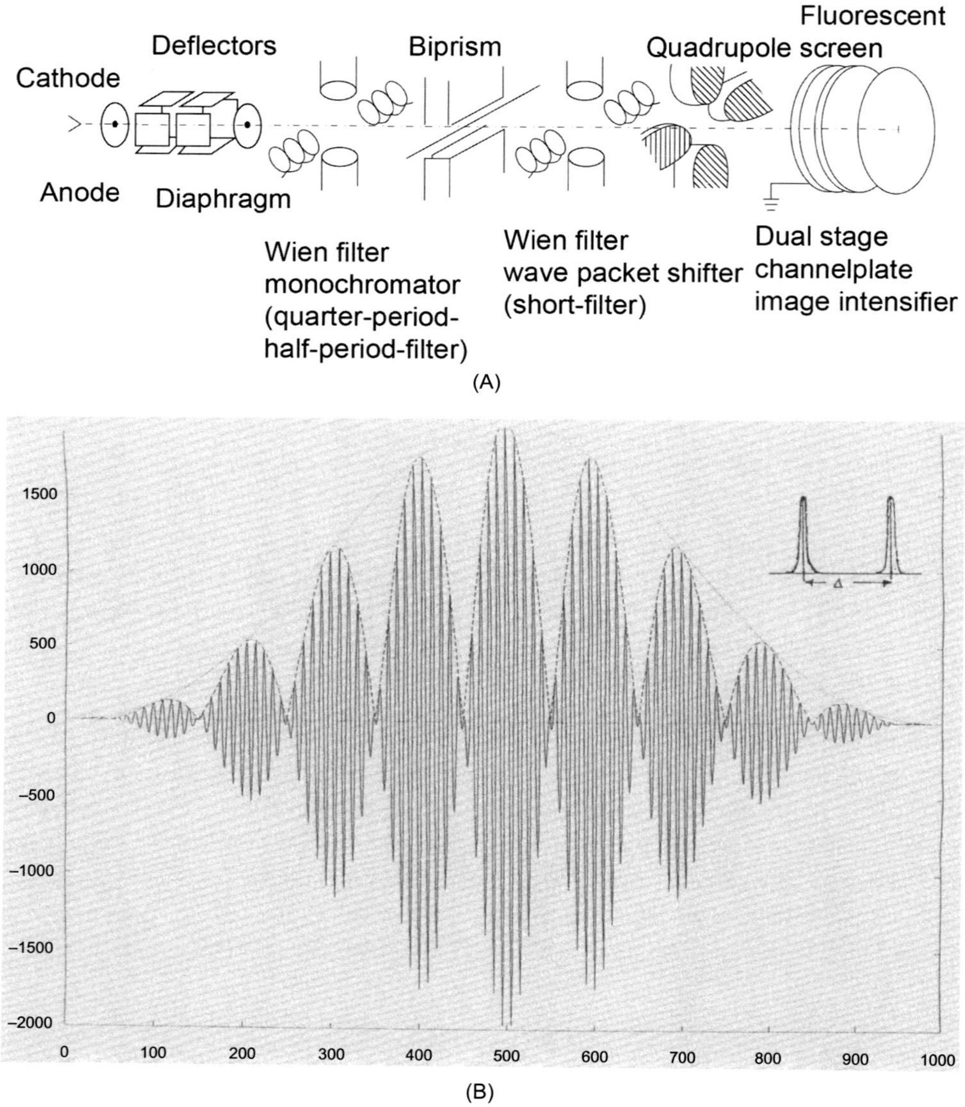

Figure 62.25

Combination of biprism and Wien filters for measuring coherence length as a function of energy spread. (A) geometry and (B), computer simulation of the interferogram generated by two (Gaussian) spectral lines separated by an energy Δ (see inset). (In the original publication, an electron interferogram is reproduced showing that gaps are indeed present but this is too dark to be reproduced here.) *After Hasselbach (1997), Courtesy Scanning Microscopy International.*

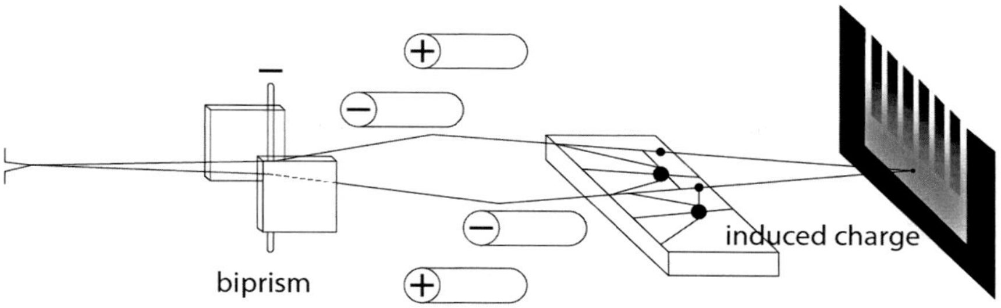

Figure 62.26
Arrangement for the measurement of decoherence. *After Hasselbach (2010), Courtesy IoP Publishing, https://doi.org/10.1088/0034-4885/73/1/016101.*

62.6.6 Decoherence

Finally, we mention a decoherence experiment in which the electron beam is separated as usual by a biprism, after which the partial beams are reconverged by means of a quadrupole. The partial beams pass close to a flat plate, in which charges are induced by the passage of the beams (Fig. 62.26). A fringe pattern is then formed and it is seen that the fringes are progressively more blurred as the distance between the plate and the partial beams is reduced (Fig. 62.27). This can be interpreted as decoherence, associated with the transition from quantum to semi-classical behaviour. For a full explanation of the phenomenon and its implications, see Sonnentag and Hasselbach (2000, 2005, 2007), Sonnentag (2006) and Hasselbach (2010). Later work suggests that the explanation of the decoherence in terms of image charges is not correct or not general enough to cover all cases (Beierle et al., 2018) and that the theory proposed by Howie is more appropriate (Howie, 2011, 2014). In the experiments of Beierle et al., a low-voltage electron beam (1.67 kV) is collimated by two slits, 25 cm apart. The resulting beam has a divergence of 61 μrad in the x-direction (perpendicular to the slits) and 120 μrad in the y-direction (parallel to the slits); the transverse (x) coherence length is about 600 nm. The beam then passes through a nanofabricated diffraction grating. A plane silicon or gold surface is then introduced in such a way that the surface is normal to the bars of the diffraction grating. The decoherence can be measured as a function of the distance between the beam and the surface in terms of the visibility of the interference fringes (Sonnentag and Hasselbach) or in terms of the transverse coherence length L,

$$L \approx \frac{\lambda}{\theta_c} \approx \frac{ad}{w}$$

in which θ_c denotes the angular spread of the collimated beam, a is the grating periodicity and d is the distance between neighbouring diffraction peaks at the detector (Beierle et al.); w denotes the full width at half-maximum of a diffraction peak. The results are very different for silicon and gold and none of the theories that have been proposed is capable of

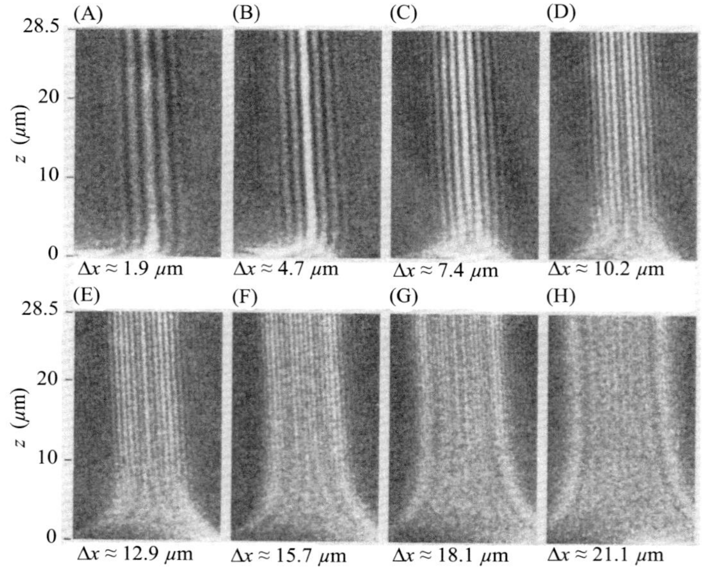

Figure 62.27
Fringe patterns obtained with the set-up shown in Fig. 62.26. The eight sets of fringes correspond to different distances between the two beams as they pass above the plate. In each case, the fringes become less sharp as the beams—plate distance z becomes smaller. *After Hasselbach (2010), Courtesy IoP Publishing, https://doi.org/10.1088/0034-4885/73/1/016101.*

explaining both findings. It seems that Howie's theory, inspired by the notion of aloof imaging, is the most promising. Recent thinking on this difficult subject is described by Keramati et al. (2020). It is not possible to reproduce the intricate material presented in their article, of which the summary and conclusions give a good idea: "Ultra-short pulsed electron beams may help the search for Hanbury Brown–Twiss antibunching signal for free electrons, and the signal depends on the spin-polarization of the ultrafast electron source. Enhanced signals and polarization dependence may be of use in distinguishing between the wanted effects due to electron spin and the unwanted effects of Coulomb repulsion between free electron pairs in experiments with pulsed electrons. The latter can lead to anti-bunching signals which are rooted in Coulombic interactions rather than spin-statistics. Therefore, a decisive observation of the Pauli exclusion principle for free electrons may be possible with spin-polarized ultrashort electron pulses, but only if the pulse duration is not much larger than the coherence time. Expressions for the Hanbury Brown–Twiss antibunching contrast, and the reduced count rate are given. Finally, a relation between our approach and decoherence theory (Joos et al., 2003; Schlossbauer, 2007, 2019; Zurek, 2003) was provided to show consistency with decoherence theory and justify the description of 'partial' coherence. We hope that our treatment of the problem assists attempts to observe the Hanbury Brown–Twiss effect for free electrons".

We also draw attention to the publications of Nicklaus and Hasselbach (1995), Hasselbach and Maier (1995, 1996, 1999), Hasselbach (1997), Maier (1997), Hasselbach and Nicklaus (1987) and Hasselbach et al. (2000, 2004), Zurek (2003), the book edited by Blanchard et al. (2000) and the books by Joos et al. (2003) and Schlosshauer (2007). A long, more recent article by Schlosshauer (2019) is very valuable.

CHAPTER 63

Holography

Electron interference and holography are intimately related, as we saw in the general introduction (Chapter 61). One major difference is that the formation of interference fringes in a two-beam interferometer can be understood without considering diffraction, whereas the reconstruction of object waves from a hologram is an intrinsic diffraction process. It is this distinction that has guided our organization of the subject material.

Before discussing in-line and off-axis holography in detail, we recall some general features of the holographic process and the associated vocabulary. The essence of holography is the formation of an interference pattern by adding a reference beam and a beam modulated by the specimen. In the case of in-line holography (where reference beam and image-forming beam coincide), the hologram may be recorded in or close to a plane conjugate to the specimen, in which case it is called a *Fresnel hologram*; alternatively, it may be recorded in some other plane, distant from an image plane, in which case it is called a *Fraunhofer hologram*. The proximity to an image plane is measured by the Fresnel number, ν, defined by

$$\nu = \frac{\pi s^2}{\lambda \Delta} \tag{63.1}$$

in which λ is the electron wavelength, Δ the distance from the hologram plane to the image plane referred back to object space and s is the size of the specimen detail of interest. For high-resolution work, $s = 0.2$ nm for example, and operation at 300 kV ($\lambda \approx 2$ pm), we have $\nu \approx 60/\Delta$ (Δ in nm). Values of ν of the order of unity or less correspond to Fraunhofer conditions, while values appreciably greater than unity are associated with Fresnel hologram formation. For values of Δ less than about 10 nm, therefore, Fresnel conditions prevail, but otherwise, we shall be involved in Fraunhofer holography. The same distinction is made in off-axis holography, introduced by Leith and Upatnieks in 1962 (see also Leith and Upatnieks, 1963, 1967). Here, image-plane holograms (Fresnel holograms with $\Delta \approx 0$) are attractive because finite source size proves to have the least effect on resolution in this case.

A further ramification is *interference holography* (Tonomura et al., 1979d, see Tonomura, 1984, 1987a), which is particularly attractive for displaying phase variations and hence thickness distributions or magnetization patterns. Here, the laser beam used in the reconstruction is split into two beams by a half-silvered mirror, one of which falls on the hologram and generates the usual reconstructed images. One of these is selected by an

Principles of Electron Optics.
DOI: https://doi.org/10.1016/B978-0-12-818979-5.00063-2

aperture and allowed to interfere with the other partial beam (Fig. 63.1A). A further refinement allows us to amplify the phase differences involved by arranging that a reconstructed image and its conjugate interfere (Fig. 63.1B). We illustrate these possibilities below. This by no means exhausts the many forms of electron holography, several of which are described in Section 63.6. For a long list, see Cowley (1992) and for many examples, see the monograph of Tonomura (1999), the collections edited by Völkl et al. (1999) and Tonomura et al. (1995) and the surveys by Midgley (2001), Lichte and Lehmann (2002) and Dunin-Borkowski et al. (2019). The use of the Wigner function to characterize holography, introduced by Lubk, is summarized in the concluding section and in Chapter 79.

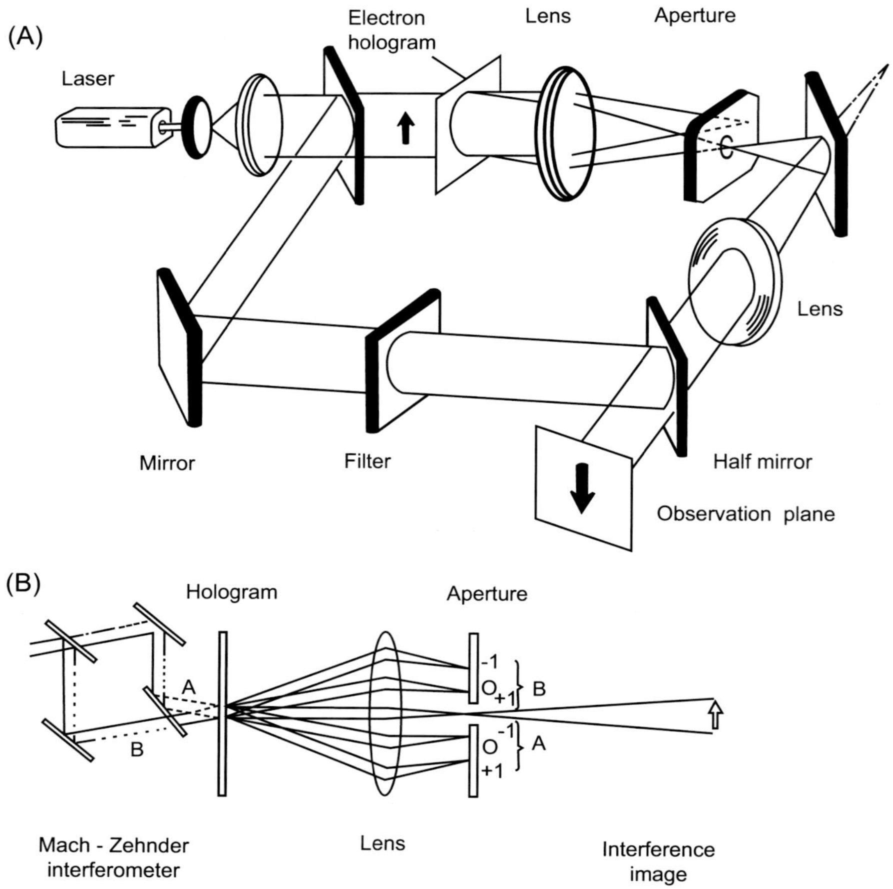

Figure 63.1
(A) Optical reconstruction system for interference microscopy. (B) Optical reconstruction system for interference microscopy with phase amplification (× 2). *After Tonomura et al. (1987), Courtesy Physical Society of Japan.*

63.1 In-Line Holography

We have already mentioned that, owing to the difficulty of separating the two images in the reconstruction step, the in-line technique lost much of its original attraction except at extremely low voltages (Section 63.6.9); nevertheless, it will be considered briefly here since it provides access to the mathematical formalism in a simple way, and furthermore, new aspects of the in-line mode have led to revived interest in it, as we shall show.

63.1.1 Hologram Recording

We start from the configuration shown in Fig. 63.2A. The axial point source S at $z = -a$ emits a monoenergetic spherical reference wave denoted $\psi_R(x, y, z)$. We adopt the Fresnel approximation and the complex wave amplitude in the hologram plane H, $(z = 0)$ is then

$$\psi_R(x, y, 0) = \frac{A}{a}\exp\left\{\frac{ik}{2a}\left(x^2 + y^2\right)\right\} \tag{63.2a}$$

Quite analogously, a point object O at $z = -b$ produces an object wave

$$\psi_o(x, y, 0) = \frac{B}{b}\exp\left\{\frac{ik}{2b}\left(x^2 + y^2\right)\right\} \tag{63.2b}$$

The intensity of the superimposed waves is then

$$\begin{aligned} J(x, y) &= |\psi_R(x, y, 0) + \psi_o(x, y, 0)|^2 \\ &= \frac{|A^2|}{a^2} + \frac{|B^2|}{b^2} + \frac{A^*B}{ab}\exp\left\{\frac{ik}{2}\left(\frac{1}{b} - \frac{1}{a}\right)\left(x^2 + y^2\right)\right\} + \text{c.c.} \end{aligned} \tag{63.3}$$

where c.c. denotes the complex conjugate of the last term. Apart from an unimportant scaling factor, this intensity pattern is recorded on the hologram. It is obvious that $J(x, y)$ is *nonlinear* in ψ_o, since a term $|\psi_o|^2 = |B|^2/b^2$ appears. When object waves emanating from many object points are superposed, such a term leads to nonlinear interactions, which make

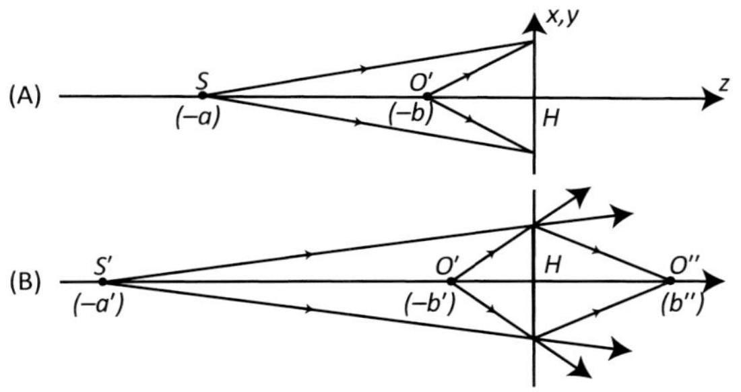

Figure 63.2
Simplified ray constructions with geometric parameters. (A) Recording step; (B) reconstruction. (In these configurations, a, b and a', b', b'' are all positive.)

direct reconstruction impossible. It is hence necessary to assume that the object waves are *weak*, which means that $|B/b|^2 \ll |A/a|^2$.

The hologram must be processed in such a way that its *transparency* $T(x, y)$ becomes proportional to $J(x, y)$:

$$\begin{aligned} T(x,y) &= T_0 J(x,y) \\ &= T_0 \frac{|A|^2}{a^2} + \frac{T_0 A^* B}{ab} \exp\left\{ \frac{\mathrm{i}k}{2f}(x^2+y^2) \right\} \\ &\quad + \frac{T_0 A B^*}{ab} \exp\left\{ -\frac{\mathrm{i}k}{2f}(x^2+y^2) \right\} \end{aligned} \tag{63.4a}$$

with

$$\frac{1}{f} := \frac{1}{b} - \frac{1}{a} > 0 \tag{63.4b}$$

63.1.2 Object Reconstruction

This step is sketched in Fig. 63.2B. The processed hologram is illuminated with a coherent wave ψ'_R emanating from an axial point source S' at $z = -a'$. The wavenumber k' may be different from that of the recording step.

On the entrance side of the hologram, we have the wave excitation

$$\psi'_R(x,y,0) = \frac{A'}{a'} \exp\left\{ \frac{\mathrm{i}k'}{2a'}(x^2+y^2) \right\} \tag{63.5}$$

and on the exit side

$$\chi(x,y) = T(x,y)\psi'_R(x,y,0) = \sum_{j=-1}^{1} \chi_j(x,y) \tag{63.6}$$

The three partial waves χ_{-1}, χ_0 and χ_1 result from the three contributions to $T(x, y)$ in (63.4a). Introducing these into (63.6) and using (63.5), we find

$$\chi_0(x,y) = \frac{T_0 |A^2| A'}{a^2 a'} \exp\left\{ \frac{\mathrm{i}k'}{2a'}(x^2+y^2) \right\} \tag{63.7a}$$

$$\chi_1(x,y) = \frac{T_0 A^* B\, A'}{aba'} \exp\left\{ \frac{\mathrm{i}}{2}\left(\frac{k'}{a'} + \frac{k}{f} \right)(x^2+y^2) \right\} \tag{63.7b}$$

and

$$\chi_{-1}(x,y) = \frac{T_0 A B^* A'}{aba'} \exp\left\{ \frac{\mathrm{i}}{2}\left(\frac{k'}{a'} - \frac{k}{f} \right)(x^2+y^2) \right\} \tag{63.7c}$$

The wave structure behind the hologram is a consequence of the diffraction in the hologram. This diffraction is described by employing χ_{-1}, χ_0 *and* χ_{+1} as *start-functions* or one-sided boundary values in the diffraction integral. In the Fresnel approximation the latter is given by

$$\psi_j'(x,y,z) = \frac{k' e^{ik'z}}{2\pi i z} \iint\limits_{-\infty}^{\infty} \chi_j(u,v) \exp\left[\frac{ik'}{2z}\left\{(x-u)^2 + (y-v)^2\right\}\right] du\, dv \qquad (63.8)$$
$$(j = -1, 0, 1; \quad z \gg D)$$

D being the diameter of the hologram. The resulting integrals, containing quadratic forms of u and v in the exponents, can all be evaluated in closed form by means of the general formula

$$\iint\limits_{-\infty}^{\infty} \exp\left\{\frac{i}{2} p\left(u^2 + v^2\right) - iq(xu + yv) + \frac{i}{2} r\left(x^2 + y^2\right)\right\} du\, dv \quad = \frac{2\pi i}{p} \exp\left\{\frac{i}{2}\left(x^2 + y^2\right)\left(r - \frac{q^2}{p}\right)\right\}$$

After some elementary calculations, we find

$$\psi_0'(x,y,z) = \frac{T_0 |A^2|}{a^2} \frac{e^{ik'z}}{z + a'} \exp\left\{\frac{ik'\left(x^2 + y^2\right)}{2(z + a')}\right\} \qquad (63.9)$$

$$\psi_1'(x,y,z) = \frac{T_0 A^* B\, A' b'}{a b a'} \frac{e^{ik'z}}{z + b'} \exp\left\{\frac{ik'\left(x^2 + y^2\right)}{2(z + b')}\right\} \qquad (63.10a)$$

with

$$\frac{1}{b'} := \frac{1}{a'} + \frac{k}{k' f} \qquad (63.10b)$$

and finally

$$\psi_{-1}'(x,y,z) = -\frac{T_0 A B^* A' b''}{a b a'} \frac{e^{ik'z}}{z - b''} \exp\left\{\frac{ik'\left(x^2 + y^2\right)}{2(z - b'')}\right\} \qquad (63.11a)$$

with

$$\frac{1}{b''} := -\frac{1}{a'} + \frac{k}{k' f} \qquad (63.11b)$$

63.1.3 Interpretation of the Results

Comparison of (63.9) with (63.5) shows that

$$\psi_0'(x,y,z) = \frac{T_0 |A|^2}{a^2} \psi_R'(x,y,z) \qquad (63.12)$$

if we generalize (63.5) for $z \neq 0$ by replacing a' by $a' + z$; ψ'_0 is hence the *reconstructed reference wave* of Fig. 63.2 and is of little importance, since this wave is not the required one and must be excluded.

The next partial wave ψ'_1 can be regarded as a *reconstructed object wave*. The corresponding object point is, however, shifted from the position O to O', as sketched in Fig. 63.2. This is a consequence of the altered illumination conditions. In fact, if we specialize to $k = k'$, $a = a'$ in (63.10a and 63.10b), we find $b = b'$ and then

$$\psi'_1(x, y, z) = \frac{T_0 A^* A'}{a^2} \psi_o(x, y, z) \tag{63.13}$$

[again by the generalization $b \rightarrow z + b$ in (63.2a and 63.2b)]. This is usually the required wave.

The third partial wave ψ'_{-1} (x, y, z) converges to an object point at a position $z = b''$ *behind* the hologram, as sketched in Fig. 63.2B. It can likewise be used as a reconstructed object wave, if some phases change their sign. This becomes easier to understand, if we again specialize to $a = a'$, $k = k'$, giving $1/b'' = 1/b - 2/a$. A further specialization is the case of parallel illumination: $a \rightarrow \infty$ and $A = A' \rightarrow \infty$ such that $C := A/a$ remains finite. We then find $b'' = b$ and hence from (63.11a):

$$\begin{aligned} \psi'_{-1}(x, y, z) &= -\frac{T_0 |C|^2 B^*}{z - b} e^{ikz} \exp\left\{\frac{ik\left(x^2 + y^2\right)}{2(z - b)}\right\} \\ &= \psi_1'^*(x, y, -z) \end{aligned} \tag{63.14}$$

The focus $z = +b$ of this wave is mirror-symmetric with respect to the original object position at $z = -b$. The presence of the factor B^* instead of B shows that the phase has reversed its sign. This wave is, therefore, often called the *conjugate wave*.

We have now shown that the method, sketched in Fig. 61.3 and briefly described in the introduction, must indeed work and we could even deal with the more general configuration shown in Fig. 63.2.

63.1.4 Focal Equations

From (63.10b) and (63.11b), we can immediately derive the focal equations

$$\frac{1}{b'} - \frac{1}{b''} = \frac{2}{a'} \tag{63.15}$$

and

$$\frac{1}{b'} + \frac{1}{b''} = \frac{2k}{k'f} =: \frac{1}{F} \tag{63.16}$$

The latter relation resembles the focal equation of a thin lens, and we may hence associate a focal length

$$F = \frac{\lambda ab}{2\lambda'(a - b)} = \frac{\lambda f}{2\lambda'} \tag{63.17}$$

with the hologram.

From this formula, it is immediately obvious that light-optical reconstruction from a hologram produced with electron waves is *not* immediately possible for technical reasons. The difference $a - b$ in the denominator cannot be made arbitrarily small, since there must be space enough to bring the object into the electron beam; we have roughly $F \approx \lambda b/2\lambda'$. Now the wavelength λ' of the laser light, used for reconstruction, is very much greater than the wavelength λ of electrons at some kilovolts. This implies that $F \ll b$, which is technically not feasible. The hologram must therefore be *highly magnified* in the electron optical system before being recorded. This situation is quite analogous to the technical conditions for the operation of an interferometer. Magnification of the hologram is equivalent to increasing the ratio λ/λ'.

The focal equations are exactly the same as for the lowest diffraction orders obtained with Fresnel *zone plates*; there must hence be a close relation between holography and zone plates (Rogers, 1950a). This relation is depicted in Fig. 63.3; the variable r is here the radial coordinate $r = (x^2 + y^2)^{1/2}$. The average transparency is $\overline{T} = T_0|A|^2 a^{-2}$ in agreement with the first term of (63.4). In this example, we have chosen the phase relations such that $\Re(A^* B) = 0$, hence

$$T(r) = \overline{T}\{1 - 2\sin(kr^2/2f)\} \tag{63.18}$$

The hologram and the zone plate have essentially the same fringe structure. The only important difference lies in the fact that the holographic transparency is *smooth*, while the zone-plate distribution has sharp edges. An important consequence is that in holography there appear just *three* different diffraction orders, while a zone plate produces an infinite (or at least very large) number of them.

We can now understand qualitatively how the reconstruction of an extended object from its hologram is possible. For this, we consider the object to be built up from a large number of very small surface elements. Each of these produces its concentric ring system on the hologram, the centre being appropriately shifted. In the reconstruction process, each ring system acts as a tiny Fresnel lens and produces three different waves. Their superposition gives the *correct* total wave, since the amplitude and phase relations are maintained by virtue of the linearization.

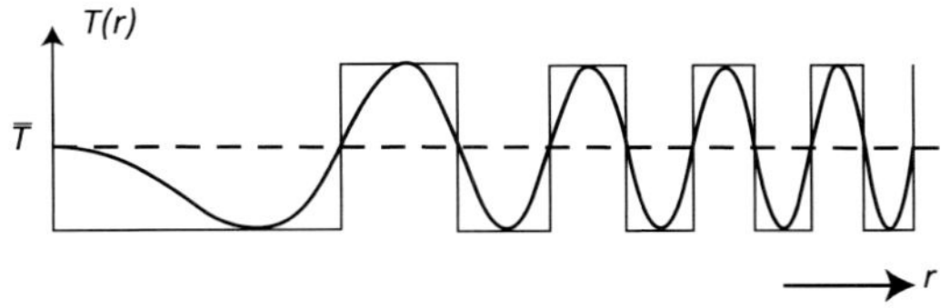

Figure 63.3
Holographic transparency distribution (smooth curve) and equivalent zone lens (steps); $\overline{T}$ is the average transparency.

It will be clear from the foregoing account and the detailed analysis of Section 66.1 that in-line electron holography and high-resolution image formation of weakly scattering specimens have much in common. We shall not pursue this interpretation of bright-field electron image formation here; the interested reader is referred to the accounts of Hanszen (1971, 1982a). An early in-line holographic reconstruction using a large enough defocus value ($\nu \approx 0.04$) to reduce the conjugate image to a blurred background behind the principal image was made by Tonomura and Watanabe in 1968 (Tonomura and Watanabe, 1968; Tonomura et al., 1968a,b; Watanabe and Tonomura, 1969). Other early attempts were made by Voronin et al. (1972) and Munch (Munch, 1975; Munch and Zeitler, 1974). This mode continues to attract interest: Latychevskaia and Fink (2009a,b, 2015), Latychevskaia et al. (2010, 2012, 2015, 2019), Koch and Lubk (2010), Koch (2014), Carlino et al. (2018), Liu et al. (2018), Huang et al. (2019), Carlino (2020) and Latychevskaia (2021).

63.2 Off-Axis Holography: Hologram Formation in a Two-Beam Interferometer

63.2.1 Expression for the Two Waves

Electron holography is nowadays mostly performed using the off-axis configuration with the aid of a two-beam interferometer. We considered fringe formation in this device in Sections 62.2–62.3, and for simplicity, we return to the arrangement shown in Fig. 62.4. Now, however, the object O may have a very fine structure and we can, therefore, no longer disregard diffraction effects within it. The reasoning becomes considerably more complicated, as we shall see.

It is convenient to express the object wave $\psi_{\overline{V}}$ emanating from the virtual source $\overline{V}$ (see Fig. 62.4), in the tilted coordinate system (x', y', z'), defined by

$$\begin{aligned} x' &= (x-d)\cos\gamma + z\sin\gamma \qquad & y' &= y \\ z' &= -(x-d)\sin\gamma + z\cos\gamma \qquad & \tan\gamma &= d/a \end{aligned} \tag{63.19}$$

The coordinates of the midpoint (M_0) of the object are then

$$\begin{aligned} z_M &= a-b \qquad & z'_M &= (a-b)\sec\gamma =: D_0 \\ x_M &= d-(a-b)\tan\gamma \qquad & x'_M &= 0 \end{aligned}$$

The object may be slightly tilted with respect to the coordinate plane $z' = D_0$ or even warped. We then describe the position of an arbitrary object point by

$$x_o = x', \quad y_o = y', \quad z_o = D_0 - h(x_o, y_o) \tag{63.20}$$

where $h(x_o, y_o)$ describes the very small vertical displacement $|h| \ll D_0$.

63.2.1.1 Determination of the Wavefunction

The wavefunction ψ_Q on the entrance side of the object can now be written as

$$\begin{aligned}\psi_Q(\boldsymbol{r}_o) &= \frac{A}{z'}\exp\left\{ikz' + \frac{ik}{2z'}\left(x'^2 + y'^2\right)\right\}\\ &= \frac{A}{D_0}\exp\left\{ikD_0 - ikh + \frac{ik}{2D_0}\left(x_o^2 + y_o^2\right)\right\}\end{aligned} \tag{63.21a}$$

The wavefunction on the exit side of the object is obtained by multiplying ψ_Q by the specimen transparency $O(x_o, y_o)$:

$$\psi_o(\boldsymbol{r}_o) = O(x_o, y_o)\psi_Q(\boldsymbol{r}_o) \tag{63.21b}$$

This function represents the boundary value for the diffraction integral. In the far zone the distance $z' - z_o$ from the object may be replaced by $z' - D_0$ in the denominator, thus neglecting the small variation $h(x_o, y_o)$. In the Fresnel approximation the diffraction integral becomes

$$\begin{aligned}\psi_o(\boldsymbol{r}) = &-\frac{ik}{2\pi(z' - D_0)}\iint_{-\infty}^{\infty}\psi_Q(\boldsymbol{r}_o)\exp\left[ik\left\{z' - D_0 + h(x_o, y_o)\right\}\right]\\ &\times \exp\left[\frac{ik}{2(z' - D_0)}\left\{(x' - x_o)^2 + (y' - y_o)^2\right\}\right]dx_o dy_o\end{aligned}$$

On substituting expressions (63.21a and 63.21b) for ψ_o *and* ψ_V into this formula, we find that the terms involving $h(x_o, y_o)$ cancel out. This is of great practical importance, as it shows that small vertical displacements do not degrade the quality of holograms.

The diffracted wave can now be written in the form

$$\psi_o(\boldsymbol{r}) = -\frac{ikA}{2\pi D_0(z' - D_0)}\exp(ikz')\iint_{-\infty}^{\infty}O(x_o, y_o)\exp\left(\frac{ik}{2D_0}R(x_o, y_o)\right)dx_o dy_o$$

$$R(x_o, y_o) := \left[\left(x_o^2 + y_o^2\right) + \frac{D_0}{z' - D_0}\left\{(x' - x_o)^2 + (y' - y_o)^2\right\}\right]$$

Before adding this wave to the reference wave ψ_R, the tilted coordinates must be eliminated. In the plane of superposition, $z = a$, (63.19) give

$$x' = x\cos\gamma, \quad y' = y, \quad z' = a\sec\gamma - x\sin\gamma$$

In the denominator, we make the approximation $z' \approx a$ and $D_0 \approx a - b$. As $\boldsymbol{r}_H$ then has the components (x, y, a), we obtain

$$\begin{aligned}\psi_o(\boldsymbol{r}_H) = &-\frac{ikA}{2\pi(a - b)b}\exp\left\{ik(a\sec\gamma - x\sin\gamma)\right\}\\ &\times \iint_{-\infty}^{\infty}O(x_o, y_o)\exp\left[\frac{ik}{2}\left\{\frac{x_o^2 + y_o^2}{a - b} + \frac{(x - x_o)^2 + (y - y_o)^2}{b}\right\}\right]dx_o dy_o\end{aligned} \tag{63.22}$$

The small-angle approximation, $\gamma \ll 1$, sin $\gamma \approx \gamma$ and cos $\gamma \approx 1$ can be made except in the term in $ka/\cos\gamma$, since $ka \gg 1$. In the same approximation the reference wave is given by

$$\psi_R(\boldsymbol{r}_H) = \frac{A}{a}\exp\left\{ik\left(a\ \sec\gamma + x\ \sin\gamma + \frac{x^2+y^2}{2a}\right)\right\} \tag{63.23}$$

63.2.2 Addition of Object Wave and Reference Wave

The intensity in the hologram plane is now given by $|\psi_o + \psi_R|^2$ but it is no longer necessary to assume that the object scatters weakly. We shall see that the diffraction pattern of the hologram separates into three distinct regions, which can in principle be made far enough apart to avoid any overlapping: (1) a central region and (2) two sidebands in which information about the amplitude and phase of the specimen is coded *linearly*. Nevertheless, high-resolution information is frequently sought from weakly scattering specimens and we shall, therefore, consider both situations.

We first consider the general case, in which the specimen is not necessarily weak, and for simplicity, we disregard the curvature of the two waves ψ_R and ψ_o. The latter are, thus, assumed to be plane waves, with wave vectors $\boldsymbol{k} = (\pm q_c/\pi,\ 0,\ k)$, where $q_c = 2\gamma/\lambda$ is known as the *carrier frequency*.

The reference wave ψ_R thus has the form

$$\psi_R(\boldsymbol{r}) = \exp\,(\mathrm{i}\pi q_c x)\mathrm{e}^{\mathrm{i}kz} \tag{63.24}$$

while the object wave is given by

$$\psi_o(\boldsymbol{r}) = A(\boldsymbol{r})\ \exp\{-\mathrm{i}\pi q_c x + \overline{\eta}\,(\boldsymbol{r})\}\mathrm{e}^{\mathrm{i}kz} \tag{63.25}$$

The functions A and $\overline{\eta}$ will in general not be exactly the same as the amplitude and phase of the specimen transparency, since (63.25) is the result of a diffraction process in the Fraunhofer approximation and phase shifts due to aberrations will hence have an effect.

In the hologram plane, where $z = a$, the intensity will be given by

$$J_H(x,y) = 1 + A^2(x,y) + 2A\ \cos\{2\pi q_c x - \overline{\eta}(x,y)\} \tag{63.26}$$

The effects of partial coherence are absent from this expression. A small modification allows us to include them to a good approximation. We find

$$J_H(x,y) = 1 + A^2(x,y) + 2A\mu\ \cos\{2\pi q_s x - \overline{\eta}(x,y) + \rho\} \tag{63.27}$$

in which

$$\mu = |\mu_s\mu_t| \quad \text{and} \quad \mu_s\mu_t =: \mu\exp(\mathrm{i}\rho) \tag{63.28}$$

(see Section 62.3) where μ_s characterizes the spatial partial coherence (finite source size) and is given by the Fourier transform of the distribution of emissive points; μ_t characterizes the temporal partial coherence (wavelength or energy spread) and is given by the Fourier transform of the energy distribution. An additional term (ρ) is sometimes included in (63.27) to represent inelastically scattered electrons.

If we consider the Fourier transform of this intensity distribution (63.26 or 63.27), we obtain three groups of terms. The unit term and $A^2(x, y)$ yield a central peak and the autocorrelation function of the object wave, which is also centred on the origin. The cosine term yields two sidebands, shifted owing to the presence of the term $2\pi q_c x$ in the argument of the cosine:

$$\begin{aligned}\mathcal{F}\{2\,A\cos(2\pi q_c x-\overline{\eta})\} &= \int A\exp(2i\pi q_c x)\exp(-i\overline{\eta})\exp(2\pi i\boldsymbol{q}\cdot\boldsymbol{r})d\boldsymbol{r}\\ &\quad+\int A\exp(-2i\pi q_c x)\exp(i\overline{\eta})\exp(2\pi i\boldsymbol{q}\cdot\boldsymbol{r})d\boldsymbol{r}\\ &= \int Ae^{-i\overline{\eta}}\exp\{2\pi i(\boldsymbol{q}+\boldsymbol{q}_c)\cdot\boldsymbol{r}\}d\boldsymbol{r}\\ &\quad+\int Ae^{i\overline{\eta}}\exp\{2\pi i(\boldsymbol{q}-\boldsymbol{q}_c)\cdot\boldsymbol{r}\}d\boldsymbol{r}\end{aligned} \tag{63.29}$$

in which $\boldsymbol{r}=(x, y)$, $\boldsymbol{q}=(q_x, q_y)$ and $\boldsymbol{q_c}=(q_c, 0)$.

In reality, the specimen will have a spatial frequency spectrum of finite extent: beyond some cut-off frequency $|\boldsymbol{q}|=\overline{q}$, the spectrum contains no useful information but only noise. Since the bandwidth of the autocorrelation function will be twice this cut-off frequency, the sidebands will not overlap the central band (Fig. 63.4) provided that the carrier frequency q_c is high enough,

$$q_c \geq 3\overline{q} \tag{63.30}$$

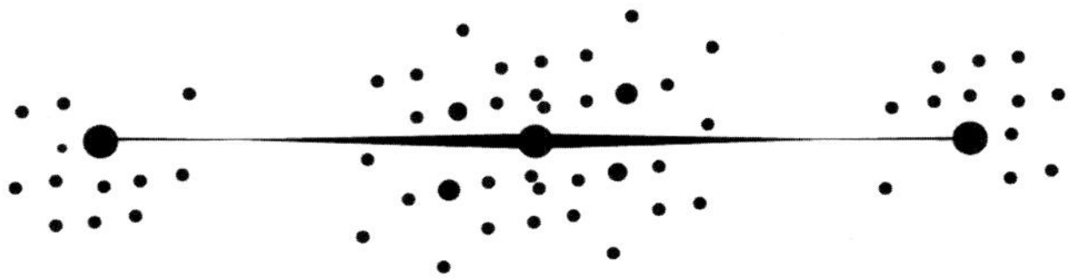

Figure 63.4
Fourier transform of a hologram showing the central autocorrelation distribution and the two sidebands. The figure demonstrates the $\boldsymbol{q}$-symmetry.

This, in turn, implies that the angle of inclination of the wavefronts must be steep enough:

$$\gamma = \frac{1}{2} q_c \lambda \geq 1.5\, \overline{q}\lambda \tag{63.31}$$

By isolating one sideband, therefore, we have direct access to the Fourier transform of $A\exp(\pm \mathrm{i}\overline{\eta})$ and should hence be able to recover the specimen phase and amplitude, perturbed by aberrations, directly. We return to this in Section 63.3, after considering the more realistic case of spherical reference and object waves.

We now form $|\psi_o + \psi_R|^2$ in which the expressions (63.22) and (63.23) are used for ψ_o and ψ_R respectively. We shall later consider the weakly scattering object and we therefore write

$$O =: 1 + S(x_o, y_o) = 1 - \sigma(x_o - y_o) + \mathrm{i}\eta(x_o, y_o) \tag{63.32}$$

For weak objects, S, σ and η will all be small.

Substituting $O = 1 + S$ in (63.22), the expression for ψ_o divides into two parts, one corresponding to the unity and the other to S. The Fresnel integral for the former can be evaluated in closed form, since

$$\iint\limits_{-\infty}^{\infty} \exp\left[\frac{ik}{2}\left\{\frac{x_o^2 + y_o^2}{a-b} + \frac{(x-x_o)^2 + (y-y_o)^2}{b}\right\}\right] dx_o dy_o \tag{63.33}$$

$$= \frac{2\pi \mathrm{i} b(a-b)}{ak} \exp\left\{\frac{\mathrm{i}k}{2a}\left(x^2 + y^2\right)\right\}$$

The resulting partial wave is then

$$\psi_V(\boldsymbol{r}_H) = \frac{A}{a}\exp\left\{\mathrm{i}k\left(a\sec\gamma - x\sin\gamma + \frac{x^2 + y^2}{2a}\right)\right\}$$

which is identical with expression (63.23) for ψ_R except that the sign of the linear term in x is reversed. This was to be expected because ψ_V represents a spherical wave that emanated from the virtual source V and passed through the object *without interaction*. The other partial wave, the scattered wave $\psi_s(\boldsymbol{r}_H)$, is obtained by replacing $O(x_o, y_o)$ in (63.22) by $S(x_o, y_o)$.

It is advantageous to combine the two strong contributions, ψ_R and ψ_V, into one background wave ψ_B $(\boldsymbol{r}_M)$; this will represent the wave pattern created by the interferometer in the absence of a specimen, its intensity being the familiar fringe pattern. In the hologram plane H, where $\boldsymbol{r}_H = (x, y, a)$, we obtain

$$\psi_B(x,y) =: \psi_R(x,y) + \psi_V(x,y)$$

$$= \frac{2A}{a}\exp\left\{\mathrm{i}k\left(a\sec\gamma + \frac{x^2 + y^2}{2a}\right)\right\}\cos xk_z \tag{63.34}$$

in which a is omitted from the arguments of the functions and we have written

$$k_x := k\sin\gamma \tag{63.35}$$

This quantity is exactly the x-component of the $\boldsymbol{k}$ vector at the midpoint of the hologram. The mean intensity is thus $\overline{J} = 2|A|^2/a^2$ and the background intensity may be written

$$J_B(x, y) = 2\overline{J}\cos^2(k_x x) \quad \forall y \tag{63.36}$$

in agreement with (62.15) for $|\gamma| \ll 1$. The large terms in ika sec γ cancel out exactly from (63.34).

The hologram intensity is given exactly by

$$J_H(x, y) = |\psi_R + \psi_V + \psi_s|^2 = |\psi_B + \psi_s|^2 \tag{63.37}$$

and we could proceed as we did in the case of plane waves and separate the various regions of the spatial spectrum of the hologram. Here, however, we consider a different situation, the weakly scattering object, for which $|S| \ll 1$. This implies that $|\psi_s| \ll |\psi_R|$ and $|\psi_s| \ll |\psi_V|$ everywhere but, near the zeros of J_B [$k_x x$ an odd multiple of $\pi/2$ in (63.22)], it will not be true that $|\psi_s|^2 \ll J_B$. Nevertheless, if we wish to obtain a hologram intensity that is always linear in ψ_s, we are obliged to omit $|\psi_s|^2$. As we saw earlier in this section, the great advantage of off-axis holography is that part of the spectrum of the hologram contains information about ψ_s and we shall show that this part can be isolated by suitable techniques. Only if we require linearity everywhere must we neglect $|\psi_s|^2$ in J_H. When we do impose this requirement, J_H takes the form

$$J_H(x, y) = J_B(x) + 2\Re\{\psi_B^*(x, y)\psi_s(x, y)\}$$

As usual, we obtain ψ_s by replacing $O(x_o, y_o)$ with $S(x_o, y_o)$ in (63.22), whereupon J_H becomes

$$J_H = 2\overline{J}\cos^2(k_x x) + \overline{J}\frac{ka}{\pi b(a-b)}\cos(k_x x)F_s \tag{63.38a}$$

in which F_s is the scattering amplitude,

$$F_s := \Im\left\{ e^{-ik_x x} \iint_{-\infty}^{\infty} S(x_o, y_o) \exp(ikL - ikW) dx_o dy_o \right\} \tag{63.38b}$$

The variable L in the exponent denotes the quadratic function

$$L := -\frac{x^2 + y^2}{2a} + \frac{x_o^2 + y_o^2}{2(a-b)} + \frac{(x - x_o)^2 + (y - y_o)^2}{2b} \tag{63.38c}$$

while W is the perturbation eikonal for the geometric aberrations.

A number of special cases of (63.38a) are of interest. We examine these briefly before turning to the various reconstruction procedures.

1. *In-line holography as a degenerate case*

When the distance $2d$ between the virtual sources V and $\overline{V}$ vanishes, $\gamma = 0$ and hence $k_x = 0$. The waves ψ_R and ψ_V coincide and form *one* coherent wave of amplitude $\hat{A} = 2A$; this is to be identified with the amplitude in Section 63.1. The midpoint of the object moves onto the optic axis. If we now set $\overline{J} = |\hat{A}|^2/a^2$, we must likewise replace $S(x_0, y_0)$ by $2S$, since the wave amplitude is twice that used earlier. Hence

$$J_H = \overline{J}\left[1 + \frac{ka}{\pi b(a-b)}\Im\left\{\iint S(x_o, y_o)\mathrm{e}^{ik(L-W)}dx_o dy_o\right\}\right] \tag{63.39}$$

where L and W are as defined above. Clearly, the separation of the hologram spectrum into regions does not occur and the advantage of off-axis operation for the reconstruction is lost.

2. *Quasi-homogeneous objects*

As in Section 62.3, we assume that the functions $\sigma(x_o, y_0)$ and $\eta(x_o, y_o)$ alter very slowly with x_o and y_o. The aberration term will be omitted here, since it has little significance.

The factor S can be taken outside the integral, but before we can do this, we must transform the quadratic function L in order to find the appropriate arguments of S. We rearrange the expression (63.38c) in such a manner that x_o and y_o appear only in quadratic terms:

$$\begin{aligned} L &= \frac{a}{2b(a-b)}\left\{\left(x_o - \frac{a-b}{a}x\right)^2 + \left(y_o - \frac{a-b}{x}y\right)^2\right\} \\ &= \frac{a}{2b(a-b)}\left\{\left(x_o - \frac{x}{M}\right)^2 + \left(y_o - \frac{y}{M}\right)^2\right\} \end{aligned}$$

with $M = a/(a-b)$, as in (62.12b). It is now obvious that the oscillations of the integrand are slow only in the vicinity of the point $(x/M, y/M)$; we may thus simplify (63.38b) to

$$F_s = \Im\left\{\mathrm{e}^{-ik_x x}S(x/M, y/M)\iint_{-\infty}^{\infty}\mathrm{e}^{ikL}dx_o dy_o\right\}$$

The remaining integral can be evaluated analytically and gives $2\pi ib(a-b)/ak$ and hence

$$F_s = \frac{2\pi b(a-b)}{ak}\Im\left\{e^{-ik_x x}S\left(\frac{x}{M}, \frac{y}{M}\right)\right\}$$

After introducing this into (63.38a), we find

$$J_H(x, y) = 2\overline{J}\cos^2 k_x x + 2\overline{J}\cos(k_x x)\Im\{\mathrm{e}^{-ik_x x}S(x/M, y/M)\}$$

or with $S = -\sigma + i\eta$:

$$J_H(x, y) = 2\overline{J}\{1 - \sigma(x/M, y/M)\}\cos^2 k_x x + \overline{J}\eta(x/M, y/M)\sin 2k_x x \tag{63.40}$$

This is in agreement with (63.36) if we linearize with respect to σ and η and identify $\overline{J}$ with $2J_R$ and k_x with $kd/a = k\gamma$. This confirms that our calculations are self-consistent.

3. *Very small objects*

The diameters of the specimens investigated in holography experiments at high resolution are very much smaller than those of the holograms finally recorded, since a high intermediate magnification is necessary to achieve the required resolution. This implies that we can ignore all terms in $x_o^2 + y_o^2$ in (63.38c). The integral in (63.38b) then reduces essentially to a Fourier integral since (63.38c) simplifies to

$$L = \frac{x^2 + y^2}{2f} - \frac{xx_o + yy_o}{b} \tag{63.41}$$

with $1/f := 1/b - 1/a$, as in (63.4b). The last term in (63.41) shows that it is convenient to define a *spatial frequency vector*

$$\boldsymbol{q} = (q_x, q_y) = \left(\frac{x}{\lambda b}, \frac{y}{\lambda b}\right) = \frac{1}{\lambda}(\theta_x, \theta_y) \tag{63.42}$$

$\theta_x = x/b$ and $\theta_y = y/b$ being essentially the *scattering angles* in the object. Eq. (63.41) can then be rewritten as

$$kL = \pi\lambda\boldsymbol{q}^2 b/M - 2\pi\boldsymbol{q}\cdot\boldsymbol{r_o} \tag{63.43}$$

Owing to the very small lateral extent of the specimen, it is sufficient to consider only those terms in the wave aberration W that depend on the angles θ_x and θ_y (isoplanatic approximation, see Chapter 65). It is convenient to express θ_x and θ_x in terms of $\boldsymbol{q}$ and to incorporate the first term in (63.43), a *defocus*, into the wave aberration. We thus write

$$w(\boldsymbol{q}) := \frac{2\pi W(\lambda\boldsymbol{q})}{\lambda} - \frac{\pi\lambda\boldsymbol{q}^2 b}{M} \tag{63.44}$$

With $k_x = k \sin\gamma \approx 2\pi\gamma/\lambda$, $x = \lambda b q_x$ (63.42) and $s := \gamma b$ we obtain

$$k_x x = 2\pi q_x s \tag{63.45}$$

s being the *off-axis* distance of the very small object detail being investigated. Putting all this together and returning to (63.38b), we see that the scattering amplitude simplifies to

$$F_s = \Im\left[\exp\{-2\pi \mathrm{i} s q_x - \mathrm{i}w(\boldsymbol{q})\}\,\tilde{S}(\boldsymbol{q})\right] \tag{63.46}$$

with the two-dimensional Fourier transform

$$\tilde{S}(\boldsymbol{q}) = \mathscr{F}^{-}(S) = \iint S(x_o, y_o)exp\{-2\pi\mathrm{i}(x_o q_x + y_o q_y)\}dx_o dy_o \tag{63.47}$$

It is now easy to express the hologram intensity J_H, given by (63.38a), in terms of q_x and q_y. We shall not pursue this, since the formulation of holography for weak objects has nowadays lost much of its earlier interest. One result of the present considerations is, however, of general interest: the conditions for the application of Fourier transforms do not need to be enforced by a special optical arrangement but are simply a consequence of the very high magnification needed for small object detail.

63.2.3 Choice of Defocus

It has been pointed out by Lichte (1990, 1991b, 1992a,b) that the best choice of defocus value for bright-field high-resolution imaging in the transmission electron microscope is not necessarily the most suitable for electron holography. The influence of the defocus value in the former case is examined in great detail in Chapter 66, where it emerges that the value known as 'Scherzer focus', $\Delta = (C_s\lambda)^{1/2}$, is particularly suitable for converting the phase variations in the electron wave into intensity variations at the image plane. Lichte argues that, in holography, it is more important to prevent unwanted exchange between amplitude and phase information and hence seeks the value of defocus for which the *amplitude* contrast transfer function (see 66.19a below) satisfies a particular condition. This function is oscillatory and is given by

$$\hat{K}_a(Q) = \cos\pi\left(DQ^2 - \frac{1}{2}Q^4\right) \tag{63.48}$$

in terms of reduced quantities (65.31 and 65.32). It first reaches zero for $Q = Q_0$, where

$$Q_0^4 - 2DQ_0^2 = 1$$

or

$$Q_0^2 = D \pm \left(D^2 + 1\right)^{1/2} \tag{63.49}$$

It passes through a minimum at Q_m, where $d\hat{K}_a(Q)/dQ = 0$, giving $Q_m^2 = D$ and hence $\hat{K}_a(Q_m) = \cos(\pi D^2/2)$. The phase-contrast transfer function (66.19b), given by

$$\hat{K}_p(Q) = \sin\pi\left(DQ^2 - \frac{1}{2}Q^4\right) \tag{63.50}$$

also passes through a minimum for this value of Q and transfer of amplitude specimen information to the amplitude variations of the wavefunction at the image and of phase information to phase variations will be greatest when the depths of the phase and amplitude minima are the same. Correspondingly, 'cross-talk', which maps specimen amplitude into image phase and vice versa, will be least. In order to find the defocus value for which this situation obtains, we equate $\hat{K}_p(Q_m)$ and $\hat{K}_a(Q_m)$:

$$\sin\frac{\pi}{2}D^2 = \cos\frac{\pi}{2}D^2 \tag{63.51}$$

which has the obvious solution $D^2 = \frac{1}{2}$. It has, therefore, been proposed that this value of defocus should be called the *Gabor focus*,[1] D_G:

$$D_G := \frac{1}{\sqrt{2}} \tag{63.52}$$

Clearly,

$$D_G = D_S/\sqrt{2} = 0.707D_S \tag{63.53}$$

where D_S is the Scherzer focus, $D_S = 1$. This value of defocus is optimal when aberration correction is not envisaged. Lichte shows that when the spherical aberration is to be cancelled, a different value is preferable, for which

$$D = 0.75Q_{\max}^2 \tag{63.54}$$

where $Q_{\max}$ is the greatest reduced spatial frequency to be included in the correction procedure. This is essentially the plane of the disc of least confusion, which we found in Section 24.3 by geometrical arguments (24.50).

63.3 Reconstruction Procedures

The numerous reconstruction techniques differ essentially in the choice of the waves that are caused to interfere. They differ too in the choice between optical and digital reconstruction but, although this choice has far-reaching practical implications, it has little bearing on the principles. The arguments on both sides are well known and will not be rehearsed here; in short, optical reconstruction is fast (once the optical setup has been chosen and aligned) and large areas can be processed with ease while numerical reconstruction requires consideration of pixel size at the detector but enables us to manipulate waves easily that would be difficult to generate optically. Today, numerical reconstruction has completely replaced optical reconstruction.[2]

The reconstruction process sets out from (63.29), the Fourier transform of the hologram intensity distribution. One of the sidebands is extracted and moved to the origin in Fourier

[1] This is not the same as the definition of Gabor focus recommended by Lichte (1991b), who gives $D_G = \{(2/\pi) \arccos 0.75\}^{1/2} \approx 0.68$. For this choice the cosine function does not fall below 0.75 before returning to 1. The above definition is suggested by Hawkes (1992).

[2] A book edited by S.H. Lee defending optical techniques against the inroads of numerical procedures was reviewed in *Nature* (**295**, 1982, 441) under the heading '*Everything e^- can do, ν can do better*'.

space. In practice, the limited number of pixels in the sampled discrete hologram can generate streaks in its transform, which make it difficult to identify the centre of the sideband exactly. This difficulty can be avoided by forming the product of the hologram with a Hann[3] window, $H(x, y)$,

$$H(x,y) := \left\{1 - \cos\left(\frac{2\pi x}{x_m}\right)\right\}\left\{1 - \cos\left(\frac{2\pi y}{y_m}\right)\right\} = \sin^2\left(\frac{\pi x}{x_m}\right)\sin^2\left(\frac{\pi y}{y_m}\right) \qquad (63.55)$$
$$0 \le x, y < x_m, y_m$$

(Nuttall, 1981). Once the sideband has been centred, the original version, before application of the window, should be used. Streaking and any distortions can be considerably attenuated with the aid of a second hologram, obtained with no specimen present; the Fourier transform of the true sideband is divided by this interference pattern and the result will be approximately free of streaks and distortions (Midgley, 2001). After inverting the resulting hologram to obtain the desired phase and amplitude distributions, one further operation is required unless the phase variations are small. When the latter are greater than 2π, the result, displayed modulo 2π, will exhibit 'phase jumps'; this phenomenon can be remedied by 'phase unwrapping' (Takeda et al., 1982; Takeda and Abe, 1996; Ghiglia and Pritt, 1998).

63.3.1 Aberration Correction

We recall that this was the purpose of Gabor's early investigations. Eqs. (63.38a), (63.38b), and (63.38c) show that the aberrations of electron microscope lenses can be removed by selecting one sideband of the hologram spectrum and cancelling the wave aberration, either with a suitably chosen glass lens (Fig. 63.5) or, more realistically, by digitizing the hologram and adding the appropriate phase shifts. Since holography is now frequently performed on aberration-corrected electron microscopes, we just record earlier attempts to put Gabor's proposal into practice: the advent of hardware aberration correctors has rendered this method of aberration correction of historic interest only.

The optical approach has been used by Tonomura et al. (1979b) and Rogers (1978, 1980). Numerical correction has been successfully demonstrated by Lichte and colleagues (Franke et al., 1987; Fu et al., 1991; Lichte, 1991a,b,c,d,e, 1993) who even included a quadratic term in q^2 of the form $\exp(cq^2)$ to reduce the attenuation due to chromatic aberration and by Harada et al. (1990a) and Ishizuka et al. (1991). A major problem with aberration correction is the need for very accurate knowledge of the aberration coefficients of the image-forming lenses. We show in Chapter 77 that it is possible, though not easy, to

[3] This window function, suggested by Julius von Hann, is universally known as a "Hanning window," presumably by confusion with the Hamming window. Use of the Hann window is sometimes known as hanning.

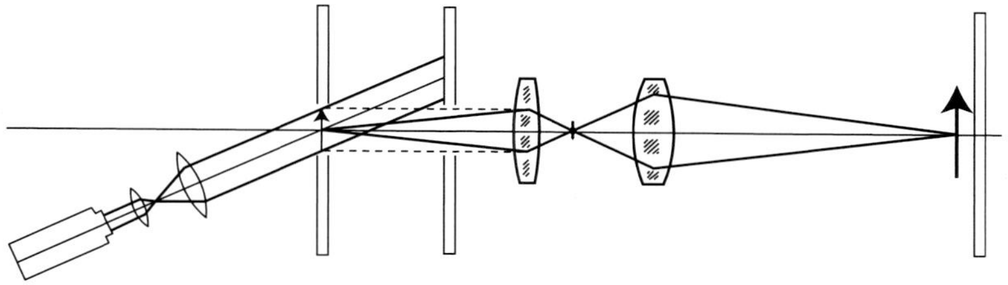

Figure 63.5
Optical reconstruction system for spherical aberration correction.

establish these quantities with the necessary precision. The theoretical studies of Hanszen et al. (1972a,b), Hanszen (1972a, 1973b) and Ade (1973) are relevant here. An ingenious way of relaxing the requirements for extremely fine sampling has been proposed by Ishizuka (1993), who has gone on to discuss correction (Ishizuka, 1994). See also Kawasaki and Rodenburg (1993) and Ishizuka (1990).

The final task is to derive the distribution of atoms in the specimen from the functions $\sigma(\boldsymbol{r}_o)$ and $\eta(\boldsymbol{r}_o)$. This question of image interpretation cannot be answered within the context of holography.

A simulation of the correction sequence calculated by Joy et al. (1993) is shown in Fig. 63.6. The images of Fig. 63.6A show the amplitude and phase of the wavefunction emerging from a silicon carbide crystal, illuminated by a coherent plane wave, as calculated by the multislice simulation programme MacTempas[4] (see Section 69.8). Fig. 63.6B shows the same quantities modified by the objective lens. In Fig. 63.6C, the hologram is shown above and its Fourier transform below; in the latter, the central disc and the two sidebands just touch: the fringe spacing is one-quarter of the atomic spacing. The final pair of figures, Fig. 63.6D, show the reconstruction, which is good but not perfect.

63.3.2 Interference Holography

If we add a plane wave to one of the reconstructed images, as illustrated in Fig. 63.1A, we can produce a contour map of specimen thickness. An example is shown in Fig. 63.7A which also illustrates the principle of phase amplification; a shearing interferogram, obtained as shown in Fig. 63.8, is also included. Here, the reconstructed image and its conjugate are brought into coincidence, to give double the phase difference. Higher

[4] Available from Total Resolution, 20 Florida Ave., Berkeley, CA 94707, United States, http://www.totalresolution.com.

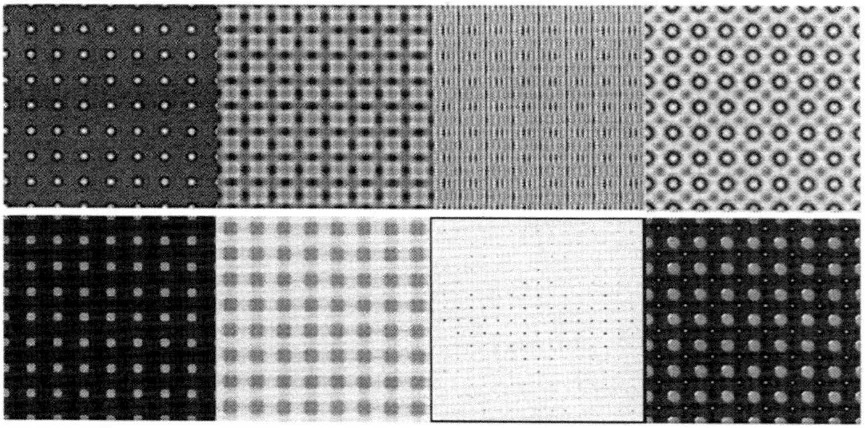

Figure 63.6
Amplitude (above) and phase (below) of the wavefunction emerging from a silicon carbide crystal: (A) simulation; (B) simulation with objective lens aberrations; (C) hologram (above) and its Fourier transform (below); and (D) reconstruction. *After Joy et al. (1993), Courtesy Elsevier.*

multiples can be obtained by using higher order diffracted beams; vivid examples are shown in Figs 63.7B and 63.7C. For extensive discussion and many examples, see Tonomura (1987a) and Völkl et al. (1999). Numerical methods were introduced into interference holography by Ohshita et al. (1990), Ru et al. (1991a,b) and Matsuda et al. (1991).

A slightly different approach to the reconstruction stage was adopted by Ade and Lauer (1988, 1990, 1991) and Lauer and Ade (1990). In their (numerical) procedure the two sidebands of (63.26) are retained and said to form a 'filtered' hologram, with the general form

$$J_H^{(f)} = b + 2A\cos\{2\pi \boldsymbol{q}\cdot\boldsymbol{r} - \overline{\eta}(\boldsymbol{x},\boldsymbol{y})\} \tag{63.56}$$

in which b is a constant bias term; the filtering consists in removing the term $A^2(x, y)$ and modifying the bias. This filtered hologram is now multiplied (Ade and Lauer, 1991) by a raised sinusoid of the form

$$s(\boldsymbol{r}) = 1 + 2B\cos(2\pi \boldsymbol{q_b}\cdot\boldsymbol{r}) \tag{63.57}$$

giving

$$\begin{aligned} J_H^{(f)}s = b &+ 2bB\cos(2\pi \boldsymbol{q}_b\cdot\boldsymbol{r}) + 2A\cos(2\pi\boldsymbol{q}\cdot\boldsymbol{r} - \overline{\eta}) \\ &+ 2AB[\cos\{2\pi(\boldsymbol{q}+\boldsymbol{q_b})\cdot\boldsymbol{r} - \overline{\eta}\} + \cos\{2\pi(\boldsymbol{q}-\boldsymbol{q_b})\cdot\boldsymbol{r} - \overline{\eta}\}] \end{aligned} \tag{63.58}$$

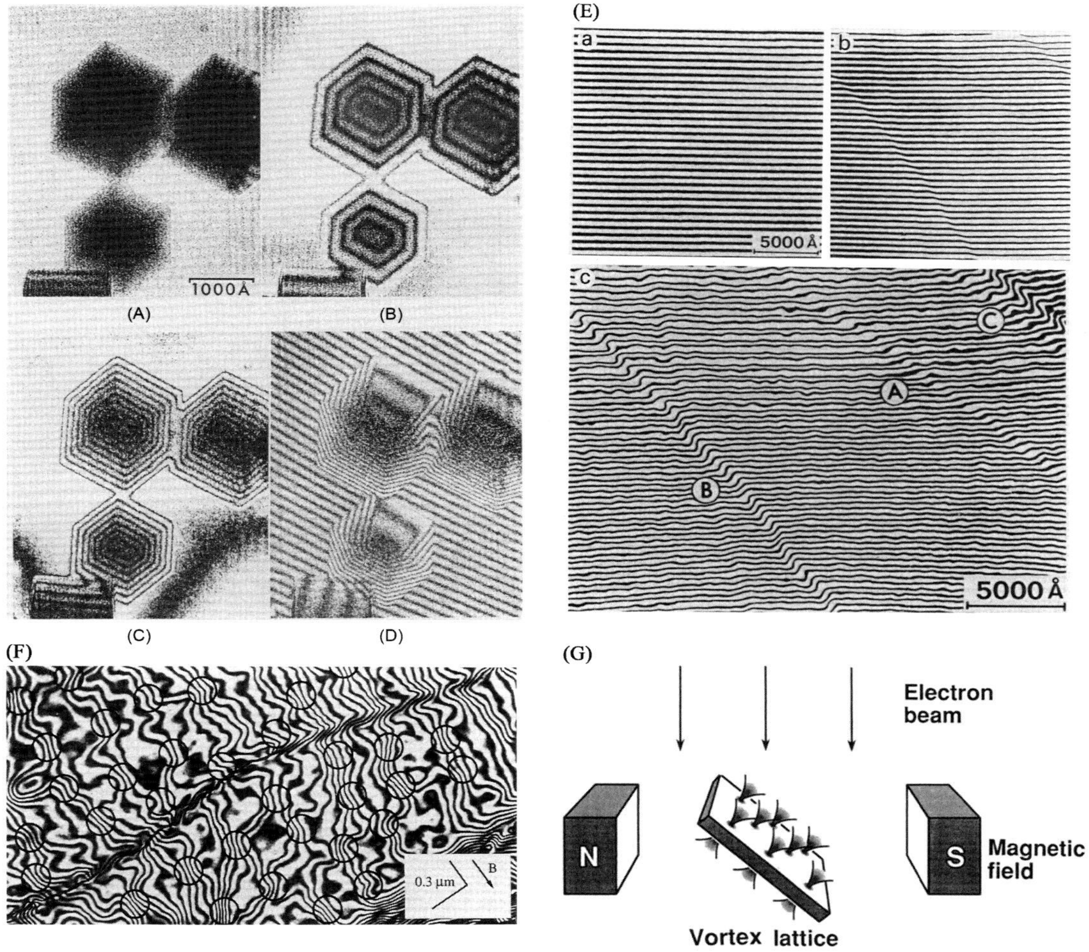

Figure 63.7

Interference micrographs of MgO particles: (A) reconstructed image; (B) contour map; (C) contour map (with phase amplification); and (D) interferogram (with plane wave at oblique incidence as in Fig. 63.8B). (E) Thin film of molybdenite. (a) Conventional interferogram. (b) x 4 phase amplification. (c) x 24 phase amplification revealing surface atomic steps. Step A corresponds to a phase shift of $2\pi/40$ and a step-height of 6.2 Å. (F) Vortex lattice in a superconducting single-crystal niobium thin film. (G) The experimental arrangement used to record Fig. 63.7F. *(A–D) After Tonomura (1984), https://doi.org/10.1093/oxfordjournals.jmicro.a050444; (E–G) After Tonomura (1994), Courtesy Elsevier.*

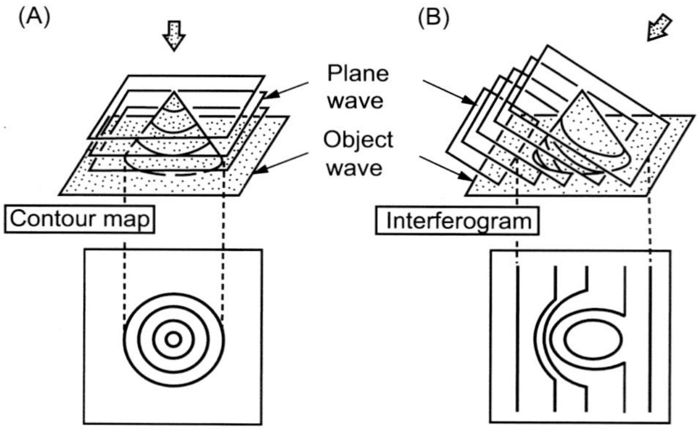

Figure 63.8
Two kinds of interference micrographs: (A) contour map obtained with the plane wave and the reconstructed object wave in the same direction; (B) interferogram obtained with the plane wave inclined to the reconstructed object wave. The latter shows whether the wavefronts have been advanced or retarded. *After Tonomura et al. (1987), Courtesy Elsevier.*

The spatial spectrum of this product consists of several terms:

$$\begin{aligned}\mathcal{F}\left(J_H^{(f)}s\right) &= b\delta(\overline{\boldsymbol{q}}) + bB\left\{\delta\left(\boldsymbol{q}-\boldsymbol{q}_s+\overline{\boldsymbol{q}}\right)+\delta\left(\boldsymbol{q}-\boldsymbol{q}_s-\overline{\boldsymbol{q}}\right)\right\}\\ &+\tilde{O}(\boldsymbol{q}+\overline{\boldsymbol{q}})+\tilde{O}^*(\boldsymbol{q}-\overline{\boldsymbol{q}})\\ &+B\left\{\tilde{O}\left(2\boldsymbol{q}-\boldsymbol{q}_s+\overline{\boldsymbol{q}}\right)+\tilde{O}^*\left(2\boldsymbol{q}-\boldsymbol{q}_s-\overline{\boldsymbol{q}}\right)\right\}\\ &+B\left\{\tilde{O}\left(\boldsymbol{q}_s+\overline{\boldsymbol{q}}\right)+\tilde{O}^*\left(\boldsymbol{q}_s-\overline{\boldsymbol{q}}\right)\right\}\end{aligned} \tag{63.59}$$

in which

$$\boldsymbol{q}_s := \boldsymbol{q} - \boldsymbol{q}_b \tag{63.60}$$

and $\tilde{O}$ is the Fourier transform of the object transparency O. A suitable choice of the difference frequency $\boldsymbol{q}_s$ brings the terms forming the last line of (63.59) close to the origin, where they can be selected and inverse Fourier transformed to give the reconstruction $r(\boldsymbol{r})$,

$$r(\boldsymbol{r}) := \mathcal{F}^{-}\left[B\left\{\tilde{O}\left(\boldsymbol{q}_s+\overline{\boldsymbol{q}}\right)+\tilde{O}^*\left(\boldsymbol{q}_s-\overline{\boldsymbol{q}}\right)\right\}\right] = 2AB\cos\left(2\pi\boldsymbol{q}_s\cdot\boldsymbol{r}-\overline{\eta}\right) \tag{63.61}$$

For visual presentation, this can either be set against a bias, b', giving

$$r(\boldsymbol{r}) + b' = b' + 2AB\cos\left(2\pi\boldsymbol{q}_s\cdot\boldsymbol{r}-\varphi\right) \tag{63.62}$$

or its modulus can be displayed, $|r(\boldsymbol{r})|$. For examples of such reconstructions, see Ade and Lauer (1990, 1991), Ade (1992) and a similar approach by Harada et al. (1990b). For earlier proposals, see Endo et al. (1979), Hanszen (1983, 1985), Hanszen and Ade (1983), Hanszen et al. (1983a) and Tonomura et al. (1985).

63.3.3 Statistical Considerations

Although the reasoning in this chapter has been expressed in terms of wavefunctions, we have said nothing about the uncertainties introduced by sampling, finiteness and the low electron dose often essential to obviate radiation damage. These questions were first studied by Ade (1980), Hanszen (1982d, 1987), Ade et al. (1984), Hanszen et al. (1984), Lauer and Hanszen (1986), Herrmann and Lichte (1986), Lichte et al. (1987, 1988), Lenz (1988), Lenz and Völkl (1990), de Ruijter and Weiss (1993) and de Ruijter (1992). We refer to their work for further details.

63.4 Holography in the Scanning Transmission Electron Microscope

With the successful incorporation of the electron biprism in the transmission electron microscope, it was natural to enquire whether the presence of such a biprism in a scanning transmission instrument would offer any new possibilities. A preliminary study by Leuthner et al. (1988, 1989, 1990) showed that the flexibility at the detector level that is characteristic of the STEM (and extensively studied in Chapter 67) can be exploited to separate the phase and amplitude of the specimen wavefunction. The experimental arrangement is shown in Fig. 63.9, where it is seen that each pixel of the object will create an interference pattern in the detector plane as the scanning probe passes over it. Clearly, this pattern can be averaged in many ways, depending on the detector geometry, as explained in connection with routine STEM operation in Chapter 67. Leuthner et al. suggested that two signals should be extracted from each pixel, one obtained by summing the incident intensity, the other obtained by modulating the incident signal by means of a grating before detection and summing. These two signals should be sufficient to yield the specimen amplitude and phase separately. Leuthner et al. also suggest an 'interferometric mode' in which the electron interference pattern is scanned over the detector in synchronism with the probe (Leuthner et al., 1989, 1990).

Shortly afterwards, STEM holography with a biprism was examined by Cowley (1990) and pursued in considerable detail by Gribelyuk and Cowley (1991, 1992a,b, 1993), Wang and Cowley (1991), Mankos et al. (1992) and by Konnert and d'Antonio (1992). We refer to those studies for a thorough analysis. There was intermittent work on STEM holography for several years but the advent of fast direct detectors, with which an entire image can be captured for each probe position, has reawakened interest in the technique, notably by Yasin et al. (2016, 2018a,b,c, d), Harvey et al. (2018), McMorran et al. (2018), McMorran (2019), Greenberg et al. (2019) and Ophus et al. (2019). A nanofabricated diffraction grating is placed between the source and the probe-forming lens (Fig. 63.10) in such a way that the diffraction pattern of the grating is formed at the specimen plane. In the case described by Harvey and Yasin, the pitch of the grating, diameter 50 μm, is 150 nm; with a 4-mrad semiconvergence angle, beams separated by 120 nm are incident on the specimen. The beam is scanned in the direction of the diffraction pattern and far-field diffraction patterns are recorded sequentially.

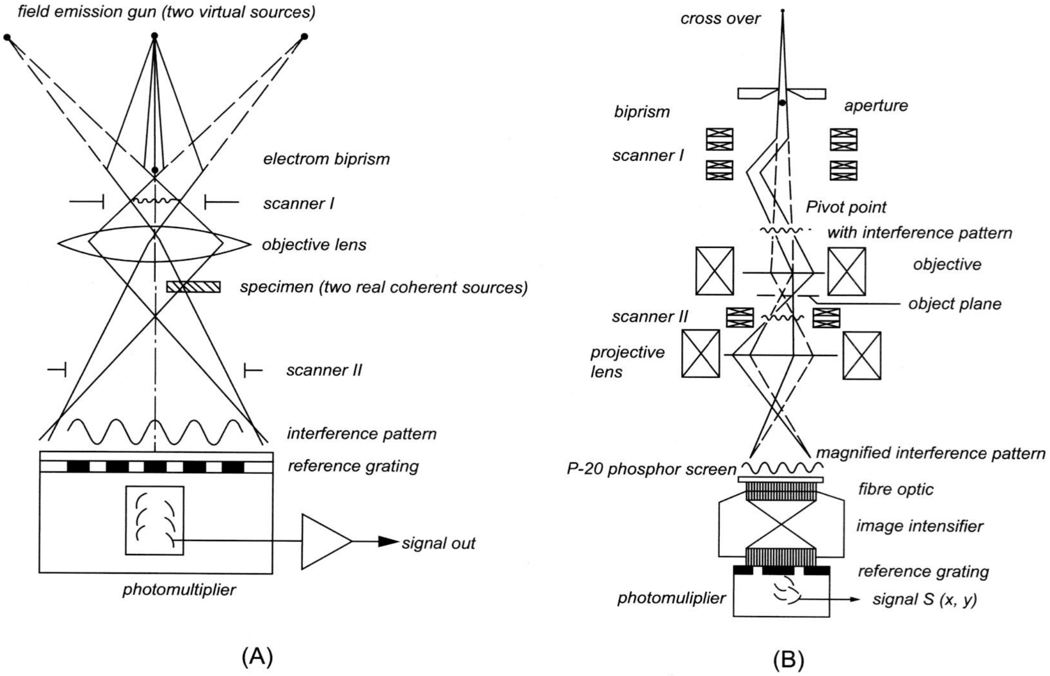

Figure 63.9
(A) Ray diagram illustrating hologram formation in the STEM. One virtual source forms a probe that explores the specimen while the other forms a probe in free space. The resulting waves interfere in the detector plane. (B) The Siemens ST100F STEM modified for holography. *After Leuthner et al. (1989), Courtesy Wiley, https://doi.org/10.1002/pssa.2211160111.*

The wavefunction incident on the specimen is now of the form

$$\psi_Q(\boldsymbol{x}) = a(\boldsymbol{x} - \boldsymbol{x}_p) = \sum_n c_n a_n(\boldsymbol{x} - \boldsymbol{x}_p - n\boldsymbol{x}_0) \tag{63.63}$$

in which $\boldsymbol{x}_p$ is the distance from the optic axis to the centre of the specimen, a_n denotes the phase and amplitude distribution of the nth diffraction order and c_n is the complex amplitude of the corresponding probe. In the work of Harvey, Yasin et al., the grating is a regular array but it could be much more elaborate, generating different phase and amplitude distributions at each probe.

The exit wave is the product of ψ_Q and the specimen transparency $t(\boldsymbol{x})$:

$$\psi_o(\boldsymbol{x}) = a(\boldsymbol{x} - \boldsymbol{x}_p)t(\boldsymbol{x}) \tag{63.64}$$

so that the far-field diffraction pattern in the detector plane is given by

$$I_d(\boldsymbol{q}) = |\Psi(\boldsymbol{q})|^2 = \{A^*(\boldsymbol{q}) \otimes T^*(\boldsymbol{q})\}\{A(\boldsymbol{q}) \otimes T(\boldsymbol{q})\} \tag{63.65}$$

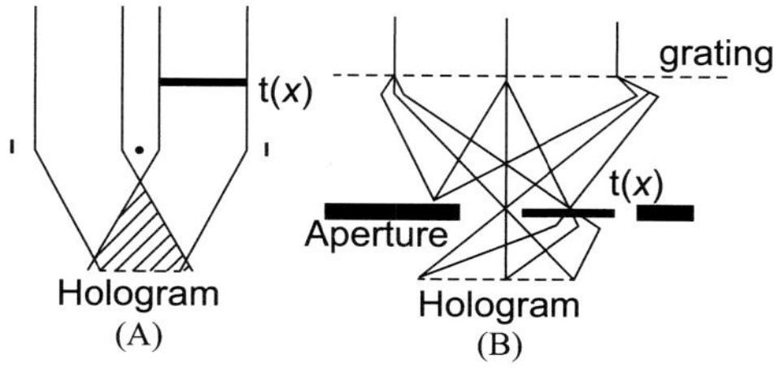

Figure 63.10
STEM holography. (A). Hologram formation with a biprism. (B) Hologram formation in STEM with a grating between the source and the probe-forming lens. *After Harvey et al. (2018), Courtesy American Physical Society, https://doi.org/10.1103/PhysRevApplied.10.061001.*

in which A and T denote the Fourier transforms of a and t. Assuming that only the lowest order diffraction peaks contribute to the signal and that only the probe $+1$ interacts with the specimen, as shown in the figure, we find

$$I_p(\boldsymbol{x}) = \sum_{n=-2}^{2} I_n(\boldsymbol{x}_p, \boldsymbol{x}) \tag{63.66a}$$

where

$$\begin{aligned}
I_2(\boldsymbol{q}) &= |c_2|^2 + 2|c_0||c_2|\cos\{4\pi\boldsymbol{q}\cdot\boldsymbol{x}_0 - \phi(0) - \phi(2\boldsymbol{x}_0) + \theta_0 - \theta_2\}\\
I_1(\boldsymbol{q}) &= |c_1|^2 + 2|c_0||c_1|\cos\{2\pi\boldsymbol{q}\cdot\boldsymbol{x}_0 + \phi(0) - \phi(\boldsymbol{x}_0) + \theta_0 - \theta_1\}\\
I_0(\boldsymbol{q}) &= |c_0|^2 + 2|c_1|[|c_{-1}|\cos\{4\pi\boldsymbol{q}\cdot\boldsymbol{x}_0 - \phi(\boldsymbol{x}_0) - \theta_1 + \theta_{-1}\}\\
&\qquad + |c_2|\cos\{2\pi\boldsymbol{q}\cdot\boldsymbol{x}_0 + \phi(1) - \phi(2\boldsymbol{x}_0) + \theta_1 - \theta_2\}]\\
I_{-1}(\boldsymbol{q}) &= |c_{-1}|^2 + 2|c_0||c_{-1}|\cos\{2\pi\boldsymbol{q}\cdot\boldsymbol{x}_0 - \phi(0) - \theta_0 + \theta_{-1}\}\\
I_{-2}(\boldsymbol{q}) &= |c_{-2}|^2 + 2|c_0||c_{-2}|\cos\{4\pi\boldsymbol{q}\cdot\boldsymbol{x}_0 - \phi(0) - \phi(2\boldsymbol{x}_0) - \theta_0 + \theta_{-2}\}
\end{aligned} \tag{63.66b}$$

The specimen transparency t can now be extracted from the set of peak values (Yasin et al., 2018a), using the relation

$$\int a_0(x)I_1(x_p, x)dx = c_o^* c_1 \left|a_0(x_p)^2\right| \otimes t^*(x_0 + x_p) \tag{63.67}$$

$a_0(x)$ is obtained by recording a reference image with no specimen present and selecting a small region around the central peak.

STEM holography generates good contrast on materials of low atomic number as well as less difficult specimens and is capable of yielding quantitative information about magnetic and electrostatic fields. Atomic resolution can be attained if the dose is sufficient.

Before leaving this topic, we should just mention that, just as bright-field TEM image formation has been interpreted in the language of in-line holography, the STEM signal in the image plane has also been discussed in the language of holography. The pioneering work of Veneklasen (1975) and the later discussion of Cowley and Walker (1981) and Lin and Cowley (1986) are examples of this. Veneklasen in particular discussed the use of structured detectors in this context.

63.5 Reflection Holography

For surface studies, a reflection mode has been investigated. For details of this technique, see Banzhof and Herrmann (1993), and for related work, see Banzhof et al. (1988, 1992), Osakabe (1992), Osakabe et al. (1988, 1989, 1993) and Takeguchi et al. (1990, 1992).

The configuration used by Banzhof, Herrmann and Osakabe to observe a surface feature such as a step is shown in Fig. 63.11. Since both the reference wave and the object wave have undergone Bragg reflection at the specimen surface, the former will not be completely unperturbed, even if it was collected from a region far from the feature of interest. For a step of height d the phase shift is given by

$$\Delta\varphi = \left\{ (2\pi m)^2 - (2kd)^2 \frac{V_c}{\Phi_0} \right\}^{1/2} \tag{63.68}$$

in which m denotes the diffraction order, $k = 2\pi/\lambda$ and V_c is the mean inner potential. The agreement with theory is reasonably good for both sets of authors. A detailed study of the technique is to be found in the work of Osakabe and colleagues, who observed the surface displacement field created by a dislocation.

63.6 Applications and Related Topics

Many applications of holography are concerned with the detection and mapping of magnetic and electrostatic fields. Very high resolution is often not needed for in many cases the fields do not change over short distances; *medium-resolution holography* is then sufficient. When higher resolution is essential, the ability to correct aberrations by means of holography was exploited in the past but this is no longer necessary; *high-resolution holography* is now best performed on an aberration-corrected microscope. The reliability of information extracted from a focal series of in-line holograms is analysed by Huang et al. (2021).

In the following sections, we first examine some common applications and then describe split-illumination holography, holographic tomography, ultrafast holography, inelastic holography, the use of multiple-biprism configurations and very low-energy holography. Electron

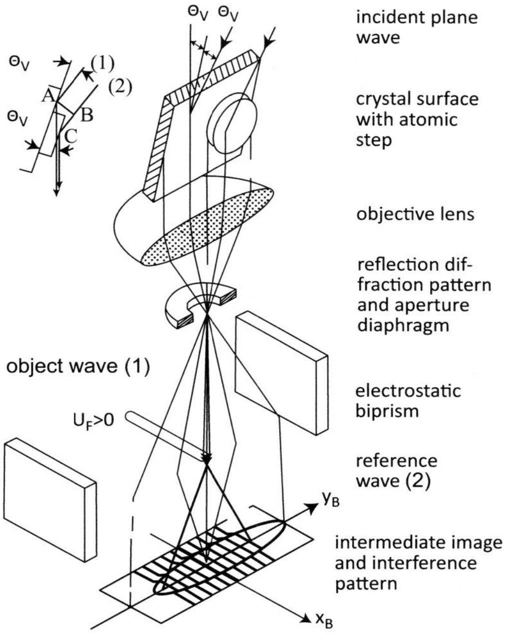

Figure 63.11
Reflection hologram formation. *After Banzhof and Herrmann (1993), Courtesy Elsevier.*

holography in the Fraunhofer region has been pursued by Harada et al. (2019a,b), who anticipate that their procedure will lead to direct phase determination in reciprocal space. A final section draws attention to a method of combating noise by 'tensor decomposition'.

For a survey, see Kawasaki et al. (2021).

63.6.1 Studies of Magnetic and Electric Fields

The reason why holography is attractive for the study of the magnetic and electrostatic properties of specimens is immediately apparent when we consider the phase shift created by an electrostatic potential $V_c(x, y, z)$ and in-plane magnetic induction $\boldsymbol{B}_\perp(x, y)$:

$$\Delta\varphi(x, y) = \frac{2\pi}{\lambda}\left(\frac{1 + 2\varepsilon\Phi_0}{2\Phi_0}\right)\int V_c(x, y, z)dz - \frac{e}{\hbar}\int \boldsymbol{B}_\perp(x, y)\cdot d\boldsymbol{S}\, dz, \quad d\boldsymbol{S} = dx\, dy \tag{63.69}$$

If V_c is constant in the z-direction, the first term collapses to

$$\Delta\varphi(x, y) = \frac{2\pi}{\lambda}\left(\frac{1 + 2\varepsilon\Phi_0}{2\Phi_0}\right)V_c(x, y)t(x, y) \approx \frac{\pi}{\lambda\Phi_0}V_c(x, y)t(x, y) \tag{63.70}$$

where t is the specimen thickness and V_c is now the inner potential. For examples of holographic measurement of electrostatic fields, see Frabboni et al. (1985, 1987), Matteucci et al. (1991, 1992, 1996, 1998, 2002), Rau et al. (1999), Cumings et al. (2002, 2008), Dunin-Borkowski et al. (2002), Lichte et al. (2002), Twitchett et al. (2002a,b, 2005), Yates (2005), Twitchett-Harrison et al. (2007, 2008a,b), Gribelyuk et al. (2002), Houben et al. (2004), Matsumoto et al. (2008) and Yamamoto et al. (2021). For many more examples, see Section 16.4 of Dunin-Borkowski et al. (2019). Lichte et al. (2002) discuss ferroelectric electron holography.

In the ideal case of a specimen of uniform thickness and composition the phase gradient is given by

$$\left(\frac{\partial\Delta\varphi}{\partial x},\frac{\partial\Delta\varphi}{\partial y}\right) = -\frac{et}{\hbar}\{B_y(x,y), -B_x(x.y)\} \quad \boldsymbol{B}_\perp = \left(B_x, B_y\right) \tag{63.71}$$

In practice, magnetic flux will often leak out of the specimen, with the result that the space through which the reference wave travels is not field-free. Simulation is then required to extract the desired field distribution.

In any experiment designed to study magnetic properties the specimen must lie in a region free of magnetic field or permeated by a field created deliberately. The objective lens of the microscope is usually switched off and magnification is provided by an auxiliary mini-lens or 'Lorentz lens', situated just below the objective (e.g. Fazzini et al., 2004; Sickmann et al., 2011). Specimen holders designed to create a controlled magnetic field are described by McVitie et al. (1995), Uhlig et al. (2000, 2003), Yi et al. (2004), Inoue et al. (2005), Cumings et al. (2008), Tsuneta et al. (2014) and Arita et al. (2014). Many examples of this application of holography can be found in the reviews by Midgley (2001), Park (2014) and Tanaka (2015) and the survey by Dunin-Borkowski et al. (2019) as well as the collections edited by Tonomura et al. (1995) and Völkl et al. (1999). Denneulin et al. (2021) have studied Néel-type skyrmions by off-axis holography.

63.6.2 Dark-Field Holography, Strain Measurement

In microelectronics, knowledge of the deformation field or strain in crystalline materials is indispensable. Methods of measuring strain by electron microscope methods have been developed, notably by Martin Hÿtch and colleagues, at first by high-resolution imaging (e.g., Hÿtch and Houdellier, 2007) and, more recently, by dark-field electron holography (Hÿtch et al., 2008, 2011, 2016, 2019; Cooper et al., 2007, 2009, 2010, 2011, 2012, 2016; Koch et al., 2010; Koch, 2014; Béché et al., 2011; Cherkashin et al., 2017; Zhang et al., 2020). Here, we concentrate on the latter; for the earlier work, see Hÿtch and Dahmen (1996), Hÿtch (1997a,b, 2001), Hÿtch et al. (1998, 2006), Hÿtch and Plamann (2001) and Hüe et al. (2008).

The principle of dark-field holography is shown in Fig. 63.12B. Fig. 63.12A is a reminder of the standard holographic arrangement, for comparison with Fig. 63.12B. Now, the entire

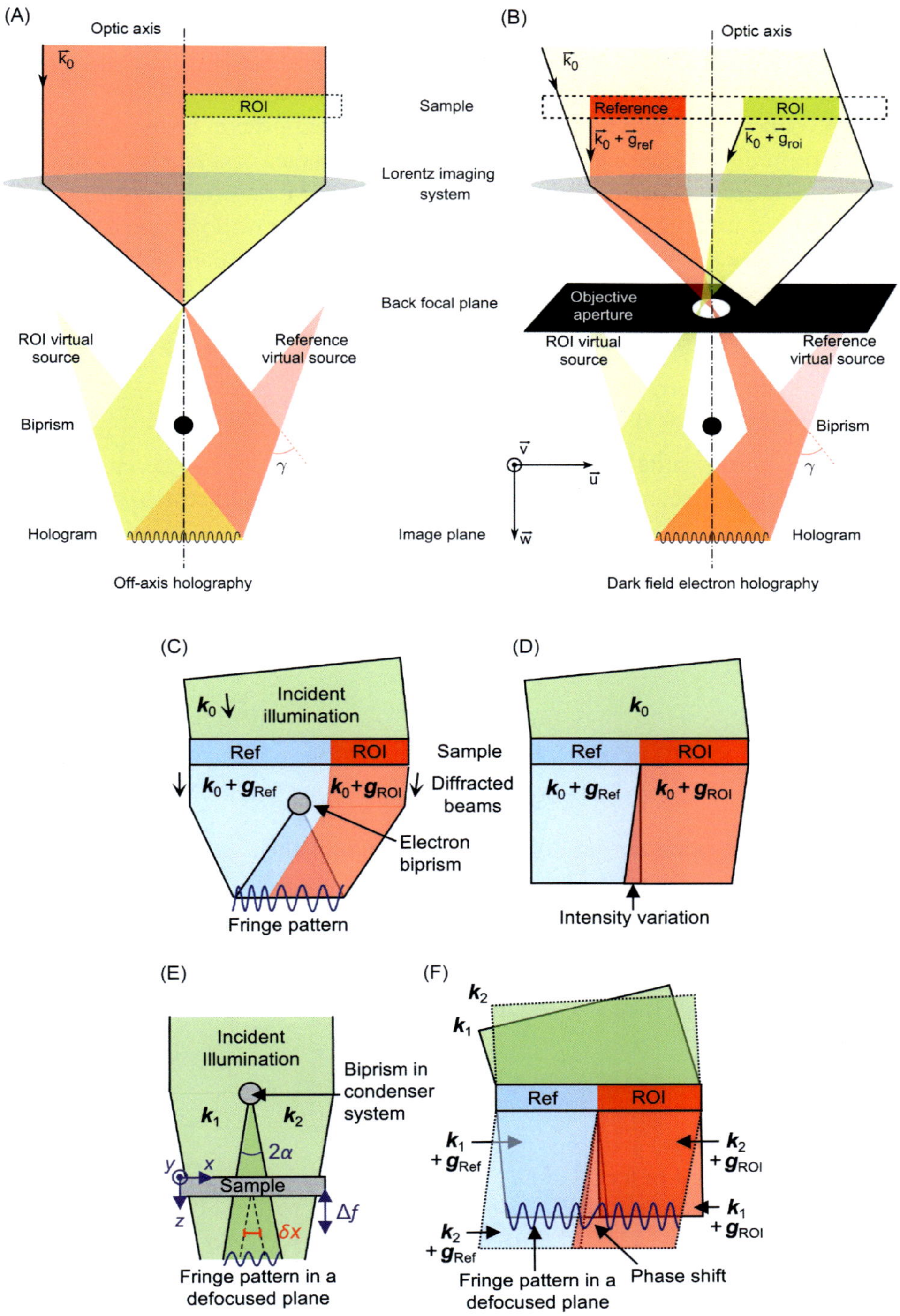

Figure 63.12

Dark-field electron holography for strain measurement: (A) standard hologram formation; (B) dark-field hologram formation; (C) off-axis dark-field arrangement; (D) in-line dark-field arrangement; (E) differential phase-contrast holography; and (F) configuration employed by Denneurlin et al. *(A and B) After Béché et al. (2011) and (C–F) after Denneulin et al. (2016), Courtesy Elsevier.*

incident beam is seen to pass through the specimen; the reference beam traverses an unstrained zone while the object wave carries information about the strain pattern. Unlike in Fig. 63.12A, an objective aperture now selects a diffracted beam corresponding to a reciprocal lattice vector $\boldsymbol{g}$ (see Section 69.3.2). The incident beam is tilted to make the $\boldsymbol{g}$ direction coincide with the optic axis; we return to this point below. The resulting phase map will provide a map of the interplanar spacing parallel to the $\boldsymbol{g}$-vector. The elements of the full strain tensor can be extracted from three such holograms obtained with different (nonplanar) $\boldsymbol{g}$-vectors as follows.

In the absence of the biprism the reference wave ψ_r and the object wave ψ_o take the form

$$\begin{aligned}\psi_r(\boldsymbol{r}_d) &= A_r \exp \mathrm{i}\left\{\left(2\pi\boldsymbol{g}_r + \boldsymbol{k}_Q\right)\cdot\boldsymbol{r}_d + \varphi_r\right\}\\ \psi_o(\boldsymbol{r}_d) &= A_o(\boldsymbol{r}_d) \exp \mathrm{i}\left\{\left(2\pi\boldsymbol{g}_o + \boldsymbol{k}_Q\right)\cdot\boldsymbol{r}_d + \varphi_o\right\}\end{aligned} \tag{63.72}$$

The biprism adds phase shifts $\pm\,\theta k_Q \boldsymbol{r}_d\cdot\boldsymbol{r}_b$ where θ is the deflection angle introduced by the biprism and $\boldsymbol{r}_b$ are unit vectors in the biprism plane. The intensity distribution in the hologram plane is then

$$J_H(\boldsymbol{r}_d) = A_o^2 + A_r^2 + 2A_rA_o\cos\left\{2k_Q\theta\boldsymbol{r}_d\cdot\boldsymbol{r}_b + 2\pi\left(\boldsymbol{g}_r - \boldsymbol{g}_o\right)\cdot\boldsymbol{r}_d + \varphi_r(\boldsymbol{r}_d) - \varphi_o(\boldsymbol{r}_d)\right\} \tag{63.73}$$

As usual, a sideband is extracted from the Fourier transform of $J_H(\boldsymbol{r}_d)$, from which we can obtain the distribution

$$\varphi(\boldsymbol{r}_d) := 2\pi\left(\boldsymbol{g}_r - \boldsymbol{g}_o\right)\cdot\boldsymbol{r}_d + \varphi_r(\boldsymbol{r}_d) - \varphi_o(\boldsymbol{r}_d) \tag{63.74}$$

and thence

$$\nabla\varphi(\boldsymbol{r}_d) = 2\pi\left(\boldsymbol{g}_r - \boldsymbol{g}_o\right) \tag{63.75}$$

The component of the strain tensor along the $\boldsymbol{g}_r$ direction is then given by

$$\varepsilon_{gg} = \frac{\nabla\varphi\cdot\boldsymbol{g}_r}{2\pi g_r^2 - \nabla\varphi\cdot\boldsymbol{g}_r} \tag{63.76}$$

In order to obtain the complete strain tensor, two more dark-field holograms are recorded and used to generate the elements of the matrix $G(\boldsymbol{r}_d)$:

$$G(\boldsymbol{r}_d) := \begin{pmatrix} g_{1x} & g_{2x} & g_{3x}\\ g_{1y} & g_{2y} & g_{3y}\\ g_{1z} & g_{2z} & g_{3z}\end{pmatrix} \tag{63.77}$$

Three matrices are derived in terms of G:

$$\begin{aligned}\mathscr{D} &= \left(G^T\right)^{-1}G_0^T - I && \text{distortion matrix}\\ \mathscr{E} &= \frac{1}{2}\left(\mathscr{D} + \mathscr{D}^T\right) && \text{strain matrix}\\ \Omega &= \frac{1}{2}\left(\mathscr{D} - \mathscr{D}^T\right) && \text{rotation matrix}\end{aligned} \tag{63.78}$$

in which I is the identity matrix and G_0 contains the $\boldsymbol{g}$-values in the reference zone. For many more details and practical examples, see the articles by Hÿtch et al. (2008) and Béché et al. (2011).

An improved approach using two biprisms is described by Denneulin and Hÿtch (2016). This differs from the two-biprism geometry of Hirayama et al. (1997) and Miyashita et al. (2004) in that the dark-field mode is employed (Fig. 63.12C). As before, we consider a plane incident wave $\psi_Q = A_Q \exp(\mathrm{i}\boldsymbol{k}_Q \cdot \boldsymbol{r})$, which interacts with the specimen and a particular $\boldsymbol{g}$-vector is selected. The two biprisms then generate four waves, each of the form $A_j \exp\,(\mathrm{i}\boldsymbol{k}_j \cdot \boldsymbol{r}_d + \varphi_j(\boldsymbol{r}_d)$ in the detector plane leading to the intensity distribution

$$
\begin{aligned}
J_H(\boldsymbol{r}_d) = {} & A_1^2 + A_2^2 + A_3^2 + A_4^2 \\
& + 2\Re[A_1A_2 \exp \mathrm{i}\left(\boldsymbol{q}_A \cdot \boldsymbol{r}_d + \varphi_1 - \varphi_2\right) \\
& + A_1A_3 \exp \mathrm{i}\left\{(\boldsymbol{q}_A + \boldsymbol{q}_B) \cdot \boldsymbol{r}_d + \varphi_1 - \varphi_3\right\} \\
& + A_1A_4 \exp \mathrm{i}\left(\boldsymbol{q}_B \cdot \boldsymbol{r}_d + \varphi_1 - \varphi_4\right) \\
& + A_2A_3 \exp \mathrm{i}\left(\boldsymbol{q}_B \cdot \boldsymbol{r}_d + \varphi_2 - \varphi_3\right) \\
& + A_2A_4 \exp \mathrm{i}\left\{(\boldsymbol{q}_B - \boldsymbol{q}_A) \cdot \boldsymbol{r}_d + \varphi_2 - \varphi_4\right\} \\
& + A_3A_4 \exp \mathrm{i}\left(-\boldsymbol{q}_A \cdot \boldsymbol{r}_d + \varphi_3 - \varphi_4\right)]
\end{aligned}
\tag{63.79}
$$

with

$$
\begin{aligned}
2\pi \boldsymbol{q}_A &= \boldsymbol{k}_1 - \boldsymbol{k}_2 = \boldsymbol{k}_4 - \boldsymbol{k}_3 \\
2\pi \boldsymbol{q}_B &= \boldsymbol{k}_1 - \boldsymbol{k}_4 = \boldsymbol{k}_2 - \boldsymbol{k}_3
\end{aligned}
\tag{63.80}
$$

Thus

$$
I(\boldsymbol{r}_d) = \sum_{q_n} \tilde{I}_{q_n} \exp\left(2\pi \mathrm{i} \boldsymbol{q}_n \cdot \boldsymbol{r}_d\right)
\tag{63.81}
$$

with

$$
\begin{aligned}
\tilde{I}_0 &= A_1^2 + A_2^2 + A_3^2 + A_4^2 \\
\tilde{I}_{q_A} &= A_1A_2 \exp \mathrm{i}\,(\varphi_1 - \varphi_2) + A_3A_4 \exp \mathrm{i}\,(\varphi_3 - \varphi_4) \\
\tilde{I}_{q_B} &= A_1A_4 \exp \mathrm{i}\,(\varphi_1 - \varphi_4) + A_2A_3 \exp \mathrm{i}\,(\varphi_2 - \varphi_3) \\
\tilde{I}_{q_A+q_B} &= A_1A_3 \exp \mathrm{i}\,(\varphi_1 - \varphi_3) \\
\tilde{I}_{q_A-q_B} &= A_2A_4 \exp \mathrm{i}\,(\varphi_2 - \varphi_4)
\end{aligned}
\tag{63.82a}
$$

This becomes much simpler when we recognize that $\varphi_1 = \varphi_o$ *and* $\varphi_2 = \varphi_3 = \varphi_4 = \varphi_r$. Setting all amplitudes to unity, we find

$$
\begin{aligned}
\tilde{I}_0 &= 4 \\
\tilde{I}_{q_A} = \tilde{I}_{q_B} &= \exp \mathrm{i}\left(\varphi_o - \varphi_r\right) + 1 = 2 \exp\left\{\mathrm{i}\left(\frac{\varphi_o - \varphi_r}{2}\right)\right\} \cos\left(\frac{\varphi_o - \varphi_r}{2}\right) \\
\tilde{I}_{q_A+q_B} &= \exp \mathrm{i}\left(\varphi_o - \varphi_r\right) \\
\tilde{I}_{q_A-q_B} &= 1
\end{aligned}
\tag{63.82b}
$$

and finally

$$J_H(\boldsymbol{r}_d) = 4 + 4\cos\left(\frac{\varphi_o - \varphi_r}{2}\right)\left\{\cos\left(2\pi\boldsymbol{q}_A\cdot\boldsymbol{r}_d + \frac{\varphi_o - \varphi_r}{2}\right) + \cos\left(2\pi\boldsymbol{q}_B\cdot\boldsymbol{r}_d + \frac{\varphi_o - \varphi_r}{2}\right)\right\}$$
$$+ 2\cos\{2\pi(\boldsymbol{q}_A + \boldsymbol{q}_B)\cdot\boldsymbol{r}_d + \varphi_o - \varphi_r\}$$
$$+ 2\cos\{2\pi(\boldsymbol{q}_A - \boldsymbol{q}_B)\cdot\boldsymbol{r}_d\} \tag{63.83}$$

(cf. Miyashita et al., 2004). The use of this expression to interpret practical patterns is discussed in full by Denneulin and Hÿtch (2016).

We have stated without comment that, for dark-field holography, the reference wave should travel along the optic axis. This is also desirable for aberration measurement (Röder and Lubk, 2015). A suitable microscope configuration for creating a tilted reference beam has been proposed and thoroughly explored by Röder et al. (2016). The following conditions must be satisfied:

1. The first biprism must be situated above the condenser lenses in order to demagnify the filament plane and hence magnify the inclination.
2. The first crossover above the biprism must be conjugate to the front focal plane of the objective lens prefield so that the illumination is parallel.
3. The biprism plane and the object plane must be conjugates.
4. The demagnification (condition 1) must be as strong as possible, to achieve a large tilt angle.
5. A cold field-emission gun is essential; since the condenser stigmators will usually be too close to the biprism to create much ellipticity, the electron emission should be rotationally symmetric.
6. Coma-free alignment is recommended to eliminate the tilt of one partial wave

We cannot reproduce here the steps that lead to the proposed solution. Instead, we reproduce figures that summarize the results. Fig. 63.13 shows the elements in the source and condenser space with which the above conditions can be satisfied. Fig. 63.14 is a schematic ray diagram for the Hitachi I2TEM at the CEMES[5], Toulouse. Fig. 63.15 shows paraxial rays passing through the objective lens for three values of the biprism filament voltage: −5, 0 and 5 V. The rays are brought to a focus at the front focal plane in order to ensure parallel illumination at the centre of the objective lens. The system is completed by two biprisms in the image space, as shown in Fig. 63.16. The first of these (B1) is conjugate to the object plane (and hence to the image plane). Two zones occupied by fringes result; in zone 1, the partial object waves interfere while in zone 2, the (partial) object and reference waves interfere. The dark bands are shadows of the biprism filaments.

The subject is reviewed briefly by Mahr et al. (2021). The ways of measuring strain have been surveyed at greater length by Pofelski (2021), whose review forms an

[5] Centre d'Elaboration de Matériaux et d'Etudes Structurales.

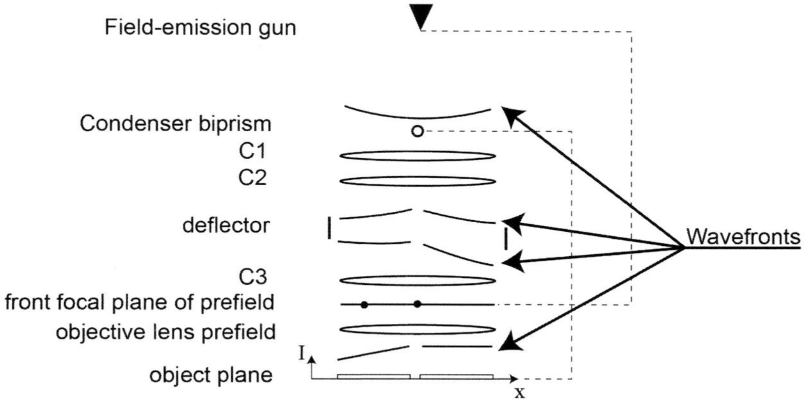

Figure 63.13
Tilted illumination. Source and condenser space. *After Röder et al. (2016), Courtesy Elsevier.*

introduction to the main theme of his work, the use of a moiré technique to extract information about strain: STEM moiré interferometry (Su and Zhu, 2010). When the scanning grid of the STEM overlaps the lattice of the crystalline specimen, moiré fringes appear. The unusual feature of this interferometric technique is the sampling rate employed. In traditional STEM, it is necessary to oversample the specimen; here, on the contrary, the scanning grid undersamples the periodic structure of the crystal, causing aliasing. The resulting moiré pattern contains valuable information about any irregularities of the crystal lattice and hence any strain pattern. The essence of the process is apparent in the identity

$$\sin(2\pi\nu_s) + \sin(2\pi\nu_c) \equiv 2\sin\{\pi(\nu_s + \nu_c)x\}\cos\{\pi(\nu_s - \nu_c)x\}$$

where ν_s *and* ν_c denote the scan frequency and the lattice frequency respectively. Cosine fringes will be seen with the difference frequency $\nu_s - \nu_c$. (Sine fringes will not be seen since their frequency is greater than the scan frequency.) The theory that permits all the elements of the strain tensor to be extracted from the moiré fringe pattern is set out in full by Pofelski. For applications of this technique, see Kim et al. (2013a,b), Haas (2017), Ishizuka et al. (2017), Naden (2018), Yamanaka et al. (2018) and Prabhakara et al. (2019). Many more references are provided by Pofelski (2021) and in earlier papers by Pofelski et al. (2021, 2017, 2019).

63.6.3 Holographic Tomography

Complete information about the distribution of magnetic and electrostatic fields in a specimen can in principle be obtained by extending the well-established methods of

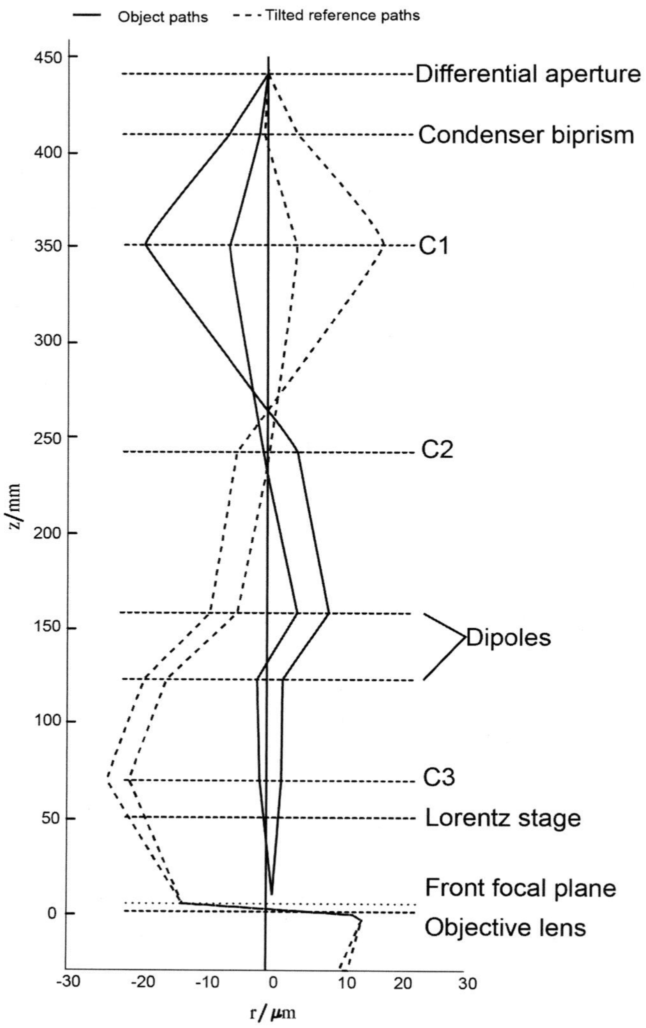

Figure 63.14
Tilted illumination. Ray diagram for an FEI 12TEM electron microscope, biprism voltage −100 V. *After Röder et al. (2016), Courtesy Elsevier.*

three-dimensional reconstruction (Chapter 75) to holography. For electrostatic fields, a tilt series of holograms is recorded at small angular intervals (1−3 degrees) up to as steep a tilt as possible (typically ±70 degrees) A tilt series of phase images is then calculated and used to create a 3D representation of the field or potential distribution. For this last step, Wolf et al. (2013a,b) have used an improved routine (W-SIRT), a hybrid between the simultaneous iterative reconstruction technique (SIRT) and weighted

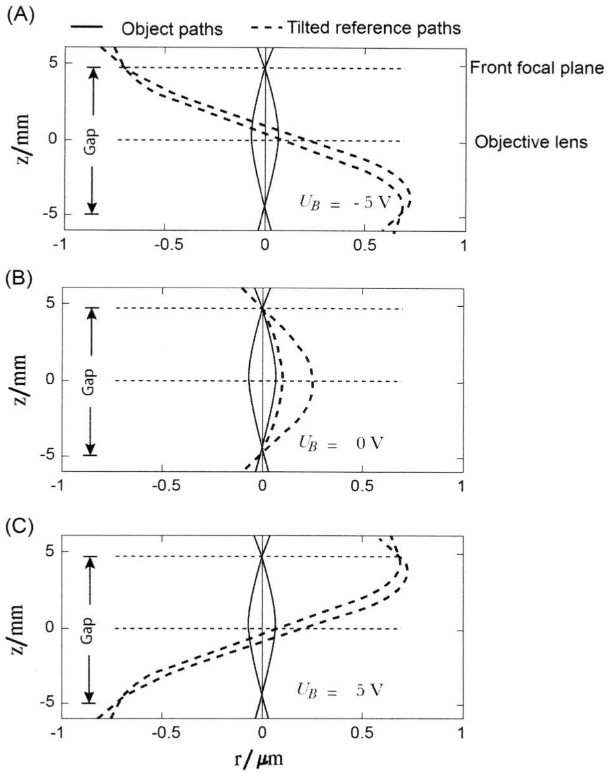

Figure 63.15
Tilted illumination. Paraxial rays passing through the objective lens for different biprism voltages U_B. The beams are focused at the front focal plane in order to provide parallel illumination at the centre of the objective. *After Röder et al. (2016), Courtesy Elsevier.*

back-projection (described in Chapter 75). These authors have also developed a software programme for automated reconstruction (Tomographic Holographic Microscope Acquisition Software, THOMAS), which speeds up the laborious sequence of steps needed: tilt series acquisition, phase tilt series calculation, unwrapping, alignment and, finally, 3-D reconstruction itself.

This work builds on earlier attempts to perform holographic tomography, notably by Lai et al. (1994a), Friedrich et al. (2004), Lade et al. (2005a,b), Twitchett-Harrison et al. (2007, 2008a,b), Fujita and Chen (2009), Wolf (2007, 2010), Wolf et al. (2008, 2010) and Tanigaki et al. (2012). Tanji and Hirayama (1997) describe a form of holographic interferometry, or shearing interferometry, from which the lateral derivative of the specimen property under study can be extracted. See also the survey by Midgley and Dunin-Borkowski (2009). Gatel et al. (2013) have shown that the charge on individual

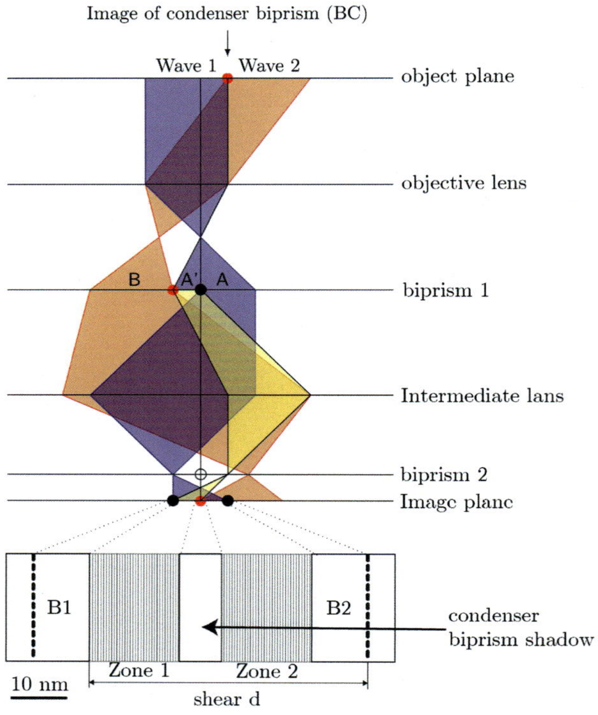

Figure 63.16
Tilted illumination. The roles of the biprisms in image space. B1: shadow of biprism 1. *After Röder et al. (2016), Courtesy Elsevier.*

nanoparticles can be counted with a precision of one elementary unit of charge by means of aberration-corrected electron holography.

Three-dimensional reconstruction of the components of the magnetic induction is proving less straightforward though very important for the study of spintronics devices, for example (Lai et al., 1994b; Dunin-Borkowski et al., 2004). Four hologram tilt series are now required, two for each in-plane component. Also, the magnetic phase shift is often much smaller than in the electrostatic case. Progress is described by Phatak et al. (2008, 2010) and Tanigaki et al. (2016) and especially by Wolf et al. (2018, 2019). The last stage of each reconstruction, in which div $\boldsymbol{B} = 0$ is used to establish the third component of $\boldsymbol{B}$, is particularly prone to errors arising from noise and artefacts. A dual-axis sample-holder capable of rotation through 360 degrees is helpful (Tsuneta et al., 2014) but the ability to rotate through a complete cycle is offset by the excessive thickness of the specimen at high tilt angles. One solution is to perform the experiments in a modern high-voltage microscope (Kawasaki et al., 2000a,b;

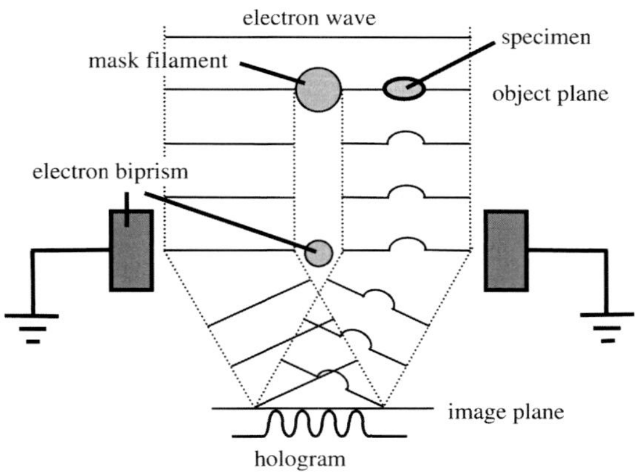

Figure 63.17
Removal of Fresnel fringes. Use of a masking filament. *After Yamamoto et al. (2004), Courtesy Elesevier.*

Tonomura, 2009[6]; Akashi et al., 2015). The improved routines described by Zhuge et al. (2017) can be expected to benefit holographic tomography.

An alternative approach, using the transport-of-intensity equation is described in Section 74.3.8 and considered by Lade et al. (2005a,b).

63.6.4 Multiple Biprisms and Split Illumination

Electrons that pass close to the biprism filament experience Fresnel diffraction. The resulting waves add to those of the reference wave and the object wave, giving rise to unwanted modulations of intensity. This effect can be considerably reduced by the simple expedient of inserting a masking filament in the specimen plane, as shown in Fig. 63.17 (Yamamoto et al., 2004). Fig. 63.18 shows that they obtained a significant improvement by this means. Nevertheless, the correction is not perfect, probably as a result of Fresnel diffraction at the masking filament.

Double-biprism interferometry was introduced by Harada and Tonomura (2004) in an attempt to control the width w of the interference pattern and the fringe spacing s independently and also eliminate the Fresnel effect. The geometry is illustrated in Fig. 63.19A–C, which shows that the filament of the second biprism is rotated relative to that of the first biprism. It was found by Harada et al. (2006) that a third biprism allows any remaining problems to be overcome – the three-biprism arrangement is

[6] This article is exceptionally well illustrated.

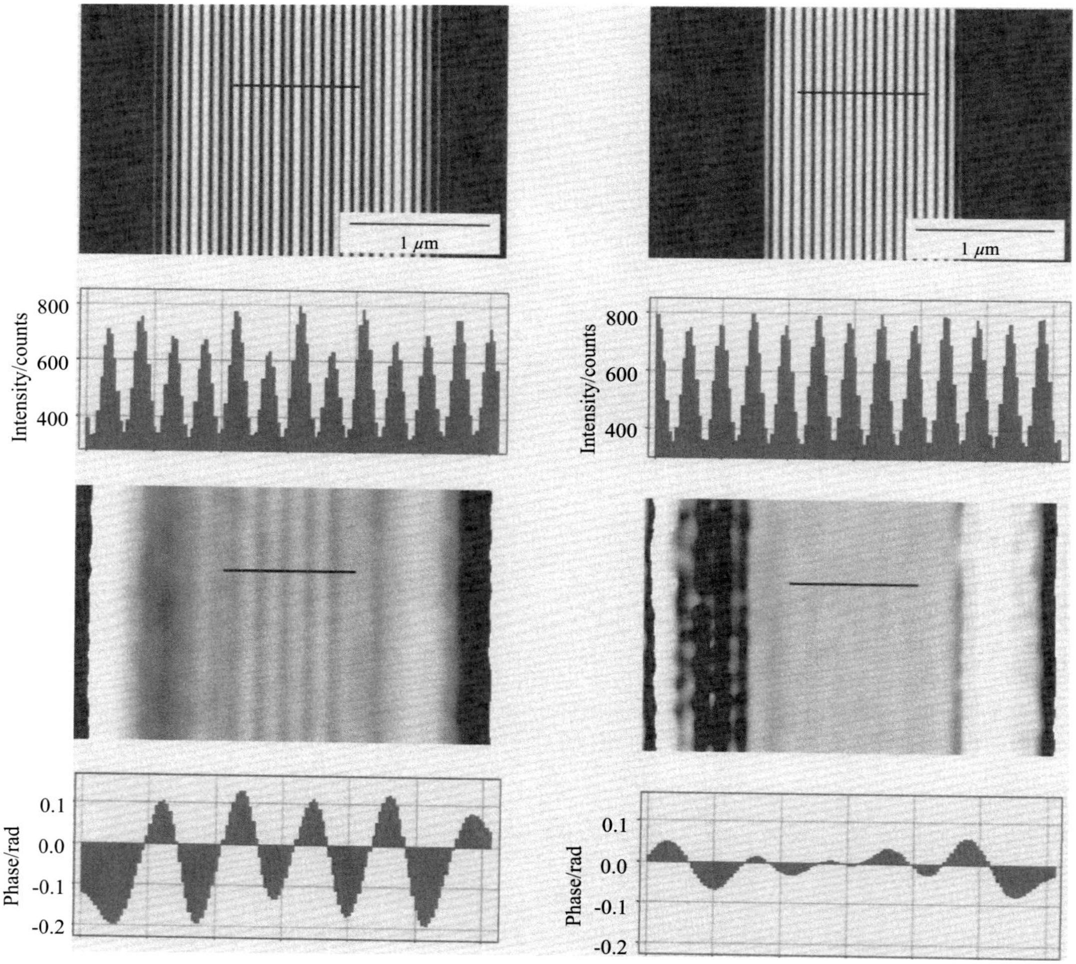

Figure 63.18

Improvement achieved by the use of the masking filament of Fig. 63.17: (left) without masking filament; (right) with masking filament; (top) fringe pattern and intensity distribution along the line shown; (bottom) phase image and phase distribution along the line shown. *After Yamamoto et al. (2004), Courtesy Elsevier.*

shown in Fig. 63.19D–G. All the parameters of interest can now be controlled independently. To see this, we reproduce Harada's formulae for the interference widths w_x and w_y, the fringe spacing s and the inclination of the fringes relative to the x-axis ϑ_o (arising from the azimuthal difference between the biprism filaments). Harada et al. also include the distance r_3 from the optic axis to the virtual source of the object or reference wave at the crossover plane. With the notation shown in Fig. 63.19D, they find

$$
\begin{aligned}
w_x &= 2\frac{1}{M_o M_1 M_2}\alpha_3 L_3 \sin\varphi_3 - \frac{d_1}{M_o \sin\varphi_1}\\
w_y &= 2\frac{1}{M_o M_1 M_2}\alpha_3 L_3 \cos\varphi_3 - \frac{d_2}{M_o M_1 \sin\varphi_1}\\
s &= \frac{1}{2}\frac{1}{M_o M_1 M_2}\frac{D_3\lambda}{r_3}\\
r_3 &= \left(r_{\cos}^2 + r_{\sin}^2\right)\\
r_{\sin} &= \frac{b_3 b_2}{a_3 a_2} D_1 \alpha_1 \sin\varphi_1 + (D_3 - L_3)\alpha_3 \sin\varphi_3\\
r_{\cos} &= \frac{b_3 b_2}{a_3 a_2} D_1 \alpha_1 \cos\varphi_1 + \frac{b_3}{a_3} D_2 \alpha_2 + (D_3 - L_3)\alpha_3 \cos\varphi_3\\
\vartheta_o &= \arctan\frac{r_{\sin}}{r_{\cos}}
\end{aligned}
\tag{63.84}
$$

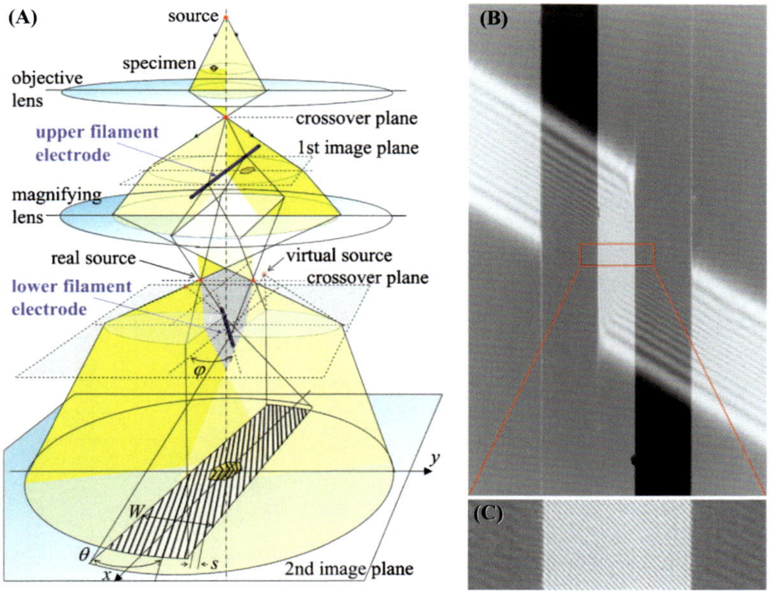

Figure 63.19

Multiple-biprism interferometry: (A–C) double-biprism geometry, wide-area view of the region of interference and enlargement in which Fresnel fringes from the lower filament are superimposed on the pattern; (D–G) triple-biprism geometry. Geometry, wide-area view and enlargements of the regions indicated (D–F). (G) Real sources (*black discs*) and virtual sources (*open stars*) in the crossover plane. *After Harada et al. (2006), Courtesy American Institute of Physics, https://doi.org/10.1063/1.2198987.*

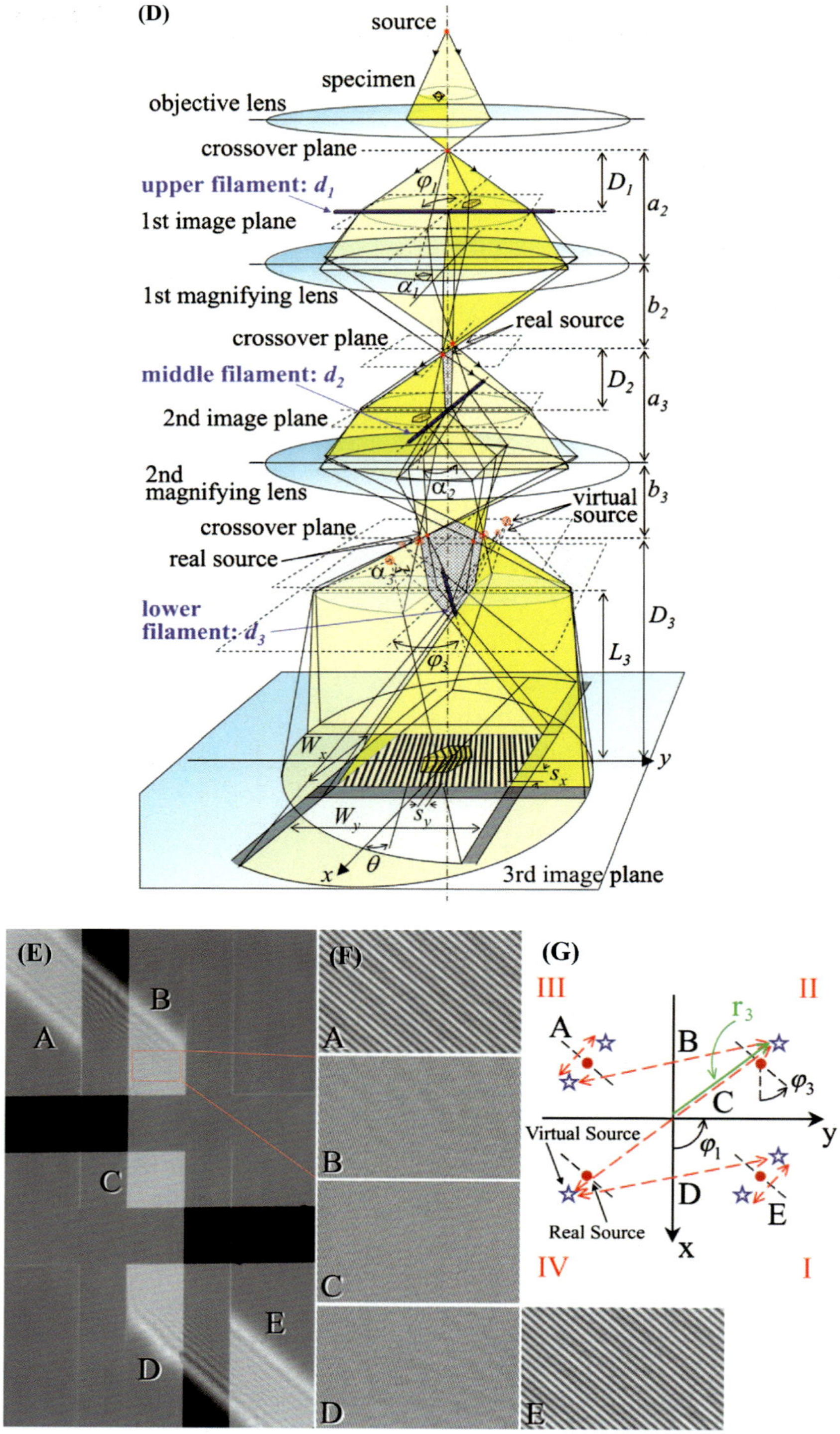

Figure 63.19
Continued.

where d_1 *and* d_2 denote the diameters of the filaments of the upper and middle biprisms; α_1, α_2 *and* α_3 are the deflections at each filament. Fig. 63.19G shows the positions of the real and virtual sources in the crossover plane. In this example the azimuthal angles of the first and third filaments are 90 and 50 degrees and the excitations are 10, 126 and 120 V. This configuration is discussed and fully illustrated by Harada et al.

Harada et al. later used the double-biprism configuration in a new attempt to perform a 'which way?' experiment, which would identify the slit through which an electron that contributed to an interference pattern had passed or confirm that this is impossible. In Harada et al. (2018a,b), the widths of the slits were different. In Harada et al. (2019a,b), the slits, drilled in a copper foil about 1.3 μm thick, were V-shaped. Later, phase modulation was added (Harada et al., 2020). Fig. 63.20A and B show the configuration employed; the results are seen in Fig. 63.20C and D.

A further development (Tanigaki et al., 2014) is an advanced form of the split-illumination electron holography tested earlier (Tanigaki et al., 2012). Now, two biprisms are present in the space between the source and the sample, as shown in Fig. 63.21A and B. The filament of the upper biprism is conjugate to the specimen plane while that of the second biprism lies on the shadow of the upper filament. Two additional biprisms are situated below the specimen; their filaments also fall within the shadow on the filament of the uppermost biprism. With this configuration, Fresnel fringes from the filaments are completely absent (Fig. 63.21C).

Yamamoto (2018) has employed the phase-shifting procedure described by Ru et al. (1991a,b) and Yamamoto et al. (2000) to study semiconductor devices. Here, several holograms are recorded with different incident beam tilts.

63.6.5 Inelastic Holography

In the foregoing sections, it was assumed that holograms were formed by interference between a reference beam and elastically scattered electrons. Inelastically scattered electrons were excluded from the theory and included in Eq. (63.27) as an undesirable background term. Nevertheless, the possibility of interference, and hence some mutual coherence, between inelastically scattered electrons has been examined, notably by Rose and colleagues (Rose, 1977, 1984; Kohl, 1983; Kohl and Rose, 1984, 1985; Müller et al., 1998; Müller and Rose, 2003) and Harscher et al. (1997).

Any coherence between electrons that have been scattered inelastically has been investigated by Lichte and Freitag (2000). For elastically scattered electrons, we have as usual (Eq. 63.27)

$$J_H(x,y) = 1 + A^2(x,y) + 2A\mu\cos\left\{2\pi q_s x - \overline{\eta}(x,y) + \rho\right\}$$

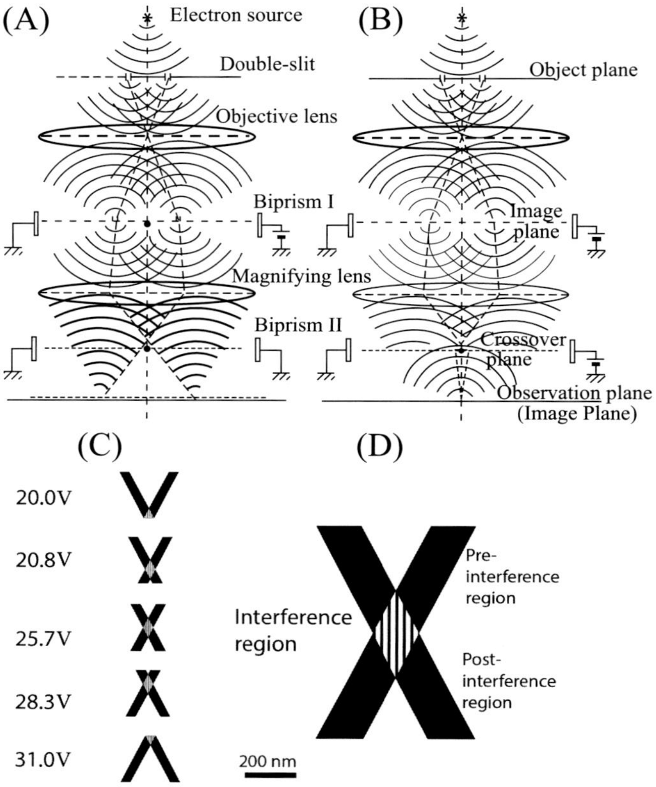

Figure 63.20

Appearance of interference patterns for a range of biprism excitations... Pattern without (A) and with (B) excitation of the lower biprism. (C) Traces of interference patterns for various values of the voltage on the lower biprism. (D). Enlargement showing that interference occurs only in the overlap zone, thus confirming that no 'which way?' information reaches the detector. *After Harada et al. (2006), Courtesy Cambridge University Press, https://doi.org/10.1017/S1431927619005452.*

As a first extension, we consider interference between electrons with a small difference in energy, κ. Eq. (63.27) now becomes

$$J_H = 1 + A^2 + 2A\mu\cos(2\pi q_0 x - \overline{\eta} + \rho + \kappa/\hbar t) \tag{63.85}$$

where t now denotes time. At a given point in the interference pattern, we therefore expect to observe beats, with a beat frequency of $\kappa/\hbar$. This has been confirmed by Möllenstedt and Lichte (1978b) and Schmid (1985). We now turn to the critical case of inelastically scattered electrons that have suffered approximately the same energy loss, studied experimentally by Lichte and Freitag. The specimen was an aluminium foil, the thickness of which was chosen to give sufficient current in the first plasmon loss (Fig. 63.22), which

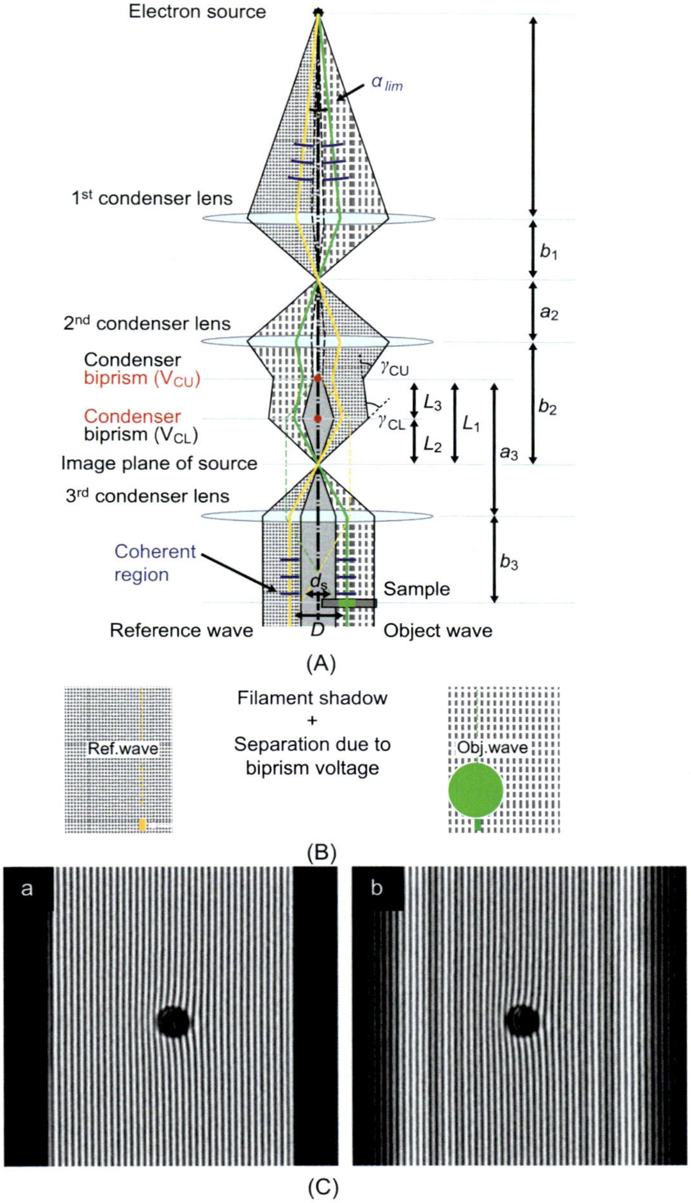

Figure 63.21

Elimination of Fresnel fringes by means of multiple biprisms: (A) microscope configuration; (B) formation of a hologram free of Fresnel fringes; for details see the original publication; (C) Electric field around a charged latex particle. (left) Hologram free of Fresnel fringes; (right) hologram from which Fresnel fringes have not been completely eliminated. *After Tanigaki et al. (2014), Courtesy Elsevier.*

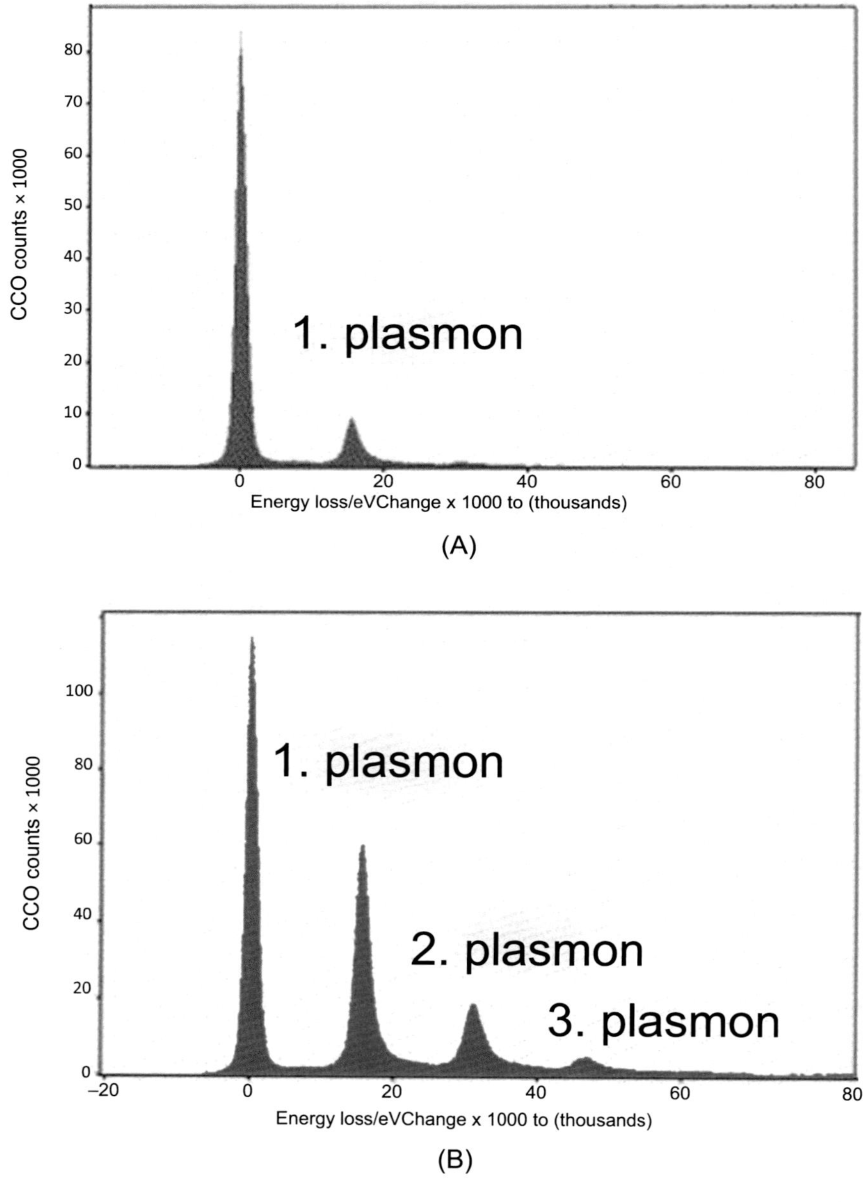

Figure 63.22

Interference and inelastic scattering. Electron energy-loss spectra showing plasmon losses in an aluminium foil. (A) Thin foil, $t/\lambda = 0.37$; (B) thicker foil, $t/\lambda = 1.29$ where λ denotes the inelastic mean free path. *After Lichte and Freitag (2000), Courtesy Elsevier.*

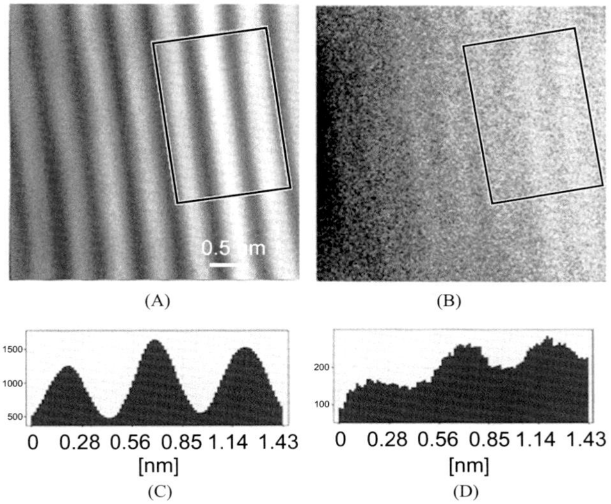

Figure 63.23

Interference and inelastic scattering. (A) Hologram obtained with elastically scattered electrons. (B) Hologram obtained with inelastically scattered electrons (first plasmon loss, see Fig. 63.22). Parts (C) and (D) show intensity profiles across the zones indicated in (A) and (B). *After Lichte and Freitag (2000), Courtesy Elsevier.*

is centred on 15.5 eV. The result is shown in Fig. 63.23, where faint fringes are seen in the pattern generated by the loss electrons; the width of the selection window was 5 eV. This is evidence of mutual coherence between inelastically scattered electrons. This work was reconsidered by Potapov et al. (2006) and complemented by Herring (2005).

In order to put these results on a sound theoretical basis, Verbeeck et al. (2008) expressed the propagation of the signal in terms of a reduced density matrix. A more recent account of this approach is given by Niermann and Lehmann (2016) and we follow their reasoning closely. For the reader's convenience, we have retained the notation of Niermann and Lehmann. In particular, the quantity $\boldsymbol{q}$ is related to that used elsewhere by $\boldsymbol{q}(\mathrm{N\ \&\ L}) \rightarrow \boldsymbol{q}/\lambda$ (this volume), which explains the apparent dimensional contradictions.

We denote the (plane) wave incident on the specimen by $\psi_Q(\boldsymbol{r}_o)$, the wave at the exit surface of the specimen by $\psi_o(\boldsymbol{r}_o)$ and in the detector plane, $\psi_d(\boldsymbol{r}_d)$. Then in standard bright-field conditions, we have

$$\tilde{\psi}_d(\boldsymbol{q}) = \tilde{\psi}_o(\boldsymbol{q})\xi(\boldsymbol{q})\exp\left(i\pi\lambda\Delta\, q^2\right) \tag{63.86}$$

where the effect of spherical aberration and objective aperture has been separated from the effect of defocus (Δ):

$$\xi(\boldsymbol{q}) = \exp\left(\frac{\mathrm{i}\pi}{2} C_s \lambda^3 q^4\right) \tag{63.87}$$

Setting $I(\boldsymbol{r}_d) = \psi(\boldsymbol{r}_d)\psi^*(\boldsymbol{r}_d)$, we have

$$\tilde{I}(\boldsymbol{q}) = \int \tilde{\psi}_o(\boldsymbol{q})\, \tilde{\psi}_o^*(\boldsymbol{q}+\bar{\boldsymbol{q}})\xi(\bar{\boldsymbol{q}})\xi^*(\boldsymbol{q}+\bar{\boldsymbol{q}})\exp\left[\mathrm{i}\pi\lambda\Delta\{\bar{q}^2 - (\boldsymbol{q}+\bar{\boldsymbol{q}})^2\}\right] d\bar{\boldsymbol{q}} \tag{63.88}$$

(Fourier transforms are indicated by a tilde.) We now extend this to the holographic situation, in which a reference wave $\psi_r(\boldsymbol{r}_d) = \psi_Q \exp\{-\mathrm{i}(2\pi \boldsymbol{g}\cdot\boldsymbol{r}_d + \vartheta)\}$ is added to the specimen wave. Here, $g\lambda$ is the angle between the reference wave and the image wave and ϑ measures any phase shift between the two waves. It will be set equal to zero later. The intensity distribution in the detector plane now becomes

$$J_H(\boldsymbol{r}_d) = \psi_d\psi_d^* + I_o + 2\Re\{\chi(\boldsymbol{r}_d)\exp\mathrm{i}\,(2\pi\boldsymbol{g}\cdot\boldsymbol{r}_d + \vartheta)\} \tag{63.89}$$

where

$$\tilde{\chi}(\boldsymbol{q}) = \psi_Q^*\tilde{\psi}_o(\boldsymbol{q})\xi(\boldsymbol{q})\exp(\mathrm{i}\pi\lambda\Delta q^2) \tag{63.90}$$

The inevitable presence of a small energy spread $\Delta E =: e\kappa$ in the incident beam is expressed by including a chromatic aberration phase shift in $\xi(\boldsymbol{q})$,

$$\bar{\xi}(\boldsymbol{q}) := \xi(\boldsymbol{q})\exp\left(\mathrm{i}\pi\lambda C_c \frac{\kappa}{\Phi_0} q^2\right) \tag{63.91}$$

We now average the expressions for $I(\boldsymbol{r}_d)$ *and* $J_H(\boldsymbol{r}_d)$ since a large number of electrons will be involved and express the result in terms of the *density matrix*,

$$\tilde{\rho}(\boldsymbol{q},\bar{\boldsymbol{q}}) := \left\langle \tilde{\psi}_o(\boldsymbol{q})\tilde{\psi}_o^*(\bar{\boldsymbol{q}})\right\rangle \tag{63.92}$$

We shall see that this collapses to a very simple form (63.95 and 63.96). In terms of the density matrix, the average of $\tilde{I}(\boldsymbol{r}_d)$ is

$$\langle\tilde{I}(\boldsymbol{q})\rangle = \int \tilde{\rho}(\bar{\boldsymbol{q}},\boldsymbol{q}+\bar{\boldsymbol{q}})T(\bar{\boldsymbol{q}},\boldsymbol{q}+\bar{\boldsymbol{q}})\exp\left[\mathrm{i}\pi\lambda\Delta\{\bar{q}^2 - (\boldsymbol{q}+\bar{\boldsymbol{q}})^2\}\right] \tag{63.93}$$

in which the transmission cross-coefficient is given by

$$T(\boldsymbol{q},\bar{\boldsymbol{q}}) := \xi(\bar{\boldsymbol{q}})\xi^*(\boldsymbol{q}+\bar{\boldsymbol{q}})\exp\left[\mathrm{i}\pi\lambda C_c\frac{\kappa}{\Phi_0}\{\bar{q}^2 - (\boldsymbol{q}+\bar{\boldsymbol{q}})^2\}\right] \tag{63.94}$$

Likewise,

$$\langle J_H(\boldsymbol{r}_d)\rangle = \langle I(\boldsymbol{r}_d)\rangle + I_0 + 2\mu\langle\chi(\boldsymbol{r}_d)\rangle\Re\exp(\mathrm{i}2\pi\boldsymbol{g}\cdot\boldsymbol{r}) \tag{63.95}$$

in which

$$\langle\tilde{\chi}(\boldsymbol{q})\rangle = \psi_Q^*\langle\tilde{\psi}_o(\boldsymbol{q})\rangle T(\boldsymbol{q},0)\exp\left(\mathrm{i}\pi\lambda\Delta q^2\right) \tag{63.96}$$

The term $\psi_Q^*\tilde{\psi}_o(\boldsymbol{q})$ also takes a simpler form (63.97). An attenuation μ is included in (63.95) to allow for various instabilities (cf. 63.27).

Suppose now that some electrons are scattered inelastically in the specimen. The quantum states of the electrons and the sample are now said to *be entangled*, which means that we must use a joint wavefunction containing the coordinates of the beam electrons and of the specimen. We denote this $\Psi(\boldsymbol{r},z,\boldsymbol{R})$; here, $\boldsymbol{R}$ is a $3N$-dimensional vector, representing the coordinates of the N elements of the specimen. The evolution of the joint wavefunction satisfies the Schrödinger equation, $\hat{H}\Psi = E\Psi$, and we write the Hamiltonian as the sum of three terms (Yoshioka, 1957):

$$\begin{aligned}\hat{H} = &-\frac{\hbar^2}{2m_0}\left(\nabla_{\boldsymbol{r}}^2 + \frac{\partial^2}{\partial z^2}\right) && \text{kinetic energy of the electron}\\ &-\gamma eV(\boldsymbol{r},z,\boldsymbol{R}) && \text{Coulomb interaction between the electron and the object}\\ &+\hat{H}_o(\boldsymbol{R}) && \text{object Hamiltonian}\end{aligned} \tag{63.97}$$

An orthonormal set of eigenstates of the object Hamiltonian $\{a_n\}$ can be obtained from

$$\hat{H}_o a_n(\boldsymbol{R}) = \mathscr{E}_n a_n(\boldsymbol{R}) \tag{63.98}$$

in terms of which $\Psi(\boldsymbol{r},z,\boldsymbol{R})$ becomes

$$\Psi(\boldsymbol{r},z,\boldsymbol{R}) = \sum_n \psi_n(\boldsymbol{r},z)\exp\left(\frac{\mathrm{i}2\pi z}{\lambda_n}\right)a_n(\boldsymbol{R}) \tag{63.99}$$

where $1/\lambda_n = K(E-\mathscr{E}_n)$ and $K(E) := \left\{2m_0E\left(1+E/2m_0c^2\right)\right\}^{1/2}/\hbar$.

Niermann and Lehmann then analyse the propagation of the electrons in terms of the density matrix and conclude that the intensity $\langle\tilde{I}(\boldsymbol{q})\rangle$ is obtained via (63.88) from the density matrix

$$\tilde{\rho}(\boldsymbol{q},\overline{\boldsymbol{q}}) = \sum_m \tilde{\psi}_{n\to m}(\boldsymbol{q})\tilde{\psi}^*_{n\to m}(\overline{\boldsymbol{q}}) \tag{63.100}$$

where the functions $\psi_{n\to m}(\boldsymbol{r})$ are the projections of the joint state onto the object states a_m when the object is initially in eigenstate a_m. This sum of products reduces to a single product only when we limit the calculation to elastic scattering, in which case

$\tilde{\rho}(\boldsymbol{q},\overline{\boldsymbol{q}}) \Rightarrow \tilde{\rho}_{elastic}(\boldsymbol{q},\overline{\boldsymbol{q}}) = \tilde{\psi}_{n\to n}(\boldsymbol{q})\psi^*_{n\to n}(\overline{\boldsymbol{q}})$. Note that when the energy spread of the incident beam is included in the calculation, an additional term appears under the summation:

$$\tilde{\rho}(\boldsymbol{q},\overline{\boldsymbol{q}}) = \sum_m \tilde{\psi}_{n\to m}(\boldsymbol{q})\tilde{\psi}^*_{n\to m}(\overline{\boldsymbol{q}})\exp\left\{\mathrm{i}\pi\lambda_0 C_c \frac{\kappa_{n\to m}}{\Phi_0}\left(q^2 - \overline{q}^2\right)\right\} \tag{63.101}$$

For the sideband of an off-axis hologram the interference terms are given by

$$\psi^*_Q\tilde{\psi}(\boldsymbol{q}) = \tilde{\psi}_{n\to n}(\boldsymbol{q})\psi^*_Q \tag{63.102}$$

This important result shows that *only elastic scattering* contributes to the fringe pattern. Any inelastic scattering will attenuate it. Niermann and Lehmann then derive expressions for these quantities after averaging over the thermal ensemble of object states and find

$$\tilde{\rho}(\boldsymbol{q},\overline{\boldsymbol{q}}) = \sum_n\sum_m p_n\tilde{\psi}_{n\to m}(\boldsymbol{q})\tilde{\psi}^*_{n\to m}(\overline{\boldsymbol{q}}), \quad p_n \propto \exp\left(-\frac{\mathscr{E}_n}{k_BT}\right) \tag{63.103}$$

k_B is Boltzmann's constant and T is the absolute temperature. For the sideband term,

$$\psi^*_Q\langle\tilde{\psi}(\boldsymbol{q})\rangle = \sum_n p_n\tilde{\psi}_{n\to n}(\boldsymbol{q})\psi^*_Q \tag{63.104}$$

Thus information obtained from the central area will be an average of squared values while that from sidebands is the average of the original values: one is an incoherent sum, the other is coherent.

63.6.6 Double- and Continuous-Exposure Off-Axis Holography

In double-exposure holography, introduced by Wahl (1975) and developed by Matteucci et al. (1991, 1996), two holograms are superposed in the recording plane. Recent progress allows us to superpose not merely two but a semicontinuous sequence of holograms; this is useful for studying specimens modified by applying a periodic potential.

We have seen that the intensity at the hologram in the standard off-axis configuration is given (in one dimension, for simplicity) by

$$J_H(x_d) = 1 + A^2(x_d) + 2A(x)\cos\{k_o x_d + \eta(x_d)\}$$

(Eq. 63.26). Suppose now that the intensity $J_H^{(T)}$ is the sum of two intensity distributions, $J_H^{(1)}$ *and* $J_H^{(2)}$. Then

$$J_H^{(T)}(x_d) = 1 + A^2(x_d) + \mu A(x_d)\left[\cos\{k_Q x_d + \eta_1(x_d)\} + \cos\{k_Q x_d + \eta_2(x_d)\}\right] \tag{63.105}$$

The usual reconstruction gives

$$\psi(x_d) = \frac{1}{2}\mu A(x_d)\exp\mathrm{i}\{\eta_1(x_d) + \eta_2(x_d)\} \tag{63.106a}$$

or setting $\eta_1 = \eta_0 + \eta, \eta_2 = \eta_0 - \eta$

$$\psi(x_d) = \mu A(x_d) \cos \eta(x_d) \exp\{\mathrm{i}\, \eta_0(x_d)\} \tag{63.106b}$$

from which we see that

$$\eta(x_d) = \arccos\left(\frac{|\psi(x_d)|}{\mu A(x_d)}\right) \tag{63.107}$$

We now apply these ideas to situations in which the properties of the specimen change either spontaneously or in response to a stimulus during the exposure time T of a hologram. The intensity recorded will now be the average of $J_H^{(T)}(x_d)$ over time:

$$\overline{J}_H^{(T)} = \frac{1}{T}\int_0^T J_H^{(T)}(x_d, t)dt = A(x_d) + \frac{\mu A(x_d)}{T}\int_0^T \cos\{k_o x_d + \eta(x_d, t)\}dt \tag{63.108}$$

which leads to

$$\psi(x_d) = \frac{\mu A(x_d)}{T}\int_0^T \exp\{\mathrm{i}\eta(x_d, t)\}dt \tag{63.109}$$

Migunov et al. (2017) consider three forms of time dependence of the object wave: a square waveform, a sinusoidal waveform and a triangular waveform. The first case corresponds to double-exposure holography. The reconstructed wavefunction is given by

$$\psi(x_d) = \mu A(x_d)\cos\eta(x_d)\exp\{\mathrm{i}\eta_0(x_d)\} \quad \text{square waveform} \tag{63.110a}$$

For the sine wave, they find

$$\psi(x_d) = \eta A(x_d) J_0(\eta(x_d))\exp\{\mathrm{i}\eta_0(x_d)\} \quad \text{sinusoidal waveform} \tag{63.110b}$$

and for the triangular case,

$$\psi(x_d) = \mu A(x_d)\frac{\sin\eta(x_d)}{\eta(x_d)}\exp(\mathrm{i}\eta_0) \quad \text{triangular waveform} \tag{63.110c}$$

The case of an arbitrary periodic waveform is also briefly considered by Migunov et al.

A simple extension of the theory permits *phase amplification*, which is required if the phase shifts involved are smaller than 2π. From Eq. (63.107), we know that

$$\cos^2\eta(x_d) = \left\{\frac{|\psi(x_d)|}{\mu A(x_d)}\right\}^2 \tag{63.111}$$

On substituting this into the expansions of $\cos(n\eta)$ as a polynomial in $\cos\eta$, we obtain the phase-amplified distribution. For $n = 4$, for example,

$$\cos(4\eta) = 8\cos^4\eta - 8\cos^2\eta + 1 \tag{63.112}$$

Volkov et al. (2013) have shown that it can be advantageous to record a second hologram after shifting the biprism by a small amount. The difference between these recordings is then used for the subsequent steps.

63.6.7 Ultrafast Holography

The rapidly expanding subject of ultrafast electron microscopy, pioneered by Oleg Bostanjoglo, was presented in Section 37.5 of volume 2. Attempts are now being made to extend this to holography. The principal difficulty arises from the small currents produced by laser-driven cold field-emission guns. The acquisition time for a hologram with adequate contrast is then long and instabilities will have a seriously adverse effect. These problems are discussed in detail by Houdellier et al. (2019), who show that the hologram quality can be improved by the image stacking technique introduced by Gatel et al. (2018) and Boureau et al. (2018). Individual holograms are acquired, using femtosecond pulses in Houdellier's experiments but, before summing, each hologram is modified to remove 'dead' pixels, which have abnormal intensities and, above all, to correct for fringe-drift arising from various forms of instability.

We draw attention to a related project to study strain dynamics in ultrafast transmission electron microscopy (Feist et al., 2018). Here, for the purposes of nanophononics, convergent-beam diffraction patterns from the edge of a single-crystal graphite membrane are generated by femtosecond pulses; from these the time dependence of the deformation can be followed.

63.6.8 Bragg Holography

The structure of nanocrystals can be reconstructed by a technique called *Bragg holography* (Latychevskaia et al., 2021). Here, the nanocrystal is illuminated by a plane wave but the diffraction pattern is imaged with a small defocus (a few micrometres at 300 kV for example). As a result, the Bragg reflected waves are separated in the detector plane and are situated at distances $\Delta\tan(2\theta_B)$ from the optic axis; Δ denotes the defocus and the angles θ_B are given by Bragg's law:

$$\sin\theta_B = \frac{n\lambda}{2d}$$

where d denotes the distance between the diffracting planes and n is an integer. The diffracted waves then interfere with the direct beam and form holograms, referred to as *Bragg images*. The entire interference pattern, incorporating all these Bragg images, is known as a *Bragg hologram*. From this, the amplitude and phase distribution of the wave can be reconstructed in the usual way. This wavefront ψ_B can be propagated back to the exit plane of the crystal and to other planes within it using (Latychevskaia and Fink, 2015)

$$\psi = \mathcal{F}^{-1}\left[\mathcal{F}(\psi_B)\exp\left\{\frac{2\pi iz}{\lambda}\left(1-\lambda^2(q_x^2+q_y^2)\right)\right\}\right]$$

thus providing enough information for three-dimensional reconstruction. Here, z is the distance between the diffraction plane and the plane in question and q_x, q_y are coordinates in the diffraction plane. Latychevskaia et al. explain how the twin image can be excluded and show examples of both biological and inorganic nanocrystals. The reconstruction step may fail if the axial resolution, $2\lambda/(NA)^2$, is not adequate (Latychevskaia, 2019); NA is the numerical aperture:

$$NA = \frac{\text{detector size}}{2L}$$

and L denotes the distance from the sample to the detector. There exist advanced reconstruction methods with which the signals from different planes along the optic axis can nevertheless be separated (Latychevskaia, 2021).

63.6.9 Holography at Very Low Electron Energy

For completeness, we mention holography at beam energies in the electronvolt range (typically 30–250 eV or even lower). Here, electrons from a very small field-emission source pass through or close to a specimen situated only a few micrometres from the emitter. At first, the technique was mostly reserved for biological specimens such as tobacco mosaic virus (Weierstall et al., 1999) and purple membrane (Spence et al., 1994), but it was subsequently employed with inorganic samples. We recall that the mean free path between scattering events increases considerably at very low energies (cf. Section 41.3.2). No lenses are used, the scattered and unscattered waves form an in-line hologram; information about the phase shift introduced by the specimen is then extracted in the usual way. This has been reasonably successful at resolutions of the order of 1–2 nm but serious difficulties remain. For background information, see Stocker et al. (1989), Fink et al. (1990, 1991), Morin et al. (1990, 1996), Kreuzer et al. (1992), Kreuzer (1995), Morin and Gargani (1993), Spence et al. (1993a,b), Fink and Schmid (1994), the chapter by Fink et al. (1995) and Beyer and Gölzhäuser (2010). For proposals for overcoming some of the obstacles arising from the twin-image and so-called biprism effect caused by a small deflection of the beam towards the centre of the specimen, see Latychevskaia et al. (2014).

In a later paper, Latychevskaia et al. (2015) draw attention to an iterative process in which information from a low-energy hologram is combined with that from a coherent diffraction pattern and claim that unambiguous phase retrieval is possible (Longchamp et al., 2013).

The coherent diffraction pattern is obtained in the same instrument but a lens is now inserted between the source and the sample to provide parallel illumination. The authors exploit the fact that the Fourier transform of the hologram is proportional to the complex-valued distribution of the scattered object wave (Latychevskaia et al., 2012). With a test object, Longchamp et al. achieved a resolution of 2 Å after 100 iterations.

We note here that Kasuya et al. (2020) have succeeded in reducing the energy width of a cold field emitter to 0.17–0.26 eV, the exact value depending on the angular current density (0.10–80 μA/sr), by replacing lanthanum hexaboride by cerium hexaboride.

The vast subject of coherent diffractive (lensless) imaging (e.g., Spence, 1992; Spence et al., 1993a,b, 1995), not covered explicitly here though present in several other sections, is surveyed by Spence (2019) and Latychevskaia (2021) and a special issue of *Journal of Optics* [**18** (5), 2016] is devoted to the subject. Lo et al. (2018) explain how dynamic processes can be studied by this technique.

63.6.10 Noise Reduction for Hologram Series

Problems arising from the noise that is inevitably present in low-dose holograms can be alleviated by tensor decomposition (Nomura et al., 2021). Here, the individual holograms of a time series are stacked to form a many-dimensional parallelepiped. In three dimensions the axes are the x- and y-axes of the individual holograms, the third axis being time. This structure is decomposed into a *core tensor* and three *factor matrices* representing the principal components of each of the axes. The core tensor indicates the level of interaction between the different components for each axis. The factor matrices are iterated and a new core tensor is generated. The latter is used to create a less noisy time series. The process is shown schematically in Fig. 63.24. The mathematical steps sketched above are given in full by Nomura et al.

63.7 Propagation and Reconstruction of the Density Matrix

For a full understanding of the behaviour of an electron beam in an electron microscope, including scattering in the specimen and division by a biprism, the density matrix is the appropriate tool. The theory of this matrix is presented by Röder and Lubk (2014), where earlier less complete studies are described (Kohl, 1983; Rose, 1984; Kohl and Rose, 1985; Dudarev et al., 1993; Schattschneider et al., 1999, 2000; Schattschneider and Lichte, 2005; Verbeeck et al., 2005; Garcia de Abajo, 2009, 2010; Rother et al., 2009; Haider et al., 2010; Howie, 2011). Here we retrace the main steps in the theory presented by Röder and Lubk.[7]

[7] There is some overlap between the theory presented here and that in Chapter 79. This is deliberate in order to make the two accounts self-contained.

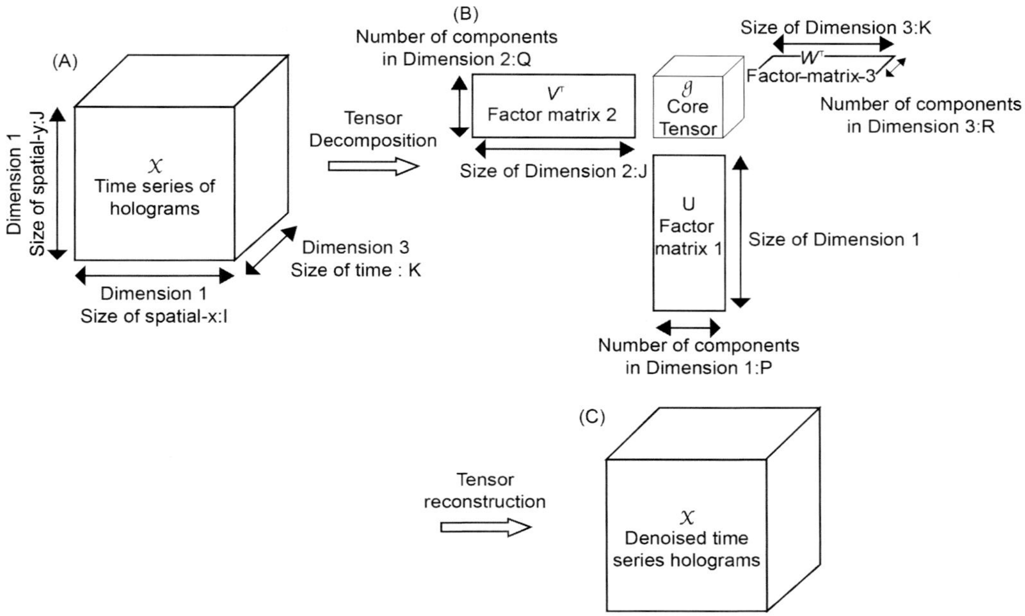

Figure 63.24

Tensor decomposition in three dimensions. (A) Stacked holograms forming a data cube. (B) Core tensor and factor matrices. (C) tensor reconstruction. The dimensions of the initial data cube are I (x-axis, 1), J (y-axis, 2) and K (time axis, 3). The numbers of components are P, Q and R, respectively. *After Nomura et al. (2021), Courtesy Oxford University Press.*

63.7.1 Introduction to the Density Matrix

As we have seen in Chapter 55, the expectation value of the current density is given by

$$\boldsymbol{j} = \frac{\hbar}{2im}\left(\psi_0^*\nabla\psi_0 - \psi_0\nabla\psi_0^*\right) = \frac{\hbar\boldsymbol{k}_0}{m}\left|\psi_0\right|^2 \tag{63.113}$$

At the specimen, the beam electron wave is *entangled* with its internal structure. We denote the ground state of the object by $\tau_o(\boldsymbol{\xi})$ and find

$$\Psi(\boldsymbol{r},\boldsymbol{\xi}) := \psi_0(\boldsymbol{r})\tau_o(\boldsymbol{\xi}) = \sum_i \psi_i(r)\tau_i(\boldsymbol{\xi}) \tag{63.114}$$

in which $\{\tau_i(\boldsymbol{\xi})\}$ form a set of orthonormal eigenfunctions of $\tau(\boldsymbol{\xi})$ and $\psi_i = \langle\tau_i|\psi\rangle$. The probability density of the beam electrons is then given by

$$\rho_s(\boldsymbol{r}) = \sum_{i,j}\psi_i(\boldsymbol{r})\psi_j^*(\boldsymbol{r})\langle\tau_j|\tau_i\rangle = \sum_i\left|\psi_i(\boldsymbol{r})\right|^2 \tag{63.115}$$

which is an (incoherent) sum of intensities. The quantity $\rho_s(\boldsymbol{r})$ is too limited to characterize the states of the beam and the specimen. The remedy is the (reduced) *density matrix*, defined by

$$\rho_s(\boldsymbol{r},\overline{\boldsymbol{r}}) := \sum_i \psi_i(\boldsymbol{r})\psi_i^*(\overline{\boldsymbol{r}}) \tag{63.116}$$

(It is 'reduced' in the sense that integration over all degrees of freedom of the object is implicit.) The diagonal elements of the density matrix are properties of a pure or mixed state while the off-diagonal elements measure the coherence of the beam electrons. The result of scattering, elastic or inelastic, on the incident electrons will be referred to as *state coherence* (suffix *s*). We shall also encounter *ensemble coherence*, in which energy spread and angular spread of the beam are incorporated. The density matrix is a representation of the density operator $\hat{\rho}$ that describes classical as well as quantum statistical properties of the beam. In the object plane, for example, we have $\rho(\boldsymbol{r},\boldsymbol{r}') = \langle \boldsymbol{r}|\hat{\rho}|\boldsymbol{r}'\rangle$. Hence

$$\begin{aligned}\rho(\boldsymbol{r},\overline{\boldsymbol{r}}) &= \int \langle \boldsymbol{r}|\boldsymbol{q}\rangle\langle \boldsymbol{q}|\hat{\rho}|\overline{\boldsymbol{q}}\rangle\langle \overline{\boldsymbol{q}}|\overline{\boldsymbol{r}}\rangle d\boldsymbol{q}d\overline{\boldsymbol{q}} \\ &= \int \rho(\boldsymbol{q},\overline{\boldsymbol{q}})\exp\{2\pi i(\boldsymbol{q}\cdot\boldsymbol{r}-\overline{\boldsymbol{q}}\cdot\overline{\boldsymbol{r}}\}d\boldsymbol{q}d\overline{\boldsymbol{q}} \\ &= \overline{\mathcal{F}}^{-1}\rho(\boldsymbol{q},\overline{\boldsymbol{q}})\end{aligned} \tag{63.117}$$

Note that this is the *definition* of the (inverse) Fourier transform $\left(\overline{\mathcal{F}},\overline{\mathcal{F}}^{-1}\right)$ that will be used in this section.

We now examine the effect of energy spread and source size on the state coherence at successive stages of the image-forming process in a transmission electron microscope equipped with a biprism.

63.7.2 The Electron Gun

Since the source plane is conjugate to the back-focal plane of the objective lens, we use the spatial frequency vectors $\boldsymbol{k},\overline{\boldsymbol{k}}$ as coordinates in the source plane. The current from each point in the emissive area is assumed to have the same energy spectrum, $f_C(E)$; the intensity is denoted $f_s(\boldsymbol{k})$. The elementary emitters are assumed to be mutually incoherent. The density matrix of the beam emitted by the source is then

$$\tilde{\rho}_Q(\boldsymbol{k},\overline{\boldsymbol{k}},E) = f_C(E)f_s(\boldsymbol{k})\delta(\boldsymbol{k}-\overline{\boldsymbol{k}}) \tag{63.118}$$

In the entrance plane of the object, this becomes

$$\rho_Q(\boldsymbol{r},\overline{\boldsymbol{r}},E) = \overline{\mathcal{F}}^{-1}\tilde{\rho}_Q(\boldsymbol{k},\overline{\boldsymbol{k}},E) = f_C(E)\tilde{f}_s(\boldsymbol{r}-\overline{\boldsymbol{r}}) \tag{63.119}$$

where $\tilde{f}_s$ denotes the inverse Fourier transform of f_s.

The degree of spatial coherence, $\mu(\boldsymbol{d})$, is given by

$$\mu(\boldsymbol{d}) = \frac{\int \rho_Q(\boldsymbol{d}, E)dE}{\int \rho_Q(0, E)dE} = f_s(\boldsymbol{d}), \quad \boldsymbol{d} = \boldsymbol{r} - \bar{\boldsymbol{r}} \tag{63.120}$$

(the van Cittert–Zernike theorem, see Section 78.5).

63.7.3 The Specimen

We make the convenient but not necessarily correct assumption that each electron in the beam incident on the specimen is scattered in the same way, leading to $\rho_s(\boldsymbol{r}, \bar{\boldsymbol{r}}, E)$. At the exit plane of the specimen, therefore, we find

$$\begin{aligned}\rho_o(\boldsymbol{r}, \bar{\boldsymbol{r}}, E) &= \int \exp\{2\pi i(\boldsymbol{q}\cdot\boldsymbol{r} - \bar{\boldsymbol{q}}\cdot\bar{\boldsymbol{r}})\} f_s(\boldsymbol{k}) f_C(E_C)\rho_s(\boldsymbol{q}-\boldsymbol{k}, \bar{\boldsymbol{q}}-\boldsymbol{k}, E - E_c)\, d\boldsymbol{k} d\, E_c\, d\boldsymbol{q} d\bar{\boldsymbol{q}} \\ &= \int \rho_Q(\boldsymbol{r} - \bar{\boldsymbol{r}}, E_C)\rho_s(\boldsymbol{r}, \bar{\boldsymbol{r}}, E - E_c) dE_C\end{aligned} \tag{63.121}$$

E_C is a measure of energy spread and E represents any energy loss associated with inelastic scattering events.

63.7.4 The Objective Lens

The aberrations of the objective lens are characterized by the transfer function, which now takes the form

$$X(\boldsymbol{q}, \bar{\boldsymbol{q}}, E) = \exp\left[-i\{\chi(\boldsymbol{q}, E) - \chi(\bar{\boldsymbol{q}}, E)\}\right] \tag{63.122}$$

The density matrix in the back-focal plane is multiplied by this function, giving

$$\rho(\boldsymbol{q}, \bar{\boldsymbol{q}}) = \int \rho(\boldsymbol{q}-\boldsymbol{k}, \bar{\boldsymbol{q}}-\boldsymbol{k}, E) f_s(\boldsymbol{k}) d\boldsymbol{k} \int f_C(E_C) X(\boldsymbol{q}, \bar{\boldsymbol{q}}, E + E_C) dE_c dE \tag{63.123}$$

in which the integration has been taken over the range of energy losses. In the back-focal plane the intensity distribution is given by

$$I(\boldsymbol{q}) = \int \rho_s(\boldsymbol{q} - \boldsymbol{q}_0, \boldsymbol{q} - q_0, E) f_s(\boldsymbol{q}_0) d\boldsymbol{q}_0 dE \tag{63.124}$$

63.7.5 The Image Plane

A further Fourier transform yields the density matrix in the image plane:

$$\rho_i(\boldsymbol{r}, \bar{\boldsymbol{r}}) = \int \exp\{2\pi i(\boldsymbol{q}\cdot\boldsymbol{r} - \bar{\boldsymbol{q}}\cdot\bar{\boldsymbol{r}})\} X(\boldsymbol{q}, \bar{\boldsymbol{q}}, E + E_C)\rho_s(\boldsymbol{q}-\boldsymbol{k}, \bar{\boldsymbol{q}}-\boldsymbol{k}, E) f_s(\boldsymbol{k}) f_C(E_C)\, d\boldsymbol{q} d\bar{\boldsymbol{q}} d\boldsymbol{k} dE dE_C \tag{63.125}$$

On writing $\boldsymbol{q}_k := \boldsymbol{q} - \boldsymbol{k}$ *and* $\overline{\boldsymbol{q}}_k := \overline{\boldsymbol{q}} - \boldsymbol{k}$, this becomes

$$\rho_i(\boldsymbol{r},\overline{\boldsymbol{r}}) = \int \exp\{2\pi i(\boldsymbol{q}_k \cdot \boldsymbol{r} - \overline{\boldsymbol{q}}_k \cdot \overline{\boldsymbol{r}})\} T(\boldsymbol{q}_k,\overline{\boldsymbol{q}}_k,\boldsymbol{r}-\overline{\boldsymbol{r}},E)\rho_s(\boldsymbol{q}_k,\overline{\boldsymbol{q}}_k,E)d\boldsymbol{q}_k d\overline{\boldsymbol{q}}_k dE \tag{63.126}$$

in which we have introduced the transmission cross-coefficient T

$$T\left(\boldsymbol{q}_k,\overline{\boldsymbol{q}}_k,\boldsymbol{d},E\right) := \int \exp(2\pi i\boldsymbol{k}\cdot\boldsymbol{d}) X\left(\boldsymbol{q}_k+\boldsymbol{k},\overline{\boldsymbol{q}}_k+\boldsymbol{k},E+E_C\right) f_s(\boldsymbol{k}) f_C(E_C) d\boldsymbol{k} dE_C \tag{63.127}$$

This very general form of the transmission cross-coefficient, which includes the dependence on $\boldsymbol{d} = \boldsymbol{r} - \overline{\boldsymbol{r}}$, enables us to calculate the effect of the spatial partial coherence of the incident beam on all the elements of the density matrix at the image plane.

The image recorded in the image plane of a transmission microscope is given by the diagonal elements of (63.126), $\rho_i(\boldsymbol{r},\boldsymbol{r})$:

$$I_i(\boldsymbol{r}) = \int \exp\{2\pi i(\boldsymbol{q}_k - \overline{\boldsymbol{q}}_k)\cdot\boldsymbol{r}\} T(\boldsymbol{q}_k,\overline{\boldsymbol{q}}_k,0,E)\rho_s(\boldsymbol{q}_k,\overline{\boldsymbol{q}}_k,E)d\boldsymbol{q}_k d\overline{\boldsymbol{q}}_k dE \tag{63.128}$$

As expected, the Fourier transform of this image intensity distribution

$$\tilde{I}_i(\boldsymbol{q}) = \int T(\boldsymbol{q}_k+\overline{\boldsymbol{q}}_k,\overline{\boldsymbol{q}}_k,0,E)\rho_s(\boldsymbol{q}_k+\overline{\boldsymbol{q}}_k,\overline{\boldsymbol{q}}_k,E)d\overline{\boldsymbol{q}}_k dE \tag{63.129}$$

involves the product of the cross-correlation of the density matrix and the transmission cross-coefficient.

(In what follows, we have dropped the suffix k on $\boldsymbol{q}$ and $\overline{\boldsymbol{q}}$.)

63.7.6 The Biprism

The biprism is placed between the back-focal plane of the objective lens and the first image plane (conjugate to the specimen), as shown in Fig. 63.25. Following Verbeeck et al. (2005, 2008) and Schattschneider and Verbeeck (2008), the effect of the biprism is represented by an operator $B(\boldsymbol{q},\overline{\boldsymbol{q}})$. This operator modulates the density matrix of the electron beam as follows:

$$B(\boldsymbol{q},\overline{\boldsymbol{q}}) = \int S\left(\frac{\boldsymbol{q}_c\cdot\boldsymbol{r}}{q_c}\right) S^*\left(\frac{\boldsymbol{q}_c\cdot\overline{\boldsymbol{r}}}{q_c}\right)\exp\{-i\pi(|\boldsymbol{q}_c\cdot\boldsymbol{r}| - |\boldsymbol{q}_c\cdot\overline{\boldsymbol{r}}|)\}\exp\{-2\pi i(\boldsymbol{q}\cdot\boldsymbol{r} - \overline{\boldsymbol{q}}\cdot\overline{\boldsymbol{r}})\} d\boldsymbol{r} d\overline{\boldsymbol{r}} \tag{63.130}$$

The function S represents the shadow cast by the biprism filament, diameter D:

$$S(x) = \begin{cases} 0 & |x| < D/2 \\ 1 & |x| > D/2 \end{cases} \tag{63.131}$$

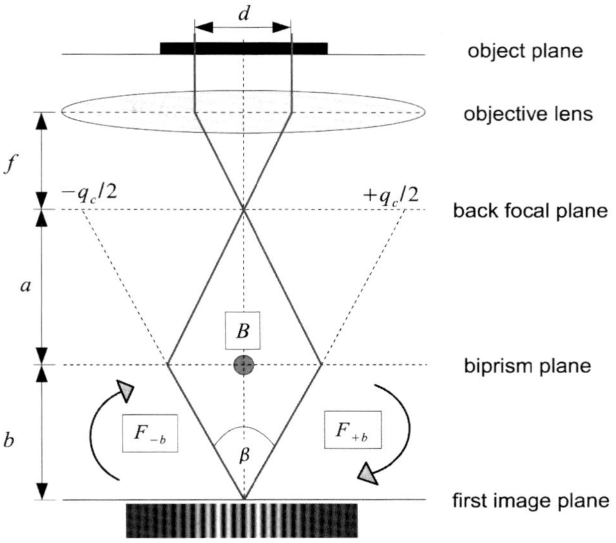

Figure 63.25
Notation used in connection with the density matrix. *After Röder and Lubk (2014), Courtesy Elsevier.*

The carrier frequency q_c is related to the angle β in Fig. 63.25 by $q_c = \beta k_0$. After some calculation, the density matrix in the image (hologram) plane is found to be

$$\begin{aligned}
\rho_h(\boldsymbol{r},\bar{\boldsymbol{r}}) &= \int \left(F^- d\boldsymbol{q}_1 d\bar{\boldsymbol{q}}_1\right) F^+ d\boldsymbol{q}_2 d\bar{\boldsymbol{q}}_2 \\
F^- &:= \rho_i\left(\boldsymbol{q}_1,\bar{\boldsymbol{q}}_1\right) F_{-b}\left(\boldsymbol{q}_1,\bar{\boldsymbol{q}}_1\right) B\left(\boldsymbol{q}_2-\boldsymbol{q}_1,\bar{\boldsymbol{q}}_2-\bar{\boldsymbol{q}}_1\right) \\
F^+ &:= F_b\left(\boldsymbol{q}_2,\bar{\boldsymbol{q}}_2\right)\exp\left\{2\pi\mathrm{i}\left(\boldsymbol{q}_2\cdot\boldsymbol{r}-\bar{\boldsymbol{q}}_2\cdot\bar{\boldsymbol{r}}\right)\right\} \\
F_b(\boldsymbol{q},\bar{\boldsymbol{q}}) &:= \exp\left\{-\mathrm{i}\pi\frac{b_o}{k_o}\left(q^2-q^2\right)\right\} \\
b_o &:= \frac{bf^2}{a(a+b)}
\end{aligned} \tag{63.132}$$

where a, b and f are also shown in Fig. 63.25. The diagonal elements of the density matrix correspond to the central zone of the hologram and the off-diagonal elements to the sidebands. On substituting (63.121) for ρ_i in (63.127), we find

$$\begin{aligned}
I_h(\boldsymbol{r}) = \rho_h(\boldsymbol{r},\boldsymbol{r}) &= \int F_r^- F_r^+ d\boldsymbol{q}_1 d\bar{\boldsymbol{q}}_1 d\boldsymbol{q}_2 d\bar{\boldsymbol{q}}_2 dE \\
F_r^- &:= \rho_s(\boldsymbol{q}_1,\boldsymbol{q}_1,E) T_h(\boldsymbol{q}_1,\bar{\boldsymbol{q}}_1,\boldsymbol{q}_2-\bar{\boldsymbol{q}}_2,E) F_{-b}(\boldsymbol{q}_1,\bar{\boldsymbol{q}}_1) B(\boldsymbol{q}_2-\boldsymbol{q}_1,\bar{\boldsymbol{q}}_2-\bar{\boldsymbol{q}}_1) \\
F_r^+ &:= F_b(\boldsymbol{q}_2,\bar{\boldsymbol{q}}_2)\exp\left\{2\pi\mathrm{i}\left(\boldsymbol{q}_2-\bar{\boldsymbol{q}}_2\right)\cdot\boldsymbol{r}\right\}
\end{aligned} \tag{63.133}$$

In this formula, T_h is the *holographic transfer cross-coefficient*, which encodes the effects of partial spatial and temporal partial coherence. When $\boldsymbol{q}_2 - \overline{\boldsymbol{q}}_2 = \boldsymbol{q}_1 - \overline{\boldsymbol{q}}_1 - \boldsymbol{q}_c$, the holographic transfer cross-coefficient can be expressed in terms of the generalized coefficient (63.127):

$$T_h(\boldsymbol{q}_1, \overline{\boldsymbol{q}}_1, \boldsymbol{q}_1 - \overline{\boldsymbol{q}}_1 - \boldsymbol{q}_c, E) = T(\boldsymbol{q}_1, \overline{\boldsymbol{q}}_1, \boldsymbol{d}, E) \qquad \boldsymbol{q}_c = \frac{k_o \boldsymbol{d}}{b_o} \tag{63.134}$$

It is convenient to re-write (63.133) in terms of the central and sideband contributions, For this, we replace the definition of $B(\boldsymbol{q}, \boldsymbol{q}')$ by the product

$$B(\boldsymbol{q}, \overline{\boldsymbol{q}}) = \{S_+(\boldsymbol{q} + \boldsymbol{q}_c/2) + S_-(\boldsymbol{q} - \boldsymbol{q}_c/2)\}\{S_-^*(\overline{\boldsymbol{q}} - \boldsymbol{q}_c/2) + S_+^*(\overline{\boldsymbol{q}} + \boldsymbol{q}_c/2)\} \tag{63.135}$$

When the biprism filament is parallel to the y-axis, the functions $S_\pm(\boldsymbol{q})$ collapse to

$$S_\pm(\boldsymbol{q}) = \frac{1}{2}\left\{\delta(q_x) \pm \frac{1}{\mathrm{i}\pi q_x}\exp(\mp \pi \mathrm{i} q_x D_o)\right\}\delta(q_y) \tag{63.136}$$

and $I_h(\boldsymbol{r})$ can be written

$$\begin{aligned} I_h(\boldsymbol{r}) &= \rho^{++}(\boldsymbol{r} + \boldsymbol{d}/2, \boldsymbol{r} + \boldsymbol{d}/2) + \rho^{--}(\boldsymbol{r} - \boldsymbol{d}/2, \boldsymbol{r} - \boldsymbol{d}/2) && \text{central band} \\ &+ \rho^{+-}(\boldsymbol{r} + \boldsymbol{d}/2, \boldsymbol{r} - \boldsymbol{d}/2)\exp(-2\pi \mathrm{i}\boldsymbol{q}_c \cdot \boldsymbol{r}) && \text{sideband} \\ &+ \rho^{-+}(\boldsymbol{r} - \boldsymbol{d}/2, \boldsymbol{r} + \boldsymbol{d}/2)\exp(2\pi \mathrm{i}\boldsymbol{q}_c \cdot \boldsymbol{r}) && \text{sideband} \end{aligned} \tag{63.137}$$

The four density matrices are now given by

$$\begin{aligned} \rho^{++}(\boldsymbol{r} + \boldsymbol{d}/2, \boldsymbol{r} + \boldsymbol{d}/2) = &\int \rho_s(\boldsymbol{q}_1, \boldsymbol{q}_1, E) T_h(\boldsymbol{q}_1, \boldsymbol{q}_1, \boldsymbol{q}_2 - \boldsymbol{q}_2, E) \\ &\times F_{-b}(\boldsymbol{q}_1, \overline{\boldsymbol{q}}_1) S_+(\boldsymbol{q}_2 - \boldsymbol{q}_1) S_+^*(\overline{\boldsymbol{q}}_2 - \overline{\boldsymbol{q}}) F_b(\boldsymbol{q}_2, \overline{\boldsymbol{q}}_2) \\ &\times \exp\{2\pi \mathrm{i}(\overline{\boldsymbol{q}}_2 \cdot (\boldsymbol{r} + \boldsymbol{d}/2) - \overline{\boldsymbol{q}}_2 \cdot (\boldsymbol{r} + \boldsymbol{d}/2))\} d\boldsymbol{q}_1 d\overline{\boldsymbol{q}}_1 d\boldsymbol{q}_2 d\overline{\boldsymbol{q}}_2 dE \end{aligned} \tag{63.138}$$

$$\begin{aligned} \rho^{--}(\boldsymbol{r} - \boldsymbol{d}/2, \boldsymbol{r} - \boldsymbol{d}/2) = &\int \rho_s(\boldsymbol{q}_1, \boldsymbol{q}_1, E) T_h(\boldsymbol{q}_1, \boldsymbol{q}_1, \boldsymbol{q}_2 - \boldsymbol{q}_2, E) \\ &\times F_{-b}(\boldsymbol{q}_1, \overline{\boldsymbol{q}}_1) S_-(\boldsymbol{q}_2 - \boldsymbol{q}_1) S_-^*(\overline{\boldsymbol{q}}_2 - \overline{\boldsymbol{q}}_1) F_b(\boldsymbol{q}_2, \boldsymbol{q}_2') \\ &\times \exp\{2\pi \mathrm{i}(\boldsymbol{q}_2 \cdot (\boldsymbol{r} - \boldsymbol{d}/2) - \overline{\boldsymbol{q}}_2 \cdot (\boldsymbol{r} - \boldsymbol{d}/2))\} d\boldsymbol{q}_1 d\overline{\boldsymbol{q}}_1 d\boldsymbol{q}_2 d\overline{\boldsymbol{q}}_2 dE \end{aligned}$$

$$\begin{aligned} \rho^{+-}(\boldsymbol{r} + \boldsymbol{d}/2, \boldsymbol{r} - \boldsymbol{d}/2) = &\int \rho_s(\boldsymbol{q}_1, \boldsymbol{q}_1, E) T_h(\boldsymbol{q}_1, \boldsymbol{q}_1, \boldsymbol{q}_2 - \boldsymbol{q}_2 - \boldsymbol{q}_c, E) \\ &\times F_{-b}(\boldsymbol{q}_1, \overline{\boldsymbol{q}}_1) S_+(\boldsymbol{q}_2 - \boldsymbol{q}_1) S_-^*(\overline{\boldsymbol{q}}_2 - \overline{\boldsymbol{q}}_1) F_b(\boldsymbol{q}_2, \overline{\boldsymbol{q}}_2) \\ &\times \exp\{2\pi \mathrm{i}(\boldsymbol{q}_2 \cdot (\boldsymbol{r} + \boldsymbol{d}/2) - \overline{\boldsymbol{q}}_2 \cdot (\boldsymbol{r} - \boldsymbol{d}/2))\} d\boldsymbol{q}_1 d\overline{\boldsymbol{q}}_1 d\boldsymbol{q}_2 d\overline{\boldsymbol{q}}_2 dE \end{aligned}$$

$$\begin{aligned} \rho^{-+}(\boldsymbol{r} - \boldsymbol{d}/2, \boldsymbol{r} + \boldsymbol{d}/2) = &\int \rho_s(\boldsymbol{q}_1, \overline{\boldsymbol{q}}_1, E) T_h(\boldsymbol{q}_1, \overline{\boldsymbol{q}}_1, \boldsymbol{q}_2 - \overline{\boldsymbol{q}}_2 + \boldsymbol{q}_c, E) \\ &\times F_{-b}(\boldsymbol{q}_1, \overline{\boldsymbol{q}}_1) S_-(\boldsymbol{q}_2 - \boldsymbol{q}_1) S_+^*(\overline{\boldsymbol{q}}_2 - \overline{\boldsymbol{q}}_1) F_b(\boldsymbol{q}_2, \overline{\boldsymbol{q}}_2) \\ &\times \exp\{2\pi \mathrm{i}(\boldsymbol{q}_2 \cdot (\boldsymbol{r} - \boldsymbol{d}/2) - \overline{\boldsymbol{q}}_2 \cdot (\boldsymbol{r} + \boldsymbol{d}/2))\} d\boldsymbol{q}_1 d\overline{\boldsymbol{q}}_1 d\boldsymbol{q}_2 d\overline{\boldsymbol{q}}_2 dE \end{aligned}$$

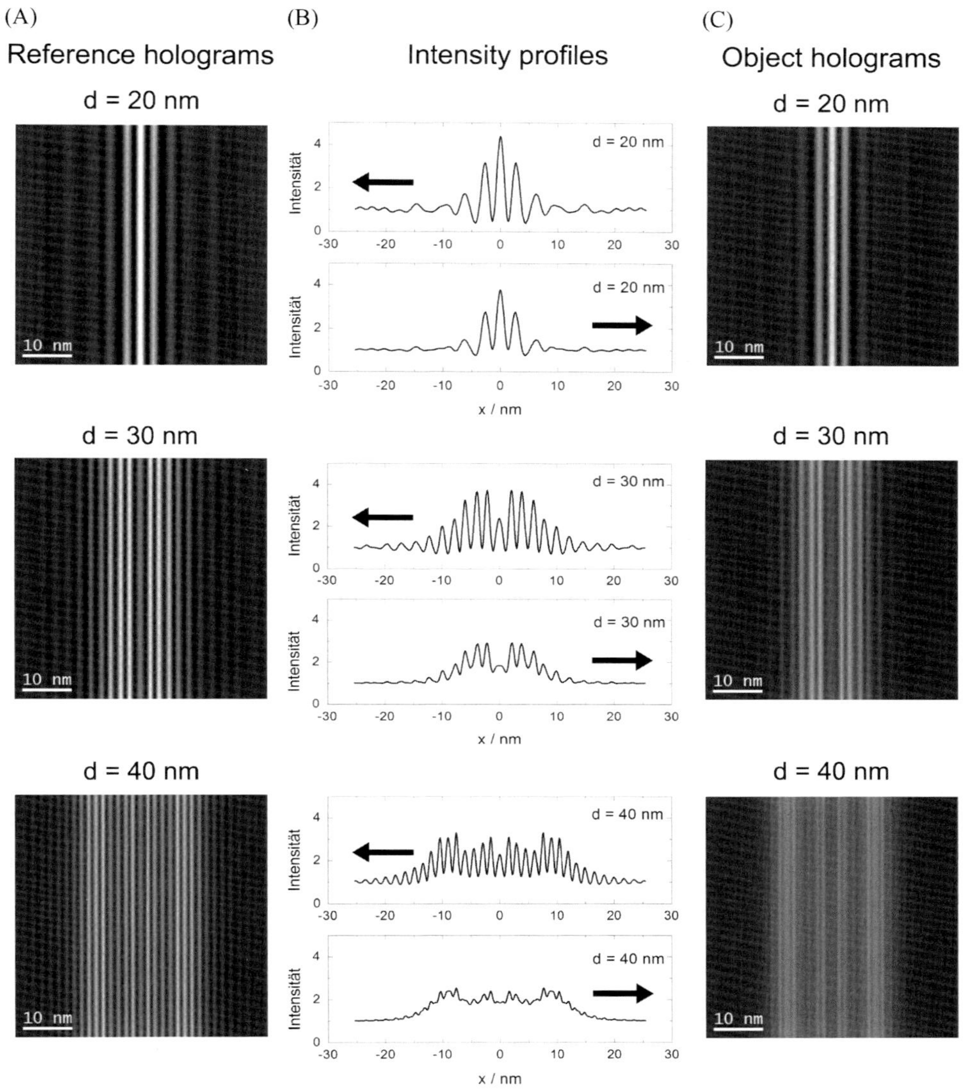

Figure 63.26
Fringe patterns calculated using the exact transfer theory (A) in the absence of any object and (C) with a homogeneous object. The corresponding profiles are shown in (B). *After Röder and Lubk (2014), Courtesy Elsevier.*

This completes the theory of information transfer into the centre and sidebands of the hologram. Röder and Lubk then describe the reconstruction process and illustrate the foregoing work with several examples (Fig. 63.26).

PART XIII

Theory of Image Formation

CHAPTER 64

General Introduction

This part is concerned with the theory of image formation in the types of transmission electron microscope (TEM) capable of very high-resolution imaging, the (conventional) transmission electron microscope (TEM, or CTEM if it is necessary to stress the fact that the conventional instrument is meant) and the scanning transmission electron microscope (STEM). Scanning electron microscopes (SEMs) of the latest generation are also capable of providing high-resolution information but the image-forming processes are very different. In this introductory chapter, we describe the principal features of these various types of microscope, limiting the account to those that affect image formation directly. In practice, a single instrument may be capable of operating in several modes: TEMs may be designed to operate as either conventional or scanning instruments; a 'dedicated' STEM (which does not provide TEM images) may furnish some of the signals routinely collected by an SEM. We disregard these aspects of instrument design.

A CTEM consists of a source, the electron gun; condenser lenses to direct the beam onto the specimen and to control the area illuminated and the angles at which the electrons are incident; an objective lens to provide the first stage of magnification; projector lenses to provide further magnification and an electron detector, connected to an image storage and processing unit (Fig. 64.1). Today, direct detectors (see Section 67.5) have replaced the fluorescent screen and photographic plates or film of the past. (Microscopists, notably in such fields as cell biology, lament that the quality of printed micrographs has deteriorated as a regrettable by-product, Ackermann, 2013, 2014). Apertures are included at several levels: in the condenser system, in the back-focal plane of the objective lens and in the plane of the first intermediate image. An electron energy-loss spectrometry unit (EELS) may be fitted below the image plane and recent instruments often include a monochromator, an aberration corrector and a biprism. There are other elements that do not concern us here (notably alignment coils and stigmators).

Two modes of operation are of everyday interest in TEM and a number of others are important in special circumstances. The two common modes are illustrated in Fig. 64.1. In bright-field imagery, a small source provides a beam of electrons, which

Principles of Electron Optics.
DOI: https://doi.org/10.1016/B978-0-12-818979-5.00064-4

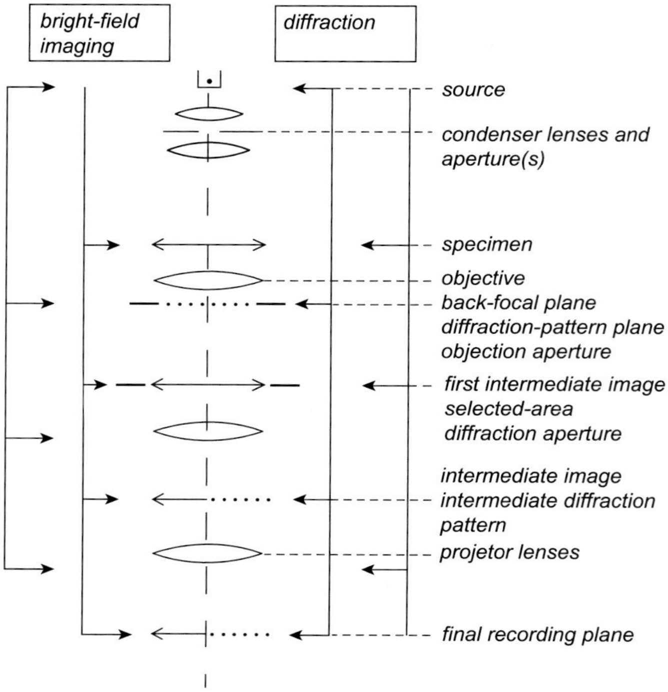

Figure 64.1
Conjugate planes in a (conventional, fixed-beam) transmission electron microscope (TEM) in the bright-field imaging mode and in the diffraction mode. Sets of conjugate planes are indicated by arrows. The condenser aperture may be regarded as conjugate to the specimen though in practice this is not always the case.

are directed onto the specimen by condenser lenses; frequently the object is immersed in the magnetic field of the objective, the prefield of which thus acts as a last condenser lens. These condensers and the associated apertures ensure that the beam covers the desired area of the specimen and that the electrons arrive with the appropriate angular spread. In much of what follows, we shall assume that the condensers have been adjusted so that the electrons are travelling parallel to the optic axis when they strike the specimen and hence correspond to a (truncated) plane incident wave. In fact, this would be possible (neglecting condenser lens aberrations) only if the source size were vanishingly small. From Eq. (16.12) of Volume 1 with $z_1 = z_{Fo}$, we see that $x_2' = -x_1/f$, where x_1 is the source width and x_2' is the corresponding beam tilt at the specimen plane; f is the focal length of the complete condenser system.

If the beam is indeed parallel to the axis when it strikes the specimen,[1] the source will be conjugate to the back-focal plane of the objective lens and to planes close to the back-focal planes of the intermediate and projector lenses if these are operating at relatively high magnification. The specimen will be conjugate to a sequence of planes beyond the objective, culminating in the final image plane, the recording plane. The first intermediate image lies in the plane of the selected-area diffraction aperture and, if all lenses are operating at relatively high magnification, these intermediate image planes will not be far from the object focal planes of successive lenses. The object plane is also frequently but not always conjugate to a condenser aperture.

In the standard diffraction mode, the strengths of the condenser and objective lenses are unaltered so that nothing in Fig. 64.1 changes down to the first intermediate image. The intermediate and projector lenses are now adjusted so that it is not this intermediate image but the objective back-focal plane that is conjugate to the final recording plane. The roles of the various planes downstream are thus exchanged. (Depending on the way in which the projector lens currents are changed, these planes may be shifted as well.)

On the basis of these simple diagrams, many other modes may be easily understood. Convergent-beam diffraction, for example, is achieved by focusing the source on the specimen. The tilted-illumination modes are obtained with the aid of coils in the space upstream from the specimen. These can be used to illuminate the object with a plane tilted beam or with a hollow-cone beam, in which the tilt angle is kept constant but the azimuth is changed so that the beam sweeps out a hollow conical volume. Tilted illumination is employed in crystal imaging and in principle improves bright-field resolution anisotropically; hollow-cone illumination should provide the same improvement isotropically but, in both cases, the gain is offset by a loss of absolute contrast. We return to this in Section 66.5. The various dark-field modes are obtained by preventing the electrons unscattered in the specimen from reaching the image. This is usually achieved by tilting the beam incident on the specimen so that the unscattered beam is intercepted by the objective aperture or by displacing the latter so that only selected scattered electrons, typically those concentrated in diffraction spots when the specimen is crystalline, reach the image. The image contrast can be improved by inserting a phase plate, analogous to the Zernike phase plate used in light microscopy, in the diffraction plane (or a plane conjugate to it). Section 66.6 is devoted to such plates.

[1] This is known as Köhler illumination (Köhler, 1893, 1899) and its merits for electron microscopy have been underlined by Probst et al. (1991a, 1991b) and Benner et al. (1990, 1991). It cannot, however, be achieved exactly when the specimen is immersed in the field of the objective (Christenson and Eades, 1986).

The STEM is a microscope in which a small probe explores the specimen and information collected by the beam electrons as they traverse each small picture-element (pixel) of the specimen is extracted by various detectors downstream from the latter. In the basic instrument, two modes of operation are routinely used but the situation is more complex than for the TEM. The microscope consists of a source (in practice only a field-emission gun is suitable); condenser lenses; sets of scanning coils; the probe-forming lens, which is commonly referred to as the objective; intermediate lenses; detectors and an electron energy analysis unit with further detectors. There are in addition alignment coils, stigmators and apertures. In both the TEM and the STEM, the column may contain an aberration corrector, a monochromator and a phase-plate holder.

In the standard imaging mode (Fig. 64.2), the source is demagnified by the condenser and objective lenses, so that crossover, a selected-area aperture and specimen are conjugates.

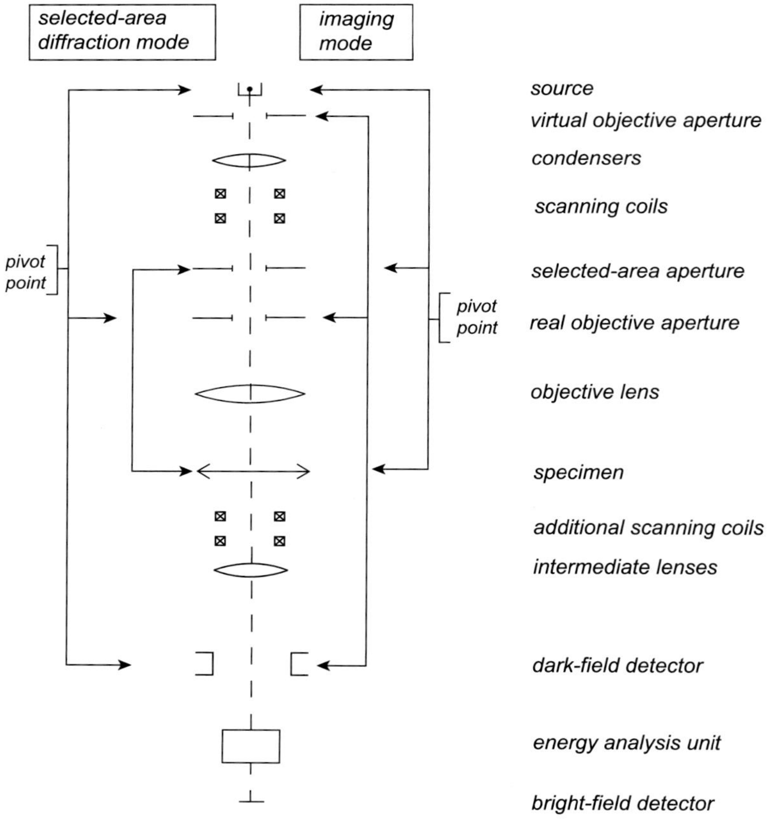

Figure 64.2
Conjugate planes in a scanning transmission electron microscope (STEM) in the bright-field imaging mode and in the diffraction mode. Sets of conjugate planes are indicated by arrows.

The pivot point of the scanning coils coincides with the front-focal plane of the objective so that, as the beam is rocked, the probe explores the specimen. This plane may also contain a real objective aperture or the latter may be replaced by a 'virtual' objective aperture in a conjugate plane closer to the source. Both of these planes will be conjugate to the plane of the dark-field detector.

In the selected-area diffraction mode, the source is no longer conjugate to the specimen but is instead conjugate to the real objective aperture (front-focal plane of the objective). By raising the pivot point of the scanning coils to the selected-area aperture plane, which is conjugate to the specimen, a parallel beam will be rocked about the axis while illuminating an (approximately) fixed area of the specimen.

The various images are thus formed sequentially and are of different kinds. Owing to the general rule that inelastically scattered electrons are deflected, on average, through smaller angles than elastically scattered electrons (see Part XIV), a crude but often very effective separation is performed by the annular dark-field detector in the imaging mode; the central opening is chosen to allow the electrons undeflected by the specimen to pass through, together with a substantial fraction of those scattered inelastically. The dark-field detector then collects mostly electrons that have been scattered elastically by each pixel in turn; the resulting dark-field signal is used to modulate the intensity of a monitor scanned in synchronism with the STEM. The electrons that pass through the opening in the dark-field detector are dispersed by a prism and can then be used to form energy-filtered images or for EELS. If a very small bright-field detector is employed, the image formation will prove to be closely analogous to that in the TEM. We shall also see that the fact that the detector geometry can be chosen freely offers new and interesting possibilities.

We note, without further comment, that the design of the lens system between the specimen and the detectors is very important, both to match the information emerging from the object to the detector geometry and, more recently, to enable the electrons scattered through large angles to be collected and exploited efficiently.

We conclude this short chapter with a reminder that the SEM, like the STEM, produces the information that is used to generate its numerous images sequentially. We shall not need to refer to its optics, which is in any case basically very simple; the problems for the designer arise from the difficulty of situating so many detectors in so confined a space. We shall, however, often have occasion to mention image processing techniques that are of particular interest for the SEM in Part XV.

Very high-resolution imaging is a major theme of the following chapters. Fig. 64.3 shows the situation in 2021.

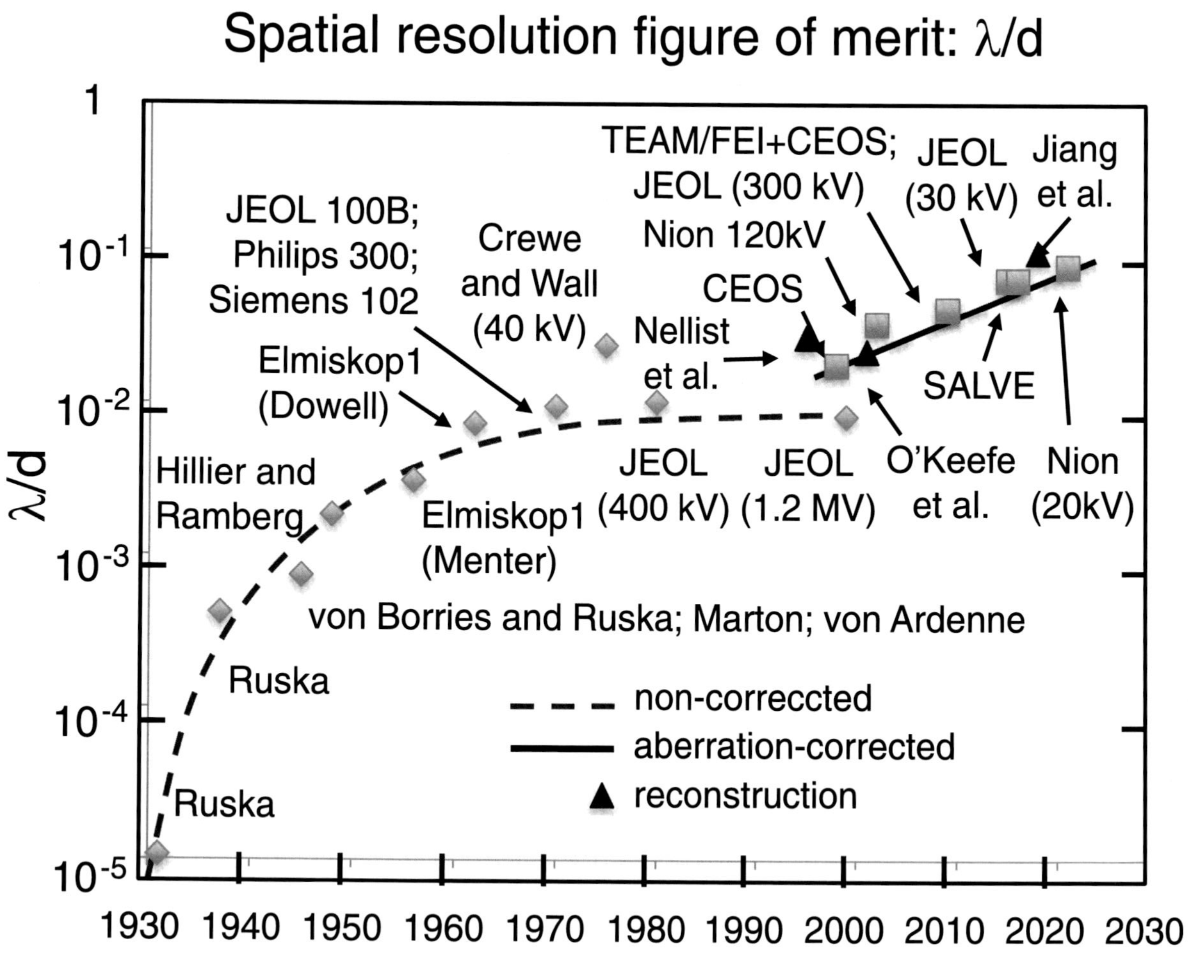

Figure 64.3

The history of resolution. The ordinate shows wavelength/spatial resolution. *After Krivanek (2021), Courtesy Cambridge University Press.*

CHAPTER 65

Fundamentals of Transfer Theory

In earlier chapters, we have developed the tools that are needed to construct a theory of image formation: a wave equation that describes electron propagation and the corresponding diffraction theory. We now reconsider electron wave propagation from a rather different standpoint. Throughout this chapter, we assume that the waves are monoenergetic; the effect of abandoning this assumption will be examined in Chapter 66.

65.1 The Integral Transformation

The most important feature of the wave equation in all its various forms is its *linearity*. This implies that, for any pair of particular solutions $\psi_1(\boldsymbol{r})$ and $\psi_2(\boldsymbol{r})$, the linear combination

$$\psi(\boldsymbol{r}) = c_1\psi_1(\boldsymbol{r}) + c_2\psi_2(\boldsymbol{r}) \tag{65.1a}$$

with arbitrary complex coefficients c_1 and c_2 is also a solution. This *superposition principle* can be cast into the following more general forms:

$$\psi(\boldsymbol{r}) = \sum_{n=1}^{N} c_n\psi_n(\boldsymbol{r}) \tag{65.1b}$$

$$\psi(\boldsymbol{r}) = \sum_{m=1}^{M}\sum_{n=1}^{N} c_{mn}\psi_{mn}(\boldsymbol{r}) \tag{65.1c}$$

$$\psi(\boldsymbol{r}) = \int_{\alpha_1}^{\alpha_2} c(\alpha)\hat{\psi}(r,\alpha)d\alpha \tag{65.2a}$$

$$\psi(\boldsymbol{r}) = \int_{\alpha_1}^{\alpha_2}\int_{\beta_1}^{\beta_2} c(\alpha,\beta)\hat{\psi}(\boldsymbol{r},\alpha,\beta)d\alpha\, d\beta \tag{65.2b}$$

The variables of integration α and β are parameters with respect to all spatial operations.

These superposition rules all refer to the same position $\boldsymbol{r}$. Our aim now is to find superposition rules that tell us how contributions from different points $\boldsymbol{r}_o$ add up to produce the wavefunction at a point $\boldsymbol{r}_i$ in the image plane.

Principles of Electron Optics.
DOI: https://doi.org/10.1016/B978-0-12-818979-5.00065-6

It proves convenient to approach this problem by dividing the object and image areas into discrete zones, as shown in Fig. 65.1, thereby introducing the first stage of digital image processing, sampling. Consider a square of side $L \times L$ in the object plane, $z = z_o$, and divide it into N^2 identical subsquares of side $h = L/N$. These are labelled with two indices m and n running from 1 to N; the coordinates of the centres of the subsquares are thus

$$x_{mn}^{(0)} = -\frac{L}{2} + \left(m - \frac{1}{2}\right)h, \quad y_{mn}^{(0)} = -\frac{L}{2} + \left(n - \frac{1}{2}\right)h \tag{65.3}$$

Next, we consider wavefunctions $\psi_{mn}(\boldsymbol{r})$ with the following values:

$$\psi_{mn}(x_o, y_o, z_o) = \begin{cases} 1 & \text{for } |x_o - x_{mn}^{(0)}| < h/2, \quad |y_o - y_{mn}^{(0)}| < h/2 \\ 0 & \text{otherwise} \end{cases} \tag{65.4}$$

Each wavefunction is thus normalized to unity within the element of area (m,n) and vanishes in the remainder of the object plane. This could be regarded as the starting condition of a wave propagation problem. Together with a reasonable assumption about the normal derivative,

$$\left(\frac{\partial \psi_{mn}}{\partial z}\right)_{\boldsymbol{r}_o} = \mathrm{i}k_{mn}\psi_{mn}(\boldsymbol{r}_o) \tag{65.5}$$

for example, we have a fully defined problem that could be solved by means of diffraction integrals. This is not, however, the most favourable way, as we shall see.

We can superimpose all these particular solutions, as in (65.1c), with weights

$$c_{mn} = \psi\left(x_{mn}^{(0)}, y_{mn}^{(0)}, z_o\right) \tag{65.6}$$

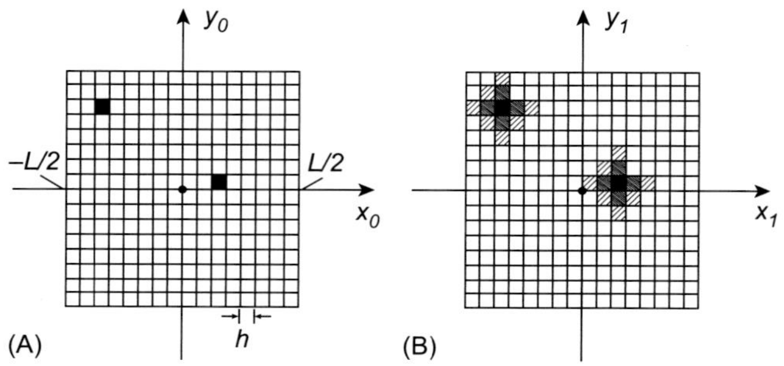

Figure 65.1
Concepts of image processing. The object (A) and the image (B) are divided into a large number (here $N^2 = 256$) of square elements of area (pixels). The image is transformed back to object size and orientation. The information relating to each object pixel is not only stored in the corresponding image pixel but is also spread over its neighbourhood.

and $M=N$, thereby obtaining an approximate solution, which is piecewise constant in the object plane. This is indeed how image processing programs operate; the value of N. is usually a power of 2, typically $N=256=2^8$ or $N=512=2^9$, since this suits both computer architectures and various widely used algorithms

The representation (65.1c) together with (65.6) is an approximate form of a *continuous* superposition of the general form (65.2b), to which it tends if we keep L fixed and let N increase indefinitely. We may then identify α and β with x_o and y_o, $\alpha_1=\beta_1=-L$ and $\alpha_2=\beta_2=L$ are the limits of integration and $c_{mn}\rightarrow\psi(x_o, y_o, z_o)$. The particular solution $\hat{\psi}$ is then the *integral kernel* $\hat{G}$ in

$$\psi(x,y,z)=\iint\hat{G}(x,y,z;x_o,y_o,z_o)\psi(x_o,y_o,z_o)dx_o\,dy_o \tag{65.7a}$$

This integral kernel can be identified with the one that appears in the diffraction theory but, in practice, (65.7a) and, its special case, (65.7b) are rarely used for calculations and serve only as a basis for the theoretical framework.

The plane $z=z_o$ is usually identified with the object plane, as we have already done, and $z=z_i$ is the image plane. Nevertheless, (65.7a) can of course be used to relate any two planes and we shall see that a form of it is needed for other pairs of planes when we consider coherence theory in Part XVI. It is often unnecessary to include z explicitly among the arguments of the wavefunction; specific planes are then indicated by a suffix, ψ_o and ψ_i for example. Eq. (65.7a) then takes the form

$$\hat{\psi}(\hat{x}_i,\hat{y}_i)=\iint_{-\infty}^{\infty}\hat{G}(\hat{x}_i,\hat{y}_i;x_o,y_o)\psi_o(x_o,y_o)dx_o\,dy_o \tag{65.7b}$$

We can extend the limits of integration formally to infinity by assuming that $\psi_o=0$ downstream from the specimen outside the region occupied by the latter.

65.2 Isoplanatism and Fourier Transforms

In certain conditions the double integral in (65.7b) takes the form of a convolution. This suggests that we should examine its Fourier transform since the latter converts convolution products into direct products. For our present purposes, we shall not consider any other approach but we note in passing that when the functions ψ_o and $\hat{G}$ have appreciably different supports (that is. $\hat{G}$ is effectively nonzero over a much smaller area than ψ_o), direct convolution may be preferable to two Fourier transforms and a direct product.

The condition in which (65.7b) reduces to a convolution is that the system is *isoplanatic*. This implies in our case that the object is so small that all off-axis lens aberrations can be ignored in the image-forming process and that only the aberrations that do not vanish for $x_o = y_o = 0$ need be retained (Fig. 65.2). This is not the original definition of isoplanatism but is a consequence of the more fundamental requirement that $\hat{G}$ may be written in the form

$$\hat{G}(\hat{x}_i, \hat{y}_i; x_o, y_o) = G(x_i - x_o, y_i - y_o) \tag{65.8}$$

which we now impose. For a careful discussion, see Dumontet (1955a). The coordinates $\hat{x}_i, \hat{y}_i$ are the natural image coordinates, measured in the same laboratory frame as x_o, y_o, while x_i, y_i are *scaled* image coordinates, related to $\hat{x}_i, \hat{y}_i$, by

$$x_i + iy_i = (\hat{x}_i + i\hat{y}_i)/M(z_i, z_o) \tag{65.9}$$

in which $M(z_i, z_o)$ is the complex magnification (see Eqs. 15.45 and 15.46 of Volume 1). We recall that

$$M(z_i, z_o) = M\exp\{i(\theta_i - \theta_o)\}$$

and that M is an algebraic quantity (negative for single-stage imaging). Eq. (65.9) means that the image is referred back to the object scale and orientation.

When $\hat{G}$ may be replaced by G (65.8 and 65.9), Eq. (65.7b) acquires the form of a *convolution integral* or *convolution product.* Such integrals are mapped into *direct products* by Fourier transformation and it is for this reason that Fourier transform techniques are ubiquitous in image analysis. The Fourier transform is not, however, the only operator that maps convolutions into a simpler type of product and we shall mention the alternatives briefly in Section 66.9.

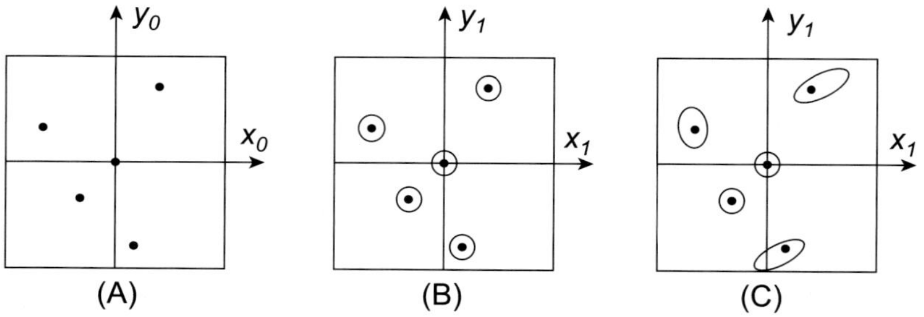

Figure 65.2
Illustration of the concept of isoplanatism. (A) Object consisting of five points. (B) Image in the isoplanatic approximation: the aberration discs are all equal in size and shape, their centres are unshifted. (C) Image with violation of the isoplanatism: the aberration discs may alter in size, shape and orientation, their centres are shifted (distortion). These effects increase with increasing distance from the optic axis (where $x_i = y_i = 0$).

The Fourier transforms are of course all two-dimensional. We first recall the one-dimensional form with 2π in the exponent, since this extends immediately to two dimensions for any pair of functions $u(x)$ and $U(q)$ satisfying the necessary mathematical requirements (essentially that the integral of their squared modulus remains finite.

We have

$$u(x) =: \mathcal{F}(U) = \int_{-\infty}^{\infty} U(q)\exp(2\pi i q x)dq \tag{65.10a}$$

with inverse[1]

$$U(q) =: \mathcal{F}^{-}(u) = \int_{-\infty}^{\infty} u(x)\exp(-2\pi i q x)dx \tag{65.10b}$$

Then for any two functions $u(x)$, $\upsilon(x)$ with transforms $U(q)$ and $V(q)$, the convolution theorem tells us that for

$$\begin{aligned} w(x) := & \int_{-\infty}^{\infty} u(x-x')\upsilon(x')dx' \equiv \int_{-\infty}^{\infty} u(x')\upsilon(x-x')dx' \\ & =: u(x) \otimes \upsilon(x) \equiv \upsilon(x) \otimes u(x) \end{aligned} \tag{65.11a}$$

or

$$w(x) =: (u \otimes \upsilon)(x) \equiv (\upsilon \otimes u)(x)$$

we have

$$W(q) = U(q)V(q) \tag{65.11b}$$

In imaging applications, in electron optics as elsewhere, all the Fourier and convolution integrals become two-dimensional. The role of the coordinate x is taken over by the position vector in the object plane, $\boldsymbol{u}_o = (x_o, y_o)$ and that of the conjugate variable q by the spatial frequency vector $\boldsymbol{q}$ with components (q_x, q_y) or occasionally (p, q) when we need to add further suffixes. Like q, q_x and q_y have the dimensions of reciprocal length. The simple product qx in the Fourier transform exponents is replaced by the scalar product $\boldsymbol{q} \cdot \boldsymbol{u}_o = q_x x_o + q_y y_o$.

It is convenient to introduce the *scaled* image coordinates (x_i, y_i), defined by (65.9) not only in the integrand of (65.7b) but also as arguments on the left-hand side. Summarizing, we write

$$\begin{aligned} \boldsymbol{u}_o &= (x_o, y_o) \\ \boldsymbol{q} &= (q_x, q_y) = (p, q) \\ \boldsymbol{u}_i &= (x_i, y_i) \end{aligned} \tag{65.12}$$

[1] The inverse Fourier transform operator is the adjoint of the direct operator but as a mnemonic aid, we denote the inverse operator by $FT^{-1}(\cdot)$ or $\mathcal{F}^{-}(\cdot)$ as a reminder that a minus sign appears in the exponent.

The convolution integral then takes the compact form

$$\begin{aligned}\psi_i(\boldsymbol{u}_i) &= \int_{-\infty}^{\infty} G(\boldsymbol{u}_i - \boldsymbol{u}_o)\psi_o(\boldsymbol{u}_o)d\boldsymbol{u}_o \\ &= (G \otimes \psi_o)(\boldsymbol{u}_i)\end{aligned} \tag{65.13}$$

in which

$$\psi(\boldsymbol{u}_i) := \hat{\psi}(\hat{\boldsymbol{u}}_i) \tag{65.14}$$

It is important to note the scale change implicit in this definition, which must not be forgotten when deriving explicit object–image relations.

Several Fourier transforms will occur so frequently that we introduce a special notation for them. In particular, we define the *object spectrum*

$$\begin{aligned}S_o(\boldsymbol{q}) &:= \mathscr{F}^-(\psi_o) \\ &= \int_{-\infty}^{\infty} \psi_o(\boldsymbol{u}_o)\exp(-2\pi i\boldsymbol{q}\cdot\boldsymbol{u}_o)d\boldsymbol{u}_o\end{aligned} \tag{65.15a}$$

the image spectrum

$$S_i(\boldsymbol{q}) := \mathscr{F}^-(\psi_i) \tag{65.15b}$$

and the instrumental wave transfer function

$$T(\boldsymbol{q}) := \mathscr{F}^-(G) \tag{65.15c}$$

From (65.13), therefore, we have

$$S_i(\boldsymbol{q}) = T(\boldsymbol{q})S_o(\boldsymbol{q}) \tag{65.16}$$

in which we have used the two-dimensional form of the convolution theorem (65.11a and 65.11b).

The intensity distribution in the image plane may now be calculated efficiently by the following sequence:

1. We define an object wavefunction $\psi_o(\boldsymbol{u}_o)$, usually in terms of a theoretical model (Part XIV), and perform the first Fourier transform (65.15a) with the aid of one of the various fast-Fourier-transform algorithms.
2. The functions T and S_o are then multiplied (65.16). The transfer function $T(\boldsymbol{q})$ is not obtained by transforming G numerically since it can be established directly in closed form by theoretical arguments, as we shall see in the next section.
3. We form the inverse of (65.15b),

$$\psi_i(\boldsymbol{u}_i) = \int_{-\infty}^{\infty} S_i(\boldsymbol{q})\exp(2\pi i q\cdot\boldsymbol{u}_i)d\boldsymbol{q} \tag{65.17}$$

and hence obtain $|\psi_i(\boldsymbol{u_i})|^2$, which is proportional to the intensity distribution in the image.

This sequence is shown in Fig. 65.3, which brings out the analogy with electronic signal processing. The transfer function $T(\boldsymbol{q})$ acts as a *complex filter*, which modifies the object spectrum prior to the recording step. Provided that $T(\boldsymbol{q})$ nowhere vanishes, a condition that will be investigated later (and shown not to be satisfied in important cases), the whole procedure from ψ_o to ψ_i is *invertible*. It is in the last step, in which $|\psi_i|^2$ is recorded, that the wave phase, $\arg(\psi_i)$, is irretrievably lost. This makes direct interpretation of recorded images unreliable whenever the phase is important. We shall see in the following chapters how this difficulty can be overcome or at least palliated. The holographic solution has already been presented in Chapter 63.

65.3 The Wave Transfer Function

The transfer calculus presented schematically in Fig. 65.3 stands or falls with the assumption (65.8), namely, that the arguments of G only appear as the differences $\boldsymbol{u_i} - \boldsymbol{u_o}$ (with the scaling of 65.9). This assumption must now be justified. An extremely useful by-product of this calculation will be a simple analytical formula for the instrumental wave transfer function, $T(\boldsymbol{q})$.

We set out from the situation shown in Fig. 59.5, where a specimen is located close to the object focal plane of an electron lens and hence highly magnified. We study the

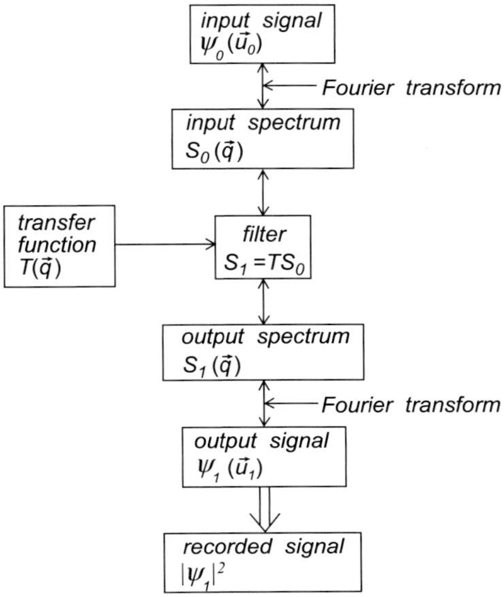

Figure 65.3
Image formation as a linear filter process with a nonlinear recording step.

wavefunction (diffraction pattern) at the asymptotic aperture (O, A), described by Eq. (59.40). The notation of the present chapter is related to that of (59.40) as follows:

$$\begin{aligned}\psi_Q(x_Q, y_Q) &\equiv \psi_o(x_o, y_o) \equiv \psi_o(\boldsymbol{u}_o)\\ \psi(\boldsymbol{r}_i) &\equiv \hat{\psi}(\hat{x}_i, \hat{y}_i)\\ (x_a, y_a) &\equiv (\hat{x}_a, \hat{y}_a) \equiv \hat{\boldsymbol{u}}_{\boldsymbol{a}}\end{aligned} \tag{65.18}$$

The use of the original (unscaled) image coordinates $\hat{\boldsymbol{u}}$ is unfavourable and we, therefore, introduce the scaled, rotated coordinates $\boldsymbol{u}_i$, with the aid of Eq. (65.9). Since $(\hat{x}_i^{(0)}, \hat{y}_i^{(0)})$ are the coordinates of the Gaussian image point, the introduction of (65.9) gives $x_i^{(0)} = x_o$, $y_i^{(0)} = y_o$. It is likewise advantageous to define *rotated* aperture coordinates,

$$x_a + \mathrm{i}y_a := (\hat{x}_a + \mathrm{i}\hat{y}_a)\exp(-\mathrm{i}\theta) \tag{65.19a}$$

and a corresponding two-dimensional vector

$$\boldsymbol{u}_a = (x_a, y_a) \tag{65.19b}$$

The scalar product in (59.40b) is invariant with respect to this rotation and the diffraction integral takes the more compact form

$$\psi_i(\boldsymbol{u}_i) = \frac{c}{b}\int_O\int_A \psi_o(\boldsymbol{u}_o)\exp(-\mathrm{i}kD)d\boldsymbol{u}_o\, d\boldsymbol{u}_a \tag{65.20a}$$

with

$$D = \overline{W}(\boldsymbol{u}_o, \boldsymbol{u}_a) - \frac{|M|}{b}\boldsymbol{u}_a \cdot (\boldsymbol{u}_i - \boldsymbol{u}_o) \tag{65.20b}$$

[In order not to overburden the notation, we have simply written $\overline{W}(\boldsymbol{u}_o, \boldsymbol{u}_a)$ here. Formally, we should have stated that $\overline{W}(\boldsymbol{r}_Q, \boldsymbol{r}_a)$ occurring in (59.40b) becomes $\overline{W}(\boldsymbol{u}_o, \hat{\boldsymbol{u}}_a)$ in our present notation and that this, in turn, becomes $\hat{W}(\boldsymbol{u}_o, \boldsymbol{u}_a)$ after making the scale change (65.19a).]

This expression must now be reconciled with Eq. (65.17). Comparison of the terms containing $\boldsymbol{u}_i$ in the exponent gives

$$2\pi\mathrm{i}\boldsymbol{q}\cdot\boldsymbol{u}_i = \frac{\mathrm{i}k|M|}{b}\boldsymbol{u}_a\cdot\boldsymbol{u}_i$$

Since this must hold for any pair of vectors $\boldsymbol{u}_a$ and $\boldsymbol{u}_i$, the spatial frequency vector $\boldsymbol{q}$ must be related to $\boldsymbol{u}_a$ by

$$\boldsymbol{q} = \frac{|M|}{\lambda b}\boldsymbol{u}_a \tag{65.21a}$$

in which we have used $k = 2\pi/\lambda$. At high magnification, $b/|M| \approx f$ and so

$$\boldsymbol{q} = \boldsymbol{u}_a/\lambda f \tag{65.21b}$$

We have $|\boldsymbol{u}_a| = r_a$ and usually $M < 0$ so that with the aid of Eq. (60.54) and the relation $g_o\lambda_o = g_i\lambda$ we can express $\boldsymbol{q}$ in terms of *angular coordinates* at the specimen. We find

$$\boldsymbol{q} = \frac{\theta_o}{\lambda_o}(\cos\varphi_o, \sin\varphi_o) \tag{65.21c}$$

and $q = |\boldsymbol{q}| = \theta_o/\lambda_o$. Formulae (65.21b) and (65.21c) have the merit of simplicity and are almost invariably used in the literature of the subject.

It must also be possible to reorganize all terms in (65.20a) and (65.20b) that depend on $\boldsymbol{u}_o$ in the form of (65.15a). Clearly, this cannot be done unless $\overline{W}$ is independent of $\boldsymbol{u}_o$ and depends only on $\boldsymbol{u}_a$:

$$\overline{W} =: W(\boldsymbol{u}_a) =: W'(\boldsymbol{q}) \tag{65.22}$$

which is the appropriate form of the *isoplanatic approximation*. Eq. (65.20a) thus becomes

$$\psi_i(\boldsymbol{u}_i) = \frac{cb\lambda^2}{M^2}\int S_o(\boldsymbol{q})\exp\{-ikW'(\boldsymbol{q})\}\exp(2\pi i\boldsymbol{q}\cdot\boldsymbol{u}_i)d\boldsymbol{q} \tag{65.23}$$

which is in agreement with (65.16) and (65.17) if we identify the transfer function $T(\boldsymbol{q})$ with the term in $W'(\boldsymbol{q})$:

$$T(\boldsymbol{q}) = T_o\exp\{-ikW'(\boldsymbol{q})\} \tag{65.24}$$

The value of the constant T_o will be determined later but we note that for a completely open diaphragm, it will simply be unity.

The exponential term in Eq. (65.24) can be cast into a more convenient form: the constant $k = 2\pi/\lambda$ refers to the image whereas the wavelength λ_0 in (65.21a), (65.21b) and (65.21c) refers to the object. We, therefore, introduce a final change of scale and define the wave aberration by

$$W'(\boldsymbol{q}) =: \frac{\lambda}{\lambda_o}W(\lambda_o\boldsymbol{q}) \equiv \frac{\lambda}{\lambda_o}W(\theta_o\cos\varphi_o, \theta_o\sin\varphi_o) \tag{65.25}$$

so that, for the completely transparent diaphragm, we obtain the standard form

$$T_L(\boldsymbol{q}) := \exp\left\{-\frac{2\pi i}{\lambda_o}W(\lambda_o\boldsymbol{q})\right\} \qquad T = T_oT_L \tag{65.26}$$

We now determine the value of T_o for this case. A pure phase factor of the form $\exp(i\alpha)$ being irrelevant, we can assume from the outset that T_o is a positive constant. From the continuity equation, it can be seen that the total (integrated) intensity,

$$I = \int |\psi_o|^2 d\boldsymbol{u}_o = \frac{\lambda_o}{\lambda}\iint |\hat{\psi}_i|^2 d\hat{x}_i\, d\hat{y}_i$$

is conserved for a wholly transparent diaphragm. The factor λ_o/λ is inconvenient and since we have already made one scale change (65.9), we introduce an intensity scale that eliminates this factor and hence normalizes the intensity integral I; we then have the very convenient relation

$$\int |\psi_o|^2 d\boldsymbol{u}_o = \int |\psi_i|^2 d\boldsymbol{u}_i = 1 \tag{65.27}$$

Parseval's theorem then tells us that

$$\int |S_o(\boldsymbol{q})|^2 dq = \int |S_i(\boldsymbol{q})|^2 d\boldsymbol{q} = 1 \tag{65.28}$$

Recalling that $S_i = T_L S_o$ and hence $|S_i| = |T_L| \cdot |S_o|$, this can only be generally true if $|T_L(\boldsymbol{q})| = 1$. We thus recover (65.26) and confirm that $T_o = 1$ for a completely open diaphragm.

We have retained the distinction between λ and λ_o for generality but, in practice, the lenses used for high-resolution imaging are almost always magnetic. We then have $\lambda = \lambda_o$ and the conservation of the norm of ψ follows directly, without an additional change of scale.

65.4 Explicit Formulae

The first few wave aberration terms that are to be considered in the isoplanatic approximation for a lens system with at worst small departures from rotational symmetry are as follows:

$$\begin{aligned}
W = {} & \frac{1}{4} C_s \theta_o^4 && \text{spherical aberration} \\
& - \frac{1}{2} \Delta_o \theta_o^2 && \text{defocus} \\
& + \theta_o (d_1 \cos\varphi_o + d_2 \sin\varphi_o) && \text{deflection axial coma} \\
& + \frac{1}{3} \theta_0^3 (c_1 \cos\varphi_o + c_2 \sin\varphi_o) && \text{cubic axial coma} \\
& + \frac{1}{2} \theta_0^2 (a_1^{(2)} \cos 2\varphi_o + a_2^{(3)} \sin 2\varphi_o) && \text{twofold axial astigmatism} \\
& + \frac{1}{3} \theta_0^3 (a_1^{(3)} \cos 3\varphi_o + a_2^{(3)} \sin 3\varphi_o) && \text{threefold axial astigmatism}
\end{aligned} \tag{65.29}$$

(cf. 60.55 and 60.72 and Table 31.1, which lists the various notations in use for these coefficients). With the aid of (65.21a), (65.21b) and (65.21c), we obtain $W(\lambda_o \boldsymbol{q})$ and hence $T_L(\boldsymbol{q})$ from (65.26).

The parasitic terms, those depending on φ_o, can be made arbitrarily small by careful alignment of the microscope, at least in principle. In practice, they are reduced as far as technically possible but the practical limits are not known beforehand, and in the past, they depended heavily on the skill and experience of the microscopist. Computer-aided methods of alignment are now usual (Section 77.3) and, even if these are not perfect, they can be expected to give reproducible results at least. Be this as it may, we can hardly hope to draw any general conclusions from these terms and we hence disregard them forthwith, though we shall occasionally return to the axial astigmatism term.

The surviving contributions are therefore the spherical aberration and the defocus, corresponding to the distance Δ_0 between the specimen and the plane conjugate to the (fixed) image plane. For these, $T_L(\boldsymbol{q})$ becomes

$$\begin{aligned}
T_L(\boldsymbol{q}) &= \exp\left(-\frac{i\pi\lambda_o^3}{2}C_s q^4 + i\pi\lambda_o\Delta_o q^2\right)\\
&= \exp\left\{-\frac{i\pi}{2}C_s\lambda_o^3(q_x^2+q_y^2)^2 + i\pi\lambda_o\Delta_o(q_x^2+q_y^2)\right\}\\
&= \exp\left\{-\frac{2\pi i}{\lambda_o}\left(\frac{1}{4}C_s\lambda_o^4 q^4 - \frac{1}{2}\Delta_o\lambda_o^2 q^2\right)\right\}\\
&= :\exp\{-i\chi(\boldsymbol{q})\}\\
\chi(\boldsymbol{q}) &:= \pi\left(\frac{1}{2}C_s\lambda_o^3 q^4 - \Delta_o\lambda_o q^2\right)
\end{aligned} \tag{65.30}$$

In aberration-corrected instruments, the fifth-order spherical aberration C_5 (and even seventh order, C_7) must be retained. It is then necessary to add the following terms to χ:

$$\pi\left(\frac{1}{3}C_5\lambda^5 q^6 + \frac{1}{4}C_7\lambda^7 q^8\right)$$

It is becoming common to regard defocus as a first-order axial or 'spherical' aberration and to denote it by C_1; here we usually retain the mnemonically superior notation Δ. The original function χ depends on three parameters, C_s, λ_o and Δ_o. It is often convenient[2] to scale distances perpendicular to the axis relative to $(C_s\lambda_o^3)^{1/4}$ and those along the axis relative to $(C_s\lambda_o)^{1/2}$, thereby reducing the number of degrees of freedom and permitting a

[2] The use of this scaling is invaluable in practice since results obtained for a particular microscope (given C_s and accelerating voltage) can be easily transferred to any other. It has been suggested (Hawkes, 1980b) that the scaling factors are given names, and this usage is occasionally adopted. If we write $(C_s\lambda)^{1/2} = :1$ scherzer $= :1$ Sch, $(C_s\lambda^3)^{1/4} = :1$ glaser $= :1$ Gl, then we can speak of a defocus of 1.5 Sch, for example or a spatial frequency cut-off at 0.5 Gl. These are of course not units in the sense that metres or volts are, since they depend on the third-order spherical aberration coefficient and the wavelength.

simpler graphical representation of the various functions that we shall be encountering. We introduce the reduced spatial frequency

$$\boldsymbol{Q}: = (C_s\lambda_o^3)^{1/4}\boldsymbol{q}, \quad Q = |\boldsymbol{Q}| \tag{65.31}$$

and the reduced defocus

$$D: = \Delta_o/(C_s\lambda_o)^{1/2} \tag{65.32}$$

The function $T_L(\boldsymbol{q})$ simplifies to

$$T_L(\boldsymbol{q}) \rightarrow \hat{T}_L(\boldsymbol{Q}) = \exp\left\{i\frac{\pi}{2}\left(2DQ^2 - Q^4\right)\right\} \tag{65.33}$$

The simplest special case, $D = 0$, is shown in Fig. 65.4; we shall see later that this is not an experimentally attractive value of D. For $Q \le 1$, the real and imaginary parts of $\hat{T}_L$ vary fairly slowly but, for $Q > 1$, they start to oscillate with increasing speed. The practical consequences of this and the behaviour of $\hat{T}_L$ for other choices of D will be examined in the next chapter.

So far, we have dealt exclusively with a completely transparent aperture plane, which has no effect on the electrons. This is unrealistic for, in all practical devices, apertures with a finite bore radius are introduced to confine the beam. What factor must we include in $T(\boldsymbol{q})$ to characterize this situation?

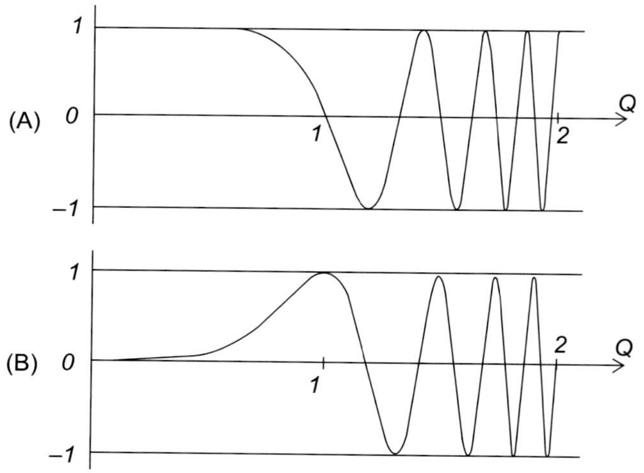

Figure 65.4
(A) Real and (B) imaginary parts of the wave transfer function for vanishing defocus, $\hat{T}_L(Q) = \exp(-i\pi Q^4/2)$. Curve (B) shows $-\Im(\hat{T}_L)$.

This factor will have the same structure as the object function given by (60.4): the laws of electron propagation that we have derived relate any two planes and we are at liberty to place the 'object' plane at the aperture and regard the aperture transparency as the 'object function'. The object coordinates (x_o, y_o) are thus to be replaced by the aperture coordinates transformed with the aid of (65.21a). We thus obtain an *aperture factor*

$$\begin{aligned} T_A(\boldsymbol{q}) &= \exp\{-\sigma_A(\boldsymbol{q}) + \mathrm{i}\eta_A(\boldsymbol{q})\} \\ &= \exp\left\{-\sigma_A\left(\frac{|M|}{\lambda b}\boldsymbol{u}_A\right) + \mathrm{i}\eta_A\left(\frac{|M|}{\lambda b}\boldsymbol{u}_A\right)\right\} \end{aligned} \tag{65.34}$$

examples of which are shown in Figs. 65.5 and 65.6. The complete transfer function is then

$$T(\boldsymbol{q}) = T_L(\boldsymbol{q})T_A(\boldsymbol{q}) \tag{65.35}$$

We have already considered the case in which no aperture is present, for which σ_A and η_A are zero everywhere and hence $T_A(\boldsymbol{q}) = 1$. The most important special case is a round hole

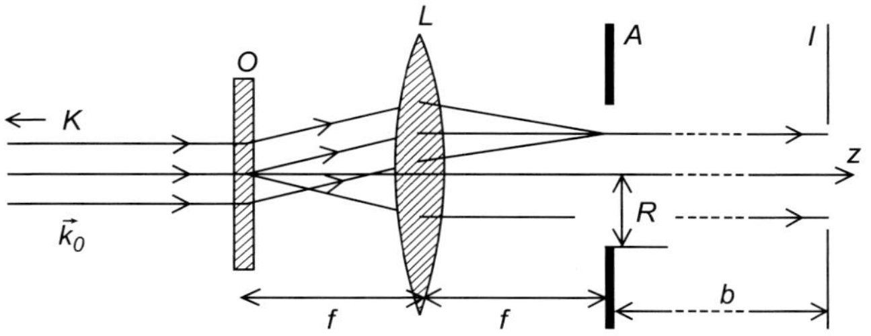

Figure 65.5
Effect of a simple hole as an aperture. *K*: condenser (far left), $\vec{k}_0$: wave vector of the incident beam, *O*: object plane, *L*: lens, *A*: aperture, *I*: image screen, far to the right, *f*: focal length of the lens, *b*: distance $\overline{AI}$, $b \gg f$, *R*: radius of the circular opening.

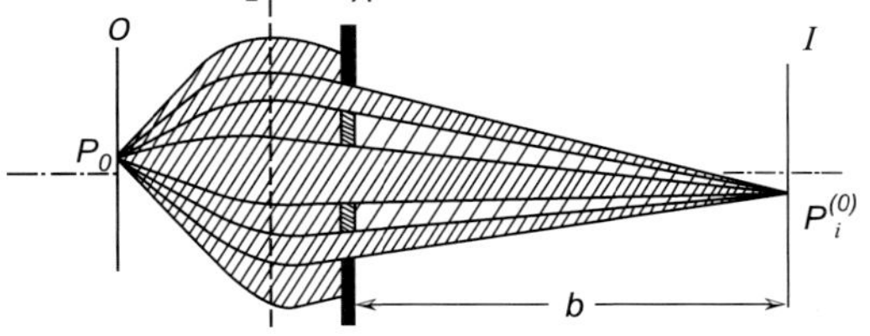

Figure 65.6
The general possibilities of beam modification by an aperture diaphragm having open zones and semitransparent zones enclosed in an opaque screen. For simplicity, only one object point P_o and the corresponding image point $P_i^{(0)}$ are considered.

in a completely opaque screen, to which the apertures in an electron microscope correspond closely. We then write

$$\begin{array}{llll} \sigma_A = 0, & \eta_A = 0 & \rightarrow T_A = 1 & : \ \text{open parts} \\ \sigma_A = \infty & \eta_A \ \text{arbitrary} & \rightarrow T_A = 0 & : \ \text{opaque parts} \end{array}$$

For a circular aperture of radius R centred on the axis, we have (Fig. 65.5)

$$T_A(\boldsymbol{q}) = \begin{cases} 1 & q \leq |M|R/\lambda b = R/\lambda f \\ 0 & q > |M|R/\lambda b = R/\lambda f \end{cases} \tag{65.36}$$

The function $T_L\ (\boldsymbol{q})$ of (65.30) is simply cut-off at the value $q_c = |M|R/\lambda b$. In Chapter 66, we shall discuss the optimum value of the cut-off value, which can be chosen freely by selecting the aperture radius appropriately.

Figs. 65.5 and 65.6 are examples of amplitude or intensity masks, consisting of zones that are either opaque or perfectly transparent to electrons. For phase objects a different type of mask is needed, as was realized by Zernike (1935), inventor of the phase plate that renders pure phase objects visible in the light microscope. This plate advances the phase of the light that has not been scattered by the specimen; the scattered light then interferes *constructively* with the unscattered beam (Fig. 65.7). Numerous attempts have been made to incorporate such plates in the electron microscope, at first to flatten the phase-contrast transfer function (Section 66.6). Today, their main role is to improve image contrast, especially when the specimen consists largely of light elements as is the case for most organic material. We examine them in detail in Section 66.6; here, we introduce the various forms that they may take.

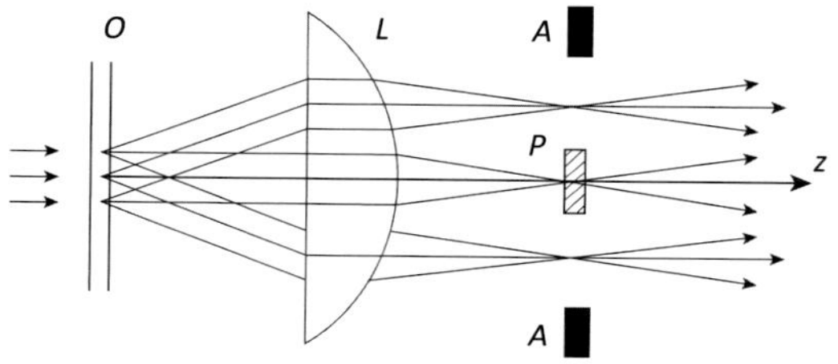

Figure 65.7

Zernike's phase-contrast procedure in light microscopy as an example of a nontrivial aperture function $T_A\ (\boldsymbol{q})$. The object O is a pure 'phase object', common in biology, which means that the specimen shifts only the phase of the incident wave and does not attenuate the amplitude of the light. The contrast plate P in the aperture plane diminishes the intensity of the central undiffracted beam and shifts its phase by $\pi/2$, thereby turning the phase differences into amplitude differences. The aperture function is $T_A(\boldsymbol{q}) = \begin{cases} -\mathrm{i}\exp(-\sigma_A) & \text{in } P \\ 1 & \text{in the open ring} \\ 0 & \text{on the opaque screen} \end{cases}$ The eyepiece is located far to the right and is not shown in the figure.

The simplest version is inspired by the Zernike plate, but unlike the latter, the phase shift is imposed on the scattered waves: a disc of uniform thickness with a central hole, situated like all these plates in the diffraction plane (usually the back focal plane) of the objective. A variant of this is the Hilbert plate, in which only half of the diffraction plane is covered by the plate, again with a (semicircular) hole or notch in the centre. This in turn may be modified so that only an annular zone of the Hilbert plate is present, the outer region being transparent – this is Buijsse's tulip plate. A hole-free Zernike plate, known as a Volta plate, is also in use; the strong unscattered beam modifies the structure of the zone through which it passes in such a way that a phase difference is again created between this central zone and the outer region.

A completely different approach, in which the physical phase plate is replaced by a local electrostatic field, was suggested by Boersch (1947c). It has the advantage that there is no unwanted scattering in the plate but it requires supporting arms that obscure part of the imaging beam, as does the field-generating element itself, typically a miniature einzel lens. Alternatively, a magnetic field can be used to modify the phase of part of the imaging beam. Such plates are known collectively as Boersch plates. A design that obscures very little of the diffraction plane consists of a very thin coaxial wire. The end of the wire is placed close to the centre of the beam, which is affected by the electric field between the central conductor and the outer conducting sheath. This device is known as a Zach phase plate. Finally, we mention a very different and potentially attractive way of creating an electric field in the path of the electron beam by means of a focused laser beam. Here, no part of the electron beam is obscured and the phase change imposed by the plate can be controlled easily.

Phase-shifting plates may also be placed upstream from the specimen, where their role is very different. They now impose a pattern on the beam incident on the specimen, and the illumination is said to be *structured*. We return to this specialized behaviour in the account of orbital angular momentum and vortex beams (Chapter 80) where we shall meet Bessel beams, Airy beams and Laguerre–Gaussian beams, the names suggested by the nature of the phase variation imprinted on the beam. In the STEM, such structuring elements make it possible to illuminate the specimen with a pattern of probes instead of a single focused probe. This is at the heart of a form of STEM ptychography.

CHAPTER 66

Image Formation in the Conventional Transmission Electron Microscope

The transfer theory established in Chapter 65 is incomplete in several respects. We have studied the linear relation between the wavefunctions in the object plane and the image plane, whereas only the intensity, the squared modulus of the wavefunction, can be detected. Furthermore, we have considered only perfectly monochromatic waves, but in practice, the illumination will always have a finite energy spectrum. Finally, we have invariably regarded the incident wave as plane or spherical. Although this is not an obvious limitation, for an appropriate set of plane waves can always be superimposed to reproduce more complicated incident wave patterns, we need to know how such a superposition interacts with the effects due to a finite energy spectrum. We now consider these various points in turn.

In the general case, the loss of the phase of the wavefunction in the recording step, which is only sensitive to $|\psi_i|^2$, makes it impossible to reconstruct the object wavefunction from its measured image intensity distribution, as we saw in Section 65.2. There are, however, situations in which reconstruction is, at least approximately, possible. Some of these involve unconventional imaging modes, of which the most successful is off-axis holography using an electron biprism; this has been described in detail in Section 63.2. A very important case is bright-field imaging in a conventional transmission electron microscope of a certain class of specimen, and we now examine this in detail. We first consider the simpler case in which the waves are assumed to be monoenergetic and plane, travelling in the axial direction. The effects of finite energy spectrum and a spread of incident directions are examined in Sections 66.2 and 66.3. At the end of Section 66.2, we show that these calculations would have been much simpler if we could have assumed at the outset that the effects were independent. In Section 66.5, we consider the representation of tilted and hollow-cone illumination in transfer theory. In Section 66.6, several types of phase plate are analysed. Transfer theory for crystalline specimens and the effect of aberration correction on transfer theory are the subjects of Sections 66.7 and 66.8. A short final section is a reminder that the Fourier transform is not the only transform that maps convolution products into direct products.

66.1 Image Contrast for Weakly Scattering Specimens

Fig. 66.1 shows a transparent foil in the object plane of a transmission electron microscope being imaged at high magnification. In this section, we assume that a *plane wave* is incident

Principles of Electron Optics.
DOI: https://doi.org/10.1016/B978-0-12-818979-5.00066-8

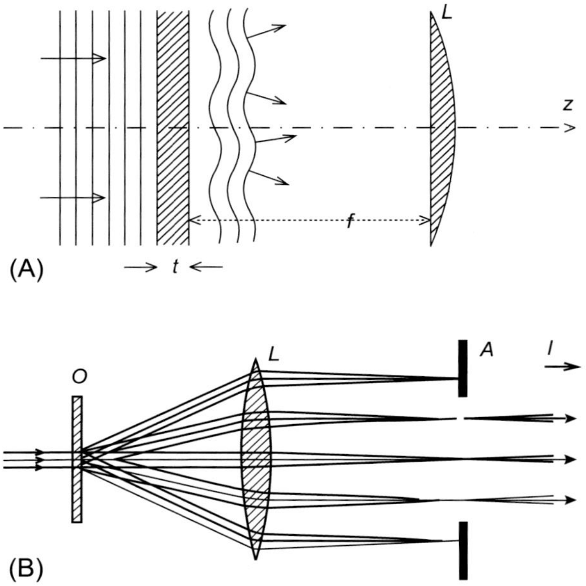

Figure 66.1
Principle of bright-field imaging, (A) Distorted waves behind a thin transparent foil located close to the front focal plane of the lens L. The effects are highly exaggerated; the thickness $t \ll f$. (B) Wave propagation as Fraunhofer diffraction, only the trajectories corresponding to a few directions are drawn. The aperture A acts selectively on the diffracted wavelets. Here only the 'zero-order' and first two 'side reflections' can pass through it.

in the axial direction from the condenser system. The Fourier transform of the wave emerging from the specimen may be pictured vividly by recalling that its squared modulus is the same as the Fraunhofer diffraction pattern of the specimen. The diffracted, locally plane, partial waves are focused in the aperture plane, where all waves travelling in the same direction are brought to a point. The aperture then selects some wavelets and excludes those beyond the spatial-frequency cut-off value (65.36).

With no loss of generality, we may normalize the amplitude of the incident plane wave to unity and set the z-coordinate origin at the entrance plane of the specimen; in this plane, the incident wave is then simply $\psi = 1$. Its passage through the thin specimen causes a decrease $\exp\{-\sigma_o(x_o, y_o)\}$ of the amplitude and a phase shift $\exp\{i\eta_o(x_o, y_o)\}$, as in Eq. (60.4). These are *total* interaction effects, obtained by integration along the classical electron paths through the specimen. More details will be found in Part XIV. A *weakly scattering specimen* is characterized by $\sigma_o \ll 1$ and $|\eta_o| \ll 1$, so that

$$\psi_o(x_o, y_o) = \exp\left(i\eta_o - \sigma_o\right) \approx 1 + i\eta_o(x_o, y_o) - \sigma_o(x_o, y_o) \tag{66.1}$$

is the wave emerging from the specimen. The 'distorted waves' of Fig. 66.1A then correspond to surfaces of equal phase and hence represent the variations of η_o. The object spectrum S_o corresponding to (66.1) is

$$S_o(\boldsymbol{q}) = \delta(\boldsymbol{q}) + \mathrm{i}\tilde{\eta}(\boldsymbol{q}) - \tilde{\sigma}(\boldsymbol{q}) \tag{66.2}$$

in which $\delta(\boldsymbol{q})$ denotes the Dirac δ-function in two dimensions, $\delta(\boldsymbol{q}) = \delta(\boldsymbol{q}_x)\delta(\boldsymbol{q}_y)$; $\tilde{\eta}$ denotes the (inverse) Fourier transform of η_o and is hence the phase shift spectrum:

$$\tilde{\eta}(\boldsymbol{q}) = \mathcal{F}^-(\eta_o) = \int_{-\infty}^{\infty} \eta_o(\boldsymbol{u}_o)\exp(-2\pi\mathrm{i}\boldsymbol{q}\cdot\boldsymbol{u}_o)d\boldsymbol{u}_o \tag{66.3}$$

while $\tilde{\sigma}$, the transform of σ_o, is the 'absorption' or amplitude spectrum,[1]

$$\tilde{\sigma}(\boldsymbol{q}) = \mathcal{F}^-(\sigma_o) = \int_{-\infty}^{\infty} \sigma_o(\boldsymbol{u}_o)\exp(-2\pi\mathrm{i}\boldsymbol{q}\cdot\boldsymbol{u}_o)d\boldsymbol{u}_o \tag{66.4}$$

For the moment, we retain the most general form of the aperture factor T_A (Eq. 65.34). The spectrum of the image wavefunction is thus

$$S_i(\boldsymbol{q}) = \exp\left\{\mathrm{i}\eta_A(\boldsymbol{q}) - \sigma_A(\boldsymbol{q}) - \frac{2\pi\mathrm{i}}{\lambda_o}W(\lambda_o\boldsymbol{q})\right\}\{\delta(\boldsymbol{q}) + \mathrm{i}\tilde{\eta}(\boldsymbol{q}) - \tilde{\sigma}(\boldsymbol{q})\} \tag{66.5}$$

On evaluating the image wavefunction $\psi_i(\mathbf{u}_i) = \mathscr{F}(S_i)$ we notice that the term in $\delta(\boldsymbol{q})$ simply gives the integrand at $\boldsymbol{q} = 0$. Since $W(0) = 0$, we obtain

$$\begin{aligned}\psi_i(\boldsymbol{u}_i) = &\exp\{\mathrm{i}\eta_A(0) - \sigma_A(0)\}\\ &+ \mathscr{F}\{(\mathrm{i}\tilde{\eta} - \tilde{\sigma})\exp(\mathrm{i}\eta_A - \sigma_A - 2\pi\mathrm{i}W/\lambda_o)\}\end{aligned} \tag{66.6}$$

This formula is still general enough to include dark-field imaging ($\sigma_A(0) \to \infty$ for central-stop dark-field) and the presence of phase-shifting devices in the aperture plane such as the Zernike phase plate. These and other examples of intervention in the back-focal plane are examined at length in Section 66.6. An essential feature of conventional bright-field imaging is that the neighbourhood of the optic axis is an *open* part of the aperture, so that the undiffracted wave, the partial wave that has passed through the specimen without any interaction, can reach the image plane unmodified. In these conditions, we have

$$\eta_A(0) = 0, \quad \sigma_A(0) = 0, \quad T_A(0) = 1 \tag{66.7}$$

[1] The physical interpretation of this term is not straightforward. In the specimens studied in the TEM, there is virtually no absorption in the sense that no incident electron is brought to a halt within the specimen. Electrons may, however, be 'lost' to the image-forming process in various ways and the absorption term is intended to account for these. For very weak specimens the amplitude term can usually be neglected but we retain it in the theory because of its importance for thicker but still fairly weak specimens.

Moreover, we now assume that the aperture is a simple circular hole centred on the optic axis (Eq. 65.36). We denote this aperture by $\mathscr{A}$

$$\mathscr{A}: q^2 = q_x^2 + q_y^2 \leq q_c^2 \tag{66.8}$$

Expression (66.6) for ψ_i now simplifies to

$$\begin{aligned}\psi_i(\boldsymbol{u}_i) &=: 1 + \mu(\boldsymbol{u}_i) \\ &= 1 + \int_{\mathscr{A}} \{\mathrm{i}\tilde{\eta}(\boldsymbol{q}) - \tilde{\sigma}(\boldsymbol{q})\}\exp\{-2\pi\mathrm{i}W(\lambda_o\boldsymbol{q})/\lambda_o\}\exp(2\pi\mathrm{i}\boldsymbol{q}\cdot\boldsymbol{u}_i)d\boldsymbol{q}\end{aligned} \tag{66.9}$$

A consequence of our earlier assumption that σ_o and $|\eta_o|$ are both much smaller than unity is that $|\mu| \ll 1$. It is this that enables us to simplify the expression for the image intensity $|\psi_i|^2$ in such a way that the *image contrast*[2] $C(\boldsymbol{u}_i)$ becomes *linearly related* to the object functions σ_o and η_o. The contrast, here defined by

$$C(\boldsymbol{u}_i) := -\frac{|\psi_i(\boldsymbol{u}_i)|^2 - 1}{|\psi_i(\boldsymbol{u}_i)|^2 + 1}, \tag{66.10a}$$

is the most direct of the various image quality measures that we shall meet. Neglecting quadratic terms in μ, we see immediately that

$$C(\boldsymbol{u}_i) = -\Re\psi_i(\boldsymbol{u}_i) \tag{66.10b}$$

The image contrast for weakly scattering specimens is clearly *very low*, which is an intrinsic drawback of the bright-field technique.

Explicit evaluation of Eq. (66.10b) with the aid of Eq. (66.9) yields two contributions, the *amplitude contrast*

$$C_a(\boldsymbol{u}_i) = \Re\int_{\mathscr{A}} \tilde{\sigma}(\boldsymbol{q})\exp\left\{2\pi\mathrm{i}\boldsymbol{q}\cdot\boldsymbol{u}_i - \frac{2\pi\mathrm{i}}{\lambda_o}W(\lambda_o\boldsymbol{q})\right\}d\boldsymbol{q} \tag{66.11a}$$

and the phase contrast

$$C_p(\boldsymbol{u}_i) = \Im\int_{\mathscr{A}} \tilde{\eta}(\boldsymbol{q})\exp\left\{2\pi\mathrm{i}\boldsymbol{q}\cdot\boldsymbol{u}_i - \frac{2\pi\mathrm{i}}{\lambda_o}W(\lambda_o\boldsymbol{q})\right\}d\boldsymbol{q} \tag{66.11b}$$

The total contrast is $C = C_a + C_p$.

In the most general case, no further progress can be made. These integral expressions are valid even for a slightly excentric or oval aperture, provided of course that the undiffracted

[2] In the literature the contrast is frequently defined as the numerator only of (66.10a). A factor 2 then appears, notably in the contrast transfer functions.

beam can pass through it. We next consider the case of a round aperture centred on the axis, for which (66.11a) and (66.11b) can be further simplified.

First, however, we reconsider the roles of σ and η and hence the relative importance of amplitude and phase contrast. We have already observed that, in the very thin specimens employed in high-resolution electron microscopy, there is no absorption and hence amplitude contrast may be expected to be absent or much weaker than phase contrast. For specimens that are thin but not excessively so, however, other phenomena appear that contribute indirectly to the 'absorption' term. Some electrons may be scattered inelastically, which means that they can no longer be included in a theory in which the wavelength is fixed; they are, therefore, 'lost' as far as the present form of the transfer theory is concerned and contribute to σ. As the specimen thickness is increased, the next term in the expansion (66.1) will become more important; this then becomes (with $|\eta_o| \gg \sigma_o$)

$$\psi_o(x_o, y_o) \approx 1 + i\eta_o - \sigma_o - \frac{1}{2}\eta_o^2 + \cdots \tag{66.1'}$$

again giving a contribution to the real part. The relative importance of these various effects is not well understood but there is experimental evidence that the 'amplitude image' can sometimes give a faithful representation of the object (Saxton, 1986, 1987).

Returning to Eqs. (66.11a) and (66.11b), we replace the factor $\exp(-2\pi i W/\lambda_o)$ by $T_L(\boldsymbol{q})$ as in Eq. (65.26), giving

$$\begin{aligned} C_a(\boldsymbol{u}_i) &= \frac{1}{2}\int_{\mathscr{A}} \tilde{\sigma}(\boldsymbol{q})T_L(\boldsymbol{q})\exp(2\pi i\boldsymbol{q}\cdot\boldsymbol{u}_i)d\boldsymbol{q} \\ &\quad + \frac{1}{2}\int_{\mathscr{A}} \tilde{\sigma}^*(\boldsymbol{q})T_L^*(\boldsymbol{q})\exp(-2\pi i\boldsymbol{q}\cdot\boldsymbol{u}_i)d\boldsymbol{q} \\ C_p(\boldsymbol{u}_i) &= \frac{1}{2i}\int_{\mathscr{A}} \tilde{\eta}(\boldsymbol{q})T_L(\boldsymbol{q})\exp(2\pi i\boldsymbol{q}\cdot\boldsymbol{u}_i)d\boldsymbol{q} \\ &\quad - \frac{1}{2i}\int_{\mathscr{A}} \tilde{\eta}^*(\boldsymbol{q})T_L^*(\boldsymbol{q})\exp(-2\pi i\boldsymbol{q}\cdot\boldsymbol{u}_i)d\boldsymbol{q} \end{aligned}$$

On changing the sign of the variable $\boldsymbol{q}$ in the second integral of each expression and using the fact that $\mathscr{A}$ is symmetric about the axis, we obtain

$$C_a(\boldsymbol{u}_i) = \frac{1}{2}\int_{\mathscr{A}} \{\tilde{\sigma}(\boldsymbol{q})T_L(\boldsymbol{q}) + \tilde{\sigma}^*(-\boldsymbol{q})T_L^*(-\boldsymbol{q})\}\exp(2\pi i\boldsymbol{q}\cdot\boldsymbol{u}_i)d\boldsymbol{q} \tag{66.12a}$$

$$C_p(\boldsymbol{u}_i) = \frac{1}{2i}\int_{\mathscr{A}} \{\tilde{\eta}(\boldsymbol{q})T_L(\boldsymbol{q}) - \tilde{\eta}^*(-\boldsymbol{q})\{T_L^*(-\boldsymbol{q})\}\exp(2\pi i\boldsymbol{q}\cdot\boldsymbol{u}_i)d\boldsymbol{q} \tag{66.12b}$$

These can be further simplified by means of various symmetry relations. First, we note that $\tilde{\sigma}^*(-\boldsymbol{q}) = \tilde{\sigma}(\boldsymbol{q})$ and $\tilde{\eta}^*(-\boldsymbol{q}) = \tilde{\eta}(\boldsymbol{q})$ since both $\sigma_o(\boldsymbol{u}_o)$ and $\eta_o(\boldsymbol{u}_o)$ are real functions. In addition, (65.30) shows that

$$T_L(-\boldsymbol{q}) = T_L(\boldsymbol{q}) \tag{66.13}$$

and consequently $T_L^*(-\boldsymbol{q}) = T_L^*(\boldsymbol{q})$. This remains true if *twofold axial astigmatism* is included but not in the general case when other terms of Eq. (65.29) are retained. With these symmetry relations, (66.12a) and (66.12b) become

$$C_a(\boldsymbol{u}_i) = \int_A \tilde{\sigma}(\boldsymbol{q})\Re T_L(\boldsymbol{q})\exp(2\pi i\boldsymbol{q}\cdot\boldsymbol{u_a})d\boldsymbol{q} \tag{66.14a}$$

$$C_p(\boldsymbol{u}_i) = \int_A \tilde{\eta}(\boldsymbol{q})\Im T_L(\boldsymbol{q})\exp(2\pi i\boldsymbol{q}\cdot\boldsymbol{u_a})d\boldsymbol{q} \tag{66.14b}$$

Each of these has the form of a Fourier transform and we, therefore, introduce their *contrast spectra*,

$$C_a(\boldsymbol{u}_i) =: \mathcal{F}(S_a), \quad S_a := \mathcal{F}^-(C_a) \tag{66.15a}$$

$$C_p(\boldsymbol{u}_i) =: \mathcal{F}(S_p), \quad S_p := \mathcal{F}^-(C_p) \tag{66.15b}$$

and the associated contrast-transfer functions,

$$K_a(\boldsymbol{q}) := \begin{cases} \Re T_L(\boldsymbol{q}) & \text{if } \boldsymbol{q} \in \mathcal{A} \\ 0 & \text{otherwise} \end{cases} \tag{66.16a}$$

$$K_p(\boldsymbol{q}) := \begin{cases} \Im T_L(\boldsymbol{q}) & \text{if } \boldsymbol{q} \in \mathcal{A} \\ 0 & \text{otherwise} \end{cases} \tag{66.16b}$$

or explicitly,

$$K_a(\boldsymbol{q}) := \begin{cases} \cos\chi(\boldsymbol{q}) & \text{if } \boldsymbol{q} \in \mathcal{A} \\ 0 & \text{otherwise} \end{cases} \tag{66.16c}$$

$$K_p(\boldsymbol{q}) := \begin{cases} -\sin\chi(\boldsymbol{q}) & \text{if } \boldsymbol{q} \in \mathcal{A} \\ 0 & \text{otherwise} \end{cases} \tag{66.16d}$$

We may then write Eqs. (66.14a) and (66.14b) in the form

$$S_a(\boldsymbol{q}) = K_a(\boldsymbol{q})\tilde{\sigma}(\boldsymbol{q}) \tag{66.17a}$$

$$S_p(\boldsymbol{q}) = K_p(\boldsymbol{q})\tilde{\eta}(\boldsymbol{q}) \tag{66.17b}$$

Recalling that (66.10b, 66.11a and 66.11b)

$$C(\boldsymbol{u}_i) = C_a(\boldsymbol{u_i}) + C_p(\boldsymbol{u_i})$$

and hence that

$$S_c(\boldsymbol{q}) = S_a(\boldsymbol{q}) + S_p(\boldsymbol{q})$$

we see that

$$S_c(\boldsymbol{q}) = K_a(\boldsymbol{q})\tilde{\sigma}(\boldsymbol{q}) + K_p(\boldsymbol{q})\tilde{\eta}(\boldsymbol{q}) \tag{66.17c}$$

Eq. (66.17c) is analogous to (65.16) but with important differences. The spectrum on the left-hand side is now given by the observed contrast and is hence a measurable quantity. There are, however, two terms on the right-hand side so that, even if the transfer functions could be inverted, a single image would not be sufficient to yield $\tilde{\sigma}$ and $\tilde{\eta}$. If σ_o is negligible, of course, this objection vanishes.

Fig. 66.2 shows the various stages of the reasoning for a weak-*phase* object, and this is to be compared with Fig. 65.3. The analogy is not exact: a minor difference is that it is no longer necessary to form the squared modulus in Fig. 66.2, since the contrast C is already the required quantity; more important is the fact that $S_p = K_p\tilde{\eta}$ is *not invertible*. We may not simply write $\tilde{\eta} = S_p/K_p$ because $K_p(\boldsymbol{q})$ has zeros and inevitably vanishes outside $\mathscr{A}$. The possibilities of object reconstruction from a recorded image are therefore limited. The same is true if we retain both the phase and the amplitude terms and record two images in different conditions, so that K_a and K_p are altered (typically by changing the defocus Δ_o). Since it is just such object reconstruction that is of paramount interest in the practical applications of electron microscopy, the theory has led us to a very unfortunate impasse! Many of the later chapters are devoted to attempts to overcome this obstacle. Electron holography (Chapter 63) and electron ptychography (Section 67.4) offer particularly attractive solutions.

We conclude this section with graphs of the phase shift χ and the contrast-transfer functions K_p and K_a in reduced coordinates for various values of the reduced defocus (D).

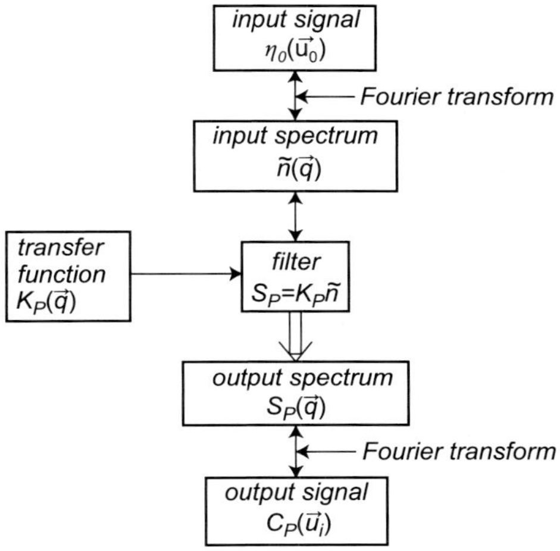

Figure 66.2
Bright-field image formation in terms of transfer functions. Only the phase term is shown.

Introducing definitions (65.31–65.33) into (66.16a), (66.16b), (66.16c) and (66.16d) and denoting the cut-off value of $\boldsymbol{q} = |\boldsymbol{q}|$ imposed by the aperture radius by $\boldsymbol{q}_c$, we write

$$Q_c = \left(C_s\lambda_o^3\right)^{1/4} q_c \tag{66.18}$$

and the transfer functions in terms of the reduced spatial frequency Q become

$$\hat{K}_a(Q)\begin{cases} \cos\pi\left(DQ^2 - \frac{1}{2}Q^4\right) & \text{for } Q \le Q_c \\ 0 & \text{for } Q > Q_c \end{cases} \tag{66.19a}$$

$$\hat{K}_p(Q)\begin{cases} \sin\pi\left(DQ^2 - \frac{1}{2}Q^4\right) & \text{for } Q \le Q_c \\ 0 & \text{for } Q > Q_c \end{cases} \tag{66.19b}$$

Fig. 66.3 shows χ/π as a function of Q for a range of values of D. Fig. 66.4A–D shows the behaviour of $K_p(Q)$ for $D = 0$, 1, $\sqrt{2}$ and $\sqrt{3}$, respectively; similar curves for $K_a(Q)$ are shown in Fig. 66.5. No cut-off value Q_c is shown. A common feature of all these curves is the onset of rapid oscillations shortly beyond the first or second zero. How are we to interpret these curves and what values of the parameters should we choose to obtain good image contrast? We defer consideration of these questions to Section 66.4.

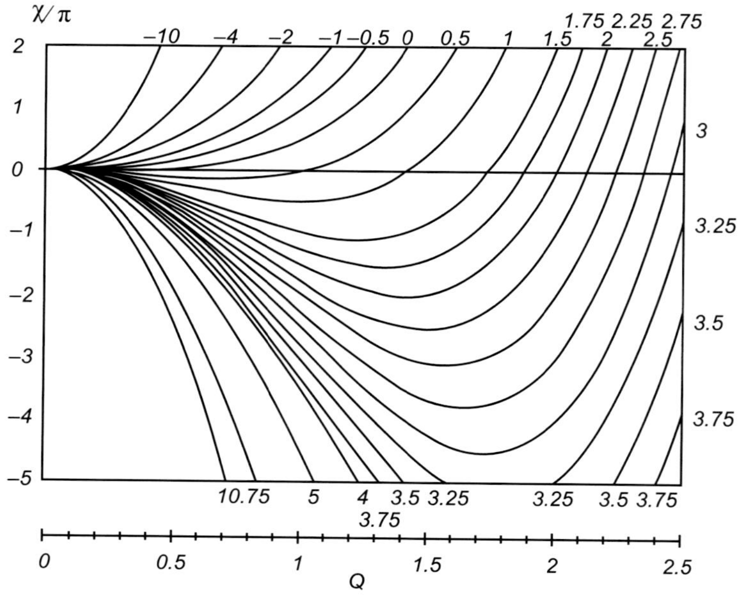

Figure 66.3
The phase shift $\chi(Q)/\pi$ as a function of Q. From (65.30) we have $\chi(Q)/\pi = Q^4/2 - DQ^2$. Each curve corresponds to the reduced defocus D indicated.

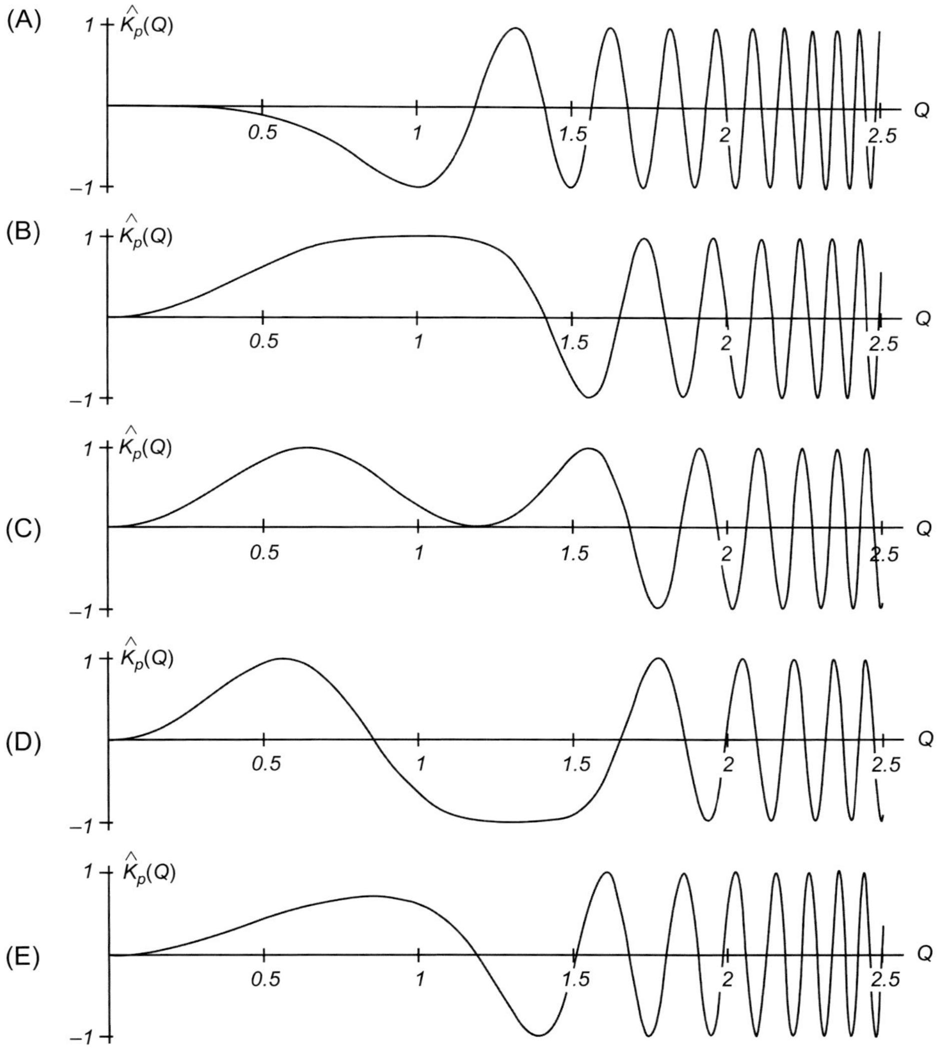

Figure 66.4
Phase contrast-transfer function $\hat{K}_p(Q)$. (A) $D = 0$. (B) $D = 1$. (C) $D = \sqrt{2}$. (*D*) $D = \sqrt{3}$. (E) $D = 1/\sqrt{2}$; this is Gabor focus, defined in Chapter 63, see (65.57).

66.1.1 Discussion

The phase and amplitude transfer functions for axial illumination were gradually developed during the 1960s, in particular by Hanszen and colleagues at the Physikalisch-Technische Bundesanstalt in Braunschweig but before tracing their work more closely, we point out that Fourier electron optics was nearly discovered by Glaser: equation (47.10) of

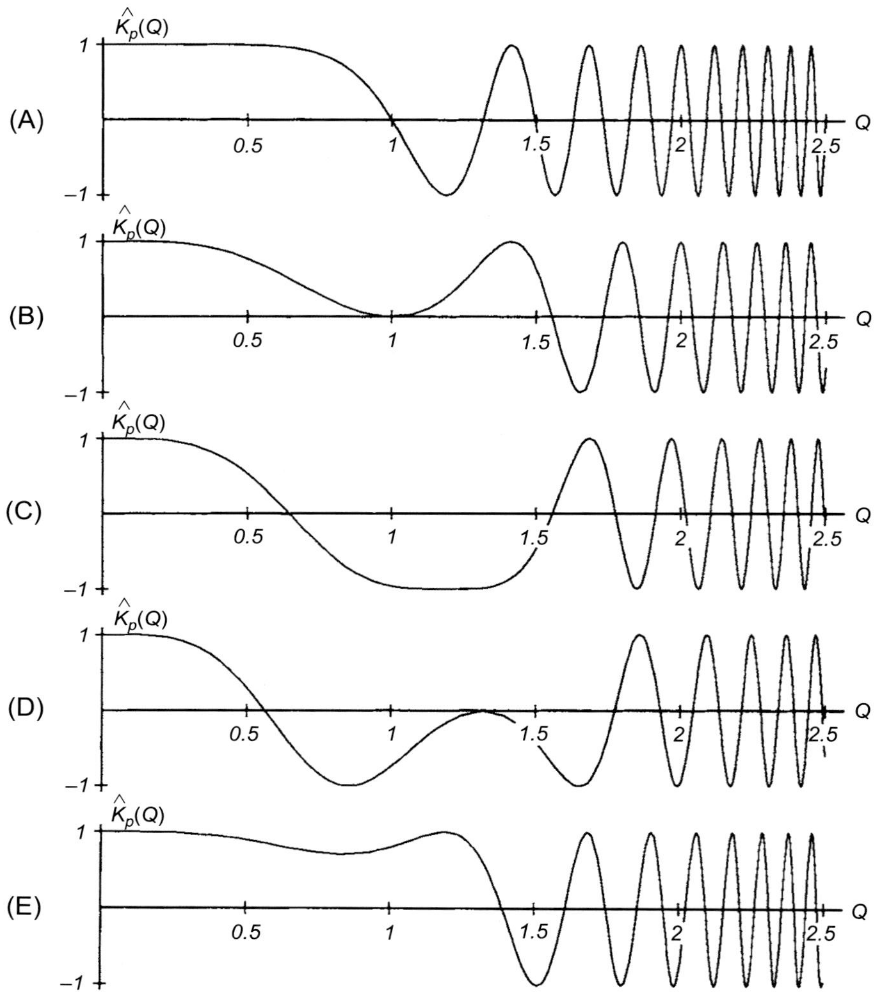

Figure 66.5
Amplitude contrast-transfer function $\hat{K}_a(Q)$. (A) $D = 0$, $\hat{K}_a = \cos(\pi Q^4/2)$. (B) $D = 1$, $\tilde{K}_a = \cos\{\pi(^2Q^2 - Q^4)/2\}$. (C) $D = \sqrt{2}$, $\hat{K}_a = \cos\{\pi(2\sqrt{2}Q^2 - Q^4)/2\}$. (D) $D = \sqrt{3}$, $\hat{K}_a = \cos\{\pi(2\sqrt{3}Q^2 - Q^4)/2\}$. (E) $D = 1/\sqrt{2}$, $\hat{K}_a = \cos\{\pi(\sqrt{2}Q^2 - Q^4)/2\}$.

Glaser (1956), which gives the image wavefunction, is already in the form of a convolution and the Fourier transform of this is at the heart of transfer theory!

The earliest discussion of contrast-transfer functions in electron optics is to be found in Hanszen et al. (1964) but the notion of such instrumental functions, although present (e.g. Eqs. 24b and 26), does not yet emerge clearly. Another paper by Hanszen et al. (1965) dealt with contrast transfer for incoherent electron imaging but the expressions for the phase and amplitude contrast-transfer functions with coherent illumination (monochromatic point source) first appear clearly in Hanszen and Morgenstern (1965) as Eqs. (32a) and (32b). The same expressions are presented more transparently in the paper by Hanszen (1966a), after which the notion of linear transfer between image contrast and object transparency (σ_o and η_o) could be considered well-established. Further publications (Hanszen, 1967, 1969, 1971) and a conference paper (Hanszen, 1966b) helped to make the theory better known but it was the dramatic optical diffraction patterns of amorphous specimens obtained by Thon (1966a, 1967, 1968a) in the University of Tübingen and presented at International and European electron microscopy conferences in Kyoto (Thon, 1966b) and Rome (Thon, 1968b) that excited widespread interest. If the specimen is a weak-phase object with a fairly 'white' spatial-frequency spectrum, so that $\tilde{\eta}$ is nearly constant for a good range of frequencies, the Fourier transform of the image contrast will show sin χ directly; the light-optical diffraction pattern of the micrograph will thus represent $\sin^2 \chi$. Thon's ring patterns (Fig. 66.6A and B), which reveal this sinusoidal variation arrestingly, were an easily grasped proof of the correctness of the approach. For an extended account, see Thon (1971) or Lenz (1971a,b), Hawkes (1973b) or Hanszen and Ade (1977). For thicker specimens a new effect occurs. As Tichelaar et al. (2020) have shown, the ring pattern is then the sum of the power spectra arising from thin slices of the film, scattering independently. The amplitudes of the Thon rings are then attenuated by a sinc function (sinc $(x) = \sin(\pi x)/\pi x$). Nodes appear in the pattern, which move to lower spatial frequencies with increasing thickness and/or decreasing accelerating voltage (Fig. 66.6C and D).

Once established, the theory was soon applied to many other imaging modes and the effects of partial source coherence were included. References are given in later sections. Hanszen and colleagues explored in very great detail the relation between in-line holography and transfer theory and their work may be traced with the aid of the fully referenced survey by Hanszen (1982). A very convenient classified bibliography of the contributions from the Physikalisch-Technische Bundesanstalt in Braunschweig has been prepared by Hanszen (1990).

The transfer functions did not, of course, emerge fully fledged in the mid-1960s; they had a long prehistory, which goes back to an early paper by Scherzer (1949) giving the wave aberration (χ of Eq. 65.30) in terms of spherical aberration and defocus. This was used on several occasions to interpret electron image formation close to the classical limit of resolution and the suggestions of Hoppe (1961, 1963) and Lenz (1963) for improving the electron image by means of zone plates, the rings of which correspond to the regions between successive zeros of sin χ, were inspired by it. Von Borries and Lenz (1956) and, more particularly, Lenz and Scheffels (1958) attempted to interpret the fine structure seen in high-resolution micrographs of supposedly amorphous specimens in the language of phase contrast but retained only the defocus term in the wave aberration, neglecting the spherical aberration (cf. Thon, 1965). Until the nature of phase contrast-transfer was fully elucidated (and indeed for some time afterwards), these fine structures were used by electron microscope manufacturers as a measure of the resolution of their instruments.

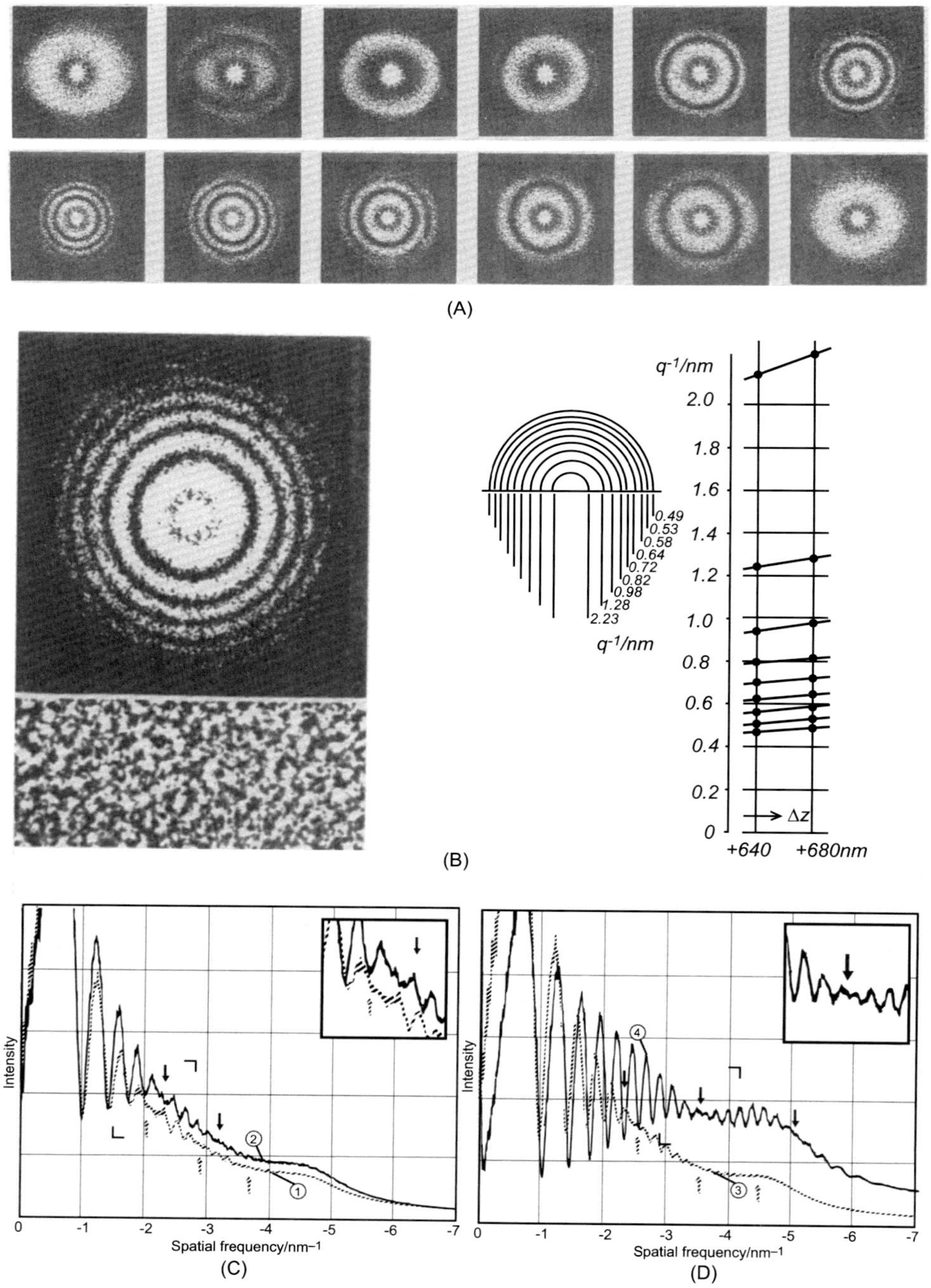

Figure 66.6

Diffractograms of a thin carbon film. (A) Focal series for values of defocus ranging from −280 to 519 nm. (B) $D = 659 \pm 19$ nm. $C_s = 4$ mm, accelerating voltage = 100 kV. (Owing to residual astigmatism D is 30 nm greater in the vertical direction.) (C) and (D) Rotational averages of power spectra (C) of images of a triple layer of carbon at various accelerating voltages. Nodes are indicated by arrows. *After Thon (1968), Courtesy Friedrich Thon. (B) After Thon (1971). (C) After Tichelaar et al. (2020), Courtesy Academic Press/Elsevier.*

During the 1970s, the use of transfer functions became common and they began to make their appearance in commercial brochures. An imaging mode employed for studying magnetic specimens that were analysed by Guigay et al. (1971) reminds us of the early explorations of Lenz and Scheffels; here the defocus has to be large and the spherical aberration can indeed be neglected. Moreover, the weak-phase constraint is considerably less stringent.

66.2 Spectral Distributions of the Illumination

The situation considered above is highly idealized in various ways: the condenser system has been assumed to furnish an ideal plane and monoenergetic wave, propagating exactly in the axial (z) direction. This is clearly unrealistic for, even ignoring the condenser lens properties, it would require a vanishingly small source emitting electrons at exactly the same energy. Electron sources do approximate this ideal but the small emitting area and the energy spread have some effect, which may be regarded as a perturbation of the ideal case so far considered. In this section, we establish formulae describing this perturbation.[3] Non-vanishing source size and non-vanishing energy spread may be regarded as *spectral effects:* the former is characterized by a spectrum of wavevectors $\boldsymbol{k} = (k_x, k_y, k_z)$, with $k_x, k_y \ll k_z$, instead of a single component $\boldsymbol{k} = (0, 0, k_z)$ and the latter by a range of values of $|\boldsymbol{k}|$.

1. *Non-vanishing source size*. The wavevector $\boldsymbol{k}$ now has small transverse components. It will be convenient to introduce the *two-dimensional* vector $\boldsymbol{\kappa}$,

$$\boldsymbol{\kappa} = \left(\kappa_x, \kappa_y\right) := \frac{1}{2\pi}\left(k_x, k_y\right) \tag{66.20}$$

   and we shall always have $\boldsymbol{\kappa}^2 = \boldsymbol{\kappa}_x^2 + \boldsymbol{\kappa}_y^2 \ll \lambda^{-2}$, since the slopes of the electron trajectories are very small, typically $\kappa\lambda \lesssim 10^{-3}$ rad. We saw in Chapter 64 how the angles at the specimen ($\boldsymbol{\theta}$) are related to the transverse dimensions of the source: for parallel illumination, $|\boldsymbol{\theta}| = r_s/f_i$ in which r_s is the crossover radius and f_i is the image focal length of the entire condenser system.
2. *Non-vanishing energy spread*. The wavenumber $k = |\boldsymbol{k}|$ now has a spectrum of non-vanishing width. The value $k_o = 2\pi/\lambda_o$ that we have used so far refers to the maximum of this spectrum. The various forms of this spectrum are discussed in Chapter 44 of Volume 2.

We recall that k is given (Eq. 56.12) by $k(\boldsymbol{r}) = \{2m_o e\hat{\Phi}(\boldsymbol{r}) + 2Em(\boldsymbol{r})\}^{1/2}/\hbar$. The specimen may well be situated within the field of a magnetic lens but there will certainly be no external electrostatic field and we may therefore replace $\hat{\Phi}$ by the constant value $\hat{U}$ and

[3] It is usual to discuss these perturbing influences in the language of partial coherence: non-vanishing source size is assimilated to partial spatial coherence and energy spread to partial temporal coherence. (The terms lateral and longitudinal coherence are also used.) Coherence is discussed more formally in Part XVI; in this chapter, we shall usually speak of source size and energy spread rather than of partial coherence.

regard $m = \gamma m_0$ as a constant. With no loss of generality, we may choose the energy scale in such a way that $E = 0$ corresponds to the maximum of the energy spectrum; $|E|$ will hence be very small and the above expression for k can be expanded as a Taylor series:

$$\begin{aligned} k_o &= \frac{2\pi}{\lambda_o} = \frac{1}{\hbar}\sqrt{2m_o e\hat{U}} \\ k &= k_o + \frac{mE}{\hbar^2 k_o} + O(E^2) \end{aligned} \tag{66.21}$$

The energy E is not very convenient as a parameter in the subsequent calculations and we, therefore, introduce its *volt equivalent* υ, $E =: e\upsilon$, whereupon the expression for k in (66.21) becomes

$$k = k_o + \frac{em}{\hbar^2 k_o}\upsilon \equiv k_o\left(1 + \frac{\gamma\upsilon}{2\hat{U}}\right) \tag{66.22}$$

where as usual (Section 2.3 of volume 1) $\gamma = 1 + 2\varepsilon U_o$, $\hat{U} = U_o(1 + \varepsilon U_o)$ and $\varepsilon = e/2m_o c^2$. The quantity υ that characterizes the energy spread is usually such that $|\upsilon| \lesssim 1$ V; see Part IX of Volume 2 for extensive discussion.

The incident electrons are now *wave packets*. It is not advantageous to use the general form for these, which would be unnecessarily complicated. Instead, we consider the *partial waves* building them up, which offer an easily understood and physically correct model. Each partial wave is given by

$$\psi_p(\boldsymbol{r}) = A(\boldsymbol{\kappa}, \upsilon)\exp\{2\pi i(x\kappa_x + y\kappa_y)\}\exp\left\{i\left(k_o + \frac{em\upsilon}{\hbar^2 k_o} - \frac{2\pi^2\kappa^2}{k_o}\right)z\right\} \tag{66.23}$$

from which the unimportant time factor $\exp(-iEt/\hbar)$ has been omitted. This expression is valid in the half-space $z \le 0$ upstream from the specimen. The amplitude A will be specified below. For the moment, we just impose the normalization

$$\iint |A(\boldsymbol{\kappa}, \upsilon)|^2 d\boldsymbol{\kappa}\, d\upsilon = 1 \tag{66.24}$$

In the entrance plane of the specimen where $z = 0$, Eq. (66.23) simplifies to

$$\psi_p(\boldsymbol{u_o}, 0) = A(\boldsymbol{\kappa}, \upsilon)\exp(2\pi i\boldsymbol{\kappa}\cdot\boldsymbol{u}_o) \tag{66.25}$$

Note that the energy does not appear in the exponential here. The specimen is assumed to be so thin that, in a first-order approximation, the phase shift η_o is unaffected by υ. At the exit surface of the specimen, therefore, the wavefunction is given by

$$\psi_o(\boldsymbol{u}_o, \boldsymbol{\kappa}, \upsilon) = \{1 + i\eta_o(\boldsymbol{u}_o)\}\exp(2\pi i\boldsymbol{\kappa}\cdot\boldsymbol{u}_o)A(\boldsymbol{\kappa}, \upsilon) \tag{66.26}$$

We shall not consider the absorption term $\sigma_o(\boldsymbol{u}_o)$ here; the calculation can be straightforwardly extended to include it if required.

The object spectrum S_o now becomes

$$\begin{aligned} S_o(\boldsymbol{q}, \boldsymbol{\kappa}, \upsilon) &= \int \psi_o(\boldsymbol{u}_o, \boldsymbol{\kappa}, \upsilon)\exp(-2\pi \mathrm{i}\boldsymbol{q}\cdot\boldsymbol{u}_o)d\boldsymbol{u}_o \\ &= A(\boldsymbol{\kappa}, \upsilon)\int\{1 + \mathrm{i}\eta(\boldsymbol{u}_o)\}\exp\{2\pi\mathrm{i}(\boldsymbol{\kappa} - \boldsymbol{q})\cdot\boldsymbol{u}_o\}d\boldsymbol{u}_o \end{aligned}$$

which, recalling (66.3) and taking into account a new effect, namely, the shift of spatial frequency from $\boldsymbol{q}$ to $\boldsymbol{q} - \boldsymbol{\kappa}$, may be written as follows:

$$S_o(\boldsymbol{q}, \boldsymbol{\kappa}, \upsilon) = A(\boldsymbol{\kappa}, \upsilon)\{\delta(\boldsymbol{q} - \boldsymbol{\kappa}) + \mathrm{i}\tilde{\eta}(\boldsymbol{q} - \boldsymbol{\kappa})\} \tag{66.27}$$

This spectrum is again transferred to the image as shown in Fig. 65.3 but the wave transfer function T now depends explicitly on υ, so that $T = T(\boldsymbol{q}, \upsilon)$. The main effect is due to the *axial chromatic aberration* C_c of the objective lens. We give the explicit formula later; in the general calculation, we simply need to recall that υ is now an argument of T.

For the image spectrum, we have

$$S_i(\boldsymbol{q}, \boldsymbol{\kappa}, \upsilon) = A(\boldsymbol{\kappa}, \upsilon)T(\boldsymbol{q}, \upsilon)\{\delta(\boldsymbol{q} - \boldsymbol{\kappa}) + \mathrm{i}\tilde{\eta}(\boldsymbol{q} - \boldsymbol{\kappa})\}$$

which is inverted to give

$$\psi_i(\boldsymbol{u}_i, \boldsymbol{\kappa}, \upsilon) = \int S_i(\boldsymbol{q}, \boldsymbol{\kappa}, \upsilon)\exp(2\pi\mathrm{i}\boldsymbol{q}\cdot\boldsymbol{u}_i)d\boldsymbol{q}.$$

The term involving the δ-function is easily integrated and we find

$$\begin{aligned} \psi_i(\boldsymbol{u}_i, \boldsymbol{\kappa}, \upsilon) &= A(\boldsymbol{\kappa}, \upsilon)T(\boldsymbol{\kappa}, \upsilon)\exp(2\pi\mathrm{i}\boldsymbol{\kappa}\cdot\boldsymbol{u}_i) \\ &+ \mathrm{i}A(\boldsymbol{\kappa}, \upsilon)\int T(\boldsymbol{q}, \upsilon)\tilde{\eta}(\boldsymbol{q} - \boldsymbol{\kappa})\exp(2\pi\mathrm{i}\boldsymbol{q}\cdot\boldsymbol{u}_i)d\boldsymbol{q} \end{aligned}$$

As usual, we neglect quadratic terms in $\tilde{\eta}$ when calculating $|\psi_i|^2$ and we also assume that $|T(\boldsymbol{\kappa}, \upsilon)|^2 = 1$. Then

$$|\psi_i(\boldsymbol{u}_i, \boldsymbol{\kappa}, \upsilon)|^2 = |A(\boldsymbol{\kappa}, \upsilon)|^2(1 - 2\Im\mu_i) \tag{66.28a}$$

where

$$\mu_i := \int T(\boldsymbol{q}, \upsilon)T^*(\boldsymbol{\kappa}, \upsilon)\tilde{\eta}(\boldsymbol{q} - \boldsymbol{\kappa})\exp\{2\pi\mathrm{i}(\boldsymbol{q} - \boldsymbol{\kappa})\cdot\boldsymbol{u}_i\}d\boldsymbol{q} \tag{66.28b}$$

This is the image intensity for a single partial wave. In order to obtain the contrast, we must integrate over the angular and energy spreads, $\boldsymbol{\kappa}$ and υ, for a definition such as (66.10a) involves the total intensity. By integrating the partial intensity (66.28a) and not the partial wave $\psi_i(\boldsymbol{u}_i, \boldsymbol{\kappa}, \upsilon)$ over $\boldsymbol{\kappa}$ and υ, we are assuming that the partial waves are completely independent, that there is no relation (on average) between electron emission at different energies and from different points on the source. We shall return to this question in Part XVI.

The integrated intensity, in which the contributions of all the partial intensities are added, is given by

$$I = \iint \left|\psi_i(\boldsymbol{u_i}, \boldsymbol{\kappa}, \upsilon)\right|^2 d\boldsymbol{\kappa} d\upsilon$$

$$= 1 - 2\iiint \left|A(\boldsymbol{\kappa}, \upsilon)\right|^2 \Im[T(q,\upsilon)T^*(\boldsymbol{\kappa},\upsilon)\tilde{\eta}(\boldsymbol{q} - \boldsymbol{\kappa})$$

$$\times \exp\{2\pi \mathrm{i}(\boldsymbol{q} - \boldsymbol{\kappa})\cdot \boldsymbol{u}_i\}]d\boldsymbol{q}\, d\upsilon\, d\boldsymbol{\kappa} \tag{66.29}$$

The image (phase) contrast is now

$$C_p(\boldsymbol{u_i}) = \frac{1 - I(\boldsymbol{u_i})}{1 + I(\boldsymbol{u_i})} \tag{66.30}$$

(cf. Eq. 66.10a) and, in the linear approximation, this can be written in the form

$$C_p(\boldsymbol{u_i}) = \Im\left[\int \left\{\iint |A(\boldsymbol{\kappa}, \upsilon)|^2 T(\boldsymbol{q} + \boldsymbol{\kappa}, \upsilon)T^*(\boldsymbol{\kappa}, \upsilon)d\upsilon\, d\boldsymbol{\kappa}\right\} \tilde{\eta}\,(\boldsymbol{q})\exp(2\pi \mathrm{i}\boldsymbol{q}\cdot \boldsymbol{u}_i)d\boldsymbol{q}\right] \tag{66.31}$$

after a change in the order of integration and a change of variable from $\boldsymbol{q}$ to $\boldsymbol{q} - \boldsymbol{\kappa}$.

We now seek to rearrange this expression in such a way that it has the form of a Fourier integral with respect to $\boldsymbol{q}$. This will enable us to write down the contrast *spectrum*, $S_c(\boldsymbol{q})$, immediately and hence to extract a contrast-transfer function $K_p(\boldsymbol{q})$. We first write out the imaginary part explicitly and make the double substitution, $\boldsymbol{q} \to -\boldsymbol{q}$ and $\boldsymbol{\kappa} \to -\boldsymbol{\kappa}$ in the final term, as in (66.12a) and (66.12b). We obtain

$$C_p(\boldsymbol{u_i}) = \frac{1}{2}\int \left\{\iint |A(\boldsymbol{\kappa}, \upsilon)|^2 T(\boldsymbol{q} + \boldsymbol{\kappa}, \upsilon)T^*(\boldsymbol{\kappa}, \upsilon)d\upsilon\, d\boldsymbol{\kappa}\right\}$$

$$\times\, \tilde{\eta}(\boldsymbol{q})\exp(2\pi \mathrm{i}\boldsymbol{q}\cdot \boldsymbol{u}_i)d\boldsymbol{q}$$

$$-\frac{1}{2\mathrm{i}}\int \left\{\iint |A(-\boldsymbol{\kappa}, \upsilon)|^2 T^*(-\boldsymbol{q} - \boldsymbol{\kappa}, \upsilon)T(-\boldsymbol{\kappa}, \upsilon)d\upsilon\, d\boldsymbol{\kappa}\right\}$$

$$\times\, \tilde{\eta}^*(-\boldsymbol{q})\exp(2\pi \mathrm{i}\boldsymbol{q}\cdot \boldsymbol{u}_i)d\boldsymbol{q}$$

We must again assume that the aperture is symmetric about the axis. Even in the presence of axial chromatic aberration, the wave transfer function T satisfies

$$T(-\bar{\boldsymbol{q}}) = T(\bar{\boldsymbol{q}}) \tag{66.32}$$

for all $\bar{\boldsymbol{q}}$, and in particular for $\bar{\boldsymbol{q}} = \boldsymbol{q} + \boldsymbol{\kappa}$ and for $\bar{\boldsymbol{q}} = \boldsymbol{\kappa}$. We must also assume that the illumination is rotationally symmetric, so that

$$\left|A(-\boldsymbol{\kappa}, \upsilon)\right|^2 = |A(\boldsymbol{\kappa}, \upsilon)|^2 \tag{66.33}$$

Finally, we know that $\tilde{\eta}^*(-\boldsymbol{q}) = \tilde{\eta}(\boldsymbol{q})$ since η is real. Bringing all these symmetries together, we find

$$C_p(\boldsymbol{u}_i) = \int_{\boldsymbol{q}} \left[\int_{\boldsymbol{\kappa}} \cdot \int_{\upsilon} |A(\boldsymbol{\kappa}, \upsilon)|^2 \frac{1}{2\mathrm{i}} \{T(\boldsymbol{q} + \boldsymbol{\kappa}, \upsilon) T^*(\boldsymbol{\kappa}, \upsilon) - \text{c.c.}\} d\upsilon \, d\boldsymbol{\kappa} \right] \times \tilde{\eta}(\boldsymbol{q}) \exp(2\pi \mathrm{i} \boldsymbol{q} \cdot \boldsymbol{u}_i) d\boldsymbol{q}$$
$$=: \int_{\boldsymbol{q}} \boldsymbol{K}(\boldsymbol{q}) \tilde{\eta}(\boldsymbol{q}) \exp(2\pi \mathrm{i} \boldsymbol{q} \cdot \boldsymbol{u}_i) d\boldsymbol{q} \qquad (66.34)$$

(c.c. signifies 'complex conjugate.') This has the desired appearance, and we see that

$$K_p(\boldsymbol{q}) = \int_{\boldsymbol{\kappa}} \cdot \int_{\upsilon} |A(\boldsymbol{\kappa}, \upsilon)|^2 \Im\{T(\boldsymbol{q} + \boldsymbol{\kappa}, \upsilon) T^*(\boldsymbol{\kappa}, \upsilon)\} d\upsilon \, d\boldsymbol{\kappa} \qquad (66.35)$$

is the generalized form of the transfer function. It should contain our earlier results (66.16a, 66.16b, 66.16c and 66.16d) as a special case. The latter corresponds to infinitely narrow spectra:

$$|A(\boldsymbol{\kappa}, \upsilon)|^2 \rightarrow \delta(\kappa_x) \delta(\kappa_y) \delta(\upsilon) \qquad (66.36)$$

which satisfies (66.24); the integration in (66.35) may then be performed, giving

$$K_p(\boldsymbol{q}) = \Im\{T(\boldsymbol{q}, 0) T^*(0, 0)\}$$
$$= \Im T(\boldsymbol{q}, 0)$$

as we expect, since $T(0,0) = 1$.

Examination of the expression (66.35) for the transfer function $K_p(\boldsymbol{q})$ shows that both instrumental parameters associated with the image-forming system (T) and parameters describing the illumination (A) are involved. We now enquire to what extent these can be separated before examining specific forms of $A(\boldsymbol{\kappa}, \upsilon)$ in Section 66.3.

For this, we must study the function $T(\boldsymbol{q}, \upsilon)$ in more detail. The associated wave aberration W including a chromatic aberration term is now

$$W = \frac{1}{4} C_s \theta_o^4 - \frac{1}{2} \Delta_o \theta_o^2 - \frac{1}{2} C_c \frac{\gamma \upsilon}{\hat{U}} \theta_o^2 \qquad (66.37)$$

in which C_c is the familiar (axial) chromatic aberration coefficient (Chapter 26) and $\hat{U}$ is again the relativistic accelerating voltage. In the relation between θ_o and $\boldsymbol{q}$, $\theta_o = \lambda_o \boldsymbol{q}$ (65.18), we use the nominal value λ_o given by (66.21) even though there is now a spread of wavelengths. This is permissible since W consists of small aberration terms and if we retained terms in $\lambda - \lambda_o$ in W we should obtain aberrations of higher order. The function $T(\boldsymbol{q}, \upsilon)$, which is the generalization of Eq. (65.30), is therefore given by

$$T(\boldsymbol{q},\upsilon)=\exp\left[\mathrm{i}\pi\left\{-\frac{1}{2}C_s\lambda_o^3\left(q_x^2+q_y^2\right)^2+\lambda_o\Delta_o\left(q_x^2+q_y^2\right)+\lambda_o C_c\gamma\frac{\upsilon}{\hat{U}}\left(q_x^2+q_y^2\right)\right\}\right] \tag{66.38}$$

and it will be convenient to write the combination $T(\boldsymbol{q}+\boldsymbol{\kappa}, \upsilon)T^*(\boldsymbol{\kappa}, \upsilon)$ in the following form:

$$T(\boldsymbol{q}+\boldsymbol{\kappa},\upsilon)T^*(\boldsymbol{\kappa},\upsilon)=:\exp\{\mathrm{i}(T_s+T_\Delta+T_\upsilon\upsilon)\} \tag{66.39}$$

with

$$\begin{aligned}
T_s&=\frac{\pi}{2}C_s\lambda_o^3\left[\{(\boldsymbol{q}+\boldsymbol{\kappa})\cdot(\boldsymbol{q}+\boldsymbol{\kappa})\}^2-(\boldsymbol{\kappa}\cdot\boldsymbol{\kappa})^2\right]\\
&=\frac{\pi}{2}C_s\lambda_o^3\{(\boldsymbol{q}\cdot\boldsymbol{q})^2+4(\boldsymbol{q}\cdot\boldsymbol{\kappa})^2+4(\boldsymbol{q}\cdot\boldsymbol{q})(\boldsymbol{q}\cdot\boldsymbol{\kappa})\\
&\quad+2(\boldsymbol{q}\cdot\boldsymbol{q})(\boldsymbol{\kappa}\cdot\boldsymbol{\kappa})+4(\boldsymbol{q}\cdot\boldsymbol{\kappa})(\boldsymbol{\kappa}\cdot\boldsymbol{\kappa})\}\\
T_\Delta&:=\pi\Delta_o\lambda_o\{(\boldsymbol{q}+\boldsymbol{\kappa})\cdot(\boldsymbol{q}+\boldsymbol{\kappa})-\boldsymbol{\kappa}\cdot\boldsymbol{\kappa}\}\\
&=\pi\Delta_o\lambda_o(\boldsymbol{q}\cdot\boldsymbol{q}+2\boldsymbol{q}\cdot\boldsymbol{\kappa})\\
T_\upsilon&:=\pi\frac{\lambda_o C_c\gamma}{\hat{U}}\{(\boldsymbol{q}+\boldsymbol{\kappa})\cdot(\boldsymbol{q}+\boldsymbol{\kappa})-\boldsymbol{\kappa}\cdot\boldsymbol{\kappa}\}=:\pi\lambda_o C_c'(\boldsymbol{q}\cdot\boldsymbol{q}+2\boldsymbol{q}\cdot\boldsymbol{\kappa})
\end{aligned} \tag{66.40}$$

and

$$C_c':=\gamma C_c/\hat{U}.$$

Returning to Eq. (66.35), we see that $K_p(\boldsymbol{q})$ can be written in the form

$$\begin{aligned}
K_p(\boldsymbol{q})&=\frac{1}{2\mathrm{i}}\iint|A(\boldsymbol{\kappa},\upsilon)|^2[\exp\{\mathrm{i}(-T_s+T_\Delta+T_\upsilon\upsilon)\}\\
&\quad-\exp\{-\mathrm{i}(-T_s+T_\Delta+T_\upsilon\upsilon)\}]d\boldsymbol{\kappa}\,d\upsilon\\
&=\frac{1}{2\mathrm{i}}\iint[|A(\boldsymbol{\kappa},\upsilon)|^2\exp\{-\mathrm{i}(T_s-T_\Delta)\}\\
&\quad-|A(\boldsymbol{\kappa},-\upsilon)|^2\exp\{\mathrm{i}(T_s-T_\Delta)\}]\exp(\mathrm{i}T_\upsilon\upsilon)d\boldsymbol{\kappa}\,d\upsilon
\end{aligned} \tag{66.41}$$

The distribution $A(\boldsymbol{\kappa},\upsilon)$ is unlikely to be exactly even in υ (though it is often assumed to be and we shall model it by an even function in the next section) and we, therefore, write

$$|A(\boldsymbol{\kappa},\upsilon)|^2=:a_e+\mathrm{i}a_o \tag{66.42}$$

in which the even and odd contributions are given by

$$a_e := \frac{1}{2}\left\{\left|A(\boldsymbol{\kappa}, \upsilon)\right|^2 + \left|A(\boldsymbol{\kappa}, -\upsilon)\right|^2\right\}$$
$$a_o := \frac{1}{2i}\left\{\left|A(\boldsymbol{\kappa}, \upsilon)\right|^2 - \left|A(\boldsymbol{\kappa}, -\upsilon)\right|^2\right\} \tag{66.43}$$

so that $K_p(\boldsymbol{q})$ becomes

$$\begin{aligned} K_p(\boldsymbol{q}) &= \frac{1}{2\mathrm{i}}\iint a_e\left[\mathrm{e}^{-\mathrm{i}(T_s-T_\Delta)} - \mathrm{e}^{\mathrm{i}(T_s-T_\Delta)}\right]\mathrm{e}^{\mathrm{i}T_\upsilon \upsilon} d\boldsymbol{\kappa}\, d\upsilon \\ &+ \frac{1}{2}\iint a_o\left[\mathrm{e}^{-\mathrm{i}(T_s-T_\Delta)} + \mathrm{e}^{\mathrm{i}(T_s-T_\Delta)}\right]\mathrm{e}^{\mathrm{i}T_\upsilon \upsilon} d\boldsymbol{\kappa}\, d\upsilon \\ &= \int\left\{\tilde{a}_e \sin(T_\Delta - T_s) + \tilde{a}_o \cos(T_\Delta - T_s)\right\}d\boldsymbol{\kappa} \end{aligned} \tag{66.44}$$

The quantities $\tilde{a_e}$ and $\tilde{a_o}$ are, with the appropriate scaling, the Fourier transforms with respect to υ of the even and odd parts of the source distribution:

$$\tilde{a}_e := \int a_e \exp(\mathrm{i}T_\upsilon \upsilon)d\upsilon, \quad \tilde{a}_o = \int a_o \exp(\mathrm{i}T_\upsilon \upsilon)d\upsilon \tag{66.45}$$

Eq. (66.44) is quite close to the representation we are seeking but requires further manipulation, particularly as $\boldsymbol{\kappa}$ appears in the arguments of $\tilde{a_e}$ and $\tilde{a_o}$. Let us now examine the terms in $T_\Delta - T_s$ in more detail.

From the definitions (66.40), we see that

$$\begin{aligned} T_\Delta - T_s &= \pi\Delta_o\lambda_o(\boldsymbol{q}\cdot\boldsymbol{q} + 2\boldsymbol{q}\cdot\boldsymbol{\kappa}) \\ &- \frac{\pi}{2}C_s\lambda_o^3\{(\boldsymbol{q}\cdot\boldsymbol{q})^2 + 4(\boldsymbol{q}\cdot\boldsymbol{\kappa})^2 + 4(\boldsymbol{q}\cdot\boldsymbol{q})(\boldsymbol{q}\cdot\boldsymbol{\kappa}) \\ &+ 2(\boldsymbol{q}\cdot\boldsymbol{q})(\boldsymbol{\kappa}\cdot\boldsymbol{\kappa}) + 4(\boldsymbol{\kappa}\cdot\boldsymbol{\kappa})(\boldsymbol{q}\cdot\boldsymbol{\kappa})\} \\ &=: -\chi(\boldsymbol{q}) + T(\boldsymbol{q}, \boldsymbol{\kappa}) \end{aligned} \tag{66.46}$$

with

$$\begin{aligned} \chi(\boldsymbol{q}) &= \frac{\pi}{2}C_s\lambda_o^3\boldsymbol{q}^4 - \pi\Delta_o\lambda_o\boldsymbol{q}^2 \\ T(\boldsymbol{q}, \boldsymbol{\kappa}) &= 2\pi\Delta_o\lambda_o\boldsymbol{q}\cdot\boldsymbol{\kappa} - \frac{\pi}{2}C_s\lambda_o^3\{4\boldsymbol{q}^2\boldsymbol{q}\cdot\boldsymbol{\kappa} + 4(\boldsymbol{q}\cdot\boldsymbol{\kappa})^2 + 2\boldsymbol{q}^2\boldsymbol{\kappa}^2 + 4\boldsymbol{\kappa}^2(\boldsymbol{q}\cdot\boldsymbol{\kappa})\} \end{aligned} \tag{66.47}$$

Hence, $K_p(\boldsymbol{q})$ may be written

$$\begin{aligned} K_p(\boldsymbol{q}) &= -\sin\chi(\boldsymbol{q})\int\left\{\tilde{a}_e\cos T(\boldsymbol{q},\boldsymbol{\kappa}) - \tilde{a}_o\sin T(\boldsymbol{q},\boldsymbol{\kappa})\right\}d\boldsymbol{\kappa} \\ &+ \cos\chi(\boldsymbol{q})\int\left\{\tilde{a}_e\sin T(\boldsymbol{q},\boldsymbol{\kappa}) + \tilde{a}_o\cos T(\boldsymbol{q},\boldsymbol{\kappa})\right\}d\boldsymbol{\kappa} \end{aligned} \tag{66.48}$$

and we have thus succeeded in showing that non-vanishing source size and energy spread lead to a *modulation* of the ideal transfer functions, sin χ and cos χ; if the energy spread is not even in υ, there will be some intermixing of the ideal functions or, in other words, a shift of the zeros. If $|A(\boldsymbol{\kappa}, \upsilon)|^2$ is even in υ, then of course a_o, $\tilde{a}_o$ vanish, leaving

$$K_p(\boldsymbol{q}) = -\sin\chi(\boldsymbol{q}) \int_{\boldsymbol{\kappa}} \tilde{a}_e \cos T(\boldsymbol{q}, \boldsymbol{\kappa}) d\boldsymbol{\kappa} + \cos\chi(\boldsymbol{q}) \int_{\boldsymbol{\kappa}} \tilde{a}_e \sin T(\boldsymbol{q}, \boldsymbol{\kappa}) d\boldsymbol{\kappa} \tag{66.49}$$

The definition of $T(\boldsymbol{q}, \boldsymbol{\kappa})$, (66.47), shows that both odd and even terms in $\boldsymbol{\kappa}$ are present and we could write $T(\boldsymbol{q}, \boldsymbol{\kappa}) =: T_e + T_o$, expand the terms in cos $T(\boldsymbol{q}, \boldsymbol{\kappa})$ and sin $T(\boldsymbol{q}, \boldsymbol{\kappa})$ and analyse the integrals further. This is fairly laborious since, even if (as has been assumed) $|A(\boldsymbol{\kappa}, \upsilon)|^2$ is even in $\boldsymbol{\kappa}$ (symmetric illumination spread), the symmetry of a_e with respect to $\boldsymbol{\kappa}$ is more complicated ($a_e(\boldsymbol{\kappa}, \boldsymbol{q}) = a_e(-\boldsymbol{\kappa}, -\boldsymbol{q})$). We shall not pursue this here. The final question concerns the interaction between energy spread, characterized by υ, and source size, measured by $\boldsymbol{\kappa}$. In what circumstances can we write the integral over $\boldsymbol{\kappa}$ (66.48 or 66.49) as the product of a term arising from $\boldsymbol{\kappa}$ and another term involving υ? As before, we assume for simplicity that $|A(\boldsymbol{\kappa}, \upsilon)|^2$ is even in υ; a more complicated reply can be found if this assumption is not made.

We first require that the roles of $\boldsymbol{\kappa}$ and υ in $A(\boldsymbol{\kappa}, \upsilon)$ be separated:

$$A(\boldsymbol{\kappa}, \upsilon) =: A_{\boldsymbol{\kappa}}(\boldsymbol{\kappa}) A_\upsilon(\upsilon) = A_{\boldsymbol{\kappa}}(\boldsymbol{\kappa}) A_\upsilon(-\upsilon) \tag{66.50}$$

Hence

$$a_e = |A_{\boldsymbol{\kappa}}(\boldsymbol{\kappa})|^2 |A_\upsilon(\upsilon)|^2, \quad a_o = 0 \tag{66.51}$$

and

$$\tilde{a}_e = |A_{\boldsymbol{\kappa}}(\boldsymbol{\kappa})|^2 \int |A_\upsilon(\upsilon)|^2 \exp\{i\pi C_c'(\boldsymbol{q}^2 + 2\boldsymbol{q}\cdot\boldsymbol{\kappa})\upsilon\} d\upsilon \tag{66.52}$$

If we can neglect $\boldsymbol{q}\cdot\boldsymbol{\kappa}$ in the exponent, then

$$\tilde{a}_e \approx |A_{\boldsymbol{\kappa}}(\boldsymbol{\kappa})|^2 t_\upsilon(\boldsymbol{q}) \tag{66.53}$$

with

$$t_\upsilon(\boldsymbol{q}) := \int |A_\upsilon(\upsilon)|^2 \exp(i\pi C_c' \boldsymbol{q}^2 \upsilon) d\upsilon \tag{66.54}$$

and as required $t_\upsilon(\boldsymbol{q})$ is independent of $\boldsymbol{\kappa}$. Thus $K_p(\boldsymbol{q})$ now reduces to

$$K_p = -\sin\{\chi(\boldsymbol{q})\} t_\upsilon(\boldsymbol{q}) t_{\boldsymbol{\kappa}}^{(c)}(\boldsymbol{q}) + \cos\{\chi(\boldsymbol{q})\} t_\upsilon(\boldsymbol{q}) t_{\boldsymbol{\kappa}}^{(s)}(\boldsymbol{q}) \tag{66.55}$$

with

$$t_{\boldsymbol{\kappa}}^{(c)}(\boldsymbol{q}) := \int_{\boldsymbol{\kappa}} |A_{\boldsymbol{\kappa}}(\boldsymbol{\kappa})|^2 \cos\{T(\boldsymbol{q}, \boldsymbol{\kappa})\} d\boldsymbol{\kappa} \tag{66.56a}$$

$$t_{\kappa}^{(s)}(\boldsymbol{q}) := \int_{\kappa} |A_{\kappa}(\boldsymbol{\kappa})|^2 \sin\{T(\boldsymbol{q}, \boldsymbol{\kappa})\} d\boldsymbol{\kappa} \tag{66.56b}$$

If $T(\boldsymbol{q}, \boldsymbol{\kappa})$ is truncated beyond the linear terms in $\boldsymbol{\kappa}$, so that

$$T(\boldsymbol{q}, \boldsymbol{\kappa}) \approx \overline{T}(\boldsymbol{q}, \boldsymbol{\kappa}) := 2\boldsymbol{q} \cdot \boldsymbol{\kappa}\left(\Delta_o \lambda_o - C_S \lambda_o^3 \boldsymbol{q}^2\right) \tag{66.57}$$

the function $t_{\kappa}^{(s)}(\boldsymbol{q})$ vanishes and

$$K_p(\boldsymbol{q}) \approx -t_{\upsilon}(\boldsymbol{q}) t_{\kappa}(\boldsymbol{q}) \sin\chi(\boldsymbol{q}) \tag{66.58}$$

with

$$t_{\kappa}(\boldsymbol{q}) := \int_{\kappa} |A_{\kappa}(\boldsymbol{\kappa})|^2 \cos\{\overline{T}(\boldsymbol{q}, \boldsymbol{\kappa})\} d\boldsymbol{\kappa} \tag{66.56c}$$

66.2.1 Discussion

The envelope functions that modulate sin $\chi(\boldsymbol{q})$ and cos $\chi(\boldsymbol{q})$ in Eq. (66.55) are both real. If we return to the discussion of Eq. (65.35) in the ideal case, we see that they occupy the same role as the aperture function, $T_A(\boldsymbol{q})$: the effect of non-vanishing source size and energy spread is analogous to that of an objective aperture with a transparency given for Eq. (66.58) by $t_{\upsilon} t_{\kappa}$. But this seems to mean that electrons are 'absorbed'! Where have they gone? Electrons have, of course, not been lost but their capacity to convey specimen information to the image has been impaired. Convenient measures of this reduction in information transfer are the image quality criteria introduced by Linfoot (1956, 1957, 1960, 1964) and discussed on several occasions (O'Neill, 1963; Franke, 1965; Frieden, 1966; Röhler, 1967; Frank et al., 1970), as pointed out by Frank (1975b), Beer et al. (1975). These are the dissimilarity, K, and the *structural resolving power*, Ξ, and two derived quantities: the *fidelity*, $\Phi := 1 - K$ and the *correlation quality* $\Psi = -(\Phi + \Xi)/2$, which is the same as the Strehl intensity ratio *(Definitionshelligkeit)* of traditional microscopy (Strehl, 1902; see Born and Wolf, 2002 Section 9.1.1). The measures K and Ξ are defined, for modulation by finite source size only, as follows:

$$K := \frac{\int |1 - t_{\kappa}(\boldsymbol{q})|^2 d\boldsymbol{q}}{\int d\boldsymbol{q}} \tag{66.59a}$$

(Frank, 1975a), in which the integration is taken over the aperture, and

$$\Xi := \frac{\int |t_{\kappa}(\boldsymbol{q})|^2 d\boldsymbol{q}}{\int d\boldsymbol{q}} \tag{66.59b}$$

The latter is interesting in that the signal-to-noise ratio at the image can be shown to remain constant as the source size is increased if the exposure, and hence the electron dose n is increased in such a way that $n^2\Xi = 1$ (signal-to-noise ratio limited by the statistics of the recording medium) or that $n^3\Xi = 1$ (ratio limited by the electron statistics, at high magnification and in minimum exposure conditions, for example). Fig. 66.7 shows the behaviour of K and Ξ as a function of source size and defocus in reduced coordinates for the Gaussian source model introduced in the next section.

The effects of non-vanishing source size and energy spread on the transfer functions were established in the early 1970s. The product representation (66.48) did not emerge from the earliest attempts to include these effects (Erickson and Klug, 1970a,b, 1971, see Erickson, 1973; Hanszen and Trepte, 1971b) but Hanszen and Trepte (1970, 1971a) did show that the effect of an energy spread (or temporal partial coherence) can be represented by an envelope. Frank (1973) noticed that the zeros of the curves of Hanszen and Trepte (1971b) and those of Hirt and Hoppe (1972), representing the effect of source size on the ideal transfer functions, remained approximately stationary for a wide range of source radii and concluded that it should be possible to derive an envelope representation for this case also. Many studies followed, notably Hahn (1973), Hahn and Seredynski (1974), Bonhomme et al. (1973), Beorchia and Bonhomme (1974), Misell (1973), Misell and Atkins (1973), Fejes (1977), Saxton (1977), Frank et al. (1978/79), Wade (1978), O'Keefe (1979), Bonnet and Bonhomme (1980), Zemlin and Schiske (1980), Lannes et al. (1981) and Krakow (1982).

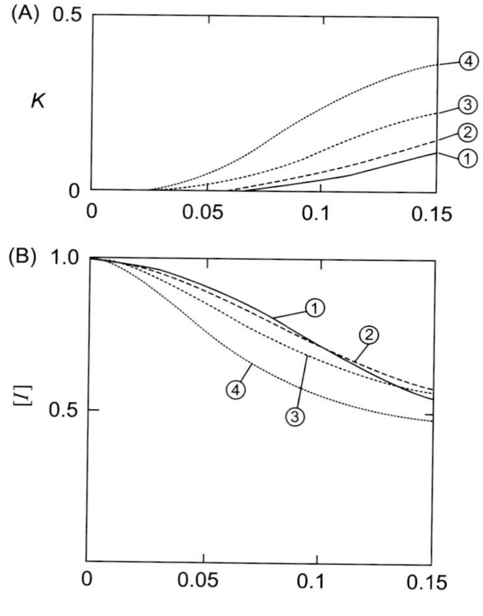

Figure 66.7
(A) Dissimilarity K, and (B) structural resolving power, Ξ *as* functions of generalized source size for four values of defocus: 1, $D = \sqrt{7}$; 2, $D = \sqrt{5}$; 3, $D = \sqrt{3}$; 4, $D = 1$.

In these early studies, source size and energy spread were studied separately but the legitimacy of simply multiplying the two envelopes was soon questioned (Frank, 1976a): was it certain that there was no interaction between the two? A careful study by Wade and Frank (1977) showed that the interaction is usually negligible for axial illumination but could become important for tilted illumination.

If the two effects can be safely assumed to be independent, the envelope representations can be obtained much more simply. For non-vanishing energy spread, for example, we associate a probability distribution $H(\Delta)$ with the variation in defocus caused by the distribution of energy. The current density at the image is thus of the form

$$\frac{dj}{j} = H(\Delta)d\Delta \quad \text{and} \quad \int H(\Delta)d\Delta = 1 \tag{66.60}$$

In the ideal case,

$$j \propto \psi_i \psi_i^* = 1 + 2\int (\tilde{\eta}\sin\chi - \tilde{\sigma}\cos\chi)\exp(2\pi i \boldsymbol{q}\cdot\boldsymbol{u}_i)d\boldsymbol{q}$$

and so

$$\frac{dj}{d\Delta} = jH(\Delta) \propto H(\Delta) + 2\int (\tilde{\eta}\sin\chi - \tilde{\sigma}\cos\chi)H(\Delta)\exp(2\pi i \boldsymbol{q}\cdot\boldsymbol{u}_i)d\boldsymbol{q}.$$

Integration over Δ then gives

$$\begin{aligned} C(\boldsymbol{u_i}) = &-\iint \tilde{\eta}\left\{\sin\chi(\Delta)H(\Delta)d\Delta\right\}\exp(2\pi i \boldsymbol{q}\cdot\boldsymbol{u}_i)d\boldsymbol{q} \\ &+ \iint \tilde{\sigma}\left\{\cos\chi(\Delta)H(\Delta)d\Delta\right\}\exp(2\pi i \boldsymbol{q}\cdot\boldsymbol{u}_i)d\boldsymbol{q} \end{aligned} \tag{66.61}$$

The defocus Δ will differ by only a small amount from the nominal defocus Δ_o:

$$\Delta = \Delta_o + \delta, \quad \delta \ll \Delta_o \tag{66.62}$$

and hence

$$\begin{aligned} \int \sin\chi(\Delta)H(\Delta)d\Delta &= \sin\chi(\Delta_o)\int H(\delta)\cos\left(\pi\lambda\boldsymbol{q}^2\delta\right)d\delta - \cos\chi(\Delta_o)\int H(\delta)\sin\left(\pi\lambda\boldsymbol{q}^2\delta\right)d\delta \\ \int \cos\chi(\Delta)H(\Delta)d\Delta &= \sin\chi(\Delta_o)\int H(\delta)\sin\left(\pi\lambda\boldsymbol{q}^2\delta\right)d\delta + \cos\chi(\Delta_o)\int H(\delta)\cos\left(\pi\lambda\boldsymbol{q}^2\delta\right)d\delta \end{aligned} \tag{66.63}$$

If $H(\Delta)$ is even, then the integrals in $\sin(\pi\lambda\boldsymbol{q}^2\delta)$ vanish, leaving

$$\begin{aligned} \int \sin\chi(\Delta)H(\Delta)d\Delta &= \sin\chi(\Delta_o)H_c\left(\lambda\boldsymbol{q}^2\right) \\ \int \cos\chi(\Delta)H(\Delta)d\Delta &= \cos\chi(\Delta_o)H_c\left(\lambda\boldsymbol{q}^2\right) \end{aligned} \tag{66.64}$$

where

$$H_c\left(\lambda \boldsymbol{q}^2\right) = \int H(\delta)\cos\left(\pi\lambda \boldsymbol{q}^2\delta\right)d\delta \tag{66.65}$$

We have thus recovered an alternative form of the envelope function (66.54); this approach is convenient if other perturbing effects (lens current fluctuations, in particular) are to be included. A similar study may be made of the envelope corresponding to the source size; see Frank (1973) or the account in Hawkes (1978c) for details. The most useful result is that for a symmetric source, the phase and amplitude contrast-transfer functions sin χ and cos χ are modulated by an envelope function of the form

$$\begin{aligned} t(\boldsymbol{q}) &\approx \int \tilde{\alpha}(\boldsymbol{\kappa})\cos\{\boldsymbol{\kappa}\cdot \mathrm{grad}\chi(\boldsymbol{q})\}d\boldsymbol{\kappa} \\ &= \alpha(\mathrm{grad}\chi(\boldsymbol{q})) \end{aligned}$$

with $\tilde{\alpha}(\boldsymbol{\kappa}) = |A(\boldsymbol{\kappa},0)|^2$.

66.3 Particular Forms of the Spectra

There are only a few forms of the function $\boldsymbol{A}(\boldsymbol{\kappa}, \upsilon)$ for which the integrations in the various approximations for $K_p(\boldsymbol{q})$ can be performed analytically. In practice, only two expressions have been examined, Gaussian distributions and top-hat functions. We shall not consider the latter here ($A(\boldsymbol{\kappa}, \upsilon) = \text{const}$, $|\boldsymbol{\kappa}| \le \kappa_o$, $|\upsilon| < \upsilon_o$, $A(\boldsymbol{\kappa}, \upsilon) = 0$ elsewhere) since the Gaussian model often matches the experimental conditions rather well. Almost all estimates of the source size and energy spread based on the measurement of the form of $K_p(\boldsymbol{q})$ have made the assumption that the distributions are Gaussian. Moreover, the rare attempts to distinguish between these two models by studying the attenuation of the transfer function suggest that, given the experimental errors and uncertainties, both are compatible with the measurements.

In the Gaussian model, we write

$$|A(\boldsymbol{\kappa}, \upsilon)|^2 = C_\upsilon \exp\left\{-C_\kappa\left(\boldsymbol{\kappa}_x^2 + \boldsymbol{\kappa}_y^2\right) - \upsilon^2/\upsilon_m^2\right\} \tag{66.66}$$

in which

$$C_\upsilon := \frac{\lambda_o^2}{\pi^{3/2}\upsilon_m\alpha_c^2}, \quad C_\kappa := \frac{\lambda_o^2}{\alpha_c^2} \tag{66.67}$$

where α_c is a measure of the aperture angle of the beam furnished by the condenser system and hence of the source size while υ_m is the mean energy spread. We see from the very outset that the contributions of $\boldsymbol{\kappa}$ and υ to $|A|^2$ are separable; we recall that this was one of the conditions for obtaining $K_p(\boldsymbol{q})$ in the product form (66.58). We note too that $|A|^2$ is even in $\boldsymbol{\kappa}$ and in υ.

Let us now reconsider the integral $\tilde{a}_e$ (66.45):

$$\begin{aligned}\tilde{\alpha}_e &= \int a_e \exp(\mathrm{i}T_\upsilon)d\upsilon \\ &= C_\upsilon \exp\left(-C_\kappa \boldsymbol{\kappa}^2\right) \int \exp\left(-\upsilon^2/\upsilon_m^2\right)\exp(\mathrm{i}T_\upsilon \upsilon)d\upsilon\end{aligned} \tag{66.68}$$

After introducing the expression for T_υ (66.40) and rearranging the exponent, we can integrate over υ, giving

$$\begin{aligned}&\int_{-\infty}^{\infty} \exp\left(-\frac{\upsilon^2}{\upsilon_m^2}\right)\exp(\mathrm{i}T_\upsilon \upsilon)d\upsilon \\ &= \pi^{1/2}\upsilon_m \exp\left(-\frac{1}{4}T_\upsilon^2 \upsilon_m^2\right) \\ &= \pi^{1/2}\upsilon_m \exp\left\{-\frac{1}{4}\pi^2 \upsilon_m^2 C_c'^2 \lambda_0^2 \left(\boldsymbol{q}^2 + 2\boldsymbol{q}\cdot\boldsymbol{\kappa}\right)^2\right\}\end{aligned} \tag{66.69}$$

We now have to evaluate

$$\int \tilde{a}_e \cos T(\boldsymbol{q}, \boldsymbol{\kappa})d\boldsymbol{\kappa} = \Re \int \tilde{a}_e \exp\{\mathrm{i}T(\boldsymbol{q}, \boldsymbol{\kappa})\}d\boldsymbol{\kappa} \tag{66.70}$$

The terms involving $\boldsymbol{\kappa}$ are of the form

$$\exp\left[-\left\{L\boldsymbol{\kappa}^2 + 2M\boldsymbol{q}\cdot\boldsymbol{\kappa} + N(\boldsymbol{q}\cdot\boldsymbol{\kappa})^2\right\}\right] \tag{66.71}$$

neglecting a small term in $\boldsymbol{\kappa}^2(\boldsymbol{q} \cdot \boldsymbol{\kappa})$; we give the explicit forms of L, M and N below. Since the result of integrating over $\boldsymbol{\kappa}$ in (66.70) is the same for all values of $|\boldsymbol{q}|$, irrespective of azimuth, we may consider the special case $\boldsymbol{q} = (\boldsymbol{q}, 0)$ without introducing any error. The integral can then be performed analytically; it takes the form

$$\int \exp\left\{-\left(L\boldsymbol{\kappa}^2 + 2M\boldsymbol{q}\kappa_x + N\boldsymbol{q}^2\kappa_x^2\right)\right\}d\boldsymbol{\kappa} \tag{66.72}$$

The general formula

$$\int_{-\infty}^{\infty} \exp\left(-ax^2 - 2bx\right)dx = \left(\pi/a\right)^{1/2}\exp\left(b^2/a^2\right) \quad \Re a > 0 \tag{66.73}$$

which, for $b=0$, reduces to $(\pi/a)^{1/2}$, enables us to integrate over κ_y immediately and (66.72) becomes

$$(\pi/L)^{1/2}\int \exp-\left\{\left(L+N\boldsymbol{q}^2\right)\boldsymbol{\kappa}_x^2+2M\boldsymbol{q}\boldsymbol{\kappa}_x\right\}d\boldsymbol{\kappa}_x$$

which, again using (66.73), is equal to

$$\left(\frac{\pi}{L}\right)^{1/2}\left(\frac{\pi}{L+N\boldsymbol{q}^2}\right)^{1/2}\exp\left(\frac{M^2\boldsymbol{q}^2}{L+N\boldsymbol{q}^2}\right) \tag{66.74}$$

The orders of magnitude of the many quantities occurring in this expression are very different and it can be replaced by an approximate formula that is still highly accurate and shows much more clearly the influence of the various parameters. In order to establish this, we examine L, M and N more closely.

$$\begin{aligned}
L &= C_{\boldsymbol{\kappa}}+\mathrm{i}\pi C_s\lambda_o^3\boldsymbol{q}^2=\frac{\lambda_o^3}{\alpha_c^2}\left(1+\mathrm{i}\pi C_s\lambda_o\alpha_c^2\boldsymbol{q}^2\right)\\
M &= \pi\lambda_o\left\{\frac{1}{2}\pi\upsilon_m^2C_c'2\lambda_o\boldsymbol{q}^2+\mathrm{i}\left(C_s\lambda_o^2\boldsymbol{q}^2-\Delta_o\right)\right\}\\
N &= \pi\lambda_o^2\left(\pi\upsilon_m^2C_c'2+2\mathrm{i}C_s\lambda_o\right)\\
L+N\boldsymbol{q}^2 &= C_{\boldsymbol{\kappa}}+3\mathrm{i}\pi C_s\lambda_o^3+\pi^2\upsilon_m^2C_c'2\lambda_o^2\boldsymbol{q}^2\\
&= \frac{\lambda_o^2}{\alpha_c^2}\left(1+3\mathrm{i}\pi C_s\lambda_o\alpha_c^2\boldsymbol{q}^2+\pi^2\upsilon_m^2C_c'2\alpha_c^2\boldsymbol{q}^2\right)
\end{aligned} \tag{66.75}$$

For typical values of the parameters ($C_s\approx10^{-3}$ m, $C_c'\approx10^{-8}$ m/V, $\lambda\approx4\times10^{-12}$ m, $\alpha_c\lesssim10^{-4}$ rad, $\upsilon_m\approx1$ V, $\boldsymbol{q}\approx10^8$ m^{-1}), the second term of L and the second and third terms of $L+Nq^2$ are found to be much smaller than the first and we therefore write

$$\begin{aligned}
L &=: \frac{\lambda_o^2}{\alpha_c^2}(1+\mathrm{i}r_1) & r_1 &:= \pi C_s\lambda_o\alpha_c^2\boldsymbol{q}^2\ll1\\
L+N\boldsymbol{q}^2 &= \frac{\lambda_o^2}{\alpha_c^2}(1+3\mathrm{i}r_1+r_2) & r_2 &:= \pi^2\upsilon_m^2C_c'2\lambda_o\alpha_c^2\boldsymbol{q}^2\ll1
\end{aligned} \tag{66.76}$$

The first two factors in (66.74) may thus be written

$$\left(\frac{\pi}{L}\right)^{1/2}\left(\frac{\pi}{L+N\boldsymbol{q}^2}\right)^{1/2}\approx\frac{\pi\alpha_c^2}{\lambda_o^2}\left(1-2\mathrm{i}r_1-\frac{1}{2}r_2\right) \tag{66.77}$$

and it will be convenient, if artificial, to regard $(1-2\mathrm{i}r_1-\frac{1}{2}r_2)$ as the first term in the expansion of an exponential, so that

$$\left(\frac{\pi}{L}\right)^{1/2}\left(\frac{\pi}{L+N\boldsymbol{q}^2}\right)^{1/2}\approx\frac{\pi\alpha_c^2}{\lambda_o^2}\exp\left(-2\mathrm{i}r_1-\frac{1}{2}r_2\right) \tag{66.78}$$

The argument of the exponential in (66.74) may be written as $M^2\boldsymbol{q}^2\alpha_c^2/\lambda_o^2$,

$$\begin{aligned}\frac{M^2\boldsymbol{q}^2\alpha_c^2}{\lambda_o^2} &= \pi^2\boldsymbol{q}^2\alpha_c^2\left\{\tfrac{1}{2}\pi\upsilon_m^2 C_c' 2\lambda_o\boldsymbol{q}^2 + \mathrm{i}\left(C_s\lambda_o^2\boldsymbol{q}^2 - \Delta_o\right)\right\}^2 \\ &\approx -\pi^2\boldsymbol{q}^2\alpha_c^2\left(C_s\lambda_o^2\boldsymbol{q}^2 - \Delta_o\right)^2\end{aligned} \tag{66.79}$$

Collecting up all the factors, we finally obtain

$$C_\upsilon\left(\pi^{1/2}\upsilon_m\right)\left(\frac{\pi\alpha_c^2}{\lambda_o^2}\right)\exp\left(-2\mathrm{i}r_1 - \frac{1}{2}r_2\right)\exp\{\mathrm{i}T(\boldsymbol{q})\}\exp\left\{-\pi^2\boldsymbol{q}^2\alpha_c^2\left(C_s\lambda_o^2\boldsymbol{q}^2 - \Delta_o\right)^2\right\}$$

But $C_\upsilon = \lambda_o^2/\pi^{3/2}\upsilon_m\alpha_c^2$ and so the phase contrast-transfer function becomes

$$\begin{aligned}K_p(\boldsymbol{q}) &= \sin\{-\chi(\boldsymbol{q}) - 2r_1\}t_\upsilon t_\kappa\exp\left(-\frac{1}{2}r_2\right) \\ &= \sin\left\{\pi\lambda_o\boldsymbol{q}^2\left(\Delta_o - \frac{1}{2}C_s\lambda_o^2\boldsymbol{q}^2 - 2C_s\alpha_c^2\right)\right\}t_\upsilon t_\kappa\exp\left(-\frac{1}{2}\pi^2\upsilon_m^2 C_c' 2\alpha_c^2\boldsymbol{q}^2\right)\end{aligned} \tag{66.80a}$$

with

$$t_\upsilon = \exp\left(-\frac{1}{4}\pi^2\upsilon_m^2 C_c' 2\lambda_0^2\boldsymbol{q}^4\right) \tag{66.81a}$$

$$t_\kappa = \exp\left\{-\pi^2\boldsymbol{q}^2\alpha_c^2\left(C_s\lambda_o^2\boldsymbol{q}^2 - \Delta_o\right)^2\right\} \tag{66.81b}$$

This expression reveals very strikingly all the features of interest of the modulation of sin $T(\boldsymbol{q})$ by the source size and energy spread: modulation by a source-size term (t_κ) and an energy-spread term (t_υ); further modulation (usually neglected and indeed usually negligible) by a mixed term $\exp(-\pi^2\upsilon_m^2 C_c' 2\alpha_c^2\boldsymbol{q}^2/2)$; and a small shift in the zeros of sin $\chi(\boldsymbol{q})$, which is likewise usually neglected.

It can be shown that a small twofold axial astigmatism, which is to be expected in practice, can be included to a good approximation simply by replacing Δ_o by an azimuth-dependent quantity:

$$\Delta_o \rightarrow \Delta_o + C_a\cos 2\varphi_o,$$

C_a being the astigmatic length constant referred back to the object (like Δ_o) and φ_o the azimuth at the exit plane of the object.

Formula (66.80a) for $K_p(\boldsymbol{q})$ contains numerous system constants. The situation can be improved slightly by the introduction of reduced quantities (Eqs. 65.31 and 65.32) but it is reasonable to look for a set of system parameters that are optimal in some well-defined sense by considering the working conditions that are advantageous from the point of view of the ideal transfer function, $\sin\chi(\boldsymbol{q})$ and then examine the influence of the envelope functions on these. This will be the subject of the next section. First, however, we extract the formulae for the phase contrast-transfer function that are most useful in practice from among the foregoing calculations and give the equivalent expressions in reduced coordinates.

1. For vanishing source size and energy spread

$$\begin{aligned} K_p(\boldsymbol{q}) &= \sin\left\{\pi \boldsymbol{q}^2\lambda\left(\Delta_o - \frac{1}{2}C_s\lambda^2\boldsymbol{q}^2\right)\right\} \\ \hat{K}_p(Q) &= \sin\left\{\pi\left(DQ^2 - \frac{1}{2}Q^4\right)\right\} \end{aligned} \tag{66.19'}$$

2. For Gaussian energy spread and angular distribution (general formula)

$$K_p(\boldsymbol{q}) = t_\upsilon(\boldsymbol{q})t_\kappa(\boldsymbol{q})\exp\left(-\frac{1}{2}\pi^2\upsilon_m^2 C_c' 2\alpha_c^2\boldsymbol{q}^2\right)\sin\left\{\pi\boldsymbol{q}^2\lambda\left(\Delta_o - \frac{1}{2}C_s\lambda^2\boldsymbol{q}^2 - 2C_s\alpha_c^2\right)\right\} \tag{66.80'}$$

$$\hat{K}_p(Q) = \hat{t}_\upsilon(Q)\hat{t}_\kappa(Q)\exp\left(-\frac{\pi}{2}B^2\alpha^2Q^2\right)\sin\left\{\pi Q^2\left(D - \frac{1}{2}Q^2 - 2\alpha^2\right)\right\} \tag{66.80''}$$

3. For Gaussian energy spread and angular distribution (approximate formula, usually adopted)

$$\begin{aligned} K_p(\boldsymbol{q}) &= t_\upsilon(\boldsymbol{q})t_\kappa(\boldsymbol{q})\sin\left\{\pi\boldsymbol{q}^2\lambda\left(\Delta_o - \frac{1}{2}C_s\lambda^2\boldsymbol{q}^2\right)\right\} \\ \hat{K}_p(Q) &= \hat{t}_\upsilon(Q)\hat{t}_\kappa(Q)\sin\left\{\pi Q^2\left(D - \frac{1}{2}Q^2\right)\right\} \end{aligned} \tag{66.82}$$

in which (Eqs. 65.31 and 65.32)

$$\begin{aligned} \boldsymbol{Q} &:= \left(C_s\lambda_0^3\right)^{1/4}\boldsymbol{q} \\ D &:= \Delta_o/(C_s\lambda_o)^{1/2} \\ \alpha &:= \alpha_c\left(C_s/\lambda_o\right)^{1/4} \end{aligned} \tag{66.83a}$$

$$B := \frac{C_c'\upsilon_m}{(C_s\lambda_o)^{1/2}} \tag{66.83b}$$

and

$$t_{\upsilon}(\boldsymbol{q}) = \exp\left(-\frac{1}{4}\pi^2\upsilon_m^2 C_c'2\lambda_o^2\boldsymbol{q}^4\right) \tag{66.81a'}$$

$$\hat{t}_{\upsilon}(Q) = \exp\left(-\frac{1}{4}\pi^2\frac{\upsilon_m^2 C_c'2}{C_s\lambda}Q^4\right) = \exp\left(-\frac{\pi^2}{4}B^2Q^4\right) \tag{66.81a''}$$

$$t_{\kappa}(\boldsymbol{q}) = \exp\left\{-\pi^2\boldsymbol{q}^2\alpha_c^2\left(C_s\lambda_o^2\boldsymbol{q}^2 - \Delta_o\right)^2\right\} \tag{66.81b'}$$

$$\hat{t}_{\kappa}(Q) = \exp\left\{-\pi^2Q^2\alpha^2\left(Q^2 - D\right)^2\right\} \tag{66.81b''}$$

We recall that

$$C_c' := \gamma C_c\hat{U}$$

66.4 Optimum Defocus and Resolution Limit

The complete expression (66.80) for the phase contrast-transfer function depends on bewilderingly many instrumental parameters. We, therefore, consider first the ideal transfer function sin $\chi(\boldsymbol{q})$, for which the influence of the spectral distributions is neglected, and then see how the latter affect it.

A common feature of all the graphs in Figs. 66.4 and 66.5 is the increasingly rapid oscillation of $\hat{K}(Q)$ for values of Q beyond about 1.25. Changes in the sign of $\hat{K}(Q)$ make the interpretation of the image complicated or even impossible. We must, therefore, choose the free parameters D (defocus of the microscope) and $\boldsymbol{q}_c$ (radius of the objective aperture) in sin $\chi(Q)$ in such a way that we obtain as wide an interval with no change of sign of K as possible.

The graphs of Fig. 66.4 suggest that curve (B), $D = 1$ with $Q_c = \sqrt{2}$, is a good choice. For this, most of the interval $0 \le Q \le \sqrt{2}$ is available and the function $\hat{K}(Q)$ has a broad maximum within it. Eq. (66.19b) now becomes

$$\hat{K}(Q) = \begin{cases} \sin\left\{\pi\left(DQ^2 - \frac{1}{2}Q^4\right)\right\} & \text{for } Q \le \sqrt{2} \\ 0 & \text{for } Q > \sqrt{2} \end{cases} \tag{66.84}$$

The optimum defocus $D = 1$ corresponds to $\Delta_{\text{opt}} = (\lambda_o C_s)^{1/2}$ (Eq. 65.28) in agreement with the value obtained in Chapter 60 (60.65), the Scherzer defocus. The cut-off $Q_c = \sqrt{2}$ corresponds to the spatial frequency $\boldsymbol{q}_c$ (66.18)

$$\boldsymbol{q}_c = \left(4/C_s\lambda_o^3\right)^{1/4} \tag{66.85a}$$

and with $\theta = \lambda_o \boldsymbol{q}$, the maximum aperture angle on the object side becomes

$$\theta_c = \lambda_o \boldsymbol{q}_c = \left(4\lambda_o / C_s\right)^{1/4} \tag{66.85b}$$

in accord with expression (60.64) for α_{opt}.

This agreement with our earlier estimates of optimal conditions shows that these are indeed realistic; reasoning based on the transfer function shows clearly what is meant by 'optimal'. We can also see that although the choice $D = 1$ is a good general value, there may well be conditions in which other choices are better. For very high-resolution work, $D = \sqrt{3}$ or even $D = \sqrt{5}$ could be preferable; with $D = \sqrt{3}$, for example, spatial frequencies in the second band would be interpreted directly and the lower resolution information discarded (or processed). The cut-off in the second band is better than $Q = \sqrt{2}$. Images have even been obtained using the information in several peaks of the oscillations, after tuning the microscope very carefully to the spatial frequency to be detected in the specimen (Hashimoto et al., 1977; Endoh and Hashimoto, 1977; Hashimoto and Endoh, 1978); this 'aberration-free focusing' is, however, rather a *tour de force* than a practical technique, for the microscopist is choosing *a priori* what is to be seen.

Returning to the case $D = 1$ (66.84), we now estimate the resolution limit with the aid of the uncertainty principle, $\Delta x \cdot \Delta p_x \geq h = 2\pi\hbar$. The result will not be overoptimistic if we identify Δx with the resolution limit $d_{\min}$ and set $\Delta p_x = hq_c$, giving

$$d_{\min} = 1/\boldsymbol{q}_c = 0.7\left(C_s \lambda_o^3\right)^{1/4} \tag{66.86}$$

This is slightly larger than $\bar{r}/|M|$ as given by (60.66) or (60.71) and is probably more realistic.

The contrast-transfer function changes quite rapidly in the neighbourhood of the optimum defocus $D = 1$, as can be seen in Fig. 66.8. If we regard a contrast of 80% as acceptable in practice, too small a value of defocus is clearly worse than too large. This is in accordance with the arguments of Section 60.5, leading to (60.62).

We now examine more closely the effects of non-vanishing source size and energy spread. Eq. (66.80) can be cast into a much simpler form by introducing the reduced spatial frequency Q (65.31) and two more dimensionless parameters (cf. 66.83a and 66.83b), A and B, which have a clear physical meaning:

$$A := \frac{2\alpha_c}{\theta_c} = \alpha_c \left(\frac{4C_s}{\lambda_o}\right)^{1/4} = \alpha\sqrt{2} \tag{66.87a}$$

is the ratio of the apertures on either side of the specimen and

$$B := \frac{1}{d_{\min}} \frac{\gamma C_c v_m \theta_c}{2\hat{U}} = \frac{\gamma C_c v_m}{\hat{U}(\lambda_o C_s)^{1/2}} = \frac{C_c{}' v_m}{(\lambda_o C_s)^{1/2}} \tag{66.87b}$$

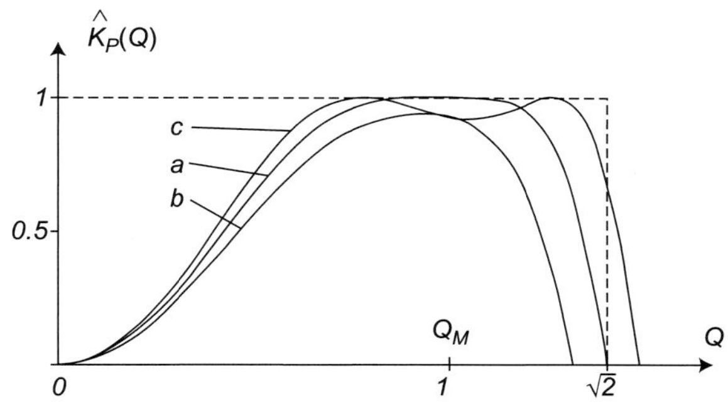

Figure 66.8
Undamped contrast-transfer function $\hat{K}_p(Q)$ for various values of the defocus D, close to $D = 1$. *Line a:* best choice $D = 1$; this has a broad maximum at $Q_M = 1$. *Line b:* too *small* a value of D, the maximum of $\hat{K}_p$ is less than unity and the zero is reached for $Q < \sqrt{2}$. *Line c:* too *large* a value of D; the curve has a local minimum at Q_M but the limiting zero is shifted beyond $Q = \sqrt{2}$.

is the ratio of the radius of the chromatic aberration disc and the resolution limit. Both parameters should be made as small as is experimentally possible in order to come close to the theoretical limit of resolution: $A \ll 1$ and $B \ll 1$.

Introducing A and B into Eq. (66.80), the transfer function takes the form

$$\hat{K}(Q) = \sin\left\{\pi Q^2\left(D - \frac{1}{2}Q^2 - A^2\right)\right\}\exp\left\{-\frac{\pi^2}{2}A^2Q^2\left(D - Q^2\right)^2\right\} \times \exp\left\{-\frac{\pi^2}{4}B^2Q^2\left(A^2 + Q^2\right)\right\} \tag{66.88a}$$

which reduces to

$$\hat{K}(Q) = \sin\left\{\pi Q^2\left(1 - \frac{1}{2}Q^2 - A^2\right)\right\}\exp\left\{-\frac{\pi^2}{2}A^2Q^2\left(1 - Q^2\right)^2\right\} \times \exp\left\{-\frac{\pi^2}{4}B^2Q^2\left(A^2 + Q^2\right)\right\} \tag{66.88b}$$

when $D = 1$. Careful inspection of these formulae shows that the *axial chromatic aberration* will in general have a *more deleterious effect* than the source size, since the exponential term involving B (energy spread) decreases monotonically with Q whereas the first exponential factor, which involves A (source size) returns to unity at $Q = D^{1/2}$; this coincides with the maximum of the sine function for $D = 1, \sqrt{3}, \sqrt{5}, \ldots$ for negligible values of A and is always close to the maximum. Beyond $Q = D^{1/2}$, this factor falls off monotonically. The separate effects of the two envelope functions are shown in Figs. 66.9 and 66.10; if B is negligible (no energy spread) and $D = 1$, we have

$$\hat{K}(Q) = \sin\left\{\pi Q^2\left(1 - A^2 - \frac{1}{2}Q^2\right)\right\}\exp\left\{-\frac{\pi^2}{2}A^2Q^2\left(1-Q^2\right)^2\right\} \qquad (66.89a)$$

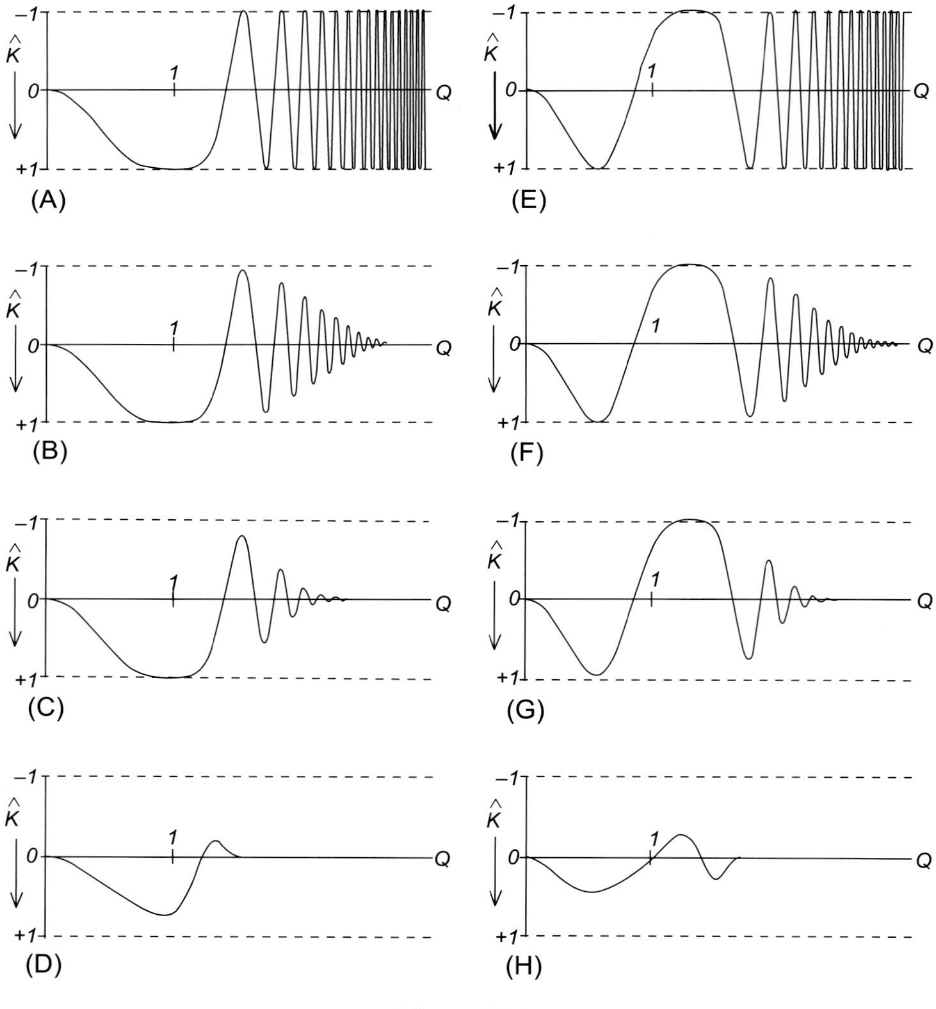

Figure 66.9
Damping effect of a non-vanishing condenser aperture $A \neq 0$, $B = 0$ on the phase contrast-transfer function for $D = 1$ (A–D) and $D = \sqrt{3}$ (E–H). (A) and (E) Undamped curve, $A = B = 0$, (B) and (F) $A = 0.05$, (C) and (G) $A = 0.1$, (D) and (H) $A = 0.5$.

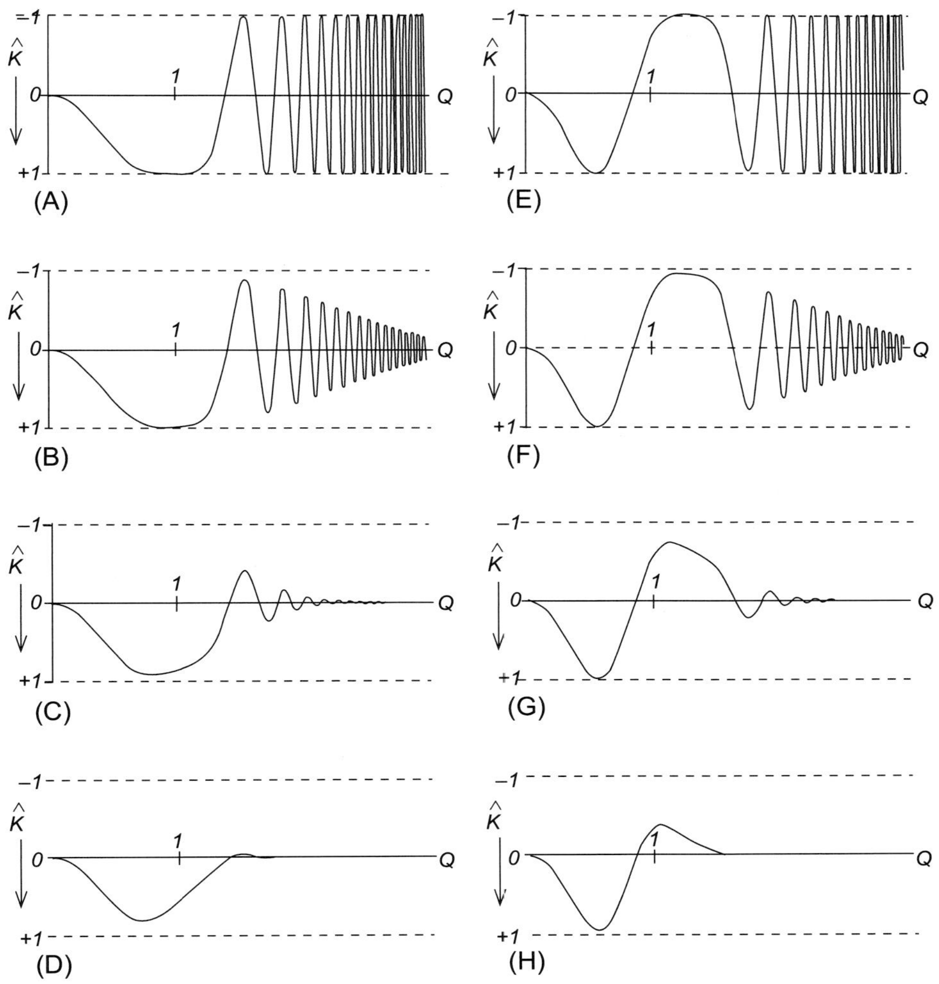

Figure 66.10

Damping effect of a non-vanishing energy spread $B \neq 0$, $A = 0$ on the phase contrast-transfer function for $D = 1$ (A–D) and $D = \sqrt{3}$ (E–H). (A) and (E) Undamped curve, $A = B = 0$, (B) and (F) $B = 0.1$, (C) and (G) $B = 0.25$, (D) and (H) $B = 0.5$.

whereas if A is set to zero (vanishing source size), we find

$$\hat{K}(Q) = \sin\left\{\pi Q^2\left(1 - \frac{1}{2}Q^2\right)\right\}\exp\left\{-\frac{\pi^2}{4}B^2Q^4\right\} \tag{66.89b}$$

These functions are plotted in Fig. 66.9A–D ($B = 0$) and Fig. 66.10A–D ($A = 0$) for a range of values of A and B. The corresponding curves for $D^2 = 3$ are shown in Figs. 66.9E–H and 66.10E–H.

For values of Q beyond $\sqrt{2}$ the damping by the exponential factors may become so strong that the oscillations of $\hat{K}(Q)$ soon become insignificant. No useful information then survives in the image. This is shown by the graphs of Fig. 66.9, which confirm that a cut-off at $Q_c = \sqrt{2}$ is a prudent choice. These curves also justify some of the approximations made in deriving (66.80) and hence (66.88a and 66.88b): the integrals were evaluated on the assumption that the upper limits were infinite and hence that $Q_c \rightarrow \infty$. Formally, therefore, we should introduce finite limits if we wish to include the effect of setting the cut-off at $Q_c = \sqrt{2}$, which would give a much more complicated result involving the Gaussian error function, Fresnel integrals and Bessel functions. The strong damping, together with the many approximations made during the derivation, shows that it is unnecessary to pursue such complicated reasoning: (66.80) will in practice be a perfectly adequate approximation.

When the ultimate resolution of the microscope is required, exceptional efforts must be made to keep A and B small and information will then survive in the image from several oscillations of the sinusoidal function. It is then usual to distinguish between the instrumental resolution and the information limit. The *instrumental resolution* indicates the capacity of the microscope to transfer spatial frequencies faithfully to the image and is usually defined to be proportional to $(C_s \lambda^3)^{1/4}$ where the constant of proportionality takes one of the values we have found earlier. The *information limit* is defined in terms of the rate of fall-off of the exponential functions, typically, the value of Q for which the product of the two exponentials in (66.88a) reaches 1/e. For careful discussion of these resolution limits, see Van Dyck (1989, 1991a,b, 1992), Van Dyck and de Jong (1992), de Jong and Van Dyck (1993), Sarikaya and Howe (1992), O'Keefe and Spence (1991), O'Keefe (1992), van den Bos (1991, 1992) and Tsuno (1993).

A convenient way of assessing the limit of information transfer has been introduced by Frank (1975a), based on his earlier work on the effect of specimen drift (1969). Two images of a specimen with a reasonably uniform spatial-frequency spectrum, the second shifted laterally with respect to the first, are superimposed. In the diffractogram the familiar ring pattern will be seen, modulated by a sinusoidal fringe pattern, the extent of which gives a good estimate of the practical resolution. The effect of source size and energy spread and the possibility of estimating these from the fringe pattern are examined in a later paper (Frank, 1976a). See Anaskin et al. (1980) and see Zemlin and Weiss (1993) for a careful examination of this idea.

The use of the spectral signal-to-noise ratio as a measure of the resolution, of correlation-averaged images in particular (though not exclusively), is very fully analysed by Unser et al. (1987a,b).

66.5 Extensions of the Theory

The foregoing sections have been devoted to the imaging of weakly scattering specimens in axial bright-field imaging conditions. Although this is by far the most widely-used mode, there are others that are also encountered and the present section will be concerned with these. In particular, we consider bright-field imaging with a tilted incident beam and its natural extension, hollow-cone imaging, introduced by Hanszen and Trepte (1971b) and recommended by Rose in 1977 (Section 66.5.1). We also devote a short section to the effect on transfer theory of aberrations, coma in particular, that render the conditions anisoplanatic (Section 66.5.2). Unusual forms of the aperture function T_A such as zone plates and phase plates are considered in Section 66.6. We then examine phase contrast for crystalline specimens that do not necessarily scatter weakly (Section 66.7). In Section 66.8, we examine the repercussions of aberration correction on the theory of contrast transfer. Finally, as promised earlier, we comment on transforms other than the Fourier transform that map convolution products into direct products (Section 66.9).

66.5.1 Tilted and Hollow-Cone Illumination

We first consider the case of a plane monochromatic illuminating wave no longer travelling in the direction of the optic axis (Hawkes, 1980a). This incident wave is now written

$$\psi = \exp(2\pi i \boldsymbol{m} \cdot \boldsymbol{x}_o) \tag{66.90}$$

and with the specimen transparency represented as before (66.1) by $1 - \sigma_o + i\eta_o$, the emergent object wave will be

$$\psi_o(x_o, y_o) = \{1 - \sigma_o(x_o, y_o) + i\eta_o(x_o, y_o)\}\exp(2\pi i \boldsymbol{m} \cdot \boldsymbol{x}_o) \tag{66.91a}$$

with spectrum

$$S_o = \delta(\boldsymbol{q} - \boldsymbol{m}) - \tilde{\sigma}(\boldsymbol{q} - \boldsymbol{m}) + i\,\tilde{\eta}(\boldsymbol{q} - \boldsymbol{m}) \tag{66.91b}$$

In the image plane, therefore,

$$S_i(\boldsymbol{q}) = T_A(\boldsymbol{q})T_L(\boldsymbol{q})\{\delta(\boldsymbol{q} - \boldsymbol{m}) - \tilde{\sigma}(\boldsymbol{q} - \boldsymbol{m}) + i\,\tilde{\eta}(\boldsymbol{q} - \boldsymbol{m})\} \tag{66.92}$$

and hence

$$\psi_i(\boldsymbol{u}_i) = \int T_A(\boldsymbol{q})T_L(\boldsymbol{q})\{\delta(\boldsymbol{q} - \boldsymbol{m}) - \tilde{\sigma}(\boldsymbol{q} - \boldsymbol{m}) + i\,\tilde{\eta}(\boldsymbol{q} - \boldsymbol{m})\}\exp(2\pi i \boldsymbol{u}_i \cdot \boldsymbol{q})d\boldsymbol{q} \tag{66.93a}$$

or with

$$p := q - m \tag{66.94}$$

$$\psi_i(\boldsymbol{u}_i) = \exp(2\pi \mathrm{i}\boldsymbol{u}_i \cdot \boldsymbol{m}) \int T_A(\boldsymbol{p}+\boldsymbol{m}) T_L(\boldsymbol{p}+\boldsymbol{m}) \{\delta(\boldsymbol{p}) - \tilde{\sigma}(\boldsymbol{p}) + \mathrm{i}\,\tilde{\eta}(\boldsymbol{p})\} \exp(2\pi \mathrm{i}\boldsymbol{u}_i \cdot \boldsymbol{p}) d\boldsymbol{p} \tag{66.93b}$$

The image contrast may again be written as the sum of an amplitude term and a phase term,

$$C(\boldsymbol{u}_i) := \frac{1 - |\psi_i|^2}{1 + |\psi_i|^2} =: C_a(\boldsymbol{u}_i) + C_p(\boldsymbol{u}_i) \tag{66.95}$$

and as before, we introduce the phase and amplitude spectra:

$$\begin{aligned} S_a = \mathcal{F}^-(C_a) &= \frac{1}{2}\tilde{\sigma}(\boldsymbol{p})\{T_A^*(\boldsymbol{m})T_L^*(\boldsymbol{m})T_A(\boldsymbol{m}+\boldsymbol{p})T_L(\boldsymbol{m}+\boldsymbol{p}) \\ &+ T_A(\boldsymbol{m})T_L(\boldsymbol{m})T_A^*(\boldsymbol{m}-\boldsymbol{p})T_L^*(\boldsymbol{m}-\boldsymbol{p})\} \\ S_p = \mathcal{F}^-(C_p) &= -\frac{1}{2}\mathrm{i}\,\tilde{\eta}(\boldsymbol{p})\{T_A^*(\boldsymbol{m})T_L^*(\boldsymbol{m})T_A(\boldsymbol{m}+\boldsymbol{p})T_L(\boldsymbol{m}+\boldsymbol{p}) \\ &- T_A(\boldsymbol{m})T_L(\boldsymbol{m})T_A^*(\boldsymbol{m}-\boldsymbol{p})T_L^*(\boldsymbol{m}-\boldsymbol{p})\} \end{aligned} \tag{66.96}$$

We must pause here to consider the limits of integration. In the presence of a real aperture, small enough to have an appreciable effect on the integration, the above expressions are not correct, since the limits for the first and second terms in C_a and C_p will be different and hence the above expressions for S_a and S_p cannot be obtained by Fourier transformation. We shall, therefore, assume that there is no physical aperture, or in practice an aperture so large that the (tilted) main beam does not come near to it and hence that the integrals can be taken to infinity. Conversely, we leave the term T_A in the formulae because, as we saw earlier, the effects of finite energy spread and source size can be incorporated in this term.

The phase and amplitude transfer functions now have the form

$$\begin{aligned} K_a(\boldsymbol{p}) = &\frac{1}{2}T_A(\boldsymbol{m})\{T_A(\boldsymbol{m}+\boldsymbol{p})T_L^*(\boldsymbol{m})T_L(\boldsymbol{m}+\boldsymbol{p}) \\ &+ T_A(\boldsymbol{m}-\boldsymbol{p})T_L(\boldsymbol{m})T_L^*(\boldsymbol{m}-\boldsymbol{p})\} \end{aligned} \tag{66.97a}$$

$$\begin{aligned} K_p(\boldsymbol{p}) = &-\frac{1}{2}\mathrm{i}T_A(\boldsymbol{m})\{T_A(\boldsymbol{m}+\boldsymbol{p})T_L^*(\boldsymbol{m})T_L(\boldsymbol{m}+\boldsymbol{p}) \\ &+ T_A(\boldsymbol{m}-\boldsymbol{p})T_L(\boldsymbol{m})T_L^*(\boldsymbol{m}-\boldsymbol{p})\} \end{aligned} \tag{66.97b}$$

in which we have taken T_A to be real. With $T_L(\boldsymbol{q}) = \exp\{-\mathrm{i}\chi(\boldsymbol{q})\}$ (Eq. 65.30), these become

$$\begin{aligned} K_a(\boldsymbol{p}) &= \frac{1}{2}T_A(\boldsymbol{m})\big[T_A(\boldsymbol{m}+\boldsymbol{p})\exp\ \mathrm{i}\{\chi(\boldsymbol{m})-\chi(\boldsymbol{m}+\boldsymbol{p})\} \\ &\quad + T_A(\boldsymbol{m}-\boldsymbol{p})\exp-\mathrm{i}\{\chi(\boldsymbol{m})-\chi(\boldsymbol{m}-\boldsymbol{p})\}\big] \\ K_p(\boldsymbol{p}) &= -\frac{1}{2}\mathrm{i}T_A(\boldsymbol{m})\big[T_A(\boldsymbol{m}+\boldsymbol{p})\exp\ \mathrm{i}\{\chi(\boldsymbol{m})-\chi(\boldsymbol{m}+\boldsymbol{p})\} \\ &\quad - T_A(\boldsymbol{m}-\boldsymbol{p})\exp-\mathrm{i}\{\chi(\boldsymbol{m})-\chi(\boldsymbol{m}-\boldsymbol{p})\}\big] \end{aligned} \tag{66.98}$$

We now consider the special case of vanishing source size and energy spread, $T_A = 1$. We introduce the reduced coordinates

$$\begin{aligned} &\boldsymbol{P} = \left(C_s\lambda^3\right)^{1/4}\boldsymbol{p} \quad \boldsymbol{M} = \left(C_s\lambda^3\right)^{1/4}\boldsymbol{m} \\ &D = \Delta/(C_s\lambda)^{1/2} \end{aligned} \tag{66.99}$$

as in (60.31) and (60.32) and rewrite the transfer functions as follows:

$$K_a(\boldsymbol{p}) \rightarrow \hat{K}_a(\boldsymbol{P}) = \exp(-2\pi\mathrm{i}\alpha)\cos\pi\beta \tag{66.100a}$$

$$K_p(\boldsymbol{p}) \rightarrow \hat{K}_p(\boldsymbol{P}) = \exp(-2\pi\mathrm{i}\alpha)\sin\pi\beta \tag{66.100b}$$

in which

$$\begin{aligned} \alpha(\boldsymbol{P}) &= \boldsymbol{P}\cdot\boldsymbol{M}\left(P^2 + M^2 - D\right) \\ \beta(\boldsymbol{P}) &= DP^2 - \frac{1}{2}\left\{P^4 + 2P^2M^2 + 4(\boldsymbol{P}\cdot\boldsymbol{M})^2\right\} \end{aligned} \tag{66.101}$$

The function $\hat{K}_p(\boldsymbol{p})$ has zeros wherever $\beta(\boldsymbol{P}) = n$ (n an integer or zero) and hence when

$$P^4 + 2\left(M^2 - D\right)P^2 + 4(\boldsymbol{P}\cdot\boldsymbol{M})^2 = 2n \tag{66.102a}$$

Setting $P^2 =: r$ and $\boldsymbol{P}\cdot\boldsymbol{M} = PM\cos\theta$, this becomes

$$r^2 + 2\left(M^2 + 2M^2\cos^2\theta - D\right)r = 2n \tag{66.102b}$$

which is the equation in polar form (r, θ) for the zeros of the phase contrast-transfer function. The form of the corresponding curves is governed by the relative magnitude of M and D. For small tilt and non-vanishing defocus the influence of $\cos\theta$ will be small and the curves resemble those for axial illumination. For large tilt and modest defocus the linear term dominates and (66.102b) resembles the polar form of the equation for an ellipse.

Returning to (66.102b), it is interesting to calculate the number of times the function returns to zero along the axes ($\theta = 0$ or π, $\theta = \pi/2$ or $3\pi/2$). For these values of θ, (66.102b) becomes

$$r^2 + 2\left(3M^3 - D\right)r = 2n$$
$$r^2 + 2\left(M^2 - D\right)r = 2n$$

and so

$$\begin{aligned} r &= D - 3M^2 \pm \left\{\left(D-3M^2\right)^2 + 2n\right\}^{1/2} \quad \theta = 0, \pi \\ r &= D - M^2 \pm \left\{\left(D-M^2\right)^2 + 2n\right\}^{1/2} \quad \theta = \pi/2,\ 3\pi/2 \end{aligned} \tag{66.103}$$

For negative values of n, therefore, r and hence P^2 may have two positive values. The curves then consist of closed loops with separate branches. Some examples are shown in Fig. 66.11.

It is, however, essential to consider the effect of source size and energy spread in the case of tilted illumination since their effect is more complicated than the simple attenuation that we have encountered for axial illumination. These effects can be included straightforwardly provided that the envelope representation (66.58) is acceptable. We consider only the phase contrast-transfer function, $K_p(\boldsymbol{p})$ of (66.97b); the amplitude term $K_a(\boldsymbol{p})$ can be analysed in a similar fashion if required.

Instead of (66.97b), we find (with 66.99)

$$\hat{K}_p(\boldsymbol{P}) = -\frac{1}{2}\mathrm{i}\left\{T_L(\boldsymbol{M}+\boldsymbol{P})T_L^*(\boldsymbol{M})t_{\boldsymbol{\kappa}}^+ t_\upsilon^+ - T_L^*(\boldsymbol{M}-\boldsymbol{P})T_L(\boldsymbol{M})t_{\boldsymbol{\kappa}}^- t_\upsilon^-\right\} \tag{66.104}$$

in which the envelopes corresponding to source size ($\boldsymbol{\kappa}$) and energy spread (υ) are given by

$$\begin{aligned} t_{\boldsymbol{\kappa}}^{\pm} &= \exp\left\{-\left(\frac{\alpha_c \tau^{\pm}}{\lambda_o}\right)^2\right\} \\ t_\upsilon^{\pm} &= \exp\left\{-\pi^2\left(P^2 \pm 2\boldsymbol{P}\cdot\boldsymbol{M}\right)^2 \upsilon_m^2/4\right\} \end{aligned} \tag{66.105}$$

in which

$$\tau^{\pm} = 2\pi\left[\left\{(\boldsymbol{M}\pm\boldsymbol{P})^2 - D\right\}(\pm\boldsymbol{P}) + \left(P^2 \pm \boldsymbol{M}\cdot\boldsymbol{P}\right)\boldsymbol{M}\right] \tag{66.106}$$

The Gaussian expressions for the source size and energy spread (66.66 and 66.67) have been used here.

The envelope $t_\upsilon^{\pm}$ is equal to unity when $P^2 \pm 2\boldsymbol{P}\cdot\ \boldsymbol{M} = 0$, that is, when

$$(P_x \pm M_x)^2\left(P_y \pm M_y\right)^2 = M_x^2 + M_y^2 \tag{66.107}$$

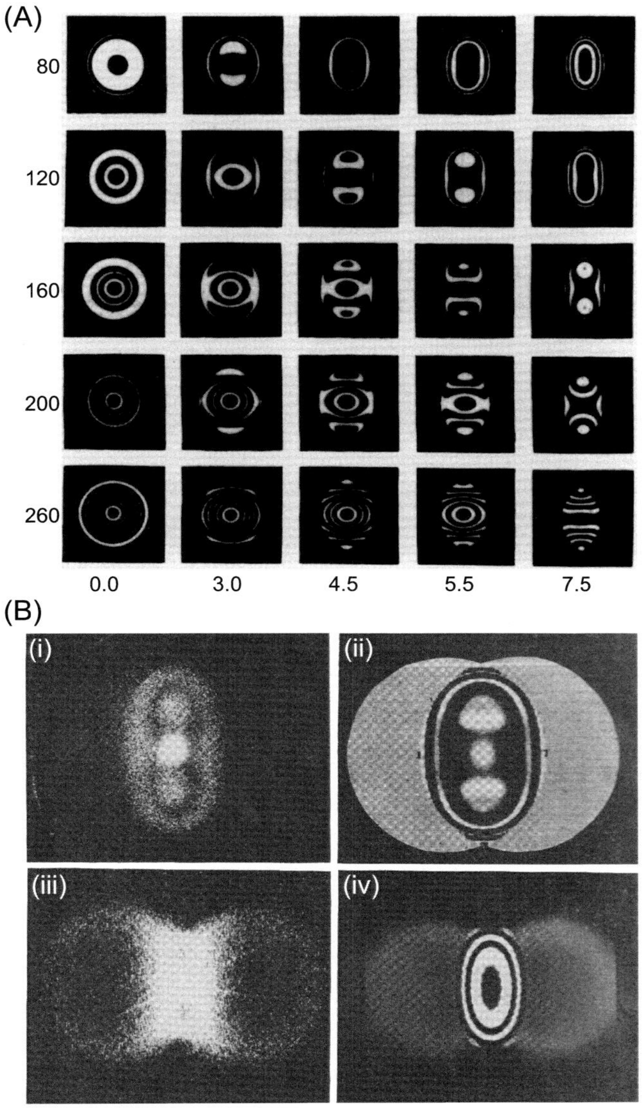

Figure 66.11

(A) Calculated diffractograms for vanishing energy spread. The ordinate is defocus in nanometres and the abscissa is beam tilt in mrad. The angular aperture of the illumination is 15 μrad, $C_s = 1.35$ mm and the accelerating voltage is 100 kV. (B) Thick and thin carbon film diffractograms. (i) Top left; Experimental diffractogram, thin film, beam tilt 5.5 mrad. (ii) Top right; Simulated diffractogram. (iii) Bottom left; Experimental diffractogram, thick film, showing the achromatic circles. (iv) Bottom right; Simulated diffractogram. *After Krakow et al. (1976), Courtesy Taylor and Francis, https://doi.org/10.1080/14786437608221929.*

These are circles of radius $|\boldsymbol{M}|$ centred on the points $(\pm M_x, \pm M_y)$ around which the energy spread has no effect: they are the *achromatic circles* described by Hoppe (1974), Hoppe et al. (1974), Willasch (1976a) and Frank (1976b).

How does the energy-spread envelope vary in the neighbourhood of an achromatic circle? The envelope falls to 1/e of its maximum value when

$$\exp\left\{-\frac{\pi^2}{2}\left(P^2+2\boldsymbol{P}\cdot\boldsymbol{M}\right)^2 D^2\right\}=1/\mathrm{e} \tag{66.108a}$$

and hence when

$$P^2 \pm 2\boldsymbol{P}\cdot\boldsymbol{M} = \pm\frac{\sqrt{2}}{D\pi} \tag{66.108b}$$

which represents circles with radii $\left(M^2 \pm \sqrt{2}/D\pi\right)^{1/2}$.

The larger the defocus D, the narrower will be the envelope around the achromatic circle (for values of $M^2 \geq \sqrt{2}/D\pi$). We shall see that the source size has an analogous effect but it is already clear that the expressions (66.100a,b) are likely to be very misleading because the envelopes $t_\upsilon^\pm$ and $t_\kappa^\pm$ may well attenuate their terms in (66.104) before these two terms overlap: we must expect three contributions to $\hat{K}_p(\boldsymbol{P})$, one corresponding to a central region in which the two terms do overlap and hence interfere and individual contributions near to $\pm\boldsymbol{M}$ for which the interference is much smaller. The effects of course vary in magnitude with the choice of parameters.

We now analyse $t_\kappa^\pm$ in a similar way. The term $\tau^\pm$ may be written

$$\tau^\pm = 2\pi\left\{\left(P^2 \pm 2\boldsymbol{P}\cdot\boldsymbol{M}\right)(\boldsymbol{M}\pm\boldsymbol{P}) \pm \left(M^2 - D\right)\boldsymbol{P}\right\} \tag{66.109}$$

which brings out the importance of the defocus value $D = M^2$. The envelopes are then symmetric about $\boldsymbol{P} = \pm\boldsymbol{M}$ in the sense that they take the same value at $\boldsymbol{P} = \boldsymbol{M} \pm \boldsymbol{\mu}$ (or $\boldsymbol{P} = -\boldsymbol{M} \pm \boldsymbol{\mu}$). The envelope function returns to unity when

$$\left(P^2 \pm 2\boldsymbol{P}\cdot\boldsymbol{M}\right)(\boldsymbol{M}\pm\boldsymbol{P}) = 0 \tag{66.110}$$

and hence at $\boldsymbol{P} = \pm\boldsymbol{M}$ and on the achromatic circle. The defocus value $D = M^2$ is, therefore, optimal in the sense that there is no attenuation around the achromatic circle and at the tilt points $(\pm M)$.

By far the most detailed study of the interaction between beam tilt, source size and energy spread is to be found in the work of Jenkins (1979), Wade and Jenkins (1978) and Jenkins and Wade (1977). Earlier work by Hoppe et al. (1975), Downing (1975), Hanszen (1976), Hanszen and Ade (1977), Wade (1976a,b), Krakow et al. (1976), Typke and Köstler (1976, 1977), Rose (1977), Krakow (1976a,b), Kiselev and Sherman (1976) and

Hoppe and Köstler (1976) is of direct relevance; see also Kunath (1979). The experimental observation by Parsons and Hoelke (1974) of wings in the optical diffractogram of images obtained with tilted illumination led to much discussion (clarified by Rose, 1977), as did the genuineness of fringes seen in images of supposedly amorphous films of germanium and silicon (see Howie et al., 1972, 1973; McFarlane, 1975; Krivanek, 1975, 1976, 1978; Krivanek and Howie, 1975; Goldfarb et al., 1975; Krivanek et al., 1976; Krakow, 1976a, b; Wade, 1976a,b; Saxton et al., 1977; Howie, 1978, 1983).

The analysis given above is only approximate but the terms neglected can be shown to be very small; for discussion of these, see McFarlane and Cochrane (1975), Wade and Jenkins (1978) and especially Jenkins (1979).

An interesting concept that emerged from this work was the *effective coherent aperture* introduced by Downing (1975) and subsequently investigated and employed by Wade (1976a,b) and Wade and Jenkins (1978). The idea here is that the integration over $\boldsymbol{\kappa}$ and v should be performed on the filtered object spectrum $S_o T_L$ instead of on the image contrast. The neglect of the quadratic terms in $|\psi_i|^2$ ensures that the results are the same, even though we seem to be adding wavefunctions and not intensities. We shall not describe this further here but simply point out that the filtered object spectrum can in this way be written as the product of the usual term $T_L(\boldsymbol{q})$, the amplitude term $T_A(\boldsymbol{q})$ defining the radius of any aperture, the (shifted) object spectrum $\tilde{\eta}(\boldsymbol{q}-\boldsymbol{m})$ and envelope functions corresponding to energy spread and source size; these have the advantage that the resulting effects at the image can be understood in terms of the behaviour of a single function for each spread function and not two for each as is the case for Eq. (66.104).

Hollow-cone illumination. The anisotropy of image formation with tilted illumination disappears if all tilt angles are present simultaneously, as they are when the electrons incident on the specimen form a hollow cone. This form of illumination was first examined by Hanszen and Trepte (1971b) and reconsidered by Rose (1977). The associated transfer functions can in principle be obtained by integrating the tilted-beam expressions over $\boldsymbol{m}$, keeping $|\boldsymbol{m}|$ constant (thin cone) or over both the angular part of $\boldsymbol{m}$ and over a small range of values of $|\boldsymbol{m}|$ as well (thick cone). The effects of source size and energy spread complicate the integrals, but not unduly, and in any case, these have to be evaluated numerically. A thorough study of hollow-cone transfer is to be found in the work of Jenkins (1979). We first reproduce his conclusions and then summarize the major contributions of Eusemann and Rose (1982) and Dinges et al. (1994).

We set out from the expression (66.104) for $\hat{K}_p(\boldsymbol{P})$, and note that the second term is the complex conjugate of the first if $\boldsymbol{P}$ is replaced by $-\boldsymbol{P}$. This means that when we integrate over $\boldsymbol{M}$, keeping the tilt-angle fixed, contributions that are themselves complex conjugates will be added and the transfer function will be real: there will be no shift in the zeros of the transfer function. Considerable insight into hollow-cone transfer can be gained by the effective aperture approach and although we confine this account to the results, we shall use this concept. It is not difficult to show that the real part of the transfer function for tilted illumination is given by

$$t_\kappa t_\upsilon \sin\left\{\frac{1}{2}\pi\left(P^2 - M^2\right)^2\right\} \tag{66.111}$$

where

$$\begin{aligned} t_\kappa &= \exp\left\{-\pi^2\frac{\left(P^2-M^2\right)P^2}{C_\kappa}\right\} \\ t_\upsilon &= \exp\left\{-\frac{1}{4}\pi^2\left(P^2-M^2\right)D^2\upsilon_m^2\right\} \end{aligned} \tag{66.112}$$

The function $\sin\frac{\pi}{2}\,(P^2 - M^2)^2$ is shown in Fig. 66.12 for the optimum defocus $M^2 = D = 1$ and in Fig. 66.13 for a series of optimum defocus values, $D = M^2 = 0.5, 0.75, \ldots, 2.75$. After rotational averaging, the bright-field hollow-cone phase contrast-transfer functions of Fig. 66.14 are obtained: for each, $D = M^2$; the reduced source size is 0.12 and the defocus corresponding to the energy spread is 0.15 Sch. As a general conclusion, we may say that hollow-cone illumination offers the possibility of extending the resolution of a given microscope beyond the cut-off for axial illumination but at the cost of lowering the contrast, which is no longer close to unity anywhere in the passband but at best 0.5.

Eusemann and Rose (1982) and Dinges et al. (1994) recognized that, in addition to defocus and angular aperture, there are two additional free parameters in hollow-cone conditions: the cone angle and the image shift. This last quantity is a measure of the distance in the Gaussian image plane between the optic axis and the corresponding point with tilted illumination. Their analysis

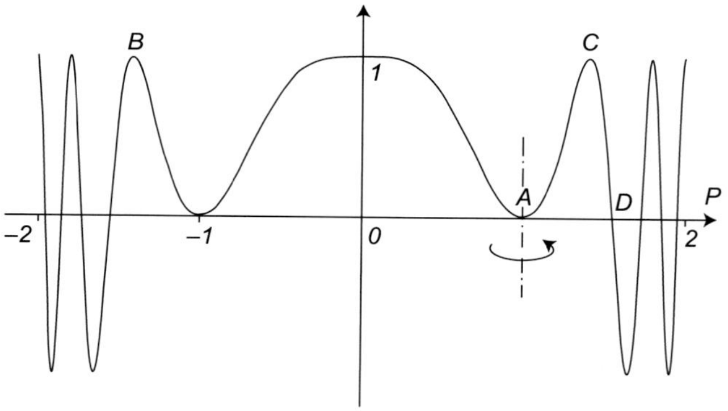

Figure 66.12
The unattenuated real part of the transfer function, $\sin\{\pi(P^2 - M^2)^2/2\}$ as a function of P for $M^2 = 1$. *After Jenkins (1979), Courtesy Kingsley Jenkins.*

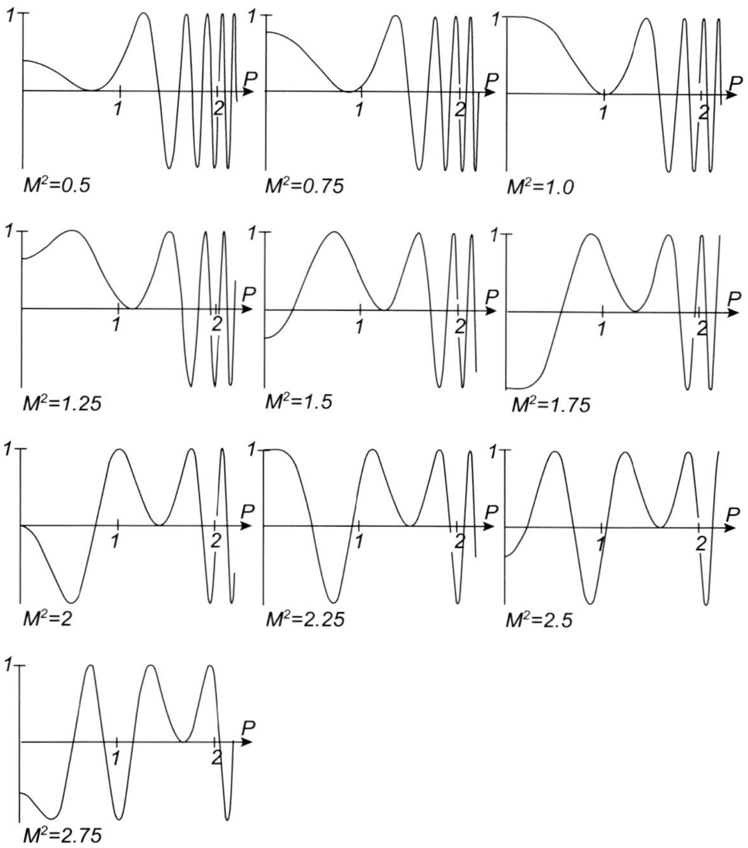

Figure 66.13
The function $\sin\{\pi(P^2 - M^2)^2/2\}$ as a function of P for the values of M^2 shown. *After Jenkins (1979), Courtesy Kingsley Jenkins.*

is too long to reproduce here; they conclude that "the "phase contrast-transfer" function for hollow-cone illumination has its first zero at a significantly higher spatial frequency than for axial illumination". Simulated images of thin layers of silicon ⟨111⟩ and YBa_2O_7 ⟨100⟩ justified this claim The gain is less obvious for thicker specimens.

For further discussion, see Hanszen and Trepte (1971b), Niehrs (1973), Krakow and Howland (1976), Kunath (1976, 1979), Krakow (1977, 1978), Rose and Fertig (1977), Rose (1977), Freeman et al. (1977, 1980), Bonnet et al. (1978), Saxton et al. (1978), Fertig and Rose (1977, 1978a,b, 1979), Saxton and Smith (1979), Kunath and Weiss (1980), Kunath et al. (1981, 1985, 1986), Zemlin et al. (1982), Balossier and Thomas (1984), Kunath and Gross (1985), Herring (1991), Wang (1994), Geipel and Mader (1996), Taya et al. (2008), Kawasaki et al. (2010) and Ma et al. (2018, light microscopy).

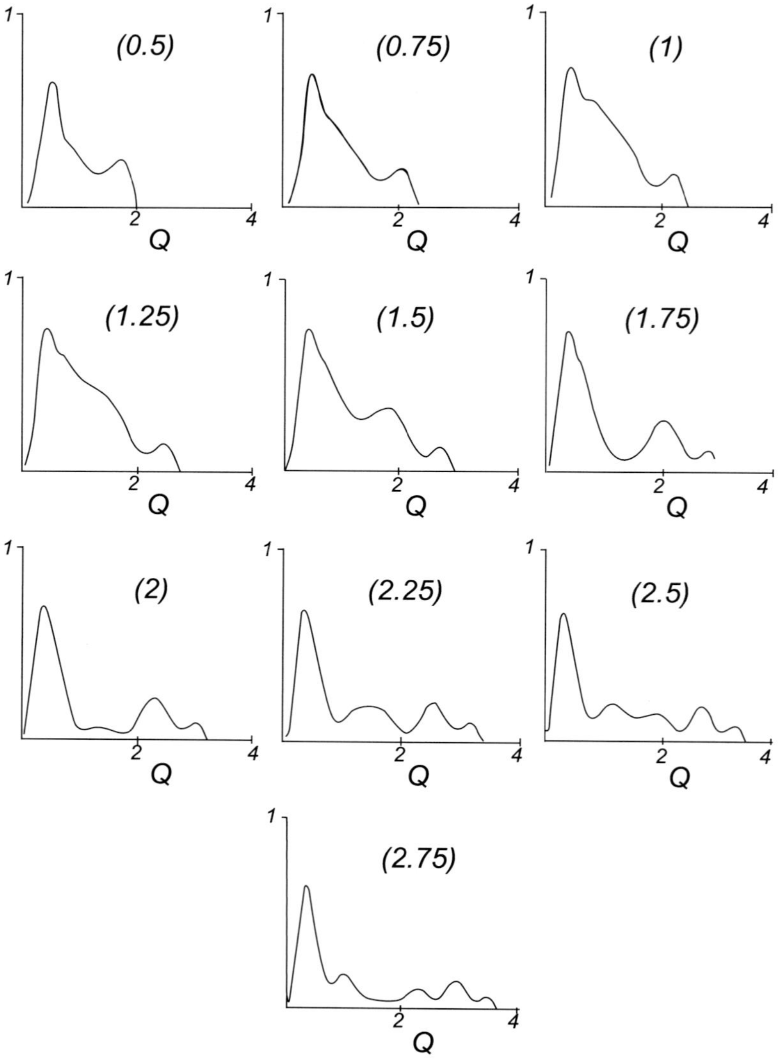

Figure 66.14
Bright-field hollow-cone transfer functions at optimum focus for various values of M^2. *After Jenkins (1979), Courtesy Kingsley Jenkins.*

Lee et al. (2017) have drawn attention to the advantage of using hollow-cone ilumimation in aberration-corrected TEM. A study of barium sulphate using the SALVE instrument (see Fig. 41.12 of Volume 2 or Kaiser, 2020) showed that both the limit of resolution and the signal-to-noise ratio were significantly better with hollow-cone than with parallel illumination; the resolution was improved from 1.07 to 0.7 Å. In a later discussion of this mode of illumination, Lee et al. (2019) showed simulated images of 2H–SiC in the traditional and hollow-cone modes. An essential feature is the use of a microscope with chromatic aberration correction. Again, the resolution at 30 kV was reduced from about 1 to 0.75 Å. Harada et al. (2019a,b) have shown that hollow-cone Foucault imaging is advantageous for the study of magnetic specimens. Here, the incident beam was tilted and rotated around the optic axis. Magnetic domains and domain walls were visible in in-focus conditions.

A procedure for simulating images and diffraction patterns in various less conventional modes is described and abundantly illustrated by Krakow (1984); this includes convergent-beam diffraction, hollow-cone bright-field and dark-field illumination and diffraction with 'virtual apertures'. In the dark-field hollow-cone mode the cone angle is so large that the direct (conical) beam is intercepted by the objective aperture. In the virtual-aperture mode the direction and intensity of the incident beam are under computer control and describe a more complicated pattern.

66.5.2 Anisoplanatism

The development of transfer theory, as we have presented it, was contingent on the convolutional structure of Eq. (65.13). This limited the geometrical aberrations that could be included to spherical aberrations of any order and axial parasitic aberrations. At very high resolution and in connection with aberration-corrector performance, we may wish to consider other aberrations, if only to show that their effect is negligible. The coma is a case in point, being the next most important third-order aberration after spherical aberration since it is quadratic in angle and linear in position in the object plane. Can we say anything about the contrast transfer in the presence of coma and perhaps other aberrations that destroy isoplanasy? We now show that a filter function analogous to $T_L(\boldsymbol{q})$ can be defined but that this varies with position in the image plane; it thus enables us to make predictions about the quality of the image but cannot be inverted (Hawkes, 1971). This is not a serious limitation as the existence of a coma-free point (Eqs. 24.134 and 36.28 of Volume 1) is exploited at high resolution. We limit the present analysis to the isotropic coma K (Section 24.4); the most highly perfected aberration correctors can also correct anisotropic coma (Fig. 41.19 and pp. 983–984 of Volume 2).

Coma contributes a term of the form $K\lambda^3\boldsymbol{q}^2\boldsymbol{q}\cdot\boldsymbol{u_o}$ to the wave aberration W (65.29). This can be forced into a form involving $\boldsymbol{u_i} - \boldsymbol{u_o}$ by writing

$$\boldsymbol{q}\cdot\boldsymbol{u}_o = -\boldsymbol{q}\cdot(\boldsymbol{u}_i - \boldsymbol{u}_o) + \boldsymbol{q}\cdot\boldsymbol{u}_i$$

so that the kernel $G(\boldsymbol{u_i}, \boldsymbol{u_o})$ in

$$\psi_i(\boldsymbol{u}_i) = \int G(\boldsymbol{u}_i, \boldsymbol{u}_o)\psi_o(\boldsymbol{u}_o)d\boldsymbol{u}_o$$

(65.7) has the form

$$G(\boldsymbol{u}_i, \boldsymbol{u}_o) = G(\boldsymbol{u}_i - \boldsymbol{u}_o; \boldsymbol{u}_i) \tag{66.113}$$

and

$$\psi_i(\boldsymbol{u}_i) = \int G(\boldsymbol{u}_i - \boldsymbol{u}_o; \boldsymbol{u}_i)\psi_o(\boldsymbol{u}_o)d\boldsymbol{u}_o$$

Setting $\boldsymbol{u}_s := \boldsymbol{u}_i - \boldsymbol{u}_o$, we have

$$\psi_i = \int G(\boldsymbol{u}_s; \boldsymbol{u}_i)\psi_o(\boldsymbol{u}_i - \boldsymbol{u}_s)d\boldsymbol{u}_s \tag{66.114}$$

This is again a convolution and hence

$$\psi_i = \int \tilde{G}(\boldsymbol{q}; u_i)S_o(\boldsymbol{q})\exp(2\pi i\boldsymbol{u}_s \cdot \boldsymbol{q})d\boldsymbol{q} \tag{66.115}$$

As usual, we write $\psi_o(\boldsymbol{u}_o) = 1 - \sigma_o + i\eta_o$, $|\sigma_o|$ and $|\eta_o| \ll 1$ and obtain, in the conditions assumed above for the isoplanatic situation,

$$\begin{aligned} C(\boldsymbol{u}_i) &= \frac{1}{2}\int \tilde{\sigma}\left\{\tilde{G}(\boldsymbol{q}; \boldsymbol{u}_i)G^{\sim *}(-\boldsymbol{q}; \boldsymbol{u}_i)\right\}\exp(-2\pi i\boldsymbol{q}\cdot\boldsymbol{u}_i)d\boldsymbol{q} \\ &- \frac{i}{2}\int \tilde{\eta}\left\{\tilde{G}(\boldsymbol{q}; \boldsymbol{u}_i)G^{\sim *}(-\boldsymbol{q}; \boldsymbol{u}_i)\right\}\exp(-2\pi i\boldsymbol{q}\cdot\boldsymbol{u}_i)d\boldsymbol{q} \end{aligned} \tag{66.116}$$

where

$$\begin{aligned} S_o(\boldsymbol{q}) &= \int \psi_o(\boldsymbol{u}_o)\exp(-2\pi i\boldsymbol{q}\cdot\boldsymbol{u}_o)d\boldsymbol{u}_o \\ \tilde{G}(\boldsymbol{q}; \boldsymbol{u}_i) &= \int G(\boldsymbol{u}_s; \boldsymbol{u}_i)\exp(-2\pi i\boldsymbol{q}\cdot u_s)du_s \end{aligned} \tag{66.117}$$

The spatial-frequency spectra of the object, characterized by $\tilde{\sigma}$ and $\tilde{\eta}$, will, therefore, be filtered in a predictable fashion but the filter functions, $\tilde{G}(\boldsymbol{q}; \boldsymbol{u}_i) \pm \tilde{G}^*(-\boldsymbol{q}; \boldsymbol{u}_i)$, now depend not only on the defocus and lens aberrations but also on the image point considered: we can still calculate how contrast is transferred from object to image but we cannot invert the transformation.

For further discussion of anisoplanatic transfer, see the later work of Hawkes (1972, 1973a), Schiske (1973) and Ade (1978).

66.6 Forms of the Aperture Function T_A: Zone Plates and Phase Plates

In addition to the standard round aperture centred on the optic axis, a number of other aperture shapes and opacities attracted sporadic attention over the past and some are now in regular use. Two roles are envisaged for these devices: either they increase the contrast of phase objects, like the Zernike phase plate of the light microscope, or, more ambitiously, they combat the adverse effects of the contrast-transfer function (Eqs. 66.16b and 66.16d). Occasionally, these roles are combined.

Phase plates were first mentioned by Hans Boersch (1947c), who described the electron-optical equivalent of the Zernike phase plate, to which we return below. The standard reference is the 1947c paper but Boersch had mentioned such a plate briefly a little earlier (Boersch, 1948), submitted in June 1946 and presented to the Austrian Academy of Sciences in October of the same year; a French translation is dated 1947 (Boersch, 1947a). Numerous attempts were made to fabricate and test such plates during the following decades (Agar et al., 1949; Locquin, 1954, 1955, 1956; Kanaya et al., 1954, 1957, 1958a,b; Kanaya and Kawakatsu, 1958; Faget et al., 1960a, b, 1962; Eisenhandler and Siegel, 1966b; Siegel et al., 1966; Badde and Reimer, 1970; Reimer and Badde,1970; Thon and Willasch, 1970, 1971a,b, 1972a,b; Tochigi et al., 1970; Müller, 1971, 1976; Müller and Rindfleisch, 1971; Parsons and Johnson, 1972; Willasch, 1973, 1975a,b; Anaskin and Ageev, 1974; Thon, 1974; Balossier et al., 1980; Laberrigue et al., 1980). When the form of the phase contrast-transfer function was established, Hoppe (1961, 1963, 1970, 1971) and Lenz (1963, 1964, 1965) considered the use of zone plates, in which the open and opaque (or phase-shifting) parts of the plate matched the positive and negative regions of the function sin χ. Attempts to make and test such plates are described by Langer and Hoppe (1966/7, 1967a,b), Möllenstedt et al. (1968) and Hoppe et al. (1970). Also see Anaskin et al. (1976) and Ageev et al. (1977).

Much recent work has been inspired by the need to increase the feeble contrast of images of unstained biological specimens. Several families of phase plates have been studied in depth: the Zernike plates consisting of a thin layer of material with a central hole, as proposed by Boersch (1947c); hole-free or Volta plates, which are unperforated Zernike plates; Hilbert plates, which impose a phase change on only half the diffraction plane apart from a hole on the optic axis; Buijsse or tulip Hilbert plates, in which only an annular zone of the plate is retained; Boersch plates in which a phase shift is created by means of an electrostatic field or more recently, a magnetic field. Zach plates are a variant of Boersch plates, in which the electrostatic field is formed at the end of a very fine coaxial cable. In another variant, which was once explored in some detail, a fine thread was stretched across the objective aperture and charge was allowed to build up on it, thereby creating a potential variation. The difficulties and uncertainties of these devices were, however, such that they have been abandoned. For details, see the very full study by Unwin (1971, see also 1970a,b, 1972, 1974) and the work of Krakow and Siegel (1975), Balossier et al. (1980) and Balossier and Bonnet (1981). Dupouy (1967) also stretched a wire across the aperture but its role was not to modify the phase but to intercept the unscattered beam. A very different way of shifting the phase of the unscattered beam exploits the Kapitza–Dirac effect (Kapitza and Dirac, 1933). Here, a laser beam travelling perpendicular to the electron beam is suitably focused and produces the desired effect.

We now examine each of these families in more detail. The subject has been reviewed by Unwin (1973), Henderson (1995), Majorovits (2002), Glaeser (2013) and in particular by

Danev and Nagayama (2010), Nagayama (2005, 2008, 2011), Danev and Baumeister (2017), Edgcombe (2017b) and Malac et al. (2021). Plates situated in the condenser space of a conventional transmission electron microscope form part of the increasingly important subject of *structured illumination*. We shall meet many examples in the context of orbital angular momentum and vortex studies in Chapter 80. Yang et al. (2020) have fabricated phase plates in the form of nanostructured membranes.

As we shall see, phase plates take many forms. In a comparative study of eight different designs, Hettler and Arenal (2021) have simulated the performance to be expected of each. Three test specimens are examined: T4 bacteriophages, spherical amorphous carbon nanoparticles and a graphene monolayer.

66.6.1 Zernike Plates

In a publication of 1935, Frits Zernike described a way of rendering phase objects visible in the light microscope by inserting a 'phase plate' in the focal plane of the objective (Zernike, 1935, 1942a,b, 1953; Köhler and Loos, 1941). The 1935 paper is the standard reference to Zernike's invention but, in fact, he mentioned it two years earlier (Zernike, 1933) at a medical congress. The thickness and refractive index of this plate were such that the phase difference between the unscattered beam, which passed through the plate, and the scattered beam was increased from $\pi/4$ to $\pi/2$, thereby creating constructive interference seen as phase contrast. A similar problem is omnipresent in transmission electron microscopy, where all specimens are phase objects, in the sense that electrons are not absorbed by the sample. Nonetheless, inelastically scattered electrons are lost in a theory of high-resolution image formation that is limited to elastic scattering and their exclusion may be seen as amplitude contrast (see Section 66.1). The image contrast is particularly weak for unstained biological specimens, notably those of interest in cryoelectron microscopy, where the design and use of phase plates for electrons has been widely studied.

Unlike the original Zernike scheme, it is the phase of the *scattered* electrons that is altered, while the unscattered central beam passes through a hole in the centre of the plate. The simple picture in which the central opening is transparent and the thickness of the plate uniform is reasonably satisfactory. Some improvement can be gained by replacing the sharp edge of the hole with a short ramp around the border of the hole (Edgcombe, 2017b). Rhinow (2016) has suggested that only an annular zone should be occupied by the plate, leaving the central hole and the periphery free; electrons scattered through larger angles, carrying high spatial-frequency information, would thus not suffer further scattering in the plate. Fig. 66.15A shows the arrangement; some key contrast-transfer functions are reproduced in Fig. 66.15B–D. The principal problems with these plates arise from the accumulation of charge and from their limited lifetime.

The phase change φ_Z created by the plate is determined by the thickness t and choice of material of which it is composed as well as the wavelength of the beam:

$$\varphi_Z = \frac{2\pi}{\lambda}(n-1)t \tag{66.118}$$

in which we have used the notion of refractive index n (Eq. 5.17 of Volume 1). Here,

$$n = \left\{ \frac{(\Phi_0 + V)(1 + \varepsilon\Phi_0 + \varepsilon V)}{\hat{\Phi}_0} \right\}^{1/2} \tag{66.119}$$

and V is the mean inner potential of the material. Values of the inner potential of several elements are listed in Table 3.1 of Reimer and Kohl (2008); for carbon, they give $V = 7.8$ V and for silicon, $V = 11.5$ V. If we consider $\varphi_Z = \pi/2$, for example, the thickness is given by

$$t = \frac{\lambda}{4(n-1)} = \frac{\hat{\Phi}_0\lambda}{2V(1+2\varepsilon\Phi_0)} \approx \frac{\Phi_0\lambda}{2V} \tag{66.120}$$

For $\Phi_0 = 100$ kV and a carbon film, we find $t = 21.8$ nm and at 300 kV, $t = 30.9$ nm. For silicon, the values are 14.8 and 20.9 nm. The value of the mean inner potential depends on

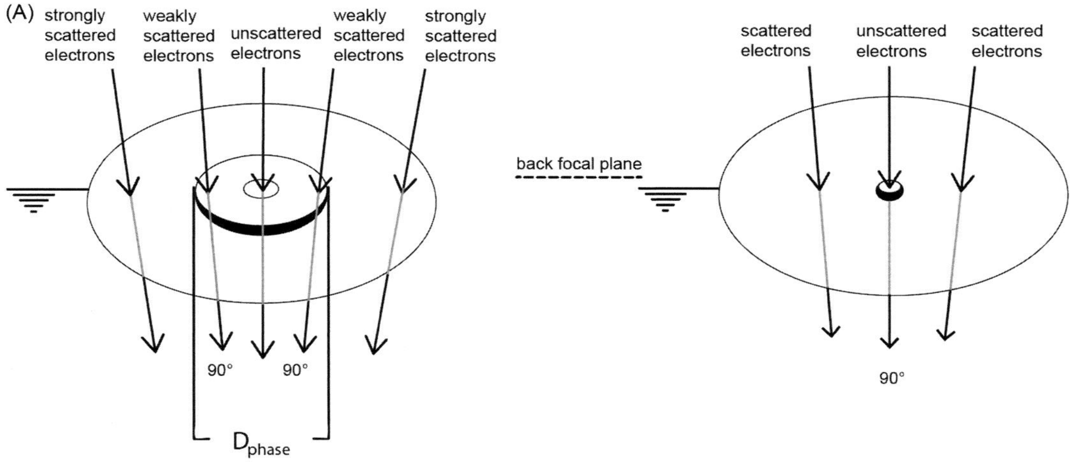

Figure 66.15

The Zernike plate. (A) The different zones of an annular Zernike plate and their roles. (B)–(D) Contrast-transfer functions. (B) No phase plate; (C) disc radius 10 μm, $\pi/2$ phase shift for spatial frequencies below 0.26 nm^{-1}; (D) disc radius 20 μm, $\pi/4$ phase shift for spatial frequencies below 0.52 nm^{-1}. In all cases, 200 kV, $f = 15$ mm, $C_s = 0.01$ mm (corrected instrument), $C_c = 7.9$ mm, $\Delta = -500$ nm. (E) Zernike plate in 'rocking mode'. (F) Images of carbon nanotubes. *Left* with a standard Zernike plate and *Right* with a gradual Zernike plate, (1–3 μm). The diameter of the central hole is 1 μm. *(A–D) After Rhinow (2016), Courtesy Elsevier. (E) After Barragán Sanz and Irsen (2019), Courtesy the authors. (F) After Obermair et al. (2019), Courtesy D. Gerthsen.*

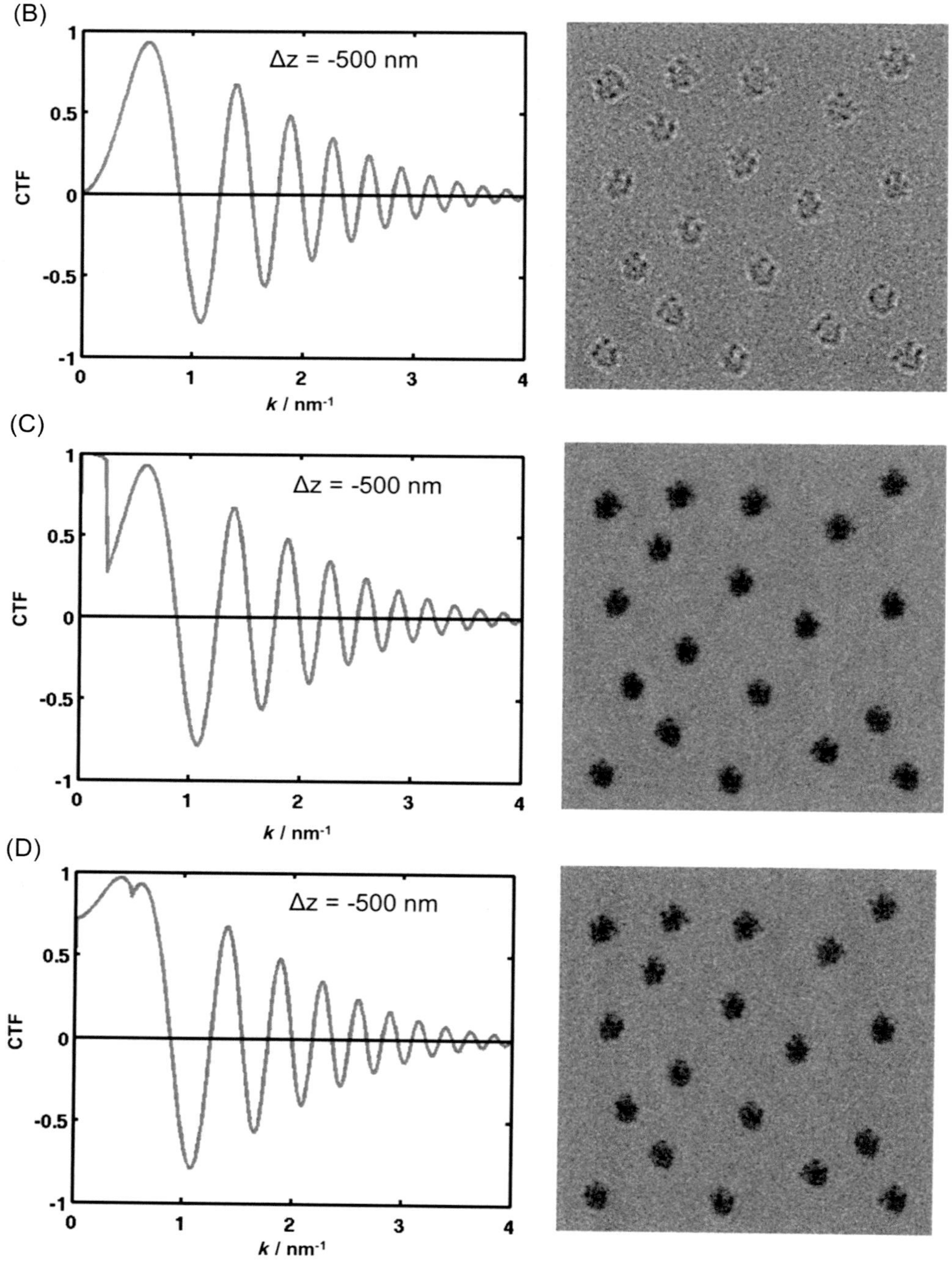

Figure 66.15
Continued.

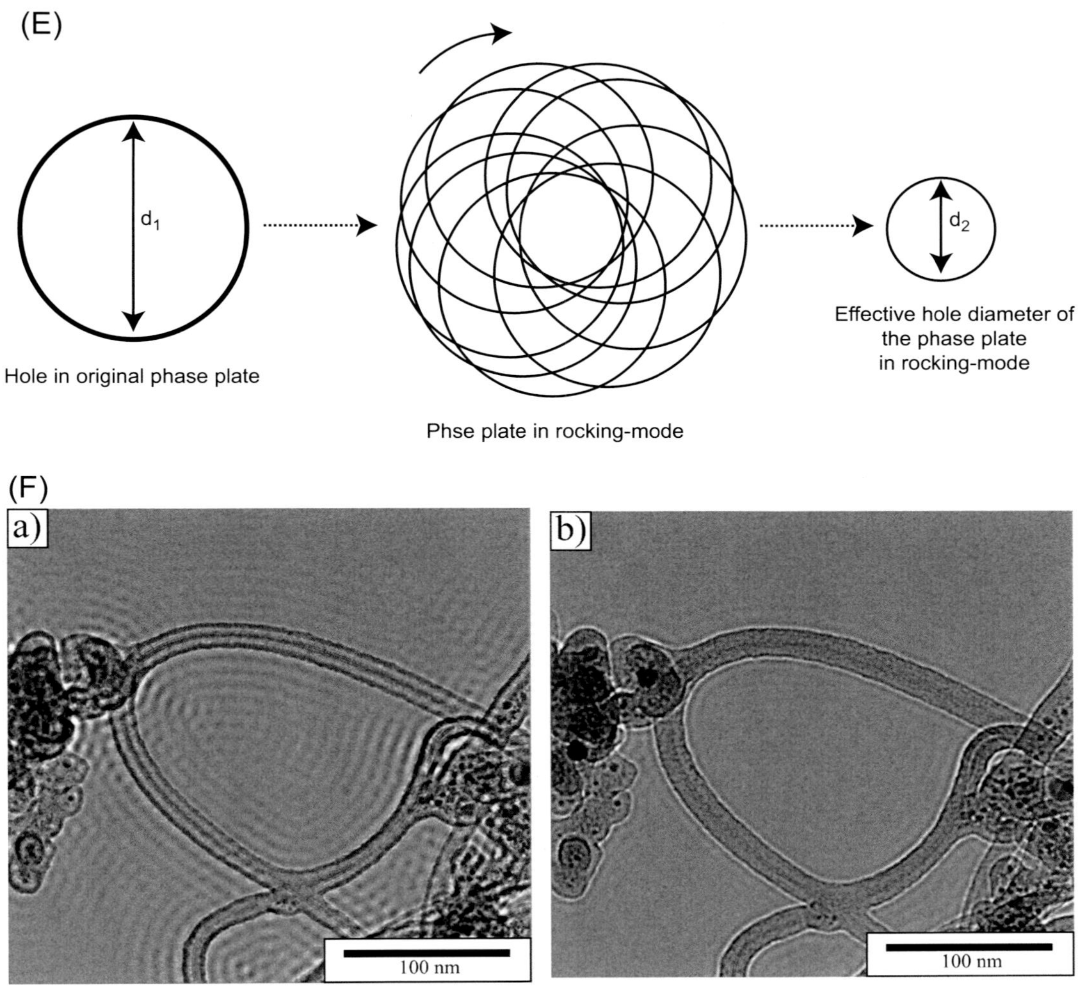

Figure 66.15
Continued.

the method used to measure it. Harscher (1999), for example, found a value of about 10.7 V for carbon by electron holography. The values for silicon include 12.1 (Wang et al., 1997, holography) and 13.5 V (Wolf et al., 2013, electron holographic tomography). For a long list and critical discussion, see Appendix B1 of Lubk (2018).

For simplicity, we examine the effect of the plate on the image of a one-dimensional phase object. The wave function emerging from the object is then exp $i\eta_o(x)$ (Eq. 66.1) and in the diffraction plane, we have

$$\tilde{\psi}_o(k) = \delta(k) + \mathcal{F}\left(e^{i\eta_o(x)} - 1\right) \tag{66.121}$$

After multiplying by $\exp(\mathrm{i}\varphi_Z)$, we obtain

$$\tilde{\psi}_i(k) = \delta(k) + \mathrm{e}^{\mathrm{i}\varphi_Z}\mathcal{F}\left(\mathrm{e}^{\mathrm{i}\eta_0(x)} - 1\right) \tag{66.122}$$

and the image intensity is hence

$$I(x, \varphi_Z) = 3 - 2\cos\varphi_Z - 2\cos\eta_o(x) + 2\cos(\varphi_Z + \eta_0(x)) \tag{66.123}$$

This is the traditional explanation of the effect of the Zernike plate but it has two weaknesses. The total current, which should be equal to unity, does not satisfy this condition unless the specimen is a weak-phase object. Also, for pure phase objects, we expect that $I(\varphi(x) + \varphi_0) = I(\varphi(x))\ \forall\varphi_0$ but this too is not true. We can remedy the situation by recognizing that the central hole has a finite size and that some of the unscattered beams may not pass through this hole (Beleggia, 2008). Eq. (66.121) should, therefore, be replaced by

$$\tilde{\psi}_o(k) = \sigma_0\delta(k) + \mathcal{F}\left(\mathrm{e}^{\mathrm{i}\eta_o(x)} - \sigma_0\right) \tag{66.124}$$

which leads to

$$I(x, \varphi_Z, \sigma_0) = 1 + 2\sigma^2 - 2\sigma^2\cos\varphi_Z - 2\sigma\cos\left(\varphi_\sigma + \eta_o(x)\right) + 2\sigma\cos\left(\varphi_\sigma + \varphi_Z - \eta_o(x)\right) \tag{66.125}$$

with $\sigma_0 = \sigma\exp(-\mathrm{i}\varphi_\sigma)$. Both conditions are now satisfied.

The sharp edge of the Zernike plate is a source of 'ringing' artefacts. Barragán Sanz and Irsen (2019) combat this problem by moving the plate in a circular path round the normal static position as mentioned by Kurth et al. (2014) (Fig, 66.15E). Their preliminary experiments confirm the benefits of this 'rocking mode'; for a full account see Barragán Sanz (2021). Alternatively, the sharp edge of the central opening can be replaced by a gradual reduction of the thickness. Obermair et al. (2019) have tested gradual phase plates in which the gradient ranged from 750 mm to 1 μm and from 1 to 3 μm. Fig. 65.15F shows that the fringing artefacts are appreciably reduced.

The phase plate remedies a weakness of imaging in which contrast is generated by defocus and spherical aberration, characterized by the contrast-transfer function (Eq. 66.16d). In the presence of the phase plate, sin (χ) is replaced by sin $(\chi - \varphi_Z)$ except in the region of the optic axis so that for $\varphi_Z = \pi/2$ we obtain $-$ cos (χ). For many more details and applications, see Danev et al. (2002, 2009, 2010, 2012, 2011), Danov et al. (2001), Danev and Nagayama (2001a,b, 2004a,b, 2008, 2010, 2011), Downing et al. (2004), Lentzen (2004), Hosokawa et al. (2005), Tosaka et al. (2005), Armbruster et al. (2008, 2010), Beleggia (2008), Gamm et al. (2008, 2010a,b), Hettler et al. (2008, 2012, 2015, 2017a,b, 2018. 2019), Malac et al. (2008), Nagayama and Danev (2008), Yamaguchi et al. (2008), Fukuda et al. (2009), Kawasaki et al. (2009), Dries et al. (2010, 2011, 2015, 2016a,b, 2018),

Hosogi et al. (2011, 2015a, b), Murata et al. (2010, 2015), Shiue and Hung (2010), Van Dyck (2010), Hall et al. (2011), Marko et al. (2011, 2013), Minoda et al. (2011, 2013, 2018a,b), Fukuda and Nagayama (2012), Iijima et al. (2012), Inayoshi et al. (2012), Qian et al. (2012), Glaeser et al. (2013), Pollard et al. (2013), Edgcombe (2014a, b, 2016), Konyuba et al. (2014a,b, 2015), Sader et al. (2014), Sannomiya et al. (2014), Shiloh et al. (2014, 2016, 2018) and Janzen et al. (2016).

Such plates have also been employed in the STEM, in which case they are inserted upstream from the probe-forming lens. See Minoda et al. (2014, 2015a,b, 2016, 2017, 2018a,b), Iijima et al. (2015), Yang et al. (2016, 2018), Hettler et al. (2017a,b), Tsubouchi et al. (2020) and Tsubouchi and Minoda (2021).

66.6.2 Hole-Free or Volta Plates

One might expect that an unperforated plate of uniform thickness would be of no interest. In reality, the strong beam, which passes through the central hole in a standard Zernike plate, causes changes in the composition of the material with the result that an advantageous phase shift between scattered and unscattered beams is perceived. This is not due to accumulated charge since the effect would require the charge to be positive. There is no conclusive explanation of the mode of action, but at first, it seemed that the impact of the primary beam caused emission of secondary electrons, leading to charging on the surfaces of the plate. Any accumulated contamination would also play a role. The situation has been made clearer by the work of Hettler et al. (2018, 2019), Pretzsch et al. (2019) and Harada et al. (2020). The latter introduce a periodic sample that splits the incident beam into three distinct well-separated beams at the hole-free plate, situated in the back-focal plane of the objective. The resulting charge distributions and their evolution with time are analysed. From this analysis, Harada et al. propose mechanisms that could explain the action of the hole-free plate. Pretzsch et al. (2019) study the phase-shifting behaviour of a hole-free plate for unscattered beams of different diameters after different lapses of time. An unexpected finding is that the phase shift introduced by the plate depends on spatial frequency; to account for this, Pretzsch et al. add a suitable term to the wave aberration function, $\Delta\varphi_{PP}(\boldsymbol{q})$ in their notation. We cannot reproduce all the details of their careful study. Fig. 66.16 together with Table 66.1 give a good idea of the nature of their conclusions. Buijsse et al. (2020) have studied the effect of a hole-free plate as a function of spatial frequency.

Hole-free plates were introduced by Malac et al. (2010), who have studied them further in Malac et al. (2012a,b, 2014a,b, 2015, 2017a,b, 2018a,b). See also Beleggia et al. (2012), Minoda et al. (2013), Buijsse et al. (2014, 2016), Danev et al. (2014a,b,c, 2015, 2017a,b), Asano et al. (2015, 2016), Fukuda et al. (2015a,b, 2017), Marko and Hsieh (2015), Hosogi et al. (2015a,b), Konyuba et al. (2015), Sader et al. (2014, 2015), Iijima et al. (2016), Mahamid et al. (2016), Marko et al. (2016), Chlanda and Locker. (2017), Hettler et al. (2017, 2018, 2019),

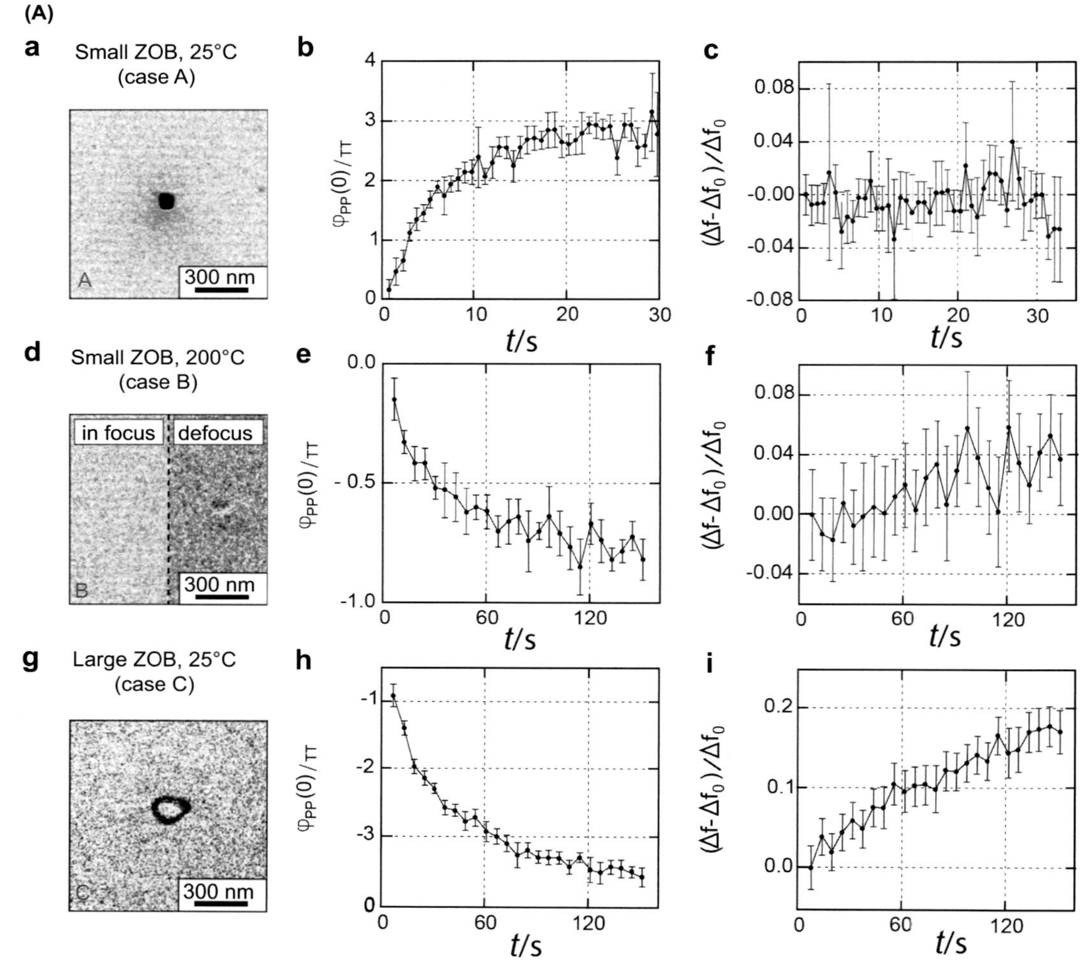

Figure 66.16

(A) Behaviour of a hole-free phase plate for the cases A, B and C defined in Table 66.1. *Left column*, TEM images of the patch on the plate. *Central column*, Phase change introduced by the plate as a function of time (t). *Right column*, Change of defocus as a function of time. (B) Sections through Thon rings for an amorphous carbon film in the presence of a hole-free plate. *Case A*, the experimental maxima and minima (*vertical bars*) coincide with the fitted values. *Case B*, a significant difference between experiment and fit is seen for low values of spatial frequency, q. *Case C*, the difference is again considerable. The dotted curves below cases B and C show the difference between the arguments of the fitted and experimental data, $\Delta\varphi_{PP}(q)$. *ZOB* = Zero-order beam.' *After Pretzsch et al. (2019), Courtesy Springer, https://doi.org/10.1186/s40679-019-0067-z.*

Sharp et al. (2017), Hayashida et al. (2018, 2019), Kotani et al. (2018a,b,c), Malac (2017a,b, 2018a,b), Obermair et al. (2018a,b), Pretzsch et al. (2018), von Loeffelholz et al. (2018), Plitzko and Baumeister (2019) and Gallagher et al. (2020).

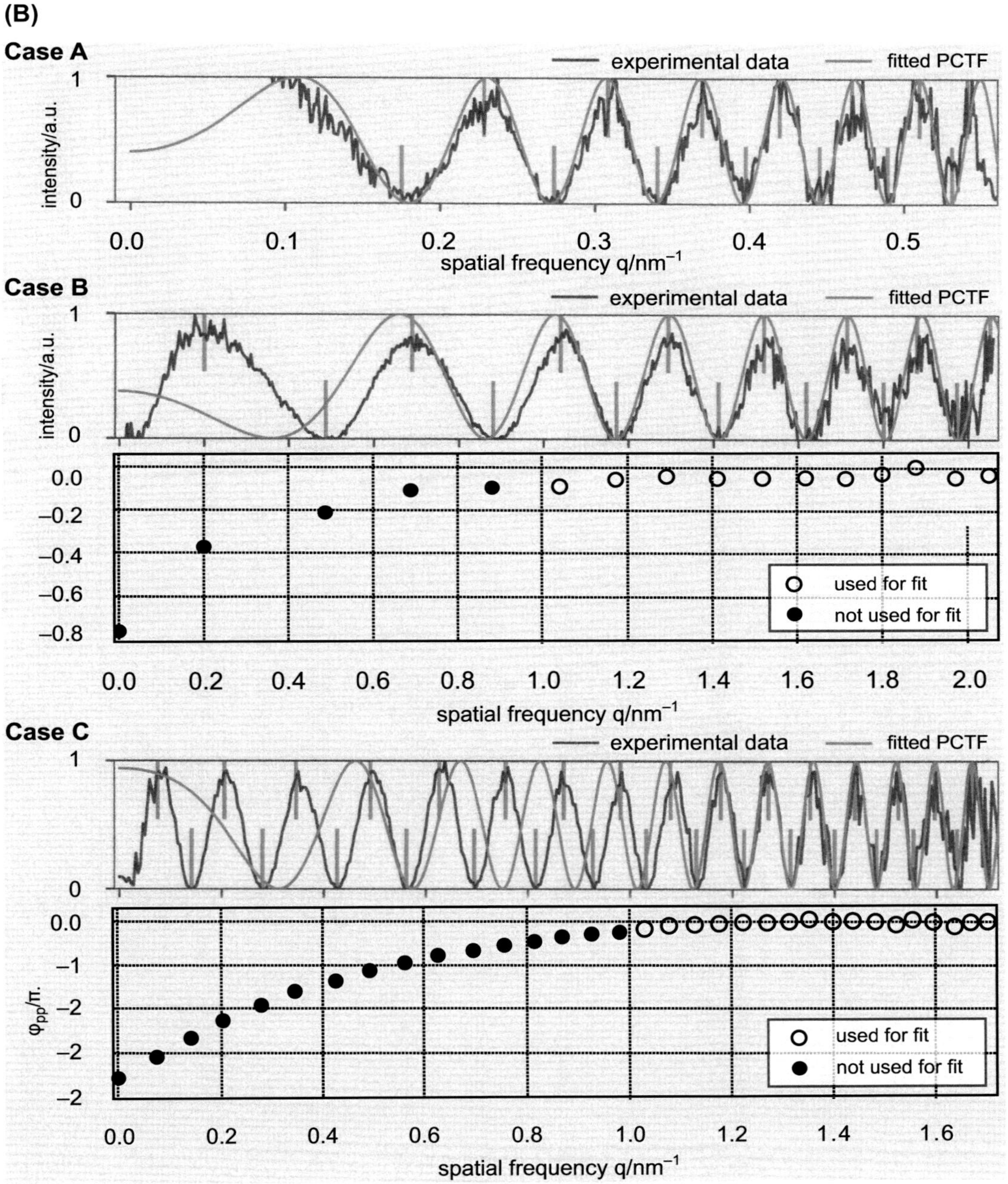

Figure 66.16
Continued.

Table 66.1: Details of the experimental conditions corresponding to Fig. 66.16.

	Case A	Case B	Case C
Diameter of zero-order beam/nm	100	100	250
Temperature of phase plate/°C	25	200	25
Phase shift and duration	3π, 30 s	0.8π, 150 s	3.5π, 150 s

66.6.3 Hilbert Plates

A Hilbert plate consists of a semicircular disc. A notch may be cut in the centre of the straight edge but the plate is commonly placed far enough off centre for the unscattered beam to pass freely. When the disc is completely opaque, it is commonly known in light microscopy as a Foucault plate (Foucault, 1858), creating schlieren contrast, a forerunner of the method attributed to Töpler (1864, 1866). Several attempts have been made to use it in electron microscopy (e.g. Downing and Siegel, 1973; Cullis and Maher, 1974, 1975; Haydon and Lemons, 1972; Andersen, 1972; Downing, 1979). See also references 300–315 in Hawkes (1978a). We shall meet semicircular opaque plates and more complicated shapes such as quadrants as a means of solving the 'phase problem' in Chapter 74. The Hilbert plate is not opaque as its role is to modify the phase of the wave incident on it. In the first attempts to use it to improve contrast, the thickness was chosen to shift the phase by π and the image was formed in the Gaussian image plane (Danev et al., 2002); Danev refers to this as differential phase-contrast microscopy. However, there are reasons to prefer a phase shift of $\pi/2$ and form the image at Scherzer focus (Koeck, 2015) as we now show.

It is assumed that the plate is situated at a distance g (gap) from the optic axis to allow the unscattered beam to pass. The plate occupies the remainder of the half-plane $\boldsymbol{q}_x > \boldsymbol{q}_g$, $\boldsymbol{q}_g = g/\lambda f$ so that its transparency $H(\boldsymbol{q})$ is described by

$$H\left(\boldsymbol{q}\right) = \begin{cases} \exp(\mathrm{i}\varphi_H) & \text{for} \quad \boldsymbol{q}_x > \boldsymbol{q}_g \\ 1 & \text{for} \quad \boldsymbol{q}_x < \boldsymbol{q}_g \end{cases} \tag{66.126}$$

For a phase object the wave function in the diffraction plane is now

$$\tilde{\psi}(\boldsymbol{q}) = \{\delta(\boldsymbol{q}) + \mathrm{i}\,\tilde{\eta}\}\mathrm{e}^{-\mathrm{i}\chi(\boldsymbol{q})}H(\boldsymbol{q}) \tag{66.127}$$

In the image plane, we find

$$I(\boldsymbol{u}) = \left\{1 + \mathrm{i}\int \tilde{\eta}\, H(\boldsymbol{q})\mathrm{e}^{-\mathrm{i}\chi(\boldsymbol{q})}\mathrm{e}^{2\pi\mathrm{i}\boldsymbol{u}\cdot\boldsymbol{q}}d\boldsymbol{u}\right\}\left\{1 - \mathrm{i}\int \tilde{\eta}\, H^*(-\boldsymbol{q})\mathrm{e}^{\mathrm{i}\chi(\boldsymbol{q})}\mathrm{e}^{2\pi\mathrm{i}\boldsymbol{u}\cdot\boldsymbol{q}}d\boldsymbol{u}\right\} \tag{66.128}$$

We note that

$$H^*\left(-\boldsymbol{q}\right) = \begin{cases} \exp(-\mathrm{i}\varphi_H) & \text{for} \quad \boldsymbol{q}_x < -\boldsymbol{q}_g \\ 1 & \text{for} \quad \boldsymbol{q}_x > -\boldsymbol{q}_g \end{cases} \tag{66.129}$$

We now consider three ranges of $\boldsymbol{q}_x$.

1. $|\boldsymbol{q}_x| < \boldsymbol{q}_g$, free space outside the plate:

$$I(\boldsymbol{u}) = 1 + 2\int \tilde{\eta}(\boldsymbol{q})\sin\chi\exp(2\pi i\boldsymbol{u}\cdot\boldsymbol{q})d\boldsymbol{q} \qquad (66.130)$$

As expected, this is the same as the expression obtained in the absence of the plate.

2. $\boldsymbol{q}_x > \boldsymbol{q}_g$, region covered by the plate:

$$I(\boldsymbol{u}) = 1 + 2\int \tilde{\eta}(\boldsymbol{q})\sin\chi\exp(2\pi i\boldsymbol{u}\cdot\boldsymbol{q})d\boldsymbol{q} \qquad (66.131)$$

3. $\boldsymbol{q}_x < -\boldsymbol{q}_g$

$$I\left(\boldsymbol{u}\right) = 1 + i\int \tilde{\eta}(\boldsymbol{q})\left(e^{-i\chi(\boldsymbol{q})} - e^{-i\pi\varphi_H}e^{i\chi(\boldsymbol{q})}\right)e^{2\pi i\mathbf{u}\cdot\boldsymbol{q}}d\boldsymbol{q} \qquad (66.132)$$

The plate examined by Danev et al. corresponds to $\varphi_H = \pi$, for which we find

$$I(\boldsymbol{u}) = 1 - 2i\int \tilde{\eta}\cos\chi(\boldsymbol{q})\mathrm{sgn}(\boldsymbol{q}_x)\exp(2\pi i\boldsymbol{u}\cdot\boldsymbol{q})d\boldsymbol{u} \qquad (66.133)$$

for $\boldsymbol{q}_x > \boldsymbol{q}_g$. For $\varphi_H = \pi/2$, the corresponding expression is

$$I(\boldsymbol{u}) = 1 + i\int \tilde{\eta}(i - \mathrm{sgn}\boldsymbol{q}_x)(\cos\chi(\boldsymbol{q}) - \sin\chi(\boldsymbol{q}))\exp(2\pi i\boldsymbol{u}\cdot\boldsymbol{q})d\boldsymbol{u} \qquad (66.134)$$

Here, we have used the signum function,

$$\mathrm{sgn}(x) = \begin{cases} 1 & x > 0 \\ 0 & x = 0 \\ -1 & x < 0 \end{cases} \qquad (66.135)$$

After eliminating the terms in sgn ($\boldsymbol{q}_x$), the contrast transfer is described by $\cos\chi$ when $\varphi_H = \pi$ or $\cos\chi - \sin\chi$ when $\varphi_H = \pi/2$. Curves D and G of Fig. 66.17 show that the contrast-transfer function falls to zero at a higher spatial frequency for a $\pi/2$ plate at Scherzer focus than for a π plate at Gaussian focus. Moreover, the thickness of the film for a $\pi/2$ plate will be half that for a π plate; inelastic scattering will therefore be reduced, an additional attraction of the $\pi/2$ plate.

Buijsse argues that the phase change should be limited to an annular zone so that high-frequency information conveyed by electrons scattered through larger angles is not affected (Buijsse et al., 2011, 2012a,b). His 'tulip plate' is shown in Fig. 66.18A. Fig. 66.18B shows Fourier transforms of a focal series of images of an amorphous carbon film. Note that the contrast-transfer function is constant behind the opaque part of the plate; the magnitude of the contrast-transfer function at the border of the semicircular pate can be matched to the

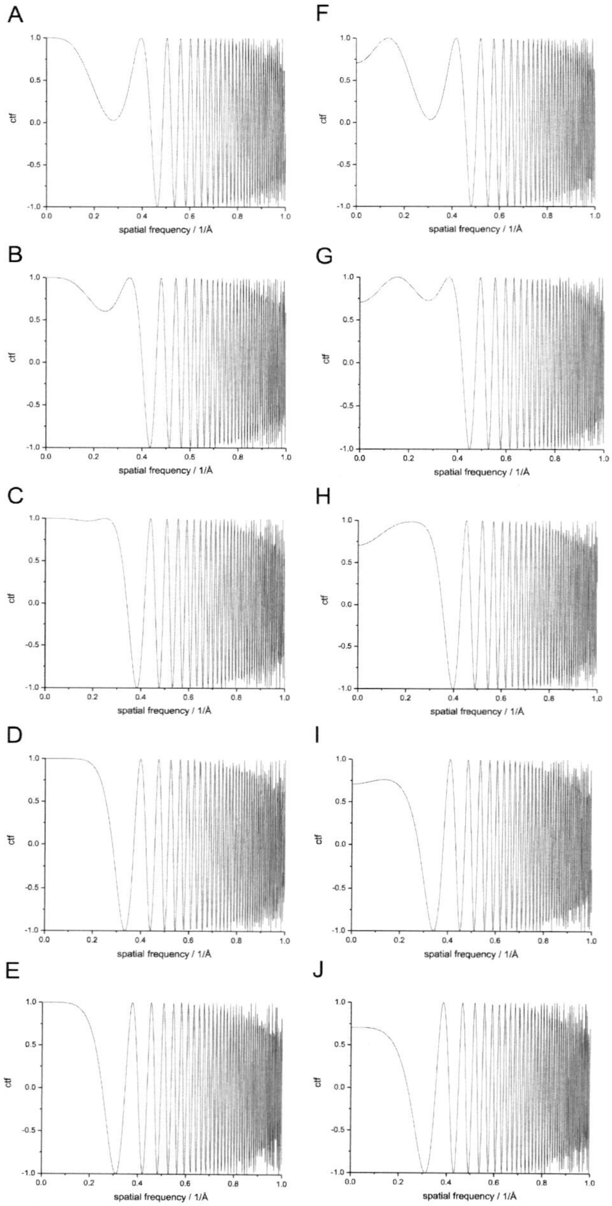

Figure 66.17

Contrast-transfer functions (outside the gap) for differential phase contrast. (A–E) left column, π plate. (F–J) right column, $\pi/2$ plate. The focus values are as follows: (A) −62.7 nm (Scherzer focus); (B) −48.5 nm; (C) −25.0 nm; (D) Gaussian plane; (E) 14.2 nm; (F) −76.9 nm (extended Scherzer focus); (G) −62.7 (Scherzer focus); (H) −39.2 nm; (I) −14.2 nm; (J) Gaussian plane. *After Koeck (2015), Courtesy Elsevier.*

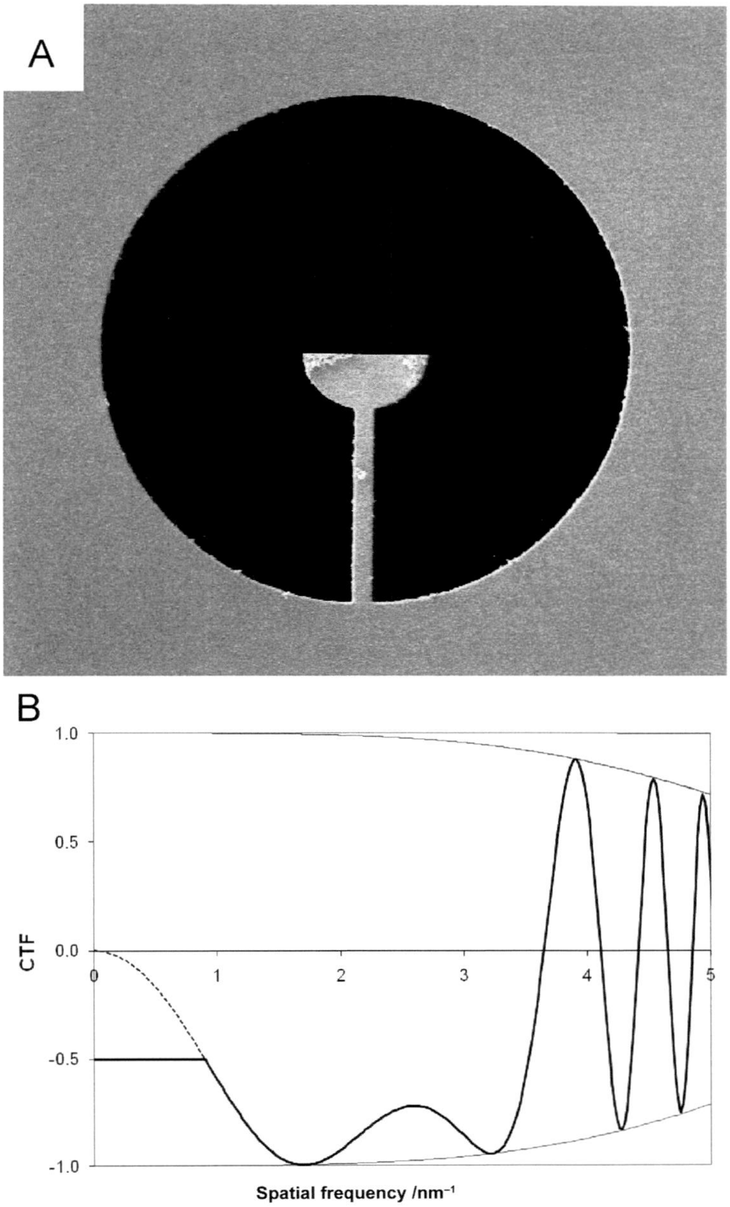

Figure 66.18

The tulip plate. (A) SEM image of the plate. (B) Contrast-transfer function at Scherzer focus, 200 kV, $C_s = C_c = 2.1$ mm, 7.5 μrad (C) Fourier transforms of a focal series of an amorphous carbon film. *Left*, close to zero defocus; *Centre*, 140 nm underfocus; *Right*, 480 nm underfocus. The magnitude of the contrast-transfer function is seen to be constant behind the opaque part of the tulip while the Thon rings are seen outside the opaque zone. (D) A modified tulip aperture. *(A) and (B) after Buijsse et al. (2011); (C) after Koeck (2017), Courtesy Elsevier.*

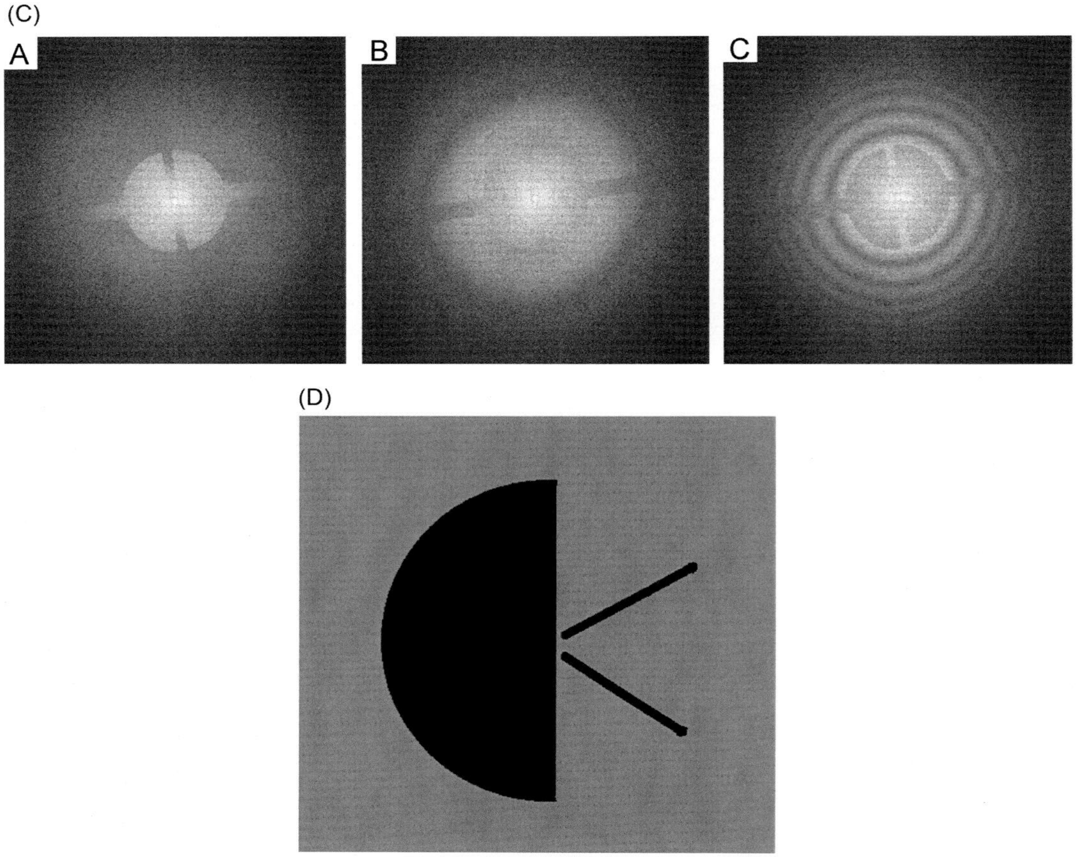

Figure 66.18
Continued.

regular function if the defocus is chosen appropriately (centre figure). A modified version is proposed by Koeck (2017), in which two slits are cut in the opaque half-plane as shown in Fig. 66.18C. Thanks to these, Thon rings can be seen in the image spectrum and thus defocus and astigmatism can be determined.

For further studies, see Buijsse et al. (2014), Dries et al. (2010, 2011, 2012a,b, 2014a,b), Danev and Nagayama (2004a,b), Nagatani and Nagayama (2011), Edgcombe (2017a), Nagayama (1999a,b) and Sugitani and Nagayama (2002, 2006).

66.6.4 Boersch Plates

In his early discussion of phase plates for the electron microscope, Boersch (1947c) proposed not only Zernike plates but also devices in which an electrostatic field imposed a

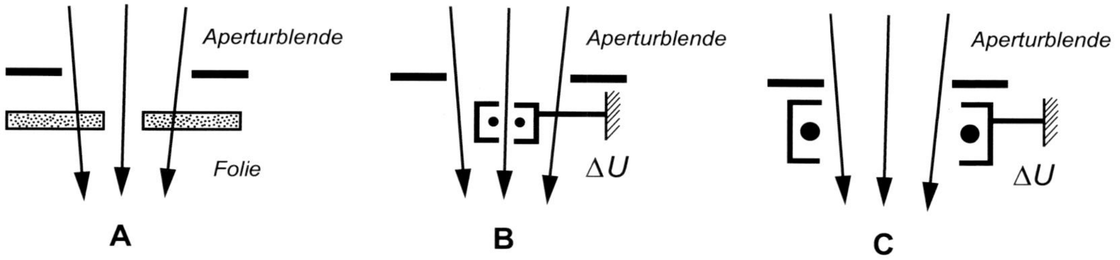

Figure 66.19
The original Boersch plate. (A) Zernike configuration; (B) Electrostatic field plate; (C) alternative electrostatic field configuration. *Aperturblende*, aperture; *Folie*, film. *After Boersch (1947c), Courtesy Zeitschrift für Naturforschung.*

phase variation on the wave function of an electron beam.[4] The latter are known as Boersch plates and were later extended to include magnetic fields. Boersch sketched two possible designs (Fig. 66.19). The idea was revived sporadically for many years, notably by Unwin (1970a,b, 1971, 1972, 1973, 1974), Krakow and Siegel (1975), Laberrigue et al. (1980), Balossier et al. (1980) and Balossier and Bonnet (1981), who stretched a conducting thread across the diffraction plane to create an electrostatic field but the results were unsatisfactory and this particular design was abandoned. The merits of Boersch's suggestion were appreciated by Matsumoto and Tonomura (1996), who analysed the properties of the phase plate shown in Fig. 66.20. This takes the form of a very small einzel lens, diameter 1 μm, ring thickness 250 nm. They showed that the integrated potential inside the ring electrode,

$$\hat{\phi}(x,y) := \int \phi(x,y,z)dz \tag{66.136a}$$

is constant inside the ring,

$$\hat{\phi} =: \hat{\phi}_0 \tag{66.136b}$$

(It is well known that the potential inside a uniformly charged sphere is constant; this is the two-dimensional analogue, cf. Glaeser and Müller, 2014.) The ensuing phase shift φ_B is given by $(2\pi/\lambda)(\hat{\phi}_0/2\Phi_0)$. For a ring of thickness 250 nm and an applied potential of 0.54 V, the phase shift is approximately $\pi/2$. In 2006 Schultheiß et al. succeeded in constructing a plate similar to the design of Matsumoto and Tonomura (Fig. 66.21). A small einzel lens surrounding the optic axis is supported by three arms, as recommended by Majorovits (2002) and Majorovits and Schröder (2002). The einzel lens consists of five layers, two outer electrodes, two insulating layers and the central 'ring' electrode to which a potential is applied via the lead shown. The outer surface of the ring electrode and the

[4] An article in French with a very similar title (Boersch, 1947a) does not mention the electrostatic plate.

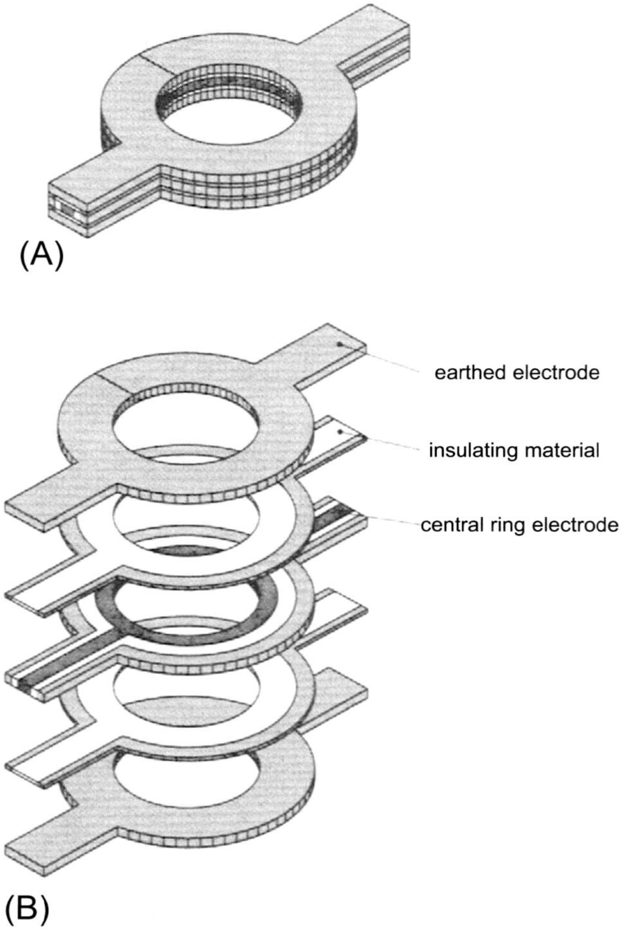

Figure 66.20
Details of a practical Boersch plate. (A) Assembled. (B) Expanded view. *After Matsumoto and Tonomura (1996), Courtesy Elsevier.*

supporting arms are coated with an insulating layer of aluminium oxide and shielded by a thin layer of gold. Thon rings (cf. Fig. 66.6A) showed that the device was capable of introducing a phase shift of $\pi/2$. See also Huang et al. (2006), Majorovits et al. (2007), Chen et al. (2008), Shiue et al. (2009), Shiue and Hung (2010), Alloyeau et al. (2010), Schultheiß (2010), Barton et al. (2011), Walter et al. (2012, 2015) and Koeck (2018).

An undesirable feature of this design is the presence of the supporting arms. The obstruction is considerably reduced in the Zach plate (Schultheiß et al., 2010a,b; Gerthsen et al., 2012; Hettler et al., 2012, 2013, 2014a,b, 2015a,b, 2016; Frindt et al., 2014;

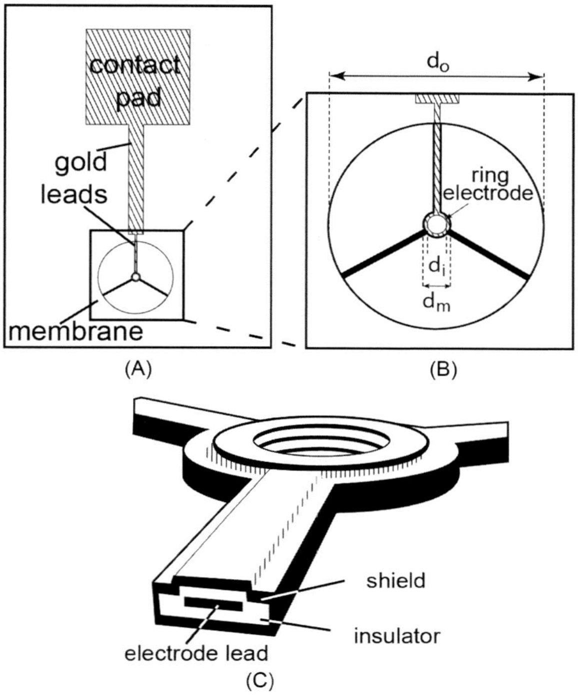

Figure 66.21
Improved design for a Boersch phase plate. (A) Gold leads linking the plate to the external control pad. (B) Close-up view and (C) perspective view of the plate. *After Schultheiß et al. (2006). Courtesy American Institute of Physics, https://doi.org/10.1063/1.2179411.*

Obermair et al., 2017, 2018a,b, 2020), in which an electrostatic field is created close to the beam by means of a thin coaxial cable, again placed in the diffraction plane (Fig. 66.22). At the extremity, an electrostatic field is created where the current in the central conductor returns via the outer sheath. A more radical solution is the anamorphotic plate (Schröder et al., 2007; Frindt et al., 2009, 2010, 2012; Rose, 2010, Section 13; Müller et al., 2012), in which the optics of aberration correctors based on quadrupoles (Section 41.2.1 of Volume 2) is exploited. In these correctors a stigmatic but highly distorted image is formed in (at least) two planes. In the first plane the distorted beam is greatly elongated in one direction (along the x-axis, say) and contracted along the y-axis – this type of imaging is said to be *anamorphotic*. In the second plane the image is elongated in the y-direction and shrunk in the x-direction. There is thus enough space in these planes to introduce devices creating electrostatic fields without obscuring the path of the electron beam. Such a device is shown in Fig. 66.23. Promising though this scheme is for the many aberration-corrected electron microscopes, it is not clear that the aspect ratio will be acceptable.

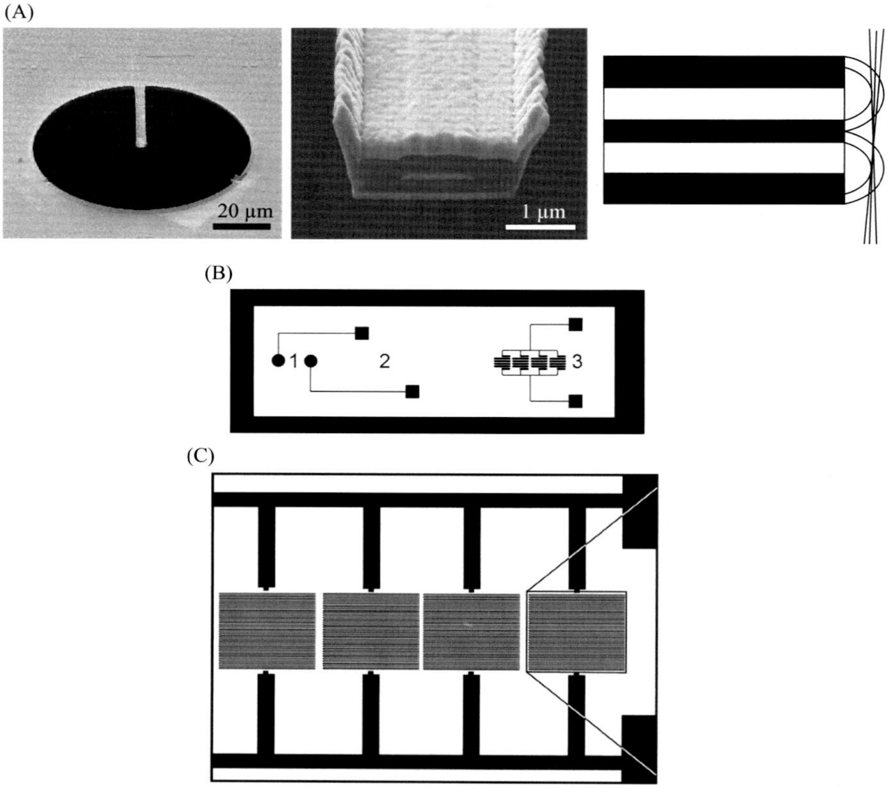

Figure 66.22

The Zach phase plate. (A) *Left*, SEM image of the complete plate; *Centre*, SEM image of the tip; *Right*, schematic cross-section showing the electrostatic field at the tip. (B) Two-phase plates (1) in their holder, contact pads for the electrode (2) and the heating device (3). (C) Details of the heater. *(A) after Schultheiß et al. (2010a), Courtesy Cambridge University Press, https://doi.org/10.1017/S1431927610055042 (B) and (C) after Hettler et al. (2012), Courtesy Cambridge University Press, https://doi.org/10.1017/S1431927612001560.*

This does not exhaust the list of electrostatic Boersch plates. Two-electrode configurations have been proposed by Cambie et al. (2007) and Koeck (2018). In the first, the field is created by a pair of concentric cylinders (Fig. 66.24A) and in the second, two rings of different radius are separated by a thin insulator (Fig. 66.24B). In a further attempt to avoid the effect of the supporting arms, Koeck (2019) examines a 'plate' consisting of two beams of electrons situated in the back-focal plane travelling normal to the optic axis and perpendicular to each other (N–S and E–W). It is the electrostatic potential associated with these beams that forms the phase plate. Koeck shows that the presence of these beams adds an approximately linear term to the lens aberration function describing spherical aberration and defocus. At Gaussian focus, the phase-contrast function is similar to that at Scherzer focus without the phase plate but is shifted towards

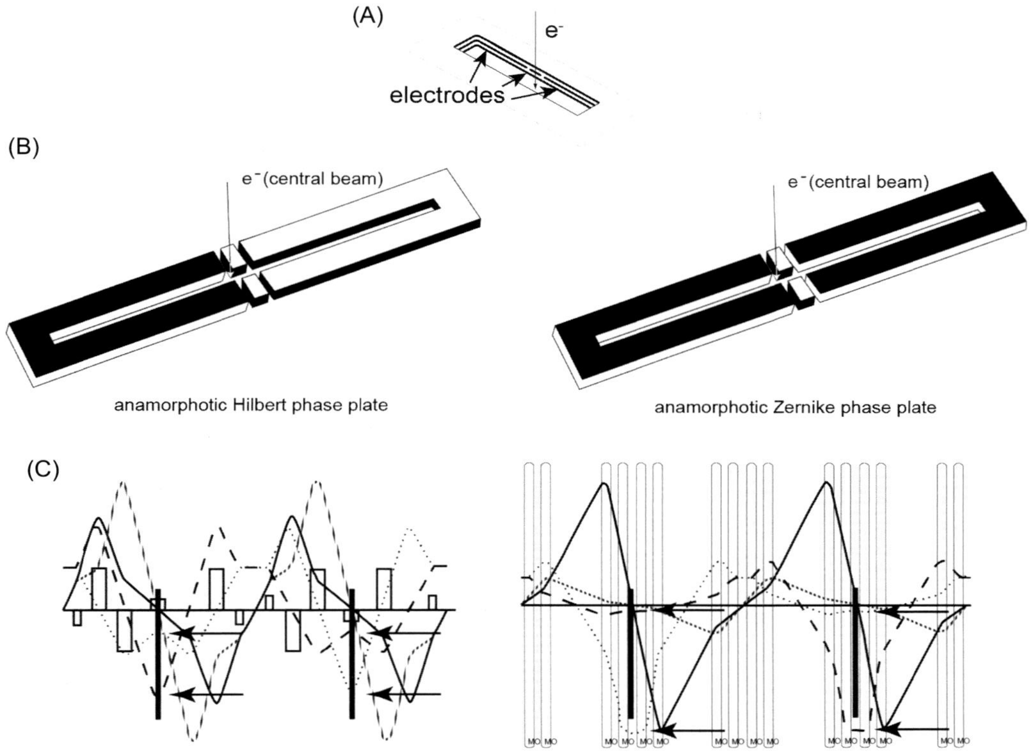

Figure 66.23
Anamorphotic phase plate. (A) Shape of the plate, 3 μm × 90 μm. (B) Excitations for Hilbert and Zernike modes. (C) Anamorphotic diffraction planes (solid black) in an aberration corrector. The arrows indicate the aspect ratio. *(A) and (C) after Schröder et al. (2007), Courtesy Cambridge University Press, https://doi.org/10.1017/S143192760708004X; (B) after Frindt et al. (2010). Courtesy Cambridge University Press, https://doi.org/10.1017/S1431927610056436.*

low spatial frequencies. A contrast-transfer function similar to that associated with large defocus in the absence of the phase plate can be achieved at a much lower defocus when the latter is activated. Koeck shows simulations of the performance that should be attainable in practice and gives estimates of the current required in the transverse beams.

Another ingenious suggestion is the contact-potential einzel lens, which does not require an external voltage source. Here, the electrodes of the lens are constructed from different metals and the desired potential distribution is generated by the contact potential between them. Perry-Houts et al. (2012) suggest a sandwich of molybdenum with a gold filling; Tamaki et al. (2013) use copper instead of molybdenum.

Although Boersch did not envisage the use of magnetic fields, devices in which the phase of an electron beam is modified by a magnetic field are also known as Boersch plates.

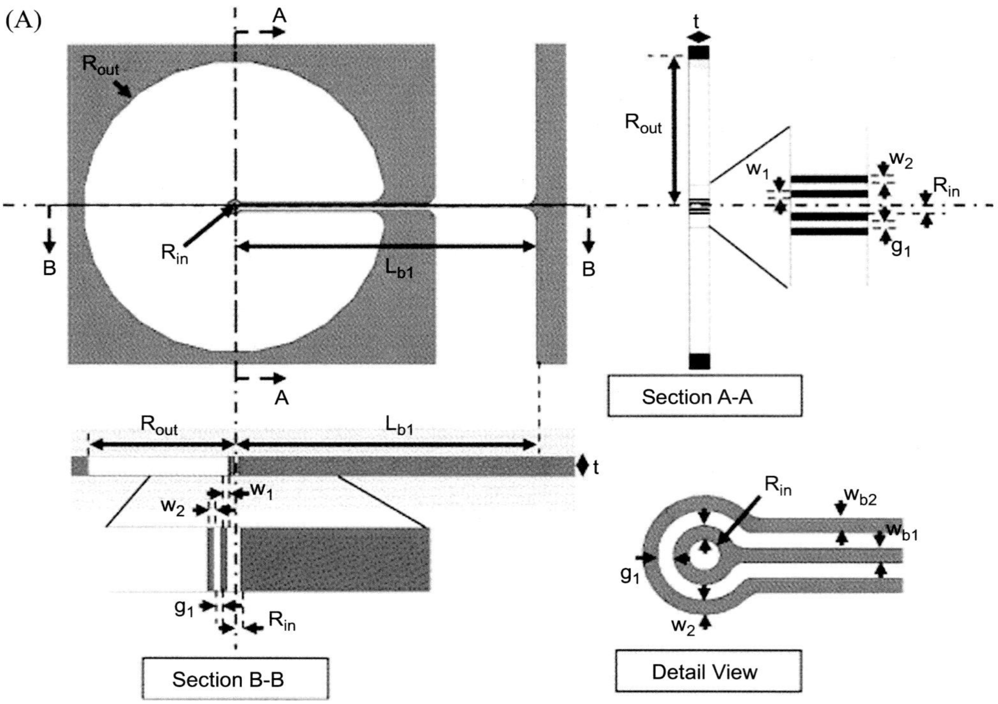

Figure 66.24

Two-electrode Boersch plates. (A) General layout with section views and a detailed sketch of the device. (B) The three layers (a–c) of the Koeck design and the complete plate (d). Metal parts are shown in black, insulating or poorly conducting parts are shaded. (C) Simulated results using the Koeck plate. (a) Object; (b) Image obtained with the plate at 60 nm underfocus; (c–e) Images obtained with no phase plate at 60, 71 and 87 nm underfocus (all close to Scherzer focus); (f) Image obtained with no phase plate at 600 nm defocus. The phase plate image represents the object faithfully and has higher contrast than (3–5). *(A) After Cambie et al. (2007). (B) and (C) After Koeck (2018). Courtesy Elsevier.*

These exploit the discovery by Ehrenberg and Siday (1949), now known as the Aharonov–Bohm effect (Sections 59.6 and 62.4), that parts of an electron beam passing on either side of a local magnetic field will experience a relative phase shift, even if the magnetic field effectively vanishes in their path. This phase difference, φ_m, is given by

$$\varphi_m = \frac{e}{\hbar}\Phi_m \qquad (66.137)$$

(Eq. 59.46) in which Φ_m denotes $\int \boldsymbol{B}\cdot d\boldsymbol{S}$ (Fig. 66.25). By introducing a magnetized ring in the diffraction plane in such a way that the unscattered beam passes inside the ring and the scattered beam outside, a phase difference will be created between the two. The technological requirements have been studied in detail by Edgcombe and colleagues, who find that such small rings can be magnetized in two ways: *vortex rings* in which the

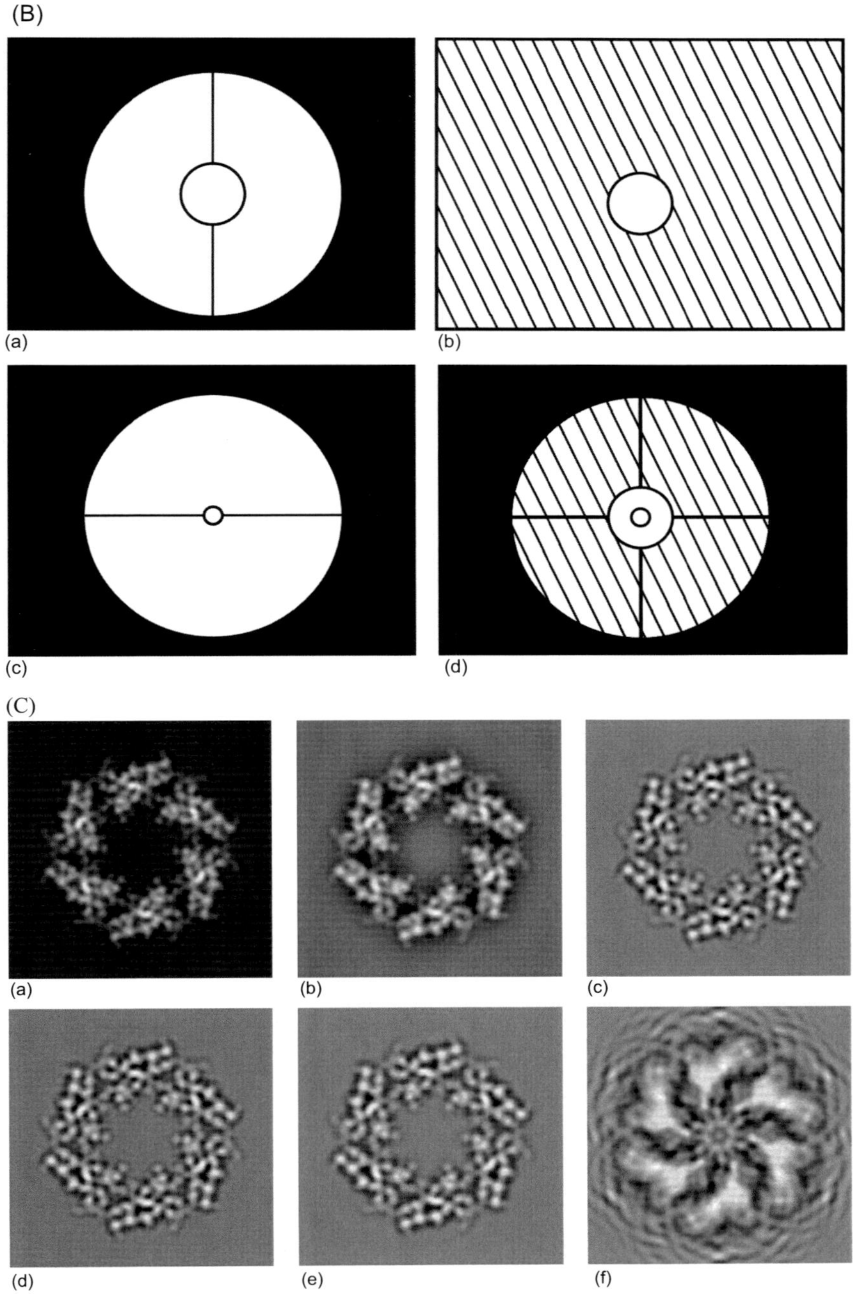

Figure 66.24
Continued.

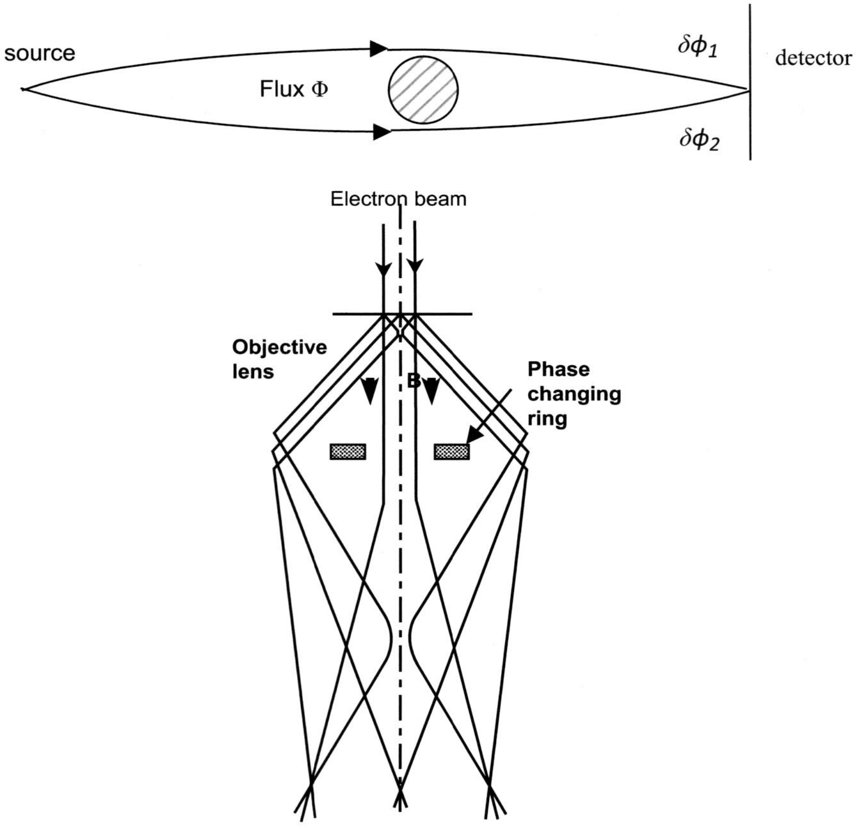

Figure 66.25
The Aharonov–Bohm effect. *Top*, origin of the effect. *Bottom*, ray diagram with the phase-shifting ring. *After Edgcombe (2010). Courtesy IoP Publishing, https://doi.org/10.1088/1742-6596/241/1/012005.*

magnetization follows the ring without interruption and *onion rings* in which the magnetization runs from a 'south pole' to a 'north pole' (Fig. 66.26). For phase plates, vortex rings are desirable and their magnetization pattern can be created and maintained in a material of high coercivity, such as cobalt (Kläui et al., 2005; Heyderman et al., 2005). In the papers of Edgcombe (2010, 2017b), Edgcombe and Loudon (2012) and Edgcombe et al. (2012), the physics of such small rings is explored and suitable dimensions are proposed.

The Aharonov–Bohm effect is also exploited in a related device, tested by Tanji et al. (2014). Here a fine platinum wire, 1 μm in diameter coated with a layer of the ferromagnetic alloy Nd–Fe–B 5 nm thick is stretched across the diffraction plane. The phase difference between the two sides of the wire was measured to be 1.5 rad (approximately $\pi/2$).

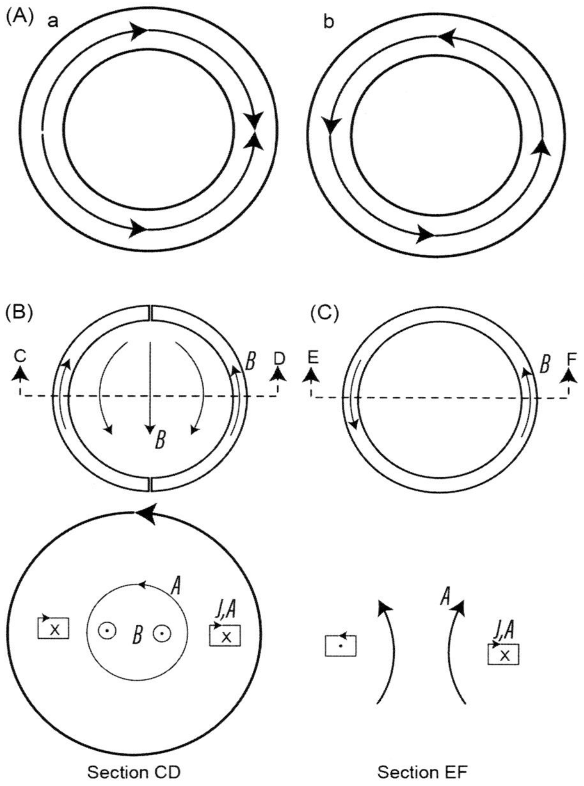

Figure 66.26
(A) Distribution of flux for rings in the onion state (a) and the vortex state (b). Flux density (B), vector potential (A) and magnetizing current in the (B) onion state and (C) vortex state. Cross ($\times$) and dot ($\cdot$) indicate that B is normal to and into the plate, respectively. *After Edgcombe et al. (2012), Courtesy Elsevier.*

An attempt to modify the wave function in the back-focal plane for crystalline specimens was envisaged by Valdrè and colleagues. Their aim was to insert a tiny coil in the aperture plane and position it around a selected diffraction spot. It was hoped that, by changing the current in the coil, information about the phase of the wave at the spot could be deduced. Brief accounts are to be found in Valdrè (1974, 1979).

66.6.5 Laser Phase Plates

Many years ago, Kapitza and Dirac (1933) showed that an electron beam should be deflected by a standing light beam (Batelaan, 2000). The effect was first observed by Freimund et al. (2001). Müller et al. (2010) have suggested that the interaction between a concentrated laser beam and an electron beam could be used to modify the phase distribution of the latter. Later developments and a proof-of-principle test are described by Schwartz et al. (2019), which we summarize here.

The interaction between the electron and light beams is caused by a repulsive ponderomotive potential,

$$V = \frac{e^2 E^2 \lambda_L^2}{16\pi mc^2} \tag{66.138}$$

in which E denote the electric field amplitude of the light wave and λ_L its wavelength. This potential is a consequence of stimulated (elastic) Compton scattering. In the scheme tested by Schwartz et al. (2019), the electrons spend such a short time in the laser beam that a power density of some tens of GW/cm^2 is needed to produce an appreciable phase shift. For this, standing waves from a continuous-wave laser are generated in an optical cavity (a nearly concentric Fabry–Perot geometry) with a wavelength of $\lambda_L = 1064$ nm as shown in Fig. 66.27. An FEI Titan microscope was modified by the addition of an extra lens beyond the objective to magnify the diffraction pattern. In this second diffraction plane the Fabry–Perot cavity and laser optics are introduced and, as shown, the waist of the laser beam coincides with the unscattered part of the electron beam. (We note in passing that it is now the phase of the unscattered beam that is changed, exactly as in Zernike's original plate.) Fig. 66.28 shows

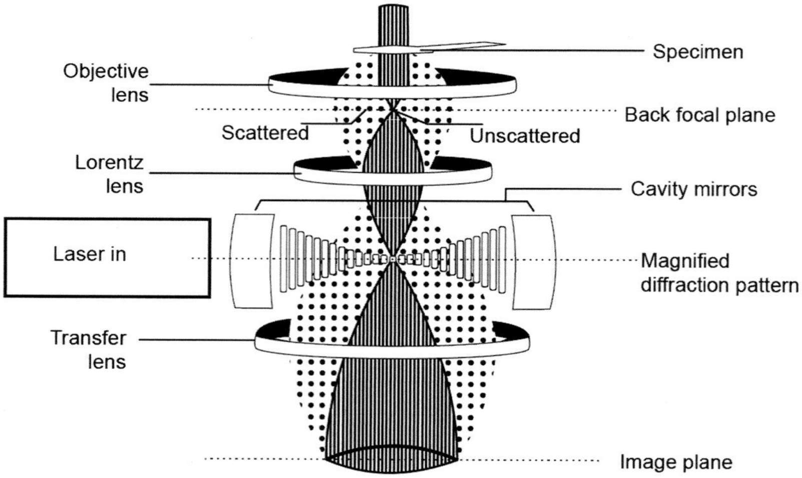

Figure 66.27
Laser-based control of the electron beam in a transmission electron microscope. The unscattered beam passes through the centre of the diffraction pattern, which is an antinode of the standing laser wave. The device thus behaves like a Zernike plate. *After Schwartz et al. (2019), Courtesy Nature, https://doi.org/10.1038/s41592-019-0552-2.*

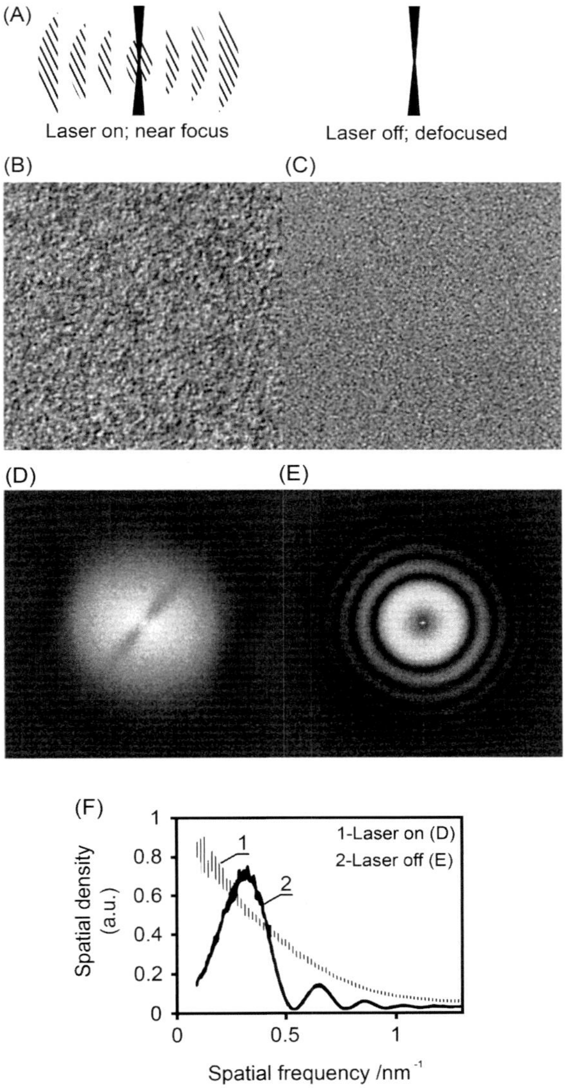

Figure 66.28

Image formation with a laser plate. (A) When the laser is on, the electron beam passes through an antinode. (B) Image of a carbon film close to focus, laser on and (D) squared modulus of the spectrum. (C) Image of a carbon film at 880 nm defocus, laser off and (E) squared modulus of the spectrum. (F) Angular averages of (D) and (E), showing that low-frequency contrast is increased when the laser phase plate is on. *After Schwartz et al. (2019), Courtesy Nature https://doi.org/10.1038/s41592-019-0552-2.*

images of an amorphous carbon film, 3 nm thick, obtained close to Gaussian focus with the laser plate (Fig. 66.28A) and at 880 nm defocus without it (Fig. 66.28B); the corresponding spectra are shown below. In Fig. 66.28D, the familiar Thon rings are seen whereas Fig. 66.28C exhibits a broad central disc crossed by a dark band arising from the laser beam, which can be understood as follows. With the laser plate present the additional phase shift is given by

$$\begin{aligned}\varphi_L &= \frac{1}{2}\varphi_0 \exp\left(-2\frac{Y^2}{1+X^2}\right)\frac{1}{(1+X^2)^{1/2}}B \\ B &= 1+\exp\left\{-2\theta^2\frac{(1+X^2)}{(\mathrm{NA})^2}\right\}\frac{1}{(1+X^2)^{1/4}}\cos\left(2\frac{X}{1+X^2}Y^2+4\frac{X}{(\mathrm{NA})^2}-\frac{3}{2}\arctan X\right)\end{aligned} \tag{66.139}$$

in which

$$Y=\frac{y}{w_0},\quad w_0=\frac{\lambda_L}{\pi(\mathrm{NA})},\quad X=\frac{x}{z_R},\quad z_R=\pi(\mathrm{NA})^2 \tag{66.140}$$

NA is the numerical aperture of the cavity mode. Here it is assumed that the electron beam propagates along the z-axis as usual and that the laser beam lies in the $x-z$ plane, intersecting the electron beam at an angle of $\pi/2-\theta,\ \theta << 1$. The reason for the dark stripe is now obvious. The contrast-transfer function is shown in Fig. 66.29.

Many more details of the experimental procedure and of the associated theory are to be found in the publications of Schwartz et al. (2017, 2019).

66.6.6 Adaptable Phase Plates

Several very different phase-plate configurations are clearly of interest and it is, therefore, natural to ask whether an adaptable plate, with which any desired geometry could be created, is possible. Three very different suggestions have been made. Okamoto (2010), in an attempt to counter the radiation-sensitivity of many biological specimens, describes a mirror-based arrangement (Fig. 66.30) in which the phase change is imposed close to the mirror surface and is defined at the pixel level. For this, a dense matrix of einzel lenses, individually excited, is required. For further discussion of this and related proposals, see Okamoto (2008, 2012, 2014, 2015), Okamoto and Nagatani (2014) and Okamoto et al. (2006). Some of these are highly original, the phase shifts being created by means of a Cooper-pair box or a SQUID.

A suggestion of Verbeeck et al. (2018a,b) is a natural extension of the electrostatic Boersch plate. The authors explain that an adaptable plate could consist of an array of individual einzel lenses and test a simple prototype consisting of only four elements (Fig. 66.31). These first results were encouraging, and a 48-pixel plate is being produced (Verbeeck, 2019, personal communication).

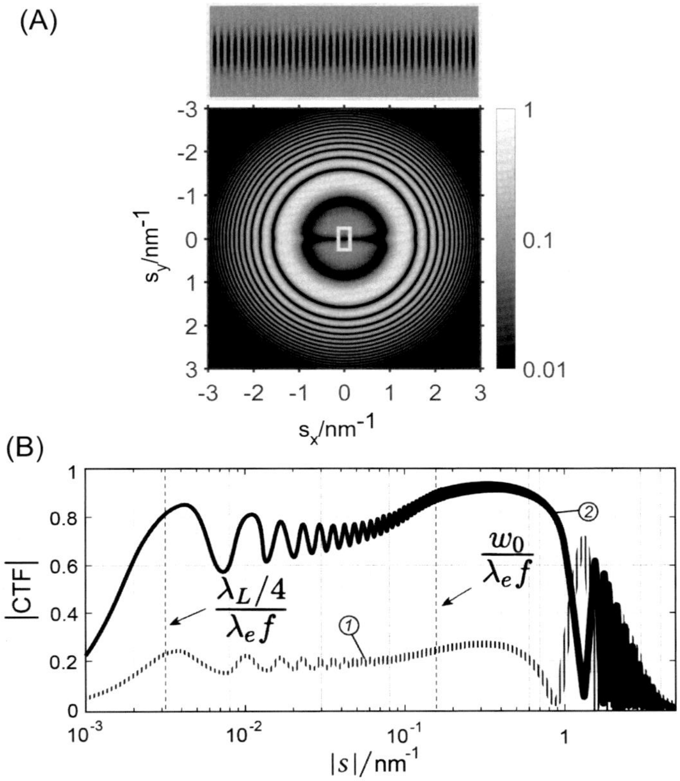

Figure 66.29

(A) Squared contrast-transfer function when the laser plate mimics a Zernike plate; an amplified view of the central box shows the standing-wave structure (B) Angularly averaged contrast-transfer function (1) for the case shown in (A) and (2) for a plate creating a $\pi/2$ phase shift. *After Schwartz et al. (2019), Courtesy Nature, https://doi.org/10.1038/s41592-019-0552-2.*

The work on multi-electron-beam systems described in Section 50.6 of Volume 2 suggests that the technical difficulties of increasing the number still further should not be insuperable.

A device consisting of many miniature electrodes clustered round the electron beam is proposed by Grillo et al. (2018); an example is shown in Fig. 66.32. The applied voltages with which a given phase distribution can be created are found as the solution of an inverse Dirichlet problem, a well-understood procedure. This approach avoids many of the difficulties with the Verbeeck and *a fortiori* the Okamoto designs. In a later paper, Ruffato et al. (2021) suggest that a very flexible type of phase plate inspired by a particular conformal transformation has many merits; see Section 80.5.3.

A programmable virtual phase plate is proposed by Krielaart and Kruit (2019). This is described at length in Chapter 80, since it is intended for use in vortex beam studies, and

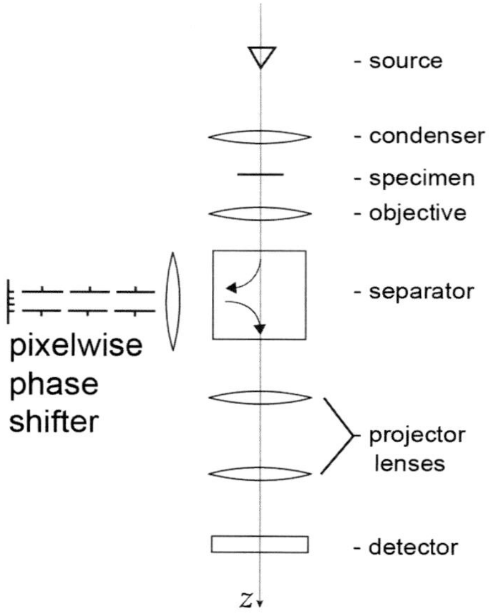

Figure 66.30
Pixelated mirror as a phase plate. The field at the mirror surface is generated by a dense matrix of electrostatic lenses which can be excited individually. *Reprinted from Okamoto (2010) with permission by the American Physical Society; https://doi.org/10.1103/PhysRevA.81.043807.*

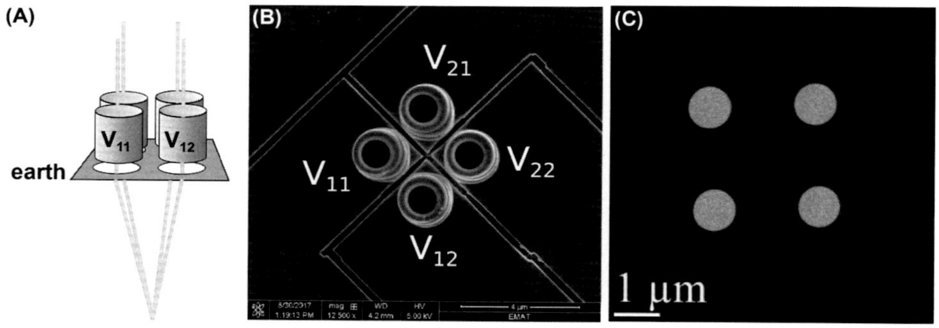

Figure 66.31
Prototype adaptable phase plate. Each of the four cylindrical electrodes alters the phase of a beamlet; the latter generate a programmable pattern in the far-field zone. (A) Array of four electrodes; (B) SEM image of a prototype; (C) TEM image. *After Verbeeck et al. (2018), Courtesy Elsevier.*

only briefly presented here. An electron beam travelling along the optic axis is brought to a focus in a plane P, the common crossover plane (Fig. 66.33). Here, it is deflected towards a second optic axis, parallel to the first, where it is re-deflected towards a lens–mirror combination. It then returns to the plane P where it is again focused and continues to a

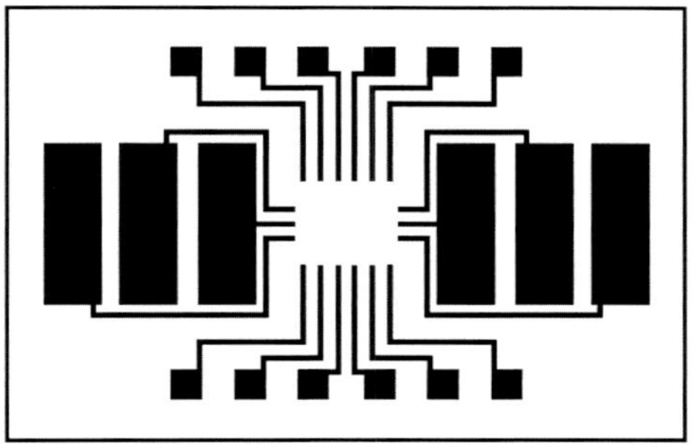

Figure 66.32
Programmable phase plate. The device consists of numerous electrodes surrounding the optic axis. Solution of the inverse Dirichlet problem yields a wide range of potential distributions. *After Grillo et al. (2018), Courtesy the authors.*

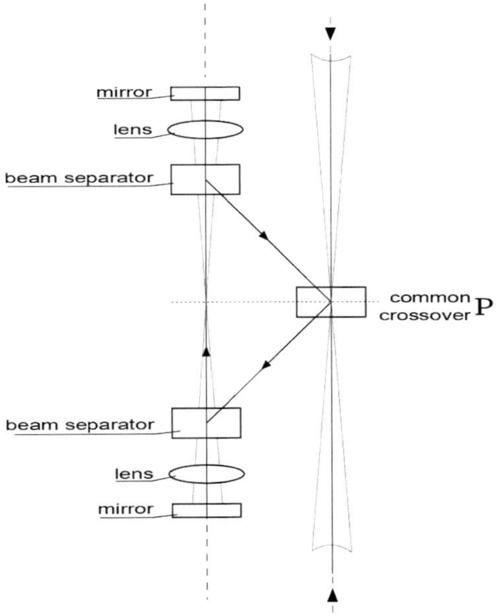

Figure 66.33
Trajectories in the creation of a virtual phase plate. *After Krielaart and Kruit (2019), Courtesy Cambridge University Press, https://doi.org/10.1017/S1431927619001193.*

second lens–mirror unit, passing through a beam separator of the Wien-filter type along the way. This second mirror impresses the desired phase pattern on the beam and reflects it to the beam separator and hence to the original optic axis. It is thus as though a phase plate had been present in the plane P but this is now a *virtual plate*.

We also mention the work of Ito et al. (1992, 1993a,b) who have fabricated very fine diffraction gratings consisting of an array of wells of different depths in a thin film. The phase change is determined by the thickness of the layer between the bottom of the well and the far surface of the film. A solid-state Fresnel lens for electron optics was later produced by this means (Ito et al., 1998).

Garcia de Abajo and Konecna (2020) have suggested that the phase distribution of electron beams in free space could be modulated by interaction with a suitably tailored light field. They describe an optical free-space electron modulator, with which a desired phase profile is imposed on the transverse wavefunction by means of 'stimulated elastic Compton scattering associated with the A^2 term in the light–electron coupling Hamiltonian' where A is the magnitude of the vector potential. The authors suggest that this could be used to correct spherical aberration but it clearly has much wider implications. Feist et al. (2021), describe a 'temporal phase plate' based on a photonic chip that exploits inelastic electron–light scattering. By employing a continuous-wave Si_3N_4 microresonator with a Q-factor greater than 10^6, very strong coupling to a continuous electron beam can be achieved. The electron beam passes close to the microresonator, which is installed in the specimen-holder of an ultrafast TEM, as in aloof interactions. Feist et al. show that when the laser is tuned to a resonance of the cavity, the energy distribution of the electron beam is broadened from a few electronvolts to about 30 eV, which shows that it has undergone a strong temporal phase modulation. They speculate that such active electron-optical configurations could act as programmable phase plates.

66.7 Transfer Theory and Crystalline Specimens

The linear transfer theory depends essentially on the assumption that the unscattered beam is considerably stronger than the scattered beam but this may well not be true for crystalline specimens, where the electrons are concentrated in spots in the back-focal plane of the objective. If several diffracted spots have intensities that are not markedly weaker than that of the central spot, interference between the diffracted waves may be significant, thereby rendering the direct interpretation of the image difficult or indeed misleading. This observation led Ishizuka (1980) to investigate the nonlinear term in the image intensity in detail, for crystalline specimens in particular, and we summarize his analysis here. The main purpose of this study is to indicate the limits that must be imposed on the credibility of the transfer theory for objects for which the weak scattering approximation may break down.

We set out from the general formula (65.16), from which we derive the spectrum of the image intensity as the weighted convolution of the spectra of ψ_i and ψ_i^*. The weighting describes the effect of non-vanishing source size and energy spread on the assumption that these can be treated separately. Eq. (65.16) tells us that

$$S_i(\boldsymbol{q}) = T(\boldsymbol{q})S_o(\boldsymbol{q}) \tag{66.141}$$

and hence

$$\begin{aligned}\mathcal{F}\left(\psi_i\psi_i^*\right) &=: S_I(\boldsymbol{q};\Delta)\\ &= \int X(\overline{\boldsymbol{q}},\boldsymbol{q};\Delta)S_o(\overline{\boldsymbol{q}}+\boldsymbol{q})S_o^*(\overline{\boldsymbol{q}})d\overline{\boldsymbol{q}}\end{aligned} \tag{66.142}$$

in which $X(\overline{\boldsymbol{q}},\boldsymbol{q};\Delta)$ is the transmission cross-coefficient, which we shall meet again in Part XVI. It involves the auto-correlation of $T(\boldsymbol{q})$, weighted by the source functions

$$H_1(\Delta) := \frac{1}{\pi^{1/2}\Delta_0}\exp\left(-\frac{\Delta^2}{\Delta_0^2}\right) \quad \text{(energy spread)} \tag{66.143a}$$

(66.60) and

$$H_2(\boldsymbol{q}) := \frac{1}{\pi \boldsymbol{q}_0^2}\exp\left(-\frac{\boldsymbol{q}^2}{\boldsymbol{q}_o^2}\right) \quad \text{(source size)} \tag{66.143b}$$

Eq. (66.143) are in conformity with (66.66 and 66.67) if we write the latter in the form

$$\left|A(\boldsymbol{\kappa},\upsilon)\right|^2 = \frac{1}{\pi^{1/2}\upsilon_m}\exp\left(-\frac{\upsilon^2}{\upsilon_m^2}\right)\times\frac{\lambda_o^2}{\pi\alpha_c^2}\exp\left(-\frac{\lambda_o^2\boldsymbol{\kappa}^2}{\alpha_c^2}\right)$$

and identify $\boldsymbol{q}_o$ with α_c/λ_o in the second factor. In the first we have chosen to characterize the effect of chromatic aberration in terms of the resulting focal shift.

Explicitly,

$$\begin{aligned}X(\overline{\boldsymbol{q}},\boldsymbol{q};\Delta) = \iint & H_2\left(\overleftrightarrow{\boldsymbol{q}}\right)H_1(\Delta)T\left(\overline{\boldsymbol{q}}+\overleftrightarrow{\boldsymbol{q}};\Delta+\overline{\boldsymbol{\Delta}}\right)\\ &\times T^*\left(\boldsymbol{q}+\overleftrightarrow{\boldsymbol{q}};\Delta+\overline{\boldsymbol{\Delta}}\right)d\overleftrightarrow{\boldsymbol{q}}\,d\overline{\boldsymbol{\Delta}}\end{aligned} \tag{66.144}$$

The object function corresponding to the spectrum $S_o(\boldsymbol{q})$ in (66.141) is now formally separated into a constant part and a variable part but the latter is now not necessarily small compared with the former (taken to be unity):

$$S_o(\boldsymbol{q}) =: \delta(\boldsymbol{q}) + s(\boldsymbol{q}) \tag{66.145}$$

After some calculation, we find that the image intensity spectrum, S_I, is composed of three parts:

$$S_I(\boldsymbol{q};\Delta) = S_I^{(0)}(\boldsymbol{q};\Delta)S_I^{(1)}(\boldsymbol{q};\Delta)S_I^{(2)}(\boldsymbol{q};\Delta) \tag{66.146}$$

The first two terms we have already met in the linear theory; it is the final term that is new and will enable us to assess the impact of the nonlinear interactions. We find

$$\begin{aligned}
S_I^{(0)}(\boldsymbol{q};\Delta) &= X(\boldsymbol{q},0;\Delta)\delta(\boldsymbol{q}) \\
S_I^{(1)}(\boldsymbol{q};\Delta) &= X(\boldsymbol{q},0;\Delta)s(\boldsymbol{q}) + X(0,-\boldsymbol{q};\Delta)s^*(-\boldsymbol{q}) \\
&\Rightarrow X(\boldsymbol{q},0;\Delta)s(\boldsymbol{q}) + X^*(\boldsymbol{q},0;\Delta)s^*(-\boldsymbol{q}) \quad (\textit{symmetric source}) \\
S_I^{(2)}(\boldsymbol{q};\Delta) &= \int X(\overline{\boldsymbol{q}}+\boldsymbol{q},\ \overline{\boldsymbol{q}};\Delta)s(\overline{\boldsymbol{q}}+\boldsymbol{q})s^*(\overline{\boldsymbol{q}})d\overline{\boldsymbol{q}}
\end{aligned} \tag{66.147}$$

A lengthy calculation, given in detail by Ishizuka, shows that the function X can be written as the product of five factors:

$$X(\overline{\boldsymbol{q}},\boldsymbol{q};\Delta) =: X_L X_p X_s X_e X_x \tag{66.148}$$

where

$$X_L := T_A(\overline{\boldsymbol{q}})T_A(\boldsymbol{q})\exp\left[-\mathrm{i}\{\chi(\overline{\boldsymbol{q}},\Delta) - \chi(\boldsymbol{q},\Delta)\}\right]$$

(effect of aperture and lens properties, $T_A T_L$)

$$\begin{aligned}
X_p(\boldsymbol{q}',\boldsymbol{q};\Delta) &:= \exp\left\{\frac{2\pi\mathrm{i}(\pi\boldsymbol{q}_o\Delta_0)^2(\chi'_\Delta - \chi_\Delta)\left(\chi'_{\boldsymbol{q}} - \chi_{\boldsymbol{q}}, \chi'_{\boldsymbol{q}\Delta} - \chi_{\boldsymbol{q}\Delta}\right)}{u}\right\} \\
X_s(\boldsymbol{q}',\boldsymbol{q};\Delta) &:= \exp\left\{-\frac{\pi\boldsymbol{q}_o^2\left(\chi'_{\boldsymbol{q}} - \chi_{\boldsymbol{q}}\right)^2}{u}\right\} \\
X_e(\boldsymbol{q}',\boldsymbol{q}) &= \exp\left\{-\frac{(\pi\Delta_0)^2(\chi'_\Delta - \chi_\Delta)^2}{u}\right\} \\
X_x(\boldsymbol{q}',\boldsymbol{q}) &= u^{-1/2}\exp\left[-\frac{(\pi^2\boldsymbol{q}_o^2\Delta_0)^2\left\{\left(\chi'_{\boldsymbol{q}} - \chi_{\boldsymbol{q}}\right)^2\left(\chi'_{\boldsymbol{q}\Delta} - \chi_{\boldsymbol{q}\Delta}\right)^2 - \left(\chi'_{\boldsymbol{q}} - \chi_{\boldsymbol{q}}, \chi'_{\boldsymbol{q}\Delta} - \chi_{\boldsymbol{q}\Delta}\right)^2\right\}}{u}\right] \\
&= u^{-1/2}\exp\left[\frac{(\pi^2\boldsymbol{q}_o^2\Delta_0)^2\left\{\left(\chi'_{\boldsymbol{q}} - \chi_{\boldsymbol{q}}\right)\left(\chi'_{\boldsymbol{q}\Delta} - \chi_{\boldsymbol{q}\Delta}\right)^2\right\}}{u}\right]
\end{aligned} \tag{66.149}$$

in which

$$u := 1 + \xi(\boldsymbol{q}' - \boldsymbol{q})^2, \quad \xi = (\pi\lambda \boldsymbol{q}_o \Delta_0)^2 \tag{66.150}$$

and

$$\begin{aligned} \chi_{\boldsymbol{q}} &= (C_s\lambda^2|\boldsymbol{q}^2| - \Delta')\lambda\boldsymbol{q} \\ \chi_{\Delta} &= -\frac{1}{2}\lambda|\boldsymbol{q}^2| \\ \chi_{\boldsymbol{q}\Delta} &= -\lambda\boldsymbol{q} \end{aligned} \tag{66.151}$$

and analogously for $\chi'_{\boldsymbol{q}}$, χ'_{Δ} and $\chi'_{\boldsymbol{q}\Delta}$.

This product representation enables us to assess in detail the various effects that may occur when the nonlinear terms cannot be neglected, a situation that is particularly likely to occur with crystals. Ishizuka gives curves and perspective plots of each term in turn and then of the cross-coefficient X. From these, it emerges that the envelopes X_e and X_s are the most important. Fig. 66.34 shows $X_e X_s$ for various defocus values for a microscope operating at 100 kV, $C_s = 0.7$ mm and $\boldsymbol{q}_o = 0.01$ Å^{-1}, $\Delta_o = 100$ Å. Clearly, the nonlinear term can have a very significant effect and Ishizuka suggests that such effects were important in the related experimental observations of Hashimoto et al. (1977), Izui et al. (1977, 1978), Sieber and Tonar (1975, 1976) and Desseaux et al. (1977).

The expressions (66.151) are in fact only first-order approximations to the correct quantities and Ishizuka has later examined the validity of this approximation (1989, 1990). We refer to these papers for further details.

Higher order coherence effects have also been studied in considerable detail by Chang (2004) and, notably in connection with rainbow lenses, by Lentzen and Thust (2005), Kimoto et al. (2013) and Kimoto (2014). When a gun-monochromator of the Wien-filter type is employed, the effective source exhibits dispersion. In consequence, a third modulation is introduced in addition to the envelope functions characterizing spatial and temporal partial coherence. It can be shown that electrons with energy $E_0 + \Delta E$ appear to impinge on the specimen with a small tilt. It is for this reason that the term *rainbow lens* is used. For a full description of the optics of this imaging mode, see Chang (2004) and for a clear description, see Kimoto (2014).

Another important effect that may be encountered when imaging crystal defects at the highest resolution is *delocalization* (Coene and Janssen, 1991). If the image is formed with very coherent illumination, so that the transfer function begins to oscillate well before the information limit is reached, the position of a crystal defect may appear to be delocalized in the image. The associated analysis is given in full by Coene and Janssen.

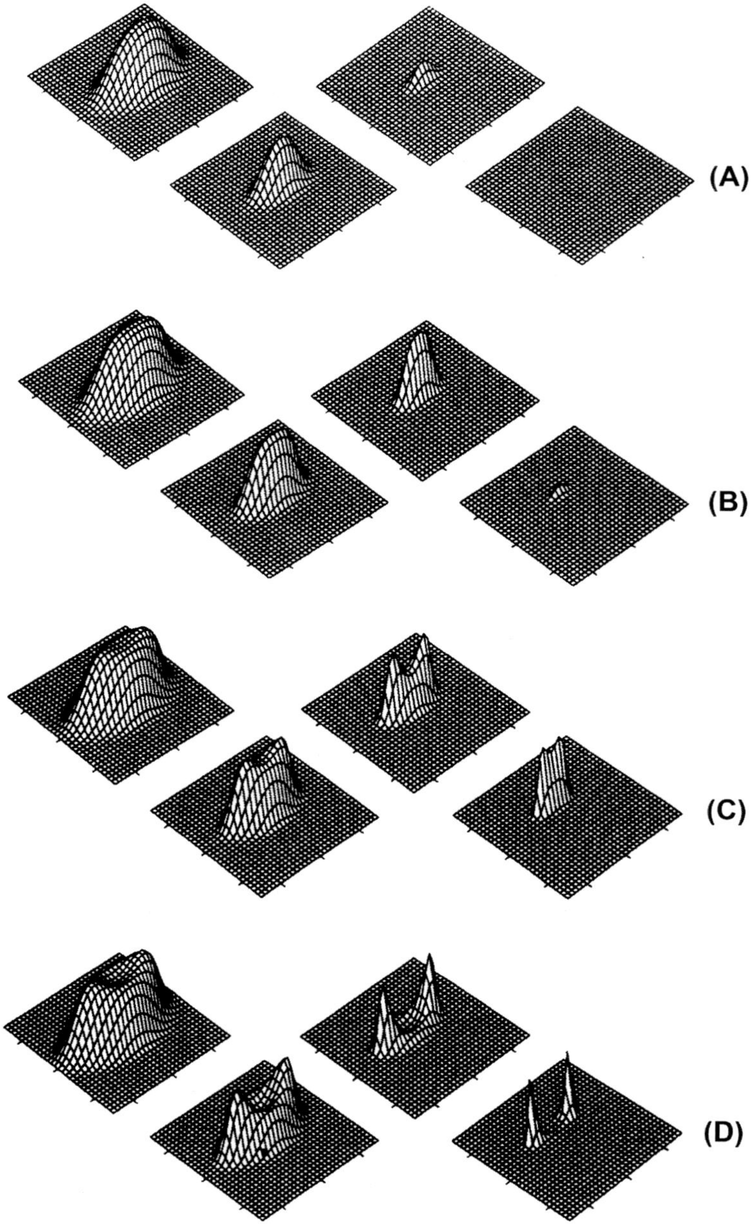

Figure 66.34
Perspective views of the overall envelope function X_eX_s for four values of the defocus. For each value, the four envelopes correspond from left to right to spatial frequencies $q_x = 1, 2, 4, 8\ \mathrm{nm}^{-1}$, $q_y = 0$. The marks on the q'_x and q'_y axes correspond to -1, 0 and 1. (A) $\Delta = -200$ nm. (B) $\Delta = 0$ nm (C) $\Delta = 200$ nm. (D) $\Delta = 400$ nm. *After Ishizuka (1980), Courtesy Elsevier.*

66.8 Contrast Transfer in Aberration-Corrected Microscopes

Since phase contrast in uncorrected instruments is created by a balance of spherical aberration C_s and defocus Δ, it seems that aberration correction will be disastrous as far as image contrast is concerned. The solution is to find a combination of third- and fifth-order spherical aberration and defocus that will generate contrast without degrading the resolution (Chang et al., 2006; Lentzen, 2008). We examine two situations. In the first, the fifth-order spherical aberration coefficient C_5 is fixed and C_s and Δ are varied; temporal partial coherence (energy spread) is excluded by Chang et al. but tacitly included by Lentzen. In the second, it is assumed that C_5 is also a free parameter and the effect of energy spread is included. These cases are studied with a critical commentary in the articles by Chang et al. and Lentzen, whom we follow closely. Different criteria are used to obtain the solution, which lead to slightly different results.

1. C_5 fixed, C_s and Δ variable

   We now include the fifth-order spherical aberration in the function $\chi(\boldsymbol{q})$ (Eq. 65.30):

$$\chi(\boldsymbol{q}) = \pi\lambda\left(-\Delta\boldsymbol{q}^2 + \frac{1}{2}C_s\lambda^2\boldsymbol{q}^4 + \frac{1}{3}\lambda^4 C_5\boldsymbol{q}^6\right) \tag{66.152}$$

   The subscript o has been omitted here. We recall that in the literature of aberration correction, C_s is frequently denoted by C_3 and the defocus by C_1 (see Table 31.1 of Vol. 1; the convention adopted for the sign of defocus should always be checked).

   Following Chang et al., we seek the conditions for which the phase shift caused by $\chi(\boldsymbol{q})$ remains between 0 and $\pi/2$ and the function $\sin\chi(\boldsymbol{q})$ remains as flat as possible. We first convert $\chi(\boldsymbol{q})$ to a more convenient form by two changes of variable. We first write

$$r := \boldsymbol{q}^2 \tag{66.153}$$

   giving

$$\chi(r) = \pi\left(-\Delta\lambda r + \frac{1}{2}C_s\lambda^3 r^2 + \frac{1}{3}\lambda^5 C_5 r^3\right) =: \boldsymbol{q}_1 r + \boldsymbol{q}_2 r^2 + \boldsymbol{q}_3 r^3 \tag{66.154}$$

   We shall need to solve a cubic equation with these coefficients and we, therefore, eliminate the quadratic term by writing

$$\hat{r} = r + \frac{\boldsymbol{q}_2}{3\boldsymbol{q}_3} \tag{66.155}$$

   which leads to

$$\frac{\chi(r)}{\boldsymbol{q}_3} := \hat{\chi}(\hat{r}) = \hat{r}^3 + \hat{\boldsymbol{q}}_1\hat{r} + \hat{\boldsymbol{q}}_4 \tag{66.156}$$

with

$$\hat{q}_1 = \frac{q_1}{q_3} - \frac{1}{3}\left(\frac{q_2}{q_3}\right)^2 \quad \text{and} \quad \hat{q}_4 = \frac{2}{27}\left(\frac{q_2}{q_3}\right)^3 - \frac{1}{3}\frac{q_2}{q_3}\frac{q_1}{q_3} \tag{66.157}$$

We now impose the desired conditions. At $\boldsymbol{q} = \overline{\boldsymbol{q}}$,

$$\chi(\overline{\boldsymbol{q}}) = \pi/4 \tag{66.158a}$$

$$\frac{d\chi}{d\boldsymbol{q}} = 0 \tag{66.158b}$$

From (66.158b), we see that

$$\hat{r}(\overline{\boldsymbol{q}}) = \pm\left(-\frac{\hat{q}_4}{3}\right)^{1/2} \tag{66.159}$$

and the best choices for
C_s and Δ are found to be

$$\begin{aligned} C_s &= -a_3(C_5^2\lambda)^{1/3} & a_3 &= 2 \times 3^{1/3} = 2.88 \\ \Delta &= -a_1\left(C_5\lambda^2\right)^{1/3} & a_1 &= \frac{3}{4} \times 9^{1/3} = 1.56 \end{aligned} \tag{66.160}$$

These expressions are not very different from those proposed by Scherzer (1970), who argued that the phase shift should be as close as possible to that of an ideal Zernike plate. He found $a_1 = 2$ and $a_3 = 3.2$. The resolution, defined in terms of the spatial frequency at the first zero of the contrast-transfer function, $\boldsymbol{q}_c$, is given by $1/\boldsymbol{q}_c = 0.68\left(C_5\lambda^5\right)^{1/6}$ (Fig. 66.35). If C_5 is negligible, we recover the familiar Scherzer focus, $\Delta = (C_s\lambda)^{1/2}$ with resolution $0.71\left(C_s\lambda^3\right)^{1/4}$ (Fig. 66.36).

Lentzen seeks the least-squares fit of the aberration function (66.152) to the value $\pi/4$; for this,

$$\iint |\chi(\boldsymbol{q}_x, \boldsymbol{q}_y) - \pi/2|^2 d\boldsymbol{q}_x d\boldsymbol{q}_y = 2\pi \int_0^{q_{max}} |\chi(\boldsymbol{q}) - \pi/2|^2 \boldsymbol{q}\, d\boldsymbol{q} \Rightarrow \min \tag{66.161}$$

in which $\boldsymbol{q}_{max}$ is determined by the temporal coherence envelope. For fixed C_5 the values of Δ and C_s are given by

$$\begin{aligned} \Delta &= -\frac{2}{\lambda \boldsymbol{q}_{max}^2} - \frac{2}{15} C_5 \lambda^4 \boldsymbol{q}_{max}^4 \\ C_s &= -\frac{10}{3\lambda^3 \boldsymbol{q}_{max}^4} - \frac{8}{9} C_5 \lambda^2 \boldsymbol{q}_{max}^2 \end{aligned} \tag{66.162}$$

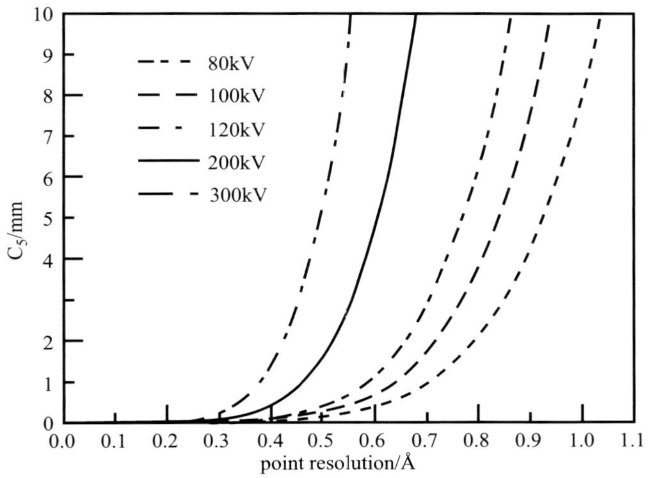

Figure 66.35
Point resolution as a function of C_5. *After Chang et al. (2006), Courtesy Elsevier.*

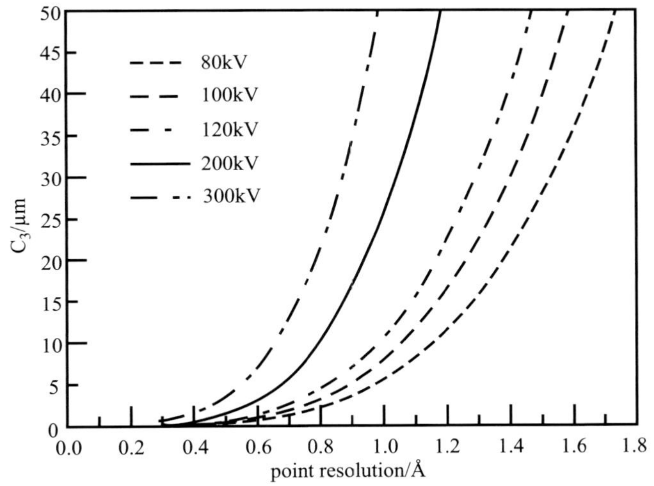

Figure 66.36
Point resolution as a function of C_s for $C_5 = 0$. *After Chang et al. (2006), Courtesy Elsevier.*

Figs. 66.37 and 66.38 show $\chi(\boldsymbol{q})$ and $\sin(2\pi\chi)$ for $\Delta = -4.9$ nm, $C_s = -10$ μm and $C_5 = 0$ and $\Delta = -5.3$ nm, $C_s = -13$ μm and $C_5 = 4$ mm, respectively.

2. C_5, C_s and Δ variable.

We have so far disregarded the effects of finite source size and finite energy spread. An aberration-corrected microscope will usually be equipped with a field-emission source, for

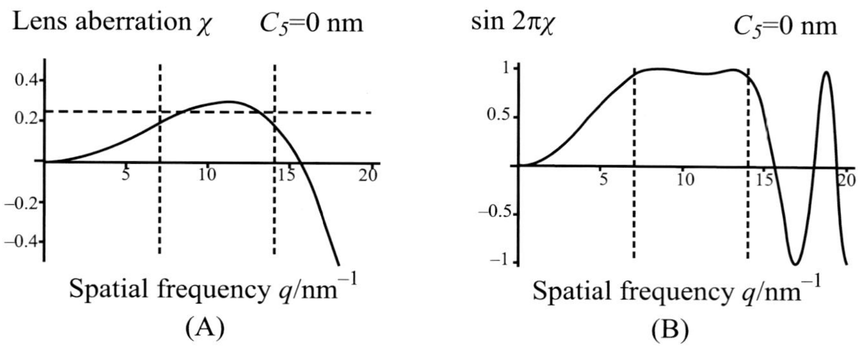

Figure 66.37

(A) Aberration function χ and (B) contrast-transfer function $\sin(2\pi\chi)$ for $C_5 = 0$, neglecting partial spatial and temporal coherence. *After Lentzen (2008), Courtesy Cambridge University Press, https://doi.org/10.1017/S1431927608080045.*

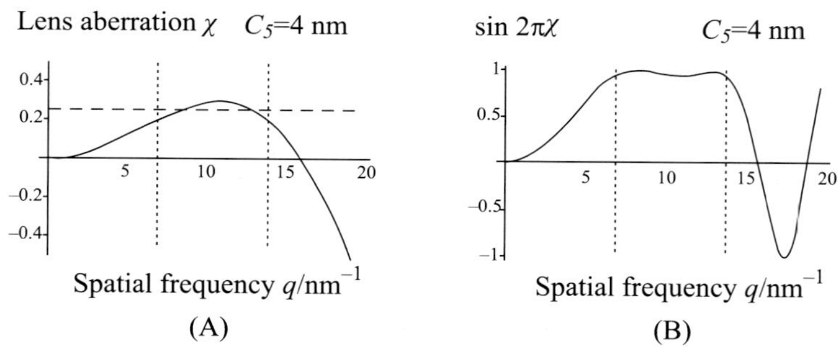

Figure 66.38

(A) Aberration function χ and (B) contrast-transfer function $\sin(2\pi\chi)$ for $C_5 = 4$ mm, neglecting partial spatial and temporal coherence. *After Lentzen (2008), Courtesy Cambridge University Press, https://doi.org/10.1017/S1431927608080045.*

which the source size is negligible. Finite energy spread is characterized by an envelope function t_υ (66.81a) and there is no point in allowing the first zero of the contrast-transfer function to fall beyond the associated information limit, defined as the $1/e^2$ threshold of the envelope function. We write the envelope function concisely as

$$t_\upsilon = \exp\left(-\frac{1}{2}t^2\lambda^2\boldsymbol{q}^4\right), \quad t^2 := \frac{1}{2}\upsilon_m^2 C_c'^2 \tag{66.163}$$

and the threshold value of $\boldsymbol{q}$, $\boldsymbol{q}_t$, is thus

$$\boldsymbol{q}_t = \left(\frac{2}{\pi\lambda t}\right)^{1/2} \tag{66.164}$$

We return to Eqs. (66.152) and (66.154) and require that the first zero of sin χ coincide with the 'chromatic' information limit, $r_t = \boldsymbol{q}_t^2$. The latter must be a degenerate root of the cubic equation which implies that $\chi(r)$ is of the form

$$\chi(r) = \boldsymbol{q}_3 r(r - r_t)^2 \tag{66.165}$$

From (66.154) and (66.165), we see that

$$\begin{aligned}\boldsymbol{q}_2 &= -2\boldsymbol{q}_3 r_t \\ \boldsymbol{q}_1 &= \boldsymbol{q}_3 r_t^2\end{aligned} \tag{66.166}$$

Condition (66.158b) gives

$$3\boldsymbol{q}_3\bar{r}^2 + 2\boldsymbol{q}_2\bar{r} + \boldsymbol{q}_1 = \boldsymbol{q}_3\left(3\bar{r}^2 - 4r_t\bar{r} + r_t^2\right) = \boldsymbol{q}_3(\bar{r} - r_t)(3\bar{r} - r_t) = 0 \tag{66.167}$$

The appropriate solution is thus

$$\bar{r} = \frac{r_t}{3} = \frac{1}{3}\left(\frac{2}{\pi\lambda t}\right)^{1/2} \tag{66.168}$$

With the aid of the condition (66.158a), we deduce that

$$\begin{aligned} C_5 &= \frac{81}{64}\frac{(\pi t)^3}{\lambda^2} = 1.266\frac{(\pi t)^3}{\lambda^2} \\ C_s &= -\frac{27}{8}\frac{(\pi t)^2}{\lambda} = -3.375\frac{(\pi t)^2}{\lambda} \\ \Delta &= -\frac{27}{16}\pi t = -1.688\pi t \end{aligned} \tag{66.169}$$

Chang et al. point out that if there is no fifth-order spherical aberration, conditions (66.158a and 66.158b) lead to $\Delta^2 = -C_s\lambda$; at the first zero, the spatial frequency is $\boldsymbol{q} = -2^{1/2}/\left(-C_s\lambda^3\right)^{1/4}$. Setting this equal to the information limit, they find

$$C_s = -\frac{(\pi t)^2}{\lambda}, \quad \Delta = -\pi t \tag{66.170}$$

These situations are illustrated in Fig. 66.39.

For this case, Lentzen finds

$$\begin{aligned} \Delta &= -\frac{15}{4\lambda\boldsymbol{q}_{\max}^2} \\ C_s &= -\frac{15}{\lambda^3\boldsymbol{q}_{\max}^4} \\ C_5 &= \frac{105}{8\lambda^5\boldsymbol{q}_{\max}^6} \end{aligned} \tag{66.171}$$

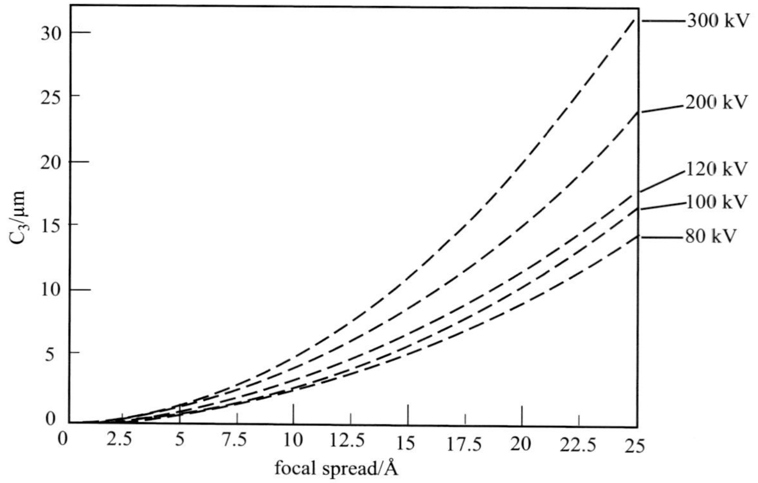

Figure 66.39
Maximum value of C_s as a function of focal spread, $C_5 = 0$. *After Chang et al. (2006), Courtesy Elsevier.*

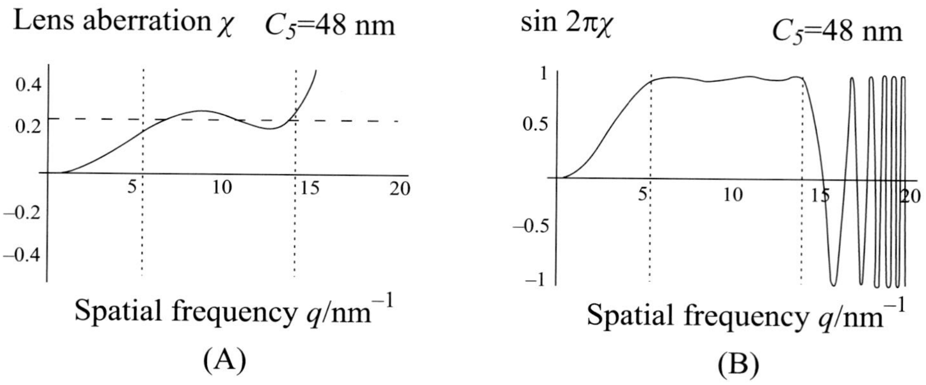

Figure 66.40
(A) Aberration function χ and (B) contrast-transfer function $\sin(2\pi\chi)$ when C_5 is variable, neglecting partial spatial and temporal coherence. The best values are now $\Delta = -9.2$ nm, $C_s = -45$ µm and $C_5 = 48$ nm. *After Lentzen (2008), Courtesy Cambridge University Press, https://doi.org/10.1017/S1431927608080045.*

Fig. 66.40 shows $\chi(\boldsymbol{q})$ and $\sin(2\pi\chi)$ for $\Delta = -9.2$ nm, $C_s = -45$ µm and $C_5 = 48$ mm. Although the coherence envelope is not included here, the information limit is again present in $\boldsymbol{q}_{\max}$.

Lentzen also explores the effect of seventh-order spherical aberration, C_7, which adds a term $(\pi/4)C_7\lambda^7(\boldsymbol{q}_x^2+\boldsymbol{q}_y^2)^4$ to χ. For a fixed value of C_7, he finds

$$\begin{aligned}
\Delta &= -\frac{15}{4\lambda \boldsymbol{q}_{\max}^2} + \frac{5C_7\lambda^6\boldsymbol{q}_{\max}^6}{112} \\
C_s &= -\frac{15}{\lambda^3\boldsymbol{q}_{\max}^4} + \frac{15C_7\lambda^4\boldsymbol{q}_{\max}^4}{28} \\
C_5 &= \frac{105}{8\lambda^5\boldsymbol{q}_{\max}^6} - \frac{45C_7\lambda^2\boldsymbol{q}_{\max}^2}{32}
\end{aligned} \tag{66.172}$$

66.9 Other Transforms Having a Convolution Theorem

Much of the theory presented in this chapter has yielded formulae of practical importance thanks to the fact that the Fourier transform maps convolution products into direct products. The Fourier transform is not alone in possessing this property and some of its rivals exhibit desirable features that it does not possess. We shall see that the principal attraction is the speed and precision of computation. Against this, we must recall that the very simple form of the filter functions (sin $\chi(\boldsymbol{q})$ and cos $\chi(\boldsymbol{q})$) is a consequence of the fact that it is the *Fourier* transform of the kernel function G (65.13) that has a simple explicit form. There is a risk that any advantage gained by applying other transforms to $C(\boldsymbol{u}_i)$ will be offset by the need to evaluate the transform of G; this remains to be investigated.

A rather general type of transform that maps convolution products into a simpler form is the polynomial transform; to understand this, some familiarity with the manipulation of rings and fields of polynomials is required. An account of this branch of algebra would lead us too far afield and we, therefore, give only a very superficial description before turning to the more easily understood number-theoretic transforms, which form a special case of the polynomial transforms.

We can see why polynomials are of interest in connection with convolutions by considering an elementary one-dimensional case,

$$y_\ell = \sum_{j=0}^{N-1} h_j x_{\ell-j} \tag{66.173}$$

in which ℓ runs from 0 to $2(N-1)$. Consider now the two polynomials in z whose coefficients are h_j and x_j:

$$H(z) := \sum_{j=0}^{N-1} h_j z^j \quad X(z) := \sum_{j=0}^{N-1} x_j z^j \tag{66.174}$$

Multiplication of H by X will give a polynomial $Y(z)$ of degree $2(N-1)$ and the coefficient of z^j in $Y(z)$ will be given by

$$y_j = \sum_{k=0}^{j} h_k x_{j-k} \tag{66.175}$$

There is thus a close relation between polynomial multiplication and convolution and this can be exploited to generate a very powerful family of transforms.

We now turn to the special case of the number-theoretic transforms and first consider the simple situation in which arithmetic is performed modulo a prime $\boldsymbol{q}$ in the Galois field GF($\boldsymbol{q}$). We again set out from Eq. (66.173) and scale so that all the coefficients h_j and x_j are integers. We now define

$$H_k := \sum_{j=0}^{N-1} h_j g^{jk} \text{ modulo } \boldsymbol{q} \tag{66.176a}$$

where g is an integer and likewise for x. We can also define an inverse

$$a_j = N^{-1} \sum_{k=0}^{N-1} H_k g^{-jk} \text{ modulo } \boldsymbol{q} \tag{66.176b}$$

with $NN^{-1} = 1$ modulo $\boldsymbol{q}$. These definitions bear a strong resemblance to those for a Fourier transform and its inverse except that an integer g replaces $\exp(2\pi i)$. Let us now examine the product of two transformed arrays and see whether its inverse has the form of a convolution. The product of the transforms of $\{h\}$ and $\{x\}$ is given by

$$H_k X_k = \sum_{j=0}^{N-1} \sum_{\ell=0}^{N-1} h_j x_\ell g^{(j+\ell)k} \text{ modulo } \boldsymbol{q} \tag{66.177a}$$

with inverse

$$a_m = N^{-1} \sum_{k=0}^{N-1} \sum_{j=0}^{N-1} \sum_{\ell=0}^{N-1} h_j x_\ell g^{(j+\ell-m)k} \text{ modulo } \boldsymbol{q} \tag{66.177b}$$

This will have the same form as (66.169), with $a_m \to y_m$, if

$$\sum_{k=0}^{N-1} g^{(j+\ell-m)k} \begin{cases} \equiv N & \text{for } j+\ell-m \equiv 0 \text{ modulo } N \\ \equiv 0 & \text{for } j+\ell-m \not\equiv 0 \text{ modulo } N \end{cases}$$

The first condition implies that we must have

$$g^N \equiv 1 \text{ modulo } \boldsymbol{q} \tag{66.178a}$$

and the second, that

$$\left(g^{j+\ell-m}-1\right)\sum_{k=0}^{N-1} g^{(j+\ell-m)} \equiv g^{N(j+\ell-m)}-1 \equiv 0 \text{ modulo } \boldsymbol{q} \tag{66.178b}$$

The integer g, which has hitherto been left undefined, must therefore be a root of unity of order N modulo $\boldsymbol{q}$, that is $g^N = 1$ modulo $\boldsymbol{q}$. For this choice of g, the transform (66.176a) maps convolution products into direct products. For given values of N and the prime $\boldsymbol{q}$, we can always find a transform provided that N divides $\boldsymbol{q} - 1$.

An analogous conclusion is reached if $\boldsymbol{q}$ is not prime. It can be shown by similar reasoning that a number-theoretic transform with the convolution mapping can be found if and only if

$$g^N \equiv 1 \text{ modulo } \boldsymbol{q} \tag{66.179a}$$

$$NN^{-1} \equiv 1 \text{ modulo } \boldsymbol{q} \tag{66.179b}$$

$g^{j+\ell-m}$ and $\boldsymbol{q}$ are mutually prime for

$$j+\ell-m = 1, 2 \ldots N-1 \tag{66.179c}$$

The various forms of the number-theoretic transform effectively correspond to different choices of the quantities g and $\boldsymbol{q}$. Thus for $\boldsymbol{q} = 2^p - 1$, p prime, the integers $\boldsymbol{q}$ are the Mersenne numbers and the transforms are likewise known as *Mersenne transforms*. Another choice is $\boldsymbol{q} = 2^p + 1$, $p = 2^t$, which means that the $\boldsymbol{q}$ coincide with the Fermat numbers and we obtain the *Fermat number transforms*. Mersenne transforms have the attraction that multiplication is unnecessary but there is no fast algorithm and word length and transform length are connected by a very restrictive relation. Fermat transforms can be calculated using addition and multiplication by a power of two for values of N up to $2^t + 2$; the connection between word length and transform length is less rigid than in the Mersenne case.

The real advantage of these transforms is only apparent when we go beyond these simple choices for $\boldsymbol{q}$ and consider the pseudo-Mersenne and pseudo-Fermat transforms. Here, the transform is defined modulo $\boldsymbol{q}_i$, where $\boldsymbol{q}_i$ is a factor of a pseudo-Mersenne number, $\boldsymbol{q} = 2^p - 1$ (p is now composite) or of a pseudo-Fermat number, $\boldsymbol{q} = 2^p + 1$ ($p \neq 2^t$). Both of these families of transforms can be extended to complex-valued arrays $\{x\}$ and $\{h\}$ and highly efficient fast algorithms have been found.

Number-theoretic transforms can therefore rival or surpass the Fourier transform in certain circumstances, depending on the transform length involved, and have the great attraction that since integers replace the exponentials of the Fourier transform, there is no rounding error. The calculations are exact. See Boussakta and Holt (1992) for an improved version.

This account can give no more than the flavour of these alternatives to the Fourier transform and we refer to specialized texts for extensive discussion, and in particular to Nussbaumer (1982), which we have followed closely, and to Hawkes (1974a,b, 1975), where they were first examined in connection with electron microscopy.

66.9.1 Other Types of Convolution

We mention in passing that the numerous discrete transforms that are commonly discussed in connection with image-processing – Walsh–Hadamard, Haar, slant are the best known – are not relevant here because they do not map convolutions of the cyclic type (66.173) into direct products. The Walsh transform does map a different type of convolution (dyadic convolution) into a simpler form but this dyadic product does not arise naturally in image formation; these transforms may, however, prove valuable for image coding (Section 71.4).

CHAPTER 67

Image Formation in the Scanning Transmission Electron Microscope

67.1 Introduction

It was realized soon after the first scanning transmission instrument (STEM) was built that its optical properties could be interpreted in terms of those of a conventional fixed-beam transmission electron microscope running backwards: a small axial STEM detector corresponds to a very small crossover in TEM; the probe-forming lens that precedes the specimen in STEM corresponds to the objective that follows it in TEM. It is harder to see what corresponds to the large image plane (fluorescent screen, film, direct detector) in the TEM, but some reflection shows that it is analogous to the large area swept out by the scanning probe *projected back to the source plane* (Zeitler and Thomson, 1970; Cowley, 1969; Krause and Rosenauer, 2017). Many of the phenomena observed in STEM operation can be easily understood by considering the corresponding TEM situation. In practice, however, there are ways of using the STEM for which the TEM analogue would be difficult to implement and which are hence not discussed in the TEM literature. We, therefore, provide a self-contained account of STEM image formation and, in particular, of the new flexibility offered at the detector level, which is readily accessible, whereas the illumination in TEM can only be practically altered in a limited number of relatively simple ways; the more complicated modifications to the incident beam that have been proposed usually involve introducing beam-shaping or phase-shifting devices into the path of the electrons in the condenser space of the TEM and some are examined in Chapter 80 in connection with vortex studies and in Chapter 63 on holography. In one STEM mode, described here, a phase plate is introduced in the illuminating beam and its properties are matched to those of the detector, in an attempt to achieve linear imaging (Ophus et al., 2016). Kahl and Rose (1995) have pointed out that the reciprocity between the STEM and TEM imaging modes can be traced back to an observation of von Helmholtz (1887).

67.2 Wave Propagation in STEM

The wave incident on the specimen is not even approximately plane, as it is in the normal imaging modes in TEM, since the source and specimen planes are conjugate and the probe-forming lens operates at a high demagnification. To a good approximation, the wave

Principles of Electron Optics.
DOI: https://doi.org/10.1016/B978-0-12-818979-5.00067-X

surface at the specimen is spherical, perturbed by the aberrations of the lenses between the source and the specimen and essentially by those of the final, probe-forming lens. The perturbation is described by the point-spread function of this lens ($\hat{G}$ or G in Chapter 65), which is itself the Fourier transform (Eq. 650.15c) of the product $T = T_A T_L$ or $T_A \exp(-i\chi)$ introduced in the discussion of image formation in TEM (Eq. 65.35). The wave incident on the specimen will be denoted by $\psi_Q\,(x_o, y_o, z_o)$ when the probe is centred on the axis, in the absence of any deflecting field therefore.

The illumination is initially assumed to be coherent (vanishing source-size, monochromatic source) after which the effect of partial coherence is examined (see also Part XVI). The intensity corresponding to this incident wave is concentrated within a very small zone, the size of which defines the pixel size and resolution of the instrument. Since probe sizes of the order of a few ångströms can be attained with uncorrected optics and substantially less than 1 Å using corrected optics, resolutions comparable with those of a TEM are routinely attained.

We assume that, as the probe is swept over the specimen surface, the wavefunction ψ_Q is translated bodily with no appreciable change of shape or direction. When the probe is no longer centred on the axis but on a point with coordinates $\boldsymbol{u}_o = \boldsymbol{\xi}$, where we have written $\boldsymbol{u}_o := (x_o, y_o)$, the incident wave will just be $\psi_Q\,(\boldsymbol{u}_o - \boldsymbol{\xi})$ therefore. For a specimen transparency $O(\boldsymbol{u}_o)$, the emergent wave will be

$$\psi_o(\boldsymbol{u}_o; \boldsymbol{\xi}) = O(\boldsymbol{u}_o)\psi_Q(\boldsymbol{u}_o - \boldsymbol{\xi}) \tag{67.1}$$

and if this is allowed to propagate freely to a detector plane distant R from the object, the wavefunction there will be

$$\hat{\psi}(\boldsymbol{u}_d; \boldsymbol{\xi}) = \int O(\boldsymbol{u}_o)\psi_Q(\boldsymbol{u}_o - \boldsymbol{\xi})\exp\left(\frac{2\pi i}{\lambda R}\boldsymbol{u}_d \cdot \boldsymbol{u}_o\right) d\boldsymbol{u}_o \tag{67.2}$$

The extensive use of the spatial frequency in Chapters 65 and 66 has accustomed us to the idea that a quantity with the dimensions of length is not always the most suitable measure of off-axial distance in wavefunctions and intensity functions. Thus in a set of conjugate planes that include the object and image planes of the TEM, the corresponding coordinates, scaled if necessary by the magnification, are usually appropriate (x_o, y_o; x_i, y_i; . . .); in the set of conjugate planes in which certain beam properties are related to those in the first set via a Fourier transform, however – these are the 'diffraction planes' – it is convenient to introduce *reciprocal* quantities with the dimension $[\text{length}]^{-1}$. The spatial frequency (65.21), which measures off-axis distance in (or close to) the plane of the objective aperture in the TEM but divided by λf, is the obvious example. The same is true in the STEM, where the detector plane is a 'reciprocal' plane relative to the specimen: at any instant, the far-field diffraction pattern of the area illuminated by the probe is formed in the detector plane. We, therefore, introduce a position coordinate in that plane that has dimension $[L]^{-1}$:

$$\boldsymbol{q}_d := \boldsymbol{u}_d/\lambda R \tag{67.3a}$$

and write

$$\psi_d(\boldsymbol{q}_d;\boldsymbol{\xi}) := \hat{\psi}(\boldsymbol{u}_d;\boldsymbol{\xi}) \tag{67.3b}$$

giving

$$\psi_d(\boldsymbol{q}_d;\boldsymbol{\xi}) = \int O(\boldsymbol{u}_o)\psi_Q(\boldsymbol{u}_o - \boldsymbol{\xi})\exp(2\pi i\boldsymbol{q}_d\cdot\boldsymbol{u}_o)d\boldsymbol{u}_o \tag{67.4}$$

The current density distribution in the detector plane, when the probe is centred on the point $\boldsymbol{u}_o = \boldsymbol{\xi}$ is denoted by $J_d(\boldsymbol{q}_d;\boldsymbol{\xi})$:

$$\begin{aligned} J_d(\boldsymbol{q}_d;\boldsymbol{\xi}) = \iint & O(\boldsymbol{u}_o)O^*(\overline{\boldsymbol{u}}_o)\psi_Q(\boldsymbol{u}_o - \boldsymbol{\xi})\psi_Q^*(\overline{\boldsymbol{u}}_o - \boldsymbol{\xi}) \\ & \times \exp\left\{2\pi i\boldsymbol{q}_d\cdot(\boldsymbol{u}_o - \overline{\boldsymbol{u}}_o)\right\}d\boldsymbol{u}_o d\overline{\boldsymbol{u}}_o \end{aligned} \tag{67.5}$$

As in Chapter 60 (Eq. 60.4), we write $O(\boldsymbol{u}_o)$ in the form

$$\begin{aligned} O(\boldsymbol{u}_o) &=: \exp\{-\sigma(\boldsymbol{u}_o) + i\eta(\boldsymbol{u}_o)\} \\ &\equiv 1 - (1 - e^{-\sigma}\cos\eta) + ie^{-\sigma}\sin\eta \\ &=: 1 - s + i\varphi \end{aligned} \tag{67.6a}$$

or alternatively,

$$O(\boldsymbol{u}_o) =: 1 - \zeta, \quad \zeta = (1 - e^{-\sigma}\cos\eta) - ie^{-\sigma}\sin\eta \tag{67.6b}$$

in which s and φ are not necessarily small; we shall make the weak-scattering approximation later on. We shall not retain the absorption term in ζ as it is a real factor and can easily be introduced in the following reasoning if required. The intensity at the detector, $J_d(\boldsymbol{q}_d;\boldsymbol{\xi})$ is now composed of three groups of terms: those independent of the specimen functions s and φ, ($J_d^{(0)}$), those linear in s or φ, ($J_d^{(1)}$) and quadratic terms, ($J_d^{(2)}$). Thus

$$J_d(\boldsymbol{q}_d;\boldsymbol{\xi}) = \sum_{j=0}^{2} J_d^{(j)}(\boldsymbol{q}_d;\boldsymbol{\xi}) \tag{67.7}$$

The first term is easily seen to be

$$J_d^{(0)}(\boldsymbol{q}_d;\boldsymbol{\xi}) = \iint \psi_Q(\boldsymbol{u}_o - \boldsymbol{\xi})\psi_Q^*(\overline{\boldsymbol{u}}_o - \boldsymbol{\xi})\exp\{2\pi i\boldsymbol{q}_d\cdot(\boldsymbol{u}_o - \overline{\boldsymbol{u}}_o)\}d\boldsymbol{u}_o d\overline{\boldsymbol{u}}_o = |T_A|^2 \tag{67.8}$$

and is independent of $\boldsymbol{\xi}$.

The second group of terms, in which ζ or s and φ appear linearly, is of greater interest. For ζ, it takes the form

$$\iint \{\zeta(\boldsymbol{u}_o) + \zeta^*(\overline{\boldsymbol{u}}_o)\}\psi_Q(\boldsymbol{u}_o - \boldsymbol{\xi})\psi_Q^*(\overline{\boldsymbol{u}}_o - \boldsymbol{\xi})\exp\{2\pi i \boldsymbol{q}_d\cdot(\boldsymbol{u}_o - \overline{\boldsymbol{u}}_o)\}d\boldsymbol{u}_o d\overline{\boldsymbol{u}}_o \tag{67.9}$$

We now examine the term in ζ in detail. On introducing the spectrum of ζ and ψ_o (66.2 and 66.3), namely

$$\tilde{\zeta}(\boldsymbol{q}_d) := \mathcal{F}^-(\zeta) = \int \zeta(\boldsymbol{u}_o)\exp(-2\pi i \boldsymbol{q}_d\cdot\boldsymbol{u}_o)d\boldsymbol{u}_o$$

and likewise for ψ_Q, the term in ζ in the above integral becomes

$$\begin{aligned}
&\iiiint\!\!\int \tilde{\zeta}(\boldsymbol{q})\tilde{\psi}_Q(\boldsymbol{m})\tilde{\psi}_Q^*(\boldsymbol{n})\exp\{\boldsymbol{q}_d\cdot\boldsymbol{u}_o - \boldsymbol{q}_d\cdot\overline{\boldsymbol{u}}_o + \boldsymbol{q}\cdot\boldsymbol{u}_o \\
&\qquad + \boldsymbol{m}\cdot(\boldsymbol{u}_o - \boldsymbol{\xi}) - \boldsymbol{n}\cdot(\overline{\boldsymbol{u}}_o - \boldsymbol{\xi})\}d\boldsymbol{u}_o\, d\overline{\boldsymbol{u}}_o\, d\boldsymbol{m}\, d\boldsymbol{n}\, d\boldsymbol{q} \\
&= \iiint \tilde{\zeta}(\boldsymbol{q})\tilde{\psi}_Q(\boldsymbol{m})\tilde{\psi}_Q^*(\boldsymbol{n})\delta(\boldsymbol{q}_d + \boldsymbol{q} + \boldsymbol{m})\delta(-\boldsymbol{q}_d - \boldsymbol{n}) \times \exp\{2\pi i\boldsymbol{\xi}\cdot(\boldsymbol{n} - \boldsymbol{m})\}d\boldsymbol{m}\, d\boldsymbol{n}\, d\boldsymbol{q} \\
&= \int \tilde{\zeta}(\boldsymbol{q})\tilde{\psi}_Q(\boldsymbol{q} + \boldsymbol{q}_d)\tilde{\psi}_Q^*(\boldsymbol{q}_d)\exp(2\pi i\boldsymbol{\xi}\cdot\boldsymbol{q})d\boldsymbol{q}
\end{aligned} \tag{67.10a}$$

in which we have assumed that $\tilde{\psi}_Q(-\boldsymbol{q}) = \tilde{\psi}_Q(\boldsymbol{q})$. An analogous calculation for the case $\zeta^*(\overline{\boldsymbol{u}}_o)$ gives

$$\int \tilde{\zeta}^*(\boldsymbol{q})\tilde{\psi}_Q(\boldsymbol{q}_d)\tilde{\psi}_Q^*(\boldsymbol{q}_d - \boldsymbol{q})\exp(2\pi i\boldsymbol{\xi}\cdot\boldsymbol{q})d\boldsymbol{q} \tag{67.10b}$$

Collecting up the four terms, we obtain in all

$$\begin{aligned}
J_d^{(1)}(\boldsymbol{q}_d;\boldsymbol{\xi}) = &-\int \tilde{s}(\boldsymbol{q})\left\{\tilde{\psi}_Q(\boldsymbol{q}_d + \boldsymbol{q})\tilde{\psi}_Q^*(\boldsymbol{q}_d) + \tilde{\psi}_Q(\boldsymbol{q}_d)\tilde{\psi}_Q^*(\boldsymbol{q}_d - \boldsymbol{q})\right\} \\
&\times \exp(2\pi i\boldsymbol{\xi}\cdot\boldsymbol{q})d\boldsymbol{q} \\
&+ i\int \tilde{\varphi}(\boldsymbol{q})\left\{\tilde{\psi}_Q(\boldsymbol{q}_d + \boldsymbol{q})\tilde{\psi}_Q^*(\boldsymbol{q}_d) - \tilde{\psi}_Q(\boldsymbol{q}_d)\tilde{\psi}_Q^*(\boldsymbol{q}_d - \boldsymbol{q})\right\} \\
&\times \exp(2\pi i\boldsymbol{\xi}\cdot\boldsymbol{q})d\boldsymbol{q}
\end{aligned} \tag{67.11a}$$

or in terms of ζ

$$\begin{aligned}
J_d^{(1)}(\boldsymbol{q}_d;\boldsymbol{\xi}) = &-\int \left\{\tilde{\zeta}(\boldsymbol{q})\tilde{\psi}_Q(\boldsymbol{q}_d + \boldsymbol{q})\tilde{\psi}_Q^*(\boldsymbol{q}_d) + \tilde{\zeta}^*(\boldsymbol{q})\psi_Q(\tilde{\boldsymbol{q}}_d\tilde{\psi}_Q^*(\boldsymbol{q}_d - \boldsymbol{q})\right\} \\
&\times \exp(2\pi i\boldsymbol{\xi}\cdot\boldsymbol{q})d\boldsymbol{q}
\end{aligned} \tag{67.11b}$$

The quadratic term is given by

$$J_d^{(2)}(\boldsymbol{q}_d;\boldsymbol{\xi}) = \left| \int \{s(\boldsymbol{u}_o) - \mathrm{i}\varphi(\boldsymbol{u}_o)\}\psi_Q(\boldsymbol{u}_o - \boldsymbol{\xi})\exp(2\pi \mathrm{i}\boldsymbol{q}_d \cdot \boldsymbol{u}_o)d\boldsymbol{u}_o \right|^2 \tag{67.12a}$$

or

$$J_d^{(2)}(\boldsymbol{q}_d;\boldsymbol{\xi}) = \left| \int \zeta(\boldsymbol{u}_o)\psi_Q(\boldsymbol{u}_o - \boldsymbol{\xi})\exp(2\pi \mathrm{i}\boldsymbol{q}_d \cdot \boldsymbol{u}_o)d\boldsymbol{u}_o \right|^2 \tag{67.12b}$$

The current density recorded per picture element will depend on the shape and response of the detector. These are characterized by means of a *detector response function, D* ($\boldsymbol{q_d}$). If the recorded current density is denoted by I_d ($\boldsymbol{\xi}$), then

$$I_d(\boldsymbol{\xi}) = \int J_d(\boldsymbol{q}_d;\boldsymbol{\xi})D(\boldsymbol{q}_d)d\boldsymbol{q}_d \tag{67.13}$$

The form of Eq. (67.5) suggests that the spatial frequency response of the detector will also be useful. We define this as in Eq. (65.15):

$$S_D(\boldsymbol{u}_d) := \mathcal{F}^-(D) = \int D(\boldsymbol{q}_d)\exp(-2\pi \mathrm{i}\boldsymbol{q}_d \cdot \boldsymbol{u}_d)d\boldsymbol{q}_d \tag{67.14}$$

(Since $\boldsymbol{q_d}$ has dimension $[\mathrm{L}]^{-1}$, $\boldsymbol{u_d}$ is a length.) The recorded current density I_d ($\boldsymbol{\xi}$) may thus be written

$$I_d(\boldsymbol{\xi}) = \iint O(\boldsymbol{u}_o)O^*(\overline{\boldsymbol{u}}_o)\psi_Q(\boldsymbol{u}_o - \boldsymbol{\xi})\psi_Q^*(\overline{\boldsymbol{u}}_o - \boldsymbol{\xi})S_D(\overline{\boldsymbol{u}}_o - \boldsymbol{u}_o)d\boldsymbol{u}_o\, d\overline{\boldsymbol{u}}_o \tag{67.15}$$

Like J_d ($\boldsymbol{q_d}$; $\boldsymbol{\xi}$), the recorded current density I_d ($\boldsymbol{\xi}$) is composed of three contributions,

$$I_d^{(j)}(\boldsymbol{\xi}) =: \int J_d^{(j)}(\boldsymbol{q}_d;\boldsymbol{\xi})D(\boldsymbol{q}_d)d\boldsymbol{q}_d \quad j = 0, 1, 2 \tag{67.16}$$

We have

$$I_d^{(0)}(\boldsymbol{\xi}) = \iint \psi_Q(\boldsymbol{u}_o)\psi_Q^*(\overline{\boldsymbol{u}}_o)S_D(\overline{\boldsymbol{u}}_o - \boldsymbol{u}_o)d\boldsymbol{u}_o\, d\overline{\boldsymbol{u}}_o$$

$$\begin{aligned} I_d^{(1)}(\boldsymbol{\xi}) = &- \iint \{s(\boldsymbol{u}_o) + s(\overline{\boldsymbol{u}}_o)\}\psi_Q(\boldsymbol{u}_o - \boldsymbol{\xi})\psi_Q^*(\overline{\boldsymbol{u}}_o - \boldsymbol{\xi})S_D(\overline{\boldsymbol{u}}_o - \boldsymbol{u}_o)d\boldsymbol{u}_o\, d\overline{\boldsymbol{u}}_o \\ &+ \mathrm{i}\iint \{\varphi(\boldsymbol{u}_o) - \varphi(\overline{\boldsymbol{u}}_o)\}\psi_Q(\boldsymbol{u}_o - \boldsymbol{\xi})\psi_Q^*(\overline{\boldsymbol{u}}_o - \boldsymbol{\xi})S_D(\overline{\boldsymbol{u}}_o - \boldsymbol{u}_o)d\boldsymbol{u}_o\, d\overline{\boldsymbol{u}}_o \end{aligned} \tag{67.17a}$$

or

$$\begin{aligned} I_d^{(1)}(\boldsymbol{\xi}) = &-\iint \tilde{s}(\boldsymbol{q})\left\{\tilde{\psi}_Q(\boldsymbol{q}_d+\boldsymbol{q})\tilde{\psi}_Q^*(\boldsymbol{q}_d)+\tilde{\psi}_Q(\boldsymbol{q}_d)\tilde{\psi}_Q^*(\boldsymbol{q}_d-\boldsymbol{q})\right\} \\ &\times D(\boldsymbol{q})\exp(2\pi i\boldsymbol{\xi}\cdot\boldsymbol{q})d\boldsymbol{q}\,d\boldsymbol{q}_d \\ &+ i\iint \tilde{\varphi}(\boldsymbol{q})\left\{\tilde{\psi}_Q(\boldsymbol{q}_d+\boldsymbol{q})\tilde{\psi}_Q^*(\boldsymbol{q}_d)-\tilde{\psi}_Q(\boldsymbol{q}_d)\tilde{\psi}_Q^*(\boldsymbol{q}_d-\boldsymbol{q})\right\} \\ &\times D(\boldsymbol{q})\exp(2\pi i\boldsymbol{\xi}\cdot\boldsymbol{q})d\boldsymbol{q}\,d\boldsymbol{q}_d \end{aligned} \tag{67.17b}$$

or again

$$\begin{aligned} I_d^{(1)}(\boldsymbol{\xi}) = &-\iiint \zeta(\boldsymbol{u}_o)\psi_Q(\boldsymbol{u}_o-\boldsymbol{\xi})\psi_Q^*(\bar{\boldsymbol{u}}_o-\boldsymbol{\xi})D(\boldsymbol{q}_d)\exp\{2\pi i\boldsymbol{q}_d\cdot(\boldsymbol{u}_o-\bar{\boldsymbol{u}}_o)\}d\boldsymbol{u}_o d\bar{\boldsymbol{u}}_o d\boldsymbol{q}_d \\ &-\iiint \zeta^*(\boldsymbol{u}_o)\psi_Q(\boldsymbol{u}_o-\boldsymbol{\xi})\psi_Q^*(\bar{\boldsymbol{u}}_o-\boldsymbol{\xi})D(\boldsymbol{q}_d)\exp\{2\pi i\boldsymbol{q}_d\cdot(\boldsymbol{u}_o-\bar{\boldsymbol{u}}_o)\}d\boldsymbol{u}_o d\bar{\boldsymbol{u}}_o d\boldsymbol{q}_d \end{aligned} \tag{67.17c}$$

The spectrum of $I_d^{(1)}$, which we denote $S_I^{(1)}(\boldsymbol{q})$, is thus

$$\begin{aligned} S_I^{(1)}(\boldsymbol{q}) &:= \mathcal{F}^-\left(I_d^{(1)}\right) \\ &= -\tilde{s}(\boldsymbol{q})K_a(\boldsymbol{q}) - i\tilde{\varphi}(\boldsymbol{q})K_p(\boldsymbol{q}) \end{aligned} \tag{67.18}$$

in which

$$K_a(\boldsymbol{q}) := \int\left\{\tilde{\psi}_Q(\boldsymbol{q}_d+\boldsymbol{q})\tilde{\psi}_Q^*(\boldsymbol{q}_d)+\tilde{\psi}_Q(\boldsymbol{q}_d)\tilde{\psi}_Q^*(\boldsymbol{q}_d-\boldsymbol{q})\right\}D(\boldsymbol{q}_d)d\boldsymbol{q}_d \tag{67.19a}$$

$$K_p(\boldsymbol{q}) := \int\left\{\tilde{\psi}_Q(\boldsymbol{q}_d)\tilde{\psi}_Q^*(\boldsymbol{q}_d-\boldsymbol{q})-\tilde{\psi}_Q(\boldsymbol{q}_d+\boldsymbol{q})\tilde{\psi}_Q^*(\boldsymbol{q}_d)\right\}D(\boldsymbol{q}_d)d\boldsymbol{q}_d \tag{67.19b}$$

$\tilde{I}_d^{(2)}$ will be small compared with this term only if the specimen scatters weakly and hence s and φ are small. For this situation the image contrast[1] is described by *transfer functions* $K_a(\boldsymbol{q})$ and $K_p(\boldsymbol{q})$, which closely resemble those encountered in the discussion of image formation in TEM.

We now return to (67.17c) and integrate the first term over $\bar{\boldsymbol{u}}_o$ and $\boldsymbol{q}_d$ and the second term over $\boldsymbol{u}_o$ and $\boldsymbol{q}_d$, which yields

$$\begin{aligned} I_d^{(1)}(\boldsymbol{\xi}) = &-\int \zeta(\boldsymbol{u}_o+\boldsymbol{\xi})\psi_Q(\boldsymbol{u}_o)\mathcal{F}\left\{D(\boldsymbol{q}_d)\mathcal{F}^{-1}\psi_Q^*\right\}d\boldsymbol{u}_o \\ &-\int \zeta^*(\bar{\boldsymbol{u}}_o+\boldsymbol{\xi})\psi_Q^*(\bar{\boldsymbol{u}}_o)\mathcal{F}^{-1}\{D(\boldsymbol{q}_d)\mathcal{F}\psi_Q\}d\bar{\boldsymbol{u}}_o \end{aligned} \tag{67.20}$$

A pattern emerges when we define a function $B(\boldsymbol{u}_o)$,

$$B(\boldsymbol{u}_o) := \psi_Q(\boldsymbol{u}_o)\mathcal{F}\left\{D(\boldsymbol{q}_d)\mathcal{F}^{-1}\psi_Q^*(\bar{\boldsymbol{u}}_o)\right\} \tag{67.21}$$

[1] Note that these transfer functions differ by a factor of two from those for the TEM because we have not defined STEM contrast in the same way, see Eq. (66.10a).

in terms of which (67.20) becomes

$$\begin{aligned} I_d^{(1)}(\boldsymbol{\xi}) &= -\int \left\{\zeta(\boldsymbol{u}_o+\boldsymbol{\xi})B(\boldsymbol{u}_o)+\zeta^*(\boldsymbol{u}_o+\boldsymbol{\xi})B^*(\boldsymbol{u}_o)\right\}d\boldsymbol{u}_o \\ &= -\frac{1}{2}\int \left\{\zeta(\boldsymbol{u}_o+\boldsymbol{\xi})+\zeta^*(\boldsymbol{u}_o+\boldsymbol{\xi})\right\}(B+B^*)d\boldsymbol{u}_o \\ &\quad -\frac{1}{2\mathrm{i}}\int \left\{\zeta(\boldsymbol{u}_o+\boldsymbol{\xi})-\zeta^*(\boldsymbol{u}_o+\boldsymbol{\xi})\right\}\mathrm{i}(B-B^*)d\boldsymbol{u}_o \end{aligned} \tag{67.22}$$

This enables us to write I_d in a form that sheds much light on the image-forming process in STEM:

$$\begin{aligned} I_d(\boldsymbol{\xi}) = &\int D(\boldsymbol{q}_d)\left|\tilde{\psi}_Q\right|^2 d\boldsymbol{q}_d I_d^{(0)}(\boldsymbol{\xi}) - \mathrm{i}\int \sin\eta(\boldsymbol{u}_o+\boldsymbol{\xi})(B-B^*)d\boldsymbol{u}_o \\ &-\int\left\{1-\cos\eta(\boldsymbol{u}_o+\boldsymbol{\xi})\right\}(B+B^*)d\boldsymbol{u}_o I_d^{(1)}(\boldsymbol{\xi}) \\ &+\iint \zeta(\boldsymbol{u}_o+\boldsymbol{\xi})\zeta^*(\overline{\boldsymbol{u}}_o+\boldsymbol{\xi})\psi_Q(\boldsymbol{u}_o)\psi_Q^*(\overline{\boldsymbol{u}}_o)S(\boldsymbol{u}_o-\overline{\boldsymbol{u}}_o)d\boldsymbol{u}_o d\overline{\boldsymbol{u}}_o I_d^{(2)}(\boldsymbol{\xi}) \end{aligned} \tag{67.23}$$

We have seen that image formation in the TEM is expressed in terms of a convolution. Eq. (67.23) shows that the STEM image involves not convolution but *cross-correlation*. The cross-correlation of two functions $f(\boldsymbol{u})$ and $g(\boldsymbol{u})$ is defined by

$$(f(\overline{\boldsymbol{u}})\odot g(\overline{\boldsymbol{u}}))(\boldsymbol{u}) := \int f^*(\overline{\boldsymbol{u}})g(\overline{\boldsymbol{u}}+\boldsymbol{u})d\overline{\boldsymbol{u}} \tag{67.24a}$$

and if the two functions are the same, this is known as their *autocorrelation*:

$$(f(\overline{\boldsymbol{u}})\odot f(\overline{\boldsymbol{u}}))(\boldsymbol{u}) = \int f^*(\overline{\boldsymbol{u}})f(\overline{\boldsymbol{u}}+\boldsymbol{u})d\overline{\boldsymbol{u}} \tag{67.24b}$$

The Fourier transform of a cross-correlation function is given by

$$\mathcal{F}^{-1}(f(\overline{\boldsymbol{u}})\odot g(\overline{\boldsymbol{u}})) = \tilde{f}^*\tilde{g} \tag{67.24c}$$

We see immediately that the terms of $I_d^{(1)}$ have the form of cross-correlations:

$$I_d^{(1)}(\boldsymbol{\xi}) = -\left\{\mathrm{i}(B-B^*)\odot\sin\eta + (B+B^*)\odot(1-\cos\eta)\right\} \tag{67.25a}$$

The final term, $I_d^{(2)}$, involves an autocorrelation. To see this, we rewrite $I_d^{(2)}$ in the form

$$\begin{aligned} I_d^{(2)}(\boldsymbol{\xi}) &= \iint \zeta(\boldsymbol{u}+\overline{\boldsymbol{u}}+\boldsymbol{\xi})\zeta^*(\boldsymbol{u}+\boldsymbol{\xi})\psi_Q(\boldsymbol{u}+\overline{\boldsymbol{u}})\psi_Q^*(u)S(\overline{\boldsymbol{u}})d\boldsymbol{u}d\overline{\boldsymbol{u}} \\ &= \int\left\{\zeta(\boldsymbol{u}+\boldsymbol{\xi})\psi_Q(\boldsymbol{u})\odot\zeta(\boldsymbol{u}+\boldsymbol{\xi})\psi_Q(\boldsymbol{u})\right\}S(\overline{\boldsymbol{u}})d\overline{\boldsymbol{u}} \end{aligned} \tag{67.25b}$$

In the following section, we shall use these expressions to analyse the various STEM imaging modes but we first introduce an approximation, studied in detail by Bosch and Lazić (2015), which simplifies digital image simulation. We assume that the source is coherent, which implies that the autocorrelation of ψ_Q is just the Fourier transform of $|T_A|^2$.

This falls rapidly to zero outside the bright-field (BF) zone. The autocorrelation function of the sample depends on its nature. For crystalline specimens, it shows peaks corresponding to the periodicities of the lattice; otherwise, it falls rapidly to a small constant value. These observations led Lazić and Bosch (2017) to write

$$\int \zeta(\boldsymbol{u}+\overline{\boldsymbol{u}}+\boldsymbol{\xi})\zeta^*(\boldsymbol{u}+\boldsymbol{\xi})\psi_Q(\boldsymbol{u}+\overline{\boldsymbol{u}})\psi_Q^*(\boldsymbol{u})d\boldsymbol{u} \approx W(\overline{\boldsymbol{u}})\int \left|\zeta(\boldsymbol{u}+\boldsymbol{\xi})\right|^2 \left|\psi_Q(\boldsymbol{u})\right|^2 d\boldsymbol{u} \qquad (67.26)$$

in which W is a window function. In its simplest form ($W^{(1)}$), it is transparent inside a small disc of radius ρ and opaque outside it. On substituting this approximate expression in Eq. (67.23), we arrive at the following expression for the three terms of $I_d(\boldsymbol{\xi})$:

$$\begin{aligned} I_d^{(0)}(\boldsymbol{\xi}) &= \int D(\boldsymbol{q}_d)\left|\tilde{\psi}_Q\right|^2 d\boldsymbol{q}_d \\ I_d^{(1)}(\boldsymbol{\xi}) &= -\mathrm{i}(M_- \odot \sin\eta)(\boldsymbol{\xi}) - (M_+ \odot (1-\cos\eta))(\boldsymbol{\xi}) \\ I_d^{(2)}(\boldsymbol{\xi}) &= 2\left(\int\limits_{|\overline{\boldsymbol{u}}_o|<\rho} S(\overline{\boldsymbol{u}}_o)d\overline{\boldsymbol{u}}_o\right)\left\{\left|\psi_Q\right|^2 \odot (1-\cos\eta)\right\}(\boldsymbol{\xi}) \end{aligned} \qquad (67.27)$$

in which $M_\pm$ denote

$$M_\pm := \psi_Q(\boldsymbol{u}_o)\mathcal{F}\left\{D(\boldsymbol{q}_d)\mathcal{F}^{-1}\left(\left(\psi_Q^*(\overline{\boldsymbol{u}}_o)\right)\right\} \pm \psi_Q^*(\boldsymbol{u}_o)\mathcal{F}^{-1}\left\{D(\boldsymbol{q}_d)\mathcal{F}\left(\left(\psi_Q(\overline{\boldsymbol{u}}_o)\right)\right\} \qquad (67.28)$$

Test calculations show that a better choice is $W^{(2)}$,

$$W^{(2)} := W^{(1)}\left(1 - \frac{|\overline{\boldsymbol{u}}|}{\rho_1}\right) \qquad (67.29)$$

In Section 67.3, we shall be discussing contrast transfer in the various STEM modes. The spatial frequency spectrum of $I_d(\boldsymbol{\xi})$, in terms of which the contrast will be analysed, is defined by

$$C(\boldsymbol{q}_d) := \mathcal{F}^{-1} I_d(\boldsymbol{\xi}) =: C^{(0)}(\boldsymbol{q}_d) + C^{(1)}(\boldsymbol{q}_d) + C^{(2)}(\boldsymbol{q}_d) \qquad (67.30)$$

in which

$$\begin{aligned} C^{(0)}(\boldsymbol{q}_d) &= \delta(\boldsymbol{q}_d)\int D(\boldsymbol{q}_d)\left|\tilde{\psi}_Q\right|^2 d\boldsymbol{q}_d \\ C^{(1)}(\boldsymbol{q}_d) &= -\mathrm{i}\left(\tilde{M}_-\mathcal{F}\sin\eta\right)(\boldsymbol{q}_d) - \left(\tilde{M}_+\mathcal{F}(1-\cos\eta)\right)(\boldsymbol{q}_d) \\ C^{(2)}(\boldsymbol{q}_d) &= \mathcal{F}^{-1}\left[\iint \{\zeta(\boldsymbol{u}+\boldsymbol{\xi})\psi_Q(\boldsymbol{u}) \odot \zeta(\boldsymbol{u}+\boldsymbol{\xi})\psi_Q(\boldsymbol{u})\}S(\overline{\boldsymbol{u}})d\overline{\boldsymbol{u}}\right] \end{aligned} \qquad (67.31)$$

or

$$C^{(2)}(\boldsymbol{q}_d) \approx 2\left(\iint\limits_{|\overline{\boldsymbol{u}}_o|<\rho} S(\overline{\boldsymbol{u}}_o)d\overline{\boldsymbol{u}}_o\right)\left\{\left(\mathcal{F}\left|\psi_Q\right|^2\right)^*\mathcal{F}(1-\cos\eta)\right\}(\boldsymbol{q}_d)$$

Typical values of ρ and ρ_1 are listed later.

We see that information about the specimen is now conveyed by *two* specimen functions, $\sin\eta$ and $(1-\cos\eta)$, each with its own transfer function. Thus image formation in STEM cannot in general be characterized by a single transfer function even for pure phase objects.

The version of the foregoing theory that appeared in the first edition has been greatly extended by Lazić and Bosch (2017). We have followed their reasoning closely.

67.3 Detector Geometry

Each of the numerous modes of operation of the STEM is associated with a particular choice of detector geometry. In the simplest cases (Fig. 67.1) the detector consists of a disc (bright-field, BF), an annulus within the bright-field disc (annular bright-field, ABF), an annulus extending everywhere outside the bright-field disc (annular dark-field, ADF) or an

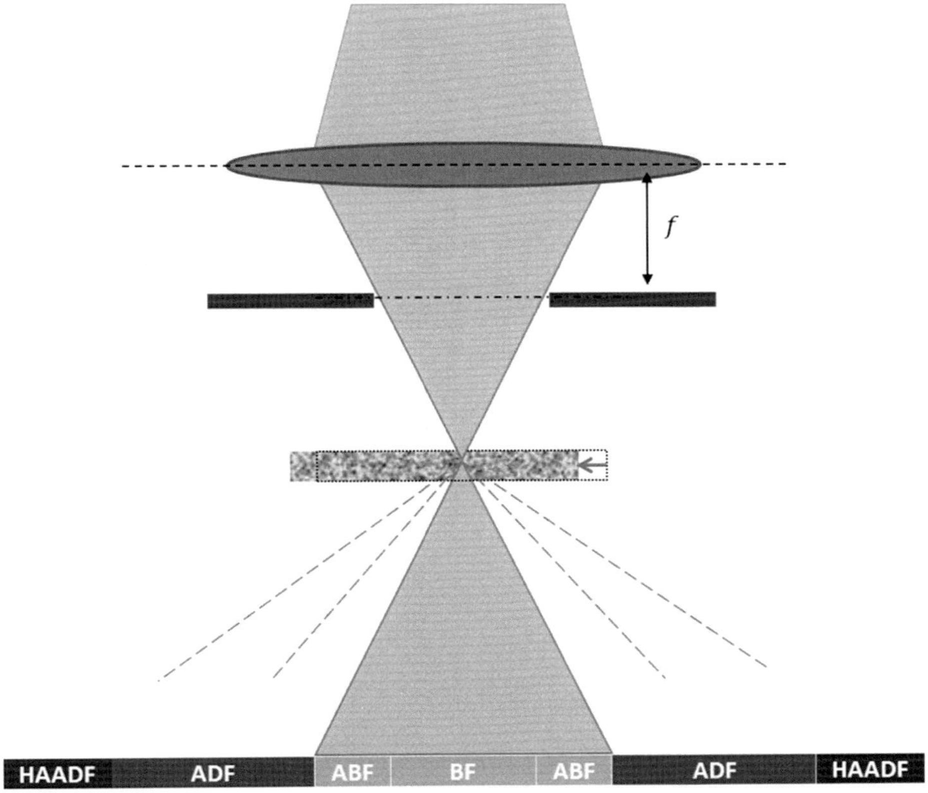

Figure 67.1
Simple detector geometries. The regions of the far-field pattern selected for bright-field, annular bright-field, annular dark-field and high-angle annular dark-field imaging. *After Lazić and Bosch (2017), Courtesy Elsevier.*

annulus covering only the outer region of the dark-field annulus (high-angle annular dark-field, HAADF). An iris aperture may be placed above the annular dark-field detector to obtain angle-resolved information (Grieb et al., 2021a,b). Next come the vector detectors, which generate two (or more) signals from each object element. The classic example of these is the family of split detectors (Fig. 67.2), in the form of quadrants or narrower segments (Cowley, 1993; Shibata et al., 2010), which may, in turn, be subdivided radially by rings; these generate differential phase contrast (DPC), which may be classified as *integrated differential phase contrast* or *differentiated differential phase contrast* (dDPC). These were originally used in studies of magnetic materials at medium resolution (McVitie and Chapman, 1990; Chapman et al., 1990) but are now capable of atomic resolution (Shibata et al., 2012; Seki et al., 2021). Importantly, the former measures the scalar electrostatic potential field of the sample and is obtained by integration of the vector image

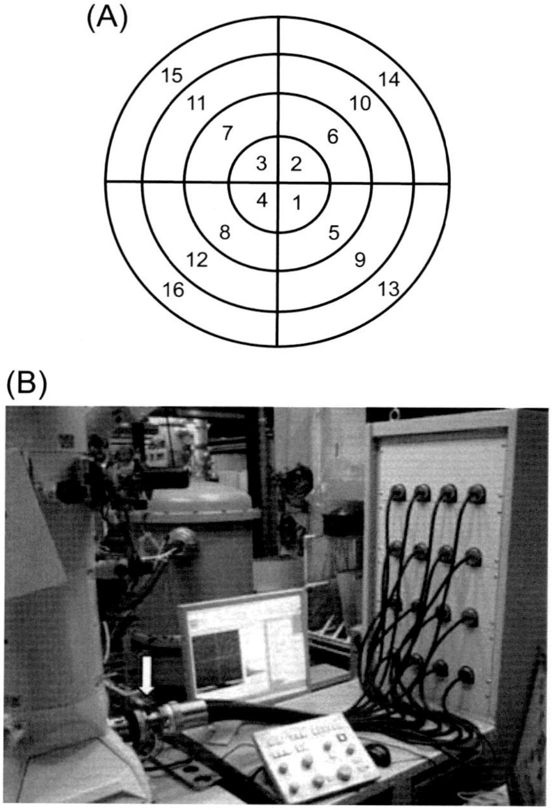

Figure 67.2
Subdivisions of a 16-zone detector. (A) Detector pattern. (B) 16 well-separated fibre bundles transmit the different signals. (C) Atomic-resolution STEM images of $SrTiO_3$ [001]. *After Shibata et al. (2010), Courtesy Oxford University Press, https://doi.org/10.1093/jmicro/dfq014.*

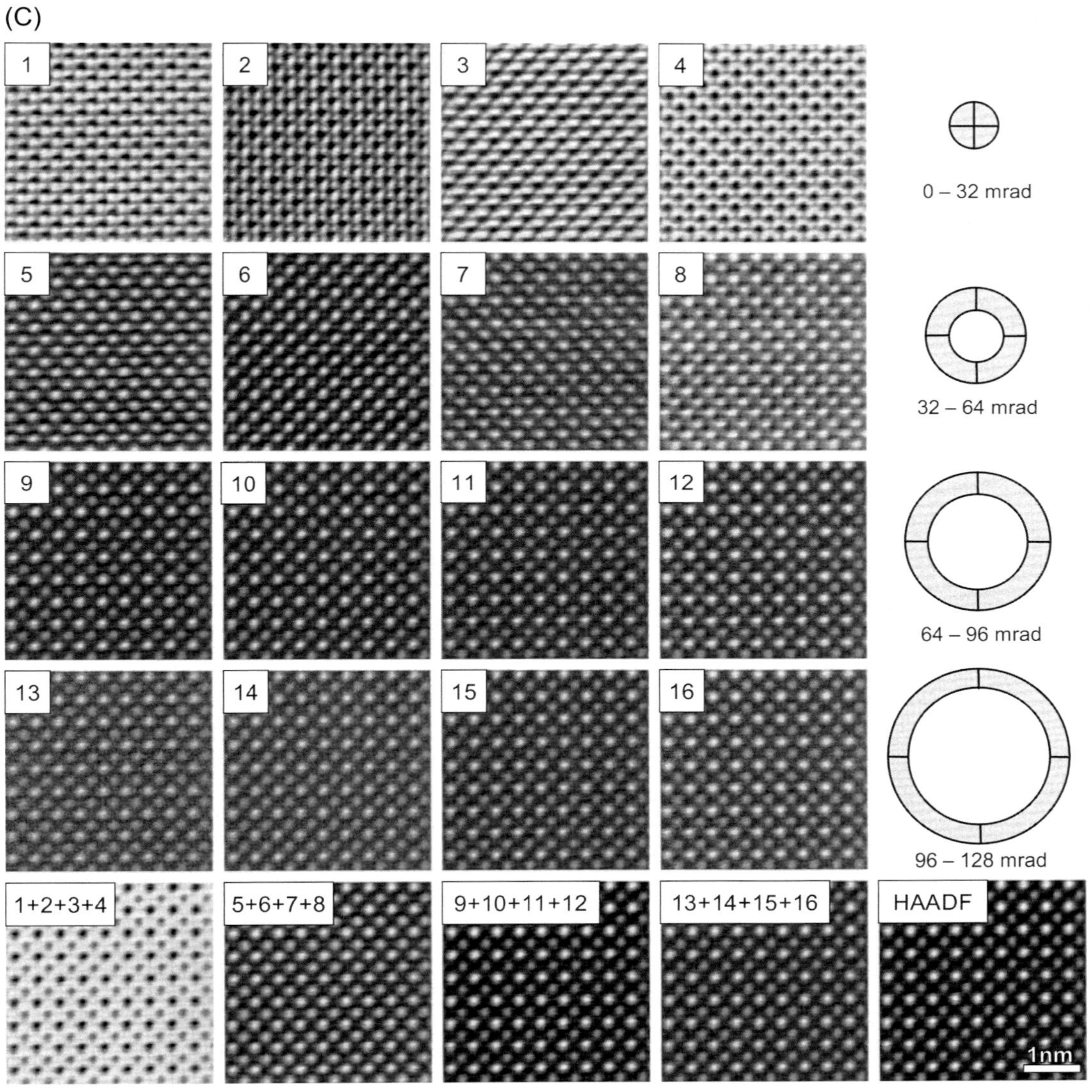

Figure 67.2
Continued.

formed by differential phase contrast (Lazić et al., 2016; Yücelen et al., 2018; cf. Winkler et al., 2017; Tanaka and Ikarashi, 2021). A particularly interesting case is the *first-moment* or *centre-of-mass detector*; this measures the displacement of the convergent-beam electron diffraction pattern, which can be linearly related to the electric field of the sample at the position of the probe (Seki et al., 2017). Still, more information is generated by pixelated detectors, of which a number have been developed (Section 67.5); these are capable of recording the current distribution from each picture element of the specimen; they may be programmed to mimic any of the shapes described above or they may generate an entire image from each object element, yielding a four-dimensional image or data set (we shall see in Section 70.2 that such an image is common in image algebra where it is known as a template). This will lead us to describe the technique known as ptychography (Section 67.4). We conclude this section with an account of confocal imaging in STEM.

67.3.1 Bright-Field Imaging

Here, the detector covers all or part of the bright central disc (Fig. 67.3):

$$D(\boldsymbol{q}_d) = \begin{cases} 1 & 0 < |\boldsymbol{q}_d| < \lambda q_d^{(b)} \\ 0 & |\boldsymbol{q}_d| > q_d^{(b)} \end{cases} \quad \text{circular disc} \tag{67.32a}$$

or

$$D(\boldsymbol{q}_d) = \begin{cases} 1 & \lambda q_d^{(b)} |\boldsymbol{q}_d| < q_d^{(b)} \\ 0 & |\boldsymbol{q}_d| > q_d^{(b)} \end{cases} \quad \text{annulus} \tag{67.32b}$$

where the label (b) is a reminder that this is the rim of the bright-field disc; $\lambda q_d^{(b)}$ is the inner radius of the annular bright-field detector; λ is a measure of the thickness of the ring and incidentally allows us to choose the size of the bright-field area captured.

The constant term $C_0(\boldsymbol{q}_d)$ is of no interest. The remaining terms consist of multiples of the spectra of sin η and of $(1-\cos\,\eta)$, which we may regard as contrast-transfer functions. Lazić and Bosch have calculated these functions for various values of λ and express the results in terms of an *effective contrast-transfer function*,

$$C_e(\boldsymbol{q}_d) := \frac{C(\boldsymbol{q}_d)}{\tilde{\eta}} \tag{67.33}$$

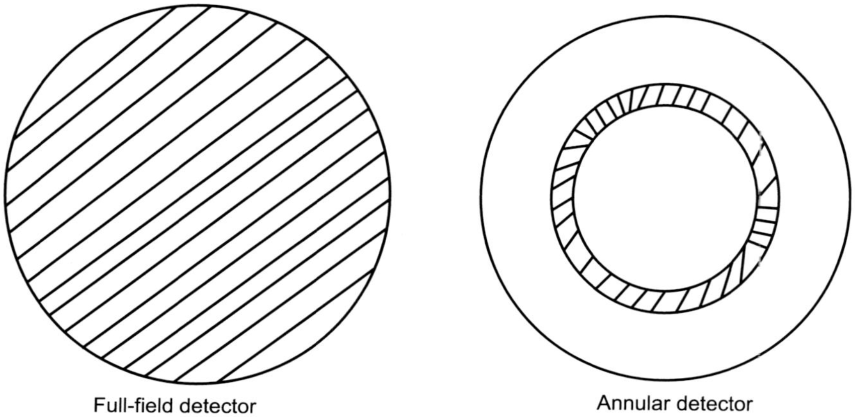

Figure 67.3
Bright-field imaging.

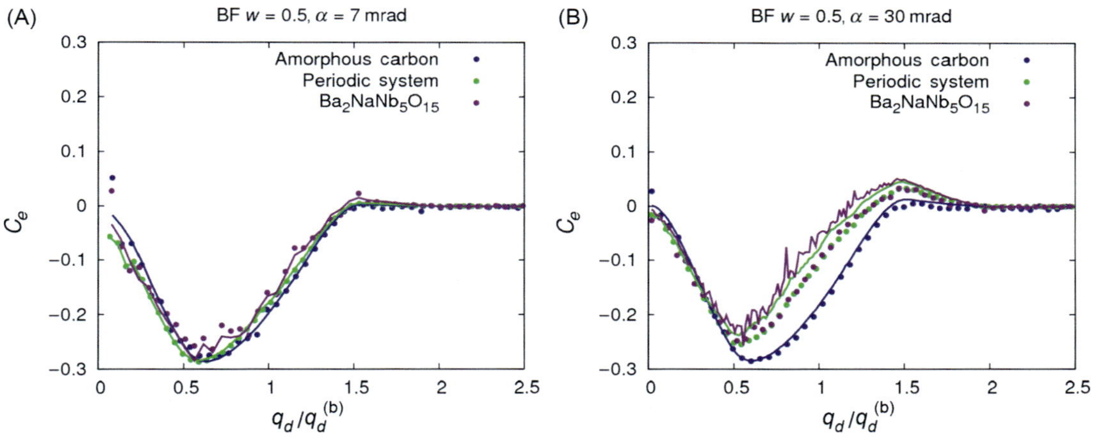

Figure 67.4
Azimuthally averaged contrast-transfer functions, bright field. (A) Uncorrected microscope, the curves coincide. (B) Corrected instrument. Three specimens are represented: (1) amorphous carbon, (2) the periodic system (see Fig. 67.6) and (3) $Ba_2NaNb_5O_{15}$. *After Lazić and Bosch (2017), Courtesy Elsevier.*

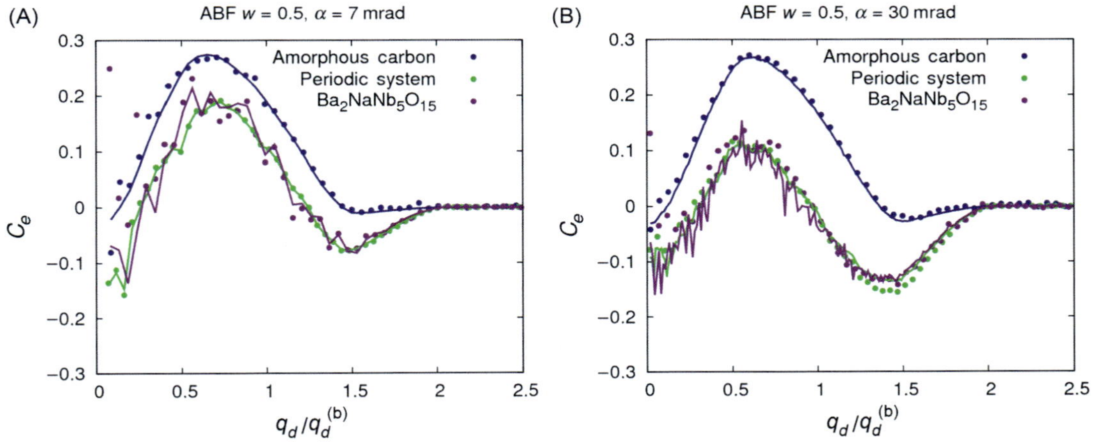

Figure 67.5
Azimuthally averaged contrast-transfer functions, annular bright field. Details as for Fig. 67.4. *After Lazić and Bosch (2017), Courtesy Elsevier.*

This is averaged around the ring or disc (this is permissible provided that only terms of even order such as defocus and spherical aberration appear in the aberration function). Their simulated results are illustrated in Fig. 67.4 (bright-field imaging) and Fig. 67.5 (annular bright-field imaging) with $\lambda = 0.5$. The left-hand column shows curves for an

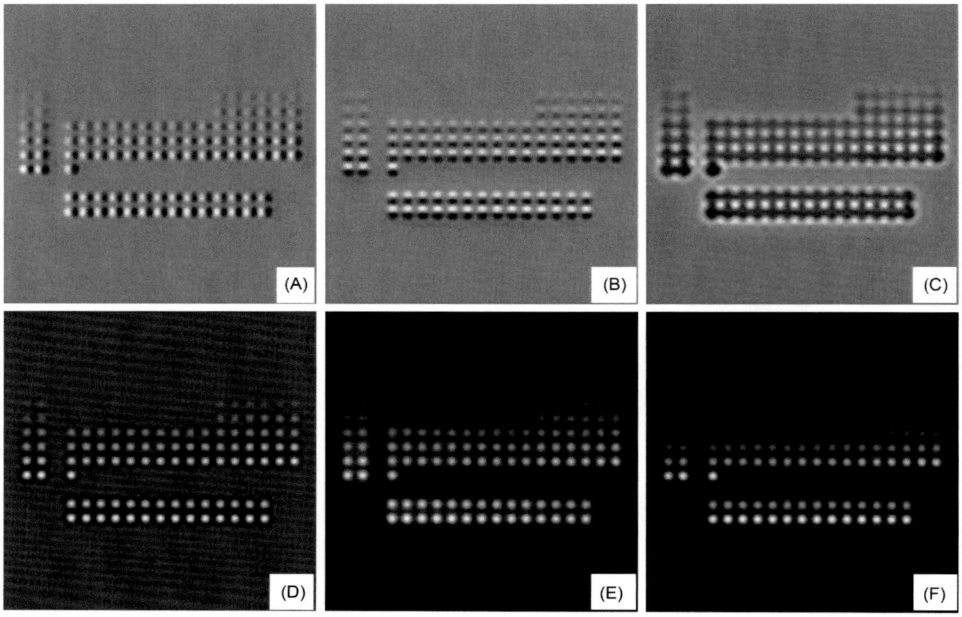

Figure 67.6
Artificial sample in which the atoms are placed as in the periodic table. Several imaging modes are shown. *After Lazić et al. (2016), Courtesy Elsevier.*

uncorrected STEM (semiangle 7 mrad) and the right-hand column for an aberration-corrected instrument (30 mrad). Three model specimens are considered: amorphous carbon, three monolayers thick (0.5 nm); a model structure consisting of single atoms with atomic numbers from 1 to 103, disposed in their positions in the periodic table (Fig. 67.6); and $Ba_2NaNb_5O_{15}$. These curves are discussed in depth by Lazić and Bosch.

If the specimen can be regarded as a weak-phase object, we obtain a simpler result: $(1 - \cos \eta)$ can be neglected and $\sin \eta \approx \eta$, so that only the first term in $C^{(1)}(\boldsymbol{q}_d)$ survives and we obtain a linear relation between object and image as in TEM.

Bright-field modes are particularly valuable for the study of very light elements such as lithium for which high-angle annular dark-field STEM is not suitable. We now summarize the approach of Ooe et al. (2019), who have established the preferred annular detector configuration for light-element imaging. First, however, we introduce the notion of an integrated phase-contrast function, introduced by Seki et al. (2018a), an idea that goes back to the work of Bonhomme and Beorchia (1983). In an aberration-corrected STEM the depth of field is not very different from the specimen thickness and the integrated phase-contrast function takes this into account. For an accelerating voltage of 200 kV and a probe-convergence angle of 22 mrad, for example, the depth of field is 9.2 nm (see Nellist, 2011,

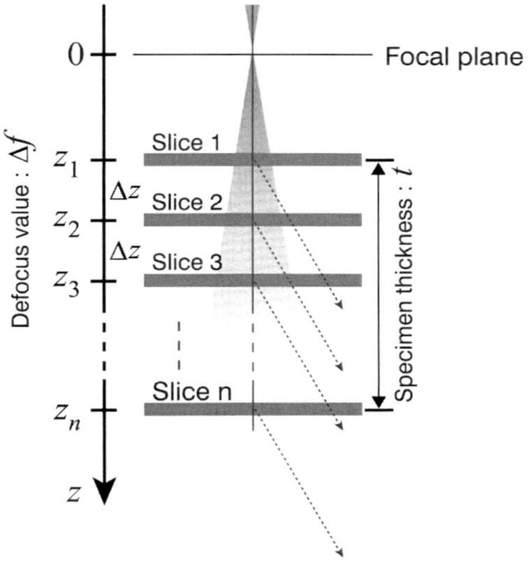

Figure 67.7
The multislice model on which the integrated phase-contrast function is based. *After Seki et al. (2018a,b), Courtesy Elsevier.*

Section 2.9). The calculation uses the multislice approach (Section 69.4) and ignores multiple scattering (Fig. 67.7). Seki et al. show that, with reasonable assumptions, the phase contrast at an annular detector is described by an *integrated phase-contrast function* of the form

$$B(\boldsymbol{q}, \Delta, t) = 2\int_D A(\boldsymbol{k}+\boldsymbol{q})A(\boldsymbol{k})\sin\left\{\chi\left(\boldsymbol{k}+\boldsymbol{q}, \Delta+\frac{1}{2}t\right) - \chi\left(\boldsymbol{k}, \Delta+\frac{1}{2}t\right)\right\}\mathrm{sinc}\left\{\frac{1}{2}\lambda t\left((\boldsymbol{k}+\boldsymbol{q})^2 - k^2\right)\right\}d\boldsymbol{k}$$

where as usual t denotes the specimen thickness. Ooe et al. have calculated this function for several annular detector geometries. Fig. 67.8A shows the three annuli studied and the central disc. Fig. 67.8B–F shows the integrated phase-contrast function for the four regions and the conventional annular bright-field result as a function of normalized spatial frequency for six specimen thicknesses. The dimensionless parameter τ is defined as

$$\tau := t\lambda k_0^2.$$

Values of the spatial frequency for typical operating conditions (120 kV, convergence semiangle 24 mrad) are shown at the top of each set of curves. Ooe et al. conclude that annulus 4 has a better information limit than the annular bright-field detector commonly used (which corresponds to annuli 3 and 4 combined). In a later paper, Ooe et al. (2021a,b) have used the signals furnished by a 16-element segmented detector to create STEM images with enhanced contrast by maximizing the signal-to-noise ratios of the signals with the aid of suitable filters. This technique,

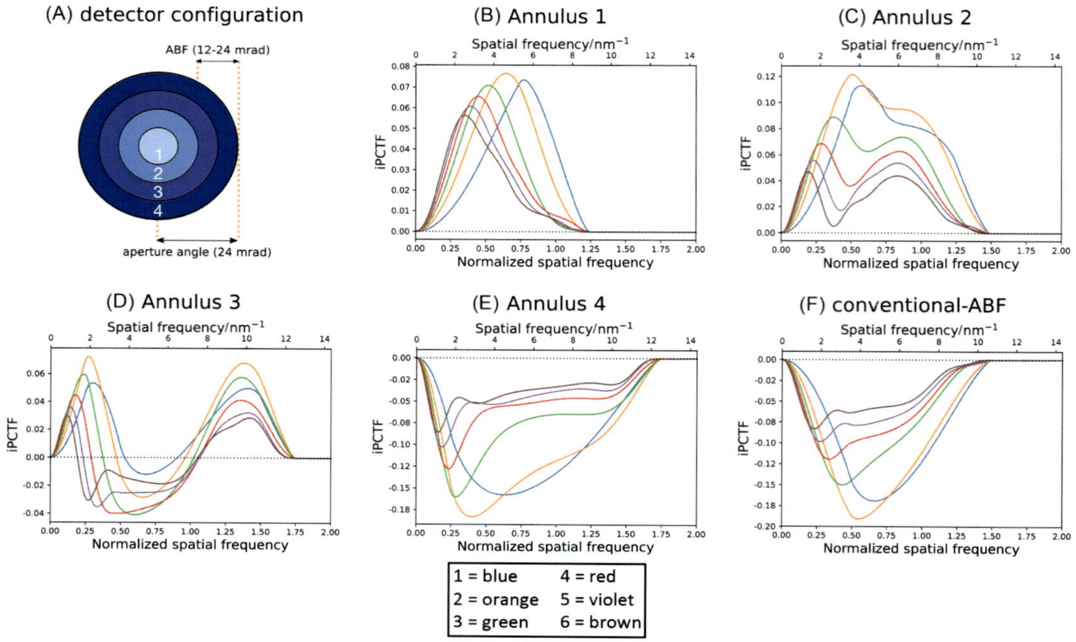

Figure 67.8
The integrated phase-contrast function. (A) The detector configuration. (B–F) The function for each annular region and for annular bright-field imaging. The lower abscissa shows the normalized spatial frequency, the upper abscissa shows the spatial frequency for a practical situation. The specimen thickness is 5 nm, $\tau = 0.91$ (1), 10 nm, $\tau = 1.71$ (2), 15 nm, $\tau = 2.61$ (3), 20 nm, $\tau = 3.41$ (4), 25 nm, $\tau = 4.31$ (5) and 30 nm. $\tau = 5.21$ (6). *After Ooe et al. (2019), Courtesy Elsevier.*

which they call *Optimum Bright-field Imaging*, is particularly attractive for radiation-sensitive specimens where low-dose operation is essential (Nakamura et al., 2021). An approximate form of the spatial frequency filter can be used to permit real-time imaging.

The resolution attainable with differential phase contrast is intimately related to the electron dose. The connection has been explored by Zweck et al. (2016), Schwarzhuber et al. (2017) and Pöllath et al. (2021). In this last paper, an 'uncertainty relation' is established between resolution and electron dose which confirms, as we should expect, that better resolution requires a higher dose. A formula is derived that relates the precision to which the first moment of a diffraction disc can be measured to the convergence angle and the number of electrons detected and hence the dose. This takes the form

$$\Delta p_x \cdot \Delta x \geq \frac{h}{2N_e^{1/2}}$$

or in practice

$$\Delta p_x = \frac{\theta h}{2\lambda N_e^{1/2}}$$

where θ is the semiangle of the illumination and N_e is the number of electrons detected:

For further discussion of these bright-field modes, see Okunishi et al. (2009), Findlay et al. (2009, 2010, 2015), Ishikawa et al. (2011), Lee et al. (2013), Kim et al. (2017) and Dinges et al. (1994).

67.3.2 Dark-Field Imaging

The only term that does not vanish when the signal outside the bright-field zone is collected is $C^{(2)}(\boldsymbol{q}_d)$ and it is immediately clear that a contrast-transfer function relates the dark-field image spectrum and the spectrum of the object characterized by $(1 - \cos\eta) \approx \eta^2/2$. For the approximation $W^{(1)}$ (Eq. 67.26b), we have

$$C^{(2)}(\boldsymbol{q}_d) = 2\left(\int_{|\overline{\boldsymbol{u}}_o|<\rho_1} S(\overline{\boldsymbol{u}}_o)d\overline{\boldsymbol{u}}_o\right)\left\{\left(\mathcal{F}|\psi_Q|^2\right)^* \mathcal{F}(1-\cos\eta)\right\}(\boldsymbol{q}_d) \tag{67.34}$$

and for a detector that collects all the beam outside the circular central (bright-field) zone up to some distance q_{d}, $q_d^{(b)} < |\boldsymbol{q}_d| < q_d^{(d)}$ this gives

$$C^{(2)}(\boldsymbol{q}_d) = 2\left\{J_0\left(2\pi q_d^{(b)}\rho\right) - J_0\left(2\pi q_d^{(d)}\rho\right)\right\}\left(\mathcal{F}|\psi_Q|^2\right)^* \tag{67.35}$$

Simulations performed by Lazić and Bosch (2017) have shown that $W^{(2)}$ (67.29) gives better results, and for this choice, (67.34) becomes

$$\begin{aligned} C^{(2)}(\boldsymbol{q}_d) = \Big[&2\left\{J_0\left(2\pi q_d^{(b)}\rho\right) - J_0\left(2\pi q_d^{(d)}\rho\right)\right\} \\ &+ \frac{\pi}{2}\left\{H_0\left(2\pi q_d^{(b)}\rho_1\right)J_1\left(2\pi q_d^{(b)}\rho_1\right) - H_0\left(2\pi q_d^{(d)}\rho_1\right)J_1\left(2\pi q_d^{(d)}\rho_1\right)\right\} \\ &- \frac{\pi}{2}\left\{H_1\left(2\pi q_d^{(b)}\rho_1\right)J_0\left(2\pi q_d^{(b)}\rho_1\right) - H_1\left(2\pi q_d^{(d)}\rho_1\right)J_0\left(2\pi q_d^{(d)}\rho_1\right)\right\}\Big]\left(\mathcal{F}|\psi_Q|^2\right)^* \end{aligned} \tag{67.36}$$

in which the functions H_0 and H_1 are the Struve functions of order 0 and 1 (Struve, 1882; Watson, 1922, Section 10.4). For careful discussion of the choice of the lengths ρ and ρ_1 (see Table 67.1) and many examples of simulated images, see the full account by Lazić and Bosch (2017).

The phase of the illuminating wave does not appear and it is, therefore, usual to say that dark-field imaging is an *incoherent* mode.

Table 67.1: Values of ρ_1.

Sample	Uncorrected probe, $\alpha = 7$ mrad ρ_1/nm	Corrected probe, $\alpha = 30$ mrad ρ_1/nm
Amorphous carbon	0.145	0.041
Periodic system	0.083	0.035
$Ba_2NaNb_5O_{15}$	0.117	0.037

The behaviour analysed above justifies a much simpler argument in which the specimen transparency is described by two terms, one of which characterizes scattering inside the bright-field cone, the other outside it:

$$O(\boldsymbol{u}_o) =: s_1 + s_2(\boldsymbol{u}_o) \tag{67.37}$$

and from (Eq. 67.6), we have

$$I_d(\boldsymbol{\xi}) = \iiint I^{(0)}(\boldsymbol{\xi})\left(1 + I^{(1)} + I^{(2)}\right) d\boldsymbol{u}_d\, d\boldsymbol{u}_o\, d\bar{\boldsymbol{u}}_o \tag{67.38}$$

in which

$$I^{(0)}(\boldsymbol{\xi}) := s_1 s_1^* \psi_Q(\boldsymbol{u}_o - \boldsymbol{\xi})\psi_Q^*(\bar{\boldsymbol{u}}_o - \boldsymbol{\xi}) D(\boldsymbol{u}_d)\exp\{2\pi i \boldsymbol{u}_d \cdot (\boldsymbol{u}_o - \bar{\boldsymbol{u}}_o)\} \tag{67.39}$$

The integral of this term vanishes since $D\,(\boldsymbol{u}_d) = 0$ in the central area occupied by s_1. The term $I^{(1)}$ has the form

$$I^{(1)} = s_2(\boldsymbol{u}_o) + s_2^*(\bar{\boldsymbol{u}}_o) \tag{67.40}$$

and the contribution to $I_d\,(\boldsymbol{\xi})$ is thus

$$I_d^{(1)}(\boldsymbol{\xi}) = \int |s_1|^2 \left|\psi_Q(\boldsymbol{u}_o - \boldsymbol{\xi})\right|^2 \{s_2(\boldsymbol{u}_o) + s_2^*(\boldsymbol{u}_o)\} d\boldsymbol{u}_o \tag{67.41}$$

The final term is quadratic in s_2:

$$I^{(2)} = s_2(\boldsymbol{u}_o)s_2^*(\boldsymbol{u}_o)$$

and

$$I_d^{(2)}(\boldsymbol{\xi}) = \int |s_1|^2 \left|\psi_Q(\boldsymbol{u}_o - \boldsymbol{\xi})\right|^2 |s_2|^2 d\boldsymbol{u}_o \tag{67.42}$$

The last term is usually dominant. Dark-field STEM image formation in these circumstances thus seems to be *incoherent* as mentioned earlier.

Examination of the dependence of the elastic and inelastic cross-sections on atomic number Z shows that, for noncrystalline material, it is useful to form the ratio of the dark-field and bright-field signals in the STEM. The resulting image exhibits *Z-contrast* and this simple combination of two signals, introduced by Crewe et al. (1970a,b), has been found beneficial for biological specimens especially.

For crystalline material, this procedure is not useful since collective diffraction effects replace the single-atom scattering that makes the ratio technique effective. In order to create Z-dependent contrast with crystalline specimens, the opening in the annular dark-field detector must be made so large that the Bragg-diffracted electrons pass through as well as the central beam, only the electrons scattered through very high angles (Rutherford scattering) being collected. In practice, the unwanted electrons are focused through the central hole by means of a lens, which also helps to bring the electrons scattered through large angles back onto the detector.

For further details of Z-contrast for noncrystalline specimens, see Isaacson et al. (1980) and for the crystalline case, see Treacy et al. (1978), Howie (1979), Pennycook (1981), Treacy (1981) and especially Pennycook (1989, 1992), Jesson and Pennycook (1990), Pennycook and Jesson (1990, 1992), Pennycook et al. (1991) and Pennycook and Boatner (1988). From among the earlier literature of ADF in ageing, we mention Loane et al. (1988, 1992), Jesson and Pennycook (1993), Hartel et al. (1996), Nellist and Pennycook (1998). Many further references are to be found in the surveys by Nellist and Pennycook (2000), Nellist et al. (2009), Watanabe (2015) and Nellist (2019).

67.3.3 Annular Differential Phase Contrast, Matched Illumination and Detector Interferometry (MIDI-STEM)

In all the foregoing imaging modes, the illuminating wave has been assumed to be spherical, perturbed by the aberrations of the probe-forming lens. A variant on this arrangement has been proposed by Ophus et al. (2016), who suggested that the presence of a phase plate (see Section 66.6) in the illumination and a suitable choice of the detector geometry could provide a linear relation between object structure and image. This was named 'matched illumination and detector interferometry' (MIDI-STEM) but Lee et al. (2019a) prefer to describe it as the 'annular differential phase contrast mode' (aDPC), a reminder that it belongs to the family of differential phase contrast modes. The phase plate consists of two zones, equal in area, between which a phase shift of $\pi/2$ is imposed by a suitable modulation of the thickness. We denote these zones $P_1(\boldsymbol{q})$ and $P_2(\boldsymbol{q})$. The wave incident on the specimen is as usual

$$\psi_Q(\boldsymbol{u}_o) = \mathcal{F}\psi(\boldsymbol{q})$$

where now

$$\psi(\boldsymbol{q}) = A(\boldsymbol{q})\{P_1(\boldsymbol{q}) + \mathrm{i}P_2(\boldsymbol{q})\}\exp(-\mathrm{i}\chi) \tag{67.43}$$

In what follows, we disregard the aberrations to simplify the calculation; they are however an essential element in the later work of Lee et al. (2019a,b).

We start from Eq. (68.25) and can show that the contrast-transfer functions $\tilde{M}_+$ and $\tilde{M}_-$ are given by

$$\begin{aligned} \tilde{M}_+ &= 2\{P_1 \odot P_1 - P_2 \odot P_2\} + 2C\{P_1 \odot P_1 + P_2 \odot P_2 + \mathrm{i}(P_1 \odot P_2 - P_2 \odot P_1)\} \\ \tilde{M}_- &= 2\{P_1 \odot P_2 + P_2 \odot P_1\} \end{aligned} \tag{67.44}$$

The constant, C is independent of probe position but does depend on the detector geometry and the specimen. In $\tilde{M}_+$, the first term will be very small and the factor C in the second term also ensures that it is small. In ideal conditions, therefore, this is a very attractive mode. Nevertheless, in its original form, it does require production and insertion of a phase plate and this disadvantage militated in favour of the vector detectors which we examine in the next section.

The technique has been reconsidered by Lee et al. (2019a,b), with the relatively low voltages (30–80 kV) of the SALVE instrument in mind (cf. Section 41.2.1 of Volume 2). Their study differs from that of Ophus et al. in two important respects. First, the roles of geometrical and chromatic aberrations are analysed as well as the effect of Johnson noise (the Uhlemann effect, Section 31.7 of Volume 1). Equally important, Lee et al. point out that it is not necessary to use a physical phase plate, with its attendant inconveniences; instead, the phase change introduced by the spherical aberration and defocus of the probe-forming lens can be chosen to mimic the zones P_1 and P_2. This, combined with a pixelated detector, makes the method much more attractive. Their study of the aberrations reveals that aberration correction is indispensable. In Figs. 67.9 and 67.10, we reproduce their curves illustrating the effect of chromatic aberration (C_c) and the spreading caused by Johnson noise. In Fig. 67.9A, where C_s ($=C_3$) and C_5 are both corrected, the effect of C_c rapidly becomes catastrophic. If C_5 is not corrected (Fig. 67.9B), some contrast survives in the worst case considered. Fig. 67.10 shows comparable curves for Johnson noise of increasing severity. The ordinate shows the mixed coherence function, $E(\boldsymbol{q}, 0)$, defined in general by

$$\begin{aligned} E(\mathbf{q}_1, \mathbf{q}_2) = & \int \frac{1}{\pi^{1/2} \upsilon_m} \exp\left\{ -\left(\frac{\upsilon}{\upsilon_m}\right)^2 \right\} \times \frac{1}{\pi\theta_0^2} \exp\left\{ -\left(\frac{\theta}{\theta_0}\right)^2 \right\} \\ & \times \exp\left[-\mathrm{i}\left\{ \chi\left(\mathbf{q}_1 + \frac{2\pi}{\lambda}\boldsymbol{\theta}, \Delta + \upsilon\right) - \chi(\mathbf{q}_1, \Delta) \right\} \right] \\ & \times \exp\left[\mathrm{i}\left\{ \chi\left(\mathbf{q}_2 + \frac{2\pi}{\lambda}\boldsymbol{\theta}, \Delta + \upsilon\right) - \chi(\mathbf{q}_2, \Delta) \right\} \right] d\upsilon d\boldsymbol{\theta} \end{aligned}$$

from which the transmission cross-coefficient and the current density can be derived straightforwardly. For a general study of noise in STEM, see Seki et al. (2018b).

Another example of structured illumination in STEM is the Fresnel zone plate (Tomita et al., 2020), situated in the condenser aperture (Fig. 67.11A and B). This improves image contrast at low spatial frequencies.

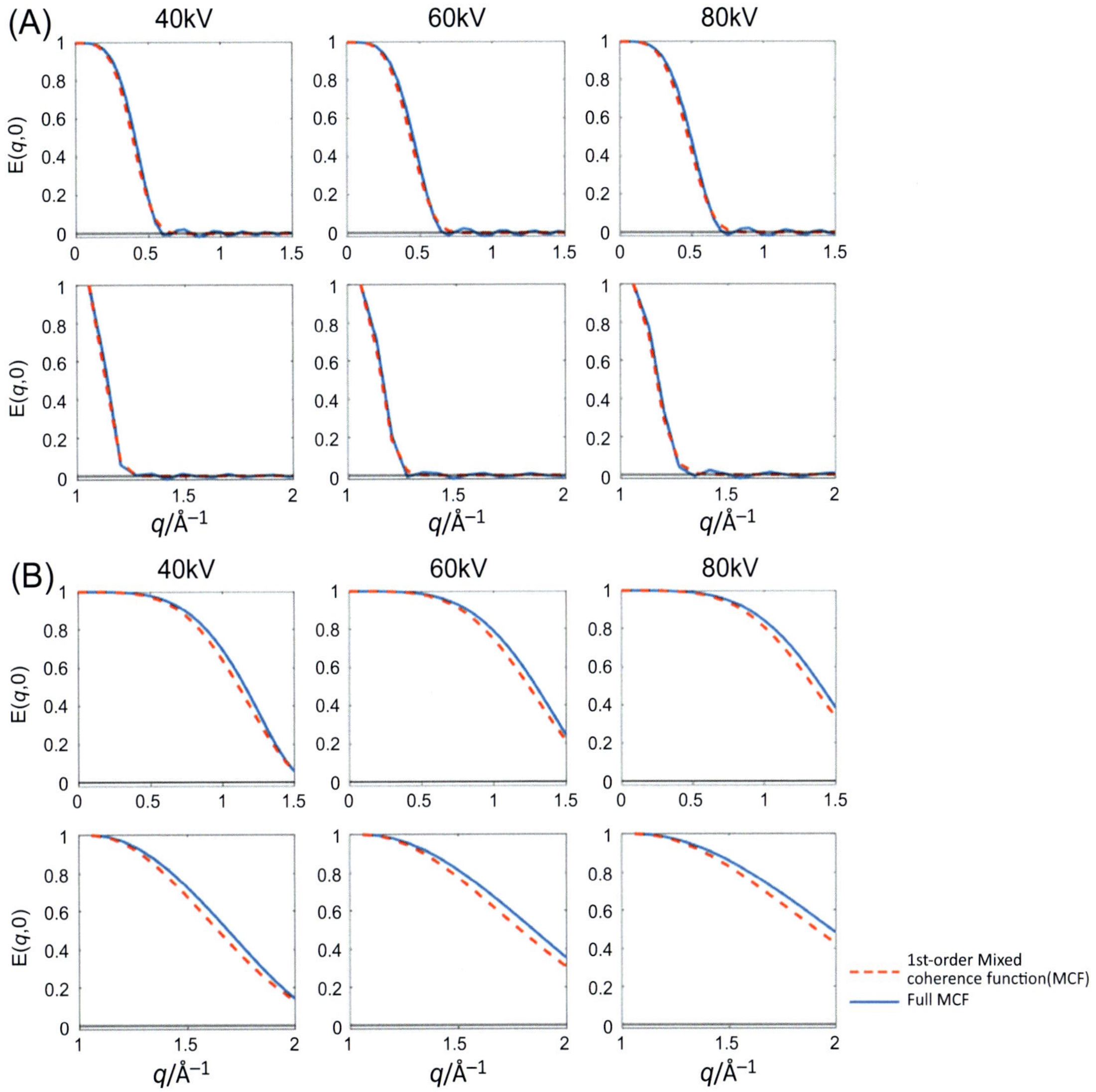

Figure 67.9
Annular differential phase contrast. The first-order and full mixed coherence functions are shown for (A) a microscope corrected for C_s and (B) a microscope corrected for C_c and C_s. *After Lee et al. (2019a,b), Courtesy Elsevier.*

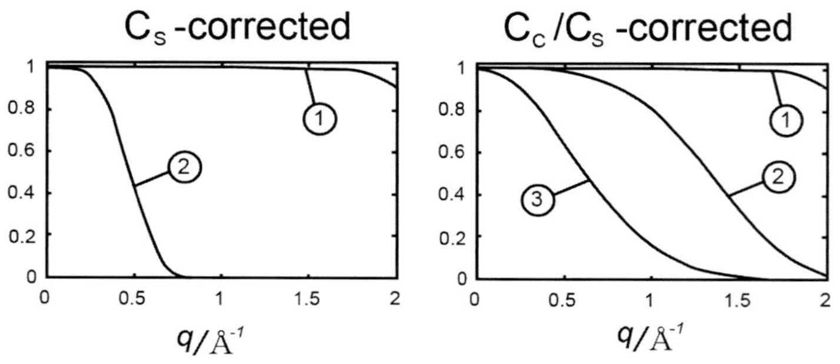

Figure 67.10
Annular differential phase contrast. Envelope functions for (A) a C_s-corrected instrument and (B) an instrument corrected for C_c and C_s at 80 kV. (1) Spatial envelope E_s, (2) Chromatic envelope E_c and (3) image-spread envelope (Uhlemann effect) E_{im}. *After Lee et al. (2019a,b), Courtesy Elsevier.*

67.3.4 Vector Detectors

Eqs. (67.17a) and (67.17b) indicate that, within the bright-field cone, the transfer of contrast to the detector plane for weakly scattering objects is similar to the mechanism of linear image formation in the TEM. This observation is at the heart of suggestions by Rose (1974a,b, 1977) to divide the bright-field detection area into zones and form a weighted sum of the signals, for specimens for which the phase component is dominant.

This was not, however, the first subdivision of the STEM detector to be proposed. In 1974 Dekkers and de Lang showed that by dividing the circular bright-field detector into two semicircular parts, the phase gradient in one direction could be detected directly; see also Dekkers et al. (1976), Dekkers and de Lang (1977, 1978a), de Lang and Dekkers (1979), Dekkers (1979) and Bouwhuis and Dekkers (1980). The natural extension to division into four quadrants to yield both components of the phase gradient was mentioned shortly after (Rose, 1975) and explored in detail by Hawkes (1978b). The theory was reconsidered in a thorough study by Landauer et al. (1995), who pointed out that the resolution of quadrant detectors is double that of a simple bright-field image (see also Landauer and Rodenburg, 1995).

The principle is easily explained with the aid of Eq. (67.14). For a detector consisting of two semicircular regions (Fig. 67.12a), we have

$$D^{(r)}(\boldsymbol{q}_d) = \begin{cases} 1 & q_{xd} > 0 \\ 0 & q_{xd} < 0 \end{cases} \quad \forall q_{yd} \quad \text{right detector}$$

$$D^{(l)}(\boldsymbol{q}_d) = \begin{cases} 1 & q_{xd} < 0 \\ 0 & q_{xd} > 0 \end{cases} \quad \forall q_{yd} \quad \text{left detector} \qquad (67.45)$$

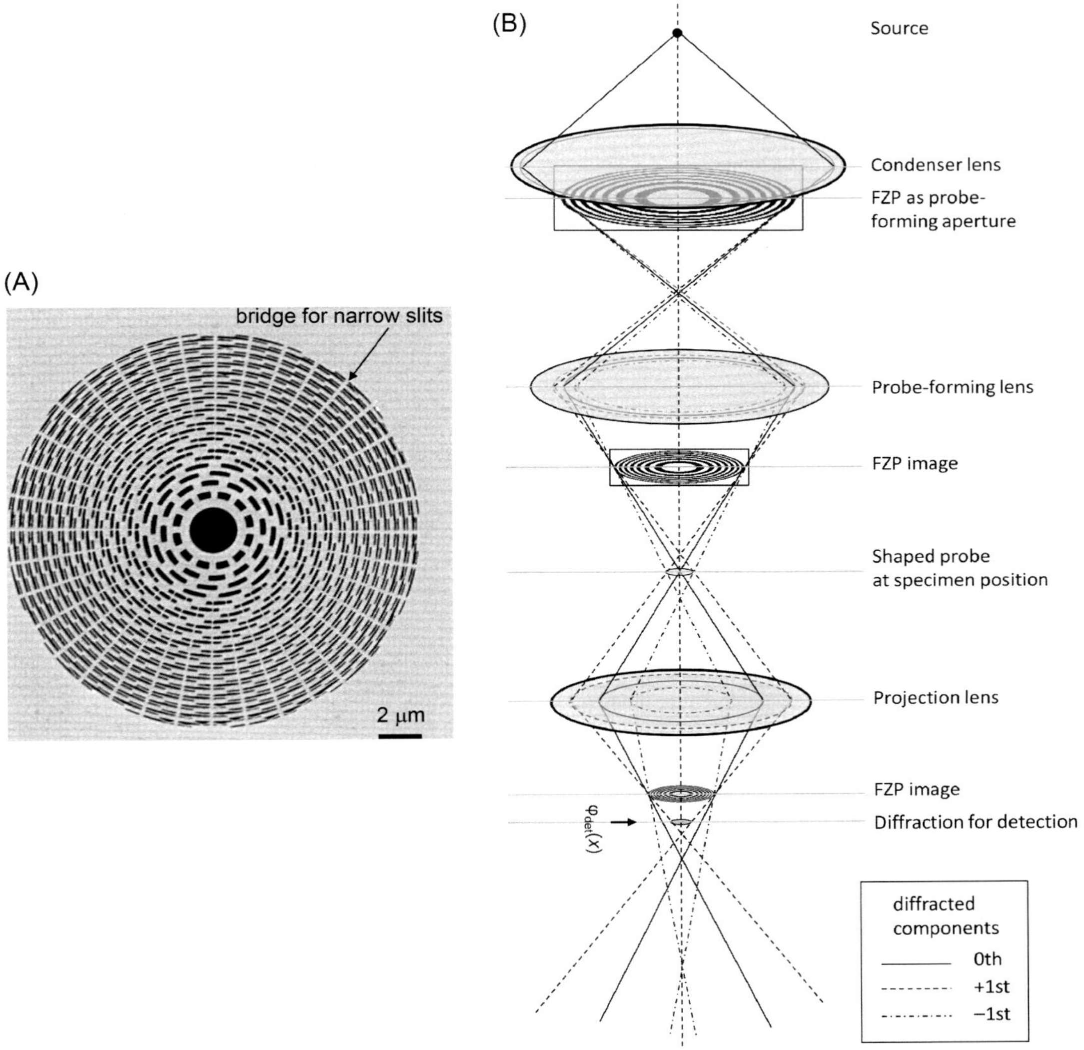

Figure 67.11
The Fresnel zone plate (FZP). (A) Scanning electron microscope image of the plate. (B) The optical system. *After Tomita et al. (2020), Courtesy Elsevier.*

with $\boldsymbol{q_d} =: (q_{xd}, q_{yd})$. Hence

$$\begin{aligned} S_D^{(r)}(\boldsymbol{u}_d) &= \frac{1}{2}\left\{\delta(x_d) - \frac{\mathrm{i}}{\pi x_d}\right\}\delta(y_d) \\ S_D^{(l)}(\boldsymbol{u}_d) &= \frac{1}{2}\left\{\delta(x_d) + \frac{\mathrm{i}}{\pi x_d}\right\}\delta(y_d) \end{aligned} \qquad (67.46)$$

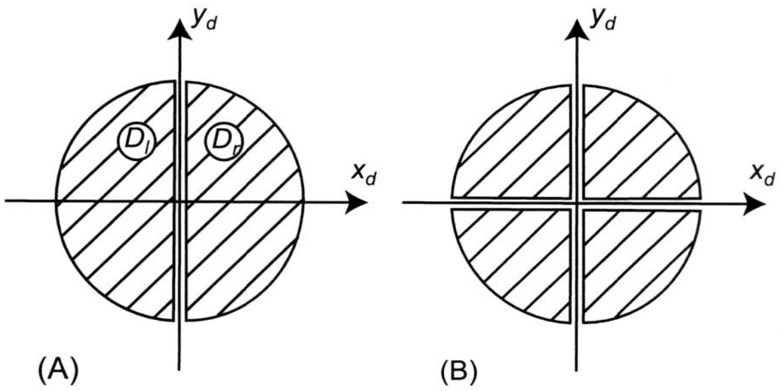

Figure 67.12
(A) The semicircular detectors of Dekkers and de Lang and (B) quadrant detectors.

A straightforward calculation set out in detail in Hawkes (1978b) shows that the sum of the two signals yields information about $s(\boldsymbol{u}_o)$ to a good approximation while the difference provides the derivative of the phase function, $\partial\varphi/\partial x_o$. Extension of the argument to a *quadrant* detector (Fig. 67.12b) shows that, as we should expect, both $\partial\varphi/\partial x$ and $\partial\varphi/\partial y$ can be extracted. Moreover, the detailed analysis of Waddell and Chapman (1979) reveals that the restriction to weak-phase variations of the specimen is not as severe as in the conventional imaging modes. Split detectors are hence very suitable for studying magnetic specimens and have been used extensively for this purpose (Chapman et al., 1978a,b, 1983, 1990; Morrison et al., 1979; Chapman and Morrison, 1983; Morrison and Chapman, 1981, 1982, 1983; Chapman, 1989; Sannomiya et al., 2004; Uhlig and Zweck, 2004; Uhlig et al., 2005; Brownlie et al., 2006; Sandweg et al., 2008; McVitie et al., 2015, 2018;Krajnak et al., 2016; Almeida et al., 2017; Nishkawa et al., 2021). Their sensitivity is examined by Schwarzhuber et al. (2017).

The split detector of Dekkers and de Lang was the first of several proposals designed to exploit the freedom of choice of detector geometry. Waddell et al. (1977, 1978) described a 'first-moment' detector, further analysed by Waddell and Chapman (1979), which involves weighting the current $J_d\ (\boldsymbol{q}_d;\ \boldsymbol{\xi})$ before adding the contributions from the points in the detector plane: $D\ (\boldsymbol{q})$ now has a more complicated variation than the simple binary response we have so far considered, $D\ (\boldsymbol{q}_d) = |\boldsymbol{q}_d|$. Schwarzhuber et al. (2018) describe a fast nonpixelated first-moment detector consisting of a duo-lateral position-sensitive diode. We return to this below.

The form of the function $K_p\ (\boldsymbol{q})$ led Rose (1974a) to investigate the signals that could be collected by subdividing the bright-field cone into complementary rings (Fig. 67.13A and B). We say no more about this here as the idea has been superseded by the more recent possibility of dividing the cone into a large number of fine rings, with which a very desirable form of the contrast-transfer function can be attained (Hammel et al., 1990).

It was soon realized, thanks largely to the work of Rose (1974a,b, 1975, 1977) and Rose and Fertig (1976), that flexible detectors, permitting a variety of forms of $D\ (\boldsymbol{q}_d)$ to be generated, were desirable and several were discussed (Cowley, 1976b; cf. Dekkers and de Lang, 1978b;

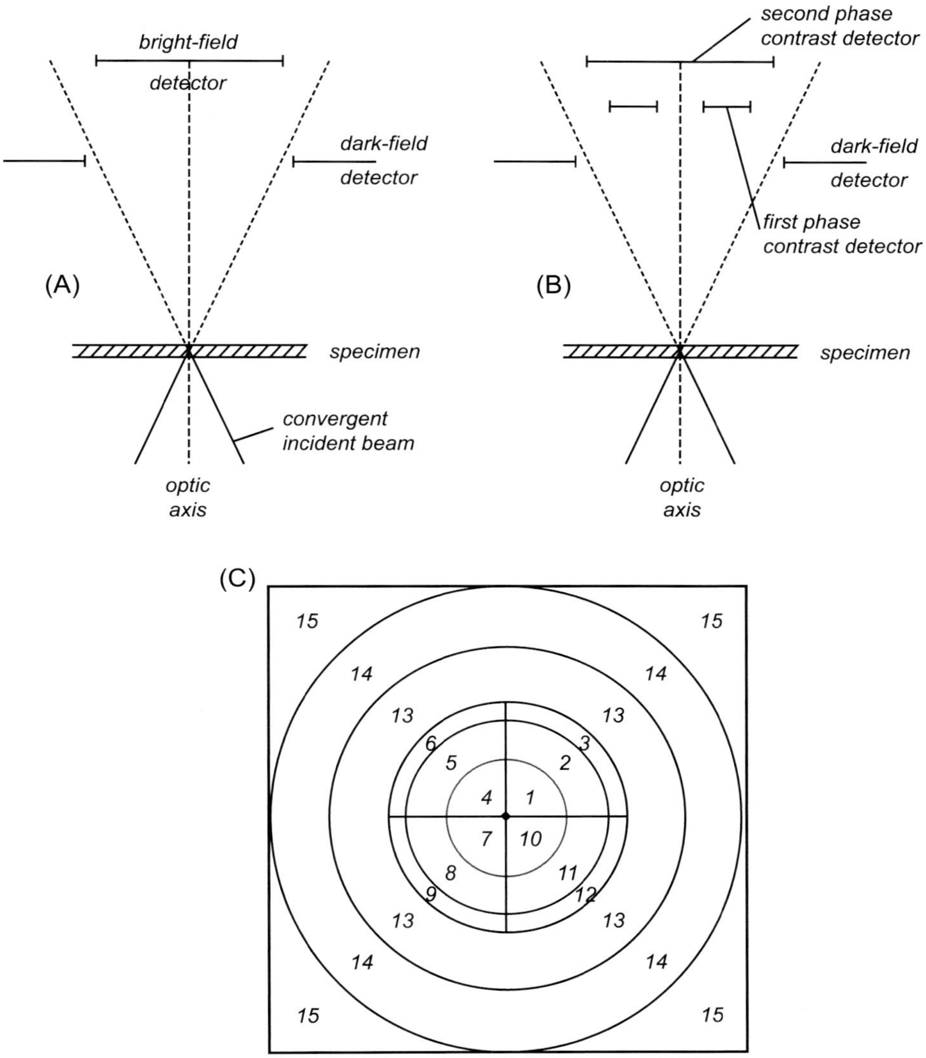

Figure 67.13

(A) Standard arrangement of detectors in STEM. (B) Subdivision of the bright-field cone to improve phase contrast. (C) Division of the detector plane with which two-ring detectors, split and quadrant detectors and other geometries can be synthesized.

Cowley and Au, 1978a,b; Lackovic et al., 1979; Burge and van Toorn, 1979, 1980; Burge et al., 1979a,b, 1982; Robinson, 1979; van Toorn and Robinson, 1980; Ward et al., 1979; Browne and Ward, 1982; Hawkes, 1980c; Cowley, 1993). Fig. 67.13C shows the subdivisions recommended by Burge and van Toorn (1980). A version with four quadrants and 32 rings was designed by Haider et al. (1988). Meanwhile, Smith and Erasmus (1982) pointed out that any desired configuration could be generated by reading the image from each pixel into computer memory and multiplying it by the appropriate pattern of zeros and ones (or weights), adding the result and storing these sums. This inevitably slowed down the acquisition, however. The idea was revived by Daberkow and Herrmann (1984, 1988) and Daberkow et al. (1993) in a more modest form, in which only a few patterns are allowed. Hammel et al. (1990) discussed the use of detectors with many rings inside the bright-field cone to combat the adverse effect of the contrast-transfer function in the high-resolution study of weakly scattering specimens; see also Wang et al. (1992). These early attempts prefigured the pixelated detectors available today.

A general study of detector configurations has been made by McCallum et al. (1995), who calculated the signals that would be collected by detector elements of arbitrary form. From this, it emerged that a detector configuration that had not been considered, a three-sector detector, is the simplest arrangement from which the complex specimen function could be extracted. Fig. 67.14 shows an example of signals from such a detector.

We now present the general theory of vector detectors (Lazić et al., 2016; Lazić and Bosch, 2017), from which the first-moment detector will emerge as a special case. We limit the discussion to antisymmetric detectors (defined below), though lower symmetries could also be envisaged. The detector function $D(\boldsymbol{q}_d)$ now has two components, D_x and D_y in $\boldsymbol{q}_d$-space and we treat it as a vector $\boldsymbol{D}(\boldsymbol{q}_d)$. Antisymmetry implies that

$$D_x(\boldsymbol{q}_d) = -D_x(-\boldsymbol{q}_d), \quad D_y(\boldsymbol{q}_d) = -D_y(-\boldsymbol{q}_d) \tag{67.47}$$

The contrast $\boldsymbol{C}(\boldsymbol{q}_d)$ is also treated as a vector, as are $\boldsymbol{M}_\pm$ (Eq. 67.28). The expression for the image spectrum (neglecting the constant term), (Eq. 67.30), now becomes

$$\boldsymbol{C}(\boldsymbol{q}_d) = \mathcal{F}(\boldsymbol{I}_d(\boldsymbol{\xi})) = -\mathrm{i}(\tilde{\boldsymbol{M}}_-\mathcal{F}\sin\eta)(\boldsymbol{q}_d) - (\tilde{\boldsymbol{M}}_+\mathcal{F}(1-\cos\eta))(\boldsymbol{q}_d) + \boldsymbol{Q} \tag{67.48}$$

in which $\boldsymbol{I}_d, \tilde{\boldsymbol{M}}_\pm$ and $\boldsymbol{Q}$ are now all vectors. We recall that

$$\boldsymbol{Q} = \mathcal{F}\left\{\iint R\mathcal{F}(\boldsymbol{D})d\boldsymbol{u}_o\right\}, \quad R(\boldsymbol{\xi}, \boldsymbol{u}_o) := \zeta(\overline{\boldsymbol{u}}_o + \boldsymbol{\xi})\psi_Q \odot \zeta(\overline{\boldsymbol{u}}_o + \boldsymbol{\xi})\psi_Q$$

We thus obtain two images of the scalar quantities $\mathcal{F}(\sin\eta)$ and $\mathcal{F}(1-\cos\eta)$, whereas we should like to obtain a single image of each of these or some simple function of them.

We can achieve this in two ways. First, we rewrite (Eq. 67.48) in the form

$$\begin{aligned}\boldsymbol{C}(\boldsymbol{q}_d) &= -\mathrm{i}\tilde{\boldsymbol{M}}_- 2\pi\mathrm{i}\boldsymbol{q}_d \cdot \frac{\boldsymbol{q}_d}{2\pi\mathrm{i}q_d^2}\mathcal{F}\sin\eta - \tilde{\boldsymbol{M}}_+ 2\pi\mathrm{i}\boldsymbol{q}_d \cdot \frac{\boldsymbol{q}_d}{2\pi\mathrm{i}q_d^2}\mathcal{F}(1-\cos\eta) + \boldsymbol{Q} \\ &= -\mathrm{i}\tilde{\boldsymbol{M}}_-^{(i)}\mathcal{F}(\mathrm{grad}\,\sin\eta) - \tilde{\boldsymbol{M}}_+^{(i)}\mathcal{F}(\mathrm{grad}(1-\cos\eta)) + \boldsymbol{Q}\end{aligned} \tag{67.49}$$

in which we have used the identity

$$\mathcal{F}(\mathrm{grad}\, f(\boldsymbol{u})) = 2\pi\mathrm{i}\boldsymbol{q}_d\mathcal{F}(f(\boldsymbol{u})) \tag{67.50}$$

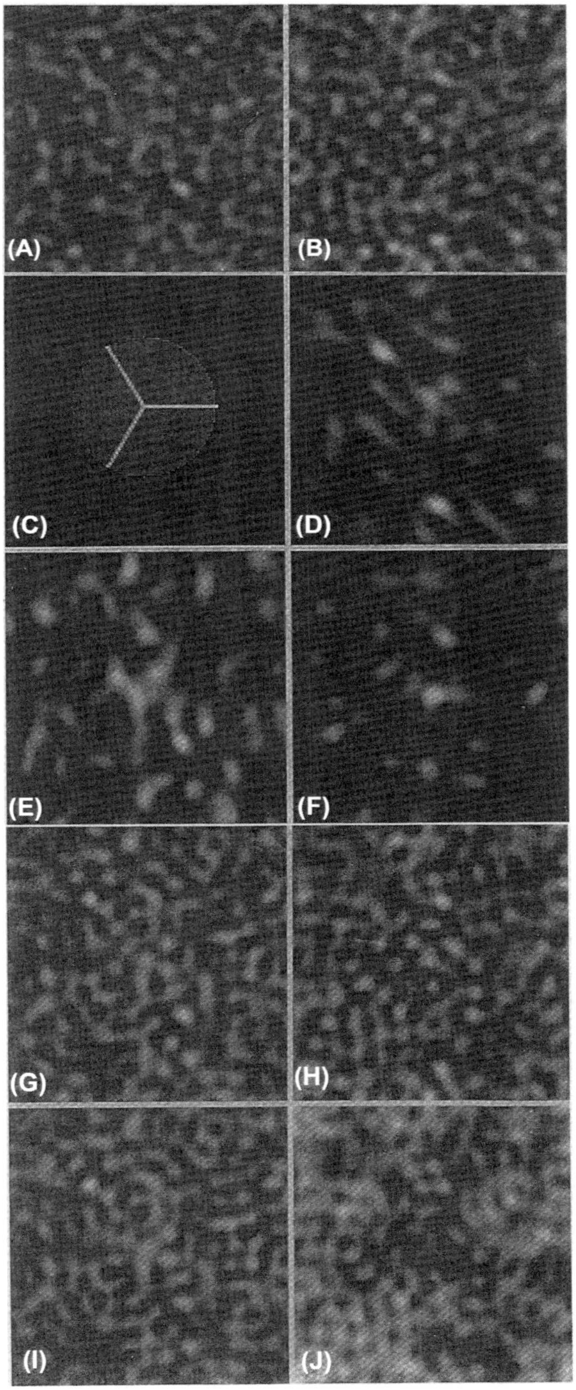

Figure 67.14

Simulation of reconstruction of a complex-valued specimen based on a three-sector detector. (A and B) Real and imaginary parts of the specimen function. (C) The detector. (D–F) Signals recorded by the upper, left and right sectors. (G and H) Specimen functions reconstructed from a noisy image, defocus = −100 nm. (I and J) Defocus = −500 nm. *After McCallum et al. (1995), Courtesy Elsevier.*

and

$$\tilde{M}_-^{(i)} = \frac{\boldsymbol{q}_d \cdot \tilde{\boldsymbol{M}}_-}{2\pi \mathrm{i} q_d^2}, \quad \tilde{M}_+^{(i)} = \frac{\boldsymbol{q}_d \cdot \tilde{\boldsymbol{M}}_+}{2\pi \mathrm{i} q_d^2} \tag{67.51}$$

Can we find a scalar image $I^{(i)}$ such that $I(\boldsymbol{\xi}) = \mathrm{grad}\, I^{(i)}(\boldsymbol{\xi})$? From (67.50), we see that

$$\begin{aligned} \mathcal{F}\left(\boldsymbol{I}^{(i)}\right) &= \frac{\boldsymbol{q}_d \cdot \mathcal{F}(\boldsymbol{I}_d)}{2\pi \mathrm{i} q_d^2} \\ &= -\mathrm{i}\left(\tilde{M}_- \mathcal{F} \sin\eta\right)(\boldsymbol{q}_d) - \left(\tilde{M}_+ \mathcal{F}(1 - \cos\eta)\right)(\boldsymbol{q}_d) + Q \end{aligned} \tag{67.52}$$

and

$$Q = \frac{\boldsymbol{\xi} \cdot \boldsymbol{Q}}{2\pi \mathrm{i} q_d^2} \tag{67.53}$$

The image $I^{(i)}$ exhibits *integrated differential phase contrast.*

At this point, we use these general formulae to examine the first-moment detector, $D(\boldsymbol{q}_d) = |\boldsymbol{q}_d|$. For this, we obtain the relation

$$\mathcal{F}\left(I^{(m)}\right) = \frac{1}{2\pi}\left\{\mathcal{F}|\psi_Q|^2\right\}^* \mathcal{F}(\eta) \tag{67.54}$$

showing that the first-moment detector (labelled (m)) is the ideal STEM detector. Its performance can be approached closely by quadrant detectors, as Lazić and Boch show in detail. We shall not reproduce their reasoning here since the growing availability of pixelated detectors means that the condition $D(\boldsymbol{q}_d) = |\boldsymbol{q}_d|$ can be satisfied closely. For this particular detector the quadratic term $\boldsymbol{Q}$ can be evaluated explicitly and it is found that

$$\begin{aligned} \boldsymbol{Q}^{(m)} = &\frac{1}{2\pi}\left\{\mathcal{F}|\psi_Q|^2\right\}^* \mathcal{F}(\eta - \sin\eta) \\ &- \frac{\boldsymbol{q}_d}{4\pi^2 q_d^2} \cdot \left[\mathcal{F}\left\{\psi_Q(\mathrm{grad}\psi_Q)^*\right\} - \psi_Q^*(\mathrm{grad}\psi_Q)\right]^* \mathcal{F}(1 - \cos\eta) \end{aligned} \tag{67.55}$$

A useful approximate expression for $\mathcal{F}(\boldsymbol{I})$ can be obtained by assuming that this expression is valid for all detector shapes, not just the first-moment detector. We then have

$$\begin{aligned} \mathcal{F}(\boldsymbol{I}) = &\ \tilde{M}_- \mathcal{F}(\eta) + N_1 \mathcal{F}(1 - \cos\eta) + N_2(\eta - \sin\eta) \\ N_1 := &\ \tilde{\boldsymbol{M}}_+ - \frac{\boldsymbol{q}_d}{4\pi^2 q_d^2} \cdot \left[F\left\{\psi_Q(\mathrm{grad}\psi_Q)^* - \psi_Q^*(\mathrm{grad}\psi_Q)\right\}\right]^* \\ N_2 := &\ \frac{1}{2\pi}\left(\mathcal{F}|\psi_Q|^2\right)^* - \tilde{M}_- \end{aligned} \tag{67.56}$$

(The terms in N_1 and N_2 vanish for the first-moment detector of course.) Simulation shows that for a four-segment detector, the terms in N_1 and N_2 are small for small values of defocus

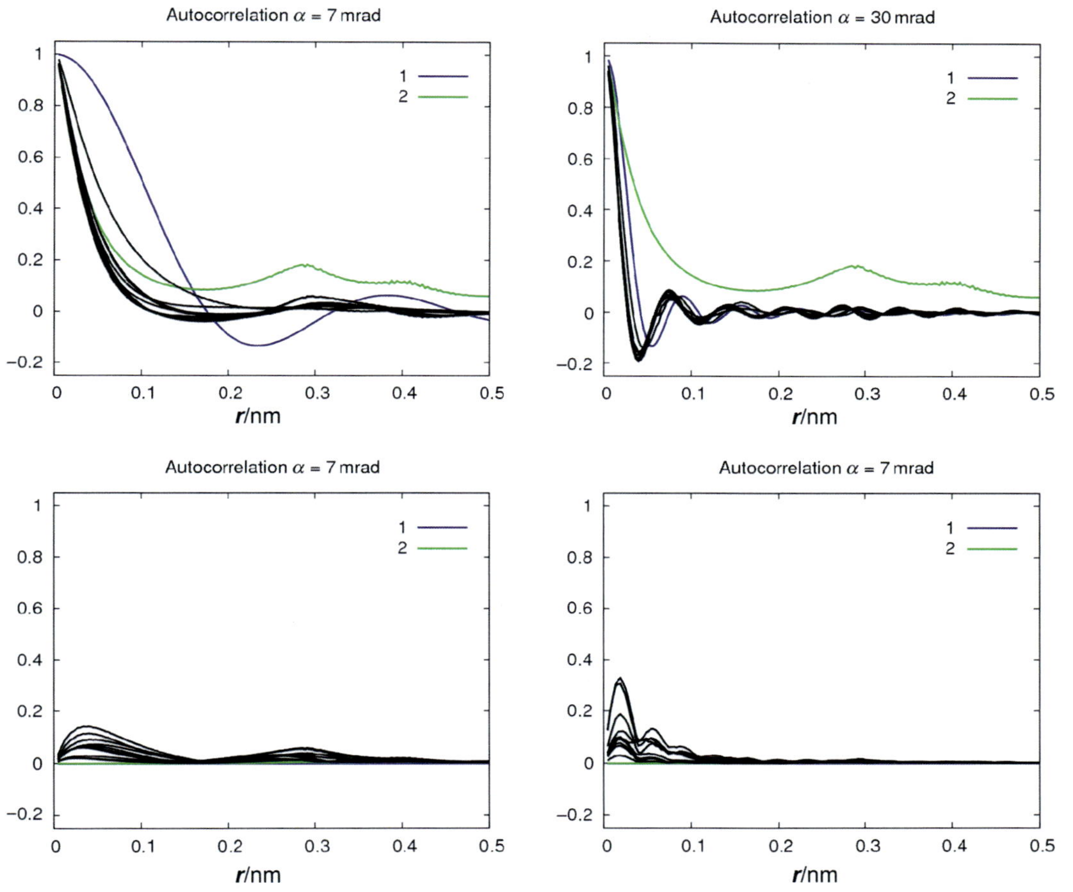

Figure 67.15
Azimuthally averaged real part (*top row*) and absolute value of the imaginary part (*bottom row*) of the normalized autocorrelation functions for different probe positions (black lines). The same functions of the probe (blue) and sample (see Fig. 67.6, green) are also shown. 1: probe, 2: sample. *After Lazić and Bosch (2017), Courtesy Elsevier.*

[measured from the disc of least confusion (Eq. 24.50 of Volume 1)]; the integrated differential phase contrast detector (Eq. 67.52) is hence an excellent approximation to the ideal first-moment configuration. Fig. 67.15 shows values of $\tilde{M}_-, N_1$ and N_2 for an uncorrected STEM (left) and an aberration-corrected instrument (right) for several values of defocus.

A second imaging mode, differentiated differential phase contrast, is obtained by considering the divergence[2] of $\boldsymbol{I}_d$,

$$I^{(\Delta)} := \text{div}\boldsymbol{I}_d \tag{67.57}$$

[2] We use the label Δ for "differentiated" to avoid confusion with d (detector).

for which

$$\mathcal{F}\left(I^{(\Delta)}\right) = 2\pi \mathrm{i}\boldsymbol{q}_d \cdot \mathcal{F}(\boldsymbol{I}_d) \tag{67.58}$$

We therefore introduce the factor $2\pi \mathrm{i}\boldsymbol{q}_d$ in (Eq. 67.56) a second time and again move this factor under the Fourier transform operator, which gives

$$\mathcal{F}\left(I^{(\Delta)}\right) = \mathrm{i}\tilde{M}_{+}\mathcal{F}\left(\nabla^2\eta\right) + N_1\mathcal{F}\left(\nabla^2(1-\cos\eta)\right) + N_2\mathcal{F}\left(\nabla^2(\eta - \sin\eta)\right) \tag{67.59}$$

As in the case of integrated differential phase contrast, the first term is dominant close to the in-focus condition and the image then gives a direct picture of $\nabla^2\eta$.

In a thorough re-examination of the integrated differential phase contrast, mode, Rose (2021) shows that, when the probe-forming lens of a STEM is corrected for both spherical and chromatic aberration and a pixelated detector is available, numerical analysis of through-focus images makes it possible to perform optical sectioning of thick objects with three-dimensional atomic resolution at about the same electron dose as would be required for a two-dimensional object, in principle at least. Li et al. (2021) have shown that the integrated differential phase contrast method can be applied to thick samples by increasing the number of detector segments. For this, it will be necessary to record Ronchigrams for each pixel and apply the detector pattern to them afterwards. Thick specimens are also studied by Mostaed et al. (2021) who combine information obtained by ptychography and annular dark-field imaging.

Hosseinnejad et al. (2021) have extended an atomic-resolution technique introduced by Goris et al. (2015) for high-angle annular dark-field STEM to integrated differential phase contrast, STEM. Here, atoms are detected iteratively and replaced by a model of the atomic potential. This can be used for both crystalline and amorphous materials.

Lopatin et al. (2016) have suggested a way of collecting the signals that a quadrant detector would record without modifying the microscope column. For magnetic specimens, for which the objective lens is usually switched off, the beam in the diffraction plane can be deflected to four lateral positions in turn (Fig. 67.16) and the resulting signals manipulated in the same way as those collected by quadrants. They refer to this as a *unitary detector.*

Information that is not routinely exploited in differential phase contrast work can be extracted from the higher moments of the signal (Löffler et al., 2021).

An eight-channel detector (annular bright-field, four quadrants, two outer rings) connected to nonmultiplexed analogue-to-digital converters, combined with unconventional scan patterns has been thoroughly studied by Seifer et al. (2021). They show that the signals can be corrected and improved in various ways, to cancel parallax for example, thereby increasing the reliability of the subsequent processing.

The importance of the degree of (spatial and temporal) coherence in image interpretation has been emphasized by Oxley and Dyck (2020).

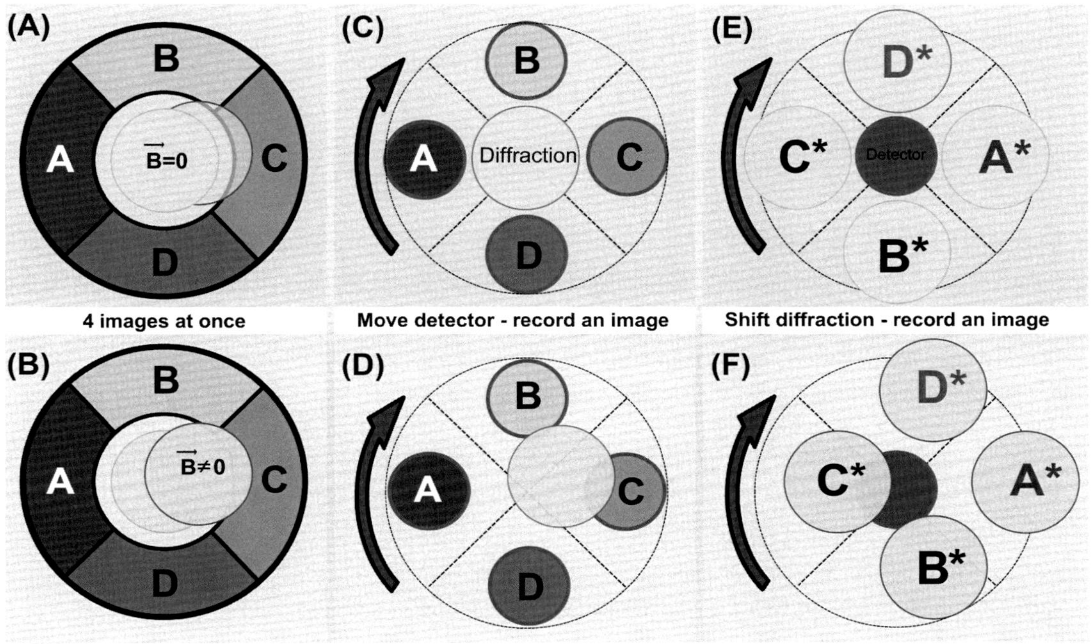

Figure 67.16
The unitary detector scheme. (A) Simple four-quadrant detector. (B) Position of the diffraction disc when a transverse magnetic field is present in the object plane. (C and D) One detector replaces each of the four quadrants in turn. (E and F) The diffraction disc is shifted around the single (unitary) detector. *After Lopatin et al. (2016), Courtesy Elsevier.*

67.3.5 Pixelated Detectors

When we are prepared to collect *all* the information for every pixel, however, a different form of the foregoing analysis is more appropriate. This involves recording not $I_d(\boldsymbol{\xi})$ but $J_d(\boldsymbol{q}_d;\boldsymbol{\xi})$ for each point $\boldsymbol{\xi}$, giving not a two-dimensional image but a four-dimensional data set. We again set out from Eq. (67.5) and consider the transform of $J_d\,(\boldsymbol{q}_d;\,\boldsymbol{\xi})$ with respect to $\boldsymbol{q}_d$ as well as to $\boldsymbol{\xi}$. We write

$$\tilde{J}_d(\boldsymbol{u}_d;\boldsymbol{x}) := \iint J_d(\boldsymbol{q}_d;\boldsymbol{\xi})\exp\{-2\pi\mathrm{i}(\boldsymbol{q}_d\cdot\boldsymbol{u}_d + \boldsymbol{\xi}\cdot\boldsymbol{x})\}d\boldsymbol{q}_d\,d\boldsymbol{\xi} \tag{67.60}$$

giving

$$\begin{aligned}\tilde{J}_d(\boldsymbol{u}_d;\boldsymbol{x}) &= \iiiint O(\boldsymbol{u}_o)O^*(\overline{\boldsymbol{u}}_o)\psi_Q(\boldsymbol{u}_o-\boldsymbol{\xi})\psi_Q^*(\overline{\boldsymbol{u}}_o-\boldsymbol{\xi})\\ &\quad\times\exp[-2\pi\mathrm{i}\{\boldsymbol{q}_d\cdot(\overline{\boldsymbol{u}}_o-\boldsymbol{u}_o+\boldsymbol{u}_d)+\boldsymbol{\xi}\cdot\boldsymbol{x}\}]d\boldsymbol{u}_o\,d\overline{\boldsymbol{u}}_o\,d\boldsymbol{q}_d\,d\boldsymbol{\xi}\\ &= \iint O(\overline{\boldsymbol{u}}_o+\boldsymbol{u}_d)O^*(\overline{\boldsymbol{u}}_o)\psi_Q(\overline{\boldsymbol{u}}_o+\boldsymbol{u}_d-\boldsymbol{\xi})\psi_Q^*(\overline{\boldsymbol{u}}_o-\boldsymbol{\xi})\\ &\quad\times\exp[-2\pi\mathrm{i}\boldsymbol{\xi}\cdot\boldsymbol{x}]d\overline{\boldsymbol{u}}_o d\boldsymbol{\xi}\end{aligned} \tag{67.61}$$

This separates very conveniently into terms in O and terms in ψ_Q if we introduce a new variable υ,

$$\upsilon := \overline{\boldsymbol{u}}_o - \boldsymbol{\xi} \tag{67.62}$$

whereupon $\tilde{J}_d$ becomes

$$\begin{aligned}\tilde{J}_d(\boldsymbol{u}_d;\boldsymbol{x}) = &\iint O(\overline{\boldsymbol{u}}_o + \boldsymbol{u}_d)O^*(\overline{\boldsymbol{u}}_o)\psi_Q\upsilon(\upsilon + \boldsymbol{u}_d)\psi^*_Q(\upsilon)\\ &\times \exp\{-2\pi i(\overline{\boldsymbol{u}}_o - \upsilon)\cdot\boldsymbol{x}\}d\overline{\boldsymbol{u}}_o\, d\upsilon\\ =:\ & A_o(\boldsymbol{u}_d,\boldsymbol{x})A_\psi(\boldsymbol{u}_d, -\boldsymbol{x})\end{aligned} \tag{67.63}$$

where

$$A_o(\boldsymbol{u}_d,\boldsymbol{x}) = \int O(\overline{\boldsymbol{u}}_o + \boldsymbol{u}_d)O^*(\overline{\boldsymbol{u}}_o)\exp(-2\pi i\overline{\boldsymbol{u}}_o\cdot\boldsymbol{x})d\overline{\boldsymbol{u}}_o \tag{67.64}$$

$$A_\psi(\boldsymbol{u}_d, -\boldsymbol{x}) = \int \psi_Q(\upsilon + \boldsymbol{u}_d)\psi^*_Q(\upsilon)\exp(2\pi i\upsilon\cdot\boldsymbol{x})d\upsilon \tag{67.65}$$

A_o and A_ψ are identical in form, apart from the change from $\overline{\boldsymbol{u}}_o$ to ψ, since υ and $\overline{\boldsymbol{u}}_o$ are essentially dummy variables:

$$A_f(\boldsymbol{u}_d,\boldsymbol{x}) = \int f(\boldsymbol{u}_d + \upsilon)f^*(\upsilon)\exp(-2\pi i\upsilon\cdot\boldsymbol{x})d\upsilon \tag{67.66}$$

The numerous methods described in Chapter 73 will yield estimates of A_o $(\boldsymbol{u}_d, \boldsymbol{x})$ and A_ψ $(\boldsymbol{u}_d, \boldsymbol{x})$ from the transformed measured distribution $\tilde{J}_d(\boldsymbol{u}_d,\boldsymbol{x})$. Fourier transformation of A_o with respect to $\boldsymbol{x}$ yields a_o,

$$a_o(\overline{\boldsymbol{u}}_o,\boldsymbol{u}_d) := O(\overline{\boldsymbol{u}}_o + \boldsymbol{u}_d)O^*(\overline{\boldsymbol{u}}_o) \tag{67.67}$$

We set the origin of $\overline{\boldsymbol{u}}_o$ at the point where $a_o(\overline{\boldsymbol{u}}_o, 0)$ is greatest and hence obtain

$$a_o(0,\boldsymbol{u}_d) = O(\boldsymbol{u}_d)O^*(0) \tag{67.68}$$

We are at liberty to choose the phase origin at a_o (0, 0) so that finally

$$O(\boldsymbol{u}_d) = \frac{a_o(0,\boldsymbol{u}_d)}{O(0)} \tag{67.69}$$

The method is thus capable of yielding the complex object transparency directly, without iteration, but at the cost of recording a huge data set; the phase distribution of the illumination can also be extracted straightforwardly from A_ψ. The electron dose that the specimen must support is clearly high if the individual diffraction patterns J_d $(\boldsymbol{q}_d;\ \boldsymbol{\xi})$ are not to be excessively noisy. This observation is at the origin of the technique known as ptychography, described in Section 67.4. The data set is still more voluminous when a focal series of 4D signals is recorded. This situation is examined by Robert et al. (2021).

67.3.6 Depth-Sectioning in STEM

We include a few paragraphs on this topic here as the theory of three-dimensional imaging of relatively thick specimens in STEM has been re-examined by Bosch and Lazić (2019); their study is based on the theory presented above.

Depth-sectioning based on (high-angle) annular dark-field STEM has a long history, recorded in the papers of Nellist and Pennycook (1999), Borisevich et al. (2006), Xin et al. (2008), Intaraprasonk et al. (2008), Nellist et al. (2008a,b), Behan et al. (2009), Xin and Muller (2009), de Jonge et al. (2010) and Nellist and Wang (2012b). Bosch and Lazić first identify the assumptions underlying earlier studies:

1. Each of the layers of the sample on which the probe is focused contributes linearly and independently (i.e. incoherently) to the convergent-beam diffraction pattern that is recorded by the detector.
2. In image simulation the contribution from each slice is computed as though the preceding and succeeding parts of the specimen did not exist.
3. Inelastic scattering can be ignored.

They obtain an expression for the contribution of a slice to the diffraction pattern when all three assumptions are justified. They then compare the results with a full multislice calculation and explain any discrepancies, notably those arising from channelling in crystalline specimens. They also provide a critical account of the notion of resolution in three dimensions.

67.3.7 The Confocal Mode

Confocal imaging is in routine use in light microscopy (Sheppard, 1987, 2003[3]; Pawley, 2006) and has now entered the electron microscopy toolbox. The basic arrangement is shown in Fig. 67.17, where it is seen that the source and specimen are now conjugate to the detector plane; the detector collects only the current arriving at a small zone around the optic axis. The lenses are shown as separate elements but in practice, the specimen is placed in the centre of the probe-forming (condenser–objective) lens. The field of the lens upstream from the specimen forms the probe, while the part downstream focuses the specimen plane onto the detector plane. Scanning coils are not shown as it is better to translate the specimen (Takeguchi et al., 2008; Hashimoto et al., 2008). The aim is to

[3] Confocal imaging has a large literature in light optics, much of which is of direct relevance to the electron optical case though not often cited. A representative selection is included in the list of references for this part.

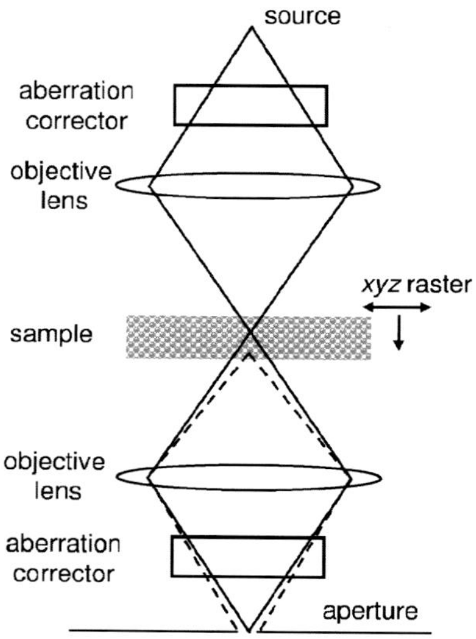

Figure 67.17
Ray diagram for the scanning confocal STEM mode. *After Cosgriff et al. (2008), Courtesy Elsevier.*

collect a set of signals from different depths within the specimen, from which a three-dimensional image can be assembled. The technique was patented by Zaluzec in 2001, after which several attempts were made to put it into practice (Frigo et al., 2002; Hallégot and Zaluzec, 2004; Zaluzec et al., 2009; see also Zaluzec, 2003, 2007), but it was only with the arrival of double-aberration-corrected STEMs that the merits of confocal scanning electron microscopy became apparent (Nellist et al., 2006, 2008a,b; Einspahr and Voyles, 2005, 2006; Nellist and Wang, 2012a,b). The angular aperture can then be much wider and the depth of field, which is inversely proportional to the square of this angle, is hence much thinner. A very detailed account is available (Cosgriff et al., 2010) and the present section is, therefore, limited to the essentials.

We shall see that when elastic scattering only is considered, the signal at the detector is very weak, contrast coming solely from the nonlinear terms in the phase variation in the specimen. For inelastic scattering the signal is better, particularly when an energy filter is present.

In the following analysis, we retain defocus Δ in the aberration function $\chi(\boldsymbol{q})$ but disregard spherical aberration, assumed to have been corrected. The wave incident on the specimen is as usual

$$\psi_Q(\boldsymbol{u}_o - \boldsymbol{\xi}, \Delta_1) = \int A(\boldsymbol{q})\exp(-\mathrm{i}\chi)\exp\{2\pi\mathrm{i}\boldsymbol{q}\cdot(\boldsymbol{u}_o - \boldsymbol{\xi})\}d\boldsymbol{q} \qquad (67.70)$$
$$\chi(\boldsymbol{q}) = \pi\lambda\Delta_1 q^2$$

(The suffix 1 is added to Δ since we shall also allow defocus in the lower lens, suffix 2.) The emergent wave, $\psi_Q O(\boldsymbol{u}_o - \boldsymbol{\xi})$, is propagated to the diffraction plane of the second lens where a defocusing term and an aperture are added; the resulting wave function is then propagated to the detector plane, where

$$\psi_d(\boldsymbol{\xi}, \Delta_1, \bar{\boldsymbol{\xi}}, \Delta_2) = \iint A_1(\boldsymbol{q})\exp(-2\pi i \boldsymbol{q}\cdot\boldsymbol{\xi})\exp(-i\chi_1)\tilde{O}(\bar{\boldsymbol{q}} - \boldsymbol{q})A_2(\bar{\boldsymbol{q}})\exp(2\pi i \bar{\boldsymbol{q}}\cdot\bar{\boldsymbol{\xi}})\exp(-i\chi_2)d\boldsymbol{q}d\bar{\boldsymbol{q}} \tag{67.71}$$

The current density at the detector aperture can be conveniently written in terms of the functions P_1 and P_2:

$$\begin{aligned} P_1(\boldsymbol{u}_o - \boldsymbol{\xi}, \Delta_1) &= \int A_1(\boldsymbol{q})\exp(-i\chi_1)\exp\{2\pi i \boldsymbol{q}\cdot(\boldsymbol{u}_o - \boldsymbol{\xi})\}d\boldsymbol{q} \\ P_2(\bar{\boldsymbol{\xi}} - \boldsymbol{u}_o, \Delta_2) &= \int A_2(\bar{\boldsymbol{q}})\exp(-i\chi_2)\exp\{2\pi i \bar{\boldsymbol{q}}\cdot(\bar{\boldsymbol{\xi}} - \boldsymbol{u}_o)\}d\bar{\boldsymbol{q}} \end{aligned} \tag{67.72}$$

and we find

$$I(\boldsymbol{\xi}, \Delta_1) = \left|O(\boldsymbol{\xi}) \otimes \{P_1(-\boldsymbol{\xi}, \Delta_1)P_2(\boldsymbol{\xi}, -\Delta_1)\}\right|^2 \tag{67.73}$$

in which we have used the fact that $\boldsymbol{u}_o$ and $\boldsymbol{u}_d$ are equal in the confocal configuration and $\Delta_2 = -\Delta_1$. The point $\boldsymbol{u}_d = \bar{\boldsymbol{\xi}}$ is conjugate to $\boldsymbol{u}_o = \boldsymbol{\xi}$.

For a weak-phase object, $O(\boldsymbol{u}_o) = 1 + i\eta(\boldsymbol{u}_o)$, the signal at a point detector on the optic axis is

$$\begin{aligned} I(0) &= \left|I^{(1)}(0) + I^{(2)}(0)\right|^2 \\ I^{(1)}(0) &= \int A_2(\bar{\boldsymbol{q}})A_1(\bar{\boldsymbol{q}})\exp\{-i(\chi_2 + \chi_1)\}d\bar{\boldsymbol{q}} \\ I^{(2)}(0) &= i\iint A_2(\bar{\boldsymbol{q}})A_1(\boldsymbol{q})\tilde{\eta}(\bar{\boldsymbol{q}} - \boldsymbol{q})\exp\{-i(\chi_2(\bar{\boldsymbol{q}}) + \chi_1(\boldsymbol{q}))\}d\boldsymbol{q}d\bar{\boldsymbol{q}} \end{aligned} \tag{67.74}$$

and this proves to be uniform, as can be seen by calculating the complex conjugate of the expression. Any contrast must come from higher order terms in the expansion of $\exp(i\eta)$.

We now consider the corresponding expression for the signal obtained when inelastically scattered electrons are considered. The derivation requires features of scattering theory that go beyond the scope of this volume and we, therefore, give only the result – the full details can be found in d'Alfonso et al. (2007, 2008). The signal collected by a point collector is now

$$I(\boldsymbol{\xi}, \Delta_1) = \iiint \phi_2^*(\boldsymbol{u}_o, t - z, \boldsymbol{\xi})\phi_2(\bar{\boldsymbol{u}}_o, t - z, \boldsymbol{\xi})\Gamma(\boldsymbol{u}_o, \bar{\boldsymbol{u}}_o, z)\phi_1^*(\boldsymbol{u}_o, z, \boldsymbol{\xi})\phi_1(\bar{\boldsymbol{u}}_o, z, \boldsymbol{\xi})d\boldsymbol{u}_o d\bar{\boldsymbol{u}}_o dz \tag{67.75}$$

in which the term $\Gamma(\boldsymbol{u}_o, \bar{\boldsymbol{u}}_o, z)$ contains the elements of the transition matrix describing the scattering. The partial waves ϕ_1 and ϕ_2 emerge during the scattering calculation.

The signal is considerably improved by energy filtering when the microscope contains an in-column analyser. Wang et al. (2014a,b) have developed a mode of operation that combines energy-loss spectroscopy (EELS) and scanning confocal microscopy. With this, quantitative chemical characterization is possible over the whole volume of the specimen.

An annular dark-field confocal mode has been proposed by Hashimoto and colleagues (Hashimoto et al., 2009a,b; Takeguchi et al., 2009, 2010; Mitsuishi et al., 2010, 2012; Zhang et al., 2012; Hamaoka et al., 2018a). Here, an annular aperture is placed in the back focal plane of the second lens. Confocal imaging in an instrument corrected for chromatic aberration as well as the geometrical aberrations, the TEAM.1 microscope, is discussed by Ercius and Xin (2012) and Xin et al. (2012, 2013a,b). A variant on the original geometry has been developed by Zheng et al. (2014a,b, 2015, 2016, 2017). The axis of the probe incident on the specimen is now tilted through an angle θ relative to the optic axis ($\theta = 26.4$ mrad in the example of Zheng et al., 2014b); θ must be greater than the semiangular convergence of the beam. As shown in Fig. 67.18, the specimen is also tilted so that the probe remains in focus during scanning. The chromatic aberration of the second lens will add an energy-dependent phase shift, $\pi\kappa\lambda\left(q_x^2 + q_y^2\right)$ where κ denotes the chromatic factor

$$\kappa = C_c \frac{\Delta E}{E_0} \tag{67.76}$$

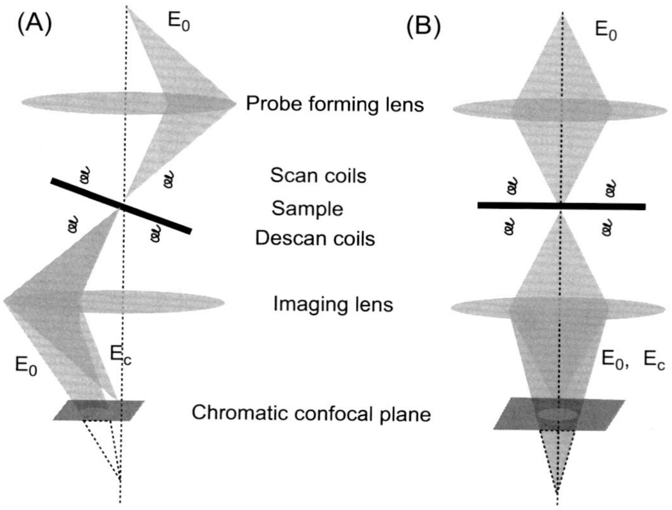

Figure 67.18
Ray diagrams for off-axis (A) and on-axis (B) scanning confocal electron microscopy. E_0, primary beam energy; E_c, 'confocal energy'. *After Zheng et al. (2014a,b), Courtesy American Physical Society, doi: 10.1103/PhysRevLett.112.166101.*

ΔE denotes the energy loss in question and E_0 is the incident beam energy. Zheng and Etheridge (2013) have shown that C_c can be measured accurately. The wavefunction at the detector plane will then be given by

$$\begin{aligned}\psi(\boldsymbol{u}_d) &= \int O(q_x - q_0, q_y)\exp(-\mathrm{i}\kappa\lambda q^2)\exp(2\pi\mathrm{i}\boldsymbol{q}\cdot\boldsymbol{u}_d)d\boldsymbol{q} \\ &= \left\{\int O(q)\exp(-\mathrm{i}\kappa\lambda q^2)\exp(2\pi\mathrm{i}\boldsymbol{q}\cdot\boldsymbol{u}_d)d\boldsymbol{q}\right\} \otimes \delta(u_x + \kappa\lambda q_0)\end{aligned} \tag{67.77}$$

The scanning intensity distribution is hence defocused as a result of the chromatic aberration and also shifted laterally by a distance $C_c\theta\kappa E/E_0$. Energy dispersion has thus been created in three dimensions. A disc placed in the back focal plane of the second lens will limit the angular range of the scattered electrons. As a result, electrons with energies close to $E_0 - \Delta E$ (the 'confocal energy', E_c) will impinge on the detector plane inside a disc that does not overlap the zero-loss disc if E_c is not too small.

The confocal image is usually composed of signals collected by a small axial detector but this is not the only possibility. Etheridge et al. (2011) have examined the signal acquired by an annular detector (Fig. 67.19), which they call an R-STEM image, and demonstrate that information difficult to obtain by other methods can be extracted from it.

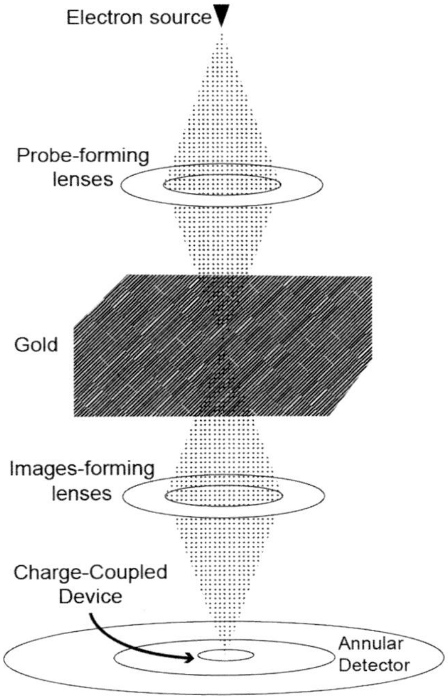

Figure 67.19

Confocal imaging with an annular detector (R-STEM). *After Etheridge et al. (2011), Courtesy American Physical Society, doi:10.1103/PhysRevLett.106.160802.*

Fast pixelated detectors make it possible to extract the confocal image or the R-STEM image of Etheridge et al. (2011) from the 4D data set or ptychogram acquired with no aperture diaphragm in the detector plane (Barnum and Williamson, 2020).

For further information about the confocal mode, see Behan et al. (2009), de Jonge et al. (2006), Hamaoka et al. (2018b), Hashimoto et al. (2009a,b, 2012), Mitsuishi et al. (2008), Mitsuishi and Takeguchi (2015), Nellist and Wang (2012a,b), Nellist et al. (2010), Takeguchi et al. (2012), van Benthem et al. (2006), Wang et al. (2009, 2010, 2011, 2012a, b, 2013, 2014a,b), Zhang et al. (2010) and Zheng et al. (2016).

67.3.8 Imaging STEM (ISTEM)

An imaging mode that combines STEM illumination with conventional TEM image formation, *Imaging STEM* or ISTEM, has been investigated by Rosenauer et al. (2014), van den Bos et al. (2017) and Krause et al. (2017). The standard image-forming optics of a conventional TEM focuses the lower surface of the specimen on the detector. The beam incident on the specimen is now a small probe, as in a STEM. The detector is set to record the images detected during a frame of the scan and then to assemble the images for each scan position into a complete image. At first sight, therefore, it seems that an incoherent image will be obtained since the individual pixels of the image have been recorded at different times. Krause et al. analyse the formation of the image in this mode and compare conventional TEM and imaging STEM images. We refer to their paper, which gives full details and, in particular, examines the assumption of complete incoherence. In a later paper (Marquardt et al., 2021), the effect of spherical aberration, astigmatism, beam tilt, contamination and shot noise on the accuracy and precision of position determination are studied. Examples are shown by Mahr et al. (2018).

67.4 Ptychography

The phase problem, arising from the fact that electron images display the electron current distribution and not the complex wavefunction, is the subject of Chapter 74. One successful solution is the technique known as ptychography; this is described here as it depends intimately on the optics of the STEM. This account is limited to the essentials: so successful has it been, notably for X-ray images, that a vast number of variants have been developed. For an extended and very lucid study of these, see Rodenburg and Maiden (2019). For background material, see Bates and Rodenburg (1989), Rodenburg (1989a,b), Rodenburg and McCallum (1991), Friedman et al. (1991), Rodenburg and Bates (1992), Plamann and Rodenburg (1992, 1998), McCallum and Rodenburg (1992, 1993a,b), Rodenburg et al. (1993), Nellist and Rodenburg (1998), Sarahan et al. (2012) and Maiden et al. (2015).

Ptychography "is a method of calculating the phase relationships between different parts of a scattered wave disturbance in a situation where only the magnitude of the wave can be measured. Its usefulness lies in its ability to obtain images without the use of lenses and hence to lead to resolution improvements and access to properties of the scattering medium [the specimen] that cannot easily be obtained from conventional imaging methods" (Rodenburg, 2008[4]).

67.4.1 Background

Just as Dennis Gabor was the founder of holography, so Walter Hoppe was the founder of ptychography. In a series of papers, Hoppe and colleagues (Hoppe, 1969a,b; Hoppe and Strube, 1969; Hegerl and Hoppe, 1970, 1972) described a way of extracting phase information from diffraction patterns, created with different illuminating waves incident on the specimen. (The word 'ptychography' first appears in the first article by Hegerl and Hoppe and is a reminder that convolution is an essential feature of the process; in German, convolution is *Faltung*, which also means 'folding' and the Greek root $\pi\tau\upsilon\xi\eta$ also means a 'fold'.) There are other diffraction-based methods of imaging (Spence, 2019) and we therefore reproduce Rodenburg's list of characteristics that are peculiar to ptychography:

1. The experimental arrangement comprises a transmission object that is illuminated by a localized field of radiation or is isolated in some way by an aperture mounted upstream of the object. Scattered radiation from this arrangement provides an interference pattern (usually, but not necessarily, a Fraunhofer diffraction pattern) at a plane where only intensity can be recorded.
2. At least two such interference patterns are recorded with the illumination function (or aperture function) changed or shifted with respect to the object function by a known amount.
3. A calculation is performed using at least two of these patterns in order to construct an estimate of the phase of all diffraction plane reflections or, equivalently, of the phase and amplitude changes that have been impressed on the incident wave in real space by the presence of the object.
4. When the change of illumination condition is a simple lateral shift (or shift of the object), then ptychography allows a large number of interference patterns (as many as required) to be processed in order to obtain an image of a nonperiodic structure of unlimited size

The analysis set out in Section 67.3.5 corresponds to a relatively advanced stage of the subject, known as Wigner distribution deconvolution. We first describe some simpler approaches.

[4] This lively account of the early history and subsequent development of ptychography by its leading exponent goes far beyond the present short text.

The objective, we recall, is to establish the phase of the electron wavefunction, given two or more versions of the convergent-beam diffraction pattern. Eq. (67.5) is repeated here for convenience:

$$J_d(\boldsymbol{q}_d;\boldsymbol{\xi}) = \iint O(\boldsymbol{u}_o)O^*(\bar{\boldsymbol{u}}_o)\psi_Q(\boldsymbol{u}_o-\boldsymbol{\xi})\psi_Q^*(\bar{\boldsymbol{u}}_o-\boldsymbol{\xi}) \times \exp\{2\pi i\boldsymbol{q}_d\cdot(\boldsymbol{u}_o-\bar{\boldsymbol{u}}_o)\}d\boldsymbol{u}_o d\bar{\boldsymbol{u}}_o \qquad (67.5')$$

Ptychography is by no means limited to crystalline specimens but the principle is more easily understood for such objects. Before embarking on a study of the general case, therefore, we examine crystals. Again, we reproduce Rodenburg's clear statement of the process:

1. A small region of the object is illuminated by a confined wave field, thus leading to the conventional crystallographic lattice points being convolved with the Fourier transform of the illumination function. As a result of this process, the amplitudes of adjacent lattice points are given the opportunity to interfere with one another. This means that if we now measure the diffracted intensity lying between the conventional diffraction spots, we can obtain an estimate of the phase difference between these amplitudes to within an uncertainty of the complex conjugate of one of these beams.
2. By shifting the illumination function (or an aperture placed in the object plane), the complex-conjugate ambiguity can be resolved. This is because the act of shifting the illumination function causes a phase ramp to be introduced across the convolving function in reciprocal space. At points between the conventional diffraction spots, the net effect is to advance the phase of one of the interfering wave components and retard the other, thus leading to a new measurement that can determine which of the two possible solutions obtained in step (1) is correct.

The sequence of measurement and calculations, described in great detail by Rodenburg (2008), is not repeated here as ptychography allows us to deal with all types of specimen, of which crystalline objects are a special case.

67.4.2 Line-Scan Measurement

Suppose that, instead of recording just two diffraction patterns, the probe is scanned along a line across the (crystalline) specimen and a set of diffraction patterns is recorded at regularly spaced positions; the spacing must be smaller than the size of the unit cell. In the first attempt to put this into practice (Nellist et al., 1995), a near-perfect image of the specimen (silicon in the [110] orientation) was reconstructed in modulus and phase. The interest of this experiment was to demonstrate that, unlike competing techniques, imperfect coherence of the illumination is not a restriction. For further work on this method, labelled

Π_{lc} by Rodenburg (the subscript lc denotes line-scanned crystal), see McCallum and Rodenburg (1993b), Plamann and Rodenburg (1998), Nellist and Rodenburg (1998) and the related publications of Spence (1978) and Spence and Cowley (1978).

67.4.3 Area-Scan Measurement (Projection Achromatic Imaging)

The illuminating spot is now scanned in both the x- and y-directions and diffraction patterns are recorded for a set of closely spaced positions of the spot. A particular Fourier component is then selected at associated detector pixels. From the Fourier transform of Eq. (67.5) with respect to ξ

$$J_d^{(p)}(\boldsymbol{q}_d, \boldsymbol{x}) = \int J_d(\boldsymbol{q}_d, \boldsymbol{\xi})\exp(2\pi i\boldsymbol{\xi}\cdot\boldsymbol{x})d\boldsymbol{\xi} \tag{67.78}$$

we select the plane in which $\boldsymbol{q}_d = \boldsymbol{x}/2$, and write

$$B(\boldsymbol{x}) = J_d^{(p)}(\boldsymbol{x}/2, \boldsymbol{x}) \tag{67.79}$$

A lengthy calculation, given in full by Rodenburg (2008), reveals that for a weak-phase specimen

$$B(\boldsymbol{x}) = |T_A|^2\{\delta(\boldsymbol{x}) + i\tilde{\eta}\} \tag{67.80}$$

The inverse Fourier transform of $B(\boldsymbol{x})$ is the *projection ptychograph*,

$$\Pi_p(\boldsymbol{u}_o) \approx |T_A|^2\eta(\boldsymbol{u}_o) \tag{67.81}$$

Tests and simulations (Rodenburg et al., 1993; Plamann and Rodenburg, 1994, 1998) show that even for thicker specimens (30 nm at 300 kV in their example), the reconstruction remains very faithful. Nevertheless, despite the many attractions of this method, it has proved very difficult to put into practice. An exception is Cowley and Winterton (2001).

67.4.4 Iterative Phase Retrieval

Iterative approaches to the phase problem have been pursued ever since Gerchberg and Saxton (1972, 1973) obtained a solution by iterating between the image and the diffraction pattern of the same object zone recorded in a TEM. The measured amplitudes in these arrays were used as constraints in each iteration (see Chapter 74). In the case of iterative ptychography (Π_π), each iteration starts with some exit wave $\psi_Q\Pi(\boldsymbol{u}_o)$. This is propagated to the detector plane, where the estimated modulus is replaced by the measured value. This, in turn, is propagated back to the object plane, where any intensity outside the illuminated zone is set to zero (another modulus constraint, here a support constraint). This yields a new estimate of the exit wave and hence the updated value of $\Pi(\boldsymbol{u}_o)$. The iteration is repeated as many times as necessary for each probe position.

In the version known as PIE (ptychographic iterative engine), the sequence is as follows. Each estimate of $\Pi^{(n+1)}(\boldsymbol{u}_o)$ is obtained from its predecessor $\Pi^{(n)}(\boldsymbol{u}_o)$ according to the rule

$$\Pi^{(n+1)}(\boldsymbol{u}_o) = \Pi^{(n)}(\boldsymbol{u}_o) + w(\boldsymbol{u}_o)\{\psi_e(\boldsymbol{u}_o, \boldsymbol{\xi}_j) - \psi_Q(\boldsymbol{u}_o - \boldsymbol{\xi}_j)\Pi^{(n)}(\boldsymbol{u}_o)\} \tag{67.82}$$

in which the weight or 'update function' w is defined by

$$w(\boldsymbol{u}_o) = \frac{|\psi_Q(\boldsymbol{u}_o - \boldsymbol{\xi}_j)|}{|\psi_{Q\max}(\boldsymbol{u}_o - \boldsymbol{\xi}_j)|} \frac{\psi_Q^*(\boldsymbol{u}_o - \boldsymbol{\xi}_j)}{|\psi_Q(\boldsymbol{u}_o - \boldsymbol{\xi}_j)|^2 + \varepsilon} \tag{67.83}$$

The label j identifies the member of the set of diffraction patterns used and ψ_e is the current estimate of the exit wavefunction at the specimen. We reproduce Rodenburg's flow diagram, which shows a typical stage (Fig. 67.20). The iteration is performed on a small set

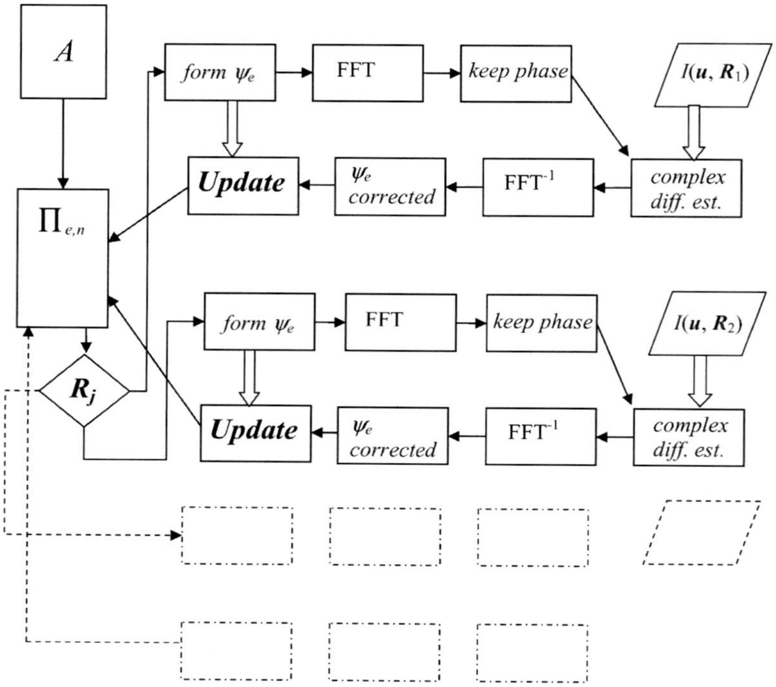

Figure 67.20
Flow diagram describing the PIE sequence, starting from the initial estimate or guess *A*. Only two illumination patterns are shown, many more can be used (*dotted boxes*). *After Rodenburg (2008) Courtesy Elsevier.*

of diffraction patterns corresponding to different probe positions. The procedure is highly insensitive to noise and has many practical merits. See Faulkner and Rodenburg (2005) and Rodenburg et al. (2007).

The weight defined by (67.83) is not the only update function that has been explored. In ePIE the weight is given by

$$w = \lambda \frac{|\psi_Q|^2}{|\psi_Q|^2_{\max}} \tag{67.84}$$

thereby removing the need for the quantity ε (familiar in Wiener filter theory, Section 73.2). This allows us to reconstruct both the specimen and the illumination distributions in amplitude and phase by simply reversing their roles in the iteration. The parameter λ is chosen according to the values of the data.

Even better is the regularized version, rPIE, for which the weight is

$$w = \frac{|\psi_Q|^2}{\lambda|\psi_Q|^2_{\max} + (1-\lambda)|\psi_Q|^2} \tag{67.85}$$

in which λ can be chosen to suit the quality of the data, as shown in Fig. 67.21. For noisy data a low weight will be preferred while for good data, a heavier weight can be used. Rodenburg and Maiden (2019) comment that in practice, $\lambda = 0.1$ provides a significantly faster convergence that PIE and ePIE.

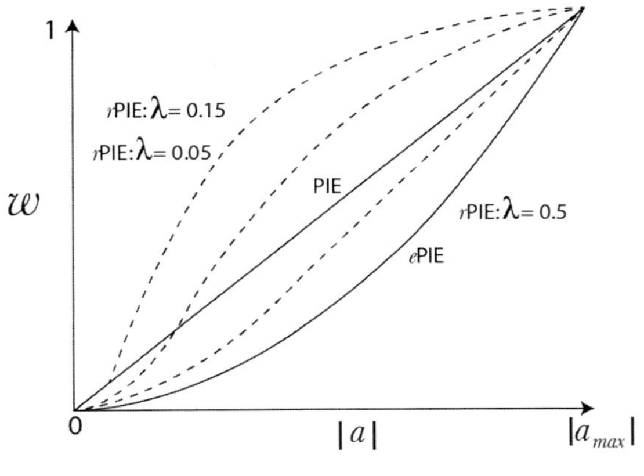

Figure 67.21
Ptychography update behaviour. The relation between probe modulus $|\alpha|$ and update strength w is shown for PIE, ePIE and three cases of rPIE ($\lambda = 0.5$, 0.15 and 0.05). *PIE*, Ptychographic iterative engine. *After Rodenburg and Maiden (2019), Courtesy Springer, https://doi.org/10.1007/978-3-030-00069-1_17.*

The rate of convergence can be further increased by introducing the notion of momentum (Maiden et al., 2017). The idea is that, after a few iterations, the process should be travelling in the right direction and can be encouraged to follow it. The basic iteration proceeds for T cycles (say five or more) after which a weighted 'momentum' (or 'velocity') term $\eta_o v^{(m)}(\boldsymbol{u}_o)$ is added to the current estimate before embarking on the next iteration. The weight η_0 is a constant, which Maiden et al. describe as the 'friction of the error landscape', and

$$v^{(n)} = v^{(n-T)} + \left(\overline{\Pi}^{(n)} - \Pi^{(n+1-T)}\right) \tag{67.86}$$

The bar in $\overline{\Pi}^{(n)}$ is a reminder that this is the estimate before the momentum term has been added. An unfortunate aspect of this rule is that the step-size in the weight must be reduced. As a result, extra parameters have to be introduced. For the regularized version (rPIE), we now have

$$w = \gamma \frac{|\psi_Q|^2}{\lambda|\psi_Q|^2_{\max} + (1-\lambda)|\psi_Q|^2} \tag{67.87}$$

and analogous factors are needed in the probe update sequence (λ'). Despite this, the improvement in speed of convergence is so impressive that this additional complication is worthwhile. Furthermore, it seems that the exact values chosen for these factors do not have a great influence. The ranges proposed by Maiden et al. are shown in Table 67.2.

The reconstruction stage of ptychography involves iteration, typically using a gradient-descent algorithm and it is always possible to arrive at a local minimum instead of the true minimum. This can be avoided if the initial estimate is not far from the correct solution and 'spectral' methods have been devised to achieve this (Candès et al., 2015). The initial estimate is now defined as the leading eigenvector of a covariance matrix ($\boldsymbol{Z}$ below), the elements of which are constructed from the measured intensities and the acquisition parameters. We express the phase-retrieval problem as

$$\boldsymbol{y} = |\boldsymbol{S}\psi|^2$$

Table 67.2: Parameter ranges for mPIE (Maiden et al., 2017).

Parameter	λ	λ'	η_{obj}	η_{probe}	γ_{obj}	γ_{probe}	T
Suggested min	0.05	0.5	0.5	0,75	0.1	0.2	10
Suggested max	0.25	5	0.9	0.99	0.5	1	100

in which $\psi \in \mathbb{C}^n$ represents the vector of the (unknown) object transmission values at each pixel. The matrix $\boldsymbol{S} \in \mathbb{C}^{p \times n}$ characterizes the sensor and p denotes the number of intensity values measured. Finally, $\boldsymbol{y}$ is the vector representing the ptychogram. The symbol $|\cdot|$ signifies the element-wise modulus operation. We now construct the covariance matrix $\boldsymbol{Z}$:

$$\boldsymbol{Z} = \boldsymbol{S}^{\dagger}\mathrm{diag}(T(\boldsymbol{y}))\boldsymbol{S}$$

The dagger denotes the Hermitian conjugate and T is a preprocessing function.

Assuming that the elements of $\boldsymbol{S}$ are independent and randomly distributed random variables, we adopt the leading eigenvector of $\boldsymbol{Z}$ as the initial estimate of ψ in the PIE.

Valzania et al. (2021) have shown that when this procedure is applied to experimental data the rate of convergence is increased threefold and the effect of noise is reduced. We refer to their paper for details of the mathematics and choice of the function T.

Kandel et al. (2019) have described an alternative version of ePIE that employs *automatic differentiation* (Griewank and Walther, 2008) to achieve the same reconstruction by a different route. The analytic derivatives that occur in ePIE are now replaced by 'automatic derivatives'. We refer to the paper by Kandel et al. for a very full and clear account of such derivatives and examples of their use and to the seminal paper of Jurling and Fienup (2014).

Konijnenberg et al. (2016) have examined the advantages of combining various PIE algorithms with the sequence described by Faulkner and Rodenburg (2004). Dwivedi et al. (2018) have suggested a way of establishing the probe position precisely.

The effect of partial coherence of the illuminating beam has been studied by Chen et al. (2020a,b). They show that, by including effects arising from the finite size of the source ('mixed-state' ptychography), the speed of data acquisition can be increased and the information limit improved for the same dose. If the original resolution is retained, the dose can be reduced by as much as 50 times (cf. Thibault and Menzel, 2013).

67.4.5 Wigner Distribution Deconvolution

This brings us to the process presented in Section 67.3.5, in which the Fourier transform of J_d with respect to both its arguments is employed. The use of this double transform was suggested to John Rodenburg by Richard Bates and has led to one of the most spectacular advances in ptychography theory. We refer to the long and fully illustrated account of the practical implementation of the numerous operations in Rodenburg and Maiden (2019). It is interesting to note that the effect of partial coherence of the source can be included very simply in the theory.

67.4.6 Defocused Ptychography

A drawback of ptychography is the heavy dose inflicted on the specimen while the large number of recordings is being collected. By defocusing the probe, so that the specimen is explored by discs, fewer probe positions are needed to reconstruct the phase distribution (Humphry et al., 2012; Wang et al., 2017). Theoretical studies suggest that the dose (high in the work of Putkunz et al., 2012; d'Alfonso et al., 2014) can be substantially reduced for a resolution equivalent to that achievable with a point probe (Pelz et al., 2017, 2018; Jiang et al., 2018). Low-dose defocused ptychography has been attempted by Song et al. (2018b, 2019). The geometry is shown in Fig. 67.22, where it is seen that the specimen is situated 80 nm above the focus, giving a disc diameter of about 4 nm at the specimen. For a probe current of 17 pA, the dose per record was 78 electrons/Å^2. We refer to their article for a discussion of their experimental results and simply reproduce their conclusions: "At low electron dose, defocused probe ptychography shows better reconstruction quality and is less reliant on extremely low beam currents than the focused-probe geometry. Thus this method has the potential to provide quantitative phase information from beam-sensitive samples such as organic crystals, biological material and frozen specimens at high resolution without damage".

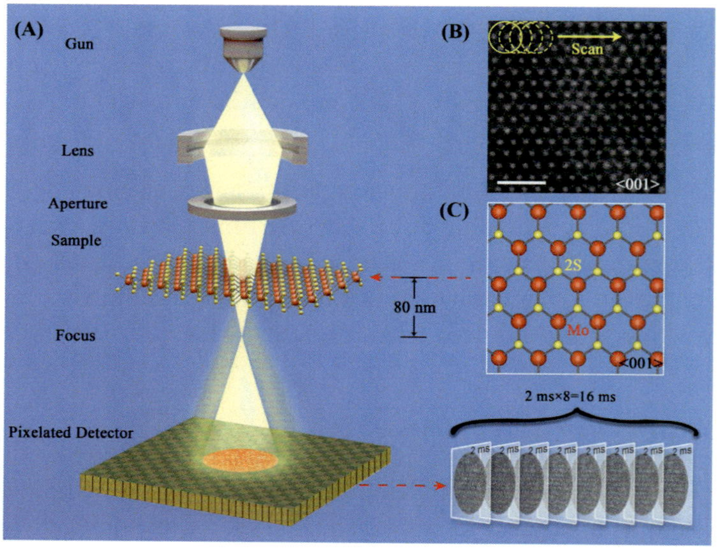

Figure 67.22
Low-dose defocused ptychography. (A) Schematic diagram. (B) High-angle annular dark-field image of molybdenum sulphide in the <001> orientation. (C) Projected atomic model (yellow = sulphur, red = molybdenum). Eight diffraction patterns (2 s each) were recorded at each probe position. *After Song et al. (2019), Courtesy SpringerNature, https://doi.org/10.1038/s41598-019-40413-z.*

67.4.7 Fourier Ptychography

The preceding sections give no more than an introduction to the subject. Many other forms of ptychography have emerged and continue to emerge, and we again refer to the survey by Rodenburg and Maiden (2019) for a clear account of these. One important variant, now known as *Fourier ptychography*, is performed in the conventional (not scanning) transmission electron microscope. It was originally described by Hoppe (1971b) and Hoppe et al. (1974, 1975) and resurfaced 20 years later (Kirkland et al., 1995, 1997); it was extended to corrected instruments by Haigh et al. (2009a,b), who recognized its relation to ptychography. It entered light microscopy in 2013 (Zheng et al., 2013; see also Zheng, 2016) and is surveyed by Konda et al. (2020).

We have mentioned the reciprocity that relates image formation in the conventional and scanning instruments and it is natural to enquire what becomes of the ptychographical operations when we transfer them to the CTEM. The answer is that there is a reciprocal equivalence which, for a weak object and purely elastic scattering, is described by Rodenberg in the Π_p schema. Collecting a series of images in the CTEM with plane-wave illumination for a set of well-defined tilt angles and azimuths can provide the same four-dimensional data set as that obtained in STEM ptychography. The reciprocal equivalence between the two approaches can be viewed as collecting all scattering angles for a single probe position in conventional ptychography compared to collecting a limited set of scattering angles but for all probe positions in the Fourier form. Experimentally, an aperture in the back focal plane (diffraction plane) or a soft aperture imposed by the partial coherence functions selects the area of the diffraction pattern that contributes to each member of the set. This method is more commonly known as *tilt-series reconstruction.* Fig. 67.23 shows a data set and the corresponding improvement in resolution resulting from the recording of an image for each value of the tilt of the illumination.

New developments continue to appear, especially in light and X-ray optics (e.g. Horstmeyer and Yang, 2014; Horstmeyer et al., 2015; Ou et al., 2015; Nguyen et al., 2018; Cheng et al., 2019; Sun et al., 2019; Aslan et al., 2020; Haas et al., 2020; Sagawa et al., 2020; O'Leary et al., 2019a,b, 2020; Zhou et al., 2020; Wei et al., 2021).

67.4.8 Multiangle Bragg Ptychography

In this variant, two degrees of freedom are included in the scan parameters: position as usual and angular scan position, corresponding to small rotations of a crystalline specimen to align it in Bragg directions. The resulting data set then contains enough information for a 3D strain-sensitive image of the specimen to be reconstructed. See Hill et al. (2018), Hruszkewycz et al. (2017), Chamard et al. (2015) and Section 4.2 of Kandel et al. (2019).

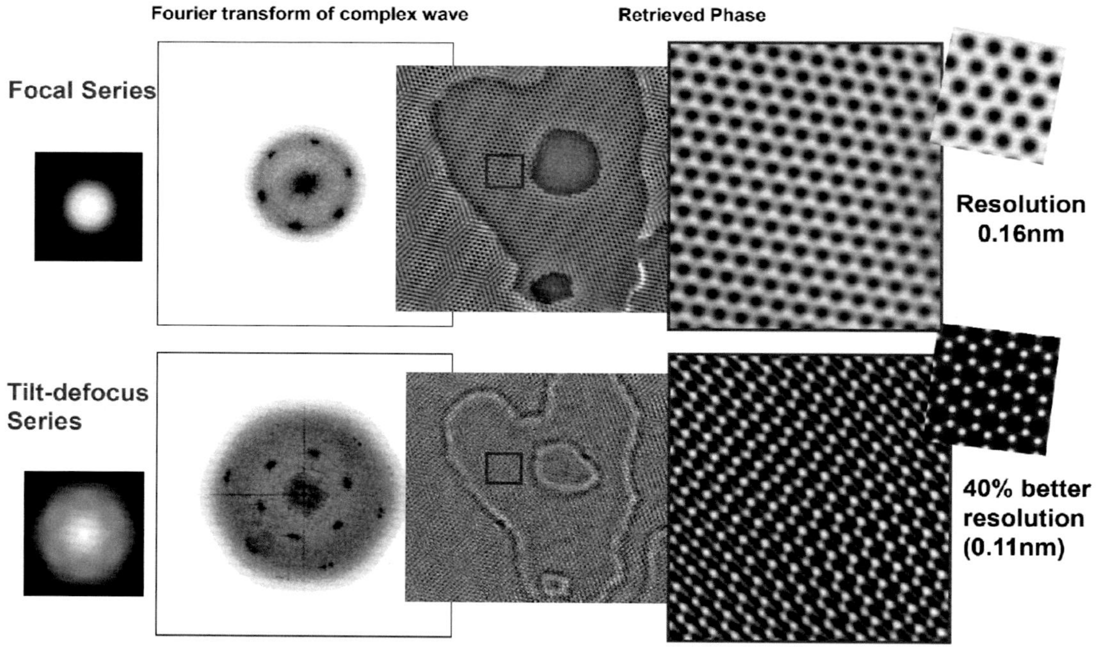

Figure 67.23
Quantitative super-resolution for graphene at 80 kV. *Courtesy Sarah Haigh.*

67.4.9 Near-Field Ptychography

The use of a small probe to generate the data for ptychographic reconstruction is not the only choice. A near-field configuration in which the specimen is explored by a wide, structured beam has advantages, of which the most striking is the small number of diffraction patterns required. In their seminal article, Allars et al. (2021) find that "as few as nine diffraction patterns are sufficient to recover a high-quality, quantitative, large field of view phase image". They also show that "the field of view can be extended to 100 μm^2 at a resolution of better than 4 nm". In this mode a parallel beam is incident on the sample. In the first plane conjugate to the specimen, a diffuser is placed. This consists of a silicon nitride disc, 50 μm in diameter, the thickness of which varies randomly. The subsequent lenses are excited in such a way that the detector records a near-field diffraction pattern; in the present example (Fig. 67.24), this is 77 cm below the Gaussian image plane, corresponding to 221 μm referred back to the object plane. The specimen is translated over a grid of positions and the image thus moves over the diffuser in small steps, respecting the overlap necessary in any ptychographic operation. Many more details and examples of phase reconstruction are given by Allars et al. The advantages and disadvantages of placing the diffuser in condenser space instead of downstream from the specimen are examined and the use of mechanical translation of the object instead of scanning is justified.

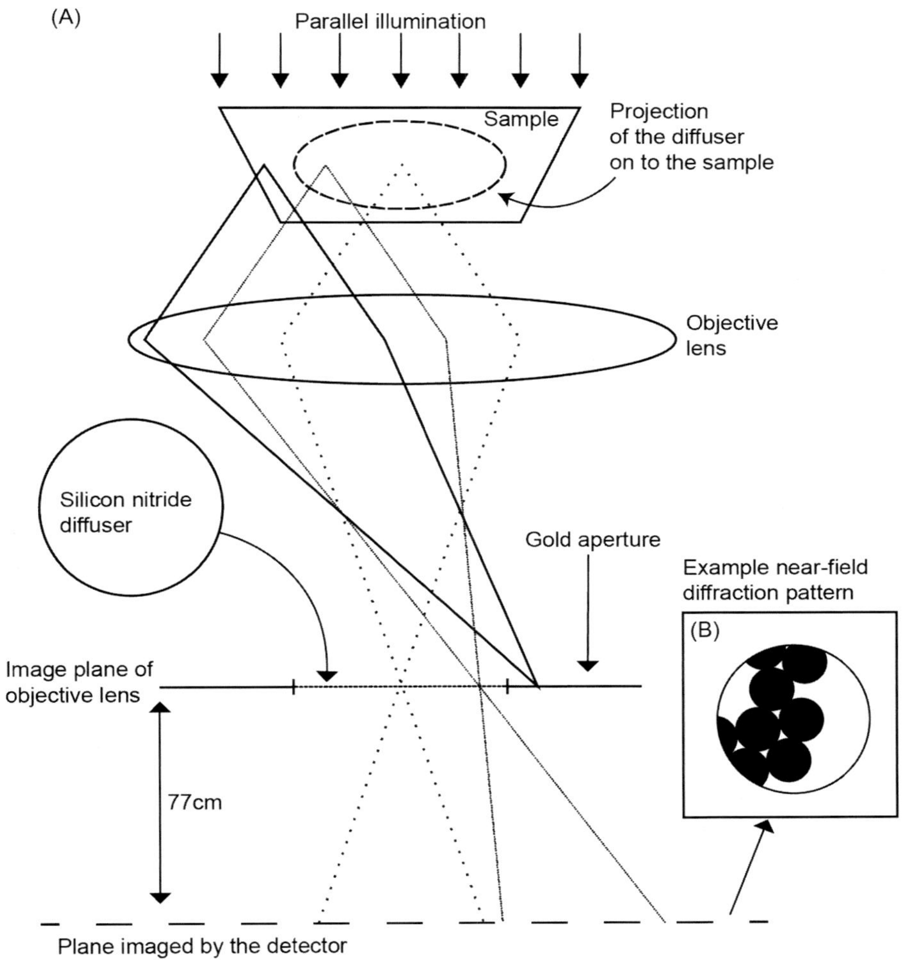

Figure 67.24
Optics of near-field ptychography. (A) Ray diagram. (B) Schematic example of a near-field diffraction pattern. *After Allars et al. (2021), Courtesy Elsevier.*

67.5 Detector Technology

The previous sections have described the various detector geometries available in STEM; for completeness, some of the key technologies underpinning these are summarized here. We note that all detectors used in electron microscopy are conveniently characterized by three measures: the modulation transfer function (MTF), which describes how an input signal is affected by the detection process and is defined as the Fourier transform of the real-space point-spread function that measures the distribution of charge generated by an incident electron; the detector quantum efficiency (DQE), which describes the loss of signal to noise in the detection process and is defined as the ratio of the squares of the output and input signals; and the radiation hardness (resistance to radiation damage), which depends critically on the thickness of the sensitive layer. These are examined closely in the opening paragraphs of the review by Clough and Kirkland (2016). Their importance for applications in electron microscopy is that the first sets a limit to the spacings that can be resolved for a given magnification, which is independent of the electron optical resolution limit, the second defines the minimum signal intensity that can be recorded at a given spatial frequency and the third imposes constraints on the semiconductor fabrication processes used. A prerequisite for quantitative STEM imaging in any mode is that the detector output is directly proportional, over the entire range of image intensities, to the time-averaged flux of electrons reaching the detector. Traditional bright-field and annular detectors are generally fabricated using $YAlO_3$ perovskite (YAP) doped with cerium as a scintillator (Baryshevsky et al., 1991; Korzhik et al., 1992; Kirkland and Thomas, 1996). This choice is based on the fact that YAP is resistant to radiation damage and its decay lifetime is ~ 30 ns (faster than the more common cerium-doped YAG(Ce) = $Y_3Al_5O_{12}$:Ce) thus ensuring that adjacent image pixel intensities do not affect one another even at short probe dwell-times. In operation the scintillator is coupled to a photomultiplier tube (PMT) and the voltage across the anode and cathode of this converts the time-averaged photon flux to a photocurrent, which is used to control the gain (image contrast). The output from the tube is then passed to a preamplifier that converts the photocurrent to a measurable voltage. Adding a constant voltage to the output (the offset of the preamplifier) changes the brightness of the image. Finally, the preamplifier output is digitized at a suitable bit depth via an analogue-to-digital (A/D) converter circuit. It is, however, important to note that even modern detectors show some nonlinearity in their response across the scintillator which must be accounted for in quantitative measurements (LeBeau and Stemmer, 2008; Findlay and LeBeau, 2013; Jones et al., 2018).

Recently, as noted in an earlier section, traditional scintillator-coupled detectors have been replaced for many applications by *pixelated direct electron detectors* operating in either an integrating or a counting mode. Although these have the significant advantage that all the imaging modes described can be computationally synthesized postacquisition from a 4D data set, their readout speeds (typically in the kHz range) fall below those achieved with conventional detectors. These detectors are closely related to those used in TEM, where they have revolutionized structural biology albeit with generally smaller pixel arrays when used in STEM (for reviews, see Clough and Kirkland, 2016; McMullan et al., 2016; Faruqi and McMullan, 2018).

Direct detectors can be broadly classified into families depending on their readout mode, integrating or counting, and their basic device structure [hybrid or monolithic active-pixel sensors (MAPSs)].

The *Medipix* series of hybrid detectors for direct detection of electrons is the longest established (Faruqi, 2001; Faruqi et al., 2003, 2005a; McMullan et al., 2007, 2009a; McMullan and Faruqi, 2008; van Gastel et al., 2009; Sikharulidze et al., 2011; Mir et al., 2016, 2017). The reason for this relatively early use of Medipix detectors for the detection of electrons is that they are based on hybrid active-pixel sensors in which the readout circuitry is protected by a sensitive layer (typically 300–500 μm thick), thereby overcoming the issue of radiation damage unless the primary electrons penetrate the sensitive layer and reach the readout electronics (Faruqi, 2001). For most hybrid detectors the pixel sizes are relatively large (55 μm for the Medipix 3 sensor) and the array size is often limited. However, this is unimportant for many STEM imaging geometries although the large pixel size and thick sensitive layer mean that the performance of this type of detector is optimal at lower voltages (Mir et al., 2017). Muller and coworkers have achieved a ptychographic reconstruction of a twisted MoS_2 bilayer with 0.39 Å resolution using an alternative hybrid electron microscope pixel array detector (EMPAD) (Jiang et al., 2018). Importantly, the larger pixel size of this detector (150 μm) significantly improves its performance at 300 kV. Hybrid Medipix detectors based on alternative substrate materials (GaAs, CdZnTe and CdTe) are also becoming available and these high atomic number substrates reduce the charge generation volume and hence have a better modulation transfer function at higher electron energies than silicon-based detectors. Thus Paton et al. (2021) have found that the use of GaAs:Cr gives a perceptibly better performance than silicon. Finally, we note that the current readout limitations can be significantly improved by using reduced bit-depth counting since the readout speed scales inversely with the counting bit depth. For STEM imaging, this is advantageous at low dose where individual pixels only record single electrons as has been recently demonstrated for ptychographic data sets recorded using binary counting (O'Leary et al., 2019a, 2020). The Timepix3 detector (a member of the Medipix family) can be used at 300 kV (van Schayck et al., 2020); these authors were then able to use a convolutional neural network (see Chapter 75) to predict the point of arrival of an electron within a pixel cluster.

The alternative structure is known as a monolithic active pixel sensor (MAPS). In such devices the readout electronics is fabricated using a CMOS[5] process within the sensitive regions of each pixel (Milazzo et al., 2005, 2010, 2011; Deptuch et al., 2007; McMullan et al., 2009b, 2014, 2016; Song et al., 2019; Chatterjee et al., 2021). Detectors of this class have smaller pixels than the hybrid detectors (typically about 10 μm) but can be fabricated with much larger array sizes (up to 8k × 8k pixels). In practice this smaller pixel size requires that the sensor be thinned to a few tens of micrometres and illuminated through the back surface of the device to avoid excessive electron scattering and consequent deterioration of the modulation transfer function. In contrast to hybrid sensors, these

[5] Complementary metal–oxide–semiconductor.

detectors have better performance at higher voltages and this, coupled with their large array sizes, makes them an excellent choice for TEM applications where they have been extensively used in low-dose imaging. Janoschka et al. (2021) demonstrate the advantage of using a fibre-coupled CMOS detector at lower energies (20–40 kV).

A disadvantage of these detectors was their relatively slow frame rate (typically not better than 1 ms corresponding to 1 Hz). In annular dark-field STEM, a dwell-time of 10 μs is routinely available. Ercius et al. (2020) describe a much faster CMOS active-pixel sensor (Johnson et al., 2018; Ciston et al., 2019). This has 576 × 576 10 μm pixels and is designed for accelerating voltages from 30 to 300 kV. The entire detector is read out at 87,000 frames/s and the resulting data rate is about 280 Gbit/s. O'Leary et al. (2019a, 2020) have shown that the detector frame rate can be considerably increased by reducing the dynamic range of the detector. In their example, only 0 and 1 were allowed, with which a frame rate of 12.5 kHz was achieved. An additional advantage of this mode is that the acquisition time is also short.

Both detector structures can operate in either integrating or counting modes. The *integrating mode* is the traditional mode of operation, in which the pixel value is simply the value read out from the analogue-to-digital converter; pixel values are directly proportional to the amount of energy deposited in the pixel in a fixed integration (exposure) time. *Counting readout* either identifies pixels in which the charge lies above a threshold value or attempts to identify individual charge clusters and assign these to pixels. For both cases the modulation transfer function and detector quantum efficiency are significantly better than for the integrating readout mode. The majority of detectors operating in this mode rely on frame-based readout, in which an entire frame is read out when an event is detected. However, for extreme frame rates, certain sensors (developed from those used in high energy physics) can operate in an event-driven mode in which only a time-stamped location of each event is read out. This reduces redundant data but requires individual frames to be assembled. Electron event-driven data recording is examined at length by Guo et al. (2020) (cf. Danev, 2020) and Jannis et al. (2022). The very high frame rates of modern cameras (e.g. Pelz et al., 2021, who mention the 87 kHz direct detector at the Lawrence Berkeley National Laboratory) make on-line compression essential. See also Datta et al. (2021). The problems arising from the vast quantities of data generated by the latest generation of detectors are also considered by Weber et al. (2020), who describe the hardware and software available for the necessary handling and processing. *Live scanning*, whereby the quality of incoming data is assessed as it is received, is also relevant in this context (Weber et al., 2021; Haas et al., 2021; Strauch et al., 2021).

Perhaps the most spectacular advantages in the use of direct detectors emerge in low-dose cryo-electron microscopy (cryo-EM) when operated in a counting mode (Amunts et al., 2014; Bammes et al., 2012; Kühlbrandt, 2014; Liao et al., 2013).

The suitability of hybrid-pixel direct detectors for energy-loss spectroscopy (EELS) is stressed by Plotkin-Swing et al. (2020), who describe the physics of the various detectors very clearly. See also Taguchi et al. (2021). Pakzad and dos Reis (2021) show that the Gatan Stela hybrid-pixel detector is valuable when low-voltage operation is required.

For measurements of the performance metrics and discussion of radiation hardness, also see Faruqi and Subramaniam (2000), Faruqi et al. (2003, 2006, 2015), Faruqi and Henderson (2007), Battaglia et al. (2009a,b,c), McMullan et al. (2009b), Guerrini et al. (2011), Milazzo et al. (2011), Faruqi and McMullan (2011, 2018), van den Broek et al. (2012), McLeod and Malac (2013), McMullan et al. (2014), Ruskin et al. (2013), Ryll et al. (2016), Susi et al. (2019) and Ishida et al. (2021). An extremely thorough study of the practical aspects of STEM imaging with a fast pixelated detector has been made by Nord et al. (2020) and Paterson et al. (2021, 2022), with particular reference to the Medipix 3 hybrid detector.

Song et al. (2018a) foresee a time when it will be possible to perform electron energy-loss spectroscopy and ptychography simultaneously, *hollow electron ptychography*; but for this, a direct detector with a central opening will be required.

In a very careful study, Hÿtch and Gatel (2021) examine the limits on phase detection in off-axis holography resulting from weaknesses of pixelated detectors.

67.6 Concluding Remarks and Historical Notes

For extensive discussion of the practical aspects of the various imaging modes in the STEM, we refer to recent texts on high-resolution microscopy, notably Spence (2013) and Zuo and Spence (2017) and above all to the collection edited by Pennycook and Nellist (2011) and a more recent account by Nellist (2019). One such mode that we have not considered here permits lattice images to be formed in the STEM (Spence and Cowley, 1978). A survey by Spence (1992), in which the close relation between this technique and other TEM and STEM operating modes is emphasized, is particularly illuminating.

The first full analysis of image formation in the STEM is to be found in the remarkable study by Zeitler and Thomson (1970), which remains invaluable. The nature of STEM imagery was investigated in great detail during the 1970s, both in research laboratories and in the development groups of the companies that produced commercial instruments (VG, Siemens and AEI) or hybrid TEM–STEM systems. The publications from Chicago, where the first STEM was built (Crewe, 1966, 1970, 1973, 1974, 1979, 1980a,b,c; Crewe and Wall, 1970a,b; Crewe et al., 1969, 1979; Crewe and Groves, 1974; Groves, 1975a,b; Beck and Crewe, 1975; Beck, 1977; Crewe and Ohtsuki, 1980, 1981a,b; Crewe and Kopf, 1980; Ohtsuki and Crewe, 1980; Thomson, 1973), Darmstadt (Rose, 1974a,b, 1975, 1977, 1978; Rose and Fertig, 1976, 1977; Fertig and Rose, 1977, 1981), Arizona (Cowley, 1969, 1970, 1975, 1976a,b,c, 1978, 1978/9, 1980; Cowley and Jap, 1976a,b; Spence and Cowley, 1978; Spence, 1978; Cowley and Au, 1978a,b, 1980;

Cowley and Spence, 1979), London (Welford, 1972; Barnett, 1973, 1974; Misell et al., 1974; Browne et al., 1975; Burge and Derome, 1976; Burge, 1977; Misell, 1977; Burge et al., 1976, 1979a,b; Burge and van Toorn, 1979, 1980) and Braunschweig (Hanszen and Ade, 1974, 1976a, 1978; Hanszen, 1974; Ade, 1977a,b) give a good idea of the way the subject developed. See also Colliex et al. (1977), Engel (1974), Engel et al. (1974), Howie (1972, 1974), Kermisch (1977), Misell et al. (1974), Reimer et al. (1975), Reimer and Hagemann (1977) and Veneklasen (1975). (Many other papers are listed in the relevant sections of Hawkes, 1978a, 1982a, 1992.)

The history of the commercial instruments – the short-lived Siemens ST100F and AEI STEM and the VG HB series – is traced by Wardell and Bovey (2009), von Harrach (2009) and Hawkes (2009). On the Siemens microscope, see Krisch et al. (1975, 1976a,b, 1977), Willasch (1975c, 1976a,b), Hubert et al. (1978) Schliepe (1978) and an anonymous account in *Microscopica Acta* 77 (1976) 455–458. For the AEI instrument, see Alderson (1971), Banbury and Bance (1973a,b), Banbury (1974), Banbury et al. (1975) and Ray et al. (1975). The VG HB5 was discussed by Wardell et al. (1973), Griffiths et al. (1973), von Harrach et al. (1974) and Wardell (1981); however, many improvements to the early version were made by users and these are reported in the conference proceedings of the period, many of which had a section on STEM, see for example Browne et al. (1975), Waddell et al. (1978), Pennycook et al. (1977), Treacy et al. (1979), Treacy and Gibson (1981). The early work on hybrid instruments is described by Thompson (1973a,b, 1975) and Kuypers et al. (1973) in the case of Philips, by Koike et al. (1974), Someya et al. (1974) and Harada et al. (1975) for JEOL and by Tochigi et al. (1974) for Akashi. The VG 300 kV STEMs are described by von Harrach et al. (1993) and von Harrach (1994, 1995, 2009).

By the 1980s the STEM was in routine use in the rather few laboratories that possessed one and only VG continued to supply such instruments commercially. Most of the manufacturers of transmission microscopes offered scanning transmission as one of the modes of operation of the TEM though it was some years before reliable field-emission sources became available. In 1996 VG ceased production of STEMs but many of their microscopes were given a new lease of life when Nion aberration-correctors were retrofitted to them (see Chapter 41 of Volume 2). The following papers give an idea of both theoretical and practical developments in the first decades: Ade (1977a), Bovey and Nicholls (1987), Brown (1981), Burge and Dainty (1976), Colliex (1985), Colliex et al. (1984, 1989), Mory and Colliex (1985), Mory et al. (1987a,b), Craven and Colliex (1977), Cowley (1970, 1975, 1976a,b,c, 1978/9, 1980, 1981, 1983, 1984, 1985, 1986, 1987, 1993), Cowley and Disko (1980), Cowley and Walker (1981), Cowley et al. (1980), Cowley and Spence (1979, 1981), Spence and Cowley (1978), Liu and Cowley (1993), Crewe (1980a, 1983, 1985), Crewe and Ohtsuki (1981a,b), Crewe and Salzman (1982), Engel (1980), Engel et al. (1974, 1976), Fertig and Rose (1977, 1981), Rose (1984), Kohl and Rose (1985), Kohl (1986), Hammel et al. (1990), Hammel and Rose (1993), Haider et al. (1988), Jones and Leonard (1978), Jones (1988), Jones and Haider (1989), Hawkes (1982b, 1985), Misell et al. (1974), Humphreys (1981), Kanaya et al. (1985, 1986), Kirkland et al. (1987),

Li and Dorignac (1986), Loane et al. (1988), Ohtsuki et al. (1979), Pennycook and Jesson (1990, 1992), Reichelt and Engel (1985, 1986), von Harrach et al. (1993) and von Harrach and Krause (1994).

In 2011 a thoroughly redesigned instrument equipped with an aberration corrector and (optional) monochromator was launched by Nion (Krivanek et al., 2008a,b; see Hawkes, 2015, for details of the gestation of this 'all-Nion' STEM). Later models incorporate an improved postcolumn imaging filter (Kahl et al., 2019); in the same year, Gatan introduced a newly designed imaging filter, the 'Continuum'.

All the earlier developments are surveyed in the masterly historical article of Pennycook (2011). The historical article by Tanaka (2015a) is particularly well illustrated. It should not be forgotten that the STEM has a prehistory: the scanning microscope built by von Ardenne (1938a,b) was in fact a scanning transmission instrument: the signal used to generate an image was collected downstream from the specimen. The same is true of a French instrument (Léauté, 1946; Hawkes and McMullan, 2004) and, even less well-known, McMullan and Smith used the transmitted signal in the 1950s (Smith, 2009).

STEM imaging techniques are in such rapid evolution that we cannot hope to cover them adequately. We draw attention to a few publications that should not be overlooked. Müller-Caspary et al. (2020) compare the merits of STEM imaging and holography for mapping electrostatic potentials. Ophus et al. (2020) explore the specimen with several probes simultaneously. Zeltmann et al. (2020) insert a mask upstream from the specimen to improve strain mapping. Oxley et al. (2020) and Findlay et al. (2020) point out that it is easy to misinterpret STEM images when the specimen is not thin.

Blackburn and McLeod (2021) compare high-resolution electron ptychography with off-axis holography. März et al. (2021) have compared several phase-retrieval algorithms including those used in electron ptychography and 'amplitude flow' (Wang et al., 2018), Mitsuishi and Sannomiya (2021) examine the effect of intermediate lens defocus on ptychography.

In a close study of noise and contrast transfer in focused-probe ptychography, O'Leary et al. (2021) show ways of minimizing the electron dose needed for reconstruction with a high signal-to-noise ratio.

Three-dimensional reconstruction based on ptychograms was considered by Maiden et al. in 2012. They pointed out that it was necessary to replace the simple representation of the specimen as a two-dimensional transparency by a multislice model (see Section 69.4.2) and this has been followed up by Heimes et al. (2021) and above all in the spectacular work of Chen et al. (2021), recalled by Muller (2021), who attained a lateral resolution that was close to the size of the atoms studied. Chen et al. describe the experimental procedure and the necessary precautions at length. We return to this subject in Chapter 75.

How far have we come in understanding quantitative STEM? (Rosenauer, 2021) makes a fitting curtain line.

CHAPTER 68

Statistical Parameter Estimation Theory

68.1 Introduction

With the unceasing miniaturization of electronic components, an exact knowledge of the positions of the atoms in the materials employed is indispensable. Certainly, electron microscope resolutions as high as 50 pm are available in favourable conditions, but resolution is not the same as precision (van den Bos and den Dekker, 2001). In this chapter, we shall see that the precision of the information about the atom positions that can be obtained by statistical methods is much better than the instrumental resolution. The importance of this precise knowledge can be appreciated from the behaviour of a layer of the superconductor $La_{1.9}Sr_{0.1}CuO_4$ grown on a substrate with a slightly different lattice constant (Locquet et al., 1998). The mismatch between the crystal structures of the layers creates strain, thereby displacing atoms close to the interface. This is sufficient to turn an insulator into a conductor.

The parameter estimation theory presented here is based on the work of the Antwerp school, especially of Dirk Van Dyck, Sandra Van Aert and Arnold Jan (Arjan) den Dekker, and on that of Adriaan van den Bos (van den Bos, 2007).

An abundantly illustrated article by Van Aert (2019), which updates an earlier account (Van Aert, 2012), surveys the subject, which was introduced in the seminal articles of den Dekker et al. (2005) and Van Aert et al. (2005). A review by De Backer et al. (2021) includes subsequent developments.

68.2 Models and Parameters

The raw data to which these statistical methods are applied are sets of images of the structure under study. We shall see in later sections that a single image may be sufficient, a valuable feature when the target is easily damaged. From this set of images, a model structure is constructed, the various features of the model being associated with and at best characterized directly by a set of *parameters,* atom positions or atom types (chemical elements), for example. Ideally, the model would be based on a first-principles calculation but often this is not possible or too complicated and a model based on a recorded image or information obtained by holography or ptychography or exit-wave reconstruction will be

Principles of Electron Optics.
DOI: https://doi.org/10.1016/B978-0-12-818979-5.00068-1

adopted instead. Reliable and very accurate estimates of the parameters are then obtained by means of the branch of probability theory known as statistical parameter estimation theory. A simple estimator is based on a least-squares minimization but this is not optimal. It is known (Section 68.4) that the variance of the estimated quantities has a lower limit, the Cramér–Rao bound, and this bound is approached asymptotically by the *maximum-likelihood estimator*, which we now describe.

The set of images that comprise the basic data will never be identical and corresponding pixel values will therefore be noisy. They are, therefore, treated as stochastic variables, which leads us to introduce their individual probability distribution function and the joint probability distribution function. (It is usual to refer to the measured quantities as *observations*, since these will not always be simple pixel values, and we follow this usage.) The set of observations is represented by a column vector $\boldsymbol{w}$,

$$\boldsymbol{w} = (w_{11}, w_{12}, \ldots, w_{KL})^T \tag{68.1}$$

with K images and L measurements in each image. The joint probability distribution of these observations is denoted $p_w(\boldsymbol{\omega})$; this is the probability that measurement of an observation will yield the value $\boldsymbol{\omega}$. It thus defines the *expectation* or mean value of each observation and its fluctuations. We write

$$\mathcal{E}(\omega_{\mathrm{kl}}) := \lambda_{\mathrm{kl}} \tag{68.2}$$

where $\mathcal{E}$ signifies 'expectation'. It is these quantities that are described by models containing parameters, θ_i; such models are denoted $f_{kl}(\boldsymbol{\theta})$; the elements of the vector $\boldsymbol{\theta}$ are the quantities to be estimated. From this, it can be seen that the notation is easier to understand if we include $\boldsymbol{\theta}$ in the joint probability function, thus $p_w(\boldsymbol{\omega}; \boldsymbol{\theta})$.

These definitions become clear when we examine a practical situation, the image formed by a (conventional or scanning) transmission electron microscope. For crystalline specimens the image intensity is concentrated around the projections of the atomic columns. It is, therefore, reasonable to model the image as a family of Gaussian functions, situated at these projections. The model that describes the intensity at some point (x_k, y_l) corresponding to the k-l pixel is then

$$f_{kl}(\boldsymbol{\theta}) = f_0 + \sum_{i=1}^{I} \sum_{m^{(i)}}^{M^{(i)}} J_{m^{(i)}} \exp\left\{ -\frac{(x_k - \overline{x}_{m^{(i)}})^2 + \left(y_l - \overline{y}_{m^{(i)}}\right)^2}{2\rho_i^2} \right\} \tag{68.3}$$

Columns of atoms of the same type are labelled i and if there are I types of atoms (chemical elements), then i runs from 1 to I. M_i denotes the number of columns containing atoms of type i; the width of the Gaussian situated on a column of type i is denoted by ρ_i. The mth column of type i is written $m^{(i)}$ and is situated at $\left(\overline{x}_{m^{(i)}}, \overline{y}_{m^{(i)}}\right)$. $J_{m^{(i)}}$ is the intensity of the $m^{(i)}$th column. The quantities to be estimated, which constitute the elements of the vector $\boldsymbol{\theta}$, are thus

$$\boldsymbol{\theta} = \left(\overline{x}_{1^{(1)}}, \overline{x}_{1^{(2)}}, \ldots, \overline{x}_{I(M^{(I)})},\ \overline{y}_{1^{(1)}}, \overline{y}_{1^{(2)}}, \ldots, \overline{y}_{I(M^{(I)})},\ \rho_1, \ldots, \rho_I,\ J_{m^{(1)}}, \ldots, J_{m^{(I)}},\ f_0\right) \tag{68.4a}$$

This is easier to appreciate if we denote the set of values of $\overline{x}$ by $\{\overline{x}\}$ and likewise for the other elements of $\boldsymbol{\theta}$ so that (68.4a) becomes

$$\boldsymbol{\theta} = (\{\overline{x}\}, \{\overline{y}\}, \{\rho\}, \{J\}, f_0) \tag{68.4b}$$

We are now ready to establish the joint probability distribution function and we assume that the pixel values are independent and that their variation can be modelled as a Poisson distribution. The probability that a measurement of an element of $\boldsymbol{\theta}$ will yield the result $\boldsymbol{\omega}$ is then

$$p_w(\boldsymbol{\omega}; \boldsymbol{\theta}) = \prod_{k=1}^{K} \prod_{l=1}^{L} \frac{\lambda_{kl}}{\omega_{kl}!} \exp(-\lambda_{kl}) \tag{68.5}$$

In practice, the observations will be approximately normally distributed (central limit theorem) with standard deviation σ and if, in addition, the variances are the same for all columns, $p_w(\boldsymbol{\omega}; \boldsymbol{\theta})$ becomes

$$p_w(\boldsymbol{\omega}; \boldsymbol{\theta}) = \prod_{k=1}^{K} \prod_{l=1}^{L} \frac{1}{\sigma\sqrt{2\pi}} \exp\left\{ -\frac{1}{2}\left(\frac{\omega_{kl} - \lambda_{kl}}{\sigma}\right)^2 \right\} \tag{68.6}$$

Since λ_{kl} has been modelled as $f_{kl}(\boldsymbol{\theta})$, the dependence of $p_w(\boldsymbol{\omega}; \boldsymbol{\theta})$ on the elements of $\boldsymbol{\theta}$ can be seen by substituting (68.3) in (68.5) or (68.6).

68.3 Estimators

The next task is to estimate the unknown parameters. It is known that the variance of the output of an unbiased estimator has a lower limit, known as the Cramér–Rao bound, to which we return in Section 68.4. Among the various estimators, the maximum-likelihood estimator has the valuable property that the variance of the results approaches the Cramér–Rao bound asymptotically as the number of observations is increased. We shall see in Section 68.5 that a large number of measured values is nevertheless not necessary, which is one of the strong points of these statistical methods: useful predictions can be obtained even for easily damaged (radiation-sensitive) material.

The maximum-likelihood estimator seeks the maximum of the joint probability distribution function in which independent variables $\{\boldsymbol{t}\}$ replace the true parameters and the measured values take the place of $\boldsymbol{\omega}$:

$$p_w(\boldsymbol{\omega}; \boldsymbol{\theta}) \rightarrow p_w(\boldsymbol{w}; \boldsymbol{t}) \tag{68.7}$$

The maximum-likelihood estimates $\hat{\boldsymbol{\theta}}$ of the parameters $\boldsymbol{\theta}$ are given by

$$\hat{\boldsymbol{\theta}} = \arg\max_{t} p_w(\boldsymbol{w};\boldsymbol{t}) = \arg\max_{t} \ln p_w(\boldsymbol{w};\boldsymbol{t}) \tag{68.8}$$

For (68.6), the maximum-likelihood estimator gives the same results as the least-squares estimator,

$$\hat{\boldsymbol{\theta}} = \arg\min_{t} \sum_{k}\sum_{l} \{w_{kl} - f_{kl}(t)\}^2 \tag{68.9}$$

Unfortunately, the number of parameters to be estimated, which is the same as the number of dimensions of the parameter space to be explored, is usually very high, especially when the specimen contains several chemical elements. A brute-force search for the optimum is likely to be trapped in a local optimum that is worse than the true optimum sought. To avoid this, it is essential to set out from a good estimate of the structure, such as a high-resolution micrograph or a high-angle annular dark-field STEM image. The role of model-based statistical estimation is then to make the information contained in the original image more *precise*, notably concerning the positions and chemical nature of the individual atoms. A software package, StatSTEM, that relieves the microscopist of the laborious programming required is available (De Backer et al., 2016, 2017c).

The theory that led to (68.8) involved the differentiation of the probability distribution function, which is permissible only if that function is (at least piecewise) continuous. This is not always the case. Estimation of the atomic number or the number of atoms in a projected column, for example, requires a different approach, based on *detection theory* (Gonnissen et al., 2014, 2016, 2017; den Dekker et al., 2013). Here, a hypothesis is made concerning the correct result and the *probability of error* is calculated on the basis of the experimental data. A binary hypothesis may be sufficient (is a given column composed of a single chemical element?) and we limit this introductory account to this situation. Van Aert (2019) discusses three such questions and we reproduce her examples.

Case (i): Which of two chemical elements, Z_0 and Z_1, is present in a given column?

Case (ii): Is a particular light atom Z_0 present or not?

Case (iii): Are there n or $n+1$ atoms in the column?

These are coded as hypotheses $\mathcal{H}_0$ and $\mathcal{H}_1$:

$$\begin{array}{llll} \text{(i)} & \mathcal{H}_0: \quad Z = Z_0 & \mathcal{H}_1: & Z = Z_1 \\ \text{(ii)} & \mathcal{H}_0: \quad Z = Z_0 & \mathcal{H}_1: & Z \in 0 \text{ (i.e., the light atom is absent)} \\ \text{(iii)} & \mathcal{H}_0: \quad n(\mathcal{H}_0) = n & \mathcal{H}_1: & n(\mathcal{H}_1) = n+1 \end{array} \tag{68.10}$$

In binary hypothesis testing, it is assumed that exactly one of the hypotheses is true. A probability $P(\mathcal{H})$ is associated with each hypothesis and

$$P(\mathcal{H}_0) + P(\mathcal{H}_1) = 1 \tag{68.11}$$

The object of the test is to minimize the probability of making the wrong choice of hypothesis. For this, we introduce a probability of error, P_e, defined by

$$P_e = P(\mathcal{H}_0|\mathcal{H}_1)P(\mathcal{H}_1) + P(\mathcal{H}_1|\mathcal{H}_0)P(\mathcal{H}_0) \tag{68.12}$$

The conditional probability of choosing $\mathcal{H}_1$ when $\mathcal{H}_0$ is true is denoted $P(\mathcal{H}_1|\mathcal{H}_0)$ and likewise for $P(\mathcal{H}_0|\mathcal{H}_1)$. Probability theory (e.g. Kay, 1998) shows that $\mathcal{H}_1$ should be chosen if

$$\frac{p_w(\boldsymbol{w};\mathcal{H}_1)}{p_w(\boldsymbol{w};\mathcal{H}_0)} > \frac{P(\mathcal{H}_0)}{P(\mathcal{H}_1)} =: p \tag{68.13}$$

Note that $p_w(\boldsymbol{w};\mathcal{H}_i), i = 1,2$ has been used to denote the conditional joint probability density $p_w(\boldsymbol{\omega};\mathcal{H}_i)$ for $\mathcal{H}_i$ to be true, evaluated with the measured values. Since $p = 1$ for equal prior probabilities, we can simplify (68.13) to

$$\frac{p_w(\boldsymbol{w};\mathcal{H}_1)}{p_w(\boldsymbol{w};\mathcal{H}_0)} > 1 \tag{68.14}$$

This quantity is known as the likelihood ratio, denoted $\Lambda(\boldsymbol{w})$. It is often written as a logarithmic inequality,

$$\ln\Lambda(\boldsymbol{w}) = \ln p_w(w;\mathcal{H}_1) - \ln p_w(\boldsymbol{w};\mathcal{H}_0) > 0 \tag{68.15}$$

For (68.5), we have

$$\ln\Lambda(\boldsymbol{w}) = \sum_{k=1}^{K}\sum_{l=1}^{L}\left\{w_{kl}\ln\left(\frac{\lambda_{kl}^{\mathcal{H}_1}}{\lambda_{kl}^{\mathcal{H}_0}}\right) - \lambda_{kl}^{\mathcal{H}_1} + \lambda_{kl}^{\mathcal{H}_0}\right\} \tag{68.16}$$

The terms in λ_{kl} represent the mean value of the intensity of pixel (k, l) for the hypothesis indicated.

Gonnissen et al. (2014), recapitulated by Van Aert (2019), have used this method to find the best annular bright-field STEM detector with which to detect the hydrogen in the rare-earth metal hydride, YH_2. A good solution for low-angle annular dark-field conditions also emerged from this study.

68.4 Fisher Information and the Cramér–Rao Bound

Given a set of observations, what is the best estimate of some particular quantity that can be derived from them? The answer depends on the estimator used, but for all unbiased estimators, there exists a lower bound on the variance of the quantity estimated. This lower limit, the Cramér–Rao bound, is established with the aid of the Fisher information matrix, $\boldsymbol{F}$, defined by

$$\boldsymbol{F} := -\mathcal{E}\left\{\frac{\partial^2 \ln p_w(\boldsymbol{w};\boldsymbol{\theta})}{\partial\theta\partial\theta^T}\right\} =: -\mathcal{E}\boldsymbol{H} \tag{68.17}$$

in which $p_w(\boldsymbol{w};\boldsymbol{\theta})$ is as before the joint probability distribution function of a set of observations $\boldsymbol{w}$ and $\boldsymbol{\theta}$ is the parameter vector with R elements. $\boldsymbol{H}$ is a Hessian $R \times R$ matrix with elements $\partial^2 \ln p_w(\boldsymbol{\omega};\boldsymbol{\theta})/\partial\boldsymbol{\theta}_\mathbf{i}\partial\boldsymbol{\theta}_\mathbf{j}$. It is known that the covariance matrix of any unbiased estimator $\hat{\boldsymbol{\theta}}$ of $\boldsymbol{\theta}$ satisfies the inequality

$$\mathrm{cov}(\hat{\boldsymbol{\theta}}) \geq \boldsymbol{F}^{-1} \tag{68.18}$$

The diagonal elements of the positive semidefinite matrix $\mathrm{cov}(\hat{\boldsymbol{\theta}}) - \boldsymbol{F}^{-1}$ cannot be negative and so the variances of the estimates $\hat{\theta}_1, \hat{\theta}_2, \ldots, \hat{\theta}_R$, which lie on the diagonal of $\mathrm{cov}(\hat{\boldsymbol{\theta}})$, cannot be smaller than the corresponding diagonal elements of $\boldsymbol{F}^{-\mathbf{1}}$:

$$\mathrm{var}(\hat{\theta}_i) \geq \boldsymbol{F}_{ii}^{-1} \tag{68.19}$$

In other words, these diagonal elements constitute lower bounds on the variances of the estimated values. These are the *Cramér–Rao lower bounds.*

Provided that certain conditions are satisfied (see Bettens et al., 1999; Van Aert et al., 2002a, 2002b, 2002c for details), the standard deviation (σ) of the estimates of the positions of the atomic columns in some target is given by

$$\sigma = \frac{\rho}{\sqrt{N}} \tag{68.20}$$

where ρ is a measure of the width of the Gaussian peaks and N is the number of atoms per column. Since ρ is proportional to the resolution given by the Rayleigh criterion or the first zero of the contrast-transfer function (Van Dyck et al., 2003), it is immediately clear that the position of a column can be established with much higher precision than the traditional resolution suggests.

68.5 Extension to Three Dimensions

In the preceding sections the arguments have been based on projections of the structures. Although methods of three-dimensional reconstruction are now well developed (Chapter 75), these are based on sets of views of the specimen, which inevitably multiplies the electron dose. For many specimens, the resulting radiation damage is not acceptable. We shall show that statistical methods can often circumvent this obstacle. In 2011 Van Aert et al. showed that the sites of individual atoms can be identified in space on the basis of projection images but again, several images were required. The same is true of the reconstructions made by Goris et al. (2015) and Xu et al. (2015). It was clearly desirable to reconstruct structure in three dimensions from a *single* projection image (Jia et al., 2014; van den Bos et al., 2016a, 2016b, 2016c; Yu et al., 2016). This led De Backer et al. (2015a, 2017a, 2017b) to propose a method in which atom counting in annular dark-field STEM (ADF-STEM) is combined with prior knowledge of the crystal structure of the sample. The 3D structure that emerges is then relaxed by minimization based on a first-principles or a Monte Carlo calculation. The method has been used successfully to study oriented attachment of lead selenide (PbSe) quantum dots in a 2D lattice, a structure easily damaged by the electron beam (Geuchies et al., 2016).

In this section, we describe the steps in the atom-counting–minimization sequence. The case considered by De Backer et al. – gold nanorods and nanodumb-bells – is examined very fully in the original publication. We note that although the aim was to reconstruct the structure from a single image, several projection images were, in fact, recorded so that the single-image result could be compared with a standard tomographic reconstruction.

The number of atoms in the atomic columns was evaluated as described earlier and leads to reliable results (De Backer et al., 2013; Van Aert et al., 2013, 2016). The process was tested on ADF STEM images of a monotype crystalline nanostructure in zone-axis orientation and the data used were the scattering cross-sections of each column. Scattering cross-sections are insensitive to incidental optical uncertainties such as astigmatism, magnification, defocus and source-size and even to small errors in specimen orientation (E et al., 2013; MacArthur et al., 2015; Martinez et al., 2014a, 2014b). Conversely, they are highly sensitive to composition and specimen thickness (De Backer et al., 2015a, 2015b). A primitive model of the structure is then assembled by placing the atoms in each column symmetrically around a central plane after which the relative heights of the columns are 'relaxed' (i.e., adjusted) first on the basis of Lenard–Jones potentials but subsequently by means of molecular dynamics calculations (described in full by De Backer et al.).

The method has been used by Altantzis et al. (2019) to quantify in 3D the evolution of facets of platinum nanoparticles in a gaseous environment. This is extremely difficult, not only because of the poor quality of the high-angle annular dark-field STEM (HAADF-STEM) images employed but also because the particles rotate under the scanning beam, thus distorting the image. These problems were overcome by the use of *deep convolutional neural network theory* (Aggarwal, 2018. Cf. Section 75.5 of Volume 4). This begins with a learning phase, in which the neural network is trained to detect undesirable features, such as distortion, and to make the appropriate corrections. The authors show that this produces satisfactory results despite the poor quality of the input data.

68.6 Use of Maximum a Posteriori Probability

The problems arising from radiation damage to fragile material and from the presence of light atoms in many nanomaterials require a different approach, which is an extended form of the maximum-likelihood procedure into which a function $g(\boldsymbol{\theta})$, known as the prior distribution or simply the *prior*, is introduced. For such specimens the signal-to-noise ratio will be low and the contrast weak. The contrast-to-noise ratio will also be low. It can be expected that different values of this ratio will be found for different atomic columns. Fatermans et al. (2019a, 2019b) argue that the integrated contrast-to-noise ratio r_{cn} defined by

$$r_{\mathrm{cn}} = \frac{\text{scattering cross-section}}{(\text{scattering cross-section} + \text{integrated background})^{1/2}} \tag{68.21}$$

in which 'integrated background' refers to the column in question, is advantageous: its connection with atom detectability is independent of the chemical element, of the incident dose and of the size of the image pixel.

The maximum a posteriori probability (MAP) rule will be used to find the most probable number of atomic columns present. The new function $g(\boldsymbol{\theta})$ describes some information about the distribution p that is available or even guessed. The estimate $\boldsymbol{\theta}$ can now be regarded as a random variable and Bayes theorem is hence applicable, leading to

$$p(\boldsymbol{\theta}|\boldsymbol{w}) = \frac{p(\boldsymbol{w}|\boldsymbol{\theta})g(\boldsymbol{\theta})}{\int p(\boldsymbol{w}|\boldsymbol{\theta})g(\boldsymbol{\theta})d\boldsymbol{\theta}} \tag{68.22}$$

From this, we deduce that the best estimate of $\boldsymbol{w}$ is given by

$$\begin{aligned}\hat{\theta} &= \arg\max_{\theta} p(\theta|\boldsymbol{w}) \\ &= \arg\max_{\theta} p(\boldsymbol{w}|\theta)g(\theta)\end{aligned} \tag{68.23}$$

This will coincide with the maximum-likelihood result if g does not depend on $\boldsymbol{\theta}$.

We now use these ideas to estimate the number of atomic columns N, typically in an image with a low contrast-to-noise ratio. Eq. (68.22) becomes

$$p(N|\boldsymbol{w}) = \frac{p(\boldsymbol{w}|N)p(N)}{p(\boldsymbol{w})} \tag{68.24}$$

$p(N|\boldsymbol{w})$ is the *posterior probability* that N atomic columns are present in the image; 'posterior' means that this is the probability found after performing the maximization. The term $p(\boldsymbol{w}|N)$ is the probability of obtaining the observations $\boldsymbol{w}$ when there are N columns. Finally, $p(N)$ is the prior ($g(\boldsymbol{\theta})$ in Eq. 68.22), which codes such knowledge as we may have about the number N. Fatermans et al. assume that all possible values of N are equally likely and since N does not appear in the denominator, we have

$$p(N|\boldsymbol{w}) \propto p(\boldsymbol{w}|N) \tag{68.25}$$

It can be shown that

$$p(\boldsymbol{w}|N) = \int p(\boldsymbol{w}|\boldsymbol{\theta}, N)p(\boldsymbol{\theta}|N)d\boldsymbol{\theta} \tag{68.26}$$

in which $p(\boldsymbol{w}|\boldsymbol{\theta}, N)$ is the probability that the observed pixel values will be $\boldsymbol{w}$ for the set of values $\boldsymbol{\theta}$ and for a particular value of N. How can we find this probability? If we assume that the uncertainty is caused by Poisson noise, then

$$p(\boldsymbol{w}|\boldsymbol{\theta}, N) = \frac{\exp\left\{-\frac{1}{2}\chi^2(\boldsymbol{\theta})\right\}}{\prod_{k=1}^{K}\prod_{l=1}^{L}\left(2\pi\sigma_{kl}^2\right)^{1/2}} \tag{68.27a}$$

in which

$$\chi^2(\boldsymbol{\theta}) = \sum_{k=1}^{K}\sum_{l=1}^{L}\frac{\left(w_{kl}-\mu_{kl}\right)^2}{\sigma_{kl}^2} \tag{68.27b}$$

Here, it is assumed that as the expectation value $f_{kl}(\boldsymbol{\theta})$ of w_{kl} increases, the Poisson distribution approaches a normal distribution with mean $\mu_{kl} = f_{kl}(\boldsymbol{\theta})$ and standard deviation $\sigma_{kl} = \sqrt{f_{kl}(\boldsymbol{\theta})}$. If we make the further assumption that

$$\sigma_{kl} = w_{kl}^{1/2}$$

and hence that the standard deviations are independent of the parameters $\boldsymbol{\theta}$, (68.27a) becomes

$$p(\boldsymbol{w}|\boldsymbol{\theta}, N) = \frac{\exp\left\{-\frac{1}{2}\chi^2(\boldsymbol{\theta})\right\}}{\prod_{k=1}^{K}\prod_{l=1}^{L}(2\pi w_{kl})^{1/2}} \tag{68.28a}$$

and now

$$\chi^2(\boldsymbol{\theta}) = \sum_{k=1}^{K}\sum_{l=1}^{L}\frac{\{w_{kl}-f_{kl}(\boldsymbol{\theta})\}^2}{w_{kl}} \tag{68.28b}$$

This is a measure of the misfit between the observations and the parametric model.

For the prior distribution, Fatermans et al. adopt

$$p(\boldsymbol{\theta}|N) = \begin{cases} \prod_{m=1}^{M}\dfrac{1}{\theta_{\max}-\theta_{\min}} & \text{for } m=1,\ldots,M; \quad \theta_{m,\min}\le\theta_m\le\theta_{m,\max} \\ 0 & \text{otherwise} \end{cases} \tag{68.29}$$

By substituting (68.28a) and (68.29) in (68.26a), an approximate expression for the posterior probability that there are N atomic columns in the image can be derived:

$$\begin{aligned} p(N|\mathbf{w}) &\propto \frac{N!(4\pi)^{2N}\exp\left(-\frac{1}{2}\chi^2_{\min}\right)\{\det(\nabla\nabla\chi^2)\}^{-1/2}}{\{(\bar{x}_{\max}-\bar{x}_{\min})(\bar{y}_{\max}-\bar{y}_{\min})(J_{\max}-J_{\min})(\rho_{\max}-\rho_{\min})\}^N} \\ \det(\nabla\nabla\chi^2) &:= \det\left(\frac{\partial^2\chi^2(\boldsymbol{\theta})}{\partial\boldsymbol{\theta}\partial\boldsymbol{\theta}^T}\right) \end{aligned} \tag{68.30}$$

In this formula, $\chi^2_{\min}$ represents $\chi^2(\hat{\boldsymbol{\theta}})$, and $\hat{\boldsymbol{\theta}}$ is the parameter vector that maximizes $p(\mathbf{w}|\boldsymbol{\theta},N)$ (68.28a). The determinant of the Hessian matrix of $\chi^2(\boldsymbol{\theta})$, $\nabla\nabla\chi^2$, is to be evaluated at $\hat{\boldsymbol{\theta}}$. Note that $\hat{\boldsymbol{\theta}}$ must lie within the support of the prior. Fatermans et al. give expressions for the posterior probability for other forms of the model.

In the remainder of their paper, Fatermans et al. discuss signal-to-noise ratio and contrast-to-noise ratio in depth and point out the advantage of using *the integrated contrast-to-noise ratio*, already mentioned (68.21). They also study ways of selecting an appropriate model structure and explain the relation between the Bayesian information criterion and the maximum a posteriori criterion. All this is illustrated with examples of practical importance. In a later paper, Fatermans et al. (2020) extend the maximum a posteriori rule to annular bright- and dark-field images recorded simultaneously.

68.7 Hidden Markov Modelling

The atomic structure of nanomaterials can change rapidly and it is, therefore, desirable to study structural *dynamics* with good temporal resolution. De wael (2021) and De wael et al. (2020a, 2020b) show that excellent results can be achieved, the temporal resolution equal to the frame time of the annular dark-field images studied, by means of hidden Markov modelling. Here, we retrace the main steps in the Markov method, referring to the original paper for many more details and in particular, for a convincing demonstration of the method applied to a platinum catalyst nanoparticle.

The Markov model employs two layers: first, the unknown quantities to be determined, which form the '*hidden*' state sequence, characterized by a stochastic tensor $\boldsymbol{H}$ and second, the measurements, of which the observed sequence, $\boldsymbol{O}$, is composed. The hidden states are the numbers of atoms in the columns of the structure. The tensor $\boldsymbol{H}$ has elements $\boldsymbol{h}(\tau_1), \boldsymbol{h}(\tau_2), \ldots, \boldsymbol{h}(\tau_T)$, where $\tau_1, \tau_2 \rightleftharpoons \tau_T$ are the times corresponding to successive frames and T is the number of frames recorded. We write $\boldsymbol{h}_i := \boldsymbol{h}(\tau_i)$. Each element $\boldsymbol{h}_i$ is itself a binary matrix: $h_{ig}^{(n)} = 1$ if and only if, at time t_i, the nth atomic column of the specimen contains g atoms. Otherwise, $h_{ig}^{(n)} = 0$. The number of atoms in each atomic column is regarded as a separate hidden state; this is important as it enables us to use the factorial hidden Markov model for which the states and the model probabilities are factorized over the atomic columns. (References to publications on Markov models are given by De wael et al. and are not repeated here.) The process is shown schematically in Fig. 68.1.

The observed sequence $\boldsymbol{O}$ is represented by a matrix, the elements of which, $\boldsymbol{o}_i$, form the observed vector at time t_i. The elements of this observed vector, $o_i^{(n)}$, are the 'scattering cross-sections' of the nth atomic column at the time in question. Here, the scattering cross-section is defined as the total intensity of the electrons scattered towards the annular detector from each atomic column, as described by Van Aert et al. (2009), E et al. (2013) or De Backer et al. (2015a, 2015b).

We now consider the factorial hidden Markov model. For this we need the joint probability density function of $\boldsymbol{H}$ and $\boldsymbol{O}$:

$$p(\boldsymbol{H}, \boldsymbol{O}; \boldsymbol{\Omega}) = p(\boldsymbol{h}_1; \boldsymbol{I}) \prod_{t=2}^{T} p(\boldsymbol{h}_t | \boldsymbol{h}_{t-1}; \boldsymbol{A}) \prod_{t'=1}^{T} p(\boldsymbol{o}_{t'} | \boldsymbol{h}_{t'}; \boldsymbol{\mu}, \sigma) \qquad (68.31)$$

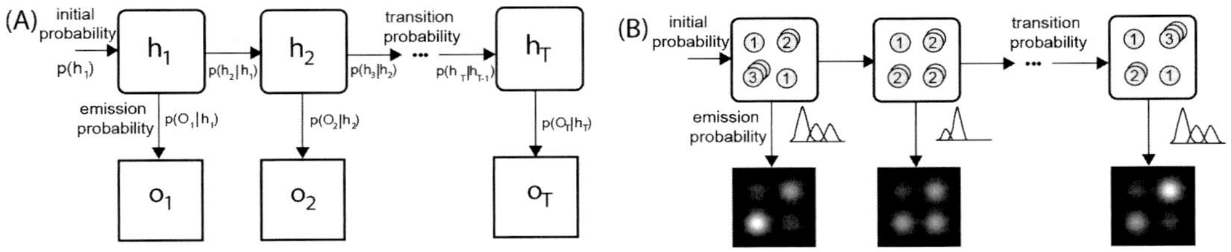

Figure 68.1
The hidden Markov model. (A) *Top row*, hidden state sequence, *bottom row*, observed sequence. The hidden states can have different possible values over time, depending on the initial and transition probabilities. (B) Atom counting. *Top row*, the number of atoms in each atomic column of the nanoparticle studied. *Bottom row*, scattering cross-sections obtained from annular dark-field STEM images. *After De wael et al. (2020a), Courtesy American Physical Society, https://doi.org/10.1103/PhysRevLett.124.106105.*

in which $\boldsymbol{\Omega} = \{\boldsymbol{I}, \boldsymbol{A}, \boldsymbol{\mu}, \sigma\}$ is the set of parameters appearing in the model. They are defined in the following paragraphs. The first term in $p(\boldsymbol{H}, \boldsymbol{O}; \boldsymbol{\Omega})$ is the initial probability distribution:

$$p(\boldsymbol{h}_1; \boldsymbol{I}) = \prod_{n=1}^{N} \prod_{g=0}^{G} t_g^{\rho} \quad \rho := h_{1g}^{(n)} \tag{68.32}$$

in which $\boldsymbol{I}$ is a vector, the elements of which are the initial probabilities that atomic columns have g atoms in the first frame; G denotes the maximum number of atoms in any column and N is the number of columns. This initializes the state sequence. The next term, the transition probability $p(\boldsymbol{h}_i | \boldsymbol{h}_{t-1}; \boldsymbol{A})$, represents changes from one frame to the next in the time series:

$$p(\boldsymbol{h}_i | \boldsymbol{h}_{t-1}; \boldsymbol{A}) = \prod_{n-1}^{N} \prod_{g_1=0}^{G} \prod_{g_2=0}^{G} A_{g_1 g_2}^{\rho_1 \rho_2}, \quad \rho_1 := h_{t-1, g_1}^{(n)}, \quad \rho_2 := h_{t, g_2}^{(n)} \tag{68.33}$$

These transition probabilities are characterized by the transition matrix, $\boldsymbol{A}$, the elements of which are the probabilities that the number of atoms in a column changes from g_1 to g_2 in successive frames. (Only nearest-neighbour frames are concerned, earlier frames have no effect.) The final term, $p(\boldsymbol{o}_t | \boldsymbol{h}_t; \boldsymbol{\mu}, \sigma)$, allows for uncertainties in the relations between the number of atoms in a column and the scattering cross-sections. These uncertainties may originate in the electron counting statistics, for example, or in instrumental instability or contamination or in factors related to the columns. These are all encompassed by random selection from a Gaussian distribution of scattering factors,

$$p(\boldsymbol{o}_t | \boldsymbol{h}_t; \boldsymbol{\mu}, \sigma) = \prod_{n=1}^{N} \prod_{g=0}^{G} \mathcal{N}(o_t^{(n)} | \mu_g, \sigma) \tag{68.34}$$

The elements of the vector $\boldsymbol{\mu}$ are the average scattering cross-sections for a column containing g atoms and σ is the width of the Gaussian distribution. The average scattering cross-sections may not be exactly equal to those found by image simulation, $\mathcal{M}_g$, and a correction factor a is included to compensate for this:

$$\mu_g =: a \mathcal{M}_g \tag{68.35}$$

We are now ready to retrieve the (hidden) state sequence. For this, we must estimate the parameters occurring in $\boldsymbol{\Omega}$:

$$\boldsymbol{\Omega} = \left(t_0, t_1, \rightleftharpoons, t_{G-1}, A_{00}, \rightleftharpoons . A_{G,G-1} a, \sigma\right)$$

Not all these parameters are independent. We can omit $G-2$ of them since $\sum_{g=0}^{G} t_g = 1$ and $\sum_{g=0}^{G} A_{jg} = 1$ for $0 \le j \le G$. The maximum-likelihood estimator enables us to obtain these parameters by maximizing the function $L(\boldsymbol{\Omega}; \boldsymbol{H}, \boldsymbol{O})$,

$$L(\boldsymbol{\Omega};\boldsymbol{H},\boldsymbol{O}) := p(\boldsymbol{h}_1)\prod_{t=2}^{T} p(\boldsymbol{h}_t|\boldsymbol{h}_{t-1})\prod_{t'=1}^{T} p(\boldsymbol{o}_{t'}|\boldsymbol{h}_{t'}) \tag{68.36}$$

The equation $\partial \log L(\boldsymbol{\Omega};\boldsymbol{H},\boldsymbol{O})/\partial\boldsymbol{\Omega} = 0$ is solved iteratively by means of the expectation-maximization (Baum–Welch) algorithm after which the hidden state sequence with the highest joint probability is extracted with the aid of the Viterbi algorithm (the individual steps in both algorithms are summarized in the Supplementary Material of De wael et al., 2020a).

The method was tested by comparing the results thus obtained with those found using the hybrid method of De wael et al. (2017). Both a model platinum nanoparticle and a real specimen were analysed (with the aid of SmartAlign, Jones et al., 2015, cf. Berkels et al., 2014). The comparison showed that the Markov method is distinctly superior to the hybrid approach. Fig. 68.2 shows the percentage of correctly counted columns as a function of electron dose per frame. Further work is included in De wael et al. (2020b).

68.8 Concluding Remarks

Other benefits of these statistical methods are appearing and many more will doubtless emerge. An important example is the study by Van Aert et al. (2019) of the balance between the knock-on damage in radiation-sensitive structures and the precision of atom-counting procedures. The authors examined the change in scattering cross-section between the frames of an image sequence. The variance v between successive frames i and j is written

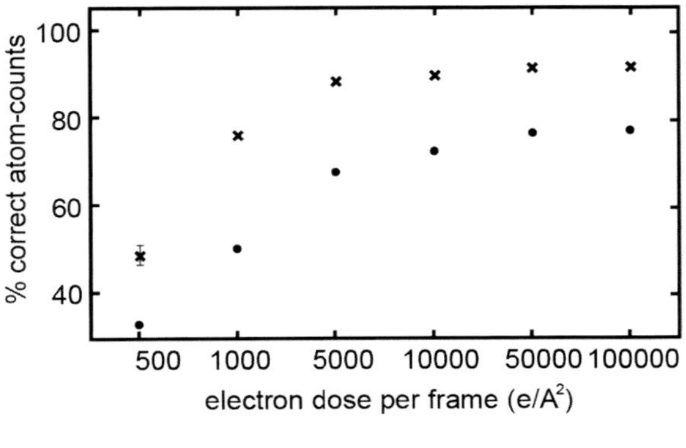

Figure 68.2

Dynamic study of the three-dimensional atomic structure of a platinum nanoparticle. Percentage of correctly counted atomic columns (95% confidence interval) as a function of the dose in each individual frame. ×: hidden Markov model, ∘: collective hybrid method. *After De wael et al. (2020a), Courtesy American Physical Society, https://doi.org/10.1103/PhysRevLett.124.106105.*

$$\begin{aligned} \upsilon^{(ij)} &= \upsilon\left(\sigma_{n,g}^{(i)} - \sigma_{n,h}^{(j)}\right) \\ &= 2\upsilon\,(\text{within}) - \upsilon\,(\text{between}) \end{aligned} \tag{68.37}$$

where $\sigma_{n.g}^{(i)}$ denotes the (column-integrated) scattering cross-section of the nth atomic column containing g atoms in frame i. Although the same columns are used in the next frame (j), the number of atoms may have changed. The variances 'within' and 'between' characterize the precision with which the scanning cross-section can be measured in a given frame and the extra contribution to the variance arising from any difference between the numbers of atoms in successive frames, $|g - h|$. Van Aert et al. analyse these variances in detail; here we reproduce only the results. If the principal damage mechanism is knock-on damage that causes displacement of an 'adatom' lying on the top surface, the variance between frames will take the form

$$\upsilon\,(\text{between}) = 2P(\varsigma)\Delta^2 \tag{68.38}$$

in which Δ is the difference between the mean integrated scattering cross-sections when $\left|g - h\right| = 1; P(\varsigma)$ is the probability of surface displacement (ς is a reminder of 'surface'). The variance 'within' has two contributions, one arising from the electron dose, $\upsilon(d)$ and the other from dose-independent scan noise.

Van Aert et al. show that $\upsilon^{(ij)}$ can be written as the sum of three terms:

$$\begin{aligned} \upsilon^{(ij)} &= 2\frac{\mu_g}{d} && \text{(Poisson noise)} \\ &+ 2\upsilon(\varsigma) && \text{(scan noise)} \\ &+ 2\sigma_{\text{sd}}\frac{N_{\text{ad}}}{N_{\text{col}}}\Delta^2 d && \text{(structural damage)} \end{aligned} \tag{68.39}$$

As before, d denotes the dose and

$$\begin{aligned} \upsilon(d) &\approx \frac{\mu_g}{d} \\ P(\varsigma) &\approx \sigma_{\text{sd}}\frac{N_{\text{ad}}}{N_{\text{col}}} \end{aligned} \tag{68.40}$$

N_{ad} is the number of 'adatoms' on the surface of the target and N_{col} is the number of atomic columns. An analytical expression is known for the knock-on cross-section, σ_{sd} (not reproduced here). All this makes it possible to estimate the dose $\hat{d}$ that should be used to balance the risk of damaging the specimen against loss of precision in atom counting. For this, we seek the minimum of $\upsilon^{(ij)}$ (68.33) and find

Table 68.1: Predicted values of the optimal electron dose (d_{opt}), atom counting reliability with and without scan noise $\left((\text{var}_d \pm \text{var}_{sn})/\delta\right)$ and probability of beam-induced surface diffusion ($\sigma_{sd}d_{opt}$) for different adatoms on a substrate and microscope operating conditions.

	A	B	A	C	D	D	E
E_{sd}/eV	1.07	0.61	0.61	0.55	1.00	1.00	0.78
E_0/keV	300	300	200	200	60	60	60
Probe semiconvergence angle/mrad	21	21	22	24	30	30	30
Detector collector range/mrad	58–190 HAADF	58–190 HAADF	51–248 HAADF	54–270 HAADF	86–190 HAADF	40–190 LAADF	40–190 LAADF
$D_{opt}/e/\text{Å}^2$	14,300	5090	4320	3380	9090	2710	18,620
Reliability, ($\text{var}_{sn} = 4.5 \times 10^{-4}\mu_g$)	0.7	0.7	0.5	0.5	0.9	0.3	1.2
Reliability, ($\text{var}_{sn} = 0$)	0.2	0.4	0.3	0.3	0.4	0.2	0.4
$\sigma_{sd}d_{opt}$ in %	14	40	24	27	16	5	17

A: Platinum on (110) platinum substrate 10 atoms thick. B: Platinum on (100) platinum substrate 10 atoms thick. C: Gold on gold substrate 10 atoms thick. D: Nitrogen on graphene. E: Hydrogen on graphene. *HAADF*, High-angle annular dark-field; *LAADF*, low-angle annular dark-field.

$$\hat{d} = \left(\frac{\mu_g}{\sigma_{sd}\left(N_{ad}/N_{col}\right)\Delta^2} \right)^{1/2} \tag{68.41}$$

Table 68.1 (Van Aert et al., 2019) shows the values predicted by this formula for several practical situations. It appears that the doses used in earlier studies (e.g. Warner et al., 2013; Yankovich et al., 2015) could be reduced by as much as 100 times and that the choice of detector geometry should also be taken into account.

Further Reading

The following publications are directly relevant to this chapter: Van Dyck et al. (1997), den Dekker et al. (1999), den Dekker et al. (2001), Van Aert et al. (2004, 2006, 2012a, 2012b, 2018b, 2021), Bals et al. (2006), Goris et al. (2012a, 2012b, 2013), De Backer et al. (2014), Koch and Van den Broek (2014), Yankovich et al. (2014), De Beenhouwer et al. (2016), Varambhia et al. (2016), Fatermans et al. (2017, 2018), Prabhat et al. (2017), Wen et al. (2019), De wael et al. (2021) and Sentürk et al. (2021).

Notes and References

The following lists of references follow the main divisions of the book. In order to avoid repetition, standard abbreviations have been adopted for several series of conference proceedings, namely the European and International conferences on electron microscopy, which have alternated every 2 years since 1954 (prior to that date they were not quite so regular); the occasional conferences on high-voltage electron microscopy; the annual meetings of the Electron Microscopy Society of America; and the biennial meetings of the Electron Microscopy and Analysis Group of the British Institute of Physics. The European and International Conferences (EUREM or EMC, and ICEM or IMC, respectively) are identified by date and place, the high-voltage conferences by date followed by HVEM and place, the American meetings by date followed by EMSA or MSA, venue and meeting number and the British meetings by date followed by EMAG and venue. The Multinational Conferences on (Electron) Microscopy and the Dreiländertagungen, now Microscopy Conferences, are labelled MCEM or MCM and MC followed by number or date and venue. HVEM refers to the High-Voltage Electron Microscopy meetings. The biennial Seminars on Recent Trends in Charged Particle Optics and Surface Physics Instrumentation ('Skalský Dvůr') are referred to by date and *Recent Trends*. Full bibliographic details of all these conference volumes are to be found at the end of this section.

Some of the lists that follow contain papers that are not cited in the text. With one exception, however, all these additional references are cited in the notes that precede the lists with some indication of their contents. The exception concerns the Preface and the introductory first chapter. This contains a very full list of books on electron optics; texts that are devoted mainly to electron microscopy are not always included, however.

Despite the length of these lists, we make no claim to completeness, and indeed, the coverage is deliberately uneven. For some topics, excellent bibliographies have been compiled and we have referred to these rather than merely repeating their contents. There are others, however, for which the literature is very scattered—phase plates and vortex studies are examples—and here we have attempted to give rather thorough coverage. We shall be most grateful to have any errors or serious omissions drawn to our attention.

* * *

Preface & Chapter 54

G. S. Agarwal, J. T. Foley and E. Wolf (1987). The radiance and phase-space representations of the cross-spectral density operator. *Opt. Commun.* **62**, 67–72.

Y. Aharonov and D. Bohm (1959). Significance of electromagnetic potentials in the quantum theory. *Phys. Rev.* **115**, 485–491.

Y. Aharonov and D. Bohm (1961). Further considerations on electromagnetic potentials in the quantum theory. *Phys. Rev.* **123**, 1511–1524.

C. C. Ahn, Ed. (2004). *Transmission Electron Energy Loss Spectrometry in Materials Science and the EELS Atlas* 2nd edn. (Wiley-VCH, Weinheim).

H. Alexander (1997). *Elektronenmikroskopie* (Teubner, Springer Fachmedien, Stuttgart, Wiesbaden).

S. Amelinckx, R. Gevers, G. Remaut and J. Van Landuyt, Eds (1970). *Modern Diffraction and Imaging Techniques in Material Science* (North-Holland, Amsterdam and London).

S. Amelinckx, D. Van Dyck, J. Van Landuyt and G. Van Tendeloo, Eds (1997). *Electron Microscopy, Principles and Fundamentals* (Wiley-VCH, Weinheim), Extracted from *Handbook of Microscopy* (3 vols), 1997.

E. Bauer (2014). *Surface Microscopy with Low Energy Electrons* (Springer, New York).

W. Baumeister and W. Vogell, Eds (1980). *Electron Microscopy at Molecular Dimensions. State of the Art and Strategies for the Future* (Springer, Berlin and New York).

D. C. Bell and N. Erdman, Eds (2013). *Low Voltage Electron Microscopy: Principles and Applications* (Wiley Chichester and RMS, Oxford).

M. Berz, K. Makino and W. Wan (2015). *An Introduction to Beam Physics* (CRC Press, Boca Raton, FL).

H. Bethge and J. Heydenreich, Eds (1982). *Elektronenmikroskopie in der Festkörperphysik* (Deutscher Verlag der Wissenschaften, Berlin and Springer, Berlin and New York).

H. Bethge and J. Heydenreich, Eds (1987). *Electron Microscopy in Solid State Physics* (Elsevier, Amsterdam, Oxford, New York and Tokyo), This is a translation with revisions of Bethge and Heidenreich (1982).

M. Blackman (1978). Preface. In *Electron Diffraction* 1927–1977 (P. J. Dobson, J. B. Pendry and C. J. Humphreys, Eds), (Institute of Physics, Bristol), Conference Series No. 41.

H. Boersch (1936a). Über das primäre und sekundäre Bild im Elektronenmikroskop. I. Eingriffe in das Beugungsbild und ihr Einfluß auf die Abbildung. *Ann. Phys. (Leipzig)* **26** (418), 631–644.

H. Boersch (1936b). Über das primäre und sekundäre Bild im Elektronenmikroskop. II. Strukturuntersuchung mittels Elektronenbeugung. *Ann. Phys. (Leipzig)* **27** (419), 75–80.

H. Boersch (1940). Fresnelsche Elektronenbeugung. *Naturwissenschaften* **28**, 709–711.

H. Boersch (1943a). Randbeugung von Elektronen. *Phys. Z.* **44**, 32–38.

H. Boersch (1943b). Randbeugung von Elektronen. *Phys. Z.* **44**, 202–211.

H. Boersch (1946a). Über die Möglichkeit der Abbildung von Atomen im Elektronenmikroskop. I. (Kontrastbildung durch elastische Streuung.). *Monatshefte Chem.* **76**, 86–92.

H. Boersch (1946b). Über die Möglichkeit der Abbildung von Atomen im Elektronenmikroskop. II. (Kontrastbildung durch unelastische Streuung.). *Monatshefte Chem.* **76**, 163–167.

H. Boersch (1947a). Sur les conditions de représentation des atomes au microscope électronique. *Publ. Inst. Rech. Sci. Tettnang* (3–4), 37–42.

H. Boersch (1947b). Über die Kontraste von Atomen im Elektronenmikroskop. *Z. Naturforsch* **2a**, 615–633.

H. Boersch (1948). Über die Möglichkeit der Abbildung von Atomen im Elektronenmikroskop. III. Kontraste von Kristallgittern und elektronenmikroskopisches Phasenkontrastverfahren. *Monatshefte Chem.* **78**, 163–171.

P. Bonhomme and A. Beorchia (1980). On-line holography of electron microscope images. EUREM-7, The Hague, **1**, 134–135.

N. Bonnet, M. Troyon and P. Gallion (1978). Possible applications of Fraunhofer holography in high resolution electron microscopy. ICEM-9, Toronto, **1**, 222–223.

M. Born and E. Wolf (1959). *Principles of Optics* (Pergamon, Oxford and New York), 7th edn (Cambridge University Press, Cambridge, 2002).

G. Botton (2008). Analytical electron microscopy. In *Science of Microscopy* (P. W. Hawkes and J. C. H. Spence, Eds), **1**, 273–405 (Springer, New York).

B. Breton, D. McMullan and K. C. A. Smith, Eds (2004). Sir Charles Oatley and the Scanning Electron Microscope. *Adv. Imaging Electron Phys.* **133**, 1–576.

Bristol. (1984). *Convergent Beam Electron Diffraction of Alloy Phases* (Adam Hilger, Bristol and Boston, MA), by the Bristol Group under the direction of John Steeds, compiled by J. Mansfield.

M. Bronsgeest (2014). *Physics of Schottky Electron Sources. Theory and Optimum Operation* (Pan Stanford, Singapore).

E. Brüche and W. Henneberg (1936). Geometrische Elektronenoptik (Entwicklung von 1934 bis Mitte 1936). *Ergebnisse Exakten Naturwiss* **15**, 365–421.

A. Bruma, Ed. (2021). *Scanning Transmission Electron Microscopy. Advanced Characterization Methods for Materials Science Applications* (CRC Press, Boca Raton, FL).

R. Brydson, Ed. (2011). *Aberration-Corrected Analytical Transmission Electron Microscopy* (Royal Microscopical Society, Oxford).

P. R. Buseck, J. M. Cowley and L. Eyring, Eds (1988). *High-Resolution Electron Microscopy and Associated Techniques* (Oxford University Press, New York and Oxford).

C. B. Carter and D. B. Williams, Eds (2016). *Transmission Electron Microscopy. Diffraction, Imaging, and Spectrometry* (Springer, New York).

W. H. Carter and E. Wolf (1977). Coherence and radiometry with quasihomogeneous planar sources. *J. Opt. Soc. Amer.* **67**, 785–796.

J. N. Chapman and A. J. Craven, Eds (1984). *Quantitative Electron Microscopy* (Scottish Universities Summer School in Physics, Edinburgh).

D. Cockayne, A. I. Kirkland, P. D. Nellist and A. Bleloch, Eds (2009). New possibilities with aberration-corrected electron microscopy. *Philos. Trans. R. Soc. Lond. A* **367**, 3631–3870.

W. Coene and D. Van Dyck (1984a). The real space method for dynamical electron diffraction calculations in high resolution electron microscopy. II. Critical analysis of the dependency on the input parameters. *Ultramicroscopy* **15**, 41–50.

W. Coene and D. Van Dyck (1984b). The real space method for dynamical electron diffraction calculations in high resolution electron microscopy. III. A computational algorithm for the electron propagation with its practical applications. *Ultramicroscopy* **15**, 287–299.

C. Colliex (1998). *La Microscopie Electronique* (Presses Universitaires de France, Paris). Translated into German as *Elektronenmikroskopie: Eine anwendungsbezogene Einführung* by H. Kohl (Wissenschaftliche Verlagsgesellschaft, 2008).

V. E. Cosslett (1947). *The Electron Microscope* (Sigma Books, London).

V. E. Cosslett (1966). *Modern Microscopy or Seeing the Very Small* (Bell, London).

J. M. Cowley (1975). *Diffraction Physics* (North-Holland, Amsterdam, New York and Oxford), 2nd edn (1981).

J. M. Cowley (1981). High energy electron diffraction in Australia. In Goodman (1981) 271–275.

J. M. Cowley, Ed. (1993). *Electron Diffraction Techniques* (Oxford University Press, Oxford and New York), 2 vols.

J. M. Cowley and A. F. Moodie (1957a). Fourier images. I. The point source. II. The out-of-focus patterns. III. Finite sources. *Proc. Phys. Soc. Lond. B* **70**, 486–496, 497–504 and 505–513.

J. M. Cowley and A. F. Moodie (1957b). The scattering of electrons by atoms and crystals. I. A new theoretical approach. *Acta Cryst.* **10**, 609–619.

J. M. Cowley and A. F. Moodie (1958). A new formulation of scalar diffraction theory for restricted aperture. *Proc. Phys. Soc. Lond.* **71**, 533–545, Corrigendum **72** (1958) 309.

J. M. Cowley and A. F. Moodie (1959). The scattering of electrons by atoms and crystals II. The effects of finite source size. III. Single-crystal diffraction patterns. *Acta Cryst.* **12**, 353–359 and 360–367.

J. M. Cowley and A. F. Moodie (1960). Fourier images. IV. The phase grating. *Proc. Phys. Soc. Lond.* **76**, 378–384.

A. V. Crewe, J. Wall and L. M. Welter (1968). A high-resolution scanning transmission electron microscope. *J. Appl. Phys.* **39**, 5861–5868.

P. Croce (1956). Etude d'une méthode de filtrage des images optiques. *Rev. Opt.* **35**, 569–589 and 642–656.

R. A. Crowther (1971). Procedures for three-dimensional reconstruction of spherical viruses by Fourier synthesis from electron micrographs. *Philos. Trans. R. Soc. Lond. B* **261**, 221–230.

R. A. Crowther, L. A. Amos, J. T. Finch, D. J. de Rosier and A. Klug (1970a). Three dimensional reconstruction of spherical viruses by Fourier synthesis from electron micrographs. *Nature* **226**, 421–425.

R. A. Crowther, D. J. DeRosier and A. Klug (1970b). The reconstruction of a three-dimensional structure from projections and its application to electron microscopy. *Proc. R. Soc. Lond. A* **317**, 319–340.

S. Czapski and O. Eppenstein (1924). *Grundzüge der Theorie der Optischen Instrumente nach Abbe* (Barth, Leipzig).

Das Übermikroskop als Forschungsmittel. Vorträge, gehalten anlaßlich der Eröffnung des Laboratoriums für Übermikroskopie der Siemens & Halske A. G., Berlin (Walter de Gruyter, Berlin, 1941).

C. J. Davisson and L. H. Germer (1927). Diffraction of electrons by a crystal of nickel. *Phys. Rev.* **30**, 705–740.

A. De Backer, J. Fatermans, A. J. den Dekker and S. Van Aert (2021). Quantitative atomic resolution electron microscopy using advanced statistical techniques. *Adv. Imaging Electron Phys.* **217**, 1–278.

L. de Broglie (1923a). Les quanta, la théorie cinétique des gaz et le principe de Fermat. *C. R. Acad. Sci. Paris* **177**, 630–632.

L. de Broglie (1923b). Ondes et quanta. *C. R. Acad. Sci. Paris* **177**, 507–510.

L. de Broglie (1923c). Quanta de lumière, diffraction et interférences. *C. R. Acad. Sci. Paris* **177**, 548–550.

L. de Broglie (1925). Recherches sur la théorie des quanta. *Ann. Phys. (Paris)* **3**, 22–128, republished in *Ann. Fond. Louis de Broglie* **17** (1992) 1–109.

L. de Broglie, Ed. (1946). *L'Optique Electronique* (Editions de la Revue d'Optique Théorique et Instrumentale, Paris).

L. de Broglie (1950). *Optique Electronique et Corpusculaire* (Hermann, Paris).

D. J. de Rosier and A. Klug (1968). Reconstruction of three dimensional structures from electron micrographs. *Nature* **217**, 130–134.

N. H. Dekkers and H. de Lang (1974). Differential phase contrast in a STEM. *Optik* **41**, 452–456.

M. M. Disko, C. C. Ahn and B. Fultz, Eds (1992). *Transmission Electron Energy Loss Spectrometry in Materials Science* (Minerals, Metals and Materials Society, Warrendale PA).

P. M. Duffieux (1940a). Analyse harmonique des images optiques. I. (Remarques sur le pouvoir de résolution). *Ann. Phys. (Paris)* **14**, 302–338.

P. M. Duffieux (1940b). Analyse harmonique des images optiques. II. (Période limite d'une pupille stigmatique quelconque). *Bull. Soc. Sci. Bretagne* **17**, 107–114.

P. M. Duffieux (1942). Analyse harmonique des images optiques. III. (Types d'incertitude). *Ann. Phys. (Paris)* **17**, 209–236.

P.-M. Duffieux (1946). *L'Intégrale de Fourier et ses Applications à l'Optique* (S.A. des Imprimeries Oberthur, Rennes), A second edition appeared in 1970 (Masson, Paris). Translated into English as *The Fourier Transform and Its Applications to Optics* (Wiley, New York and Chichester, 1983).

P. M. Duffieux (1970). Comment j'ai pris contact avec la transformation de Fourier. In *Applications de l'Holographie* (J.-C. Viénot, J. Bulabois and J. Pasteur, Eds), xviii–xx. (Université de Besançon, Besançon).

R. F. Egerton (1996). *Electron Energy-Loss Spectroscopy in the Electron Microscope* (3rd edn) (Springer, New York), 1st edn (1986), 2nd edn (1996) (Plenum, New York).

R. F. Egerton (2005). *Physical Principles of Electron Microscopy. An Introduction to TEM, SEM, and AEM* (Springer, New York).

W. Ehrenberg and R. E. Siday (1949). The refractive index in electron optics and the principles of dynamics. *Proc. Phys. Soc. Lond.* **B62**, 8–21.

Elektronenmikroskopie. eine anwendungsbezogene Einführung (Wissenschaftliche Verlagsgesellschaft, Darmstadt, 2008).

W. Elsasser (1925). Bemerkungen zur Quantenmechanik freier Elektronen. *Naturwissenschaften* **13**, 711.

R. Erni (2015). *Aberration-Corrected Imaging in Transmission Electron Microscopy. An Introduction* (2nd edn) (Imperial College Press, London), 1st edn (2010).

F. Ernst and M. Rühle, Eds (2003). *High-Resolution Imaging and Spectroscopy of Materials* (Springer, Berlin).

M. Fournier (2009). Electron microscopy in Second World War Delft. In *Scientific Research in World War II. What Scientists Did in the War* (A. Maas and H. Hooijmaijers, Eds), 77–95 (Routledge, London and New York).

J. Frank (1973). The envelope of electron microscopic transfer functions for partially coherent illumination. *Optik* **38**, 519–536.

J. Frank, Ed. (1992). *Electron Tomography. Three-Dimensional Imaging with the Transmission Electron Microscope* (Plenum, New York and London), 2nd edn, *Electron Tomography: Methods for Three-Dimensional Visualization of Structures in the Cell* (Springer, New York and Heidelberg, 2006).

J. Frank and M. van Heel (1980). Intelligent averaging of single molecules using computer alignment and correspondence analysis. I. The basic method. EUREM-7, The Hague, **2**, 690–691.

R. Frisch and O. Stern (1933). Beugung von Materiestrahlen. In *Handbuch der Physik* (H. Geiger and K. Scheel, Eds), **24**. Pt. 1. 313–354 (Springer, Berlin).

J. R. Fryer (1979). *The Chemical Applications of Transmission Electron Microscopy* (Academic Press, London and New York).

J. R. Fryer and D. L. Dorset, Eds (1991). *Electron Crystallography of Organic Molecules* (Kluwer, Dordrecht, Boston, MA and London).

B. Fultz and J. Howe (2001). *Transmission Electron Microscopy and Diffractometry* (Springer, Berlin), 2nd edn (2002), 3rd edn (2008), 4th edn (2013).

D. Gabor (1945). *The Electron Microscope* (Hulton, London), 2nd edn (Chemical Publishing Co., Brooklyn, 1948, see Gabor, 1948b).

D. Gabor (1948a). A new microscopic principle. *Nature* **161**, 777–778.

D. Gabor (1948b). *The Electron Microscope. Its Development, Present Performance and Future Possibilities* (Chemical Publishing Co, Brooklyn, NY).

D. Gabor (1968). Preface to Marton (1968).

P. Gallion, M. Troyon and A. Beorchia (1975). Formation des images électroniques en holographie de Fraunhofer en ligne. Expressions des fonctions de transfert du contraste. *Opt. Acta* **22**, 731–743.

H. R. Gelderblom and D. H. Krüger (2014). Helmut Ruska (1908–1973): his role in the evolution of electron microscopy in the life sciences, and especially virology. *Adv. Imaging Electron Phys.* **182**, 1–94.

H. Gelderblom and H. Schwarz (2022). Rudolf Rühle and the Bosch electron microscope: another early commercial instrument. *Adv. Imaging Electron Phys.* **221** (in press).

R. W. Gerchberg and W. O. Saxton (1972). A practical algorithm for the determination of phase from image and diffraction plane pictures. *Optik* **35**, 237–246.

R. W. Gerchberg and W. O. Saxton (1973). Wave phase from image and diffraction plane pictures. In *Image Processing and Computer-Aided Design in Electron Optics* (P. W. Hawkes, Ed.), 66–81 (Academic Press, London and New York).

W. Glaser (1943). Bildentstehung und Auflösungsvermögen des Elektronenmikroskops vom Standpunkt der Wellenmechanik. *Z. Phys.* **121**, 647–666, additional note *ibid.* **125** (1949) 541.

W. Glaser (1949). Auflösungsvermögen und Grenzvergrößerung des magnetischen Übermikroskops in ihrer Abhängigkeit von Voltgeschwindigkeit und magnetischer Feldstärke. *Acta Phys. Austriaca* **3**, 38–51.

W. Glaser (1950a). Zur wellenmechanischen Theorie der elektronenoptischen Abbildung. *Sitzungsber. Öst. Akad. Wiss., Mathem.-Naturw. Klasse, Abt. IIa* **159**, 297–360.

W. Glaser (1950b). Etude en théorie ondulatoire de la répartition des intensités dans les images électroniques. ICEM-2, Paris, **1**, 63–72 and 164.

W. Glaser (1952). *Grundlagen der Elektronenoptik* (Springer, Vienna).

W. Glaser (1956). Elektronen- und Ionenoptik. In *Handbuch der Physik* (S. Flügge, Ed.), **33**, 123–395 (Springer, Berlin).

W. Glaser and G. Braun (1954–1955). Zur wellenmechanischen Theorie der elektronenoptischen Abbildung. I, II. *Acta Phys. Austriac.* **9**, 41–74 and 267–296.

W. Glaser and P. Schiske (1953). Elektronenoptische Abbildung auf Grund der Wellenmechanik. I, II. *Ann. Phys. (Leipzig)* **12** (447), 240–266 and 267–280.

T. Gonen and B. L. Nannenga, Eds (2021). *CryoEM* (Humana Press, New York).

P. Goodman, Ed. (1981). *Fifty Years of Electron Diffraction* (Reidel, Dordrecht, Boston, MA and London).

T. R. Groves (2015). *Charged Particle Optics Theory. An Introduction* (CRC Press, Boca Raton, FL).

H. Grümm and P. Schiske (1996). Reminiscences of Walter Glaser. *Adv. Imaging Electron Phys.* **96**, 59–66.

M. E. Haine and T. Mulvey (1952). The formation of the diffraction image with electrons in the Gabor diffraction microscope. *J. Opt. Soc. Amer.* **42**, 763–773.

C. E. Hall (1953). *Introduction to Electron Microscopy* (McGraw-Hill, New York and London), 2nd edn (1966).

T. W. Hansen and J. B. Wagner, Eds (2016). *Controlled Atmosphere Transmission Electron Microscopy. Principles and Practice* (Springer, Cham).

K.-J. Hanszen (1966). Generalisierte Angaben über Phasenkontrast- und Amplitudenkontrast-Übertragungsfunktionen für elektronenmikroskopische Objektive. *Z. Angew. Phys.* **20**, 427–435.

K.-J. Hanszen (1971). Optical transfer theory of the electron microscope: fundamental principles and applications. *Adv. Opt. Electron Microsc.* **4**, 1–84.

K.-J. Hanszen (1973). Contrast transfer and image processing. In *Image Processing and Computer-Aided Design in Electron Optics* (P. W. Hawkes, Ed.), 16–53 (Academic Press, London and New York).

K.-J. Hanszen (1982). Holography in electron microscopy. *Adv. Electron Electron Phys.* **59**, 1–77.

K.-J. Hanszen and B. Morgenstern (1965). Die Phasenkontrast- und Amplitudenkontrast-Übertragung des elektronenmikroskopischen Objektivs. *Z. Angew. Phys.* **19**, 215–227.

K.-J. Hanszen and L. Trepte (1971). Der Einfluß von Strom- und Spannungsschwankungen, sowie der Energiebreite der Strahlelektronen auf Kontrastübertragung und Auflösung des Elektronenmikroskops. *Optik* **32**, 519–538.

R. G. Hart (1968). Electron microscopy of unstained biological material: the polytropic montage. *Science* **159**, 1464–1467.

H. Hashimoto (1954). Study of thin crystalline films by universal electron diffraction microscope. *J. Phys. Soc. Jpn.* **9**, 150–161.

H. Hashimoto, A. Howie and M. J. Whelan (1960). Anomalous electron absorption effects in metal foils. *Philos. Mag.* **5**, 967–974.

H. Hashimoto, M. Mannami and T. Naiki (1961a). Dynamical theory of electron diffraction for the electron microscopic image of crystal lattices. I. Images of single crystals. *Philos. Trans. R. Soc. Lond.* **253**, 459–489.

H. Hashimoto, M. Mannami and T. Naiki (1961b). Dynamical theory of electron diffraction for the electron microscopic image of crystal lattices. II. Image of superposed crystals (moiré pattern). *Philos. Trans. R. Soc. Lond.* **253**, 490–516.

H. Hashimoto, A. Howie and M. J. Whelan (1962). Anomalous electron absorption effects in metal foils: theory and comparison with experiment. *Proc. R. Soc. Lond. A* **269**, 80–103.

K. Hattar and K. L. Jungjohann (2021). Possibility of an integrated transmission electron microscope: enabling complex in-situ experiments. *J. Mater. Sci.* **56**, 5309–5320.

P. W. Hawkes, Ed. (1973). *Image Processing and Computer-Aided Design in Electron Optics* (Academic Press, London and New York).

P. W. Hawkes, Ed. (1980). *Computer Processing of Electron Microscope Images* (Springer, Berlin and New York).

P. W. Hawkes, Ed. (1985). *The Beginnings of Electron Microscopy. Advances in Electronics and Electron Physics Elsevier* (Suppl. 16); reissued as *Advances in Imaging and Electron Physics* **220** (2021) and **221** (2022).

P. W. Hawkes (1995). A lifetime in electron optics: a biographical sketch of Tom Mulvey with a bibliography. *J. Microsc. (Oxford)* **179**, 97–104.

P. W. Hawkes, Ed. (2008). *Aberration-Corrected Electron Microscopy. Advances in Imaging and Electron Physics* **153**, 1–538.

P. W. Hawkes (2009a). Two commercial STEMs: the Siemens ST100F and the AEI STEM-1. *Adv. Imaging Electron Phys.* **159**, 187–219.

P. W. Hawkes, Ed. (2009b). *Cold Field Emission and the Scanning Transmission Electron Microscope. Advances in Imaging and Electron Physics* **159**, 1–418.

P. W. Hawkes and D. McMullan (2004). A forgotten French scanning electron microscope and a forgotten text on electron optics. *Proc. R. Microsc. Soc.* **39**, 285–290.

P. W. Hawkes and J. C. H. Spence, Eds (2007). *Science of Microscopy* (Springer, New York), Corrected printing 2008.

P. W. Hawkes and J. C. H. Spence, Eds (2019). *Handbook of Microscopy* (Springer, New York, Cham).

P. W. Hawkes and U. Valdrè, Eds (1990). *Biophysical Electron Microscopy* (Academic Press, London and New York).

P. Hayman (1974). *Théorie Dynamique de la Microscopie et Diffraction Electroniques* (Presses Universitaires de France, Paris).

A. K. Head, P. Humble, L. M. Clarebrough, A. J. Morton and C. T. Forwood (1973). *Computed Electron Micrographs and Defect Identification* (North-Holland, Amsterdam and London, American Elsevier, New York). Volume 7 of *Defects in Crystalline Solids* (S. Amelinckx, R. Gevers and J. Nihoul, Eds).

D. W. O. Heddle (2000). *Electrostatic Lens Systems* (Institute of Physics Publishing, Bristol & Philadelphia, PA).

R. D. Heidenreich (1949). Electron microscope and diffraction study of metal crystal textures by means of thin sections. *J. Appl. Phys.* **20**, 993–1010.

R. D. Heidenreich (1951). Electron transmission through thin metal sections with application to self-recovery in cold worked aluminum. *Bell Syst. Tech. J.* **30**, 867–887.

R. D. Heidenreich (1964). *Fundamentals of Transmission Electron Microscopy* (Wiley–Interscience, New York and London).

T. Hibi (1956). Pointed filament (I). Its production and its applications. *J. Electronmicrosc.* **4**, 10–15.

T. Hibi (1985). My recollection of the early history of our work on electron optics and the electron microscope. In *The Beginnings of Electron Microscopy*, 297–315. *Adv. Electron Electron Phys.* (Suppl. 16). Reissued as *Advances in Imaging and Electron Physics* **221** (2022).

J. Hillier (1940). Fresnel diffraction of electrons as a contour phenomenon in electron supermicroscope images. *Phys. Rev.* **58**, 842.

J. Hillier (1941). A discussion of the fundamental limit of performance of an electron microscope. *Phys. Rev.* **60**, 743–745.

P. B. Hirsch (1980). Direct observations of dislocations by transmission electron microscopy: recollections of the period 1946–56. *Proc. R. Soc. Lond. A* **371**, 160–164.

P. B. Hirsch (1986). Direct observations of moving dislocations: reflections on the thirtieth anniversary of the first recorded observations of moving dislocations by transmission electron microscopy. *Mater. Sci. Eng.* **84**, 1–10.

P. B. Hirsch, R. W. Horne and M. J. Whelan (1956). Direct observations of the arrangement and motion of dislocations in aluminium. *Philos. Mag.* **1**, 677–684.

P. B. Hirsch, A. Howie and M. J. Whelan (1960). A kinematical theory of diffraction contrast of electron transmission microscope images of dislocations and other defects. *Philos. Trans. R. Soc. Lond.* **A252**, 499–529.

P. B. Hirsch, A. Howie, R. B. Nicholson, D. W. Pashley and M. J. Whelan (1965). *Electron Microscopy of Thin Crystals* (Butterworths, London), reprinted, with an additional chapter, by Krieger, New York (1977).

D. B. Holt and D. C. Joy, Eds (1989). *SEM Microcharacterization of Semiconductors* (Academic Press, London and San Diego, CA).

D. B. Holt, M. D. Muir, P. R. Grant and I. M. Boswarva, Eds (1974). *Quantitative Scanning Electron Microscopy* (Academic Press, London and New York).

W. Hoppe (1961). Ein neuer Weg zur Erhöhung des Aulösungsvermögens des Elektronenmikroskops. *Naturwissenschaften* **48**, 736–737.

W. Hoppe (1983). Electron diffraction with the transmission electron microscope as a phase-determining diffractometer – from spatial frequency filtering to the three-dimensional structure analysis of ribosomes. *Angew. Chem.* **95**, 465–494, *Internat. Ed.* in English **22**, 456–485.

W. Hoppe, R. Langer, G. Knesch and C. Poppe (1968). Protein-Kristallstrukturanalyse mit Elektronenstrahlen. *Naturwissenschaften* **55**, 333–336.

W. Hoppe, J. Gaßmann, N. Hunsmann, H. J. Schramm and M. Sturm (1974). Three-dimensional reconstruction of individual negatively stained yeast fatty-acid synthetase molecules from tilt series in the electron microscope. *Hoppe–Seyler's Z. Physiol. Chem.* **355**, 1483–1487.

W. Hoppe, H. J. Schramm, M. Sturm, N. Hunsmann and J. Gaßmann (1976). Three-dimensional electron microscopy of individual biological objects. I. Methods. II. Test calculations. III. Experimental results on yeast fatty acid synthetase. *Z. Naturforsch* **31a**, 645–655, 1370–1379 and 1380–1390.

A. Howie and U. Valdrè, Eds (1988). *Surface and Interface Characterization by Electron Optical Methods* (Plenum, New York and London), Vol. 191 of the NATO ASI Series B.

A. Howie and M. J. Whelan (1960a). Numerical solution of the dynamical equations of electron diffraction for several simultaneous reflections. EUREM-2, Delft, **1**, 181–185.

A. Howie and M.J. Whelan (1960b). The dynamical theory of diffraction from dislocations. EUREM-2, Delft, **1**, 194–198.

A. Howie and M. J. Whelan (1961). Diffraction contrast of electron microscope images of crystal lattice defects. II. The development of a dynamical theory. *Proc. R. Soc. Lond. A* **263**, 217–237.

A. Howie and M. J. Whelan (1962). Diffraction contrast of electron microscope images of crystal lattice defects. III. Results and experimental confirmation of the dynamical theory of dislocation image contrast. *Proc. R. Soc. Lond. A* **267**, 206–230.

Z.-y. Hua and G.-x. Gu (1993). *Electron Optics* (Fudan University Press, Shanghai) [in Chinese].

L.-y. Huang and X.-p. Liu (1991). *Electron Microscopy and Electron Optics* (Beijing Science Press, Beijing) [in Chinese].

K. Ishizuka and N. Uyeda (1977). A new theoretical and practical approach to the multislice method. *Acta Cryst.* **A33**, 740–749.

C. Jacobsen (2019). *X-Ray Optics* (Cambridge University Press, Cambridge).

F. H. Karrenberg (2019). *In the Light of the Electron Microscope, in the Shadow of the Nobel Prize* (BOD–Books on Demand, Norderstadt).

A. Khursheed (2011). *Scanning Electron Microscope Optics and Spectrometers* (World Scientific, Singapore).

A. Khursheed (2021). *Secondary Electron Energy Spectroscopy in the Scanning Electron Microscope* (World Scientific, Singapore).

S. Kikuchi (1928a). Diffraction of cathode rays by mica. *Proc. Imp. Acad. Jpn.* **4**, 271–274.

S. Kikuchi (1928b). Further study on the diffraction of cathode rays by mica. *Proc. Imp. Acad. Jpn.* **4**, 275–278.

S. Kikuchi (1928c). Diffraction of cathode rays by mica, Part III. Part IV. *Proc. Imp. Acad. Jpn.* **4**, 354–356 and 471–474.

A. I. Kirkland and S. Haigh, Eds (2015). *Nanocharacterisation* 2nd edn. (Royal Society of Chemistry, Cambridge).

A. I. Kirkland and J. L. Hutchison, Eds (2007). *Nanocharacterisation* (Royal Society of Chemistry, Cambridge).

J. K. Koehler, Ed. (1978). Advanced Techniques in Biological Electron Microscopy *I, II, III* (Springer, Berlin and New York).

W. Kossel (1941). Zur Struktur der im konvergenten Elektronenbündel auftretenden Interferenzbilder. *Ann. Phys. (Leipzig)* **40** (432), 17–38.

W. Kossel and G. Möllenstedt (1938). Elektroneninterferenzen in konvergentem Bündel. *Naturwissenschaften* **26**, 660–661.

W. Kossel and G. Möllenstedt (1939). Elektroneninterferenzen im konvergenten Bündel. *Ann. Phys. (Leipzig)* **36** (428), 113–140.

W. Kossel and G. Möllenstedt (1942). Dynamische Anomalie von Elektroneninterferenzen. *Ann. Phys. (Leipzig)* **42** (434), 287–293, *Berichtigung* **51** (443) (1943) 508.

W. Krakow and M. O'Keefe (1989). *Computer Simulation of Electron Microscope Diffraction and Images* (Minerals, Metals and Materials Society, Warrendale, PA).

W. Krakow, D. A. Smith and L. W. Hobbs, Eds (1984). *Electron Microscopy of Materials* (North-Holland, New York, Amsterdam and Oxford).

A. Lannes and J.-P. Pérez (1983). *Optique de Fourier en Microscopie Electronique* (Masson, Paris and New York).

P. K. Larsen and P. J. Dobson, Eds (1988). *Reflection High-Energy Electron Diffraction and Reflection Electron Imaging of Surfaces* (Plenum, New York and London), Vol. 188 of the NATO ASI Series B.

E. N. Leith and J. Upatnieks (1962). Reconstructed wavefronts and communication theory. *J. Opt. Soc. Amer.* **52**, 1123–1130.

E. N. Leith and J. Upatnieks (1963). Wavefront reconstruction with continuous-tone objects. *J. Opt. Soc. Amer.* **53**, 1377–1381.

F. Lenz (1953). Zur Einfachstreuung schneller Elektronen in kleine Winkel. *Naturwissenschaften* **39**, 265.

F. Lenz (1954). Zur Streuung mittelschneller Elektronen in kleinste Winkel. *Z. Naturforsch* **9a**, 185–204.

K. G. Lickfeld (1979). *Elektronenmikroskopie. Eine Einführung in die Grundlagen der Durchstrahlungs-Elektronenmikroskopie und ihrer Präparationstechniken* (Eugen Ulmer, Stuttgart).

H. Liebl (2008). *Applied Charged Particle Optics* (Springer, Berlin).

D. F. Lynch and M. A. O'Keefe (1972). *n*-Beam lattice images. II. Methods of calculation. *Acta Cryst. A* **28**, 536–548.

V. M. Lyuk'yanovich (1960). *Elektronnaya Mikroskopiya v Fiziko-Khimicheskikh Issledovaniyakh. Metodika i Primenenie* (Izd. Akad. Nauk SSSR, Moscow).

A. Maas and H. Hooijmaijers, Eds (2009). *Scientific Research in World War II. What Scientists Did in the War* (Routledge, London & New York).

N. C. MacDonald (1968). Computer-controlled scanning electron microscopy. EMSA 26, New Orleans, LA, 362–363.

N. C. MacDonald (1969). Quantitative scanning electron microscopy: solid state applications. *Scanning Electron Microsc.*, 431–437.

C. H. MacGillavry (1940a). Diffraction of convergent electron beams. *Nature* **145**, 189–190.

C. H. MacGillavry (1940b). Zur Prüfung der dynamischen Theorie der Elektronenbeugung am Kristallgitter. *Physica* **7**, 329–343.

E. W. Marchand and E. Wolf (1972a). Generalized radiometry for radiation from partially coherent sources. *Opt. Commun.* **6**, 305–308.

E. W. Marchand and E. Wolf (1972b). Angular correlation and the far-zone behavior of partially coherent fields. *J. Opt. Soc. Amer.* **62**, 379–385.

E. W. Marchand and E. Wolf (1974). Radiometry with sources in any state of coherence. *J. Opt. Soc. Amer.* **64**, 1219–1226.

A. Maréchal and P. Croce (1953). Un filtre de fréquences spatiales pour l'amélioration du contraste des images optiques. *C.R. Acad. Sci. Paris* **237**, 607–609.

A. Maréchal and M. Françon (1960). *Diffraction. Structure des Images* (Editions de la Revue d'Optique Théorique et Instrumentale, Paris).

M. Marko and H. Rose (2010). The contributions of Otto Scherzer (1909–1982) to the development of the electron microscope. *Microsc. Microanal.* **16**, 366–374.

L. Marton (1934a). La microscopie électronique des objets biologiques. *Bull. Acad. Roy. Belgique (Cl. Sci.)* **20**, 439–446.

L. Marton (1934b). Le microscope électronique; premiers essais d'application à la biologie. *Ann. Bull. Soc. Roy. Sci. Med. Nature Brux.* **Nos** 5–6, 92–106.

L. Marton (1936). Quelques considérations concernant le pouvoir séparateur en microscopie électronique. *Physica* **3**, 959–967.

L. Marton (1952). Electron interferometer. *Phys. Rev.* **85**, 1057–1058.

L. Marton (1968). *Early History of the Electron Microscope* (San Francisco Press, San Francisco, CA), 2nd edn (1994).

L. Marton and L. I. Schiff (1941). Determination of object thickness in electron microscopy. *J. Appl. Phys.* **12**, 759–765.

L. Marton, J. A. Simpson and J. A. Suddeth (1953). Electron beam interferometer. *Phys. Rev.* **90**, 490–491.

J. W. Menter (1956). The direct study by electron microscopy of crystal lattices and their imperfections. *Proc. R. Soc. Lond. A* **236**, 119–135.

P. G. Merli and M. V. Antisari, Eds (1992). *Electron Microscopy in Materials Science* (World Scientific, Singapore).

M. K. Miller and R. G. Forbes (2014). *Atom-Probe Tomography: The Local Electrode Atom Probe* (Springer, New York).

D. L. Misell (1978). *Image Analysis, Enhancement and Interpretation* (North-Holland, Amsterdam, New York and Oxford). *Practical Methods in Electron Microscopy* (A. Glauert, Ed.) vol. 7.

D. L. Misell and E. B. Brown (1987). *Electron Diffraction: An Introduction for Biologists* (Elsevier, Amsterdam, New York and Oxford). *Practical Methods in Electron Microscopy* (A. Glauert, Ed.) vol. 12.

G. Möllenstedt (1941). Messungen an den Interferenzerscheinungen im konvergenten Elektronenbündel. *Ann. Phys. (Leipzig)* **40** (432), 39–65.

G. Möllenstedt (1989). My early work on convergent-beam electron diffraction. *Phys. Status Solidi A* **116**, 13–22.

G. Möllenstedt (1991). The invention of the electron Fresnel interference biprism. *Adv. Opt. Electron Microsc.* **12**, 1–23.

G. Möllenstedt and H. Düker (1955). Fresnelscher Interferenzversuch mit einem Biprisma für Elektronenwellen. *Naturwissenschaften* **42**, 41.

G. Möllenstedt and H. Düker (1956). Beobachtungen und Messungen an Biprisma-Interferenzen mit Elektronenwellen. *Z. Phys.* **145**, 377–397.

G. Möllenstedt and H. Wahl (1968). Elektronenholographie und Rekonstruktion mit Laserlicht. *Naturwissenschaften* **55**, 340–341.

F. Müller (2009). The birth of a modern instrument and its development during World War II: electron microscopy in Germany from the 1930s to 1945. In *Scientific Research in World War II. What Scientists Did in the War* (D. Maas and H. Hooijmaijers, Eds), 121–146 (Routledge, London & New York).

F. Müller (2021). *Jenseits des Lichts. Siemens, AEG und die Anfänge der Elektronenmikroskopie in Deutschland* (Wallstein, Göttingen). Deutsches Musem Abhandlungen und Berichte, Neue Folge, Band 35.

T. Mulvey, Ed. (1996). *The Growth of Electron Microscopy. Adv. Imaging Electron Phys.* **96**, xix–xxxii and 1–885.

J. Munch (1975). Experimental electron holography. *Optik* **43**, 79–99.

S. Nishikawa and S. Kikuchi (1928). The diffraction of cathode rays by calcite. *Proc. Imp. Acad. Jpn.* **4**, 475–477.

M. A. O'Keefe (1973). *n*-beam lattice images. IV. Computed two-dimensional images. *Acta Cryst.* **A29**, 389–401.

J. Orloff, Ed. (1997). *Handbook of Charged Particle Optics* (CRC Press, Boca Raton, FL), 2nd edn (2009).

S. J. Pennycook and P. D. Nellist, Eds (2011). *Scanning Transmission Electron Microscopy Imaging and Analysis* (Springer, New York).

M. Peshkin and A. Tonomura (1989). *The Aharonov–Bohm Effect* (Springer, Berlin, New York and London), Lecture Notes in Physics, vol. 340.

J. Picht (1939). *Einführung in die Theorie der Elektronenoptik* (Barth, Leipzig), 2nd edn (1957), 3rd edn (1963).

J. Picht (1945). *Was Wir über Elektronen Wissen* (Carl Winter, Universitätsverlag, Heidelberg).

J. Picht and J. Heydenreich (1966). *Einführung in die Elektronenmikroskopie* (Verlag Technik, Berlin).

Z. G. Pinsker (1949). *Diffraktsia Elektronov* (Izd. Akad. Nauk SSSR, Moscow and Leningrad), Translated into English as *Electron Diffraction* (Butterworths, London, 1953).

L. Qing (1995). *Zur Frühgeschichte des Elektronenmikroskops* (GNT-Verlag, Stuttgart).

D. Quaglino, E. Falcieri, M. Catalano, A. Diaspro, A. Montone, P. Mengucci and C. Pellicciari, Eds (2006). *1956–2006. Cinquanta Anni di Microscopia in Italia tra Storia, Progresso ed Innovazione* (Società Italiana di Scienze Microscopiche.

N. Rasmussen (1997). *Picture Control. The Electron Microscope and the Transformation of Biology in America, 1940–1960* (Stanford University Press, Stanford).

Lord Rayleigh (1881). On copying diffraction-gratings, and on some phenomena connected therewith. *Philos. Mag.* **11**, 196–205.

L. Reimer (1984). *Transmission Electron Microscopy: Physics of image formation and microanalysis* (Springer, Berlin and New York), 2nd edn (1989), 3rd edn (1993), 4th edn (1997), 5th edn (2008) by L. Reimer and H. Kohl.

L. Reimer (1985). *Scanning Electron Microscopy. Physics of Image Formation and Microanalysis* (Springer, Berlin), 2nd edn (1998).

L. Reimer (1993). *Image Formation in the Low-Voltage Scanning Electron Microscope* (SPIE Optical Engineering Press, Bellingham, WA).

L. Reimer, Ed. (1995). *Energy-Filtering Transmission Electron Microscopy* (Springer, Berlin & New York).
L. Reimer and H. Kohl (2008). *Transmission Electron Microscopy. Physics of Image Formation* (4th edn of Reimer, 1984) (Springer, New York).
L. Reimer and G. Pfefferkorn (1973). *Raster-Elektronenmikroskopie* (Springer, Berlin and New York), 2nd edn (1977).
H. Rose (1974). Phase contrast in scanning transmission electron microscopy. *Optik* **39**, 416–436.
H. Rose (1983). Otto Scherzer gestorben. *Optik* **63**, 185–188.
H. Rose (2012). *Geometrical Charged-Particle Optics* (2nd edn) (Springer, Berlin), 1st edn (2009).
E. Ruedl and U. Valdrè, Eds (1976). *Electron Microscopy in Materials Science, Parts III and IV* (Commission of the European Communities, Luxemburg).
R. Rühle (1949). *Das Elektronenmikroskop* (Curt E. Schwab, Stuttgart).
E. Ruska (1934). Über Fortschritte im Bau und in der Leistung des magnetischen Elektronenmikroskops. *Z. Phys.* **87**, 580–602.
E. Ruska (1979). Die Frühe Entwicklung der Elektronenlinsen und der Elektronenmikroskopie. *Acta Hist. Leopoldina,* Nr 12, 136 pp. Translated into English as *The Early Development of Electron Lenses and Electron Microscopy* by T. Mulvey (1980) (Hirzel, Stuttgart); also published as *Microsc. Acta* (Suppl. 5) 140 pp.
J. V. Sanders and P. Goodman (1981). Fourier images and the multi-slice approach. In Goodman (1981), 281–283.
W. O. Saxton (1978). Computer techniques for image processing in electron microscopy. *Adv. Electron Electron Phys.* (Suppl. 10). Reissued as *Adv. Imaging Electron Phys.* **214** (2020).
P. Schattschneider, Ed. (2012). *Linear and Chiral Dichroism in the Electron Microscope* (Pan Stanford, Singapore).
O. Scherzer (1949). The theoretical resolution limit of the electron microscope. *J. Appl. Phys.* **20**, 20–29.
L. I. Schiff (1942a). Atomic images with the electron microscope. *Phys. Rev.* **61**, 391.
L. I. Schiff (1942b). Ultimate resolving power of the electron microscope. *Phys. Rev.* **61**, 721–722.
P. Schiske (1968). Zur Frage der Bildrekonstruktion durch Fokusreihen. EUREM-4, Rome, **1**, 145–146.
P. Schiske (1973). Image processing using additional statistical information about the object. In *Image Processing and Computer-Aided Design in Electron Optics* (P. W. Hawkes, Ed.), 82–90 (Academic Press, London and New York).
P. F. Schmidt (2017). *Praxis der Rasterelektronenmikroskopie und Mikrobereichanalyse* (Expert Verlag).
B. M. Siegel and D. R. Beaman, Eds (1975). *Physical Aspects of Electron Microscopy and Microbeam Analysis* (Wiley, New York and London).
K. C. A. Smith (2013). Electron microscopy at Cambridge University with Charles Oatley and Ellis Cosslett: some reminiscences and recollections. *Adv. Imaging Electron Phys.* **177**, 189–277.
J. C. H. Spence (1980). *Experimental High-Resolution Electron Microscopy* (Oxford University Press, Oxford), 2nd edn (1988), 3rd edn (2003).
J. C. H. Spence (2013). *High-Resolution Electron Microscopy* (Oxford University Press, Oxford), 4th edn of Spence (1980).
J. C. H. Spence and J. M. Zuo (1992). *Electron Microdiffraction* (Plenum, New York and London).
C. Süsskind (1985). L. L. Marton, 1901–1979. In *The Beginnings of Electron Microscopy* (P. W. Hawkes, Ed.) 501–523. *Adv. Electron Electron Phys.* (Suppl. 16), Reissued as *Adv. Imaging Electron Phys.* **221** (2022).
J. L. Synge (1954). *Geometrical Mechanics and de Broglie Waves* (Cambridge University Press, Cambridge).
H. F. Talbot (1836). Facts relating to optical science, No. IV *Philos. Mag.* **9**, 401–407.
N. Tanaka, Ed. (2015). *Scanning Transmission Electron Microscopy of Nanomaterials. Basics of Imaging and Analysis* (Imperial College Press, London).
M. Tanaka and M. Terauchi (1985). *Convergent-Beam Electron Diffraction* (JEOL, Tokyo).
M. Tanaka, M. Terauchi and T. Kaneyama (1988). *Convergent-Beam Electron Diffraction II* (JEOL, Tokyo).
T.-t. Tang (1996). *Advanced Electron Optics* (Beijing Institute of Technology Press, Beijing), [in Chinese].
G. P. Thomson (1927). The diffraction of cathode rays by thin films of platinum. *Nature* **120**, 802.

G. P. Thomson (1928). Experiments on the diffraction of cathode rays. *Proc. R. Soc. Lond. A* **117**, 600–609.

G. P. Thomson and A. Reid (1927). Diffraction of cathode rays by a thin film. *Nature* **119**, 890.

F. Thon (1966a). On the defocusing dependence of phase contrast in electron microscopical images. *Z. Naturforsch.* **21a**, 476–478.

F. Thon (1966b). Imaging properties of the electron microscope near the theoretical limit of resolution. ICEM-6, Kyoto, **1**, 23–24.

H. Tomita, T. Matsuda and T. Komoda (1970a). Electron microholography by two-beam method. ICEM-7, Grenoble, **1**, 151–152.

H. Tomita, T. Matsuda and T. Komoda (1970b). Electron microholography by two-beam method. *Jpn. J. Appl. Phys.* **9**, 719.

H. Tomita, T. Matsuda and T. Komoda (1972). Off-axis electron micro-holography. *Jpn. J. Appl. Phys.* **11**, 143–149.

A. Tonomura (1969). Electron beam holography. *J. Electron Microsc.* **18**, 77–78.

A. Tonomura (1998). *The Quantum World Unveiled by Electron Waves* (World Scientific, Singapore).

A. Tonomura (1999). *Electron Holography,* 2nd edn (Springer, Berlin). 1st edn (1993).

A. Tonomura and H. Watanabe (1968). Electron beam holography, [in Japanese] *Nihon Butsuri Gakkai-Shi [Proc. Phys. Soc. Jpn.]* **23**, 683–684.

A. Tonomura, A. Fukuhara, H. Watanabe and T. Komoda (1968). Optical reconstruction of image from Fraunhofer electron-hologram. *Jpn. J. Appl. Phys.* **7**, 295.

A. Tonomura, N. Osakabe, T. Matsuda, T. Kawasaki, J. Endo, S. Yano and H. Yamada (1986). Evidence for Aharonov–Bohm effect with magnetic field completely shielded from electron wave. *Phys. Rev. Lett.* **56**, 792–795.

M. Troyon, P. Gallion and A. Laberrigue (1976). In-line Fraunhofer holography: advantages of field emission gun. EUREM-6, Jerusalem, **1**, 344–345.

S. Uhlemann and M. Haider (1998). Residual wave aberrations in the first spherical aberration corrected transmission electron microscope. *Ultramicroscopy* **72**, 109–119.

K. Ura (1979). *Denshikogaku [Electron Optics]* (Kyoritsu, Tokyo).

B. K. Vainshtein (1956). *Strukturnaya Elektronografiya* (Izd. Akad. Nauk SSSR, Moscow and Leningrad), Translated into English as *Structure Analysis by Electron Diffraction* (Pergamon, Oxford and New York, 1964).

U. Valdrè, Ed. (1971). *Electron Microscopy in Material Science* (Academic Press, New York and London).

U. Valdrè and E. Ruedl, Eds (1976). *Electron Microscopy in Materials Science, Parts I and II* (Commission of the European Communities, Luxembourg).

D. Van Dyck (1980). Fast computational procedures for the simulation of structure images in complex or disordered crystals: a new approach. *J. Microsc. (Oxford)* **119**, 141–152.

D. Van Dyck and W. Coene (1984). The real space method for dynamical electron diffraction calculations in high resolution electron microscopy. I. Principles of the method. *Ultramicroscopy* **15**, 29–40.

M. van Heel and J. Frank (1980). Intelligent averaging of single molecules using computer alignment and correspondence analysis. II. Localization of image features. EUREM-7, The Hague, **2**, 692–693.

M. van Heel and J. Frank (1981). Use of multivariate statistics in analysing the images of biological macromolecules. *Ultramicroscopy* **6**, 187–194.

G. Van Tendeloo, D. Van Dyck and S. J. Pennycook, Eds (2012). *Handbook of Nanoscopy* (Wiley-VCH, Weinheim), 2 vols.

B. von Borries (1949). *Die Übermikroskope. Einführung, Untersuchung ihrer Grenzen und Abriß ihrer Ergebnisse* (Editio Cantor, Aulendorf/Württ).

H. S. von Harrach (2009). Development of the 300-kV Vacuum Generators STEM (1985–1996). *Adv. Imaging Electron Phys.* **159**, 287–323.

R. H. Wade and J. Frank (1977). Electron microscope transfer functions for partially coherent axial illumination and chromatic defocus spread. *Optik* **49**, 81–92.

H. Wahl (1974). Experimentelle Ermittlung der komplexen Amplitudentransmission nach Betrag und Phase beliebiger elektronenmikroskopischer Objekte mittels der Off-Axis- Bildebenenholographie. *Optik* **39**, 585–588.

H. Wahl (1975). Bildebenenholographie mit Elektronen. Habilitationsschrift, Tübingen.

A. Walther (1968). Radiometry and coherence. *J. Opt. Soc. Amer.* **58**, 1256–1259.
I. R. M. Wardell and P. E. Bovey (2009). A history of Vacuum Generators 100-kV scanning transmission electron microscope. *Adv. Imaging Electron Phys.* **159**, 221–285.
H. Watanabe and A. Tonomura (1969). Electron beam holography, [in Japanese] *Nihon Kessho Gakkai-Shi [J. Crystallogr. Soc. Jpn.]* **11**, 23–25.
H. Weisel (1910). Über die nach Fresnelscher Art beobachteten Beugungserscheinungen der Gitter. *Ann. Phys. (Leipzig)* **33** (338), 995–1031.
O. C. Wells (1974). *Scanning Electron Microscopy* (McGraw-Hill, New York).
E. W. White, H. A. McKinstry and G. G. Johnson (1968). Computer processing of SEM images. *Scanning Electron Microsc., 95–103.*
D. B. Williams and C. B. Carter (2009). *Transmission Electron Microscopy. A Textbook for Materials Science* (Springer, New York), 1st edn (1996).
M. Wolfke (1913). Über die Abbildung eines Gitters außerhalb der Einstellebene. *Ann. Phys. (Leipzig)* **40** (345), 194–200.
C. Wolpers (1991). Electron microscopy in Berlin 1928–1945. *Adv. Electron Electron Phys.* **81**, 211–229.
E. Zeitler and M. G. R. Thomson (1970). Scanning transmission electron microscopy. *Optik* **31**, 258–280 and 359–366.
J. M. Zuo and J. C. H. Spence (2017). *Advanced Transmission Electron Microscopy. Imaging and Diffraction in Nanoscience* (Springer, New York).
V. K. Zworykin, G. A. Morton, E. G. Ramberg, J. Hillier and E. E. Vance (1945). *Electron Optics and the Electron Microscope* (Wiley, New York and Chapman and Hall, London).

Part XI, Chapters 55–60

Other studies on wave propagation and diffraction have been made by Bremmer (1951a,b) and by Komrska, whose work is recapitulated in a long review (Komrska, 1971).

We recall here that the paper by Ehrenberg and Siday in which the "Aharonov–Bohm" effect was first noticed was primarily concerned with the definition of the refractive index in electron optics. Their observation provoked a response by Glaser (1951a,b) and further comment by Ehrenberg and Siday (1951).

* * *

M. Abramowitz and I. A. Stegun, Eds (1965). *Handbook of Mathematical Functions* (Dover, New York).
Y. Aharonov and D. Bohm (1959). Significance of electromagnetic potentials in the quantum theory. *Phys. Rev.* **115**, 485–491.
Y. Aharonov and D. Bohm (1961). Further considerations on electromagnetic potentials in the quantum theory. *Phys. Rev.* **123**, 1511–1524.
J. Babinet (1837). Memoires d'optique météorologique. *C. R. Acad. Sci. Paris* **4**, 638–648.
H. Boersch (1951). Über die Gültigkeit des Babinetschen Theorems. *Z. Phys.* **131**, 78–81.
H. Boersch (1968). Differenzfrequenzen von Elektronenwellen. *Z. Phys.* **215**, 28–33.
H. Boersch, H. Hamisch, K. Grohmann and D. Wohlleben (1961). Experimenteller Nachweis der Phasenschiebung von Elektronenwellen durch das magnetische Vektorpotential. *Z. Phys.* **165**, 79–93.
H. Boersch, H. Hamisch and K. Grohmann (1962). Experimenteller Nachweis der Phasenschiebung von Elektronenwellen durch das magnetische Vektorpotential. II. *Z. Phys.* **169**, 263–272.
H. Boersch, K. Grohmann, H. Hamisch, B. Lischke and D. Wohlleben (1981). On the Aharonov–Bohm effect. *Lett. Nuovo Cim.* **30**, 257–258.
M. Born and E. Wolf (2002). *Principles of Optics* (7th edn) (Cambridge University Press, Cambridge), 1st edn (Pergamon, Oxford and New York, 1959).
H. Bremmer (1951a). On the theory of optical images affected by artificial influences in the focal plane. *Physica* **17**, 63–70.
H. Bremmer (1951b). On a phase-contrast theory of electron-optical image formation. *Natl. Bur. Stand. Circular* **527**, 145–158, Published 1954.

R. Castañeda (2016). Spectrum of classes of point emitters of electromagnetic wave fields. *J. Opt. Soc. Amer. A* **33**, 1769–1776.

R. Castañeda (2017). Discreteness of the real point emitters as a physical condition for diffraction. *J. Opt. Soc. Amer. A* **34**, 184–192.

R. Castañeda and G. Matteucci (2017). New physical principle for interference of light and material particles. *Adv. Imaging Electron Phys.* **204**, 1–37.

R. Castañeda and G. Matteucci (2019). Modelo geométrico para interferencia y difracción con ondas particulas/ Geometric model for interference and diffraction with waves and particles. *Rev. Acad. Colombiana Cienc. Exactas, Fis. y. Naturales* **43**, 177–192, Open access in Spanish and English.

R. Castañeda, G. Matteucci and R. Capelli (2016a). Interference of light and of material particles: A departure from the superposition principle. *Adv. Imaging Electron Phys.* **197**, 1–43.

R. Castañeda, G. Matteucci and R. Capelli (2016b). Quantum interference without wave-particle duality. *J. Mod. Phys.* **7**, 375–389.

R. Castañeda, P. Bedoya and G. Matteucci (2021). Non-locality and geometric potential provide the phenomenology of the double-hole single massive particle and light interference. *Phys. Scr.* **96**, 125036, 20 pp.

R. Castañeda, P. Bedoya and G. Matteucci (2022). Diffractive description of the magnetic Aharonov–Bohm effect. *In preparation*.

R. G. Chambers (1960). Shift of an electron interference pattern by enclosed magnetic flux. *Phys. Rev. Lett.* **5**, 3–5.

R. Danev and K. Nagayama (2001). Transmission electron microscopy with Zernike phase plate. *Ultramicroscopy* **88**, 243–252.

R. W. Ditchburn (1963). *Light* (2nd edn) (Blackie, London), See Section 6.46.

E. Durand (1953). Le principe de Huygens et la diffraction de l'électron en théorie de Dirac. *C. R. Acad. Sci. Paris* **236**, 1337–1339.

C. Dwyer (2005a). Multislice theory of fast electron scattering incorporating atomic inner-shell ionization. *Ultramicroscopy* **104**, 141–151.

C. Dwyer (2005b). Relativistic effects in atomic inner-shell ionization by a focused electron probe. *Phys. Rev. B* **72**, 144102.

C. Dwyer and J. S. Barnard (2006). Relativistic effects in core-loss electron diffraction. *Phys. Rev. B* **74**, 064106.

W. Ehrenberg and R. E. Siday (1949). The refractive index in electron optics and the principles of dynamics. *Proc. Phys. Soc. Lond. B* **62**, 8–21.

W. Ehrenberg and R. E. Siday (1951). The refractive index in electron optics. *Proc. Phys. Soc. Lond. B* **64**, 1088–1089.

H. A. Ferwerda, B. J. Hoenders and C. H. Slump (1986a). Fully relativistic treatment of electron-optical image formation based on the Dirac equation. *Opt. Acta* **33**, 145–157.

H. A. Ferwerda, B. J. Hoenders and C. H. Slump (1986b). The fully relativistic foundation of linear transfer theory in electron optics based on the Dirac equation. *Opt. Acta* **33**, 159–183.

H. A. Fowler, L. Marton, J. A. Simpson and J. A. Suddeth (1961). Electron interferometer studies of iron whiskers. *J. Appl. Phys.* **32**, 1153–1155.

M. Frigge (2011). Numerische Berechnung von Wirkungsquerschnitten im Zentralfeldmodell. Dissertation, Münster.

W. Glaser (1943). Bildentstehung und Auflösungsvermögen des Elektronenmikroskops vom Standpunkt der Wellenmechanik. *Z. Phys.* **121**, 647–666.

W. Glaser (1950a). Zur wellenmechanischen Theorie der elektronenoptischen Abbildung. *Sitzungsber. Öst. Akad. Wiss., Math.-Naturwiss. Kl., Abt. IIa* **159**, 297–360.

W. Glaser (1950b). Etude en théorie ondulatoire de la répartition des intensités dans les images électroniques. ICEM-2, Paris, **1**, 63–72 and 164.

W. Glaser (1951a). The refractive index of electron optics and its connection with the Routhian function. *Proc. Phys. Soc. Lond. B* **64**, 114–118.

W. Glaser (1951b). The refractive index in electron optics. *Proc. Phys. Soc. Lond. B* **64**, 1089.

W. Glaser (1951c). Fundamental problems of theoretical electron optics. *Natl. Bur. Stand. Circular* **527**, 111–126, Published 1954.
W. Glaser (1952). *Grundlagen der Elektronenoptik* (Springer, Vienna).
W. Glaser (1953). Über die Bewegung eines "Wellenpakets" in einer Elektronenlinse. *Öst. Ing.-Arch* **7**, 144–152.
W. Glaser (1954). Bildentstehung im Elektronenmikroskop. *Optik* **11**, 101–117.
W. Glaser (1956). Elektronen- und Ionenoptik. *Handbuch der Physik* **33**, 123–395.
W. Glaser and G. Braun (1954). Zur wellenmechanischen Theorie der elektronenoptischen Abbildung. I. *Acta Phys. Austriaca* **9**, 41–74.
W. Glaser and G. Braun (1955). Zur wellenmechanischen Theorie der elektronenoptischen Abbildung. II. *Acta Phys. Austriaca* **9**, 267–296.
W. Glaser and P. Schiske (1953). Elektronenoptische Abbildung auf Grund der Wellenmechanik. I. II. *Ann. Phys. (Leipzig)* **12** (447), 240–266 and 267–280.
P. Grivet (1965). *Electron Optics* (Pergamon, Oxford and New York), 2nd edn (1972).
W. Hoppe (1961). Ein neuer Weg zur Erhöhung des Auflösungsvermögens des Elektronenmikroskops. *Naturwissenschaften* **48**, 736–737.
W. Hoppe (1963). Fresnelsche Zonenkorrekturplatten für das Elektronenmikroskop. *Optik* **20**, 599–606.
R. Hosemann and D. Joerchel (1954). Die notwendige Korrektion am Babinetschen Theorem. *Z. Phys.* **138**, 209–221.
A. Howie (1983). Problems of interpretation in high resolution electron microscopy. *J. Microsc. (Oxford)* **129**, 239–251.
A. Ishizuka, K. Mitsuishi and K. Ishizuka (2018). Direct observation of curvature of the wave surface in transmission electron microscope using transport intensity equation. *Ultramicroscopy* **194**, 7–14.
R. Jagannathan (1990). Quantum theory of electron lenses based on the Dirac equation. *Phys. Rev.* **A 42**, 6674–6689; corrigendum, *ibid.* **A 44** (1991) 7856.
R. Jagannathan and S. A. Khan (1996). Quantum theory of the optics of charged particles. *Adv. Imaging Electron Phys.* **97**, 257–358.
R. Jagannathan and S. A. Khan (2019). *Quantum Mechanics of Charged Particle Beam Optics* (CRC Press, Boca Raton, FL).
R. Jagannathan, R. Simon, E. C. G. Sudarshan and N. Mukunda (1989). Quantum theory of magnetic electron lenses based on the Dirac equation. *Phys. Lett. A* **134**, 457–464.
E. Kamke (1977). *Differentialgleichungen, Lösungsmethoden und Lösungen* (Teubner, Stuttgart).
E. Kasper (1973). Eine Verallgemeinerung der Schrödingerschen Wellenmechanik in relativistische Gültigkeitsbereiche. *Z. Naturforsch.* **28a**, 216–221.
S. A. Khan (2006). The Foldy–Wouthuysen transformation technique in optics. *Optik* **117**, 481–488.
S. A. Khan (2008). The Foldy–Wouthuysen transformation technique in optics. *Adv. Imaging Electron Phys.* **152**, 49–78.
S. A. Khan (2017). Quantum methodologies in Maxwell optics. *Adv. Imaging Electron Phys.* **201**, 57–135.
S. A. Khan (2018). E.C.G. Sudarshan and the quantum mechanics of charged-particle beam optics. *Curr. Sci.* **115**, 1813–1814.
S. A. Khan and R. Jagannathan (1995). Quantum mechanics of charged-particle beam transport through magnetic lenses. *Phys. Rev. E* **51**, 2510–2515.
S. A. Khan and R. Jagannathan (2020). Quantum mechanics of bending of a nonrelativistic charged particle beam by a dipole magnet. *Optik* **206**, 163626, 11 pp.
S. A. Khan and R. Jagannathan (2021). Quantum mechanics of round magnetic electron lenses with Glaser and power law models of $B(z)$. *Optik* **229**, 166303, 23 pp.
R. Knippelmayer (1996). Relativistische Betrachtung der unelastischen Streuung von Elektronen für Anwendungen in der Transmissionselektronenmikroskopie. Dissertation, Münster.
R. Knippelmayer, P. Wahlbring and H. Kohl (1997). Relativistic ionisation cross sections for use in microanalysis. *Ultramicroscopy* **68**, 25–41.

J. Komrska (1971). Scalar diffraction theory in electron optics. *Adv. Electron Electron Phys.* **30**, 139–234.

J. Komrska and M. Lenc (1972). Wave mechanical approach to magnetic lenses. EUREM-5, Manchester, 78–79.

E. Krimmel (1960). Kohärente Teilung eines Elektronenstrahls durch Magnetfelder. *Z. Phys.* **158**, 35–38.

E. Krimmel (1961). Elektronen-Interferenzen in der Umgebung der Brennlinie einer magnetischen Quadrupollinse. *Z. Phys.* **163**, 339–355.

F. Lenz (1961). Eine Erweiterung des Cornu-Diagramms auf den Fall einer teildurchlässigen und phasenschiebenden Halbebene. *Z. Phys.* **164**, 425–427.

F. Lenz (1962). Zur Phasenschiebung von Elektronenwellen im feldfreien Raum durch Potentiale. *Naturwissenschaften* **49**, 82.

F. Lenz (1963). Zonenplatten zur Öffnungsfehlerkorrektur und zur Kontrasterhöhung. *Z. Phys.* **172**, 498–502.

F. Lenz (1964). Dimensionierung und Anordnung von Hoppe-Platten und Kontrastplatten in starken Magnetlinsen. *Optik* **21**, 489–493.

F. Lenz (1965). The influence of lens imperfections on image formation. *Lab. Invest.* **14**, 808–818.

F. Lenz and E. Krimmel (1961). Die allgemeine Intensitätsverteilung in der kohärent ausgeleuchteten Umgebung einer Kaustikfläche. *Z. Phys.* **163**, 356–362.

F. Lenz and E. Krimmel (1963). Bilder lokaler Aufladungen im Elektronen-Spiegelmikroskop. *Z. Phys.* **175**, 235–241.

F. Lenz and A. P. Wilska (1966). Electron optical systems with annular apertures and with corrected spherical aberration. *Optik* **24**, 383–396.

H. Lipson and K. Walkley (1968). On the validity of Babinet's principle for Fraunhofer diffraction. *Opt. Acta* **15**, 83–91.

R. F. Loane, P. Xu and J. Silcox (1991). Thermal vibrations in convergent-beam electron diffraction. *Acta Cryst. A* **47**, 267–278.

S. Löffler, A.-L. Hamon, D. Aubry and P. Schattschneider (2020). A quantum propagator for electrons in a round magnetic lens. *Adv. Imaging Electron Phys.* **215**, 89–105.

A. Lubk (2018). Holography and tomography with electrons. *Adv. Imaging Electron Phys.* **206**, 1–318.

S. Majert and H. Kohl (2019). Simulation of atomically resolved elemental maps with a multislice algorithm for relativistic electrons. *Adv. Imaging Electron Phys.* **211**, 1–120.

G. Möllenstedt and W. Bayh (1962). Messung der kontinuierlichen Phasenschiebung von Elektronenwellen im kraftfeldfreien Raum durch das magnetische Vektorpotential einer Luftspule. *Naturwissenschaften* **49**, 81–82.

G. Möllenstedt, K. H. von Grote and C. Jönsson (1963). Production of Fresnel zone plates for extreme ultraviolet and soft radiation. In *X-Ray Optics and X-Ray Microanalysis* (H. H. Pattee, V. E. Cosslett and A. Engström, Eds), 73–79 (Academic Press, New York and London).

F. W. J. Olver, D. W. Lozier, R. F. Boisvert and C. W. Clark (2010). *NIST Handbook of Mathematical Functions* (Cambridge University Press, Cambridge).

Phan–Van–Loc. (1953). Sur le principe de Huygens en théorie de l'électron de Dirac. *C.R. Acad. Sci. Paris* **237**, 649–651.

Phan–Van–Loc. (1954). Principe de Huygens en théorie de l'électron de Dirac et intégrales de contour. *C.R. Acad. Sci. Paris* **238**, 2494–2496.

Phan–Van–Loc. (1955). Diffraction des ondes ψ_n de l'électron de Dirac. *Ann. Fac. Sci. Univ. Toulouse* **18**, 178–192.

Phan–Van–Loc. (1958a). Interprétation physique de l'expression mathématique du principe de Huygens en théorie de l'électron de Dirac. *Cah. de. Phys.* **12** (97), 327–340.

Phan–Van–Loc. (1958b). Une nouvelle manière d'établir l'expression mathématique du principe de Huygens en théorie de l'électron de Dirac. *C.R. Acad. Sci. Paris* **246**, 388–390.

Phan–Van–Loc (1960). Principes de Huygens en Théorie de l'Electron de Dirac. Thèse, Toulouse.

F. Pokroppa (1999). Berechnung von relativistischen Wirkungsquerschnitten im Rahmen des Zentralfeldmodells für Anwendungen in der Elektronenenergieverlustspektroskopie. Dissertation, Münster.

O. Rang (1964). Zur Eichtransformation der Elektronenwelle. *Optik* **21**, 59–65.

O. Rang (1977). Is the electron wavelength an observable? *Ultramicroscopy* **2**, 149–151.

L. Reimer (1998). *Scanning Electron Microscopy* (Springer, Berlin and New York), 1st edn (1985).

L. Reimer and H. Kohl (2008). *Transmission Electron Microscopy* (Springer, New York).
H. Rose (2012). *Geometrical Charged-Particle Optics* (Springer, Berlin).
A. Rother and K. Scheerschmidt (2009). Relativistic effects in elastic scattering of electrons in TEM. *Ultramicroscopy* **109**, 154–160.
A. Rubinowicz (1934). Über das Kirchhoffsche Beugungsproblem für Elektronenwellen. *Acta Phys. Pol.* **3**, 143–163.
A. Rubinowicz (1957). *Die Beugungswelle in der Kirchhoffschen Theorie der Beugung* (Pańtswowe Wydawnictwo Naukowe, Warsaw).
A. Rubinowicz (1963). Beugungswellen verschiedener Felder im Falle beliebiger einfallender Wellen. *Acta Phys. Pol.* **23**, 727–744.
A. Rubinowicz (1965). The Miyamoto–Wolf diffraction wave. *Prog. Opt.* **4**, 199–240, See Part II, Section 3: Diffraction wave in the Kirchhoff theory of Dirac-electron waves.
A. Rubinowicz (1966). *Die Beugungswelle in der Kirchhoffschen Theorie der Beugung* (2nd edn) (PWN, Warsaw and Springer, Berlin and New York).
S. Samarin, O. Artamonov and J. Williams (2018). *Spin-Polarized Two-Electron Spectroscopy of Surfaces* (Springer Nature, Switzerland).
W. O. Saxton, M. A. O'Keefe, D. J. H. Cockayne and M. Wilkens (1983). Sign conventions in electron diffraction and imaging. *Ultramicroscopy* **12**, 75–77, Corrigendum *ibid.* **13**, 349–350.
O. Scherzer (1949). The theoretical resolution limit of the electron microscope. *J. Appl. Phys.* **20**, 20–29.
A. Sommerfeld (1954). *Optics* (Academic Press, New York).
J. C. H. Spence (2013). *High-Resolution Electron Microscopy* (Oxford University Press, Oxford).
K. Strehl (1902). Ueber Luftschlieren und Zonenfehler. *Z. Instrumentenkde* **22**, 213–217.
A. Tonomura, T. Matsuda, R. Suzuki, A. Fukuhara, N. Osakabe, H. Umezaki, J. Endo, K. Shinagawa, Y. Sugita and H. Fujiwara (1982). Observation of Aharonov–Bohm effect by electron holography. *Phys. Rev. Lett.* **48**, 1443–1446.
A. Tonomura, H. Umezaki, T. Matsuda, N. Osakabe, J. Endo and Y. Sugita (1984). Electron holography, Aharonov–Bohm effect and flux quantization. In *Proceedings of the International Symposium on Foundations of Quantum Mechanics in the Light of New Technology*, Tokyo 1983 (S. Kamefuchi, H. Ezawa, Y. Murayama, M. Namiki, S. Nomura, Y. Ohnuki and T. Yajima, Eds), 20–28 (Physical Society of Japan, Tokyo).
D. B. Williams and C. B. Carter (2009). *Transmission Electron Microscopy,* 2nd edn (Springer, New York).

Part XII, Chapters 61–63

The references cited in this part represent only a small fraction of the voluminous literature of electron interferometry and holography. We have, therefore, grouped here further papers, of a general nature or dealing with aspects of the subject not considered here. More recent publications are cited in the text.

Electron interferometry. Instrumental aspects are considered by Buhl (1961b), Costa et al. (1989), Harada et al. (1988), Hibi and Takahashi (1963), Kawasaki et al. (1990a,b, 1992a), Matteucci (1978), Matteucci et al. (1982b), Merli et al. (1974), Möllenstedt and Krimmel (1964), Ogai et al. (1991), Pozzi (1977, 1983a,b, 1992), Rau et al. (1991a,b), Ru et al. (1991a, 1992a,b), Schaal (1971), Stumpp (1984), Stumpp et al. (1984) and Tanji et al. (1991).

For more details on electron interferometry, refer Donati et al. (1973), Faget et al. (1960), Fagot et al. (1961), Fert and Faget (1958), Fert et al. (1962a,b), Komrska (1971, 1975), Komrska and Lenc (1970), Komrska and Vlachová (1973), Komrska et al. (1964a,b, 1967), Fischer and Lischke (1967), Gabor (1956), Hasselbach (1992), Lenz (1965), Lichte (1984b, 1986d), Matsuda et al. (1978), Matteucci et al. (1979), Merli et al. (1976a), Möllenstedt (1960, 1962, 1987), Möllenstedt and Bayh (1961a,b), Möllenstedt and Lenz (1962), Möllenstedt and Lichte (1978a,b, 1979), Pozzi (1975, 1980b), Rang (1953), Rogers (1970), Stroke (1967), Takeda and Ru (1985), Tonomura et al. (1978c, 1989), Fu et al. (1987) and Yatagai et al. (1987). The light shed on the concept of electron phase by interference experiments is discussed notably by Möllenstedt (1988) and Möllenstedt and Lichte (1989).

Electron holography. The following papers are concerned with some aspect of the technique: Ade (1982, 1994), Allard et al. (1992), Anaskin and Stoyanova (1972), Boersch (1967), Bonhomme and Beorchia (1980), Bonnet et al. (1978), Boseck et al. (1986), Buhl (1961a), Chen et al. (1987, 1993), Cowley (1991), Endo and Tonomura (1990), Endo et al. (1986, 1989), Estrada et al. (1991), Gabor (1968/9), Gabor et al. (1965), Gajdardziska-Josifovska et al. (1993), Greenaway and Huiser (1976), Hanszen (1969, 1970a, 1980, 1982c, 1984, 1986b) Hanszen and Ade (1976a,b, 1977, 1984), Hanszen and Lauer (1980), Hanszen et al. (1980, 1981, 1982, 1985, 1986), Harada and Shimizu (1991), Herrmann et al. (1978), Laberrigue et al. (1980), Lannes (1978, 1980, 1982), Lauer (1982, 1984a,b), Lauer and Ade (1992), Lichte (1982, 1984a, 1985, 1986a,b,c, 1989, 1991c,d,e, 1992a,b,c), Lichte and Völkl (1988, 1991), Lichte et al. (1992), Matteucci et al. (1982c), Menzel et al. (1973), Plass and Marks (1992), Pozzi (1992, 1993), Pozzi and Prola (1987), Saxon (1972a,b), Tonomura (1969, 1986b, 1987a-c, 1989, 1990, 1991a,b, 1992d), Tonomura et al. (1978a,b, 1979a, 1979c), Völkl and Lichte (1990a,b), Wade (1974, 1975, 1980), Weingärtner et al. (1969a,b, 1970, 1971), Weiss et al. (1993), Zeitler (1979) and Zhang and Joy (1991).

Some unconventional forms of holography involving electrons are discussed by Bartell (1972, 1975), Bartell and Johnson (1977), Bartell and Ritz (1974) and Gabor (1980); by Bates and Lewitt (1975); by Garcia (1989); by Fink et al. (1991), Fink and Kreuzer (1992) and Spence et al. (1992); by Lichte and Hornstein (1982); by Tong et al. (1991); by Tonomura and Matsuda (1980); by Saldin (1991); and by Qian et al. (1993), Scheinfein et al. (1993) and Spence and Qian (1992).

For further discussion of the relation between contrast-transfer theory and holography, see the following papers, from Hanszen's group: Hanszen (1970b, 1971, 1972b, 1973a, 1974, 1976), Hanszen and Ade (1974), Hanszen et al. (1983b) and the historical accounts by Hanszen (1990a,b).

The most numerous applications of holography are concerned with magnetic specimens. The following list contains a representative selection of this work: Arii et al. (1981), Fabbri et al. (1987), Frost and Lichte (1988), Fukuhara et al. (1983), Hasegawa et al. (1989), Lau and Pozzi (1978), Matsuda et al. (1982), Matteucci et al. (1982a, 1988a,c), Matteucci and Muccini (1992), Migliori and Pozzi (1992), Olivei (1969), Osakabe et al. (1983), Pozzi (1980a), Pozzi and Missiroli (1973), Ru et al. (1991b), Tonomura (1972, 1983, 1987b, 1992b), Tonomura et al. (1980a,b, 1982b,c,d), Wahl and Lau (1979), Yoshida et al. (1983). Trapped flux in superconductors and fluxons are particularly examined by Boersch and Lischke (1970a,b), Boersch et al. (1974), Harada et al. (1992), Lischke (1969, 1970), Matsuda et al. (1989, 1990), Migliori and Pozzi (1991), Migliori et al. (1993), Tonomura et al. (1984, 1990) and Wahl (1968/9, 1970a,b). Ferroelectrics have also been studied by Zhang et al. (1992, 1993a,b).

In an unpublished holographic display of scanning electron microscope images was proposed by Kulick et al. (1987). In an unpublished exchange of letters between Dr H. Börsch [sic] in Tettnang and Dr D. Gabor in Rugby, Boersch comments on Gabor's note in Nature (1948) and draws attention to his article in the *Zeitschrift für Technische Physik* (1938). He assumes (rightly) that Gabor has not seen this article. Boersch's letter is dated 5 July 1948 and Gabor's reply, 18 July 1948.

* * *

G. Ade (1973). Der Einfluss der Bildfehler dritter Ordnung auf die elektronenmikroskopische Abbildung und die Korrektur dieser Fehler durch holographische Rekonstruktion. Dissertation, Braunschweig and PTB-Bericht APh–3.

G. Ade (1980). Effects of film nonlinearities on reconstructed images in electron microscopy. EUREM-7, The Hague, **1**, 138–139.

G. Ade (1982). Limitation of resolution in electron microscopical Fourier transform holography. ICEM-10, Hamburg, **1**, 425–426.

G. Ade (1986). Dynamical phase shifts of transmitted waves behind a monocrystal and their holographic visualization. ICEM-11, Kyoto, **1**, 687–688.

G. Ade (1988). On the digital reconstruction of the complex object function in electron off-axis holography. EUREM-9, York, **1**, 201–202.

G. Ade (1992). Phase distribution and differential phase images of the object obtained by digital processing of off-axis holograms. EUREM-10, Granada, **1**, 489–490.

G. Ade (1994). Digital techniques in electron off-axis holography. *Adv. Electron Electron Phys.* **89**, 1–51.

G. Ade (1997). A digital method for noise reduction in holographic reconstructions and electron microscopical images. *Scanning Microsc.* **11**, 375–378.
G. Ade and R. Lauer (1988). Nearly online digital reconstruction of off-axis electron holograms. EUREM-9, York, **1**, 203–204.
G. Ade and R. Lauer (1990). Digital phase amplification in electron off-axis holography. ICEM-12, Seattle, WA, **1**, 232–233.
G. Ade and R. Lauer (1991). Digital interferometric reconstruction of the object wave from off-axis electron holograms. *Optik* **88**, 103–108.
G. Ade and R. Lauer (1992a). Rekonstruktion der Objektphase und Bestimmung von atomaren Dickenstufen durch digitale Verarbeitung von Off-Axis-Elektronenhologrammen. *PTB–Mitt* **102**, 181–187.
G. Ade and R. Lauer (1992b). Digital phase determination and amplification techniques in electron off-axis holography. *Optik* **91**, 5–10.
G. Ade, K.-J. Hanszen and R. Lauer (1984). Image properties and noise in off-axis holography and Fourier holography. EUREM-8, Budapest, **1**, 283–284.
A. Agarwal, C.-s Kim, R. Hobbs, D. Van Dyck and K. K. Berggren (2017). A nanofabricated, monolithic, path-separated electron interferometer. *Sci. Rep.* **7**, 1677, 15 pp.
Y. Aharonov and D. Bohm (1959). Significance of electromagnetic potentials in the quantum theory. *Phys. Rev.* **115**, 485–491.
Y. Aharonov and D. Bohm (1961). Further considerations on electromagnetic potentials in the quantum theory. *Phys. Rev.* **123**, 1511–1524.
Y. Aharonov and G. Carmi (1973). Quantum aspects of the equivalence principle. *Found. Phys.* **3**, 493–498.
T. Akashi, Y. Takahashi, T. Tanigaki, T. Shimakura, T. Kawasaki, T. Furutsu, H. Shinada, H. Müller, M. Haider, N. Osakabe and A. Tonomura (2015). Aberration corrected 1.2-MV cold field-emission transmission electron microscope with a sub-50-pm resolution. *Appl. Phys. Lett.* **106**, 074101.
L. F. Allard, T. A. Nolan, D. C. Joy and H. Hashimoto (1992). Digital imaging for high-resolution electron holography. EMSA 50, Boston, MA, 944–945.
I. F. Anaskin and I. G. Stoyanova (1967). Three-beam interference of electrons with the help of an electrostatic prism. *Dokl. Akad. Nauk. SSSR* **174**, 56–59, *Sov. Phys. Dokl.* **12**, 447–450.
I. Anaskin and I. Stoyanova (1968a). Multi-beam interference of the electrons by the splitting of wave front. EUREM-4, Rome, **1**, 149–150.
I. Anaskin and I. G. Stoyanova (1968b). Concerning the determination of the mean internal potential in an electron interference microscope. *Zh. Eksp. Teor. Fiz.* **54**, 1687–1689, *J. Exp. Theor. Phys.* **27**, 904–905.
I. F. Anaskin and I. G. Stoyanova (1968c). Quality of the interference image in an electron interferometer. *Radiotekh. Elektron.* **13**, 913–920, *Radio Eng. Electron Phys.* **13**, 789–794.
I. F. Anaskin and I. G. Stoyanova (1968d). Investigation of three-beam interference of electrons obtained with the aid of an electrostatic prism. *Radiotekh. Elektron.* **13**, 1031–1040, *Radio Eng. Electron Phys.* **13**, 895–902.
I. F. Anaskin and I. G. Stoyanova (1972). Resolution limit of reconstruction methods in electron microscopy. EUREM-5, Manchester, 636–637.
I. F. Anaskin, I. G. Stoyanova and A. F. Chyapas (1966). An electron interference microscope and electron interferometer based on the UEMV-100 electron microscope. *Izv. Akad. Nauk. SSSR (Ser. Fiz.)* **30**, 766–768, *Bull. Acad. Sci. USSR (Phys. Ser.)* **30**, 793–796.
I. F. Anaskin, I. G. Stoyanova and M. D. Shpagina (1968). Multibeam interference of electrons achieved by splitting the electron wavefront. *Izv. Akad. Nauk. SSSR* **32**, 1016–1021, *Bull. Acad. Sci. USSR* **32**, 941–946.
T. Arii, K. Mihama, T. Matsuda and A. Tonomura (1981). Observation of magnetic structure for polyhedral fine iron particles by Lorentz microscopy and electron holography. *J. Electron Microsc.* **30**, 121–127.
M. Arita, R. Tokuda, K. Hamada and Y. Takahashi (2014). Development of TEM holder generating in-plane magnetic field used for in-situ TEM observation. *Mater. Trans.* **55**, 403–409.
R. Bach, D. Pope, S.-h Liou and H. Batelaan (2013). Controlled double-slit electron diffraction. *New J. Phys.* **15**, 033018, 7 pp.

S. Bals, S. Van Aert, G. Van Tendeloo and D. Ávila-Brande (2006). Statistical estimation of atomic positions from exit wave reconstruction with a precision in the picometer range. *Phys. Rev. Lett.* **96**, 096106, 4 pp.

S. Bals, M. Casavola, M. A. van Huis, S. Van Aert, K. J. Batenburg, G. Van Tendeloo and D. Vanmaekelbergh (2011). Three-dimensional atomic imaging of colloidal core–shell nanocrystals. *Nano Lett.* **11**, 3420–3424.

S. Bals, S. Van Aert, C. P. Romero, K. Lauwaet, M. J. Van Bael, B. Schoeters, B. Partoens, E. Yücelen, P. Lievens and G. Van Tendeloo (2012). Atomic scale dynamics of ultrasmall germanium clusters. *Nature Commun.* **3**, 897, 6 pp.

P. Banerjee, C. Roy, S. K. De, A. J. Santos, F. M. Morales and S. Bhattacharyya (2020). Atomically resolved tomographic reconstruction of nanoparticles from single projection: influence of amorphous carbon support. *Ultramicroscopy* **221**, 113177, 15 pp.

H. Banzhof and K.-H. Herrmann (1993). Reflection electron holography. *Ultramicroscopy* **48**, 475–481.

H. Banzhof, K.-H. Herrmann and H. Lichte (1988). Reflexion electron interferometry of single crystal surfaces. EUREM-9, York, **1**, 263–264.

H. Banzhof, K.-H. Herrmann and H. Lichte (1992). Reflection electron microscopy and interferometry of atomic steps on gold and platinum single crystal surfaces. *Microsc. Res. Tech.* **20**, 450–456.

L. S. Bartell (1972). Determination of electron distributions by electron diffraction. *Trans. Amer. Cryst. Assoc.* **8**, 37–57.

L. S. Bartell (1975). Images of gas atoms by electron holography. I. Theory. II. Experiment and comparison with theory. *Optik* **43**, 373–390 and 403–418.

L. S. Bartell and R. D. Johnson (1977). Molecular images by electron-wave holography. *Nature* **268**, 707–708.

L. S. Bartell and C. L. Ritz (1974). Atomic images by electron-wave holography. *Science* **185**, 1163–1165.

H. Batelaan (2007). Illuminating the Kapitza–Dirac effect with electron matter optics. *Rev. Mod. Phys.* **79**, 929–941.

H. Batelaan and M. Becker (2015). Dispersionless forces and the Aharonov–Bohm effect. *Europhys. Lett.* **112**, 40006.

H. Batelaan and S. McGregor (2014). Do dispersionless forces exist? In *In Memory of Akira Tonomura, Physicist and Electron Microscopist* (K. Fujikawa and Y. A. Ono, Eds), 122–129 (World Scientific, Singapore).

H. Batelaan and A. Tonomura (2009). The Aharonov–Bohm effects: variations on a subtle theme. *Phys. Today* **62**, 38–43.

R. H. T. Bates and R. M. Lewitt (1975). Crysto-holography. *Optik* **44**, 1–16.

H. Bauer, F. Hasselbach, H. Kiesel, U. Maier and A. Schäfer (1996). New experiments in charged particle interferometry. In *Quantum Interferometry* (F. de Martini, G. Denardo and Y. Shiy, Eds), 349–360 (VCH, Weinheim).

W. Bayh (1962). Messung der kontinuierlichen Phasenschiebung von Elektronenwellen im kraftfeldfreien Raum durch das magnetische Vektorpotential einer Wolfram-Wendel. *Z. Phys.* **169**, 492–510.

G. Baym and K. Shen (2014). Hanbury Brown–Twiss interferometry with electrons: Coulomb vs. quantum statistics. In *In Memory of Akira Tonomura, Physicist and Electron Microscopist* (K. Fujikawa and Y. A. Ono, Eds), 201–210 (World Scientific, Singapore).

A. Béché, J. L. Rouvière, J. P. Barnes and D. Cooper (2011). Dark field electron holography for strain measurement. *Ultramicroscopy* **111**, 227–238.

M. Becker, G. Guzzinati, A. Béché, J. Verbeeck and H. Batelaan (2019). Asymmetry and non-dispersivity in the Aharonov–Bohm effect. *Nature Commun.* **10**, 1700, 10 pp.

P. J. Beierle, L. Zhang and H. Batelaan (2018). Experimental test of decoherence theory using electron matter waves. *New J. Phys.* **20**, 113030, 12 pp.

M. V. Berry (1999). Aharonov–Bohm beam deflection: Shelankov's formula, exact solution, asymptotics and an optical analogue. *J. Phys. A: Math. Gen.* **32**, 5627–5642.

A. Beyer and A. Gölzhäuser (2010). Low energy electron point source microscopy: beyond imaging. *J. Phys.: Cond. Mater.* **22**, 343001t.

L. Biberman, N. Sushkin and V. Fabrikant (1949). Diffraction of electrons moving in succession. *Dokl. Akad. Nauk. SSSR* **66**, 185–186.

P. Blanchard, E. Joos, D. Giulini, C. Kiefer, and I.-O. Stamatescu, Eds (2000). *Decoherence: Theoretical, Experimental, and Conceptual Problems* (Springer, Berlin). Lecture Notes in Physics, vol. 538.

R. Blanco (1999). On a hypothetical explanation of the Aharonov–Bohm effect. *Found. Phys.* **29**, 693–720.

H. Boersch (1938). Zur Bilderzeugung im Mikroskop. *Z. Tech. Phys.* **19**, 337–338.

H. Boersch (1967). Holographie und Elektronenoptik. *Phys. Blätt* **23**, 393–404.

H. Boersch and B. Lischke (1970a). Direkte Beobachtung einzelner magnetischer Flußquanten in supraleitenden Hohlzylindern. I. *Z. Phys.* **237**, 449–468.

H. Boersch and B. Lischke (1970b). Electron interferometric measurements of quantized magnetic flux trapped in superconducting tubes. ICEM-7, Grenoble, **1**, 69–70.

H. Boersch, H. J. Hamisch, D. Wohlleben and K. Grohmann (1960). Antiparallele Weißsche Bereiche als Biprisma für Elektroneninterferenzen. *Z. Phys.* **159**, 397–404.

H. Boersch, H. Hamisch, K. Grohmann and D. Wohlleben (1961). Experimentelle Nachweis der Phasenschiebung von Elektronenwellen durch das magnetische Vektorpotential. *Z. Phys.* **165**, 79–93.

H. Boersch, H. J. Hamisch, K. Grohmann and D. Wohlleben (1962a). Antiparallele Weißsche Bereiche als Biprisma für Elektroneninterferenzen. II. *Z. Phys.* **167**, 72–82.

H. Boersch, H. Hamisch and K. Grohmann (1962b). Experimentelle Nachweis der Phasenschiebung von Elektronenwellen durch das magnetische Vektorpotential. II. *Z. Phys.* **169**, 263–272.

H. Boersch, B. Lischke and H. Söllig (1974). Dynamics of single flux lines in superconducting films by vortex microscopy experiments. *Phys. Stat. Sol. (b)* **61**, 215–222.

P. Bonhomme and A. Beorchia (1980). On-line holography of electron microscope images. EUREM-7, The Hague, **1**, 134–135.

N. Bonnet, M. Troyon and P. Gallion (1978). Possible applications of Fraunhofer holography in high resolution electron microscopy. ICEM-9, Toronto, **1**, 222–223.

P. G. Borzjak, J. A. Kuljupin, S. A. Nepijko, A. P. Ostranitsa and V. G. Shamonja (1977). On new possibilities of electron interference. *Dopovidi Akad. Nauk. Ukr. RSR, Ser. A* (10), 920–922 [in Ukrainian; Note, names transliterated by authors].

S. Boseck, H. Block, B. Schmidt and E. Reuber (1986). Use of synthetic holograms in coherent image processing for high resolution micrographs of a CTEM. ICEM-11, Kyoto, **1**, 683–684.

V. Boureau, R. McLeod, B. Mayall and D. Cooper (2018). Off-axis electron holography combining summation of hologram series with double-exposure phase-shifting: theory and application. *Ultramicroscopy* **193**, 52–63.

T. H. Boyer (1972). Misinterpretation of the Aharonov-Bohm effect. *Amer. J. Phys.* **40**, 56–59.

T. H. Boyer (1987). The Aharonov–Bohm effect as a classical electromagnetic-lag effect: an electrostatic analogue and possible experimental test. *Nuovo Cim.* **100B**, 685–701.

T. H. Boyer (2000a). Does the Aharonov–Bohm effect exist? *Found. Phys.* **30**, 893–905.

T. H. Boyer (2000b). Classical electromagnetism and the Aharonov–Bohm phase shift. *Found. Phys.* **30**, 907–932.

T. H. Boyer (2002a). Classical electromagnetic interaction of a point charge and a magnetic moment: considerations related to the Aharonov–Bohm phase shift. *Found. Phys.* **32**, 1–39.

T. H. Boyer (2002b). Semiclassical explanation of the Matteucci–Pozzi and Aharonov–Bohm phase shifts. *Found. Phys.* **32**, 41–49.

T. H. Boyer (2008). Comment on experiments related to the Aharonov–Bohm phase shift. *Found. Phys.* **38**, 498–505.

T. H. Boyer (2015). Classical interaction of a magnet and a point charge: the Shockley–James paradox. *Phys. Rev. E* **91**, 013201, 11 pp.

K.-J. Braun (1972). Untersuchung der Kohärenzeigenschaften von Elektronenwellen mit dem Elektroneninterferometer. Diplomarbeit, Tübingen.

W. Brünger (1968). Feldemissionskathode zur kohärenten Beleuchtung des Elektronen-Biprismas. *Naturwissenschaften* **55**, 295.

W. Brünger (1972). Elektroneninterferometer mit Feldemissionskathode zur Messung von Kontaktpotentialdifferenzen im Ultrahochvakuum. *Z. Phys.* **250**, 263–272.

W. Brünger and M. Klein (1977). Contact potential difference between Au and Ag measured by electron interferometry. *Surf. Sci.* **62**, 317–320.

R. Buhl (1958). Elektronen-Interferenz-Mikroskopie. ICEM-4, Berlin, **1**, 233–234.

R. Buhl (1959). Interferenzmikroskopie mit Elektronenwellen. *Z. Phys.* **155**, 395–412.

R. Buhl (1961a). Vefahren zur Kompensation von magnetischen Wechselfeldern in Elektronen-Interferometern. *Z. Angew. Phys.* **13**, 232–235.

R. Buhl (1961b). Elektronen-Dreistrahlinterferenzen mit elektrostatischen Biprismen. *Naturwissenschaften* **48**, 298–299.

B. F. Buxton, G. M. Rackham and J. W. Steeds (1978). The dynamical theory of a double crystal electron interferometer. ICEM-9, Toronto, **1**, 188–189.

A. Caprez and H. Batelaan (2009). Feynman's relativistic electrodynamics paradox and the Aharonov–Bohm effect. *Found. Phys.* **39**, 295–306.

A. Caprez, B. Barwick and H. Batelaan (2007). Macroscopic test of the Aharonov–Bohm effect. *Phys. Rev. Lett.* **99**, 20401, 4 pp.

R. Carles, O. Pujol and J.-P. Pérez (2021). The ESAB effect and the physical meaning of the vector potential. *Adv. Imaging Electron Phys.* **218**, 181–194.

E. Carlino (2020). In-line holography in transmission electron microscopy for the atomic resolution imaging of single particles of radiation-sensitive matter. *Materials* **13**, 1413, 19 pp.

E. Carlino, F. Scattarella, L. de Caro, C. Giannini, D. Siliqi, A. Colombo and D. E. Galli (2018). Coherent diffraction imaging in transmission electron microscopy for atomic resolution quantitative studies of the matter. *Materials* **11**, 2323, 14 pp.

R. G. Chambers (1960). Shift of an electron interference pattern by enclosed magnetic flux. *Phys. Rev. Lett.* **5**, 3–5.

J. W. Chen, G. Matteucci, G. F. Missiroli and G. Pozzi (1987). Contour maps from double exposure electron holograms, SIME, *Microscopia Elettronica* **8** (2, Suppl.), 259–260.

J. Chen, T. Hirayama, G. Lai, T. Tanji, K. Ishizuka and A. Tonomura (1993). Real-time electron-holographic interference microscopy with a liquid-crystal spatial light modulator. *Opt. Lett.* **18**, 1887–1889.

N. Cherkashin, T. Denneulin and M. J. Hÿtch (2017). Electron microscopy by specimen design: application to strain measurements. *Sci. Rep.* **7**, 12384, 13 pp.

S. Chung, D. J. Smith and M. R. McCartney (2007). Determination of the inelastic mean-free-path and mean inner potential for AlAs and GaAs using off-axis electron holography and convergent beam electron diffraction. *Microsc. Microanal.* **13**, 329–335.

W. Coene, G. Janssen, M. Op de Beeck and D. Van Dyck (1992). Phase retrieval through focus variation for ultra-resolution in field-emission transmission electron microscopy. *Phys. Rev. Lett.* **69**, 3743–3746.

D. Cooper, R. Truche, P. Rivallin, J.-M. Hartmann, F. Laugier, F. Bertin, A. Chabli and J.-L. Rouvière (2007). Medium resolution off-axis electron holography with millivolt sensitivity. *Appl. Phys. Lett.* **91**, 143501.

D. Cooper, J.-P. Barnes, J.-M. Hartmann, A. Béché and J.-L. Rouvière (2009). Dark field electron holography for quantitative strain measurements with nanometer-scale spatial resolution. *Appl. Phys. Lett.* **95**, 053501.

D. Cooper, A. Béché, J.-M. Hartmann, V. Carron and J.-L. Rouvière (2010). Strain evolution during the silicidation of nanometer-scale SiGe semiconductor devices studied by dark field electron holography. *Appl. Phys. Lett.* **96**, 113508.

D. Cooper, J.-L. Rouvière, A. Béché, S. Kadkhogazadeh, E. S. Semenova, K. Yvind and R. E. Dunin-Borkowski (2011). Quantitative strain mapping of InAs/InP quantum dots with 1 nm spatial resolution using dark field electron holography. *Appl. Phys. Lett.* **99**, 261911.

D. Cooper, C. Le Royer, A. Béché and J.-L. Rouvière (2012). Strain mapping for the silicon-on-insulator generation of semiconductor devices by high-angle annular dark field scanning electron transmission microscopy. *Appl. Phys. Lett.* **100**, 233121.

D. Cooper, T. Denneulin, N. Bernier, A. Béché and J.-L. Rouvière (2016). Strain mapping of semiconductor specimens with nm-scale resolution in a transmission electron microscope. *Micron* **80**, 145–165.

A. Costa, G. Matteucci and S. Patuelli (1989). Asta porta-biprisma per un Philips EM 400T, SIME, *Microscopia Elettronica* **10** (2, Suppl.), 267–268.

J. M. Cowley (1990). High resolution side-band holography with a STEM instrument. *Ultramicroscopy* **34**, 293–297.

J. M. Cowley (1991). Alternative approaches to ultra-high resolution imaging. EMSA 49, San Jose, CA, 650–651.

J. M. Cowley (1992). Twenty forms of electron holography. *Ultramicroscopy* **41**, 335–348.

J. M. Cowley and D. J. Walker (1981). Reconstruction from in-line holograms by digital processing. *Ultramicroscopy* **6**, 71–75.

A. V. Crewe and J. Saxon (1971). Electron holography and the correction of spherical aberration. EMSA 29, Boston, MA, 12–13.

J. Cumings, A. Zettl, M. R. McCartney and J. C. H. Spence (2002). Electron holography of field-emitting carbon nanotubes. *Phys. Rev. Lett.* **88**, 056804, 4 pp.

J. Cumings, E. Olsson, A. K. Petford-Long and Y. Zhu (2008). Electric and magnetic phenomena studied by in situ transmission electron microscopy. *MRS Bull.* **33**, 101–106.

I. Daberkow, W. Dreher, Q. Fu, H. Lichte and E. Völkl (1988). Demands at an image digitizer for numerical processing of electron holograms. EUREM-9, York, **1**, 193–194.

L. de Caro, F. Scattarella and E. Carlino (2016). Determination of the projected atomic potential by deconvolution of the auto-correlation function of TEM electron nano-diffraction patterns. *Crystals* **6**, 141, 19 pp.

F. de Martini, G. Denardo and Y. Shiy, Eds (1996). *Quantum Interferometry* (VCH, Weinheim).

W. J. de Ruijter (1992). Quantitative high-resolution electron microscopy and holography. Dissertation, Delft.

W. J. de Ruijter and J. K. Weiss (1993). Detection limits in quantitative off-axis electron holography. *Ultramicroscopy* **50**, 269–283.

T. Denneulin and M. Hÿtch (2016). Four-wave dark-field electron holography for imaging strain fields. *J. Phys. D: Appl. Phys* **49**, 244003, 8 pp.

T. Denneulin, J. Caron, M. Hoffmann, M. Lin, H. K. Tan, A. Kovács, S. Blügel and R. E. Dunin-Borkowski (2021). Off-axis electron holography of Néel-type skyrmions in multilayers of heavy metals and ferromagnets. *Ultramicroscopy* **220**, 113155, 10 pp.

Y. Ding, Y. Liu, K. C. Pradel, Y. Bando, N. Fukata and Z. L. Wang (2015). Quantifying mean inner potential of ZnO nanowires by off-axis electron holography. *Micron* **78**, 67–72.

O. Donati, G. F. Missiroli and G. Pozzi (1973). An experiment with electron interference. *Amer. J. Phys.* **41**, 639–644.

V. Drahoš and A. Delong (1963). Úprava prozařovacího elektronového mikroskopu pro interferenční elektronovou mikroskopii. *Čs. Čas. Fys.* **13**, 278–286.

V. Drahoš and A. Delong (1964). The source width and its influence on interference phenomena in a Fresnel electron bi-prism. *Opt. Acta* **11**, 173–181.

R. Dronyak, K. S. Liang, Y. P. Stetsko, T.-k. Lee, C.-k. Feng, J.-s. Tsai and F.-r. Chen (2009). Electron diffractive imaging of nano-objects using a guided method with a dynamic support. *Appl. Phys. Lett.* **95**, 111908.

M. Duchamp, O. Girard, G. Pozzi, H. Soltner, F. Winkler, R. Speen, R. E. Dunin-Borkowski and D. Cooper (2018). Fine electron biprism on a Si-on-insulator chip for off-axis electron holography. *Ultramicroscopy* **185**, 81–89.

S. L. Dudarev, L.-m. Peng and M. J. Whelan (1993). Correlations in space and time and dynamical diffraction of high-energy electrons by crystals. *Phys. Rev. B* **48**, 13408–13429.

H. Düker (1955). Lichtstarke Interferenzen mit einem Biprisma für Elektronenwellen. *Z. Naturforsch.* **10a**, 256.

R. E. Dunin-Borkowski, A. C. Twitchett and P. A. Midgley (2002). The determination and interpretation of electrically active charge density profiles at reverse biased p-n junctions from electron holograms. *Microsc. Microanal.* **8** (Suppl. 2), 42–43.

R. E. Dunin-Borkowski, T. Kasama, A. Wei, S. L. Tripp, M. Hÿtch, E. Snoeck, R. J. Harrison and A. Putnis (2004). Off-axis electron holography of magnetic nanowires and chains, rings, and planar arrays of magnetic nanoparticles. *Microsc. Res. Tech.* **64**, 390–402.

R. E. Dunin-Borkowski, A. Kovács, T. Kasama, M. R. McCartney and D. J. Smith (2019). Electron holography. In *Handbook of Microscopy* (P. W. Hawkes and J. C. H. Spence, Eds), 767–818 (Springer-Nature, Cham).

M. Durand, J. Faget, J. Ferré and C. Fert (1958). Mesure interférométrique du potentiel interne du graphite. *C. R. Acad. Sci. Paris* **247**, 590–593.

J. Dyson (1950). The optical synthesizer for the gabor diffraction microscope. ICEM-2, Paris, **1**, 126–128.

W. Ehrenberg and R. E. Siday (1949). The refractive index in electron optics and the principles of dynamics. *Proc. Phys. Soc. Lond. B* **62**, 8–21.

J. Endo and A. Tonomura (1990). Applications of electron holography to materials science. *Mater. Trans. Jpn. Inst. Met.* **31**, 551–560.

J. Endo, T. Matsuda and A. Tonomura (1979). Interference electron microscopy by means of holography. *Jpn. J. Appl. Phys.* **18**, 2291–2294.

J. Endo, N. Osakabe, T. Matsuda, S. Yano, H. Yamada, T. Kawasaki and A. Tonomura (1986). Development of low temperature specimen stage and its applications to electron holographic measurement. EMSA 44, Albuquerque, 614–615.

J. Endo, T. Kawasaki, T. Masuda and A. Tonomura (1989). Development of 350 kV holography electron microscope equipped with magnetic type field-emission gun. EMSA 47, San Antonio, TX, 104–105.

F. F. M. Estrada, G. F. Missiroli and E. Nichelatti (1991). Amplificazione digitale di fase di ologrammi elettronici in doppia esposizione, SIME, *Microscopia Elettronica* **12** (2, Suppl.), 411–412.

R. Fabbri, G. Matteucci and G. Pozzi (1987). Observation of magnetic domains in thin films by electron holography with a commercial TEM with FEG. *Phys. Status Solidi A* **102**, K127–K129.

J. Faget (1961). Interférences des ondes électroniques. Application à une méthode de microscopie électronique interférentielle. *Rev. Opt.* **40**, 347–381.

J. Faget and C. Fert (1956). Franges de diffraction et d'interférences en optique électronique: diffraction de Fresnel, trous d'Young, biprisme de Fresnel. *C.R. Acad. Sci. Paris* **243**, 2028–2029.

J. Faget and C. Fert (1957a). Diffraction et interférences en optique électronique. *Cahiers de Physique* **11**, 285–296 [Cahier No. 83].

J. Faget and C. Fert (1957b). Microscopie interférentielle et mesure de la différence de phase introduite par une lame en optique électronique. *C. R. Acad. Sci. Paris* **244**, 2368–2371.

J. Faget, J. Ferré and C. Fert (1958). Le biprisme de Fresnel en optique électronique: influence de la largeur de la source; effet d'une tension périodique appliquée sur le fil du biprisme. *C.R. Acad. Sci. Paris* **246**, 1404–1407.

J. Faget, M. Fagot and C. Fert (1960). Microscopie électronique en éclairage cohérent: microscopie interférentielle, contraste de défocalisation, strioscopie et contraste de phase. EUREM-2, Delft, **1**, 18–24.

M. Fagot, J. Ferré and C. Fert (1961). Quelques observations sur la formation de l'image en microscopie électronique. *C.R. Acad. Sci. Paris* **252**, 3766–3768.

P. F. Fazzini, P. G. Merli and G. Pozzi (2004). Electron microscope calibration for the Lorentz mode. *Ultramicroscopy* **99**, 201–209.

P. F. Fazzini, L. Ortolani, G. Pozzi and F. Ubaldi (2006). Interference electron microscopy of one-dimensional electron-optical phase objects. *Ultramicroscopy* **106**, 620–629.

A. Feist, N. Rubiano da Silva, W. Liang, C. Ropers and S. Schäfer (2018). Nanoscale diffractive probing of strain dynamics in ultrafast transmission electron microscopy. *Struct. Dyn.* **5**, 014302, 13 pp.

A. Feltynowski (1963). Beobachtung von Elektronen-Biprisma-Interferenzen im Elmiskop. *Z. Angew. Phys.* **15**, 312–315.

C. Fert (1961). Interférences, diffraction en optique électronique et leurs applications à la microscopie. In *Traité de Microscopie Electronique* (C. Magnan, Ed.), **1**, 333–390 (Hermann, Paris).

C. Fert (1962). The conditions of image-formation in electron optics. Electron microscopy with coherent and incoherent illumination. *J. Electron Microsc.* **11**, 1–9.

C. Fert and J. Faget (1958). Interférométrie électronique et microscopie électronique interférentielle. ICEM-4, Berlin, **1**, 234–239.

C. Fert, J. Faget, M. Fagot and J. Ferré (1962a). Un microscope électronique interférentiel. *J. Microsc. (Paris)* **1**, 1–12.

C. Fert, J. Faget, M. Fagot and J. Ferré (1962b). Effets de diffraction et d'interférence dans les images électroniques; formation de l'image en optique électronique. *J. Phys. Soc. Jpn.* **17** (Suppl. B–II), 186–190.

H.-W. Fink and H. J. Kreuzer (1992). Reply to Spence et al. (1992). *Phys. Rev. Lett.* **68**, 3257.

H.-W. Fink and H. Schmid (1994). Electron and ion microscopy without lenses. In *Nanostructures and Quantum Effects* (H. Sakaki and H. Noge, Eds), 17–27 (Springer, Berlin).

H.-W. Fink, W. Stocker and H. Schmid (1990). Holography with low-energy electrons. *Phys. Rev. Lett.* **65**, 1204–1206.

H.-W. Fink, H. Schmid, H. J. Kreuzer and A. Wierbicki (1991). Atomic resolution in lensless low-energy electron holography. *Phys. Rev. Lett.* **67**, 1543–1546, cf. Spence et al. (1992).

H.-W. Fink, H. J. Kreuzer and H. Schmid (1995). State of the art of low-energy electron holography. In *Electron Holography* (A. Tonomura, L. F. Allard, G. Pozzi, D. C. Joy and Y. A. Ono, Eds), 257–266 (Elsevier, Amsterdam).

D. Fischer and B. Lischke (1967). Biprismainterferenzen mit langsamen Elektronen. *Z. Phys.* **205**, 458–464.

H. A. Fowler, L. Marton, J. A. Simpson and J. A. Suddeth (1961). Electron interferometric studies of iron whiskers. *J. Appl. Phys.* **32**, 1153–1155.

S. Frabboni (2011). Two and three slit electron interference and diffraction experiments. *Amer. J. Phys.* **79**, 615.

S. Frabboni, G. Matteucci, G. Pozzi and M. Vanzi (1985). Electron holographic observations of the electrostatic field associated with thin reverse-biased p-n junctions. *Phys. Rev. Lett.* **55**, 2196–2199.

S. Frabboni, G. Matteucci, G. F. Missiroli, G. Pozzi, G. Pizzochero and M. Vanzi (1986). Electron holography of p-n junctions. ICEM-11, Kyoto, **1**, 685–686.

S. Frabboni, G. Matteucci and G. Pozzi (1987). Observation of electrostatic fields by electron holography: the case of reverse-biased p-n junctions. *Ultramicroscopy* **23**, 29–38.

S. Frabboni, G. C. Gazzadi and G. Pozzi (2007). Young's double-slit interference experiment with electrons. *Amer. J. Phys.* **75**, 1053–1055.

S. Frabboni, C. Frigeri, G. C. Gazzadi and G. Pozzi (2010). Four slits interference and diffraction experiments. *Ultramicroscopy* **110**, 483–487.

S. Frabboni, A. Gabrielli, G. C. Gazzadi, F. Giorgi, G. Matteucci, G. Pozzi, N. S. Cesari, M. Villa and A. Zoccoli (2012). The Young-Feynman two-slits experiment with single electrons: build-up of the interference pattern and arrival-time distribution using a fast-readout pixel detector. *Ultramicroscopy* **116**, 73–76.

S. Frabboni, G. C. Gazzadi, V. Grillo and G. Pozzi (2015). Elastic and inelastic electrons in the double-slit experiment: a variant of Feynman's which-way set-up. *Ultramicroscopy* **154**, 49–56.

J. C. Francken (1967). Analogical methods for resolving Laplace's and Poisson's equations. In *Focusing of Charged Particles* (A. Septier, Ed.), **1**, 101–162 (Academic Press, New York & London).

F. J. Franke, K.-H. Herrmann and H. Lichte (1986). Off-axis electron holography with digital image reconstruction. ICEM-11, Kyoto, **1**, 677–678.

F. J. Franke, K.-H. Hermann and H. Lichte (1987). Numerical reconstruction of the electron object wave from an electron hologram including the correction of aberrations. In Hawkes et al. (1988) 59–67.

F. Frémont, A. Hajaji, R. O. Barrachina and J.-Y. Chesnel (2008). Interférences de type Young avec une source à un seul électron. *C. R. Phys.* **9**, 469–475.

H. Friedrich, M. R. McCartney and P. R. Buseck (2004). Electron holographic tomography – challenge and opportunity. *Microsc. Microanal.* **10** (Suppl. 2), 1174–1175.

B. Frost and H. Lichte (1988). Electron holography for magnetic microstructures. EUREM-9, York, **2**, 267–268.

S. Fu, J. W. Chen, Z. Wang and H. Cao (1987). Experimental investigation of electron interference and electron holography. *Optik* **76**, 45–47.

Q. Fu, H. Lichte and E. Völkl (1991). Correction of aberrations of an electron microscope by means of electron holography. *Phys. Rev. Lett.* **67**, 2319–2322.

K. Fujikawa and Y. A. Ono, Eds (1996). *Proceedings of the 5th International Symposium on Foundations of Quantum Mechanics in the Light of New Technology* (Elsevier/North Holland, Amsterdam).

K. Fujikawa and Y. A. Ono, Eds (2014). *In Memory of Akira Tonomura, Physicist and Electron Microscopist* (World Scientific, Singapore).

T. Fujita and M. Chen (2009). Quantitative electron holographic tomography for a spherical object. *J. Electron Microsc.* **58**, 301–304.

A. Fukuhara, K. Shinagawa, A. Tonomura and H. Fujiwara (1983). Electron holography and magnetic specimens. *Phys. Rev. B* **27**, 1839–1843.

D. Gabor (1948). A new microscope principle. *Nature* **161**, 777–778.

D. Gabor (1949a). Problems and prospects of electron diffraction microscopy. ICEM-1, Delft, 55–59.

D. Gabor (1949b). Microscopy by reconstructed wave-fronts. *Proc. R. Soc. Lond. A* **197**, 454–487.

D. Gabor (1950). Generalized schemes of diffraction microscopy. ICEM-2, Paris, **1**, 129–137.

D. Gabor (1951a). Microscopy by reconstructed wave fronts. II. *Proc. Phys. Soc. Lond. B* **64**, 449–469.

D. Gabor (1951b). Progress in microscopy by reconstructed wavefronts. *Natl Bur. Stand. Circular* **527**, 237–245 Published 1954.

D. Gabor (1956). Theory of electron interference experiments. *Rev. Mod. Phys.* **28**, 260–276.

D. Gabor (1968). The outlook for holography. *Optik* **28**, 437–441.

D. Gabor (1980). Heavy atom holography in electron microscopy. *Isr. J. Technol.* **18**, 209–213.

D. Gabor, G. W. Stroke, R. Restrick, A. Funkhouser and D. Brumm (1965). Optical image synthesis (complex amplitude addition and subtraction) by holographic [sic] Fourier transformation. *Phys. Lett.* **18**, 116–118.

A. Gabrielli, F. Giorgi, N. S. Cesari, M. Villa, A. Zoccoli, G. Matteucci, G. Pozzi, S. Frabboni and G. C. Gazzadi (2011). Application of a HEPE-oriented 4096-MAPS to time analysis of single electron distribution in a two-slits interference experiment. *J. Instrum.* **6**, C12029, 7 pp.

A. Gabrielli, F. M. Giorgi, N. Semprini, M. Villa, A. Zoccoli, G. Matteucci, G. Pozzi, S. Frabboni and G. C. Gazzadi (2013). A 4096-pixel MAPS detector used to investigate the single-electron distribution in a Young–Feynman two-slit interference experiment. *Nucl. Instrum. Meth. Phys. Res. A* **699**, 47–50.

M. Gajdardziska–Josifovska, M. R. McCartney, W. J. de Ruijter, D. J. Smith, J. K. Weiss and J. M. Zuo (1993). Accurate measurements of mean inner potential of crystal wedges using digital electron holograms. *Ultramicroscopy* **50**, 285–299.

Z.-F. Gan, S. Ahn, H. Yu, D. J. Smith and M. R. McCartney (2015). Measurement of mean inner potential and inelastic mean free path of ZnO nanowires and nanosheet. *Mater. Res. Express* **2**, 105003.

N. Garcia (1989). Electron emission from small sources and its influence in electron holography and interferometry, SIME, *Microscopia Elettronica* **10** (2, Suppl.), 255–256.

F. J. Garcia de Abajo (2009). Photons and electrons team up. *Nature* **462**, 861.

F. J. Garcia de Abajo (2010). Optical excitations in electron microscopy. *Rev. Mod. Phys.* **82**, 209–275.

C. Gatel, A. Lubk, G. Pozzi, E. Snoeck and M. Hÿtch (2013). Counting elementary charges on nanoparticles by electron holography. *Phys. Rev. Lett.* **111**, 025501, 5 pp.

C. Gatel, J. Dupuy, F. Houdellier and M. Hÿtch (2018). Unlimited acquisition time in electron holography by automated feedback control of transmission electron microscope. *Appl. Phys. Lett.* **113**, 133102, 5 pp.

D. C. Ghiglia and M. D. Pritt (1998). *Two-Dimensional Phase Unwrapping: Theory, Algorithms, and Software* (Wiley, Chichester).

A. H. Greenaway and A. M. J. Huiser (1976). A new motivation for off-axis holography in electron microscopy. *Optik* **45**, 295–300.

A. Greenberg, F. Yasin, C. Johnson and B. McMorran (2019). Lorentz implementation of STEM holography. *Microsc. Microanal.* **25** (Suppl. 2), 96–97.

M. A. Gribelyuk and J. M. Cowley (1991). Computer analysis of side-band holography in STEM. EMSA 49, San Jose, CA, 684–685.

M. A. Gribelyuk and J. M. Cowley (1992a). Determination of imaging conditions in electron holography. EUREM-10, Granada, **1**, 649–650.

M. A. Gribelyuk and J. M. Cowley (1992b). Computer analysis of side-band holography in STEM. *Ultramicroscopy* **45**, 103–113.
M. A. Gribelyuk and J. M. Cowley (1993). Determination of experimental imaging conditions for off-axis transmission electron holography. *Ultramicroscopy* **50**, 29–40.
M. A. Gribelyuk, M. R. McCartney, J. Li, C. S. Murthy, P. Ronsheim, B. Doris, J. S. McMurray, S. Hegde and D. J. Smith (2002). Mapping of electrostatic potential in deep submicron CMOS devices by electron holography. *Phys. Rev. Lett.* **89**, 025502, 4 pp.
G. Gronniger, B. Barwick, H. Batelaan, T. A. Savas, D. Pritchard and A. D. Cronin (2005). Electron diffraction from free-standing, metal-coated transmission gratings. *Appl. Phys. Lett.* **87**, 124104.
G. Gronniger, B. Barwick and H. Batelaan (2006). A three-grating electron interferometer. *New J. Phys.* **8**, 224.
A. Günther, A. Rembold, G. Schütz and A. Stibor (2015). Multifrequency perturbations in matter-wave interferometry. *Phys. Rev. A* **92**, 053607, 7 pp.
B. Haas (2017). Development of quantitative diffraction- and imaging-based techniques for scanning transmission electron microscopy. Dissertation, Grenoble-Alpes.
M. Haider, P. Hartel, H. Müller and S. Uhlemann (2010). Information transfer in a TEM corrected for spherical and chromatic aberration. *Microsc. Microanal.* **16**, 393–408.
M. E. Haine and J. Dyson (1950). A modification to Gabor's proposed diffraction microscope. *Nature* **166**, 315–316.
M. E. Haine and T. Mulvey (1950). Initial results in the practical realisation of Gabor's diffraction microscope. ICEM-2, Paris, **1**, 120–125.
M. E. Haine and T. Mulvey (1951). Problems in the realization of diffraction microscopy with electrons. *Natl Bur. Stand. Circular* **527**, 247–250, Published 1954.
M. E. Haine and T. Mulvey (1952). The formation of the diffraction image with electrons in the Gabor diffraction microscope. *J. Opt. Soc. Amer.* **42**, 763–773.
R. Hanbury Brown and R. Q. Twiss (1954). A new type of interferometer for use in radio astronomy. *Philos. Mag.* **45**, 663–682.
R. Hanbury Brown and R. Q. Twiss (1956). Correlation between photons in two coherent beams of light. *Nature* **177**, 27–29.
B. A. M. Hansson, N. Machida, K. Furuya, L.-E. Wernersson and L. Samuelson (2000). Simulation of interference patterns in solid-state biprism devices. *Solid-State Electron* **44**, 1275–1280.
K.-J. Hanszen (1969). Einseitenband-Holographie. *Z. Naturforsch.* **24a**, 1849.
K.-J. Hanszen (1970a). In-line holographic reconstruction methods in electron microscopy. ICEM-7, Grenoble, **1**, 21–22.
K.-J. Hanszen (1970b). Holographische Rekonstruktionsverfahren in der Elektronenmikroskopie und ihre kontrastübertragungstheoretische Deutung. A. In-line Fresnel-Holographie. *Optik* **32**, 74–90.
K.-J. Hanszen (1971). Optical transfer theory of the electron microscope: fundamental principles and applications. *Adv. Opt. Electron Microsc.* **4**, 1–84.
K.-J. Hanszen (1972a). Die Wellenaberration einer korrigierten Linse in Abhängigkeit von Defokussierung und Pupillenlage – Ein Beitrag zur Festlegung der Rekonstruktionsbedingungen für elektronenmikroskopische Aufnahmen. *Optik* **35**, 431–444.
K.-J. Hanszen (1972b). In-line-holographische Erfahrungen mit Radialgittern als Testobjekten in lichtoptischen Modellanordnungen für das Elektronenmikroskop. *Optik* **36**, 41–54.
K.-J. Hanszen (1973a). Contrast transfer and image processing. In *Image Processing and Computer-Aided Design in Electron Optics* (P. W. Hawkes, Ed.), 16–53 Academic Press, London and New York).
K.-J. Hanszen (1973b). Neuere theoretische Erkenntnisse und praktische Erfahrungen über die holographische Rekonstruktion elektronenmikroskopischer Aufnahmen. PTB-Bericht Aph–4, 30 pp.
K.-J. Hanszen (1974). Die übertragungstheoretische Behandlung des Einflusses der Bildfehler erster Ordnung auf die holographische Rekonstruktion elektronenmikroskopischer Aufnahmen. *Optik* **39**, 520–542.
K.-J. Hanszen (1976). In-line holography in electron microscopy using tilted illumination without an objective aperture stop. EUREM-6, Jerusalem, **1**, 95–96.

K.-J. Hanszen (1980). Experience and results obtained in electron microscopical holography by using a reference beam in the light optical reconstruction step. EUREM-7, The Hague, **1**, 136–137.
K.-J. Hanszen (1982a). Holography in electron microscopy. *Adv. Electron Electron Phys.* **59**, 1–77.
K.-J. Hanszen (1982b). A simple reconstruction device for the interferometric detection of the object phase recorded in off-axis image field holograms. *Ultramicroscopy* **9**, 159–166.
K.-J. Hanszen (1982c). Anwendungen und Grenzen der elektronenmikroskopischen Holographie, unter besonderer Berücksichtigung der Rauscheinflüsse die von Objektträgerfolie und Hologrammplatte ausgehen. PTB-Bericht APh–17, 34 pp.
K.-J. Hanszen (1982d). Phase contrast improvement and residual noise problems in electron microscopical holography. ICEM-10, Hamburg, **1**, 423–424.
K.-J. Hanszen (1983). Interferometrische Verarbeitung von elektronenmikroskopischen Hologrammen. C. Untersuchungen über die Besonderheiten der Differentialinterferometrie unter Einbeziehung des Rauschens. *Optik* **65**, 153–177.
K.-J. Hanszen (1984). Phase-contrast imaging surpassing the background noise in holographic reconstructions of weak objects. EUREM-8, Budapest, **1**, 279–280.
K.-J. Hanszen (1985). Quantitative interferometrical evaluation of electron holograms of weak phase objects, with special regard to in-line holography. *Optik* **71**, 155–162.
K.-J. Hanszen (1986a). Methods of off-axis electron holography and investigations of the phase structure in crystals. *J. Phys. D: Appl. Phys* **19**, 373–395.
K.-J. Hanszen (1986b). Recent developments and prospective trends in electron holography. In *Proceedings of the International Symposium on Electron Optics, ISEOB* (J.-y. Ximen, Ed.), 66–69 (Institute of Electronics, Academia Sinica, Beijing).
K.-J. Hanszen (1987). The noise in in-line Fraunhofer electron holography. *Optik* **77**, 57–61.
K.-J. Hanszen (1990a). 40 Jahre elektronenoptische Forschung in der Physikalisch-Technischen Bundsanstalt. PTB-Bericht APh–33 (2 parts).
K.-J. Hanszen (1990b). 40 Jahre Forschung und Entwicklung auf dem Gebiet der Elektronenoptik in der PTB. *PTB-Mitt* **100**, 363–368.
K.-J. Hanszen and G. Ade (1974). Problems and results of the optical transfer theory and of reconstruction methods in electron microscopy. PTB-Bericht APh–5, 68 pp.
K.-J. Hanszen and G. Ade (1976a). A consistent Fourier optical representation of CEM and STEM imaging and holographic reconstruction. EUREM-6, Jerusalem, **1**, 446–447.
K.-J. Hanszen and G. Ade (1976b). Aspects of some image reconstructions and holographic methods in electron microscopy. PTB-Bericht APh–10, 38 pp.
K.-J. Hanszen and G. Ade (1977). A consistent Fourier optical representation of image formation in the conventional fixed beam electron microscope, in the scanning transmission electron microscope and of holographic reconstruction. PTB-Bericht APh–11, 31pp.
K.-J. Hanszen and G. Ade (1983). Interferometrische Verarbeitung von elektronenmikroskopischen Hologrammen. A. Theoretische Behandlung unter Einschluß des photographischen Rauschens. *Optik* **63**, 247–264.
K.-J. Hanszen and G. Ade (1984). The speckled background in reconstructions of electron off-axis holograms occurring when very small apertures are used in the diffraction plane of the reconstruction device. *Optik* **68**, 81–95.
K.-J. Hanszen and R. Lauer (1980). Holographic phase determination of strong objects. EUREM-7, The Hague, **1**, 140–141.
K.-J. Hanszen, G. Ade and R. Lauer (1972a). Genauere Angaben über sphärische Längsaberration, Verzeichnung in der Pupillenebene und über die Wellenaberration von Elektronenlinsen. *Optik* **35**, 567–590.
K.-J. Hanszen, R. Lauer and G. Ade (1972b). Eine Beziehung zwischen der Abhängigkeit des Öffnungsfehlers von der Objektlage und der Koma, ausgedrückt in der Terminologie der Elektronenopik. *Optik* **36**, 156–159.

K.-J. Hanszen, R. Lauer and G. Ade (1980). Discussions on the possibilities and limitations of in-line and off-axis holography in electron microscopy. PTB-Bericht APh–15, 38 pp.
K.-J. Hanszen, R. Lauer and G. Ade (1981). Elektronenmikroskopische Holographie. PTB-Bericht APh–16, 35 pp.
K.-J. Hanszen, R. Lauer and G. Ade (1982). Recent developments in electron microscopical holography. PTB-Bericht APh–19, 37 pp.
K.-J. Hanszen, R. Lauer and G. Ade (1983a). Interferometrische Verarbeitung von elektronenmikroskopischen Hologrammen. B. Praktische Erfahrungen mit den verschiedenen Verfahren. *Optik* **63**, 285–303.
K.-J. Hanszen, R. Lauer and G. Ade (1983b). Recent investigations in electron holography and electron transfer theory. PTB-Bericht APh–20, 35 pp.
K.-J., Hanszen, G. Ade and R. Lauer (1984). Image properties and noise in electron holography. PTB-Bericht APh–23, 31 pp.
K.-J. Hanszen, R. Lauer and G. Ade (1985). Beiträge zur Nachverarbeitung elektronenmikroskopischer Aufnahmen und zur Elektronenholographie. PTB-Bericht APh–25, 56 pp.
K.-J. Hanszen, R. Lauer and G. Ade (1986). Prospective trends in electron holography; influence of the noise on reconstructions of off-axis holograms and dynamical phase effects in perfect crystals. PTB-Bericht APh–30, 49 pp.
K. Harada (2021). Interference and interferometry in electron holography. *Microscopy* **70**, 3–16.
K. Harada and R. Shimizu (1991). A new FFT method for numerical reconstruction in electron holography. *J. Electron Microsc.* **40**, 92–96.
K. Harada and A. Tonomura (2004). Double-biprism electron interferometry. *Appl. Phys. Lett.* **84**, 3229–3231.
K. Harada, H. Endoh and R. Shimizu (1988). Anti-contamination electron biprism for electron holography. *J. Electron Microsc.* **37**, 199–201.
K. Harada, K. Ogai and R. Shimizu (1990a). Aberration correction by electron holography using numerical reconstruction method. *J. Electron Microsc.* **39**, 465–469.
K. Harada, K. Ogai and R. Shimizu (1990b). The fringe scanning method as numerical reconstruction for electron holography. *J. Electron Microsc.* **39**, 470–476.
K. Harada, T. Matsuda, J. Bonevich, M. Igarashi, S. Kondo, G. Pozzi, U. Kawabe and A. Tonomura (1992). Real-time observation of vortex lattices in a superconductor by electron microscopy. *Nature* **360**, 51–53.
K. Harada, T. Matsuda, A. Tonomura, T. Akashi and Y. Togawa (2006). Triple-biprism electron interferometry. *J. Appl. Phys.* **99**, 113502, 7 pp.
K. Harada, T. Kodama, T. Akashi and Y. Takahashi (2019a). Double-slit electron interference experiment with zero propagation distance using electron biprism. *Microsc. Microanal.* **25** (Suppl. 2), 944–945.
K. Harada, K. Niitsu, K. Shimada, T. Kodama, T. Akashi, Y. A. Ono, D. Shindo, H. Shinada and S. Mori (2019b). Electron holography on Fraunhofer diffraction. *Microscopy* **68**, 254–260.
K. Harada, Y. Takahashi, T. Akashi, Y. Ono, T. Kodama and S. Mori (2020). Double-slit electron interference experiment with phase modulation. *Microsc. Microanal.* **26** (Suppl. 2), 2152–2153.
J. H. Harris and M. D. Semon (1980). A review of the Aharonov-Carmi thought experiment concerning the inertial and electromagnetic vector potentials. *Found. Phys.* **10**, 151–162.
A. Harscher (1999). Elektronenholographie biologischer Objekte: Grundlagen und Anwendungsbeispiele. Dissertation, Tübingen.
A. Harscher, H. Lichte and J. Mayer (1997). Interference experiments with energy filtered electrons. *Ultramicroscopy* **69**, 201–209.
T. R. Harvey, F. S. Yasin, J. J. Chess, J. S. Pierce, R. M. S. dos Reis, V. B. Özdöl, P. Ercius, J. Ciston, W. Feng, N. A. Kotov, B. J. McMorran and C. Ophus (2018). Interpretable and efficient interferometric contrast in scanning transmission electron microscopy with a diffraction-grating beam splitter. *Phys. Rev. Appl.* **10**, 061001, 7 pp.
T. Harvey, F. Yasin, J. Chess, C. Ophus and B. McMorran (2019). Holography in scanning transmission electron microscopy. MC-2019, Berlin, 398–399.

S. Hasegawa, T. Kawasaki, J. Endo, A. Tonomura, Y. Honda, M. Futamoto, K. Yoshida, F. Kugiya and M. Koizumi (1989). Sensitivity-enhanced electron holography and its application to magnetic recording investigations. *J. Appl. Phys.* **65**, 2000–2004.

F. Hasselbach (1988). A ruggedized miniature UHV electron biprism interferometer for new fundamental experiments and applications. *Z. Phys.* **B71**, 443–449.

F. Hasselbach (1992). Recent contributions of electron interferometry to wave-particle duality. In *Wave–Particle Duality* (F. Selleri, Ed.), 109–125 Plenum, New York and London).

F. Hasselbach (1997). Selected topics in charged particle interferometry. *Scanning Microsc.* **11**, 345–366.

F. Hasselbach (2010). Progress in electron- and ion-interferometry. *Rep. Prog. Phys.* **73**, 016101, 43 pp.

F. Hasselbach and U. Maier (1995). A biprism interferometer. *Optik* **100** (Suppl. 6), 7.

F. Hasselbach and U. Maier (1996). A biprism interferometer for ions and some of its future applications. In *Quantum Coherence and Decoherence* (K. Fujiwara and Y. A. Ono, Eds), 69–72 (North-Holland, Amsterdam).

F. Hasselbach and U. Maier (1999). Biprism interferences of He^+-ions. In *Quantum Coherence and Decoherence* (Y.A. Ono and K. Fujiwara, Eds), 299–302 (North-Holland, Amsterdam).

F. Hasselbach and M. Nicklaus (1986). An electron interferometer for the demonstration of the rotational phase shift of charged particle waves (Sagnac effect). ICEM-11, Kyoto, **1**, 691–692.

F. Hasselbach and M. Nicklaus (1987). An electron interferometer for the demonstration of the rotational phase shift of charged particle waves (Sagnac effect). In *Proceedings of the International Symposium on Electron Optics, ISEOB* (J.-y. Ximen, Ed.), 70–73 (Institute of Electronics, Academia Sinica, Beijing).

F. Hasselbach and M. Nicklaus (1988). An electron optical Sagnac experiment. *Physica B* **151**, 230–234.

F. Hasselbach and M. Nicklaus (1990). Phase shift of electron waves in a rotating frame of reference. ICEM-12, Seattle, WA, **1**, 212–213.

F. Hasselbach and M. Nicklaus (1993). Sagnac experiment with electrons: observation of the rotational phase shift of electron waves in vacuum. *Phys. Rev. A* **48**, 143–151.

F. Hasselbach and A. Schäfer (1990). Interferometric (Fourier-spectroscopic) measurement of electron energy distributions. ICEM-12, Seattle, WA, **2**, 10–11.

F. Hasselbach, A. Schäfer and H. Wachendorfer (1995). Interferometric measurement of charged particle spectra (Fourier-spectroscopy). *Nucl. Instrum. Meth, Phys. Res. A* **363**, 232–238.

F. Hasselbach, H. Kiesel and P. Sonnentag (2000). Exploration of the fundamentals of quantum mechanics by charged particle interferomentry. In *Decoherence: Theoretical, Experimental, and Conceptual Problems* (P. Blanchard, E. Joos, D. Giulini, C. Kiefer and I.-O. Stamatescu, Eds), 201–213 (Springer, Berlin).

F. Hasselbach, H. Kiesel and P. Sonnentag (2004). Electron interferometry: interferences between two electrons and a precision method of measuring decoherence. *Ann. Fond, Louis de. Broglie* **29** (Suppl. 1), 857–872.

P. W. Hawkes (1978). Coherence and electron optics. *Adv. Opt. Electron Microsc.* **7**, 101–184.

P. W. Hawkes (1980). Preface. In *Computer Processing of Electron Microscope Images* (P. W. Hawkes, Ed.), v (Springer, Berlin and New York).

P. W. Hawkes (1992). An alternative definition of Gabor focus. *Ultramicroscopy* **41**, 441.

P. W. Hawkes, F. P. Ottensmeyer, W. O. Saxton and A. Rosenfeld, Eds (1988). *Image and Signal Processing in Electron Microscopy. Proceedings of the 6th Pfefferkorn Conference, Niagara Falls, 1987* (Scanning Microscopy International, AMF O'Hare/Chicago, IL). *Scanning Microsc.* (Suppl. 2).

P. W. Hawkes, W. O. Saxton and M. A. O'Keefe, Eds (1992). *Signal and Image Processing in Microscopy and Microanalysis. Proceedings of the 10th Pfefferkorn Conference, Cambridge 1991* (Scanning Microscopy Int., Chicago, IL). Scanning Microsc. (Suppl. 6), dated 1992, published 1994.

R. M. Herman (1992). Classical origins of the Aharonov–Bohm effect. *Found. Phys.* **22**, 713–725.

R. A. Herring (2005). Energy-filtered electron-diffracted beam holography. *Ultramicroscopy* **104**, 261–270.

R. A. Herring and T. Tanji (1993). CBED + EBI holography (CBED + EBH) of materials structures. MSA 51, Cincinnati, OH, 1086–1087.

R. A. Herring, T. Tanji and A. Tonomura (1992). Determination of mean inner potential and thickness in wedge shaped specimens by interferometry and electron holography. EMSA 50, Boston, MA, 990–991.

R. A. Herring, G. Pozzi, T. Tanji and A. Tonomura (1993a). Realization of a mixed type of interferometry using convergent-beam electron diffraction and an electron biprism. *Ultramicroscopy* **50**, 94–100.
R. A. Herring, G. Pozzi, T. Tanji and A. Tonomura (1993b). Convergent beam electron diffraction interferometry using an electron biprism (CBED + EBI). MSA 51, Cincinnati, OH, 1056–1057.
R. A. Herring, G. Pozzi, T. Tanji and A. Tonomura (1995). Interferometry using convergent electron diffracted beams plus an electron biprism (CBED + EBI). *Ultramicroscopy* **60**, 153–169.
K.-H. Herrmann and H. Lichte (1986). Quantum noise limitations in high resolution off axis electron holography. ICEM-11, Kyoto, **1**, 679–680.
K.-H. Herrmann, D. Krahl, A. Kübler and V. Rindfleisch (1969). Comparative investigations of various TV image intensifiers at the electron microscope. *Siemens Rev., X-Ray Microscopy No. 3* **36**, 6–10.
K.-H. Herrmann, D. Krahl, A. Kübler and V. Rindfleisch (1971). Image recording with semiconductor detectors and video amplification devices. In *Electron Microscopy in Material Science* (U. Valdrè, Ed.), 236–272 Academic Press, New York & London).
K. H. Herrmann, D. Krahl and V. Rindfleisch (1972). Use of TV image intensifiers in electron microscopy. *Siemens Forsch. Entwickl. Ber.* **1**, 167–178.
K.-H. Herrmann, E. Reuber and P. Schiske (1978). A simple way for producing holographic filters suitable for image improvement. ICEM-9, Toronto, **1**, 226–227.
T. Hibi (1956). Pointed filament. (I). Its production and its applications. *J. Electronmicrosc.* **4**, 10–15.
T. Hibi and S. Takahashi (1963). Electron interference microscope. *J. Electron Microsc.* **12**, 129–133.
T. Hibi and S. Takahashi (1969). Relation between coherence of electron beam and contrast of electron image. *Z. Angew. Phys.* **27**, 132–138.
T. Hibi and K. Yada (1976). Electron interference microscope. In *Principles and Techniques of Electron Microscopy* (M. A. Hayat, Ed.), **6**, 312–343 (Van Nostrand Reinhold, New York and London).
T. Hirayama (1999). Interference of three electron waves and its application to direct visualization of electric fields. *Mater. Charact.* **42**, 193–200.
T. Hirayama, G. Lai, T. Tanji, N. Tanaka and A. Tonomura (1997). Interference of three electron waves by two biprisms and its application to direct visualization of electromagnetic fields in small regions. *J. Appl. Phys.* **82**, 522–527.
T. Hirayama, K. Yamamoto, K. Miyashita and T. Saito (2005). Amplitude-division three-electron-wave interference for observing pure phase objects having low spatial frequency. *J. Electron Microsc.* **54**, 51–55.
H. Hoffmann and C. Jönsson (1965). Elektroneninterferometrische Bestimmung der mittleren inneren Potentiale von Al, Cu and Ge unter Verwendung eines neuen Präparationsverfahrens. *Z. Phys.* **182**, 360–365.
D. Home and F. Selleri (1992). The Aharonov–Bohm effect from the point of view of local realism. In *Wave-Particle Duality* (F. Selleri, Ed.), 127–137 (Plenum, New York and London).
L. Houben, M. Luysberg and T. Brammer (2004). Illumination effects in holographic imaging of the electrostatic potential of defects and *pn* junctions in transmission electron microscopy. *Phys. Rev. B* **70**, 165313, 8 pp.
F. Houdellier, F. Röder and E. Snoeck (2015). Development of splitting convergent beam electron diffraction (SCBED). *Ultramicroscopy* **159**, 59–66.
F. Houdellier, G. M. Caruso, S. Weber, M. J. Hÿtch, C. Gatel and A. Arbouet (2019). Optimization of off-axis electron holography performed with femtosecond electron pulses. *Ultramicroscopy* **202**, 26–32.
A. Howie (2011). Mechanisms of decoherence in electron microscopy. *Ultramicroscopy* **111**, 761–767.
A. Howie (2014). Addressing Coulomb's singularity, nanoparticle recoil and Johnson's noise. *J. Phys: Conf. Ser* **522**, 012001, 8 pp (EMAG 2013).
M. R. S. Huang, A. Eljarrat and C. T. Koch (2019). Influence of the experimental setup on inline electron holography reconstructions. MC-2019, Berlin, 407–408.
M. R. S. Huang, A. Eljarrat and C. T. Koch (2021). Quantifying the data quality of focal series for inline electron holography. *Ultramicroscopy* **231**, 113264, 8 pp.
F. Hüe, M. Hÿtch, F. Houdellier, E. Snoeck and A. Claverie (2008). Strain mapping in MOSFETS by high-resolution electron microscopy and electron holography. *Mater. Sci. Eng. B* **154–155**, 221–224.

M. Hÿtch (1997a). Analysis of variations in structure from high resolution electron microscope images by combining real space and Fourier space information. *Microsc. Microanal. Microstruct.* **8**, 41–57.

M. Hÿtch (1997b). Geometric phase analysis of high resolution electron microscope images. *Scanning Microsc.* **11**, 53–66.

M. J. Hÿtch (2001). Measurement of displacement and strain by high-resolution transmission electron microscopy. In *Stress and Strain in Epitaxy: Theoretical Concepts, Measurements and Applications* (M. Hanbücken and J.-P. Deville, Eds), 201–219 (Elsevier, Amsterdam).

M. J. Hÿtch and U. Dahmen (1996). Measurement of displacements at defects in Al grain boundaries by holographic reconstruction of geometric phase. EUREM-11, Dublin, **1**, 388–389.

M. Hÿtch and F. Houdellier (2007). Mapping stress and strain in nanostructures by high-resolution transmission electron microscopy. *Microelectron. Eng.* **84**, 460–463.

M. Hÿtch and T. Plamann (2001). Imaging conditions for reliable measurement of displacement and strain in high-resolution electron microscopy. *Ultramicroscopy* **87**, 199–212.

M. Hÿtch, E. Snoeck and R. Kilaas (1998). Quantitative measurement of displacement and strain fields from HREM micrographs. *Ultramicroscopy* **74**, 131–146.

M. Hÿtch, J.-L. Puteaux and J. Thibault (2006). Stress and strain around grain-boundary dislocations measured by high-resolution electron microscopy. *Philos. Mag.* **86**, 4641–4656.

M. Hÿtch, F. Houdellier, F. Hüe and E. Snoeck (2008). Nanoscale holographic interferometry for strain measurements in electronic devices. *Nature* **453**, 1086–1089.

M. J. Hÿtch, N. Cherkashin, S. Reboh, F. Houdellier and A. Claverie (2011). Strain mapping in layers and devices by electron holography. *Phys. Status Solidi A* **208**, 580–583.

M. Hÿtch, C. Gatel, A. Ishizuka and K. Ishizuka (2016). Mapping 2D strain components from STEM moiré fringes. EMC-16, Lyon, **1**, 515–516.

M. Hÿtch, N. Cherkashin, A. Ishizuka and K. Ishizuka (2019). Extending geometric phase analysis (GPA). MC-2019, Berlin, 514.

M. Inoue, H. Tomita, M. Naruse, Z. Akase, Y. Murakami and D. Shindo (2005). Development of a magnetizing stage for in situ observations with electron holography and Lorentz microscopy. *J. Electron Microsc.* **54**, 509–513.

S. Ishioka and K. Fujikawa, Eds (2006). *Proceedings of the 8th International Symposium on Foundations of Quantum Mechanics in the Light of New Technology* (World Scientific, Singapore).

S. Ishioka and K. Fujikawa, Eds (2009). *Proceedings of the 9th International Symposium on Foundations of Quantum Mechanics in the Light of New Technology* (World Scientific, Singapore).

K. Ishizuka (1990). New form of transmission cross coefficient for high-resolution imaging. ICEM-12, Seattle, WA, **1**, 60–61.

K. Ishizuka (1993). Optimized sampling scheme for off-axis holography. *Ultramicroscopy* **52**, 1–5.

K. Ishizuka (1994). Coma-free alignment of a high-resolution electron microscope with threefold astigmatism. *Ultramicroscopy* **55**, 407–418.

K. Ishizuka, T. Tanji and A. Tonomura (1991). Atomic resolution electron holography. In Hawkes et al. (1992), 423–432.

A. Ishizuka, M. Hÿtch and K. Ishizuka (2017). STEM moiré analysis for 2D strain measurements. *J. Electron Microsc.* **66**, 217–221.

C. Johnson, A. Turner, F. J. Garcia de Abajo and B. McMorran (2021). 2-grating inelastic free electron interferometry. *Microsc. Microanal.* **27** (Suppl. 1), 1474–1477.

C. Jönsson (1961). Elektroneninterferenzen an mehreren künstlich hergestellten Feinspalten. *Z. Phys.* **161**, 454–474. Translated into English as Electron diffraction at multiple slits. *Amer. J. Phys.* **42** (1974) 4–11.

C. Jönsson, H. Hoffmann and G. Möllenstedt (1965). Messung des mittleren inneren Potentials von Be im Elektronen-Interferometer. *Phys. Kond. Mater.* **3**, 193–199.

E. Joos, H. D. Zeh, C. Kiefer, D. J. W. Giulini, J. Kupsch and I. O. Stamatescu (2003). *Decoherence and the Appearance of a Classical World in Quantum Theory* (Springer, Berlin).

D. C. Joy, Y.-s. Zhang, X. Zhang, T. Hashimoto, R. D. Bunn, L. Allard and T. A. Nolan (1993). Practical aspects of electron holography. *Ultramicroscopy* **51**, 1–14.

S. Kamefuchi, H. Ezawa, Y. Murayama, M. Namiki, S. Nomura, Y. Ohnuki and T. Yajima, Eds (1984). *Proceedings of the International Symposium on Foundations of Quantum Mechanics in the Light of New Technology, Tokyo 1983* (Physical Society of Japan, Tokyo).

E. Kasper (1992). On the geometrical aberrations in electrostatic biprisms. *Optik* **92**, 45–47.

K. Kasuya, T. Kusunoki, T. Hashizume, T. Ohshima, S. Katagiri, Y. Sakai and N. Arai (2020). Monochromatic electron emission from CeB_6 (310) cold field emitter. *Appl. Phys. Lett.* **117**, 213103.

T. Kawasaki and J. M. Rodenburg (1993). Deconvolving lens transfer functions in electron holograms. *Ultramicroscopy* **52**, 248–252.

T. Kawasaki, J. Endo, T. Matsuda and A. Tonomura (1990a). Development of a 350-kV holography electron microscope and its applications. ICEM-12, Seattle, WA, **1**, 222–223.

T. Kawasaki, T. Matsuda, J. Endo and A. Tonomura (1990b). Observation of a 0.055 nm spacing lattice image in gold using a field emission electron microscope. *Jpn. J. Appl. Phys.* **29**, L508–L510.

T. Kawasaki, Q. X. Ru, T. Matsuda, Y. Bando and A. Tonomura (1991a). High-resolution holography observation of $H\text{-}Nb_2O_5$. *Jpn. J. Appl. Phys.* **30**, L1830–L1832.

T. Kawasaki, Q. X. Ru, T. Matsuda and A. Tonomura (1991b). Electron holography of crystal structure image. EMAG, Bristol, 483–486.

T. Kawasaki, G. Pozzi and A. Tonomura (1992a). Three-beam electron interference experiments. EUREM-10, Granada, **1**, 651–652.

T. Kawasaki, Q. X. Ru and A. Tonomura (1992b). High-resolution holography observations of InP. EUREM-10, Granada, **1**, 653–654.

T. Kawasaki, I. Matsui, T. Yoshida, T. Katsuta, S. Hayashi, T. Onai, T. Furutsu, K. Myochin, M. Numata, H. Mogaki, M. Gorai, T. Akashi, O. Kamimura, T. Matsuda, N. Osakabe, A. Tonomura and K. Kitazawa (2000a). Development of a 1 MV field-emission transmission electron microscope. *J. Electron Microsc.* **49**, 711–718.

T. Kawasaki, T. Yoshida, T. Matsuda, N. Osakabe, A. Tonomura, K. Matsui and K. Kitazawa (2000b). Fine crystal lattice fringes observed using a transmission electron microscope with 1 MeV coherent electron waves. *Appl. Phys. Lett.* **76**, 1342–1344.

T. Kawasaki, Y. Takahashi and T. Tanigaki (2021). Holography: application to high-resolution imaging. *Microscopy* **70**, 39–46.

J. P. Keating and J. M. Robbins (2001). Force and impulse from an Aharonov–Bohm flux line. *J. Phys. A: Math. Gen.* **34**, 807–827.

A. Keller (1956). Estudios acerca de una possibilidad para medir las posiciones de fase de los planos reticulares reflectores de distintos órdenes de un cristal. *Scientia* (Valparaiso) **23**, 1936.

M. Keller (1958). Biprisma-Interferometer für Elektronenwellen und seine Anwendung zur Messung von inneren Potentialen. ICEM-4, Berlin, **1**, 230–232.

M. Keller (1961a). Ein Biprisma-Interferometer für Elektronenwellen und seine Anwendung. *Z. Phys.* **164**, 274–291.

M. Keller (1961b). Elektronen-Zweistrahlinterferenzen mit hohem Gangunterschied. *Z. Phys.* **164**, 292–294.

A. Keller (1962). Vorschlag eines Elektronen-Interferometers zur Bestimmung von Kristallphasen. *Optik* **19**, 117–121.

S. Keramati, E. R. Jones, J. Armstrong and H. Batelaan (2020). Partially coherent quantum degenerate electron matter waves. *Adv. Imaging Electron Phys.* **213**, 3–26.

E. Kerschbaumer (1967). Ein Biprisma-Interferometer für 100 keV Elektronen und seine Anwendung. *Z. Phys.* **201**, 200–208.

H. Kiesel (2000). Nachweis von Korrelationen der Ankünftszeiten von freien Fermiteilchen: electron-antibunching. Dissertation, Tübingen.

H. Kiesel, A. Renz and F. Hasselbach (2002). Observation of Hanbury Brown–Twiss anticorrelations for free electrons. *Nature* **418**, 392–394.

S. Kim, Y. Kondo, K. Lee, G. Byun, J. J. Kim, S. Lee and K. Lee (2013a). Quantitative measurement of strain field in strained-channel-transistor arrays by scanning moiré fringe imaging. *Appl. Phys. Lett.* **103**, 033523.

S. Kim, S. Lee, Y. Kondo, K. Lee, G. Byun, S. Lee and K. Lee (2013b). Strained hetero interfaces in Si/SiGe/SiGe/SiGe multi-layers studied by scanning moiré fringe imaging. *J. Appl. Phys.* **114**, 053518.

S.-I. Kobayashi, H. Ezawa, Y. Murayama and S. Nomura, Eds (1990). *Proceedings of the 3rd International Symposium on Foundations of Quantum Mechanics in the Light of New Technology* (Physical Society of Japan, Tokyo).

D. H. Kobe, V. C. Aguilera-Navarro and R. M. Ricotta (1992). Asymmetry of the Aharonov–Bohm diffraction pattern and Ehrenfest's theorem. *Phys. Rev. A* **45**, 6192–6197.

C. T. Koch (2014). Towards full-resolution inline electron holography. *Micron* **63**, 69–75.

C. T. Koch and A. Lubk (2010). Off-axis and inline electron holography: A quantitative comparison. *Ultramicroscopy* **110**, 460–471.

T. Kodama, N. Osakabe and A. Tonomura (2011). Correlation in a coherent electron beam. *Phys. Rev. A* **83**, 063616, 9 pp.

H. Kohl (1983). Image formation by inelastically scattered electrons: image of a surface plasmon. *Ultramicroscopy* **11**, 53–65.

H. Kohl and H. Rose (1984). Delocalization of electronic excitations as a function of spectrometer acceptance angle. EUREM-6, Budapest, **1**, 445–446.

H. Kohl and H. Rose (1985). Theory of image formation by inelastically scattered electrons in the electron microscope. *Adv. Electron Electron Phys.* **65**, 173–227.

J. Komrska (1971). Scalar diffraction theory in electron optics. *Adv. Electron Electron Phys.* **30**, 139–234.

J. Komrska (1975). Ohyb an interference elektronů. *Čs. Čas. Fyz. A* **25**, 1–13.

J. Komrska and M. Lenc (1970). The wave mechanical interpretation of the interference phenomena produced by an electrostatic biprism. ICEM-7, Grenoble, **1**, 67–68.

J. Komrska and B. Vlachová (1973). Justification of the model for electron interference produced by an electron biprism. *Opt. Acta* **20**, 207–215.

J. Komrska, V. Drahoš and A. Delong (1964a). Interpretation of interference phenomena in Fresnel's electron biprism. EUREM-3, Prague, **A**, 1–2.

J. Komrska, V. Drahoš and A. Delong (1964b). The application of Fresnel fringes to the determination of the local filament diameter in an electron biprism. *Czech. J. Phys. B* **14**, 753–756.

J. Komrska, V. Drahoš and A. Delong (1964c). Fresnel diffraction of electrons by a filament. *Opt. Acta* **11**, 145–157.

J. Komrska, V. Drahoš and A. Delong (1967). Intensity distributions in electron interference phenomena produced by an electrostatic bi-prism. *Opt. Acta* **14**, 147–167.

J. Konnert and P. d'Antonio (1992). Simulation and interpretation of STEM side-band holograms. *Ultramicroscopy* **45**, 281–290.

H. J. Kreuzer (1995). Low energy electron point source microscopy. *Micron* **26**, 503–509.

H. J. Kreuzer, K. Nakamura, A. Wierzbicki, H.-W. Fink and H. Schmid (1992). Theory of the point source electron microscope. *Ultramicroscopy* **45**, 381–403.

E. Krimmel (1960). Kohärente Teilung eines Elektronenstrahls durch Magnetfelder. *Z. Phys.* **158**, 35–38.

E. Krimmel (1961). Elektronen-Inteferenzen in der Umgebung der Brennlinie einer magnetischen Quadrupollinse. *Z. Phys.* **163**, 339–355.

E. Krimmel, G. Möllenstedt and W. Rothemund (1964). Measurement of contact potential differences by electron interferometry. *Appl. Phys. Lett.* **5**, 209–210.

J. H. Kulick, M. D. Ball, S. A. Benton, C. Gallerneault, E. Krantz, L. Wilson and S. Whitmarsh (1987). Holographic imaging of SEM data. EMAG, Manchester, 213–216.

Y. A. Kulyupin, S. A. Nepijko, N. N. Sedov and V. G. Shamonya (1978). Use of interference microscopy to measure electric field distributions. *Optik* **52**, 101–109.

C. G. Kuper (1980). Electromagnetic potentials in quantum mechanics: a proposed test of the Aharonov–Bohm effect. *Phys. Lett. A* **79**, 413–416.

A. Laberrigue, G. Balossier, A. Beorchia, P. Bonhomme, N. Bonnet and M. Troyon (1980). Traitement direct en M.E.T. Déconvolution holographique à l'aide de l'Electrotitus. Utilisation d'un diaphragme de phase de type électrostatique. *J. Microsc. Spectrosc. Electron* **5**, 655–664.

S. J. Lade, D. Paganin and M. J. Morgan (2005a). 3-D vector tomography of Doppler-transformed fields by filtered-backprojection. *Opt. Commun.* **253**, 382–391.

S. J. Lade, D. Paganin and M. J. Morgan (2005b). Electron tomography of electromagnetic fields, potentials and sources. *Opt. Commun.* **253**, 392–400.

G. Lai, T. Hirayama, A. Fukuhara, K. Ishizuka, T. Tanji and A. Tonomura (1994a). Three-dimensional reconstruction of magnetic vector fields using electron-holographic interferometry. *J. Appl. Phys.* **75**, 4593–4598.

G. Lai, T. Hirayama, K. Ishizuka, T. Tanji and A. Tonomura (1994b). Three-dimensional reconstruction of electric-potential distribution in electron-holographic interferometry. *Appl. Opt.* **33**, 829–833.

W. Langbein (1958). Elektroneninterferometrische Messung des inneren Potentials von Kohlenstoff-Folien. *Naturwissenschaften* **45**, 510–511.

A. Lannes (1978). On a generalized view of holography in relation to the phase problem in electron microscopy. ICEM-9, Toronto, **1**, 228–229.

A. Lannes (1980). Abstract holography. *J. Math. Anal. Appl.* **74**, 530–559.

A. Lannes (1982). Reconstruction d'un objet à partir de ses projections, analyse des principes de l'holographie classique et itérative. *J. Opt. (Paris)* **13**, 27–39.

T. Latychevskaia (2019). Iterative phase retrieval for digital holography: tutorial. *J. Opt. Soc. Amer. A* **36**, D31–D40.

T. Latychevskaia and H. W. Fink (2009a). Practical algorithms for simulation and reconstruction of digital in-line holograms. *Appl. Opt.* **54**, 2424–2434.

T. Latychevskaia and H. W. Fink (2009b). Simultaneous reconstruction of phase and amplitude contrast from a single holographic record. *Opt. Express* **17**, 10697–10705.

T. Latychevskaia and H.-W. Fink (2015). Reconstruction of purely absorbing, absorbing and phase-shifting, and strong phase-shifting objects from their single-shot in-line holograms. *Appl. Opt.* **54**, 3925–3932.

T. Latychevskaia, P. Formanek, C. T. Koch and A. Lubk (2010). Off-axis and inline electron holography: experimental comparison. *Ultramicroscopy* **110**, 472–482.

T. Latychevskaia, J.-N. Longchamp and H.-W. Fink (2012). When holography meets coherent diffraction imaging. *Opt. Express* **20**, 28871–28892.

T. Latychevskaia, J.-N. Longchamp, C. Escher and H.-W. Fink (2014). On artefact-free reconstruction of low-energy (30–250 eV) electron holograms. *Ultramicroscopy* **145**, 22–27.

T. Latychevskaia, J.-N. Longchamp, C. Escher and H.-W. Fink (2015). Holography and coherent diffraction with low-energy electrons: A route towards structural biology at the single molecule level. *Ultramicroscopy* **159**, 395–402.

T. Latychevskaia, C. Cassidy and T. Shintake (2021). Bragg holography of nano-crystals. *Ultramicroscopy* **230**, 113376, 8 pp.

B. Lau and G. Pozzi (1978). Off-axis electron micro-holography of magnetic domain walls. *Optik* **51**, 287–296.

R. Lauer (1982). Electron microscopy and holography employing coherent quasi-cylindrical illumination waves. ICEM-10, Hamburg, **1**, 427–428.

R. Lauer (1984a). Fourier-Holographie mit Elektronen. *Optik* **66**, 159–174.

R. Lauer (1984b). Interferometry by electron Fourier holography. *Optik* **67**, 291–293.

R. Lauer and G. Ade (1990). Visualisation and quantitative evaluation of small thickness and phase steps from electron holograms by digital image processing. ICEM-12, Seattle, WA, **1**, 230–231.

R. Lauer and G. Ade (1992). Direct phase evaluation from single off-axis holograms. EUREM-10, Granada, **1**, 523–524.

R. Lauer and K.-J. Hanszen (1986). Quantitative determination of noise in reconstructions of off-axis electron holograms. ICEM-11, Kyoto, **1**, 681–682.

R. Lauer and K. Lickfeld (1988). Electron holographic investigation of the cutting edge of glass knives for ultramicrotomy. EUREM-9, York, **3**, 351–352.

E. N. Leith and J. Upatnieks (1962). Reconstructed wavefronts and communication theory. *J. Opt. Soc. Amer.* **52**, 1123–1130.

E. N. Leith and J. Upatnieks (1963). Wavefront reconstruction with continuous-tone objects. *J. Opt. Soc. Amer.* **53**, 1377–1381.

E. N. Leith and J. Upatnieks (1967). Recent advances in holography. *Prog. Opt.* **6**, 1–52.
F. Lenz (1962). Zur Phasenschiebung von Elektronenwellen im feldfreien Raum durch Potentiale. *Phys. Blätt* **18**, 305–307.
F. Lenz (1965). Kann man Biprisma-Interferenzstreifen zur Messung des elektronenmikroskopischen Auflösungsvermögen verwenden? *Optik* **22**, 270–288.
F. Lenz (1972). Path length differences in electron interferometers using mirrors. *Z. Phys.* **249**, 462–464.
F. Lenz (1987). Coherence and contrast in electron microscopy and electron interferometry. In *Proceedings of the 2nd International Symposium on Foundations of Quantum Mechanics, Tokyo 1986* (M. Namiki, Y. Ohnuki, Y. Murayama and S. Nomura, Eds), 112–116 (Physical Society of Japan, Tokyo).
F. Lenz (1988). Statistics of phase and contrast determination in electron holograms. *Optik* **79**, 13–14.
F. Lenz and E. Völkl (1990). Stochastic limitations to phase and contrast determinations in electron holography. ICEM-12, Seattle, WA, **1**, 228–229.
F. Lenz and G. Wohland (1984). Effect of chromatic aberration and partial coherence on the interference pattern of an electron biprism interferometer. *Optik* **67**, 315–329.
T. Leuthner, H. Lichte and K.-H. Herrmann (1988). STEM-holography using an electron biprism. EUREM-9, York, **1**, 177–178.
T. Leuthner, H. Lichte and K.-H. Herrmann (1989). STEM-holography using the electron biprism. *Phys. Status Solidi A* **116**, 113–121.
T. Leuthner, H. Lichte, K.-H. Herrmann and J. Sum (1990). STEM holography. ICEM-12, Seattle, WA, **1**, 224–225.
F. Li, M. R. McCartney, R. E. Dunin-Borkowski and D. J. Smith (1999). Determination of mean inner potential of germanium using off-axis electron holography. *Acta Cryst. A* **55**, 652–658.
H. Lichte (1979). Ein Elektronen-Auflicht-Interferenzmikroskop zur Präzisionsmessung von Unebenheiten und Potentialunterschieden auf Oberflächen. *PTB-Mitt*, 229–236.
H. Lichte (1980a). Ein Elektronen-Auflicht-Interferenzmikroskop zur Präzisionsmessung von Unebenheiten und Potentialunterschieden auf Oberflächen. *Optik* **57**, 35–67.
H. Lichte (1980b). Application of the electron mirror interference microscope for measuring the height of steps on cleavage faces of single crystals. EUREM-7, The Hague, **1**, 30–31.
H. Lichte (1982). Electron holography. ICEM-10, Hamburg, **1**, 411–418.
H. Lichte (1984a). High resolution imaging by off axis electron holography. EUREM-8, Budapest, **1**, 281–282.
H. Lichte (1984b). Coherent electron optical experiments using an electron mirror. In *Proceedings of the International Symposium on Foundations of Quantum Mechanics in the Light of New Technology, Tokyo 1983* (S. Kamefuchi, H. Ezawa, Y. Murayama, M. Namiki, S. Nomura, Y. Ohnuki, et al., Eds), 29–38 (Physical Society of Japan, Tokyo).
H. Lichte (1985). Electron biprism interference fringes of 0.08 nm spacing for high resolution electron holography. *Optik* **70**, 176–177.
H. Lichte (1986a). Electron holography approaching atomic resolution. *Ultramicroscopy* **20**, 293–304.
H. Lichte (1986b). Electron off axis holography of atomic structures. ICEM-11, Kyoto, **1**, 675–676.
H. Lichte (1986c). Principles of off axis electron holography and its applications to objects in atomic dimensions. In *Proceedings of the International Symposium on Electron Optics, ISEOB* (J.-y. Ximen, Ed.) 150–153 (Institute of Electronics, Academia Sinica, Beijing).
H. Lichte (1986d). Electron interferometry applied to objects of atomic dimensions. *Ann. N. Y. Acad. Sci.* **480**, 175–189.
H. Lichte (1988). Electron holography of atomic structures – prospect and limitations. *Physica B* **151**, 214–222.
H. Lichte (1989). Is electron holography at atomic dimensions going to meet the expectations of Dennis Gabor? EMSA 47, San Antonio, TX, 2–3.
H. Lichte (1990). Electron holography improving transmission electron microscopy. ICEM-12, Seattle, WA, **1**, 208–209.
H. Lichte (1991a). Electron image plane off-axis holography of atomic structures. *Adv. Opt. Electron Microsc.* **12**, 25–91.

H. Lichte (1991b). Optimum focus for taking electron holograms. *Ultramicroscopy* **38**, 13–22.
H. Lichte (1991c). Electron holography opens up wave optics in an electron microscope. EMAG, Bristol, 465–471.
H. Lichte (1991d). High-resolution electron holography. EMSA 49, San Jose, CA, 494–495.
H. Lichte (1991e). Holography – just another method of image processing? In Hawkes et al. (1992), 433–440.
H. Lichte (1992a). Electron holography – state of the art in Tübingen. EUREM-10, Granada, **1**, 637–641.
H. Lichte (1992b). High resolution electron holography – state of the art in Tübingen. APEM-5, Beijing, **1**, 234–237.
H. Lichte (1992c). Electron holography. I. Can electron holography reach 0.1 nm resolution? *Ultramicroscopy* **47**, 223–230.
H. Lichte (1993). Parameters for high-resolution electron holography. *Ultramicroscopy* **51**, 15–20.
H. Lichte and B. Freitag (2000). Inelastic electron holography. *Ultramicroscopy* **81**, 177–186.
H. Lichte and R. Hornstein (1982). Electron interferometric measurement of the enhancement of the phase modulation in low voltage electron holography. ICEM-10, Hamburg, **1**, 431–432.
H. Lichte and G. Möllenstedt (1977). An attempt to investigate the quality of plane surfaces using an electron mirror interference microscope. In *Eighth Int. Cong. X-Ray Optics and Microanalysis* (D. R. Beaman, R. E. Ogilvie and D. B. Wittry, Eds), 211–212 (Pendell, Midland MI).
H. Lichte and G. Möllenstedt (1979). Measurement of the roughness of supersmooth surfaces using an electron mirror interference microscope. *J. Phys. E: Sci. Instrum.* **12**, 941–944.
H. Lichte and E. Völkl (1988). Vignetting in high resolution electron holography. EUREM-9, York, **1**, 191–192.
H. Lichte and E. Völkl (1991). Electron holography in material science. EMSA 49, San Jose, CA, 670–671.
H. Lichte and U. Weierstall (1988). Can electron holography improve EM imaging of biological objects? EUREM-9, York, **3**, 325–326.
H. Lichte, G. Möllenstedt and H. Wahl (1972). A Michelson interferometer using electron waves. *Z. Phys.* **249**, 456–461.
H. Lichte, K.-H. Herrmann and F. Lenz (1987). Electron noise in off-axis image plane holography. *Optik* **77**, 135–140.
H. Lichte, K.-H. Herrmann and F. Lenz (1988). Electron noise in image-plane off-axis holography. EUREM-9, York, **1**, 189–190.
H. Lichte, E. Völkl and K. Scheerschmidt (1992). Electron holography. II. First steps of high resolution electron holography into materials science. *Ultramicroscopy* **47**, 231–240.
H. Lichte, M. Reibold, K. Brand and M. Lehmann (2002). Ferroelectric electron holography. *Ultramicroscopy* **93**, 199–212.
J. A. Lin and J. M. Cowley (1986). Reconstruction from in-line electron holograms by digital processing. *Ultramicroscopy* **19**, 179–190.
B. Lischke (1969). Direct observation of quantized magnetic flux in a superconducting hollow cylinder with an electron interferometer. *Phys. Rev. Lett.* **22**, 1366–1368.
B. Lischke (1970a). Bestimmung des Fluxoidquants in supraleitenden Hohlzylindern. II. *Z. Phys.* **237**, 469–474, Part I is Boersch and Lischke (1970a).
B. Lischke (1970b). Direkte Beobachtung einzelner magnetischer Flußquanten in supraleitenden Hohlzylindern. III. *Z. Phys.* **239**, 360–378.
S. Liu, T. Shi, Q.-y. Chen, A. Zhai, H. Tian, Z.-j. Ding, J.-x. Li and Z. Zhang (2018). Towards quantitative mapping of the charge distribution along a nanowire by in-line electron holography. *Ultramicroscopy* **194**, 126–132.
Y. H. Lo, L. Zhao, M. Gallagher-Jones, A. Rana, J. J. Lodico, W. Xiao, B. C. Regan and J. Miao (2018). In situ coherent diffractive imaging. *Nature Commun.* **9**, 1826, 10 pp.
E. V. Loewenstein (1966). The history and current status of Fourier transform spectroscopy. *Appl. Opt.* **5**, 845–854.
J.-N. Longchamp, T. Latychevskaia, C. Escher and H.-W. Fink (2013). Graphene unit cell imaging by holographic coherent diffraction. *Phys. Rev. Lett.* **110**, 255501, 4 pp.

C. Mahr, K. Müller-Caspary, T. Grieb, F. F. Krause, M. Schowalter and A. Rosenauer (2021). Accurate measurement of strain at interfaces in 4D-STEM: a comparison of various methods. *Ultramicroscopy* **221**, 113196, 10 pp.

U. Maier (1997). Ein Biprisma-Interferometer für Ionen. Dissertation, Tübingen.

G. B. Malykin (1997). Earlier studies of the Sagnac effect. *Usp. Fiz. Nauk.* **167**, 337–342, *Phys Usp.* **40**, 317–321.

G. B. Malykin (2000). The Sagnac effect: correct and incorrect explanations. *Usp. Fiz. Nauk.* **170**, 1325–1349, *Phys. Usp.* **43**, 1229–1252.

G. B. Malykin (2002). Sagnac effect on a rotating frame of reference. Relativistic Zeno paradox. *Usp. Fiz. Nauk.* **172**, 969–970, *Phys. Usp.* **45**, 907–909.

M. Mankos, S. Y. Wang, J. K. Weiss and J. M. Cowley (1992). New detection system for HAADF and holography in STEM. EMSA 50, Boston, MA, 102–103.

L. Marton (1952). Electron interferometer. *Phys. Rev.* **85**, 1057–1058.

L. Marton (1954). Electron interference phenomena. ICEM-3, London, 272–279.

L. Marton, J. A. Simpson and J. A. Suddeth (1953). Electron beam interferometer. *Phys. Rev.* **90**, 490–491.

L. Marton, J. A. Simpson and J. A. Suddeth (1954). An electron interferometer. *Rev. Sci. Instrum.* **25**, 1099–1104.

T. Matsuda, A. Tonomura and T. Komoda (1978). Observation of lattice images with a field emission electron microscope. *Jpn. J. Appl. Phys.* **17**, 2073–2074.

T. Matsuda, A. Tonomura, R. Suzuki, J. Endo, N. Osakabe, H. Umezaki, H. Tanabe, Y. Sugita and H. Fujiwara (1982). Observation of microscopic distribution of magnetic fields by electron holography. *J. Appl. Phys.* **53**, 5444–5446.

T. Matsuda, S. Hasegawa, M. Igarashi, T. Kobayashi, M. Naito, H. Kajiyama, J. Endo, N. Osakabe and A. Tonomura (1989). Magnetic field observation of a single flux quantum by electron-holographic interferometry. *Phys. Rev. Lett.* **62**, 2519–2522.

T. Matsuda, S. Hasegawa, J. Endo, N. Osakabe, A. Tonomura and R. Aoki (1990). Observation of single magnetic-flux quanta using electron holography. ICEM-12, Seattle, WA, **1**, 210–211.

T. Matsuda, A. Fukuhara, T. Yoshida, S. Hasegawa, A. Tonomura and Q. Ru (1991). Computer reconstruction from electron holograms and observation of fluxon dynamics. *Phys. Rev. Lett.* **66**, 457–460.

S. Matsui, K.-I. Nakamatsu, K. Yamamoto and T. Hirayama (2009). Fabrication of fine electron biprism filament by focused-ion-beam chemical-vapor-deposition. *Microsc. Microanal.* **15** (Suppl. 2), 316–317.

T. Matsumoto, N. Osakabe, J. Endo, T. Matsuda and A. Tonomura (1991). Electron holography of frozen-hydrated biological specimen. EMSA 49, San Jose, CA, 686–687.

T. Matsumoto, M. Koguchi, K. Suzuki, H. Nishimura, Y. Motoyoshi and N. Wada (2008). Ferroelectric 90° domain structure in a thin film of $BaTiO_3$ fine ceramics observed by 300 kV electron holography. *Appl. Phys. Lett.* **92**, 072902.

G. Matteucci (1978). On the use of a Wollaston wire in a Möllenstedt–Düker electron biprism. *J. Microsc. Spectrosc. Electron* **3**, 69–71.

G. Matteucci (1990). Electron wavelike behavior: A historical and experimental introduction. *Amer. J. Phys.* **58**, 1143–1147.

G. Matteucci (2007). Quantum effects of electric fields and potentials on electron motion: an introduction to theoretical and practical aspects. *Eur. J. Phys.* **28**, 625–634.

G. Matteucci (2011). On the presentation of wave phenomena of electrons with the Young–Feynman experiment. *Eur. J. Phys.* **32**, 733–738.

G. Matteucci (2013). Interference with electrons – from thought to real experiments. *Proc. SPIE* **8785**, CF1–CF9.

G. Matteucci and C. Beeli (1998). An experiment on electron wave–particle duality including a Planck constant measurement. *Amer. J. Phys* **66**, 1055–1059.

G. Matteucci and M. Muccini (1992). Electron holographic simulations of the field of ferromagnetic tips. EUREM-10, Granada, **1**, 657–658.

G. Matteucci and G. Pozzi (1978). Two further experiments on electron interference. *Amer. J. Phys.* **46**, 619–623.
G. Matteucci and G. Pozzi (1980). A “mixed” type electron interferometer. *Ultramicroscopy* **5**, 219–222.
G. Matteucci and G. Pozzi (1985). New diffraction experiment on the electrostatic Aharonov–Bohm effect. *Phys. Rev. Lett.* **54**, 2469–2472.
G. Matteucci, G. F. Missiroli, G. Pozzi, P. G. Merli and I. Vecchi (1979). Interference electron microscopy in thin film investigations. *Thin Solid Films* **62**, 5–17.
G. Matteucci, P. G. Merli, G. F. Missiroli, G. Pozzi, M. Vanzi and I. Vecchi (1980). Conversion of a standard electron microscope into an interference microscope. *Ultramicroscopy* **5**, 380.
G. Matteucci, G. F. Missiroli and G. Pozzi (1981). Amplitude division electron interferometry. *Ultramicroscopy* **6**, 109–114.
G. Matteucci, G.F. Missiroli and G. Pozzi (1982a). A new holographic method using a high resolution electron microscope. ICEM-10, Hamburg, **1**, 429–430.
G. Matteucci, G. F. Missiroli and G. Pozzi (1982b). A “mixed” type electron interferometer. II. *Ultramicroscopy* **7**, 277–286.
G. Matteucci, G. F. Missiroli and G. Pozzi (1982c). The realization of amplitude division electron interferometry demonstrated on ferromagnetic domain walls. *Ultramicroscopy* **7**, 295–298.
G. Matteucci, G. F. Missiroli and G. Pozzi (1982d). A new off-axis Fresnel holographic method in transmission electron microscopy: an application on the mapping of ferromagnetic domains. III. *Ultramicroscopy* **8**, 403–409.
G. Matteucci, G. F. Missiroli and G. Pozzi (1982e). A new electrostatic phase-shifting effect. *Ultramicroscopy* **10**, 247–252.
G. Matteucci, G. Pozzi and M. Vanzi (1982f). Olografia con elettroni. *Giornale di Fisica* **23**, 17–32.
G. Matteucci, G. F. Missiroli and G. Pozzi (1984). Interferometric and holographic techniques in transmission electron microscopy for the observation of magnetic domain structures. *IEEE Trans.* **MAG–20**, 1870–1875.
G. Matteucci, G. F. Missiroli, E. Nichelatti, J. W. Chen and G. Pozzi (1987). Electron holography of electrostatic fields of spherical charged particles, SIME, *Microscopia Elettronica* **8** (2, Suppl.), 257–258.
G. Matteucci, G. F. Missiroli, J. W. Chen and G. Pozzi (1988a). Mapping of microelectric and magnetic fields with double-exposure electron holography. *Appl. Phys. Lett.* **52**, 176–178.
G. Matteucci, A. Migliori, G. Pozzi and M. Vanzi (1988b). Image simulation of holographic contour maps of reverse-biased p-n junctions. EUREM-9, York, **1**, 195–196.
G. Matteucci, G.F. Missiroli, E. Nichelatti and G. Pozzi (1988c). On the holographic mapping of strong magnetic leakage fields. EUREM-9, York, 2, 265–266.
G. Matteucci, A. Migliori, G. F. Missiroli, E. Nichelatti, G. Pozzi and M. Vanzi (1989). Olografia elettronica di campi a lungo range, SIME, *Microscopia Elettronica* **10** (2, Suppl.), 259–260.
G. Matteucci, G. F. Missiroli, E. Nichelatti, A. Migliori, M. Vanzi and G. Pozzi (1991). Electron holography of long-range electric and magnetic fields. *J. Appl. Phys.* **69**, 1835–1842.
G. Matteucci, F. F. Medina and G. Pozzi (1992a). Electron-optical analysis of the electrostatic Aharonov–Bohm effect. *Ultramicroscopy* **41**, 255–268.
G. Matteucci, G. F. Missiroli, M. Muccini and G. Pozzi (1992b). Electron holography in the study of the electrostatic fields: the case of charged microtips. *Ultramicroscopy* **45**, 77–83.
G. Matteucci, G. F. Missiroli and G. Pozzi (1996). Electron holography of electrostatic fields. *J. Electron Microsc.* **45**, 27–35.
G. Matteucci, G. F. Missiroli and G. Pozzi (1997). Simulations of electron holograms of long range electrostatic field. *Scanning Microsc.* **11**, 367–374.
G. Matteucci, G. F. Missiroli and G. Pozzi (1998). Electron holography of long-range electrostatic fields. *Adv. Imaging Electron Phys.* **99**, 171–240.
G. Matteucci, G. F. Missiroli and G. Pozzi (2002). Electron holography of long-range electrostatic fields. *Adv. Imaging Electron Phys.* **122**, 173–249.
G. Matteucci, D. Iencinella and C. Beeli (2003). The Aharonov–Bohm phase shift and Boyer’s critical considerations: new experimental result but still an open subject? *Found. Phys.* **33**, 577–590.

G. Matteucci, A. Migliori, F. Medina and R. Castaneda (2009). An experiment on the particle–wave nature of electrons. *Eur. J. Phys.* **30**, 217–226.

G. Matteucci, R. Castaneda, S. Serna, F. Medina and J. Garcia-Sucerquia (2010a). Discovering the puzzling behaviour of electrons with the Grimaldi–Young experiment. *Eur. J. Phys.* **31**, 347–356.

G. Matteucci, L. Ferrari and A. Migliori (2010b). The Heisenberg uncertainty principle demonstrated with an electron diffraction experiment. *Eur. J. Phys.* **31**, 1287–1293.

G. Matteucci, M. Pezzi, G. Pozzi, G. L. Alberghi, F. Giorgi, A. Gabrielli, N. S. Cesari, M. Villa, A. Zoccoli, S. Frabboni and G. C. Gazzadi (2013). Build-up of interference patterns with single electrons. *Eur. J. Phys.* **34**, 511–517.

B. J. McMorran (2019). Seeing with phase: interferometric electron microscopy for magnetic materials and biological specimens. *Microsc. Microanal.* **25** (Suppl. 2), 1210–1211.

B. J. McMorran, T. R. Harvey, C. Ophus, J. Pierce and F. Yasin (2018). Demonstration of STEM holography using diffraction gratings. *Microsc. Microanal.* **24** (Suppl. 1), 200–201.

S. McVitie, J. N. Chapman, L. Zhou, L. J. Heyderman and W. A. P. Nicholson (1995). In-situ magnetising experiments using coherent magnetic imaging in TEM. *J. Magn. Magn. Mater.* **148**, 232–236.

F. F. Medina and G. Pozzi (1990). Spatial coherence of anisotropic and astigmatic sources in interference electron microscopy and holography. *J. Opt. Soc. Amer. A* **7**, 1027–1033.

C. Menu and D. Evrard (1971). Interférence en optique électronique. Etude de la visibilité des franges. *C.R. Acad. Sci. Paris B* **273**, 309–312.

E. Menzel, W. Mirandé and I. Weingärtner (1973). *Fourier-Optik und Holographie* (Springer, Vienna and New York), especially Section 10.9–a.

P. G. Merli and G. Pozzi (1978). Stationary phase approximation of defocused images of p-n junctions. *Optik* **51**, 39–48.

P. G. Merli, G. F. Missiroli and G. Pozzi (1974). Electron interferometry with the Elmiskop 101 electron microscope. *J. Phys. E: Sci. Instrum.* **7**, 729–732.

P. G. Merli, G. F. Missiroli and G. Pozzi (1976a). On the statistical aspect of electron interference phenomena. *Amer. J. Phys.* **44**, 306–307.

P. G. Merli, G. F. Missiroli and G. Pozzi (1976b). Recent results on the observation of the microelectric field of reverse-biased p-n junctions by interference electron microscopy. EUREM-6, Jerusalem, **1**, 478–479.

P. G. Merli, G. F. Missiroli and G. Pozzi (2003). L'esperimento di interferenza degli elettroni singoli. *Il Nuovo Saggiatore (Bull. Soc. Ital. Fis.)* **39**, 37–40.

B. M. Mertens and P. Kruit (1997). Element-specific image acquisition by sampling Fourier components. *Scanning Microsc.* **11**, 241–250.

B.M. Mertens and P. Kruit (1999). Results of a pilot experiment on direct phase determination of diffracted beams in TEM. EMAG, Sheffield, 133–136.

B. M. Mertens, M. H. F. Overwijk and P. Kruit (1999). Off-axis holography with a crystal beam splitter. *Ultramicroscopy* **77**, 1–11.

S. Meuret (2020a). Intensity interferometry experiment: photon bunching in cathodoluminescence. *Adv. Imaging Electron Phys.* **215**, 1–45.

S. Meuret (2020b). Applications of photon bunching in cathodoluminescence. *Adv. Imaging Electron Phys.* **215**, 47–87.

S. Meuret, Y. Auad, L. Tizei, H. C. Chang, F. Houdellier, M. Kociak and A. Arbouet (2020). Time-resolved cathodoluminescence in a transmission electron microscope applied to NV centers in diamond. *Microsc. Microanal.* **26** (Suppl. 2), 2022–2023.

A. A. Michelson (1891a). Visibility of interference-fringes in the focus of a telescope. *Philos. Mag.* **31**, 256–259.

A. A. Michelson (1891b). On the application of interference-methods to spectroscopic measurements. I. *Philos. Mag.* **31**, 336–346.

P. A. Midgley (2001). An introduction to off-axis electron holography. *Micron* **32**, 167–184.

P. A. Midgley and R. E. Dunin-Borkowski (2009). Electron tomography and holography in materials science. *Nature Mater.* **8**, 271–280.

A. Migliori and G. Pozzi (1991). Simulations of electron holography observations of the magnetic field of superconducting flux lines, SIME, *Microscopia Elettronica* **12** (2, Suppl.), 407–408.

A. Migliori and G. Pozzi (1992). Computer simulations of electron holographic contour maps of superconducting flux lines. *Ultramicroscopy* **41**, 169–179.

A. Migliori, G. Pozzi and A. Tonomura (1993). Computer simulations of electron holographic contour maps of superconducting flux lines. II. The case of tilted specimens. *Ultramicroscopy* **49**, 87–94.

V. Migunov, A. London, M. Farle and R. E. Dunin-Borkowski (2015). Model-independent measurement of the charge density distribution along an Fe atom probe needle using off-axis electron holography without mean inner potential effects. *J. Appl. Phys.* **117**, 134301.

V. Migunov, C. Dwyer, C. B. Boothroyd, G. Pozzi and R. E. Dunin-Borkowski (2017). Prospects for quantitative and time-resolved double and continuous exposure off-axis electron holography. *Ultramicroscopy* **178**, 48–61.

W. Mirandé, I. Weingärtner and E. Menzel (1969). Holographie bei Teilkohärenz. II. Fourier-Holographie. *Optik* **29**, 537–548.

G. F. Missiroli, G. Pozzi and U. Valdrè (1981). Electron interferometry and interference electron microscopy. *J. Phys. E: Sci. Instrum.* **14**, 649–671.

G. F. Missiroli, M. Muccini and G. Pozzi (1991). Olografia elettronica di campi elettrostatici generati da micropunte, SIME, *Microscopia Elettronica* **12** (2, Suppl.), 409–410.

K. Miyashita, K. Yamamoto, T. Hirayama and T. Tanji (2004). Direct observation of electrostatic microfields by four-electron-wave interference using two electron biprisms. *J. Electron Microsc.* **53**, 577–582.

G. Möllenstedt (1960). Aktuelle Probleme der Elektronenmikroskopie. EUREM-2, Delft, **1**, 1–17.

G. Möllenstedt (1962). Beiträge zur experimentellen Elektronenoptik. ICEM-5, Philadelphia, PA, post-deadline paper.

G. Möllenstedt (1987a). Basic concepts of electron interferometry. In *Proceedings of the International Symposium on Electron Optics, ISEOB* (J.-y. Ximen, Ed.), 171–176 (Institute of Electronics, Academia Sinica, Beijing).

G. Möllenstedt (1987b). How I got an idea of electron beam interference and developed it [in Japanese]. *Denshikenbikyo* **21**, 190–200.

G. Möllenstedt (1988). Some remarks on the quantum mechanics of the electron. *Physica B* **151**, 201–205.

G. Möllenstedt (1991). The invention of the electron Fresnel interference biprism. *Adv. Opt. Electron Microsc.* **12**, 1–23.

G. Möllenstedt and W. Bayh (1961a). Ein verbessertes Elektronen-Interferometer für weitgetrennte Bündel. *Phys. Verhandl* **12**, 142.

G. Möllenstedt and W. Bayh (1961b). Elektronen-Biprisma-Interferenzen mit weit getrennten kohärenten Teilbündeln. *Naturwissenschaften* **48**, 400.

G. Möllenstedt and W. Bayh (1962a). Messung der kontinuierlichen Phasenschiebung von Elektronenwellen im kraftfeldfreien Raum durch das magnetische Vektorpotential einer Luftspule. *Naturwissenschaften* **49**, 81–82.

G. Möllenstedt and W. Bayh (1962b). Kontinuierliche Phasenschiebung von Elektronenwellen im kraftfeldfreien Raum durch das magnetische Vektorpotential eines Solenoids. *Phys. Blätt* **18**, 299–305.

G. Möllenstedt and R. Buhl (1957). Ein Elektronen-Interferenz-Mikroskop. *Phys. Blätt* **13**, 357–360.

G. Möllenstedt and H. Düker (1955). Fresnelscher Interferenzversuch mit einem Biprisma für Elektronenwellen. *Naturwissenschaften* **42**, 41.

G. Möllenstedt and H. Düker (1956). Beobachtungen und Messungen an Biprisma-Interferenzen mit Elektronenwellen. *Z. Phys.* **145**, 377–397.

G. Möllenstedt and C. Jönsson (1959). Elektronen-Mehrfachinterferenzen an regelmäßig hergestellten Feinspalten. *Z. Phys.* **155**, 472–474.

G. Möllenstedt and M. Keller (1957). Elektroneninterferometrische Messung des inneren Potentials. *Z. Phys.* **148**, 34–37.

G. Möllenstedt and E. Krimmel (1964). Über ein Elektronen-Weitwinkel-Interferometer mit neuartigem Strahlengang. *Mikroskopie* **19**, 29.

G. Möllenstedt and F. Lenz (1962). Some electron interference experiments and their theoretical interpretation. *J. Phys. Soc. Jpn.* **17** (Suppl. B–II), 183–186.

G. Möllenstedt and H. Lichte (1978a). Young–Fresnelscher Interferenzversuch mit zwei nebeneinander stehenden Spiegeln für Elektronenwellen. *Optik* **51**, 423–428.

G. Möllenstedt and H. Lichte (1978b). Doppler shift of electron waves. ICEM-9, Toronto, **1**, 178–179.

G. Möllenstedt and H. Lichte (1979). Electron interferometry. In *Neutron Interferometry* (U. Bonse and H. Rauch, Eds), 363–388 (Clarendon, Oxford).

G. Möllenstedt and H. Lichte (1989). Einige Bemerkungen zur Phase der Elektronenwelle in Interferometrie und Holographie. *Nova Acta Leopoldina* **64** (274), 57–62.

G. Möllenstedt and H. Wahl (1968). Elektronenholographie und Rekonstruktion mit Laserlicht. *Naturwissenschaften* **55**, 340–341.

G. Möllenstedt and G. Wohland (1980). Direct interferometric measurement of the coherence length of an electron wave packet using a Wien filter. EUREM-7, The Hague, **1**, 28–29.

W. Moreau and D. K. Ross (1994). Complementary electric Aharonov-Bohm effect. *Phys. Rev. A* **49**, 4348–4352.

R. Morin (1999). Computer simulation and object reconstruction in low-energy off-axis electron holography. *Ultramicroscopy* **76**, 1–12.

R. Morin and A. Gargani (1993). Ultra-low-energy-electron projection holograms. *Phys. Rev. B* **48**, 6643–6645.

R. Morin, A. Gargani and F. Bel (1990). A simple UHV electron projection microscopy. *Microsc. Microanal. Microstruct.* **1**, 289–297.

R. Morin, M. Pitaval and E. Vicario (1996). Low energy off-axis holography in electron microscopy. *Phys. Rev. Lett.* **76**, 3979–3982.

H. Müller and H. Rose (2003). Electron scattering. In *High-Resolution Imaging and Spectroscopy of Materials* (F. Ernst and M. Rühle, Eds), 9–68 (Springer, Berlin).

H. Müller, H. Rose and P. Schorsch (1998). A coherence function approach to image simulation. *J. Microsc. (Oxford)* **190**, 73–88.

E. Müller, P. Kruse, D. Gerthsen, M. Schowalter and A. Rosenauer (2005). Measurement of the mean inner potential of ZnO nanorods by transmission electron holography. *Appl. Phys. Lett.* **86**, 154108.

J. Munch (1975). Experimental electron holography. *Optik* **43**, 79–99.

J. Munch and E. Zeitler (1974). Interference experiments with a field emission electron microscope. EMSA 32, St Louis, MO, 386–387.

A. B. Naden, K. J. O'Shea and D. A. MacLaren (2018). Evaluation of crystallographic strain, rotation and defects in functional oxides by the moiré effect in scanning transmission electron microscopy. *Nanotechnology* **29**, 165704.

K.-I. Nakamatsu, K. Yamamoto, T. Hirayama and S. Matsui (2008). Fabrication of fine electron biprism filament by free-space-nanowiring technique of focused-ion-beam + chemical vapor deposition for accurate off-axis electron holography. *Appl. Phys. Espress* **1**, 117004.

M. Namiki, Y. Ohnuki, Y. Murayama and S. Nomura, Eds (1987). *Proceedings of the 2nd International Symposium Foundations of Quantum Mechanics in the Light of New Technology, Tokyo 1986* (Physical Society of Japan, Tokyo).

R. Neutze and F. Hasselbach (1998). Sagnac experiment with electrons: reanalysis of a rotationally induced phase shift for charged particles. *Phys. Rev. A* **58**, 557–565.

M. Nicklaus (1989). Ein Sagnac-Experiment mit Elektronenwellen. (Dissertation, Tübingen.

M. Nicklaus and F. Hasselbach (1993). Wien filter: a wave-packet-shifting device for restoring longitudinal coherence in charged-matter-wave interferometers. *Phys. Rev. A* **48**, 152–160.

M. Nicklaus and F. Hasselbach (1995). Experiments on the influence of electro-magnetic and gravito-inertial potentials and fields on the quantum mechanical phase of matter waves. *Ann. N. Y. Acad. Sci.* **755**, 877–879.

T. Niermann and M. Lehmann (2016). Holographic focal series: differences between inline and off-axis electron holography at atomic resolution. *J. Phys. D: Appl. Phys* **49**, 194002, 14 pp.

Y. Nomura, K. Yamamoto, S. Anada, T. Hirayama, E. Igaki and K. Saitoh (2021). Denoising of series electron holograms using tensor decomposition. *Microscopy* **70**, 255–264.

A. Nuttall (1981). Some windows with very good sidelobe behavior. *IEEE Trans. Acoust. Speech Sig. Proc.* **29**, 84–91.

L. O'Raifeartaigh, N. Straumann and A. Wipf (1991). On the origin of the Aharonov–Bohm effect. *Comments Nucl. Part. Phys.* **20**, 15–22.

K. Ogai, Y. Kimura, R. Shimizu, K. Ishibashi, Y. Aoyagi and S. Namba (1991). Microfabricated submicron Al-filament biprism as applied to electron holography. *Jpn. J. Appl. Phys.* **30**, 3272–3276.

A. Ohshita, H. Teraoka and H. Tomita (1984). Production of central electrode of electron biprism using ion beam thinning technique. *Ultramicroscopy* **12**, 247–250.

A. Ohshita, H. Minamide, Y. Saito and H. Tomita (1990). Holographic interference electron microscopy by numerical reconstruction method. ICEM-12, Seattle, WA, **1**, 220–221.

M. Ohtsuki and E. Zeitler (1977). Young's experiment with electrons. *Ultramicroscopy* **2**, 147–148.

A. Olivei (1969). Holography and interferometry in electron Lorentz microscopy. *Optik* **30**, 27–43.

Y. A. Ono and K. Fujikawa, Eds (1999). *Proceedings of the 6th International Symposium on Foundations of Quantum Mechanics in the Light of New Technology* (Elsevier, New York & Oxford).

Y. A. Ono and K. Fujikawa, Eds (2002). *Proceedings of the 7th International Symposium on Foundations of Quantum Mechanics in the Light of New Technology* (World Scientific, Singapore).

C. Ophus, T. R. Harvey, F. S. Yasin, H. G. Brown, P. M. Pelz, B. H. Savitzky, J. Ciston and B. J. McMorran (2019). Advanced phase reconstruction methods enabled by four-dimensional scanning transmission electron microscopy. *Microsc. Microanal.* **25** (Suppl. 2), 10–11.

N. Osakabe (1992). Observation of surfaces by reflection electron holography. *Microsc. Res. Tech.* **20**, 457–462.

N. Osakabe (2014). How the test of Aharonov-Bohm effect was initiated at Hitachi laboratory. In *In Memory of Akira Tonomura, Physicist and Electron Microscopist* (K. Fujikawa and Y. A. Ono, Eds), 62–73 (World Scientific, Singapore).

N. Osakabe, K. Yoshida, Y. Horiuchi, T. Matsuda, H. Tanabe, T. Okuwaki, J. Endo, H. Fujiwara and A. Tonomura (1983). Observation of recorded magnetization pattern by electron holography. *Appl. Phys. Lett.* **42**, 746–748.

N. Osakabe, T. Matsuda, T. Kawasaki, J. Endo, A. Tonomura, S. Yano and H. Yamada (1986). Experimental confirmation of Aharonov–Bohm effect using a toroidal magnetic field confined by a superconductor. *Phys. Rev. A* **34**, 815–822.

N. Osakabe, T. Matsuda, J. Endo and A. Tonomura (1988). Observation of atomic steps by reflection electron holography. *Jpn. J. Appl. Phys.* **27**, L1772–L1774.

N. Osakabe, J. Endo, T. Matsuda, A. Tonomura and A. Fukuhara (1989). Observation of surface undulation due to single-atomic shear of a dislocation by reflection-electron holography. *Phys. Rev. Lett.* **62**, 2969–2972.

N. Osakabe, T. Matsuda, J. Endo and A. Tonomura (1993). Reflection electron holographic observation of surface displacement field. *Ultramicroscopy* **48**, 483–488.

N. Osakabe, T. Kodama, J. Endo, A. Tonomura, K. Ohbayashi, T. Urakami, S. Ohsuka, H. Tsuchiya and Y. Tsuchiya (1995). Fast and precise electron counting system for the observation of quantum mechanical electron intensity correlation. *Nucl. Instrum. Meth. Phys. Res. A* **365**, 585–587.

H. S. Park (2014). Nanomagnetism visualized by electron holography. In *In Memory of Akira Tonomura, Physicist and Electron Microscopist* (K. Fujikawa and Y. A. Ono, Eds), 180–191 (World Scientific, Singapore).

M. Peshkin (1999). Force-free interactions and nondispersive phase shifts in interferometry. *Found. Phys.* **29**, 481–489.

M. Peshkin and A. Tonomura (1989). *The Aharonov–Bohm Effect* (Springer, Berlin and New York), Lecture Notes in Physics, vol 340.

C. Phatak, M. Beleggia and M. de Graef (2008). Vector field electron tomography of magnetic materials: theoretical development. *Ultramicroscopy* **108**, 503–513.

C. Phatak, A. K. Petford-Long and M. de Graef (2010). Three-dimensional study of the vector potential of magnetic structures. *Phys. Rev. Lett.* **104**, 253901, 4 pp.

R. Plass and L. D. Marks (1992). Holographic contrast transfer theory. EMSA 50, Boston, MA, 984–985.

A. Pofelski (2021). Strain characterization methods in transmission electron microscopy. *Adv. Imaging Electron Phys.* **219**.

A. Pofelski, S. Y. Woo, B. H. Le, X. Liu, S. Zhao, Z. Mi, S. Löffler and G. Botton (2018). 2D strain mapping using scanning transmission electron microscopy moiré interferometry and geometrical phase analysis. *Ultramicroscopy* **187**, 1–12.

A. Pofelski, S. Ghanad-Tavakoli, D. A. Thompson and G. Botton (2020). Sampling optimization of moiré geometrical phase analysis for strain characterization in scanning transmission electron microscopy. *Ultramicroscopy* **209**, 112858, 13 pp.

A. Pofelski, V. Whabi, S. Ghanad-Tavakoli and G. Botton (2021). Assessment of the strain depth sensitivity of moiré sampling scanning transmission electron microscopy geometrical phase analysis through a comparison with dark-field electron holography. *Ultramicroscopy* **223**, 113225, 7 pp.

A. Pooch (2018). Coherence and decoherence studies in electron matter wave interferometry. Dissertation, Tübingen.

A. Pooch, M. Seidling, M. Layer, A. Rembold and A. Stibor (2017). A compact electron matter wave interferometer for sensor technology. *Appl. Phys. Lett.* **110**, 223108.

A. Pooch, M. Seidling, N. Kerker, R. Röpke, A. Rembold, W. T. Chang, I. S. Hwang and A. Stibor (2018). Coherent properties of a tunable low-energy electron-matter-wave source. *Phys. Rev. A* **97**, 013611, 7 pp.

E. J. Post (1967). Sagnac effect. *Rev. Mod. Phys.* **39**, 475–493.

P. L. Potapov, H. Lichte, J. Verbeeck and D. Van Dyck (2006). Experiments on inelastic electron holography. *Ultramicroscopy* **106**, 1012–1018.

G. Pozzi (1975). Asymptotic approximation of the image wavefunction in interference electron microscopy. *Optik* **42**, 97–102.

G. Pozzi (1977). Off-axis image electron holography: a proposal. *Optik* **47**, 105–107.

G. Pozzi (1980a). Theoretical interpretation of shadow and interference electron microscopy images of magnetic domain walls. EUREM-7, The Hague, **1**, pp 32–33.

G. Pozzi (1980b). Asymptotic approximation of the image wavefunction in interference electron microscopy. II. Extension to the biprism edges. *Optik* **56**, 243–250.

G. Pozzi (1983a). Off-axis image electron holography with a mixed type interferometer. *Optik* **63**, 227–238.

G. Pozzi (1983b). Amplitude division off-axis Fresnel holography in transmission electron microscopy. *Optik* **66**, 91–100.

G. Pozzi (1992). Electron holography. In *Electron Microscopy in Materials Science* (P. G. Merli and M. V. Antisari, Eds), 269–278 (World Scientific, Singapore, River Edge NJ and London).

G. Pozzi (1993). Recent developments in electron holography, SIME, *Microscopia Elettronica* **14** (2, Suppl.), 3–6.

G. Pozzi (2013). *Microscopia e Olografia con Elettroni* (Bononia University Press, Bologna).

G. Pozzi (2016). Particles and waves in electron optics and microscopy. *Adv. Imaging Electron Phys.* **194**, 1–303.

G. Pozzi and G. F. Missiroli (1973). Interference microscopy of magnetic domains. *J. Microsc. (Paris)* **18**, 103–108.

G. Pozzi and R. Prola (1987). Coerenza ellittica in olografia elettronica, SIME, *Microscopia Elettronica* **8** (2, Suppl.), 255–256.

G. Pozzi and M. Vanzi (1982). Interpretation of electron interference images of reverse-biased p-n junctions. *Optik* **60**, 175–180.

V. Prabhakara, D. Jannis, A. Béché, H. Bender and J. Verbeeck (2019). Strain measurement in semiconductor FinFET devices using a novel moiré demodulation technique. *Semicon. Sci. Technol.* **35**, 034002.

V. Prabhakara, D. Jannis, G. Guzzinati, A. Béché, H. Bender and J. Verbeeck (2020). HAADF-STEM block-scanning strategy for local measurement of strain at the nanoscale. *Ultramicroscopy* **219**, 113099, 10 pp.

W. Qian, M. R. Scheinfein and J. C. H. Spence (1993). Electron optical properties of nanometer field emission electron sources. *Appl. Phys. Lett.* **62**, 315–317.

G. M. Rackham, J. E. Loveluck and J. W. Steeds (1978). A double-crystal electron interferometer. In *Electron Diffraction,* 1927–1977 (P. J. Dobson, J. B. Pendry and C. J. Humphteys, Eds), 435–440 (Institute of Physics, Bristol), IoP Conference Series No. 41.

O. Rang (1953). Fern-Interferenzen von Elektronenwellen. *Z. Phys.* **136**, 465–479.

W. D. Rau, H. Lichte, E. Völkl and U. Weierstall (1991a). Real-time reconstruction of electron-off-axis holograms recorded by means of a high pixel CCD camera. EMSA 49, San Jose, CA, 680–681.
W. D. Rau, H. Lichte, E. Völkl and U. Weierstall (1991b). Real-time reconstruction of electron-off-axis holograms recorded with a high pixel CCD camera. *J. Computer-Assist. Microsc.* **3**, 51–63.
W. D. Rau, P. Schwander, F. H. Baumann, W. Höppner and A. Ourmazd (1999). Two-dimensional mapping of the electrostatic potential in transistors by electron holography. *Phys. Rev. Lett.* **82**, 2614–2617.
A. Rembold (2017). Second-order correlation analysis of multifrequence dephasing in single-particle interferomentry. Dissertation, Tübingen.
A. Rembold, G. Schütz, W. T. Chang, A. Stefanov, A. Pooch, I. S. Hwang, A. Günther and A. Stibor (2014). Correction of dephasing oscillations in matter-wave interferometry. *Phys. Rev. A* **89**, 033635, 5 pp.
A. Rembold, G. Schütz, R. Röpke, W. T. Chang, I. S. Hwang, A. Günther and A. Stibor (2017a). Vibrational dephasing in matter-wave interferometers. *New J. Phys.* **19**, 033009, 13 pp.
A. Rembold, R. Röpke, G. Schütz, J. Fortágh, A. Stibor and A. Günther (2017b). Second-order correlations in single-particle interferometry. *New J. Phys.* **19**, 103029, 26 pp.
J. M. Rodenburg (2001). A simple model of holography and some enhanced resolution methods in electron microscopy. *Ultramicroscopy* **87**, 105–121.
F. Röder and A. Lubk (2014). Transfer and reconstruction of the density matrix in off-axis electron holography. *Ultramicroscopy* **146**, 103–116.
F. Röder and A. Lubk (2015). A proposal for the holographic correction of incoherent aberrations by tilted reference waves. *Ultramicroscopy* **152**, 63–74.
F. Röder, F. Houdellier, T. Denneulin, E. Snoeck and M. Hÿtch (2016). Realization of a tilted reference wave for electron holography by means of a condenser biprism. *Ultramicroscopy* **161**, 23–40ß.
P. Rodgers (2002). The double-slit experiment. *Phys. World* **15** (9), 15–16.
G. L. Rogers (1950a). Gabor diffraction microscopy: the hologram as a generalized zone-plate. *Nature* **166**, 237.
G. L. Rogers (1950b). The black and white hologram. *Nature* **166**, 1027.
G. L. Rogers (1952). Experiments in diffraction microscopy. *Proc. R. Soc. Edinb. A* **63**, 193–221.
G. L. Rogers (1970). The 'equivalent interferometer' in holography. *Opt. Acta* **17**, 527–538.
J. Rogers (1978). The design and use of an optical model of the electron microscope. In *Optica Hoy y Mañana* (J. Bescos, A. Hidalgo, L. Plaza and J. Santamaria, Eds), 235–238 (Sociedad Española de Optica, Madrid).
J. Rogers (1980). Electron holography. In *Imaging Processes and Coherence in Physics* (M. Schlenker, M. Fink, J. P. Goedgebuer, C. Malgrange, J. C. Viénot and R. H. Wade, Eds), 365–370 (Springer, Berlin and New York).
R. Röpke, N. Kerker and A. Stibor (2020). Data transmission by quantum matter wave modulation. *New J. Phys.* **23**, 023038, 11 pp.
H. Rose (1977). Nonstandard imaging methods in electron microscopy. *Ultramicroscopy* **2**, 251–267.
H. Rose (1984). Information transfer in transmission electron microscopy. *Ultramicroscopy* **15**, 173–192.
R. Rosa (2012). The Merli–Missiroli–Pozzi two-slit electron-interference experiment. *Phys. Perspect.* **14**, 178–195.
A. Rother and K. Scheerschmidt (2009). Relativistic effects in elastic scattering of electrons in TEM. *Ultramicroscopy* **109**, 154–160.
A. Rother, T. Gemming and H. Lichte (2009). The statistics of the thermal motion of the atoms during imaging process in transmission electron microscopy and related techniques. *Ultramicroscopy* **109**, 139–146.
Q. Ru (1994). Digital analysis of Young's interference fringes. *Ultramicroscopy* **55**, 15–17.
Q. Ru (1995). Amplitude-division electron holography. In *Electron Holography* (A. Tonomura, L. F. Allard, G. Pozzi, D. C. Joy and J. A. Ono, Eds), 343–353 (Elsevier, Amsterdam).
Q. Ru, J. Endo, T. Tanji and A. Tonomura (1991a). Phase-shifting electron holography by beam tilting. *Appl. Phys. Lett.* **59**, 2372–2374.
Q. Ru, T. Matsuda, A. Fukuhara and A. Tonomura (1991b). Digital extraction of the magnetic-flux distribution from an electron interferogram. *J. Opt. Soc. Amer. A* **8**, 1739–1745.
Q. Ru, T. Hirayama, J. Endo and A. Tonomura (1992a). Hologram-shifting method for high-speed electron hologram reconstruction. *Jpn. J. Appl. Phys.* **31**, 1919–1921.

Q. Ru, T. Hirayama, J. Endo and A. Tonomura (1992b). Approaches to real-time electron holography. EUREM-10, Granada, **1**, 661–662.

Q. Ru, N. Osakabe, J. Endo and A. Tonomura (1994a). Electron holography available in a non-biprism transmission electron microscope. *Ultramicroscopy* **53**, 1–7.

Q. Ru, G. Lai, K. Aoyama, J. Endo and A. Tonomura (1994b). Principle and application of phase-shifting electron holography. *Ultramicroscopy* **55**, 209–220.

H. Rubens and R. W. Wood (1911). Focal isolation of long heat-waves. *Philos. Mag.* **21**, 249–261.

D. K. Saldin (1991). Atomic resolution electron holography – a realization of Gabor's dream? *Adv. Mater.* **3**, 159–161.

A. S. Sanz and S. Miret-Artés (2012). *A Trajectory Description of Quantum Processes. I. Fundamentals* (Springer, Berlin).

A. S. Sanz and S. Miret-Artés (2014). *A Trajectory Description of Quantum Processes. II. Applications* (Springer, Berlin).

G. Saxon (1972a). Division of wavefront side-band Fresnel holography with electrons. *Optik* **35**, 195–210.

G. Saxon (1972b). The compensation of magnetic lens wavefront aberrations in side-band holography with electrons. *Optik* **35**, 359–375.

G. Schaal (1971). Ein Biprismainterferometer für 300 keV-Elektronen. *Z. Phys.* **241**, 65–81.

G. Schaal, C. Jönsson and E. F. Krimmel (1966). Weitgetrennte kohärente Elektronen-Wellenzüge und Messung des Magnetflusses $\phi_0 = h/e$. *Optik* **24**, 529–538.

P. Schattschneider and H. Lichte (2005). Correlation and the density-matrix approach to inelastic electron holography in solid state plasmas. *Phys. Rev. B* **71**, 045130, 9 pp.

P. Schattschneider and J. Verbeeck (2008). Fringe contrast in inelastic LACBED holography. *Ultramicroscopy* **108**, 407–414.

P. Schattschneider, M. Nelhiebel and B. Jouffrey (1999). Density matrix of inelastically scattered fast electrons. *Phys. Rev. B* **59**, 10959–10969.

P. Schattschneider, M. Nelhiebel, H. Souchay and B. Jouffrey (2000). The physical significance of the mixed dynamic form factor. *Micron* **31**, 333–345.

K. Scheerschmidt (1997). Direct retrieval of object information from diffracted electron waves. *Scanning Microsc.* **11**, 455–465.

M. R. Scheinfein, W. Qian and J. C. H. Spence (1993). Aberrations of emission cathodes: nanometer diameter field-emission electron sources. *J. Appl. Phys.* **73**, 2057–2068.

M. A. Schlosshauer (2007). *Decoherence and the Quantum-to-Classical Transition* (Springer, Berlin).

M. A. Schlosshauer (2019). Quantum decoherence. *Phys. Reports* **831**, 1–57.

H. Schmid (1984). Coherence length measurement by producing extremely high phase shifts. EUREM-8, Budapest, **1**, 285–286.

H. Schmid (1985). Ein Elektronen-Interferometer mit 300 μm weit getrennten kohärenten Teilbündeln zur Erzeugung hoher Gangunterschiede und Messung der Phasenschiebung durch das magnetische Vektorpotential bei metallisch abgeschirmtem Magnetfluss. Dissertation, Tübingen.

G. Schütz (2018). Optimierung und Characterisierung eines Biprisma-Interferometers für die Aharonov–Bohm Physik. Dissertation, Tübingen.

G. Schütz, A. Rembold, A. Pooch, S. Meier, P. Schneeweiss, A. Rauschenbeutel, A. Günther, W. T. Chang, I. S. Hwang and A. Stibor (2014). Biprism electron interferometry with a single atom tip source. *Ultramicroscopy* **141**, 9–15.

G. Schütz, A. Rembold, A. Pooch, H. Prochel and A. Stibor (2015). Effective beam separation schemes for the measurement of the electric Aharonov–Bohm effect in an ion interferometer. *Ultramicroscopy* **158**, 65–73.

F. Selleri, Ed. (1992). *Wave–Particle Duality* (Plenum, New York and London).

M. D. Semon (1982). Experimental verification of an Aharonov–Bohm effect in rotating reference frames. *Found. Phys.* **12**, 49–57.

M. D. Semon and J. R. Taylor (1986). A dynamical formulation of the Aharonov–Bohm effect. In *Fundamental Questions in Quantum Mechanics* (L. M. Roth and A. Inomata, Eds), 191–198 (Gordon and Breach, New York).

M. D. Semon and J. R. Taylor (1987a). Expectation values in the Aharonov–Bohm effect. *Nuovo Cim. B* **97**, 25–49.

M. D. Semon and J. R. Taylor (1987b). Expectation values in the Aharonov–Bohm effect. II. *Nuovo Cim. B* **100**, 389–401.

M. D. Semon and J. R. Taylor (1988). The Aharonov–Bohm effect: still a thought-provoking experiment. *Found. Phys.* **18**, 731–740.

M. D. Semon and J. R. Taylor (1994). Comment on "Asymmetry of the Aharonov–Bohm diffraction pattern and Ehrenfest's theorem". *Phys. Rev. A* **50**, 1954–1958.

A. Septier (1959). Bipartition d'un faisceau de particules par un biprisme électrostatique. *C.R. Acad. Sci. Paris* **249**, 662–664.

A. I. Shelankov (1998). Magnetic force exerted by the Aharonov–Bohm line. *Europhys. Lett.* **43**, 623–628.

J. Sickmann, P. Formánek, M. Linck, U. Muehle and H. Lichte (2011). Imaging modes for potential mapping in semiconductor devices by electron holography with improved lateral resolution. *Ultramicroscopy* **111**, 290–302.

M. P. Silverman (1987a). Distinctive quantum features of electron intensity correlation interferometry. *Nuovo Cim, B* **97**, 200–219.

M. P. Silverman (1987b). Fermion ensembles that show statistical bunching. *Phys. Lett. A* **124**, 27–31.

M. P. Silverman (1987c). On the feasibility of observing electron antibunching in a field-emission beam. *Phys. Lett. A* **120**, 442–446.

M. P. Silverman (1987d). Second-order temporal and spatial coherence of thermal electrons. *Nuovo Cim, B* **99**, 227–245.

M. P. Silverman (1988). Quantum interference effects on fermion clustering in a fermion interferometer. *Physica B* **151**, 291–297.

M. P. Silverman (2008). *Quantum Superposition. Counterintuitive Consequences of Coherence, Entanglement, and Interference* (Springer, Berlin).

G. Simonsohn and E. Weihreter (1979). The double-slit experiment with single-photoelecton detection. *Optik*, 199–208.

J. A. Simpson (1954). The theory of the three-crystal electron interferometer. *Rev. Sci. Instrum.* **25**, 1105–1109.

J. A. Simpson (1956). Electron interference experiments. *Rev. Mod. Phys.* **28**, 254–260.

F. Sonier (1968). Application d'une cathode à pointe å la réalisation d'un interféromètre électronique à lentilles magnétiques. *C.R. Acad. Sci. Paris B* **267**, 187–190.

F. Sonier (1970). Détermination de la valeur du potentiel interne de MgO en microscopie interférentielle. *C.R. Acad. Sci. Paris B* **270**, 1536–1539.

F. Sonier (1971). Microscopie électronique interférentielle. *J. Microsc. (Paris)* **12**, 17–32.

P. Sonnentag (2006). Ein Experiment zur kontrollierten Dekohärenz in einem Elektronen-Biprisma-Interferometer. Dissertation, Tübingen.

P. Sonnentag and F. Hasselbach (2005). Decoherence of electron waves due to induced charges moving through a nearby resistive material. *Brazil. J. Phys.* **35**, 385–390.

P. Sonnentag and F. Hasselbach (2007). Measurement of decoherence of electron waves and visualization of the quantum-classical transition. *Phys. Rev. Lett.* **98**, 200402, 4 pp.

P. Sonnentag, H. Kiesel and F. Hasselbach (2000). Visibility spectroscopy with electron waves using a Wien filter: higher order corrections. *Micron* **31**, 451–456.

R. Speidel and D. Kurz (1977). Richtstrahlwertmessungen an einem Strahlerzeugungssystem mit Feldemissionskathode. *Optik* **49**, 173–185.

J. C. H. Spence (1992). Convergent-beam nano-diffraction, in-line holography and coherent shadow imaging. *Optik* **92**, 57–68.

J. C. H. Spence (2009). Electron interferometry. In *Compendium of Quantum Physics* (D. Greenberger, K. Hentschel and F. Weinert, Eds), 188–195 (Springer, Berlin).

J. C. H. Spence (2019). Diffractive imaging of single particles. In *Handbook of Microscopy* (P. W. Hawkes and J. C. H. Spence, Eds), 1009–1036 (Springer-Nature, Cham).

J. C. H. Spence and W. Qian (1992). Transmission-electron Fourier imaging of crystal lattices using low-voltage field-emission sources: theory. *Phys. Rev. B* **45**, 10271–10279.

J. C. H. Spence and J. M. Zuo (1997). Does electron holography energy-filter? *Ultramicroscopy* **69**, 185–190.

J. C. H. Spence, J. M. Zuo and W. Qian (1992). Comment on "Atomic resolution in lensless low-energy electron holography". *Phys. Rev. Lett.* **68**, 3256, See Fink et al. (1991).

J. C. H. Spence, W. Qian and A. J. Melmed (1993a). Experimental low-voltage point-projection microscopy and its possibilities. *Ultramicroscopy* **52**, 473–477.

J. C. H. Spence, J. M. Cowley and J. M. Zuo (1993b). Comment on "Electron holographic study of ferroelectric domain walls". *Appl. Phys. Lett.* **62**, 2446–2447, See Zhang et al. (1993a).

J. C. H. Spence, W. Qian and X. Zhang (1994). Contrast and radiation damage in point-projection electron imaging of purple membrane at 100 V. *Ultramicroscopy* **55**, 19–23.

J. C. H. Spence, X. Zhang and W. Qian (1995). On the reconstruction of low voltage point projection holograms. In *Electron Holography* (A. Tonomura, L. F. Allard, G. Pozzi, D. C. Joy and J. A. Ono, Eds), 267–276 (Elsevier, Amsterdam).

W. Stocker, H.-W. Fink and R. Morin (1989). Low-energy electron and ion projection microscopy. *Ultramicroscopy* **31**, 379–384.

I. Stoyanova and I. Anaskin (1968). The method of determination of incoherent part of wave function forming the image in electron microscope. EUREM-4, Rome, **1**, 161.

I. Stoyanova and I. Anaskin (1972). *Fizicheskie Osnovy Metodov Prosvechivayushchei Elektronnoi Mikroskopii* (Nauka, Moscow).

G. W. Stroke (1967). Interference effects in electron microscopy: some theoretical considerations. In *Record IEEE 9th Annual Symposium on Electron, Ion and Laser Beam Technology* (R. F. W. Pease, Ed.), 287–294 (San Francisco Press, San Francisco, CA).

G. W. Stroke and M. Halioua (1972a). Attainment of diffraction-limited imaging in high-resolution electron microscopy by 'a posteriori' holographic image sharpening. I. *Optik* **35**, 50–65.

G. W. Stroke and M. Halioua (1972b). Image deblurring by holographic deconvolution with partially-coherent low-contrast objects and application to electron microscopy, II. *Optik* **35**, 489–508.

G. W. Stroke and M. Halioua (1973). Image improvement in high-resolution electron microscopy with coherent illumination (low-contrast objects) using holographic image-deblurring deconvolution, III, Part A, Theory. Part B, Experimental implementations and applications to electron beam image holography. *Optik* **37**, 192–203 and 249–264.

G. W. Stroke, M. Halioua, A. J. Saffir and D. J. Evins (1971a). High-resolution enhancement in scanning electron microscopy by *a posteriori* holographic image processing. *Scanning Electron Microsc.* (I), 57–64.

G. W. Stroke, M. Halioua, A. J. Saffir and D. J. Evins (1971b). Resolution enhancement in electron microscopy by a posteriori holographic image deblurring. EMSA 29, Boston, MA, 92–93.

G. W. Stroke, M. Halioua and A. J. Saffir (1973). Imaging improvements in high-resolution electron microscopy by 'a posteriori' holographic image deblurring. *Phys. Lett. A* **44**, 115–117.

G. W. Stroke, M. Halioua, F. Thon and D. Willasch (1974). Image improvement in high-resolution electron microscopy using holographic image deconvolution. *Optik* **41**, 319–343.

G. W. Stroke, M. Halioua, F. Thon and D. H. Willasch (1977). Image improvement and three-dimensional reconstruction using holographic image processing. *Proc. IEEE* **65**, 39–62.

H. Stumpp (1984). Ein 1,25 MeV-Elektronenbeschleuniger für Beugung und Interferometrie mit Elektronenwellen. *Optik* **68**, 193–207.

H. Stumpp, H. Lichte and G. Möllenstedt (1984). Biprism interferences with 1 MeV electrons from a monofile van de Graaff generator. *Optik* **68**, 147–152.

D. Su and Y. Zhu (2010). Scanning moiré fringe imaging by scanning transmission electron microscopy. *Ultramicroscopy* **110**, 229–233.

A. Subbarao (1991). On the singularities in electron-holography. *Optik* **89**, 91.

M. Takeda and T. Abe (1996). Phase unwrapping by a maximum cross-amplitude spanning tree algorithm: a comparative study. *Opt. Eng.* **35**, 2345–2351.

M. Takeda and Q.-S. Ru (1985). Computer-based highly sensitive electron-wave interferometry. *Appl. Opt.* **24**, 3068–3071.

M. Takeda, H. Ina and S. Kobayashi (1982). Fourier-transform method of fringe-pattern analysis for computer-based topography and interferometry. *J. Opt. Soc. Amer.* **72**, 156–160.

M. Takeguchi, K. Harada and R. Shimizu (1990). Observation of GaAs (110) surface defect by reflection electron holography. *J. Electron Microsc.* **39**, 269–272.

M. Takeguchi, C. Hanqing, K. Shibata and R. Shimizu (1992). Construction of UHV-reflection electron holographic microscope. APEM-5, Beijing, **1**, 238–239.

N. Tanaka (2015). Historical survey of the development of STEM instruments. In *Scanning Transmission Electron Microscopy of Materials* (N. Tanaka, Ed.), 9–38 (Imperial College Press, London).

T. Tanigaki, S. Aizawa, T. Suzuki and A. Tonomura (2012). Three-dimensional reconstructions of electrostatic potential distributions with 1.5-nm resolution using off-axis electron holography. *J. Electron Microsc.* **61**, 77–84.

T. Tanigaki, S. Aizawa, H. S. Park, T. Matsuda, K. Harada and D. Shindo (2014). Advanced split-illumination electron holography without Fresnel fringes. *Ultramicroscopy* **137**, 7–11.

T. Tanigaki, K. Harada, Y. Murakami, K. Niitsu, T. Akashi, Y. Takahashi, A. Sugawara and D. Shindo (2016). New trend in electron holography. *J. Phys. D: Appl. Phys* **49**, 244001, 13 pp.

T. Tanji and T. Hirayama (1997). Differential microscopy in off-axis transmission electron microscope holography. *Scanning Microsc.* **11**, 417–425.

T. Tanji, K. Urata and K. Ishizuka (1991). High-resolution electron holography of MgO. EMSA 49, San Jose, CA, 672–673.

T. Tanji, S. Manabe, K. Yamamoto and T. Hirayama (1999). Electron differential microscopy using an electron trapezoidal prism. *Ultramicroscopy* **75**, 197–202.

A. H. Tavabi, V. Migunov, C. Dwyer, R. E. Dunin-Borkowski and G. Pozzi (2015). Tunable caustic phenomena in electron wavefields. *Ultramicroscopy* **157**, 57–64.

A. H. Tavabi, M. Duchamp, R.E. Dunin-Borkowski and G. Pozzi (2016). Double crystal interference experiments. EMC-16, Lyon, **1**, 711–712.

A. H. Tavabi, M. Duchamp, V. Grillo, R. E. Dunin-Borkowski and G. Pozzi (2017). New experiments with a double crystal electron interferometer. *Eur. Phys. J. Appl. Phys.* **78**, 10701, 3 pp.

A. Thust, M. Lentzen and K. Urban (1997). The use of stochastic algorithms for phase retrieval in high resolution transmission electron microscopy. *Scanning Microsc.* **11**, 437–454.

L. H. G. Tizei and M. Kociak (2017). Quantum nanooptics in the electron microscope. *Adv. Imaging Electron Phys.* **199**, 185–235.

H. Tomita and M. Savelli (1968). Mesure du potentiel interne de MgO par microscopie interférentielle. *C. R. Acad. Sci. Paris B* **267**, 580–583.

H. Tomita, T. Matsuda and T. Komoda (1970a). Electron microholography by two-beam method. *Jpn. J. Appl. Phys.* **9**, 719.

H. Tomita, T. Matsuda and T. Komoda (1970b). Electron microholography by two-beam method. ICEM-7, Grenoble, **1**, 151–152.

H. Tomita, T. Matsuda and T. Komoda (1972). Off-axis electron micro-holography. *Jpn. J. Appl. Phys.* **11**, 143–149.

S. Y. Tong, H. Li and H. Huang (1991). Energy extension in three-dimensional atomic imaging by electron-emission holography. *Phys. Rev. Lett.* **67**, 3102–3105.

A. Tonomura (1969). Electron beam holography. *J. Electron Microsc.* **18**, 77–78.

A. Tonomura (1972). The electron interference method for magnetization measurement of thin films. *Jpn. J. Appl. Phys.* **11**, 493–502.

A. Tonomura (1983). Observation of magnetic domain structure in thin ferromagnetic films by electron holography. *J. Magnetism Magnet. Mater.* **31–34**, 963–969.

A. Tonomura (1984). Applications of electron holography using a field-emission electron microscope. *J. Electron Microsc.* **33**, 101–115.

A. Tonomura (1986a). Electron holography. *Prog. Opt.* **23**, 183–220.
A. Tonomura (1986b). Electron holography. ICEM-11, Kyoto, **1**, 9–14.
A. Tonomura (1987a). Applications of electron holography. *Rev. Mod. Phys.* **59**, 639–669.
A. Tonomura (1987b). Electron holography to image magnetic domains. *J. Appl. Phys.* **61**, 4297–4302.
A. Tonomura (1987c). Holographic image reconstruction in electron microscopy, SIME, *Microscopia Elettronica* **8** (2, Suppl.), 151–154.
A. Tonomura (1987d). Electron holography and fundamental physics, SIME, *Microscopia Elettronica* **8** (2, Suppl.), 321–324.
A. Tonomura (1989). Present and future of electron holography. *J. Electron Microsc.* **38**, S43–S50.
A. Tonomura (1990). Electron holography: a new view of the microscopic. *Phys. Today* **43** (4), 22–29.
A. Tonomura (1991a). Applications of electron holography to interference microscopy. EMSA 49, San Jose, CA, 676–677.
A. Tonomura (1991b). Recent developments of electron holography, SIME, *Microscopia Elettronica* **12** (2, Suppl.), 403–406.
A. Tonomura (1992a). Electron-holographic interference microscopy. *Adv. Phys.* **41**, 59–103.
A. Tonomura (1992b). Electron holography of magnetic materials and observation of flux-line dynamics. *Ultramicroscopy* **47**, 419–424.
A. Tonomura (1992c). Experiments on the Aharonov–Bohm effect. In *Wave-Particle Duality* (F. Selleri, Ed.), 291–299 (Plenum, New York and London).
A. Tonomura (1992d). Recent applications of electron holography. APEM-5, Beijing, **1**, 230–233.
A. Tonomura (1994). Recent advances in electron phase microscopy. *Surf. Sci. Rep.* **20**, 317–364.
A. Tonomura (1999). *Electron Holography,* (2nd edn) (Springer, Berlin and New York), 1st edn (1993). Springer Series in Optical Sciences, **70**.
A. Tonomura (2009). Development of electron phase microscopes. *Nucl. Instrum. Meth, Phys. Res. A* **601**, 203–212.
A. Tonomura and T. Matsuda (1980). Phase-contrast microscopy by electron holography. EUREM-7, The Hague, **1**, 24–25.
A. Tonomura and H. Watanabe (1968). Electron beam holography [in Japanese]. *Nihon Butsuri Gakkai-Shi [Proc. Phys. Soc. Jpn.]* **23**, 683–684.
A. Tonomura, A. Fukuhara, H. Watanabe and T. Komoda (1968a). Optical reconstruction of image from Fraunhofer electron-hologram. *Jpn. J. Appl. Phys.* **7**, 295.
A. Tonomura, A. Fukuhara, H. Watanabe and T. Komoda (1968b). Optical reconstruction of image from Fraunhofer electron holograms. EUREM-4, Rome, **1**, 277–278.
A. Tonomura, T. Matsuda and T. Komoda (1978a). Off-axis electron holography by field emission electron microscope. ICEM-9, Toronto, **1**, 224–225 and 670.
A. Tonomura, T. Matsuda and T. Komoda (1978c). Two beam interference with field emission electron beam. *Jpn. J. Appl. Phys.* **17**, 1137–1138.
A. Tonomura, T. Matsuda and J. Endo (1979a). High resolution electron holography with field emission electron microscope. *Jpn. J. Appl. Phys.* **18**, 9–14.
A. Tonomura, T. Matsuda and J. Endo (1979b). Spherical-aberration correction of an electron lens by holography. *Jpn. J. Appl. Phys.* **18**, 1373–1377.
A. Tonomura, T. Matsuda and J. Endo (1979c). Electron beam holography. *Denshikenbikyo* **14**, 47–52.
A. Tonomura, J. Endo and T. Matsuda (1979d). An application of electron holography to interference microscopy. *Optik* **53**, 143–146.
A. Tonomura, T. Matsuda, J. Endo, T. Arii and K. Mihama (1980a). Direct observation of fine structure of magnetic domain walls by electron holography. *Phys. Rev. Lett.* **44**, 1430–1433.
A. Tonomura, T. Matsuda, J. Endo, T. Arii and K. Mihama (1980b). Direct observation of magnetic domain structures of ferromagnetic particles by electron holography. EUREM-7, The Hague, **1**, 22–23.
A. Tonomura, T. Matsuda, R. Suzuki, A. Fukuhara, N. Osakabe, H. Umezaki, J. Endo, K. Shinagawa, Y. Sugita and H. Fujiwara (1982a). Observation of Aharonov–Bohm effect by electron holography. *Phys. Rev. Lett.* **48**, 1443–1446.

A. Tonomura, T. Matsuda, H. Tanabe, N. Osakabe, J. Endo, A. Fukuhara, K. Shinagawa and H. Fujiwara (1982b). Electron holography technique for investigating thin ferromagnetic films. *Phys. Rev. B* **25**, 6799–6804.

A. Tonomura, T. Matsuda, H. Tanabe, N. Osakabe and H. Fujiwara (1982c). Magnetic domain observation of thin ferromagnetic films by electron holography. ICEM-10, Hamburg, **1**, 419–420.

A. Tonomura, T. Matsuda, J. Endo, R. Suzuki, N. Osakabe, H. Umezaki, Y. Sugita and H. Fujiwara (1982d). Observation of micro-magnetic field by electron holography. ICEM-10, Hamburg, **1**, 421–422.

A. Tonomura, H. Umezaki, T. Matsuda, N. Osakabe, J. Endo and Y. Sugita (1983). Is magnetic flux quantized in a toroidal ferromagnet? *Phys. Rev. Lett.* **51**, 331–334.

A. Tonomura, H. Umezaki, T. Matsuda, N. Osakabe, J. Endo and Y. Sugita (1984). Electron holography, Aharonov–Bohm effect and flux quantization. In *Proceedings of the International Symposium on Foundations of Quantum Mechanics in the Light of New Technology*, Tokyo 1983 (S. Kamefuchi, H. Ezawa, Y. Murayama, M. Namiki, S. Nomura, Y. Ohnuki and T. Yajima, Eds), 20–28 (Physical Society of Japan, Tokyo).

A. Tonomura, T. Matsuda, T. Kawasaki, J. Endo and N. Osakabe (1985). Sensitivity-enhanced electron-holographic interferometry and thickness-measurement applications at atomic scale. *Phys. Rev. Lett.* **54**, 60–62.

A. Tonomura, N. Osakabe, T. Matsuda, T. Kawasaki, J. Endo, S. Yano and H. Yamada (1986). Evidence for Aharonov–Bohm effect with magnetic field completely shielded from electron wave. *Phys. Rev. Lett.* **56**, 792–795.

A. Tonomura, S. Yano, N. Osakabe, T. Matsuda, H. Yamada, T. Kawasaki and J. Endo (1987). Proofs of the Aharonov–Bohm effect with completely shielded magnetic field. In *Proceedings of the 2nd International Symposium on Foundations of Quantum Mechanics*, Tokyo 1986 (M. Namiki, Y. Ohnuki, Y. Murayama and S. Nomura, Eds), 97–105 (Physical Society of Japan, Tokyo).

A. Tonomura, J. Endo, T. Matsuda, T. Kawasaki and H. Ezawa (1989). Demonstration of single-electron buildup of an interference pattern. *Amer. J. Phys.* **57**, 117–120.

A. Tonomura, T. Matsuda, S. Hasegawa, M. Igarashi, T. Kobayashi, M. Naito, M. Kajiyama, J. Endo, N. Osakabe and R. Aoki (1990). Electron-interferometric observation of magnetic flux quanta using the Aharonov–Bohm effect. In *Proceedings of the 3rd International Symposium on Foundations of Quantum Mechanics*, Tokyo 1989 (S.-I. Kobayashi, H. Ezawa, Y. Murayama and S. Nomura, Eds), 15–24 (Physical Society of Japan, Tokyo).

A. Tonomura, L. F. Allard, G. Pozzi, D. C. Joy and Y. A. Ono, Eds (1995). *Electron Holography* (Elsevier, Amsterdam).

S. Trépout (2019). Tomographic collection of block-based sparse STEM images: practical implementation and impact on the quality of the 3D reconstructed volume. *Materials* **12**, 2281, 14 pp.

M. Tsukada, S.-I. Kobayashi, S. Kurihara and S. Nomura, Eds (1992). *Proceedings of the 4th International Symposium on Foundations of Quantum Mechanics in the Light of New Technology* (Japanese Society of Applied Physics, Tokyo).

R. Tsuneta, H. Kashima, T. Iwane, K. Harada and M. Koguchi (2014). Dual-axis 360° rotation specimen holder for analysis of three-dimensional magnetic structures. *Microscopy* **63**, 469–473.

R. Q. Twiss, A. G. Little and R. Hanbury Brown (1957). Correlation between photons, in coherent beams of light, detected by a coincidence counting technique. *Nature* **180**, 324–326.

A. C. Twitchett, R. E. Dunin-Borkowski and P. A. Midgley (2002a). Quantitative electron holography of biased semiconductor devices. *Phys. Rev. Lett.* **88**, 238302, 4 pp.

A. C. Twitchett, R. E. Dunin-Borkowski and P. A. Midgley (2002b). Quantitative examination of reverse-biased semiconductor devices using off-axis electron holography. *Microsc. Microanal.* **8** (Suppl. 2), 518–519.

A. C. Twitchett, R. E. Dunin-Borkowski, R. J. Halifax and R. F. Broom (2005). Off-axis electron holography of unbiased and reverse-biased focused ion beam milled Si *p-n* junctions. *Microsc. Microanal.* **11**, 66–78.

A. C. Twitchett-Harrison, T. J. V. Yates, S. B. Newcomb, R. E. Dunin-Borkowski and P. A. Midgley (2007). High-resolution three-dimensional mapping of semiconductor dopant potentials. *Nano Lett.* **7**, 2020–2023.

A. C. Twitchett-Harrison, T. J. V. Yates, R. E. Dunin-Borkowski and P. A. Midgley (2008a). Quantitative electron holographic tomography for the 3D characterisation of semiconductor device structures. *Ultramicroscopy* **108**, 1401–1407.

A. C. Twitchett-Harrison, R. E. Dunin-Borkowski and P. A. Midgley (2008b). Mapping the electrical properties of semiconductor junctions—the electron holographic approach. *Scanning* **30**, 299–309.

T. Uhlig, M. Heumann, M. Schneider, H. Hoffmann and J. Zweck (2000). Construction and characterisation of a TEM specimen holder for in situ application of magnetic in-plane fields. EUREM-12, Brno, **3**, I439–I440.

T. Uhlig, M. Heumann and J. Zweck (2003). Development of a specimen holder for in situ generation of pure in-plane magnetic fields in a transmission electron microscope. *Ultramicroscopy* **94**, 193–196.

U. Valdrè (1974). Recenti sviluppi della microscopia elettronica a contrasto di fase controllato, su immagini in fuoco. In *Atti del Convegno del Gruppo di Strumentazione e Tecniche Nonbiologiche della Società Italiana di Microscopia Elettronica* (P. G. Merli, A. Armigliato and L. Pedulli, Eds), 37–56 (CLUE, Bologna).

U. Valdrè (1979). Electron microscope stage design and applications. *J. Microsc. (Oxford)* **117**, 55–75.

G. Vanasse and H. Sakai (1967). Fourier spectroscopy. *Prog. Opt.* **6**, 259–330.

M. Vanzi (1981). Electron interferometry with a Siemens Elmiskop 102 for measuring electromagnetic microfields. *Optik* **58**, 103–124.

L. H. Veneklasen (1975). On line holographic imaging in the scanning transmission electron microscope. *Optik* **44**, 447–468.

J. Verbeeck, D. Van Dyck, H. Lichte, P. L. Potapov and P. Schattschneider (2005). Plasmon holographic experiments: theoretical framework. *Ultramicroscopy* **102**, 239–255.

J. Verbeeck, G. Bertoni and P. Schattschneider (2008). The Fresnel effect of a defocused biprism on the fringes in inelastic holography. *Ultramicroscopy* **108**, 263–269.

E. Völkl (1991). Höchstauflösende Elektronenholographie. Dissertation, Tübingen.

E. Völkl and H. Lichte (1990a). Point resolution below 1 Å by means of electron holography? ICEM-12, Seattle, WA, **1**, 226–227.

E. Völkl and H. Lichte (1990b). Electron holograms for subångström point resolution. *Ultramicroscopy* **32**, 177–180.

E. Völkl, L. F. Allard and D. C. Joy, Eds (1999). *Introduction to Electron Holography* (Kluwer/Plenum, New York).

V. V. Volkov, M. G. Han and Y. Zhu (2013). Double-resolution electron holography with simple Fourier transform of fringe-shifted holograms. *Ultramicroscopy* **134**, 175–184.

Yu. M. Voronin, I. P. Demenchenok, A. V. Mokhnatkin and R. Yu. Khaitlina (1972). Electron microholography by the one-beam method. *Izv. Akad. Nauk. SSSR (Ser. Fiz.)* **36**, 1293–1296, *Bull. Acad. Sci. USSR (Phys. Ser.)* **36**, 1154–1156.

R. H. Wade (1974). Spectral analysis of holograms and reconstructed images. *Optik* **40**, 201–216.

R. H. Wade (1975). A unified view of holographic schemes. EMAG, Bristol, 197–200.

R. H. Wade (1980). Holographic methods in electron microscopy. In *Computer Processing of Electron Microscope Images* (P. W. Hawkes, Ed.), 223–255. (Academic Press, London and New York).

H. Wahl (1968). Zur elektroneninterferometrischen Vermessung des in einem supraleitenden Hohlzylinder eingefrorenen Magnetflusses. *Optik* **28**, 417–420.

H. Wahl (1970a). Electron optics with highly coherent electron waves. *Ber. Bunsen-Ges. Phys. Chem.* **74**, 1142–1148.

H. Wahl (1970b). Elektroneninterferometrische Messung von quantisierten Magneteinflüssen in supraleitenden Hohlzylinder. I, II. *Optik* **30**, 508–520 and 577–589.

H. Wahl (1974). Experimentelle Ermittlung der komplexen Amplitudentransmission nach Betrag und Phase beliebiger elektronenmikroskopischer Objekte mittels der Off-Axis-Bildebenenholographie. *Optik* **39**, 585–588.

H. Wahl (1975). Bildebenenholographie mit Elektronen. Habilitationsschrift, Tübingen, 63 pp.

H. Wahl and B. Lau (1979). Theoretische Analyse des Verfahrens, die Feldverteilung in dünnen magnetischen Schichten durch lichtholographische Auswertung elektronen-interferenzmikroskopischer Aufnahmen zu veranschaulichen. *Optik* **54**, 27–36.

A. Walstad (2010). A critical reexamination of the electrostatic Aharonov–Bohm effect. *Int. J. Theor. Phys.* **49**, 2929–2934.

A. Walstad (2017). Further considerations regarding the Aharonov—Bohm effect and the wavefunction of the entire system. *Int. J. Theor. Phys.* **56**, 965—970.

Z. L. Wang (1993). Electron reflection, diffraction and imaging of bulk crystal surfaces in TEM and STEM. *Rep. Prog. Phys.* **56**, 997—1065.

R.-f. Wang (2015). A possible interplay between electron beams and magnetic fluxes in the Aharonov—Bohm effect. *Front. Phys.* **10**, 358—363.

S.-y. Wang and J. M. Cowley (1991). Probe-shifting method in in-line holography. EMSA 49, San Jose, CA, 682—683.

Y. C. Wang, T. M. Chou, M. Libera and T. F. Kelly (1997). Transmission electron holography of silicon nanospheres with surface oxide layers. *Appl. Phys. Lett.* **70**, 1296—1298.

Y. C. Wang, T. M. Chou, M. Libera, E. Voelkl and B. G. Frost (1998). Measurement of polystyrene mean inner potential by transmission electron holography of latex spheres. *Microsc. Microanal.* **4**, 146—157.

Y.-g. Wang, H.-r. Liu, Q.-b. Yang and Z. Zhang (2003). Determination of inelastic mean free path by electron holography along with electron dynamic calculation. *Chin. Phys. Lett.* **20**, 888—890.

M. Wanner, D. Bach, D. Gerthsen, R. Werner and B. Tesche (2006). Electron holography of thin amorphous carbon films: measurement of the mean inner potential and a thickness-independent phase shift. *Ultramicroscopy* **106**, 341—345.

H. Watanabe and A. Tonomura (1969). Electron beam holography [in Japanese]. *Nihon Kessho Gakkai-Shi [J. Crystallogr. Soc. Jpn.]* **11**, 23—25.

U. Weierstall, J. C. H. Spence, M. Stevens and K. H. Downing (1999). Point-projection electron imaging of tobacco mosaic virus at 40 eV electron energy. *Micron* **30**, 335—338.

I. Weingärtner, W. Mirandé and E. Menzel (1969a). Holographie bei Teilkohärenz. I. Fresnel-Holographie. *Optik* **29**, 87—104, Part II is Mirandé et al. (1969).

I. Weingärtner, W. Mirandé and E. Menzel (1969b). Enhancement of resolution in electron microscopy by image holography. *Optik* **30**, 318—322.

I. Weingärtner, W. Mirandé and E. Menzel (1970). Holographie bei Teilkohärenz. III. Theorie zur Bildebenen-Holographie. *Optik* **31**, 335—353.

I. Weingärtner, W. Mirandé and E. Menzel (1971). Verbesserung der Auflösung im Elektronenmikroskop durch Bildebenen-Holographie. *Ann. Phys. (Leipzig)* **26**, 289—301.

J. K. Weiss, W. J. de Ruijter, M. Gajdardziska-Josifovska, D. J. Smith, E. Völkl and H. Lichte (1991). Applications of electron holography to multilayer interfaces. EMSA 49, San Jose, CA, 674—675.

J. K. Weiss, W. J. de Ruijter, M. Gajdardziska-Josifovska, M. R. McCartney and D. J. Smith (1993). Applications of electron holography to the study of interfaces. *Ultramicroscopy* **50**, 301—311.

F. G. Werner and D. R. Brill (1960). Significance of electromagnetic potentials in the quantum theory in the interpretation of electron interferometer fringe observations. *Phys. Rev. Lett.* **4**, 344—346.

G. Wohland (1981). Messung der Kohärenzlänge von Electronen im Elektroneninterferometer mit Wien-Filter. Dissertation, Tübingen.

D. Wolf (2007). Towards quantitative electron-holographic tomography. *Microsc. Microanal.* **13** (Suppl. 3), 112—113.

D. Wolf (2010). Elektronen-Holographische Tomographie zur 3D-Abbildung von elektrostatischen Potentialen in Nanostrukturen. Dissertation, Dresden.

D. Wolf, A. Lenk and H. Lichte (2008). Three-dimensional mapping of nanostructures with electron-holographic tomography. EMC-14, Aachen, **1**, 339—340.

D. Wolf, A. Lubk, H. Lichte and H. Friedrich (2010). Towards automated electron holographic tomography for 3D mapping of electrostatic potentials. *Ultramicroscopy* **110**, 390—399.

D. Wolf, A. Lubk, A. Lenk, S. Sturm and H. Lichte (2013a). Tomographic investigation of Fermi level pinning at focused ion beam milled semiconductor surfaces. *Appl. Phys. Lett.* **103**, 264104.

D. Wolf, A. Lubk, F. Röder and H. Lichte (2013b). Electron holographic tomography. *Curr. Opin. Solid State Mater. Sci.* **17**, 126—134.

D. Wolf, R. Hübner, T. Niermann, S. Sturm, P. Prete, N. Lovergine, B. Büchner and A. Lubk (2018). Three-dimensional composition and electric potential mapping of III—V core—multishell nanowires by correlative STEM and holographic tomography. *Nano Lett.* **18**, 4777—4784.

D. Wolf, N. Bizière, S. Sturm, D. Reyes, T. Wade, T. Niermann, J. Krehl, B. Warot-Fonrose, B. Büchner, E. Snoeck, C. Gatel and A. Lubk (2019). Holographic vector field electron tomography of three-dimensional nanomagnets. *Commun. Phys.* **2**, 87.

K. Yada, K. Shibata and T. Hibi (1973). A high resolution electron interference microscope and its application to the measurement of mean inner potential. *J. Electron Microsc.* **22**, 223–230.

J. Yamaguchi, M. Shibano and T. Saito (1994). Immuno-scanning electron microscopic study of cytoskeletons and actin-binding proteins on phagocytosis of zymosans in mouse macrophages by using double marking method. ICEM-13, Paris, **3A**, 43–44.

K. Yamamoto (2018). High precision phase-shifting electron holography with multiple biprisms for GaN semiconductor devices. *Microsc. Microanal.* **24** (Suppl. 1), 1554–1555.

K. Yamamoto, I. Kawajiri, T. Tanji, M. Hibino and T. Hirayama (2000). High precision phase-shifting electron holography. *J. Electron Microsc.* **49**, 31–39.

K. Yamamoto, T. Hirayama and T. Tanji (2004). Off-axis electron holography without Fresnel fringes. *Ultramicroscopy* **101**, 265–269.

K. Yamamoto, S. Anada, T. Sato, N. Yoshimoto and T. Hirayama (2021). Phase-shifting electron holography for accurate measurement of potential distributions in organic and inorganic semiconductors. *Microscopy* **70**, 24–38.

J. Yamanaka, M. Shirakura, C. Yamamoto, K. Sato, T. Yamada, K. Hara, K. Arimoto, K. Nakagawa, A. Ishizuka and K. Ishizuka (2018). Feasibility study to evaluate lattice-space changing of a step-graded SiGe/Si (110) using STEM moiré. *J. Mater. Sci. Chem. Eng.* **6**, 8–15.

Y. Y. Yang, D. Cooper, J.-L. Rouvière, C. E. Murray, N. Bernier and J. Bruley (2015). Nanoscale strain distributions in embedded SiGe semiconductor devices revealed by precession electron diffraction and dual lens dark field electron holography. *Appl. Phys. Lett.* **106**, 042104.

F. S. Yasin, T. R. Harvey, J. J. Chess, J. S. Pierce and B. J. McMorran (2016). Development of STEM-holography. *Microsc. Microanal.* **22** (Suppl. 3), 506–507.

F. S. Yasin, K. Harada, D. Shindo, H. Shinada, B. J. McMorran and T. Tanigaki (2018a). A tunable path-separated electron interferometer with an amplitude-dividing grating beamsplitter. *Appl. Phys. Lett.* **113**, 233102, 5 pp.

F. S. Yasin, T. R. Harvey, J. J. Chess, J. S. Pierce and B. J. McMorran (2018b). Path-separated electron interferometry in a scanning transmission electron microscope. *J. Phys. D: Appl. Phys* **51**, 205104, 6 pp.

F. S. Yasin, T. R. Harvey, J. J. Chess, J. S. Pierce, C. Ophus, P. Ercius and B. J. McMorran (2018c). Probing light atoms at subnanometer resolution: realization of scanning transmission electron microscope holography. *Nano Lett.* **18**, 7118–7123.

F. Yasin, T. Harvey, J. Pierce, C. Ophus, P. Ercius, K. Harada, D. Shindo, T. Tanigaki, H. Shinada, B. McMorran and J. J. Chess (2018d). 4D STEM holography with an amplitude-division diffraction grating. IMC-19, Sydney.

T. Yatagai, K. Ohmura, S. Iwasaki, S. Hasegawa, J. Endo and A. Tonomura (1987). Quantitative phase analysis in electron holographic interferometry. *Appl. Opt.* **26**, 377–382.

T. J. V. Yates (2005). The development of electron tomography for nanoscale materials science applications. Dissertation, Cambridge.

G. Yi, W. A. P. Nicholson, C. K. Lim, J. N. Chapman, S. McVitie and C. D. W. Wilkinson (2004). A new design of specimen stage for in situ magnetising experiments in the transmission electron microscope. *Ultramicroscopy* **99**, 65–72.

K. Yoshida, T. Okuwaki, N. Osakabe, H. Tanabe, Y. Horiuchi, T. Matsuda, K. Shinagawa, A. Tonomura and H. Fujiwara (1983). Observation of recorded magnetization patterns by electron holography. *IEEE Trans.* **MAG–19**, 1600–1604.

A. Zeilinger (1986). Generalized Aharonov–Bohm experiments with neutrons. In *Fundamental Aspects of Quantum Theory* (V. Gorini and A. Frigerio, Eds), 311–318 (Plenum, New York).

E. Zeitler (1979). Electron holography. EMSA 37, San Antonio, TX, 376–379.

X. Zhang and D. Joy (1991). Divergent beam holography with a 200 keV FE TEM. EMSA 49, San Jose, CA, 678–679.

X. Zhang, T. Hashimoto and D. C. Joy (1992). Electron holographic study of ferroelectric domain walls. *Appl. Phys. Lett.* **60**, 784–786.

X. Zhang, T. Hashimoto and D. C. Joy (1993a). Response to "Comment on 'Electron holographic study of ferroelectric domain walls'". *Appl. Phys. Lett.* **62**, 2447. Cf. Spence et al. (1993).
X. Zhang, D. C. Joy, Y. Zhang, T. Hashimoto, L. Allard and T. A. Nolan (1993b). Electron holography techniques for study of ferroelectric domain walls. *Ultramicroscopy* **51**, 21–30.
R. Zhang, S. Zeltmann, C. Ophus, B. Savitzky, T. Pekin, E. Rothchild, K. Bustillo, M. Asta, D. Chrzan and A. Minor (2020). Imaging short-range order and extracting 3-D strain tensor using energy-filtered 4D-STEM techniques. *Microsc. Microanal.* **26** (Suppl. 2), 936–938.
F. Zhou (2001). The principle of a double crystal electron interferometer. *J. Electron Microsc.* **50**, 371–376.
X. Zhuge, H. Jinnai, R. E. Dunin-Borkowki, V. Migunov, S. Bals, P. Cool, A.-J. Bons and K. J. Batenburg (2017). Automated discrete electron tomography—towards routine high-fidelity reconstruction of nanomaterial. *Ultramicroscopy* **175**, 87–96.
J.-M. Zuo (2019). Electron nanodiffraction. In *Handbook of Microscopy* (P. W. Hawkes and J. C. H. Spence, Eds), 905–969 (Springer-Nature, Cham).
J. M. Zuo, M. Gao, J. Tao, B. Q. Li, R. Twesten and I. Petrov (2004). Coherent nano-area electron diffraction. *Microsc. Res. Tech.* **64**, 347–355.
W. H. Zurek (2003). Decoherence, einselection, and the quantum origins of the classical. *Rev. Mod. Phys.* **75**, 715–775.

Part XIII, Chapters 64–68

The literature on the detection and detectability of single atoms is not examined here but we do list the early comments of Eisenhandler and Siegel (1966a) and of Zeitler and Thomson (1969), forerunners of much later work. A paper by van Dorsten et al. (1968) considers phase and amplitude. Incoherent transfer is studied by van Heel (1978).

As mentioned in Chapter 67, confocal microscopy has been extensively studied in light microscopy and many of these publications are of interest for electrons. We include a small selection mainly by Colin Sheppard and Tim Wilson, pioneers in this domain: Sheppard (1977, 1989), Sheppard and Choudhury (1977), Sheppard and Wilson (1979a,b, 1980a,b, 1981, 1986), Ash (1980), Sheppard et al. (1983), Sheppard and Hamilton (1984), Wilson and Harrison (1984), Wilson and Carlini (1989), Wilson (1990), Sheppard and Cogswell (1990a,b), Sheppard and Gu (1991a,b, 1992, Juskaitis and Wilson (1994) and Gu (1996).

The articles by Weßels et al. (2022) and Sagawa et al. (2022) are relevant for Chapters 66 and 67, respectively.

* * *

H.-W. Ackermann (2013). The sad and sorry state of phage electron microscopy. MC-2013, Regensburg, **2**, 557–558.
H.-W. Ackermann (2014). Sad state of phage electron microscopy. Please shoot the messenger. *Microorganisms* **2**, 1–10.
G. Ade (1977a). On the incoherent imaging in the scanning transmission electron microscope (STEM). *Optik* **49**, 113–116.
G. Ade (1977b). Probleme der Kontrastübertragung bei der Verarbeitung von Differenzsignalen. PTB-Bericht APh–12, 21 pp.
G. Ade (1978). Erweiterung der Kontrastübertragungstheorie auf nichtisoplanatische Abbildungen. *Optik* **50**, 143–162.
A. W. Agar, R. S. M. Revell and R. A. Scott (1949). A preliminary report on attempts to realise a phase contrast electron microscope. ICEM-1, Delft, 52–54.
E. V. Ageev, I. F. Anaskin and P. A. Stoyanov (1977). Experimental analysis of image correction methods for electron microscopy. *Izv. Akad. Nauk. SSSR (Ser. Fiz.)* **41**, 1447–1451, *Bull. Acad. Sci. USSR (Phys. Ser.)* **41** (7), 116–119.
C. C. Aggarwal (2018). *Neural Networks and Deep Learning* (Springer Nature, Cham).
R. H. Alderson (1971). High resolution scanning transmission electron microscopy. *AEI Electron Microscope News* **2** (10), 2 pp.
F. Allars, P.-H. Lu, M. Kruth, R. E. Dunin-Borkowski, J. M. Rodenburg and A. M. Maiden (2021). Efficient large field of view electron phase imaging using near-field electron ptychography with a diffuser. *Ultramicroscopy* **231**, 113257, 9 pp.

D. Alloyeau, W. K. Hsieh, E. H. Anderson, L. Hilken, G. Benner, X. Meng, F. R. Chen and C. Kisielowski (2010). Imaging of soft and hard materials using a Boersch phase plate in a transmission electron microscope. *Ultramicroscopy* **110**, 563–570.

T. P. Almeida, R. Temple, J. Massey, K. Fallon, D. McGrouther, T. Moore, C. H. Marrows and S. McVitie (2017). Quantitative TEM imaging of the magnetostructural and phase transitions in FeRh thin film systems. *Sci. Rep.* **7**, 17835, 11 pp.

T. Altantzis, I. Lobato, A. De Backer, A. Béché, Y. Zhang, S. Basak, M. Porcu, Q. Xu, A. Sánchez-Iglesias, L. M. Liz-Marzán, G. Van Tendeloo, S. Van Aert and S. Bals (2019). Three-dimensional quantification of the facet evolution of Pt nanoparticles in a variable gaseous environment. *Nano Lett.* **19**, 477–481.

A. Amunts, A. Brown, X.-c. Bai, J. L. Llácer, T. Hussain, P. Emsley, F. Long, G. Murshudov, S. H. W. Scheres and V. Ramakrishnan (2014). Structure of the yeast mitochondrial large ribosomal subunit. *Science* **343**, 1485–1489.

I. F. Anaskin and E. V. Ageev (1974). Frequency–contrast characteristics of an electron microscope with a zone plate. *Izv. Akad. Nauk. SSSR (Ser. Fiz.)* **38**, 1389–1392, *Bull. Acad. Sci. USSR (Phys. Ser.)* **38** (7), 26–29.

I. F. Anaskin, E. V. Ageev, P. A. Stoyanov and V. V. Moseev (1976). Correction of electron microscope images by means of holographic filters. *Prib. Tekh. Eksp.* (1), 188–190, *Instrum. Exp. Tech.* **19**, 228–229.

I. F. Anaskin, E. V. Ageev and P. A. Stoyanov (1980). Correlation diffraction patterns – a means of determining electron microscope resolution. *Prib. Tekh. Eksp.* (6), 168–170, *Instrum. Exp. Tech.* **23**, 1500–1502.

W. H. J. Andersen (1972). Phase contrast enhancement by single sideband modulation transfer. EUREM-5, Manchester, 396–397.

B. L. Armbruster, J. Brink, H. Furukawa, T. Isabell, M. Kawasaki and M. Kersker (2008). New frontiers in cryo-electron tomography with Zernike phase contrast imaging for transmission electron microscopy. *Microsc. Microanal.* **14** (Suppl. 2), 1072–1073.

B. L. Armbruster, J. Brink, R. Danev, T. C. Isabell, M. Kawasaki, M. Marko, S. Motoki and K. Nagayama (2010). Hardware considerations to optimize Zernike phase contrast TEM for cryo-tomography and single particle data acquisition. *Microsc. Microanal.* **16** (Suppl. 2), 554–555.

S. Asano, Y. Fukuda, F. Beck, A. Aufderheide, F. Förster, R. Danev and W. Baumeister (2015). A molecular census of 26S proteasomes in intact neurons. *Science* **347**, 439–442.

S. Asano, B. D. Engel and W. Baumeister (2016). In situ cryo-electron tomography: a post-reductionist approach to structural biology. *J. Mol. Biol.* **428**, 332–343.

E. A. Ash, Ed. (1980). *Scanned Image Microscopy* (Academic Press, London).

K. Atsuzawa, N. Usuda, A. Nakazawa, M. Fukasawa, R. Danev, S. Sugitani and K. Nagayama (2009). High-contrast imaging of plastic-embedded tissues by phase contrast electron microscopy. *J. Electron Microsc.* **58**, 35–45.

R. Bach, D. Pope, S.-h. Liou and H. Batelaan (2013). Controlled double-slit electron diffraction. *New J. Phys.* **15**, 033018, 7 pp.

H. G. Badde and L. Reimer (1970). Der Einfluß einer streuenden Phasenplatte auf das elektronenmikroskopische Bild. *Z. Naturforsch.* **25a**, 760–765.

G. Balossier and N. Bonnet (1981). Use of an electrostatic phase plate in TEM. Transmission electron microscopy: improvement of phase and topographical contrast. *Optik* **58**, 361–376.

G. Balossier and X. Thomas (1984). Kit informatisé générateur de faisceau conique annulaire en M.E.T. *J. Microsc. Spectrosc. Electron* **9**, 343–350.

G. Balossier, N. Bonnet, D. Génotel and A. Laberrigue (1980). Use of an electrostatic phase plate in TEM. Improvement of phase and topographical contrast. EUREM-7, The Hague, **1**, 26–27.

S. Bals, S. Van Aert, G. Van Tendeloo and D. Ávila-Brande (2006). Statistical estimation of atomic positions from exit wave reconstruction with a precision in the picometer range. *Phys. Rev. Lett.* **96**, 096106, 4 pp.

S. Bals, M. Casavola, M. A. van Huis, S. Van Aert, K. J. Batenburg, G. Van Tendeloo and D. Vanmaekelbergh (2011). Three-dimensional atomic imaging of colloidal core–shell nanocrystals. *Nano Lett.* **11**, 3420–3424.

S. Bals, S. Van Aert, C. P. Romero, K. Lauwaet, M. J. Van Bael, B. Schoeters, B. Patoens, E. Yücelen, P. Lievens and G. Van Tendeloo (2012). Atomic scale dynamics of ultrasmall germanium clusters. *Nature Commun.* **3**, 897.

B. E. Bammes, R. H. Rochat, J. Jakana, D.-h Chen and W. Chiu (2012). Direct electron detection yields cryo-EM reconstructions at resolutions beyond 3/4Nyquist frequency. *J. Struct. Biol.* **177**, 589–601.

J. R. Banbury (1974). Operation of an experimental high voltage field emission STEM. ICEM-8, Canberra, **1**, 44–45.

J. R. Banbury and U. R. Bance (1973a). A prototype field emission scanning transmission electron microscope. EMAG, Scanning Electron Microscopy, Newcastle, 164–168.

J. R. Banbury and U. R. Bance (1973b). A prototype field-emission scanning transmission electron microscope. EMSA, New Orleans, LA, **31**, 296–297.

J. R. Banbury, I. W. Drummond and I. L. F. Ray (1975). Performance and applications of a high resolution field emission STEM. EMSA, Las Vegas, NV, **33**, 112–113.

M. E. Barnett (1973). The reciprocity theorem and the equivalence of conventional and scanning transmission microscopes. *Optik* **38**, 585–588.

M. E. Barnett (1974). Image formation in optical and electron transmission microscopy. *J. Microsc. (Oxford)* **102**, 1–28.

K. Barragán Sanz (2021). Implementation and improvement of metal-film phase plates for single-particle cryo-EM. Dissertation, Bonn.

K. Barragán Sanz and S. Irsen (2019). The rocking phase plate – another step towards improved stability. MC-2019, Berlin, 693–694.

B. Barton and R. Schröder (2007). Electron tomography of unstained cell sections using a half-plane (Hilbert) phase plate. *Microsc. Microanal.* **13** (Suppl. 2), 1312–1313, also (Suppl. 3), 106–107.

B. Barton, D. Rhinow, A. Walter, R. Schröder, G. Benner, E. Majorovits, M. Matijevic, H. Niebel, H. Müller, M. Haider, M. Lacher, S. Schmitz, P. Holik and W. Kühlbrandt (2011). In-focus electron microscopy of frozen-hydrated biological samples with a Boersch phase plate. *Ultramicroscopy* **111**, 1696–1705.

V. G. Baryshevsky, M. V. Korzhik, V. I. Moroz, V. B. Pavlenko and A. A. Fyodorov (1991). $YAlO_3$:Ce-fast-acting scintillators for detection of ionizing radiation. *Nucl. Instrum. Meth, Phys. Res. B* **58**, 291–293.

R. H. T. Bates and J. M. Rodenburg (1989). Sub-ångström transmission microscopy: a Fourier transform algorithm for microdiffraction plane intensity information. *Ultramicroscopy* **31**, 303–307.

M. Battaglia, D. Contarato, P. Denes, D. Doering, P. Giubilato, T. S. Kim, S. Mattiazzo, V. Radmilovic and S. Zalusky (2009a). A rad-hard CMOS active pixel sensor for electron microscopy. *Nucl. Instrum. Meth, Phys. Res. A* **598**, 642–649.

M. Battaglia, D. Contarato, P. Denes, D. Doering and V. Radmilovic (2009b). CMOS pixel sensor response to low energy electrons in transmission electron microscopy. *Nucl. Instrum. Meth, Phys. Res. A* **605**, 350–352.

M. Battaglia, D. Contarato, P. Denes and P. Giubilato (2009c). Cluster imaging with a direct detection CMOS pixel sensor in transmission electron microscopy. *Nucl. Instrum. Meth, Phys. Res. A* **608**, 363–365.

V. Beck (1977). Asymmetrical dark field detectors in the STEM. *Ultramicroscopy* **2**, 351–360.

V. Beck and A. V. Crewe (1975). High resolution imaging properties of the STEM. *Ultramicroscopy* **1**, 137–144.

M. Beer, J. Frank, K.-J. Hanszen, E. Kellenberger and R. C. Williams (1975). The possibilities and prospects of obtaining high-resolution information (below 30Å) on biological material using the electron microscope. Some comments and reports inspired by the EMBO workshop held at Gais, Switzerland, October 1973. *Q. Rev. Biophys.* **7**, 211–238.

G. Behan, E. C. Cosgriff, A. I. Kirkland and P. D. Nellist (2009). Three-dimensional imaging by optical sectioning in the aberration-corrected scanning transmission electron microscope. *Philos. Trans. R. Soc. A* **367**, 3825–3844.

M. Beleggia (2008). A formula for the image intensity of phase objects in Zernike mode. *Ultramicroscopy* **108**, 953–958.

M. Beleggia, P. F. Fazzini and G. Pozzi (2003). A Fourier approach to field and electron-optical phase shifts calculations. *Ultramicroscopy* **96**, 93–103.

M. Beleggia, M. Malac, R.F. Egerton and M. Kawasaki (2012). Imaging with a hole-free phase plate. IMC-15, Manchester, **2**, 499–500.

G. Benner, J. Bihr and M. Prinz (1990). A new illumination system for an analytical transmission electron microscope using a condenser objective lens. ICEM, Seattle, WA, **1**, 138–139.

G. Benner, J. Bihr and E. Weimer (1991). Advantages of Koehler illumination for selected area diffraction. EMSA, San Jose, CA, **49**, 1008–1009.

A. Beorchia and P. Bonhomme (1974). Experimental studies of some dampings of electron microscope phase contrast transfer function. *Optik* **39**, 437–442.

B. Berkels, P. Binev, D. A. Blom, W. Dahmen, R. Sharpley and T. Vogt (2014). Optimized imaging using non-rigid registration. *Ultramicroscopy* **138**, 46–56.

E. Bettens, D. Van Dyck, A. J. den Dekker, J. Sijbers and A. van den Bos (1999). Model-based two-object resolution from observations having counting statistics. *Ultramicroscopy* **77**, 37–48.

A. M. Blackburn and R. A. McLeod (2021). Practical implementation of high-resolution electron ptychography and comparison with off-axis electron holography. *Microscopy* **70**, 131–147.

H. Boersch (1947a). Contrastes donnés par les atomes au microscope électronique. *Publ. de l'Institut de Recherches Scientifiques, Tettnang* **5**, 14–29.

H. Boersch (1947b). Sur les conditions de représentation des atomes au microscope électronique. *Publ. de l'Institut de Recherches Scientifiques, Tettnang* **3/4**, 37–42.

H. Boersch (1947c). Über die Kontraste von Atomen im Elektronenmikroskop. *Z. Naturforsch.* **2a**, 615–633.

H. Boersch (1948). Über die Möglichkeit der Abbildung von Atomen im Elektronenmikroskop. III. *Monatshefte Chem.* **78**, 163–171.

P. Bonhomme and A. Beorchia (1983). The specimen thickness effect upon the electron microscope image contrast transfer of amorphous objects. *J. Phys. D: Appl. Phys* **16**, 705–713.

P. Bonhomme, A. Beorchia and N. Bonnet (1973). Influence de l'ouverture du faisceau éclairant l'échantillon, sur le transfert du contraste de phase. *C. R. Acad. Sci. Paris B* **277**, 83–86.

N. Bonnet and P. Bonhomme (1980). On the shape of the partially coherent attenuation envelope in bright field transmission electron microscopy. *Optik* **56**, 353–362.

N. Bonnet, A. Beorchia and P. Bonhomme (1978). Fonctions de transfert en microscopie électronique avec éclairage conique annulaire. *J. Microsc. Spectrosc. Electron* **3**, 497–511.

A. Y. Borisevich, A. R. Lupini, T. S. and S. J. Pennycook (2006). Depth sectioning of aligned crystals with the aberration-corrected scanning transmission electron microscope. *J. Electron Microsc.* **55**, 7–12.

M. Born and E. Wolf (2002). *Principles of Optics,* 7th edn (Cambridge University Press, Cambridge), 1st edn (1959) (Pergamon, Oxford and New York).

E. G. T. Bosch and I. Lazić (2015). Analysis of HR-STEM theory for thin specimens. *Ultramicroscopy* **158**, 59–62.

E. G. T. Bosch and I. Lazić (2019). Analysis of depth-sectioning STEM for thick samples and 3D imaging. *Ultramicroscopy* **207**, 112831, 22 pp.

S. Boussakta and A. G. J. Holt (1992). New number theoretic transform. *Electron Lett.* **28**, 1683–1684.

G. Bouwhuis and N. H. Dekkers (1980). Ultramicroscopy in scanning microscopes. *Optik* **56**, 233–242.

P. E. Bovey and A. W. Nicholls (1987). Recent advances in high spatial resolution STEM. *Microscopia Elettronica* **8** (Pt 2 (Suppl.)), 247–250.

T. B. Britton, R. N. Clough, A. I. Kirkland, G. Moldovan and A. J. Wilkinson (2013). Direct detection of electron backscatter diffraction patterns. *Phys. Rev. Lett.* **111**, 065506.

L. M. Brown (1981). Scanning transmission electron microscopy: microanalysis for the microelectronic age. *J. Phys. F: Met. Phys* **11**, 1–26.

H. G. Brown, Z. Chen, M. Weyland, C. Ophus, J. Ciston, L. J. Allen and S. D. Findlay (2018). Structure retrieval at atomic resolution in the presence of multiple scattering of the electron probe. *Phys. Rev. Lett.* **121**, 266102, 6 pp.

M. T. Browne and J. F. L. Ward (1982). Detectors for STEM, and the measurement of their detective quantum efficiency. *Ultramicroscopy* **7**, 249–262.

M. T. Browne, S. Lackovic and R. E. Burge (1975). Instrumentation and recording for the Vacuum Generators HB5 STEM instrument. EMAG, Bristol, 27–30.

N. D. Browning, M. F. Chisholm and S. J. Pennycook (1993). Atomic-resolution chemical analysis using a scanning transmission electron microscope. *Nature* **366**, 143–146.

C. Brownlie, S. McVitie, J. N. Chapman and C. D. W. Wilkinson (2006). Lorentz microscopy studies of domain wall trap structures. *J. Appl. Phys.* **100**, 033902.

B. Buijsse, F. M. H. M. van Laarhoven, A. K. Schmid, R. Cambie, S. Cabrini, J. Jin and R. M. Glaeser (2011). Design of a hybrid double-sideband/single-sideband (schlieren) objective aperture suitable for electron microscopy. *Ultramicroscopy* **111**, 1688–1695.

B. Buijsse, F. M. H. M. van Laarhoven, A. K. Schmid, R. Cambie, S. Cabrini, J. Jin and R. M. Glaeser (2012a). A 'tulip aperture' providing in-focus phase-contrast in TEM. EMC-15, Manchester, **2**, 509–510.

B. Buijsse, F. M. H. M. van Laarhoven, A. K. Schmid, R. Cambie, S. Cabrini, J. Jin and R. M. Glaeser (2012b). A 'tulip aperture' providing in-focus phase-contrast. *Microsc. Microanal.* **18** (Suppl. 2), 472–473.

B. Buijsse, G. van Duinen, K. Sader and R. Danev (2014). Challenges in phase plate product development. *Microsc. Microanal.* **20** (Suppl. 3), 218–219.

B. Buijsse, R. Danev, K. Sader and S. Welsch (2016). Optimizing the FEI Volta phase plate for efficient and artefact-free data acquisition. *Microsc. Microanal.* **22** (Suppl. 3), 58–59.

B. Buijsse, P. Trompenaars, V. Altin, R. Danev and R. M. Glaeser (2020). Spectral DQE of the Volta phase plate. *Ultramicroscopy* **218**, 113079, 6 pp.

R. E. Burge (1977). Scanning transmission electron microscopy at high resolution. In *Analytical and Quantitative Methods in Microscopy* (G. A. Meek and H. Y. Elder, Eds), 171–191 (University Press, Cambridge).

R. E. Burge and A. F. Clark (1981). STEM multiple images, the Karhunen–Loeve transform and data compression. EMAG, Cambridge, 315–320.

R. E. Burge and J. C. Dainty (1976). Partially coherent image formation in the scanning transmission electron microscope (STEM). *Optik* **46**, 229–240.

R. E. Burge and M. Derome (1976). *Z*-discrimination by real-space measurements in dark field S.T.E.M. *Optik* **46**, 161–182.

R. E. Burge and P. van Toorn (1979). Multi-signal detection and processing in STEM. EMAG, Brighton, 249–252.

R. E. Burge and P. van Toorn (1980). Multiple images and image processing in STEM. *Scanning Electron Microsc.* (I), 81–91.

R. E. Burge, M. T. Browne, J.C. Dainty, M. Derome and S. Lackovic (1976). High resolution with STEM. EUREM-6, Jerusalem, **1**, 442–443.

R. E. Burge, M. T. Browne, S. Lackovic and J. F. L. Ward (1979a). STEM imaging at high resolution: the influence of detector geometry. *Scanning Electron Microsc.* (I), 127–136.

R. E. Burge, M. T. Browne, J. C. Dainty, S. Lackovic, D. Robinson and J. Ward (1979b). In *Machine-Aided Image Analysis 1978* (W. E. Gardner, Ed.), 107–113 (Institute of Physics, Bristol), Conference Series **44**.

R. E. Burge, M. T. Browne, P. Charalambous, A. Clark and J. K. Wu (1982). Multiple signals in STEM. *J. Microsc. (Oxford)* **127**, 47–60.

J. Bürger, T. Riedl and J. K. N. Lindner (2020). Influence of lens aberrations, specimen thickness and tilt on differential phase contrast STEM images. *Ultramicroscopy* **219**, 113118, 14 pp.

R. Cambie, K. H. Downing, D. Typke, R. M. Glaeser and J. Jin (2007). Design of a microfabricated two-electrode phase-contrast element suitable for electron microscopy. *Ultramicroscopy* **107**, 329–339.

J. Candès, X. Li and M. Soltanokotabi (2015). Phase retrieval via Wirtinger flow: theory and algorithms. *IEEE Trans. Inf. Theory* **61**, 1985–2007.

V. Chamard, M. Allain, P. Godard, A. Talneau, G. Patriarche and M. Burghammer (2015). Strain in a silicon-on-insulator nanostructure revealed by 3D x-ray Bragg ptychography. *Sci. Rep.* **5**, 9827, 8 pp.

L.-y. Chang (2004). Studies of indirect methods in high-resolution electron microscopy. Dissertation, Cambridge.

L.-y. Chang, A. I. Kirkland and J. M. Titchmarsh (2006). On the importance of fifth-order spherical aberration for a fully corrected electron microscope. *Ultramicroscopy* **106**, 301–306.

J. N. Chapman (1989). High resolution imaging of magnetic structures in the transmission electron microscope. *Mater. Sci. Eng. B* **3**, 355–358.

J. N. Chapman and G. R. Morrison (1983). Quantitative determination of magnetisation distribution in domains and domain walls by scanning transmission electron microscopy. *J. Magnetism Magnet. Mater.* **35**, 254–260.

J. N. Chapman, P. E. Batson, E. M. Waddell and R. P. Ferrier (1978a). The direct determination of magnetic domain wall profiles by differential phase contrast electron microscopy. *Ultramicroscopy* **3**, 203–214.

J. N. Chapman, P. E. Batson, E. M. Waddell and R. P. Ferrier (1978b). The direct determination of magnetic domain wall profiles using a split detector STEM. ICEM-9, Toronto, **1**, 466–467.

J. N. Chapman, G. R. Morrison, J. P. Jakubovics and R. A. Taylor (1983). Investigations of micromagnetic structures by STEM. EMAG, Guildford, 197–200.

J. N. Chapman, I. R. McFadyen and S. McVitie (1990). Modified differential phase contrast Lorentz microscopy for improved imaging of magnetic structures. *IEEE Trans. Magnetics* **26**, 1506–1511.

D. Chatterjee, J. Wei, A. Kvit, B. Bammes, B. Levin, R. Bilhorn and P. Voyles (2021). An ultrafast direct electron camera for 4D STEM. *Microsc. Microanal.* **27** (Suppl. 1), 1004–1006.

J. H. Chen, K. Urban, B. Kabius, M. Lentzen, J. Jansen and H. W. Zandbergen (2002). Atomic imaging in aberration-corrected HRTEM with application to Al alloys. *Microsc. Microanal.* **8** (Suppl. 2), 468–469.

K.-f. Chen, C.-s. Chang, J. Shiue, Y. Hwu, W.-h. Chang, J.-j. Kai and F.-r. Chen (2008). Study of mean absorption potential using Lenz model: toward quantification of phase contrast from an electrostatic phase plate. *Micron* **39**, 749–756.

Z. Chen, Y. Jiang, M. Odstrcil, Y. Han, G. Fuchs and D. Muller (2020a). Efficient phase-contrast imaging via mixed-state electron ptychography: from crystal structures to electromagnetic fields. *Microsc. Microanal.* **26** (Suppl. 2), 26–28.

Z. Chen, M. Odstrcil, Y. Jiang, Y. Han, M.-h. Chiu, L.-j. Li and D. A. Muller (2020b). Mixed-state electron ptychography enables sub-angstrom resolution imaging with picometer precision at low dose. *Nature Commun.* **11**, 2994, 10 pp.

Y. F. Cheng, M. Strachan, Z. Weiss, M. Deb, D. Carone and V. Ganapati (2019). Illumination pattern design with deep learning for single-shot Fourier ptychographic microscopy. *Opt. Express* **27**, 644–656.

W. Chiu, C. J.-y. Fu, H. Khant and S. Motoki (2015). Zernike phase plate configuration at intermediate lens position on JEM2200FS. *Microsc. Microanal.* **21** (Suppl. 3), 2143–2144.

P. Chlanda and J. K. Locker (2017). The sleeping beauty kissed awake: new methods in electron microscopy to study cellular membranes. *Biochem. J.* **474**, 1041–1053.

K. K. Christenson and J. A. Eades (1986). On "parallel" illumination in the transmission electron microscope. *Ultramicroscopy* **19**, 191–194.

J. Ciston, I. J. Johnson, B. R. Draney, P. Ercius, E. Fong, A. Goldschmidt, J. M. Joseph, J. R. Lee, A. Mueller, C. Ophus, A. Selvarajan, D. E. Skinner, T. Stezelberger, C. S. Tindall, A. M. Minor and P. Denes (2019). The 4D camera: very high speed electron counting for 4D-STEM. *Microsc. Microanal.* **25** (Suppl. 1), 1930–1931.

R. Clough and A. I. Kirkland (2016). Direct digital electron detectors. *Adv. Imaging Electron Phys.* **198**, 1–42.

W. Coene and A. J. E. M. Janssen (1991). Image delocalisation and high resolution transmission electron microscopic imaging with a field emission gun. *Scanning Microsc.* (Suppl. 6, published 1994), 379–403; Proceedings of the 10th Pfefferkorn Conference, Cambridge, 1991.

W. Coene, G. Janssen, M. Op de Beeck and D. Van Dyck (1992). Phase retrieval through focus variation for ultra-resolution in field-emission transmission electron microscopy. *Phys. Rev. Lett.* **69**, 3743–3746.

C. J. Cogswell and C. J. R. Sheppard (1992). Confocal differential interference contrast (DIC) microscopy: including a theoretical analysis of conventional and confocal DIC imaging. *J. Microsc. (Oxford)* **165**, 81–101.

C. Colliex (1985). L'instrument STEM – une introduction au microscope électronique à balayage en transmission. *J. Microsc. Spectrosc. Electron* **10**, 313–332.

C. Colliex, A. J. Craven and C. J. Wilson (1977). Fresnel fringes in STEM. *Ultramicroscopy* **2**, 327–335.

C. Colliex, C. Jeanguillaume and C. Mory (1984). Unconventional modes for STEM imaging of biological structures. *J. Ultrastruct. Res.* **88**, 177–206.

C. Colliex, C. Mory, A. L. Olins and M. Tencé (1989). Energy filtered STEM imaging of thick biological sections. *J. Microsc. (Oxford)* **153**, 1–21.

E. C. Cosgriff, P. D. Nellist, A. J. d'Alfonso, S. D. Findlay, G. Behan, P. Wang, L. J. Allen and A. I. Kirkland (2010). Image contrast in aberration-corrected scanning confocal electron microscopy. *Adv. Imaging Electron Phys.* **162**, 45–76.

J. M. Cowley (1969). Image contrast in a transmission scanning electron microscope. *Appl. Phys. Lett.* **15**, 58–59.

J. M. Cowley (1970). High-voltage transmission scanning electron microscopy. *J. Appl. Cryst.* **3**, 49–58.

J. M. Cowley (1975). Coherent and incoherent imaging in the scanning transmission electron microscope. *J. Phys. D.: Appl. Phys* **8**, L77–L79.

J. M. Cowley (1976a). The extension of scanning transmission electron microscopy by use of diffraction information. *Ultramicroscopy* **1**, 255–262.

J. M. Cowley (1976b). Scanning transmission electron microscopy of thin specimens. *Ultramicroscopy* **2**, 3–16, cf. Dekkers and de Lang (1978b).

J.M. Cowley (1976c). Image contrast for dark-field STEM. EMSA, Miami Beach, FL, **34**, 466–467.

J. M. Cowley (1978). Electron microdiffraction. *Adv. Electron Electron Phys.* **46**, 1–53.

J. M. Cowley (1978/9). High resolution studies of crystals using STEM. *Chem. Scr.* **14**, 33–38.

J. M. Cowley (1980). Interference effects in a STEM instrument. *Micron* **11**, 229–233.

J. M. Cowley (1981). Coherent interference effects in SIEM and CBED. *Ultramicroscopy* **7**, 19–26.

J. M. Cowley (1983). The STEM approach to the imaging of surfaces and small particles. *J. Microsc. (Oxford)* **129**, 253–261.

J. M. Cowley (1984). STEM imaging of thick specimens with off-axis detectors. *J. Electron Microsc. Tech.* **1**, 83–94.

J. M. Cowley (1985). A new detector system for the HB5 STEM instrument. EMSA, Louisville, KY, **43**, 134–135.

J. M. Cowley (1986). Electron diffraction phenomena observed with a high resolution STEM instrument. *J. Electron Microsc. Tech.* **3**, 25–44.

J. M. Cowley (1987). High resolution electron microscopy. *Annu. Rev. Phys. Chem.* **38**, 57–88.

J. M. Cowley (1993). Configured detectors for STEM imaging of thin specimens. *Ultramicroscopy* **49**, 4–13.

J. M. Cowley and A. Y. Au (1978a). Image signals and detector configurations for STEM. *Scanning Electron Microsc.* (I), 53–60.

J. M. Cowley and A. Y. Au (1978b). Bright-field image contrast and resolution in STEM and CTEM. ICEM-9, Toronto, **1**, 172–173.

J. M. Cowley and M. M. Disko (1980). Fresnel diffraction in a coherent convergent electron beam. *Ultramicroscopy* **5**, 469–477.

J. M. Cowley and B. K. Jap (1976a). The use of diffraction information to augment STEM imaging. *Scanning Electron Microsc.* (I), 377–384 and 344.

J. M. Cowley and B. K. Jap (1976b). The use of diffraction pattern information in STEM. EMSA, Miami Beach, FL, **34**, 460–461.

J. M. Cowley and J. C. H. Spence (1979). Innovative imaging and microdiffraction in STEM. *Ultramicroscopy* **3**, 433–438.

J. M. Cowley and J. C. H. Spence (1981). Convergent beam electron microdiffraction from small crystals. *Ultramicroscopy* **6**, 359–366.

J. M. Cowley and D. J. Walker (1981). Reconstruction from in-line holograms by digital processing. *Ultramicroscopy* **6**, 71–75.

J. M. Cowley and J. Winterton (2001). Ultra-high-resolution electron microscopy of carbon nanotube walls. *Phys. Rev. Lett.* **87**, 016101, 4 pp.

J. M. Cowley, M. Strahm and J. H. Butler (1980). Recording and processing STEM images. *Micron* **11**, 285–286.

A. J. Craven and C. Colliex (1977). High resolution energy filtered images in STEM. *J. Microsc. Spectrosc. Electron* **2**, 511–522.

A. V. Crewe (1966). Scanning electron microscopes: is high resolution possible? *Science* **154**, 729–738.

A. V. Crewe (1970). The current state of high resolution scanning electron microscopy. *Q. Rev. Biophys.* **3**, 137–175.

A. V. Crewe (1971). High resolution scanning microscopy of biological specimens. *Philos. Trans. R. Soc. Lond. B* **261**, 61–70.

A. V. Crewe (1973). Production of electron probes using a field emission source. *Prog. Opt.* **11**, 223–246.

A. V. Crewe (1974). Scanning transmission electron microscopy. *J. Microsc. (Oxford)* **100**, 247–259.

A. V. Crewe (1978). Direct imaging of single atoms and molecules using the STEM. *Chem. Scr.* **14**, 17–20.

A. V. Crewe (1979). Development of high resolution STEM and its future. *J. Electron Microsc.* **28**, S9–S16.

A. V. Crewe (1980a). The physics of the high-resolution scanning microscope. *Rep. Prog. Phys.* **43**, 621–639.

A. V. Crewe (1980b). Imaging in scanning microscopes. *Ultramicroscopy* **5**, 131–138.

A. V. Crewe (1980c). Theory of optimal scanning in the STEM. EMSA, San Francisco, CA, **38**, 60–61.

A. V. Crewe (1983). High-resolution scanning transmission electron microscopy. *Science* **221**, 325–330.

A. V. Crewe (1985). Towards the ultimate scanning electron microscope. *Scanning Electron Microsc.* (II), 467–472.

A. V. Crewe and T. Groves (1974). Thick specimens in the CEM and STEM. I. Contrast. *J. Appl. Phys.* **45**, 3662–3672.

A. V. Crewe and D. Kopf (1980). The use of non-optimal apertures in the STEM. *Optik* **55**, 325–327.

A. V. Crewe and M. Ohtsuki (1980). The use of optimal scanning in the STEM. EUREM-7, The Hague, **2**, 588–589.

A.V. Crewe and M. Ohtsuki (1981a). Digital processing of STEM images. EMSA, Atlanta, GA, **39**, 236–237.

A. V. Crewe and M. Ohtsuki (1981b). Optimal scanning and image processing with the STEM. *Ultramicroscopy* **7**, 13–18.

A. V. Crewe and D. B. Salzman (1982). On the optimum resolution for a corrected STEM. *Ultramicroscopy* **9**, 373–377.

A. V. Crewe and J. S. Wall (1970a). A scanning microscope with 5Å resolution. *J. Mol. Biol.* **48**, 375–393.

A. V. Crewe and J. S. Wall (1970b). Contrast in a high resolution scanning transmission electron microscope. *Optik* **30**, 461–474.

A. V. Crewe, M. Isaacson and D. Johnson (1969). A simple scanning electron microscope. *Rev. Sci. Instrum.* **40**, 241–246.

A. V. Crewe, J. Wall and J. Langmore (1970a). Visibility of single atoms. *Science* **168**, 1338–1340.

A. V. Crewe, J. Langmore, J. Wall and M. Beer (1970b). Single atom contrast in a scanning microscope. EMSA, Houston, TX, **28**, 250–251.

A. V. Crewe, M. S. Isaacson and E. Zeitler (1979). Progress in scanning transmission electron microscopy at the University of Chicago. *Adv. Structure Res. Diffraction Methods* **7**, 23–48.

A. G. Cullis and D. M. Maher (1974). High-resolution topographical imaging by direct transmission electron microscopy. *Philos. Mag.* **30**, 447–451.

A. G. Cullis and D. M. Maher (1975). Topographical contrast in the transmission electron microscope. *Ultramicroscopy* **1**, 97–112.

A. J. d'Alfonso, S. D. Findlay, M. P. Oxley, S. J. Pennycook, K. van Benthem and L. J. Allen (2007). Depth sectioning in scanning transmission electron microscopy based on core-loss spectroscopy. *Ultramicroscopy* **108**, 17–28.

A. J. d'Alfonso, E. C. Cosgriff, S. D. Findlay, G. Behan, A. I. Kirkland, P. D. Nellist and L. J. Allen (2008). Three-dimensional imaging in double aberration-corrected scanning confocal electron microscopy, Part II: Inelastic scattering. *Ultramicroscopy* **108**, 1567–1578.

A. J. d'Alfonso, A. J. Morgan, A. W. C. Yan, P. Wang, H. Sawada, A. I. Kirkland and L. J. Allen (2014). Deterministic electron ptychography at atomic resolution. *Phys. Rev. B* **89**, 064101, 10 pp.

I. Daberkow and K.-H. Herrmann (1984). Computer-controlled angle-resolving recording system in STEM. EUREM-8, Budapest, **1**, 115–116.

I. Daberkow and K.-H. Herrmann (1988). Application of an angle-resolving recording system in STEM. EUREM-9, York, **1**, 125–126.

I. Daberkow, K.-H. Herrmann and F. Lenz (1993). A configurable angle-resolving detector system in STEM. *Ultramicroscopy* **50**, 75–82.

W. Dai, C. Fu, D. Raytcheva, J. Flanagan, H. A. Khant, X. Liu, R. H. Rochat, C. Haase-Pettingell, J. Piret, S. J. Ludtke, K. Nagayama, M. F. Schmid, J. A. King and W. Chiu (2013). Visualizing virus assembly intermediates inside marine cyanobacteria. *Nature* **502**, 707–710.

W. Dai, C. Fu, H. A. Khant, S. J. Ludtke, M. F. Schmid and W. Chiu (2014). Zernike phase-contrast electron cryotomography applied to marine cyanobacteria infected with cyanophages. *Nature Protocols* **9**, 2630–2642.

R. Danev (2020). Electrons receive individual treatment with electron-event representation. *IUCrJ* **7**, 780–781.

R. Danev and W. Baumeister (2017). Expanding the boundaries of cryo-EM with phase plates. *Curr. Opin. Struct. Biol.* **46**, 87–94.

R. Danev and K. Nagayama (2001a). Complex observation in electron microscopy. II. Direct visualization of phases and amplitudes of exit wave functions. *J. Phys. Soc. Jpn.* **70**, 696–702.

R. Danev and K. Nagayama (2001b). Transmission electron microscopy with a Zernike phase plate. *Ultramicroscopy* **88**, 243–252.

R. Danev and K. Nagayama (2004). Complex observation in electron microscopy. IV. Reconstruction of complex object wave from conventional and half plane phase plate image pair. *J. Phys. Soc. Jpn.* **73**, 2718–2724.

R. Danev and K. Nagayama (2008). Single particle analysis based on Zernike phase contrast transmission electron microscopy. *J. Struct. Biol.* **161**, 211–218.

R. Danev and K. Nagayama (2010). Phase plates for transmission electron microscopy. *Methods Enzymol.* **481**, 343–369.

R. Danev and K. Nagayama (2011). Optimizing the phase shift and the cut-on periodicity of phase plates for TEM. *Ultramicroscopy* **111**, 1305–1315.

R. Danev, H. Okawara, N. Usuda, K. Kametani and K. Nagayama (2002). A novel phase-contrast transmission electron microscopy [sic] producing high-contrast topographic images of weak objects. *J. Biol. Phys.* **28**, 627–635.

R. Danev, R. M. Glaeser and K. Nagayama (2009). Practical factors affecting the performance of a thin-film phase plate for transmission electron microscopy. *Ultramicroscopy* **109**, 312–325.

R. Danev, S. Kanamaru, M. Marko and K. Nagayama (2010). Zernike phase contrast cryo-electron microscopy. *J. Struct. Biol.* **171**, 174–181.

R. Danev, R. M. Glaeser and B. Buijsse (2012). Properties and behavior of amorphous carbon films related to phase plate applications. *Microsc. Microanal.* **18** (Suppl. 2), 482–483.

R. Danev, B. Buijsse, Y. Fukuda, M. Khoshouei, J. Plitzko and W. Baumeister (2014a). Electron cryo-tomography with a new type of phase plate. IMC-18, Prague, 2806.

R. Danev, B. Buijsse, Y. Fukuda, M. Khoshouei, J. Plitzko and W. Baumeister (2014b). Automated cryo-tomography and single particle analysis with a phase plate. *Microsc. Microanal.* **20** (Suppl. 3), 206–207.

R. Danev, B. Buijsse, M. Khoshouei, J. M. Plitzko and W. Baumeister (2014c). Volta potential phase plate for in-focus phase contrast transmission electron microscopy. *Proc. Natl. Acad. Sci. U.S.A.* **111**, 15635–15640.

R. Danev, B. Buijsse, M. Khoshouei, Y. Fukuda and W. Baumeister (2015). Practical aspects and usage tips for the Volta phase plate. *Microsc. Microanal.* **21** (Suppl. 3), 1391–1392.

R. Danev, M. Khoshouei and W. Baumeister (2017a). Single particle analysis applications of the Volta phase plate. *Microscopy* **66** (Suppl. 1), i9.

R. Danev, D. Tegunov and W. Baumeister (2017b). Using the Volta phase plate with defocus for cryo-EM single particle analysis. *eLife* **6**, e23006.

K. Danov, R. Danev and K. Nagayama (2001). Electric charging of thin films measured using the contrast transfer function. *Ultramicroscopy* **87**, 45–54.

A. Datta, K. F. Ng, D. Balakrishnan, M. Ding, S. W. Chee, Y. Ban, J. Shi and N. D. Loh (2021). A data reduction and compression description for high throughput time-resolved electron microscopy. *Nature Commun.* **12**, 664, 15 pp.

A. De Backer, G. T. Martinez, A. Rosenauer and S. Van Aert (2013). Atom counting in HAADF STEM using a statistical model-based approach: methodology, possibilities, and inherent limitations. *Ultramicroscopy* **134**, 23–33.

A. De Backer, G. T. Martinez, K. E. MacArthur, L. Jones, A. Béché, P. D. Nellist and S. Van Aert (2014). Possibilities and limitations for atom counting using quantitative ADF STEM. IMC-18, Prague, 2253.

A. De Backer, A. De wael, J. Gonnissen and S. Van Aert (2015a). Optimal experimental design for nano-particle atom-counting from high-resolution STEM images. *Ultramicroscopy* **151**, 46–55.

A. De Backer, G. T. Martinez, K. E. MacArthur, L. Jones, A. Béché, P. D. Nellist and S. Van Aert (2015b). Dose limited reliability of quantitative annular dark field scanning transmission electron microscopy for nano-particle atom-counting. *Ultramicroscopy* **151**, 56–61.

A. De Backer, K. H. W. van den Bos, W. Van den Broek, J. Sijbers and S. Van Aert (2016). StatSTEM: an efficient approach for accurate and precise model-based quantification of atomic resolution electron microscopy images. *Ultramicroscopy* **171**, 104–116.

A. De Backer, L. Jones, I. Lobato, T. Altantzis, B. Goris, P. D. Nellist, S. Bals and S. Van Aert (2017a). Three-dimensional atomic models from a single projection using *Z*-contrast imaging: verification by electron tomography and opportunities. *Nanoscale* **9**, 8791–8798.

A. De Backer, L. Jones, I. Lobato, T. Altantzis, P. D. Nellist, S. Bals and S. Van Aert (2017b). Reconstructing three-dimensional atomic details of nanostructures from single Z-contrast electron microscopy images. MC-2017, Lausanne, 725–726.

A. De Backer, K.H.W. van den Bos, W. van den Broek, J. Sijbers and S. Van Aert (2017c). StatSTEM – an efficient approach for accurate and precise model-based quantification of atomic resolution electron microscopy images. MC-2017, Lausanne, 662–663.

A. De Backer, L. Jones, I. Lobato, T. Altantzis, P.D. Nellist, S. Bals and S. Van Aert (2018). Three-dimensional atomic models from a single projection using z-contrast imaging: verification by electron tomography and opportunities. IMC-19, Sydney.

A. De Backer, J. Fatermans, A. J. den Dekker and S. Van Aert (2021). Quantitative atomic resolution electron microscopy using advanced statistical techniques. *Adv. Imaging Electron Phys.* **217**, 1–278.

J. De Beenhouwer, I. Lobato, D. Van Dyck, S. Van Aert and J. Sijbers (2016). Direct estimation of 3D atom positions of simulated Au nanoparticles in HAADF STEM. EMC-16, Lyon, **1**, 61–62.

A. F. de Jong and D. Van Dyck (1993). Ultimate resolution and information in electron microscopy. II. The information limit of transmission electron microscopes. *Ultramicroscopy* **49**, 66–80.

N. de Jonge, A. R. Lupini, K. van Benthem, A. Y. Borisevich and S. J. Pennycook (2006). Depth-related contrast in aberration-corrected confocal STEM. *Microsc. Microanal.* **12** (Suppl. 2), 1574–1575.

N. de Jonge, R. Sougrat, B. M. Northan and S. J. Pennycook (2010). Three-dimensional scanning transmission electron microscopy of biological specimens. *Microsc. Microanal.* **16**, 54–63.

H. de Lang and N. H. Dekkers (1979). A posteriori focussing of a STEM using shadow image processing. *Optik* **53**, 353–365.

A. De wael (2021). Model-based quantitative scanning transmission electron microscopy for measuring dynamic structural changes at the atomic scale. Dissertation, Antwerp.

A. De wael, A. De Backer, L. Jones, P. D. Nellist and S. Van Aert (2017). Hybrid statistics-simulations based method for atom-counting from ADF STEM images. *Ultramicroscopy* **177**, 69–77.

A. De wael, A. de Backer, L. Jones, A. Varambhia, P. D. Nellist and S. Van Aert (2020a). Measuring dynamic structural changes of nanoparticles at the atomic scale using scanning transmission electron microscopy. *Phys. Rev. Lett.* **124**, 106105, 7 pp.

A. De wael, A. De Backer and S. Van Aert (2020b). Hidden Markov model for atom-counting from sequential ADF STEM images: methodology, possibilities and limitations. *Ultramicroscopy* **219**, 113131, 14 pp.

A. De Wael, A. De Backer, I. Lobato and S. Van Aert (2021). Modelling ADF STEM images using elliptical Gaussian peaks and its effects on the quantification of structure parameters in the presence of sample tilt. *Ultramicroscopy* **230**, 113391, 8 pp.

N. H. Dekkers (1979). Object wave reconstruction in STEM. *Optik* **53**, 131–142.

N. H. Dekkers and H. de Lang (1974). Differential phase contrast in a STEM. *Optik* **41**, 452–456.

N. H. Dekkers and H. de Lang (1977). A detection method for producing phase and amplitude images simultaneously in a scanning transmission electron microscope. *Philips Tech. Rev.* **37** (1), 1–9.

N. H. Dekkers and H. de Lang (1978a). A calculation of bright field single-atom images in STEM with half plane detectors. *Optik* **51**, 83–92.

N. H. Dekkers and H. de Lang (1978b). Comment on "Scanning transmission electron microscopy of thin specimens" by J.M. Cowley. *Ultramicroscopy* **3**, 101–102.

N. H. Dekkers, H. de Lang and K. van der Mast (1976). Field emission STEM on a Philips EM 400 with a new detection system for phase and amplitude contrast. *J. Microsc. Spectrosc. Electron* **1**, 511–512.

A. J. den Dekker (1999). How to optimize the design of a quantitative HREM experiment so as to attain the highest precision. *J. Microsc. (Oxford)* **194**, 95–104.

A. J. den Dekker, S. Van Aert, D. Van Dyck, A. van den Bos and P. Geuens (2001). Does a monochromator improve the precision in quantitative HRTEM? *Ultramicroscopy* **89**, 275–290.

A. J. den Dekker, S. Van Aert, A. van den Bos and D. Van Dyck (2005). Maximum likelihood estimation of structure parameters from high resolution electron microscopy images. Part I: A theoretical framework. *Ultramicroscopy* **104**, 83–106, Part II is Van Aert et al. (2005).

A. J. den Dekker, J. Gonnissen, A. De Backer, J. Sijbers and S. Van Aert (2013). Estimation of unknown structure parameters from high-resolution (S)TEM images: what are the limits? *Ultramicroscopy* **134**, 34–43.

G. Deptuch, A. Besson, P. Rehak, M. Szelezniak, J. Wall, M. Winter and Y. Zhu (2007). Direct electron imaging in electron microscopy with monolithic active pixel sensors. *Ultramicroscopy* **107**, 674–684.

J. Desseaux, A. Renault and A. Bourret (1977). Multi-beam lattice images from germanium oriented in (011). *Philos. Mag.* **35**, 357–372.

C. Dinges, H. Kohl and H. Rose (1994). High-resolution imaging of crystalline objects by hollow-cone illumination. *Ultramicroscopy* **55**, 91–100.

K. H. Downing (1975). Note on transfer functions in electron microscopy with tilted illumination. *Optik* **43**, 199–203.

K. H. Downing (1979). Possibilities of heavy atom discrimination using single-sideband techniques. *Ultramicroscopy* **4**, 13–31.

K. H. Downing and B. M. Siegel (1973). Phase shift determination in single-sideband holography. *Optik* **38**, 21–28.

K. H. Downing, M. R. McCartney and R. M. Glaeser (2004). Experimental characterization and mitigation of specimen charging on thin films with one conducting layer. *Microsc. Microanal.* **10**, 783–789.

M. Dries, B. Gamm, K. Schultheiß, A. Rosenauer, R. Schröder and D. Gerthsen (2010). Object-wave reconstruction by carbon film-based Zernike- and Hilbert-phase plate microscopy: a theoretical study not restricted to weak-phase objects. *Microsc. Microanal.* **16** (Suppl. 2), 552–553.

M. Dries, K. Schultheiß, B. Gamm, A. Rosenauer, R. R. Schröder and D. Gerthsen (2011). Object-wave reconstruction by carbon film-based Zernike- and Hilbert-phase plate microscopy: a theoretical study not restricted to weak-phase objects. *Ultramicroscopy* **111**, 159–168.

M. Dries, B. Gamm, S. Hettler, E. Müller, W. Send, D. Gerthsen and A. Rosenauer (2012a). A nanocrystalline Hilbert-phase plate for phase-contrast transmission electron microscopy. EMC-15, Manchester, **2**, 399–400.

M. Dries, B. Gamm, S. Hettler, E. Müller, W. Send, D. Gerthsen and A. Rosenauer (2012b). A nanocrystalline Hilbert-phase plate for phase-contrast transmission electron microscopy of amorphous objects. *Microsc. Microanal.* **18** (Suppl. 2), 496–497.

M. Dries, S. Hettler, B. Gamm, E. Müller, W. Send, D. Gerthsen, K. Müller and A. Rosenauer (2014a). A nanocrystalline Hilbert phase-plate for phase-contrast transmission electron microscopy. *Microsc. Microanal.* **20** (Suppl. 3), 236–237.

M. Dries, S. Hettler, B. Gamm, E. Müller, W. Send, K. Müller, A. Rosenauer and D. Gerthsen (2014b). A nanocrystalline Hilbert phase-plate for phase-contrast transmission electron microscopy. *Ultramicroscopy* **139**, 29–37.

M. Dries, S. Hettler, T. Schulze, W. Send, E. Müller, R. Schneider, D. Gerthsen, Y. Luo and K. Samwer (2015). Thin-film-based phase plates for transmission electron microscopy fabricated from metallic glasses. *Microsc. Microanal.* **21** (Suppl. 3), 1575–1576.

M. Dries, S. Hettler, T. Schulze, W. Send, E. Müller, R. Schneider, Y. Luo and K. Samwer (2016a). Thin-film phase plates for transmission electron microscopy fabricated from metallic glasses. *Microsc. Microanal.* **22**, 955–963.

M. Dries, R. Janzen, T. Schulze, J. Schundelmeier, S. Hettler, U. Golla-Schindler, B. Jaud, U. Kaiser and D. Gerthsen (2016b). The role of secondary electron emission in the charging of thin-film phase plates. *Microsc. Microanal.* **22** (Suppl. 3), 64–65.

M. Dries, M. Obermair, S. Hettler, P. Hermann, K. Seemann, F. Seifried, S. Ulrich, R. Fischer and D. Gerthsen (2018). Oxide-free aC/$Zr_{0.65}Cu_{0.275}$/aC phase plates for transmission electron microscopy. *Ultramicroscopy* **189**, 39–45.

P. Dumontet (1955a). Sur la correspondance objet–image en optique. *Opt. Acta* **2**, 53–63.

P. Dumontet (1955b). Imagerie en cohérence partielle. *Publ. Sci. Univ. Alger B* **1**, 33–44.

P. Dumontet (1956). La correspondance objet–image en optique et possibilités de corriger certains défauts liés à la diffraction. *Publ. Sci. Univ. Alger B* **2**, 151–179 and 203–294.

G. Dupouy (1967). Contrast improvement in electron microscope images of amorphous objects. *J. Electron Microsc.* **16**, 5–16.

P. Dwivedi, A. P. Konijnenberg, S. F. Pereira and H. P. Urbach (2018). Lateral position correction in ptychography using the gradient of intensity patterns. *Ultramicroscopy* **192**, 29–36.

H. E, K. E. MacArthur, T. J. Pennycook, E. Okunishi, A. J. D'Alfonso, N. R. Lugg, L. J. Allen and P. D. Nellist (2013). Probe integrated scattering cross sections in the analysis of atomic resolution HAADF STEM images. *Ultramicroscopy* **133**, 109–119.

C. J. Edgcombe (2010). A phase plate for transmission electron microscopy using the Aharonov–Bohm effect. *J. Phys. Conf. Ser.* **241**, 012005, 4 pp (EMAG 2009, Sheffield).

C. J. Edgcombe (2012). The positioning of thin-film magnetic rings as phase plates for transmission electron microscopy. IMC-15, Manchester, **2**, 489–490.

C. J. Edgcombe (2014a). Effect of a phase plate on TEM imaging. IMC-18, Prague, 2476.

C. J. Edgcombe (2014b). Imaging of weak phase objects by a Zernike phase plate. *Ultramicroscopy* **136**, 154–159.

C. J. Edgcombe (2016). Imaging by Zernike phase plates in the TEM. *Ultramicroscopy* **167**, 57–63.

C. J. Edgcombe (2017a). Imaging with straight-edge phase plates in the TEM. *Ultramicroscopy* **182**, 124–130.

C. J. Edgcombe (2017b). Phase plates for transmission electron microscopy. *Adv. Imaging Electron Phys.* **200**, 61–102.

C. J. Edgcombe and J. C. Loudon (2012). Use of Aharonov–Bohm effect and chirality control in magnetic phase plates for transmission microscopy. *J. Phys. Conf. Ser.* **371**, 012006, 4 pp (EMAG 2011, Birmingham).

C. J. Edgcombe, A. Ionescu, J. C. Loudon, A. M. Blackburn, H. Kurebayashi and C. H. W. Barnes (2012). Characterisation of ferromagnetic rings for Zernike phase plates using the Aharonov–Bohm effect. *Ultramicroscopy* **120**, 78–85.

W. Ehrenberg and R. E. Siday (1949). The refractive index in electron optics and the principles of dynamics. *Proc. Phys. Soc. Lond. B* **62**, 8–21.

J. J. Einspahr and P. M. Voyles (2005). Confocal scanning transmission electron microscopy: theoretical analysis of three-dimensional imaging. *Microsc. Microanal.* **11** (Suppl. 2), 320–321.

J. J. Einspahr and P. M. Voyles (2006). Prospects for 3D, nanometer-resolution imaging by confocal STEM. *Ultramicroscopy* **106**, 1041–1052.

C. B. Eisenhandler and B. M. Siegel (1966a). Imaging of single atoms with the electron microscope by phase contrast. *J. Appl. Phys.* **37**, 1613–1620.

C. B. Eisenhandler and B. M. Siegel (1966b). A zone-plate aperture for enhancing resolution in phase-contrast electron microscopy. *Appl. Phys. Lett.* **8**, 258–260.

H. Endoh and H. Hashimoto (1977). Electron channelling in metal crystal lattices and high resolution electron microscopy at high voltage. HVEM, Kyoto, 293–296.

A. Engel (1974). The principle of reciprocity and its application to conventional and scanning dark field electron microscopy. *Optik* **41**, 117–126.

A. Engel (1980). Current state of biological scanning transmission electron microscopy. In *Electron Microscopy at Molecular Dimensions* (W. Baumeister and W. Vogell, Eds), 170–178 (Springer, Berlin).

A. Engel, J. W. Wiggins and D. C. Woodruff (1974). A comparison of calculated images generated by six modes of transmission electron microscopy. *J. Appl. Phys.* **45**, 2739–2747.

A. Engel, J. Dubochet and E. Kellenberger (1976). Some progress in the use of a scanning transmission electron microscope for the observation of biomacromolecules. *J. Ultrastruct. Res.* **57**, 322–330.

P. Ercius and N. L. Xin (2012). Quantitative confocal sectioning in double-corrected STEM utilizing electron energy loss spectroscopy and post-specimen C_c correction. *Microsc. Microanal.* **18** (Suppl. 2), 1026–1027.

P. Ercius, I. Johnson, H. Brown, P. Pelz, S.-l. Hsu, B. Draney, E. Fong, A. Goldschmidt, J. Joseph, J. Lee, J. Ciston, C. Ophus, M. Scott, A. Selvarajan, D. Paul, D. Skinner, M. Hanwell, C. Harris, P. Avery, T. Stezelberger, C. Tindall, R. Ramesh, A. Minor and P. Denes (2020). The 4D Camera – a 87 kHz frame-rate detector for counted 4D-STEM experiments. *Microsc. Microanal.* **26** (Suppl. 2), 1896–1897.

H. P. Erickson (1973). The Fourier transform of an electron micrograph – first order and second order theory of image formation. *Adv. Opt. Electron Microsc.* **5**, 163–199.

H. P. Erickson and A. Klug (1970a). The Fourier transform of an electron micrograph: effects of defocussing and aberrations, and implications for the use of underfocus contrast enhancement. *Ber. Bunsen-Ges. Phys. Chem.* **74**, 1129–1137.

H. P. Erickson and A. Klug (1970b). Phase contrast electron microscopy and compensation of aberrations by Fourier image processing. EMSA, Houston, TX, **28**, 248–249.

H. P. Erickson and A. Klug (1971). Measurement and compensation of defocusing and aberrations by Fourier processing of electron micrographs. *Philos. Trans. R. Soc. Lond. B* **261**, 105–118.

J. Etheridge, S. Lazar, C. Dwyer and G. A. Botton (2011). Imaging high-energy electrons propagating in a crystal. *Phys. Rev. Lett.* **106**, 160802, 4 pp.

J. Faget, J. Ferré and C. Fert (1960a). Contraste de phase en microscopie électronique. *C. R. Acad. Sci. Paris* **251**, 526–528.

J. Faget, M. Fagot and C. Fert (1960b). Microscopie électronique en éclairage cohérent: microscopie interférentielle, contraste de défocalisation, strioscopie et contraste de phase. EUREM-2, Delft, **1**, 18–24.

J. Faget, M. Fagot, J. Ferré and C. Fert (1962). Microscopie électronique à contraste de phase. ICEM-5, Philadelphia, PA, **1**, A–7.

A. R. Faruqi (2001). Prospects for hybrid pixel detectors in electron microscopy. *Nucl. Instrum. Meth, Phys. Res. A* **466**, 146–154.

A. R. Faruqi and R. Henderson (2007). Electronic detectors for electron microscopy. *Curr. Opin. Struct. Biol.* **17**, 549–555.

A. R. Faruqi and G. McMullan (2011). Electronic detectors for electron microscopy. *Q. Rev. Biophys.* **44**, 357–390.

A. R. Faruqi and G. McMullan (2018). Direct imaging detectors for electron microscopy. *Nucl. Instrum. Meth, Phys. Res. A* **878**, 180–190.

A. R. Faruqi and S. Subramaniam (2000). CCD detectors in high-resolution biological electron microscopy. *Q. Rev. Biophys.* **33**, 1–27.

A. R. Faruqi, D. M. Cattermole, R. Henderson, B. Mikulec and C. Raeburn (2003). Evaluation of a hybrid pixel detector for electron microscopy. *Ultramicroscopy* **94**, 263–276.

A. R. Faruqi, R. Henderson and L. Tlustos (2005a). Noiseless direct detection of electrons in Medipix2 for electron microscopy. *Nucl. Instrum. Meth, Phys. Res. A* **546**, 160–163.

A. R. Faruqi, R. Henderson, M. Pryddetch, P. Allport and A. Evans (2005b). Direct single electron detection with a CMOS detector for electron microscopy. *Nucl. Instrum. Meth, Phys. Res. A* **546**, 170–175.

A. R. Faruqi, R. Henderson and J. Holmes (2006). Radiation damage studies on STAR250 CMOS sensor at 300keV for electron microscopy. *Nucl. Instrum. Meth, Phys. Res. A* **565**, 139–143.

A. R. Faruqi, R. Henderson and G. McMullan (2015). Progress and development of direct detectors for electron cryomicroscopy. *Adv. Imaging Electron Phys.* **190**, 103–141.

J. Fatermans, K. Müller-Caspary, A. J. den Dekker and S. Van Aert (2017). Detection of atomic columns from noisy STEM images. MC-2017, Lausanne, 445–446.

J. Fatermans, A. J. den Dekker, K. Müller-Caspary, I. Lobato, C. O'Leary, P. D. Nellist and S. Van Aert (2018). Single atom detection from low contrast-to-noise ratio electron microscopy images. *Phys. Rev. Lett.* **121**, 056101, 6 pp.

J. Fatermans, S. Van Aert and A. J. den Dekker (2019a). The maximum a posteriori probability rule for atom column detection from HAADF STEM images. *Ultramicroscopy* **201**, 81–91.

J. Fatermans, A. J. den Dekker, C. M. O'Leary, P. D. Nellist and S. Van Aert (2019b). Atom column detection from STEM images using the maximum a posteriori probability rule. MC-2019, Berlin, 515–516.

J. Fatermans, A. J. den Dekker, K. Müller-Caspary, N. Gauquelin, J. Verbeeck and S. Van Aert (2020). Atom column detection from simultaneously acquired ABF and ADF STEM images. *Ultramicroscopy* **219**, 113046, 10 pp.

H. M. Faulkner and J. Rodenburg (2004). Movable aperture lensless transmission microscopy: a novel phase retrieval algorithm. *Phys. Rev. Lett.* **93**, 023903, 4 pp.

H. M. L. Faulkner and J. M. Rodenburg (2005). Error tolerance of an iterative phase retrieval algorithm for moveable illumination microscopy. *Ultramicroscopy* **103**, 153–164.

A. Feist, S. V. Yalunin, S. Schäfer and C. Ropers (2020). High-purity free-electron momentum states prepared by three-dimensional optical phase modulation. *Phys. Rev. Res.* **2**, 043227, 7 pp.

A. Feist, A. S. Raja, J. W. Henke, J. Liu, G. Arend, G. Huang, F. J. Kappert, R. N. Wang, J. Pan, O. Kfir, T. Kippenberg and C. Ropers (2021). Continuous-wave photonic chip-based temporal phase plates for electron microscopy. MC-2021, Vienna (virtual), 478–479.

P. L. Fejes (1977). Approximations for the calculation of high-resolution electron-microscope images of thin films. *Acta Cryst.* **A33**, 109–113.

J. Fertig and H. Rose (1977). A reflection on partial coherence in electron microscopy. *Ultramicroscopy* **2**, 269–279.

J. Fertig and H. Rose (1978a). Image of a sulfur atom: fact or artifact? *Optik* **51**, 213–220.

J. Fertig and H. Rose (1978b). Computer simulation of dark-field imaging as a tool for image interpretation. ICEM-9, Toronto, **1**, 238–239.

J. Fertig and H. Rose (1979). On the theory of image formation in the electron microscope. II. *Optik* **54**, 165–191, Part I is Rose (1975).

J. Fertig and H. Rose (1981). Resolution and contrast of crystalline objects in high-resolution scanning transmission electron microscopy. *Optik* **59**, 165–191.

S. D. Findlay and J. M. LeBeau (2013). Detector non-uniformity in scanning transmission electron microscopy. *Ultramicroscopy* **124**, 52–60.

S. Findlay, N. Shibata, H. Sawada, E. Okunishi, Y. Kondo, T. Yamamoto and Y. Ikuhara (2009). Robust atomic resolution imaging of light elements using scanning transmission electron microscopy. *Appl. Phys. Lett.* **95**, 191913.

S. D. Findlay, N. Shibata, H. Sawada, E. Okunishi, Y. Kondo and Y. Ikuhara (2010). Dynamics of annular bright field imaging in scanning transmission electron microscopy. *Ultramicroscopy* **110**, 903–923.

S. D. Findlay, N. Shibata and Y. Ikuhara (2015). Theory for annular bright field STEM imaging. In *Scanning Transmission Electron Microscopy of Materials* (N. Tanaka, Ed.), 217–230 (Imperial College Press, London).

S. Findlay, L. Allen, H. Brown, Z. Chen, J. Ciston, Y. Ikuhara, R. Ishikawa, C. Ophus, G. Sánchez-Santolino, T. Seki, N. Shibata and M. Weyland (2020). Phase-contrast-based structure retrieval methods in atomic resolution scanning transmission electron microscopy – when they hold and when they don't. *Microsc. Microanal.* **26** (Suppl. 2), 442–443.

L. Foucault (1858). Description des procédés employés pour reconnaître la configuration des surfaces optiques. *C. R. Acad. Sci. Paris* **47**, 958–959.

J. Frank (1969). Nachweis von Objektbewegungen im lichtoptischen Diffraktogramm von elektronenmikroskopischen Aufnahmen. *Optik* **30**, 171–180.

J. Frank (1973). The envelope of electron microscopic transfer functions for partially coherent illumination. *Optik* **38**, 519–536.

J. Frank (1975a). A practical resolution criterion in optics and electron microscopy. *Optik* **43**, 25–34.

J. Frank (1975b). Partial coherence and efficient use of electrons in bright field electron microscopy. *Optik* **43**, 103–109.

J. Frank (1976a). Determination of source size and energy spread from electron micrographs using the method of Young's fringes. *Optik* **44**, 379–391.

J. Frank (1976b). Phase contrast and partial coherence. EUREM-6, Jerusalem, **1**, 97–98.

J. Frank, P. Bußler, R. Langer and W. Hoppe (1970). Einige Erfahrungen mit der rechnerischen Analyse und Synthese von elektronenmikroskopischen Bildern hoher Auflösung. *Ber. Bunsen-Ges. Phys. Chem.* **74**, 1105–1115.

J. Frank, S. C. McFarlane and K. H. Downing (1978). A note on the effect of illumination aperture and defocus spread in brightfield electron microscopy. *Optik* **52**, 49–60.

G. Franke (1965). Bildgütekriterien. *Optik* **23**, 20–25.

L. A. Freeman, A. Howie and M. M. J. Treacy (1977). Bright field and hollow cone dark field electron microscopy of palladium catalysts. *J. Microsc. (Oxford)* **111**, 165–178.

L. A. Freeman, A. Howie and A. B. Mistry (1980). Comparison of hollow cone and axial bright field electron microscope imaging techniques. *J. Microsc. (Oxford)* **119**, 3–18.

D. L. Freimund and H. Batelaan (2002). Bragg scattering of free electrons using the Kapitza–Dirac effect. *Phys. Rev. Lett.* **89**, 283602, 4 pp.

D. L. Freimund, K. Aflatooni and H. Batelaan (2001). Observation of the Kapitza–Dirac effect. *Nature* **413**, 142–143.

B. R. Frieden (1966). Image evaluation by use of the sampling theorem. *J. Opt. Soc. Amer.* **56**, 1355–1362.

S. L. Friedman, J. M. Rodenburg and B. C. McCallum (1991). Phase reconstruction imaging in scanning transmission microscopy via the microdiffraction plane. EMAG, Bristol, 491–494.

S. P. Frigo, Z. H. Levine and N. J. Zaluzec (2002). Submicron imaging of buried integrated circuit structures using scanning confocal electron microscopy. *Appl. Phys. Lett.* **81**, 2112–2124.

N. Frindt, B. Gamm, M. Dries, K. Schultheiß, D. Gerthsen and R. R. Schröder (2009). Simulating Hilbert phase contrast produced by an anamorphotic electrostatic phase plate. MC-2009, Graz, 61–62.

N. Frindt, K. Schultheiß, B. Gamm, M. Dries, J. Zach, D. Gerthsen and R. Schröder (2010). The way to an ideal matter-free Zernike and Hilbert TEM phase plate: anamorphotic design and first experimental verification in isotropic optics. *Microsc. Microanal.* **16** (Suppl. 2), 518–519.

N. Frindt, M. Oster, R. Schröder, S. Hettler, B. Gamm, M. Dries, D. Gerthsen and K. Schultheiß (2012). Tunable phase contrast of vitrified macromolecular complexes by an obstruction minimized electrostatic phase plate. *Microsc. Microanal.* **18** (Suppl. 2), 468–469.

N. Frindt, M. Oster, S. Hettler, B. Gamm, L. Dieterle, W. Kowalsky, D. Gerthsen and R. R. Schröder (2014). In-focus electrostatic Zach phase plate imaging for transmission electron microscopy with tunable phase contrast of frozen hydrated biological samples. *Microsc. Microanal.* **20**, 175–183.

Y. Fukuda and K. Nagayama (2012). Zernike phase contrast cryo-electron tomography of whole mounted frozen cells. *J. Struct. Biol.* **177**, 484–489.

Y. Fukuda, Y. Fukazawa, R. Danev, R. Shigamoto and K. Nagayama (2009). Tuning of the Zernike phase-plate for visualization of detailed ultrastructure in complex biological specimens. *J. Struct. Biol.* **168**, 476–484.

Y. Fukuda, F. Beck, R. Danev, I. Nagy and W. Baumeister (2015a). Electron cryo-tomography of *Thermoplasma acidophilum* with Volta phase plate. *Microscopy* **64** (Suppl. 1), i69.

Y. Fukuda, U. Laughs, V. Lučić, W. Baumeister and R. Danev (2015b). Electron cryotomography of vitrified cells with a Volta phase plate. *J. Struct. Biol.* **190**, 143–154.

T. Fukuda, F. Beck and W. Baumeister (2017). In situ structural studies of macro molecular complexes in cells by cryo-electron tomography with Volta phase plate. *Microscopy* **66** (Suppl. 1), i9.

M. Furuhata, R. Danev, K. Nagayama, Y. Yamada, H. Kawakami, K. Toma, Y. Hattori and Y. Matanai (2008). Decaarginine–PEG–artificial lipid/DNA complex for gene delivery: nanostructure and transfection efficiency. *J. Nanosci. Nanotechnol.* **8**, 2308–2315.

J. Gallagher, N. Gulati and A. Harris (2020). Phase-plate cryo-electron tomography facilitates the identification of influenza virus condensed core structures. *Microsc. Microanal.* **26** (Suppl. 2), 1308–1310.

B. Gamm, K. Schultheiß, D. Gerthsen and R. R. Schröder (2008). Effect of a physical phase plate on contrast transfer in an aberration-corrected transmission electron microscope. *Ultramicroscopy* **108**, 878–884.

B. Gamm, M. Dries, K. Schultheiß, H. Blank, A. Rosenauer, R. Schröder and D. Gerthsen (2010a). Wave-function reconstruction by phase-plate transmission electron microscopy. *Microsc. Microanal.* **16** (Suppl. 2), 538–539.

B. Gamm, M. Dries, K. Schultheiß, H. Blank, A. Rosenauer, R. R. Schröder and D. Gerthsen (2010b). Object wave reconstruction by phase-plate transmission electron microscopy. *Ultramicroscopy* **110**, 807–814.

F. J. Garcia de Abajo and A. Konečná (2021). Optical modulation of electron beams in free space. *Phys. Rev. Lett.* **126**, 123901, 6 pp.

T. Geipel and W. Mader (1996). Practical aspects of hollow-cone imaging. *Ultramicroscopy* **63**, 65–74.

R. W. Gerchberg and W. O. Saxton (1972). A practical algorithm for the determination of phase from image and diffraction plane pictures. *Optik* **35**, 237–246.

R. W. Gerchberg and W. O. Saxton (1973). Wave phase from image and diffraction plane pictures. In *Image Processing and Computer-Aided Design in Electron Optics* (P. W. Hawkes, Ed.), 66–81 (Academic Press, London and New York).

D. Gerthsen, S. Hettler, M. Dries, B. Gamm, K. Schultheiß, N. Frindt, R. Schröder and J. Zach (2012). Electrostatic Zach phase plates: optimization of properties and applications. *Microsc. Microanal.* **18** (Suppl. 2), 466–467.

J. J. Geuchies, C. van Overbeek, W. H. Evers, B. Goris, A. De Backer, A. P. Gantapara, F. T. Rabouw, J. Hilhorst, J. L. Peters, O. Konovalov, A. V. Petukhov, M. Dijkstra, L. D. A. Siebbels, S. Van Aert, S. Bals and D. Vanmaekelbergh (2016). In situ study of the formation mechanism of two-dimensional superlattices from PbSe nanocrystals. *Nature Mater.* **15**, 1248–1254.

R. M. Glaeser (2013). Methods for imaging weak-phase objects in electron microscopy. *Rev. Sci. Instrum.* **84**, 111101.

R. M. Glaeser and H. Müller (2014). Generalization of the Matsumoto–Tonomura approximation for the phase shift within an open aperture. *Ultramicroscopy* **138**, 1–3.

R. M. Glaeser, S. Sassolini, R. Cambie, J. Jin, S. Cabrini, A. Schmid, R. Danev, B. Buijsse, R. Csencsits, K. H. Downing, D. M. Larson, D. Typke and R. C. Han (2013). Minimizing electrostatic charging of an aperture used to produce in-focus contrast in the TEM. *Ultramicroscopy* **135**, 6–15.

W. Glaser (1956). Elektronen- und Ionenoptik. *Handbuch der Physik* **33**, 123–395.

W. Goldfarb, W. Krakow, D. Ast and B. M. Siegel (1975). Imaging of amorphous objects by tilted beam bright field illumination. EMSA, Las Vegas, NV, **33**, 186–187.

J. Gonnissen, A. De Backer, A. J. den Dekker, G. T. Martinez, A. Rosenauer, J. Sijbers and S. Van Aert (2014). Optimal experimental design for the detection of light atoms from high-resolution scanning transmission electron microscopy images. *Appl. Phys. Lett.* **105**, 063116, 5 pp.

J. Gonnissen, A. De Backer, A. J. den Dekker, J. Sijbers and S. Van Aert (2016). Detecting and locating light atoms from high-resolution STEM images: the quest for a single optimal design. *Ultramicroscopy* **170**, 128–138.

J. Gonnissen, A. De Backer, A. J. den Dekker, J. Sijbers and S. Van Aert (2017). Atom-counting in high resolution electron microscopy: TEM or STEM – that's the question. *Ultramicroscopy* **174**, 112–120.

B. Goris, S. Bals, W. Van den Broek, E. Carbó-Argibay, S. Gómez-Graña, L. M. Liz-Marzán and G. Van Tendeloo (2012a). Atomic-scale determination of surface facets in gold nanorods. *Nature Mater.* **11**, 930–935.

B. Goris, W. Van den Broek, K. J. Batenburg, H. H. Mezerji and S. Bals (2012b). Electron tomography based on a total variation minimization reconstruction technique. *Ultramicroscopy* **113**, 120–130.

B. Goris, A. De Backer, S. Van Aert, S. Gómez-Graña, L. M. Liz-Marzán, G. Van Tendeloo and S. Bals (2013). Three-dimensional elemental mapping at the atomic scale in bimetallic nanocrystals. *Nano Lett.* **13**, 4236–4241.

B. Goris, J. De Beenhouwer, A. De Backer, D. Zanage, K. J. Batenburg, A. Sánchez-Iglesias, L. M. Liz-Marzán, S. Van Aert, S. Bals, J. Sijbers and G. Van Tendeloo (2015). Measuring lattice strain in three dimensions through electron microscopy. *Nano Lett.* **15**, 6996–7001.

T. Grieb, F. F. Krause, K. Müller-Caspary, S. Firoozabadi, C. Mahr, M. Schowalter, A. Beyer, O. Oppermann, K. Volz and A. Rosenauer (2021a). Angle-resolved STEM using an iris aperture: scattering contributions and sources of error for the quantitative analysis in Si. *Ultramicroscopy* **221**, 113175, 11 pp.

T. Grieb, F. F. Krause, K. Müller-Caspary, R. Ritz, M. Simson, J. Schörmann, C. Mahr, J. Müßener, M. Schowalter, H. Soltau, M. Eickhoff and A. Rosenauer (2021b). 4D-STEM at interfaces to GaN: centre-of-mass approach & NBED-disc detection. *Ultramicroscopy* **228**, 113321, 13 pp.

A. Griewank and A. Walther (2008). *Differentiation,* 2nd edn (SIAM, Philadelphia, PA).

B. W. Griffiths, A. V. Jones and I. R. M. Wardell (1973). An ultra-high vacuum scanning electron microscope with Auger analysis facilities. EMAG, Newcastle-upon-Tyne, 42–45.

V. Grillo, P. Lu, H. Soltner, A. Roncaglia, A.H. Tavabi, P. Rosi, C. Menozzi, R. Balboni, S. Frabboni, G. C. Gazzadi, G. Pozzi, R. E. Dunin-Borkowki and E. Karimi (2018). Towards a programmable phase plate for electrons. IMC-19, Sydney.

T. Groves (1975a). Thick specimens in the CEM and STEM. Resolution and image formation. *Ultramicroscopy* **1**, 15–31.

T. Groves (1975b). Plural scattering and thick specimens in transmission electron microscopy. *Ultramicroscopy* **1**, 170–172.

L. Grünewald, D. Gerthsen and S. Hettler (2018). Generation of non-diffracting Bessel beams with amorphous phase masks. IMC-19, Sydney.

M. Gu (1996). *Principles of Three-Dimensional Imaging in Confocal Microscopes* (World Scientific, Singapore).

R. C. Guerrero-Ferreira and E. R. Wright (2014). Zernike phase contrast cryo-electron tomography of whole bacterial cells. *J. Struct. Biol.* **185**, 129–133.

N. Guerrini, R. Turchetta, G. van Hoften, R. Henderson, G. McMullan and A. R. Faruqi (2011). A high frame rate, 16 million pixels, radiation hard CMOS sensor. *J. Instrument.* **6**, C03003, 9 pp.

J. P. Guigay, R. H. Wade and C. Delpla (1971). Optical diffraction of Lorentz microscope images. EMAG, Cambridge, 238–239.

H. Guo, E. Franken, Y. Deng, S. Benlekbir, G. S. Lezcano, B. Janssen, L. Yu, Z. A. Ripstein, Y. Z. Tana and J. L. Rubinstein (2020). Electron-event representation data enable efficient cryoEM file storage with full preservation of spatial and temporal resolution. *IUCrJ* **7**, 860–869.

B. Haas, A. Mittelberger, C. Meyer, B. Plotkin-Swing, N. Dellby, O. Krivanek, T. Lovejoy and C. T. Koch (2021). 4D-STEM live-processing at 15’000 detector images per second for efficient series acquisition. MC-2021, Vienna (virtual), 587–588.

M. Hahn (1973). Theoretische und experimentelle Untersuchungen zum Nachweis von Einzelatomen mit Durchstrahlungs-Elektronenmikroskopen. Dissertation, Düsseldorf.

M. Hahn and J. Seredynski (1974). Imaging single atoms with limited spatial and temporal coherence. ICEM-8, Canberra, **1**, 234–235.

M. Haider, C. Boulin and A. Epstein (1988). A versatile multichannel STEM phase-contrast detector. EUREM-9, York, **1**, 123–124.

M. Haider, A. Epstein, P. Jarron and C. Boulin (1994). A versatile, software configurable, multichannel STEM detector for angle-resolved imaging. *Ultramicroscopy* **54**, 41–59.

S. J. Haigh, H. Sawada and A. I. Kirkland (2009a). Atomic structure imaging beyond conventional resolution limits in the transmission electron microscope. *Phys. Rev. Lett.* **103**, 126101, 4 pp.

S. J. Haigh, H. Sawada and A. I. Kirkland (2009b). Optimal tilt magnitude determination for aberration-corrected super resolution exit wave function reconstruction. *Philos. Trans. R. Soc. Lond. A* **367**, 3755–3771.

R. J. Hall, E. Nogales and R. M. Glaeser (2011). Accurate modeling of single-particle cryo-EM images quantitates the benefits expected from using Zernike phase contrast. *J. Struct. Biol.* **174**, 468–475.

P. Hallégot and N. J. Zaluzec (2004). Scanning confocal electron microscopy of thick biological materials. *Microsc. Microanal.* **10** (Suppl. 2), 1290–1291.

T. Hamaoka, A. Hashimoto, K. Mitsuishi and M. Takeguchi (2018a). 4D-data acquisition in scanning confocal electron microscopy for depth-sectioned imaging. *J. Surf. Sci. Nanotechnol.* **16**, 247–252.

T. Hamaoka, C.-y. Jao and M. Takeguchi (2018b). Annular dark-field scanning confocal electron microscopy studied using multislice simulations. *Microscopy* **67**, 232–243.

M. Hammel and H. Rose (1993). Resolution and optimum conditions for dark-field STEM and CTEM imaging. *Ultramicroscopy* **49**, 81–86.

M. Hammel, H. Kohl and H. Rose (1990). Information enhancement in the STEM by employing a multiple-signals imaging procedure. ICEM-12, Seattle, WA, **1**, 120–121.

K.-J. Hanszen (1966a). Generalisierte Angaben über die Phasenkontrast- und Amplitudenkontrast-Übertragungsfunktionen für elektronenmikroskopische Objektive. *Z. Angew. Phys.* **20**, 427–435.

K.-J. Hanszen (1966b). The phase-contrast transfer-functions of the electron microscopic objective. ICEM-6, Kyoto, **1**, 39–40.

K.-J. Hanszen (1967). Neue Erkenntnisse über Auflösung und Kontrast im elektronenmikroskopischen Bild. *Naturwissenschaften* **54**, 125–133.

K.-J. Hanszen (1969). Problems of image interpretation in electron microscopy with linear and nonlinear transfer. *Z. Angew. Phys.* **27**, 125–131.

K.-J. Hanszen (1971). The optical transfer theory of the electron microscope: fundamental principles and applications. *Adv. Opt. Electron Microsc.* **4**, 1–84.

K.-J. Hanszen (1974). The relevance of dark field illumination in conventional and scanning transmission electron microscopy. PTB-Bericht APh–7, 26 pp.

K.-J. Hanszen (1976). In-line holography in electron microscopy using tilted illumination without an aperture stop. EUREM-6, Jerusalem, **1**, 95–96.

K.-J. Hanszen (1982). Holography in electron microscopy. *Adv. Electron Electron Phys.* **59**, 1–77.

K.-J. Hanszen (1990). 40 Jahre elektronenoptische Forschung in der Physikalisch-Technischen Bundesanstalt. Teil I: Aus der Geschichte des Laboratoriums für Elektronenoptik; Teil II: Kurzberichte aus den PTB-Mitteilungen; Veröffentlichungs- und Vortragslisten. PTB-Bericht Aph–33.

K.-J. Hanszen and G. Ade (1974). The relevance of dark field illumination in conventional and scanning electron microscopy. ICEM-8, Canberra, **1**, 196–197.

K.-J. Hanszen and G. Ade (1976a). A consistent Fourier optical representation of CEM and STEM imaging and holographic reconstruction. EUREM-6, Jerusalem, **1**, 446–447.

K.-J. Hanszen and G. Ade (1976b). Linearität als objektunabhängige Apparateeigenschaft in der Elektronenmikroskopie. *Optik* **44**, 237–249.

K.-J. Hanszen and G. Ade (1977). A consistent Fourier optical representation of image formation in the conventional fixed beam electron microscope, in the scanning transmission electron microscope and of holographic reconstruction. PTB-Bericht APh–11, 31 pp.

K.-J. Hanszen and G. Ade (1978). Phase contrast transfer with different imaging modes in electron microscopy. *Optik* **51**, 119–126.

K.-J. Hanszen and B. Morgenstern (1965). Die Phasenkontrast- und Amplitudenkontrast-Übertragung des elektronenmikroskopischen Objektivs. *Z. Angew. Phys.* **19**, 215–227.

K.-J. Hanszen and L. Trepte (1970). The influence of energy width and of voltage and current fluctuations on contrast transfer and resolution in electron microscopy. ICEM-7, Grenoble, **1**, 45–46.

K.-J. Hanszen and L. Trepte (1971a). Der Einfluß von Strom- und Spannungsschwankungen, sowie der Energiebreite der Strahlelektronen auf Kontrastübertragung und Auflösung des Elektronenmikroskops. *Optik* **32**, 519–538.

K.-J. Hanszen and L. Trepte (1971b). Die Kontrastübertragung im Elektronenmikroskop bei partiell kohärenter Beleuchtung. Teil A: Ringkondensor; Teil B: Ausgedehnte scheibenförmige Strahlquelle und Zweiseitenband-Übertragung. *Optik* **33**, 166–198.

K.-J. Hanszen, B. Morgenstern and K.-J. Rosenbruch (1964). Aussagen der optischen Übertragungstheorie über Auflösung und Kontrast im elektronenmikroskopischen Bild. *Z. Angew. Phys.* **16**, 477–486.

K.-J. Hanszen, K.-J. Rosenbruch and F.-A. Sunder-Plassmann (1965). Die Kontrastübertragung des elektronenmikroskopischen Objektivs bei inkohärenter Beleuchtung und Verwendung einer ringförmigen Apertur. *Z. Angew. Phys.* **18**, 345–350.

Y. Harada, N. Tamura and T. Goto (1975). Application of field emission electron gun to STEM and SEM. EMAG, Bristol, 15–18.

K. Harada, A. Kawaguchi, A. Kotani, Y. Fujibayashi, K. Shimada and S. Mori (2019a). Hollow-cone Foucault imaging. *Kenbikyo* **44** (Suppl. 1), 104.

K. Harada, A. Kawaguchi, A. Kotani, Y. Fujibayashi, K. Shimada and S. Mori (2019b). Hollow-cone Foucault imaging for magnetic structure observations. MC-2019, Berlin, 412.

K. Harada, K. Niitsu, T. Kodama, T. Akashi, Y. A. Ono, D. Shindo, H. Shinada and S. Mori (2019c). Electron Fraunhofer holography. *Kenbikyo* **44** (Suppl. 1), 104.

K. Harada, M. Malac, M. Hayashida, K. Niitsu, K. Shimada, D. Homeniuk and M. Beleggia (2020). Toward the quantitative interpretation of hole-free phase plate images in a transmission electron microscope. *Ultramicroscopy* **209**, 112875, 9 pp.

A. Harscher (1999). Elektronenholographie biologischer Objekte. Dissertation, Tübingen.

P. Hartel, H. Rose and C. Dinges (1996). Conditions and reasons for incoherent imaging in STEM. *Ultramicroscopy* **63**, 93–114.

H. Hashimoto and H. Endoh (1978). In *Electron Diffraction 1927–1977* (P. J. Dobson, J. B. Pendry and C. J. Humphreys, Eds), 188–194 (Institute of Physics, Bristol), Conference Series **41**.

H. Hashimoto, H. Endoh, T. Tanji, A. Ono and E. Watanabe (1977). Direct observation of fine structure within images of atoms in crystals by transmission electron microscopy. *J. Phys. Soc. Jpn.* **42**, 1073–1074.

A. Hashimoto, M. Takeguchi, M. Shimojo, K. Mitsuishi, M. Tanaka and K. Furuya (2008). Development of stage-scanning system for confocal scanning transmission electron microscopy. *e-J Surf. Sci. Nanotechnol.* **6**, 111–114.

A. Hashimoto, M. Shimojo, K. Mitsuishi and M. Takeguchi (2009a). Three-dimensional imaging of carbon nanostructures by scanning confocal electron microscopy. *J. Appl. Phys.* **106**, 086101.

A. Hashimoto, M. Shimojo, K. Mitsuishi and M. Takeguchi (2009b). Three-dimensional observation of carbon nanostructures with confocal scanning transmission electron microscopy. *Microsc. Microanal.* **15** (Suppl. 2), 636–637.

A. Hashimoto, P. Wang, M. Shimojo, K. Mitsuishi, P. D. Nellist, A. I. Kirkland and M. Takeguchi (2012). Three-dimensional analysis of nanoparticles on carbon support using aberration-corrected scanning confocal electron microscopy. *Appl. Phys. Lett.* **101**, 253108.

P. W. Hawkes (1971). To what extent is isoplanatism a restriction in contrast transfer theory? EMAG, Cambridge, 230–232.

P. W. Hawkes (1972). The effect of coma or astigmatism on the contrast transfer function. EUREM-5, Manchester, 398–399.

P. W. Hawkes (1973a). The effect of coma, astigmatism and field curvature on the electron optical transfer function. I. Coma; II. Astigmatism and field curvature. *Optik* **37**, 366–375 and 376–384.

P. W. Hawkes (1973b). Introduction to electron optical transfer theory. In *Image Processing and Computer-Aided Design in Electron Optics* (P. W. Hawkes, Ed.), 2–14 (Academic Press, London and New York).

P. W. Hawkes (1974a). Transforms for discrete electron image processing and filtering. *Optik* **40**, 539–556.

P. W. Hawkes (1974b). An alternative theory of electron image formation avoiding Fourier transforms. ICEM-8, Canberra, **1**, 202–203.

P. W. Hawkes (1975). Mapping two-dimensional convolution products into direct products in finite (Galois) fields. *Optik* **42**, 433–438.

P. W. Hawkes (1978a). Electron image processing: a survey. *Computer Graph. Image Proc.* **8**, 406–442.

P. W. Hawkes (1978b). Half-plane apertures in TEM, split detectors in STEM and ptychography. *J. Opt. (Paris)* **9**, 235–241.

P. W. Hawkes (1978c). Coherence in electron optics. *Adv. Opt. Electron Microsc.* **7**, 101–184.

P. W. Hawkes (1980a). Electron microscope transfer functions in closed form with tilted illumination. *Optik* **55**, 207–212.

P. W. Hawkes (1980b). Units and conventions in electron microscopy, for use in *Ultramicroscopy*. *Ultramicroscopy* **5**, 67–70.

P. W. Hawkes (1980c). Improvements in STEM imaging by special probe and detector shaping techniques. *Scanning Electron Microsc.* (I), 93–98.

P. W. Hawkes (1982a). Electron image processing: 1978–1980. *Comput. Graph. Image Proc.* **18**, 58–96.

P. W. Hawkes (1982b). Is the STEM a ptychograph? *Ultramicroscopy* **9**, 27–30.

P. W. Hawkes (1985). The transfer function of the STEM; its relation to detector strategy. *J. Microsc. Spectrosc. Electron* **10**, 395–398.

P. W. Hawkes (1992). Electron image processing: 1981–1990. *J. Computer-Assist. Microsc.* **4**, 1–72 and **5** (1993) 255.

P. W. Hawkes (2009). Two commercial STEMs: the Siemens ST100F and the AEI STEM-1. *Adv. Imaging Electron Phys.* **159**, 187–219.

P. W. Hawkes (2015). The correction of electron lens aberrations. *Ultramicroscopy* **156**, A1–A64.

P. W. Hawkes and D. McMullan (2004). A forgotten French scanning electron microscope and a forgotten text on electron optics. *Proc. R. Microsc. Soc.* **39**, 285–290.

M. Hayashida, A. M. Najarian, R. McCleery and M. Malac (2018). Hole free phase plate electron tomography in material sciences. *Microsc. Microanal.* **24** (Suppl. 1), 2224–2225.

M. Hayashida, K. Cui, A. Morteza Najarian, R. McCreery, N. Jehanathan, C. Pawlowicz, S. Motoki, M. Kawasaki, Y. Konyuba and M. Malac (2019). Hole free phase plate tomography for materials science samples. *Micron* **116**, 54–60.

G. B. Haydon and R. A. Lemons (1972). Optical shadowing in the electron microscope. *J. Microsc. (Oxford)* **95**, 483–491.

R. Hegerl and W. Hoppe (1970). Dynamische Theorie der Kristallstrukturanalyse durch Elektronenbeugung im inhomogenen Primärstrahlwellenfeld. *Ber. Bunsen-Ges. Phys. Chem.* **74**, 1148–1154.

R. Hegerl and W. Hoppe (1972). Phase evaluation in generalized diffraction. EUREM-5, Manchester, 628–629.

D. Heimes, A. Beyer-Leser, S.S. Firoozabadi, J. Belz and K. Volz (2021). Combining 4DSTEM and multislice simulations to quantify long-range electric fields. MC-2021, Vienna (virtual), 600.

R. Henderson (1995). The potential and limitations of neutrons, electrons and x-rays for atomic resolution microscopy of unstained biological molecules. *Quart. Rev. Biophys.* **28**, 171–193.

R. Henderson (2015). Overview and future of single particle electron cryomicroscopy. *Arch. Biochem. Biophys.* **581**, 19–24.

R. A. Herring (1991). Application of hollow-cone illumination to high-resolution electron microscopy. EMSA, San Jose, CA, **49**, 692–693.

S. Hettler and R. Arenal (2021). Comparative image simulations for phase-plate transmission electron microscopy. *Ultramicroscopy* **227**, 113319, 12 pp.

S. Hettler, B. Gamm, M. Dries, N. Frindt, R. R. Schröder and D. Gerthsen (2012). Improving fabrication and application of Zach phase plates for phase-contrast transmission electron microscopy. *Microsc. Microanal.* **18**, 1010–1015.

S. Hettler, J. Wagner, M. Dries, N. Frindt, R. R. Schröder and D. Gerthsen (2013). Electrostatic Zach phase plate for transmission electron microscopy. MC-2013, Regensburg, **1**, 95–96.

S. Hettler, J. Wagner, M. Dries, M. Oster, R. Schröder and D. Gerthsen (2014a). Application of Zach phase plates for phase-contrast transmission electron microscopy: status and future experiments. *Microsc. Microanal.* **20** (Suppl. 3), 214–215.

S. Hettler, J. Wagner, M. Dries and D. Gerthsen (2014b). Inelastic phase contrast using electrostatic Zach phase plates. *Microsc. Microanal.* **20** (Suppl. 3), 216–217.

S. Hettler, J. Wagner, M. Dries, M. Oster, C. Wacker, R. R. Schröder and D. Gerthsen (2015a). On the role of inelastic scattering in phase-plate transmission electron microscopy. *Ultramicroscopy* **155**, 27–41.

S. Hettler, M. Dries, T. Schulze, M. Oster, C. Wacker, R. Schröder and D. Gerthsen (2015b). High-resolution transmission electron microscopy with Zach phase plate. *Microsc. Microanal.* **21** (Suppl. 3), 1581–1582.

S. Hettler, M. Dries, M. Oster, R. Schröder and D. Gerthsen (2016). High-resolution transmission electron microscopy with an electrostatic phase plate. *New J. Phys.* **18**, 053005, 7 pp.

S. Hettler, M. Dries, P. Hermann, M. Obermair, D. Gerthsen and M. Malac (2017a). Carbon contamination in scanning transmission electron microscopy and its impact on phase-plate applications. *Micron* **96**, 38–47.

S. Hettler, P. Hermann, M. Dries, M. Obermair, D. Gerthsen and M. Malac (2017b). Contamination and charging of amorphous thin films suitable as phase plates for phase-contrast transmission electron microscopy. *Microsc. Microanal.* **23** (Suppl. 1), 830–831.

S. Hettler, E. Kano, M. Dries, D. Gerthsen, L. Pfaffmann, M. Bruns, M. Beleggia and M. Malac (2018). Charging of carbon thin films in scanning and phase-plate transmission electron microscopy. *Ultramicroscopy* **184**, 252–266.

S. Hettler, J. Onoda, R. Wolkow, J. Pitters and M. Malac (2019). Charging of electron beam irradiated amorphous carbon thin films at liquid nitrogen temperature. *Ultramicroscopy* **196**, 161–166.

L. J. Heyderman, M. Kläui, R. Schäublin, U. Rüdiger, C. A. F. Vaz, J. A. C. Bland and C. David (2005). Fabrication of magnetic ring structures for Lorentz electron microscopy. *J. Magn. Magn. Mater.* **290–291**, 86–89.

M. O. Hill, I. Calvo-Almazan, M. Allain, M. V. Holt, A. Ulvestad, J. Treu, G. Koblmüller, C. Huang, X. Huang, H. Yan, E. Nazaretski, Y. S. Chu, G. B. Stephenson, V. Chamard, L. J. Lauhon and S. O. Hruszkewycz (2018). Measuring three-dimensional strain and structural defects in a single InGaAs nanowire using coherent x-ray multiangle Bragg projection ptychography. *Nano Lett.* **18**, 811–819.

A. Hirt and W. Hoppe (1972). Experiences with an electron microscopical installation for high resolution. EUREM-5, Manchester, 12–13.

W. Hoppe (1961). Ein neuer Weg zur Erhöhung des Auflösungsvermögens des Elektronenmikroskops. *Naturwissenschaften* **48**, 736–737.

W. Hoppe (1963). Fresnelsche Zonenkorrekturplatten für das Elektronenmikroskop. *Optik* **20**, 599–606.

W. Hoppe (1969a). Beugung im inhomogenen Primärstrahlwellenfeld. I. Prinzip einer Phasenmessung von Elektronenbeugungsinterferenzen. *Acta Cryst. A* **25**, 495–501.

W. Hoppe (1969b). Beugung im inhomogenen Primärstrahlwellenfeld. III. Amplituden- und Phasenbestimmung bei unperiodischen Objekten. *Acta Cryst. A* **25**, 508–514.

W. Hoppe (1970). Principles of structure analysis at high resolution using conventional electron microscopes and computers. *Ber. Bunsen-Ges. Phys. Chem.* **74**, 1090–1100.

W. Hoppe (1971a). Use of zone correction plates and other techniques for structure determination of aperiodic objects at atomic resolution using a conventional electron microscope. *Philos. Trans. R. Soc. Lond. B* **261**, 71–94.

W. Hoppe (1971b). Zur "Abbildung" komplexer Bildfunktionen in der Elektronenmikroskopie. *Z. Naturforsch.* **26a**, 1155–1168.

W. Hoppe (1974a). Towards three-dimensional "electron microscopy" at atomic resolution. *Naturwissenschaften* **61**, 239–249.

W. Hoppe (1974b). High resolution studies using computerized image reconstruction methods. ICEM-8, Canberra, **1**, 240–241.

W. Hoppe (1982). Trace structure analysis, ptychography, phase tomography. *Ultramicroscopy* **10**, 187–198.

W. Hoppe and D. Köstler (1976). Experimental results in high resolution electron microscopy using the tilt image reconstruction method. EUREM-6, Jerusalem, **1**, 99–104.

W. Hoppe and G. Strube (1969). Beugung im inhomogenen Primärstrahlwellenfeld. II. Lichtoptische Analogieversuche zur Phasenmessung von Gitterinterferenzen. *Acta Cryst. A* **25**, 502–507.

W. Hoppe, R. Langer and F. Thon (1970). Verfahren zur Rekonstruktion komplexer Bildfunktionen in der Elektronenmikroskopie. *Optik* **30**, 538–545.

W. Hoppe, D. Köstler and P. Sieber (1974). Light diffractograms of electron micrographs at tilted illumination. *Z. Naturforsch.* **29a**, 1933–1934.

W. Hoppe, D. Köstler, D. Typke and N. Hunsmann (1975). Kontrastübertragung für die Hellfeld-Bildrekonstruktion mit gekippter Beleuchtung in der Elektronenmikroskopie. *Optik* **42**, 43–56.

R. Horstmeyer and C. Yang (2014). A phase space model of Fourier ptychographic microscopy. *Opt. Express* **22**, 338–358.

R. Horstmeyer, R. Y. Chen, X. Ou, B. Ames, J. A. Tropp and C. Yang (2015). Solving ptychography with a convex relaxation. *New J. Phys.* **17**, 053044, 14 pp.

N. Hosogi, H. Shigematsu, H. Terashima, M. Homma and K. Nagayama (2011). Zernike phase contrast cryo-electron tomography of sodium-driven flagellar hook-based bodies from *Vibrio alginolyticus*. *J. Struct. Biol.* **173**, 67–76.

N. Hosogi, H. Iijima, Y. Konyuba and A. Sen (2015a). Cryo-TEM applications with Zernike and hole-free phase plate. *Microscopy* **64** (Suppl. 1), i126.

N. Hosogi, A. Sen and H. Iijima (2015b). Comparison of cryo TEM images obtained with Zernike and hole-free phase plates. *Microsc. Microanal.* **21** (Suppl. 3), 1389–1390.

F. Hosokawa, R. Danev, Y. Arai and K. Nagayama (2005). Transfer doublet and an elaborated phase plate holder for 120 kV electron-phase microscope. *J. Electron Microsc.* **54**, 317–324.

S. Hosseinnejad, E. G. T. Bosch, H. Kohr, I. Lazić, V. Zharinov, E. Franken, J. Sijbers and J. D. Beenhouwer (2021). 3D atomic resolution tomography from iDPC-STEM images using multiple atom model prior. MC-2021, Vienna (virtual), 544–545.

A. Howie (1972). The reciprocity theorem in electron microscopy and diffraction. EUREM-5, Manchester, 408–413.

A. Howie (1974). Theory of diffraction contrast effects in the scanning electron microscope. In *Quantitative Scanning Electron Microscopy* (D. B. Holt, M. D. Muir, P. R. Grant and I. M. Boswarva, Eds), 183–211 (Academic Press, London and New York).

A. Howie (1978). High resolution electron microscopy of amorphous thin films. *J. Non-Cryst. Sol.* **31**, 41–55.

A. Howie (1979). Image contrast and localized signal selection techniques. *J. Microsc. (Oxford)* **117**, 11–23.

A. Howie (1983). Problems of interpretation in high resolution electron microscopy. *J. Microsc. (Oxford)* **129**, 239–251.

A. Howie, O. Krivanek and M. L. Rudee (1972). Electron microscopy of "amorphous" materials. EUREM-5, Manchester, 450–451.

A. Howie, O. L. Krivanek and M. L. Rudee (1973). Interpretation of electron micrographs and diffraction patterns of amorphous materials. *Philos. Mag.* **27**, 235–255.

S. O. Hruszkewycz, M. Allain, M. V. Holt, C. E. Murray, J. R. Holt, P. H. Fuoss and V. Chamard (2017). High-resolution three-dimensional structural microscopy by single-angle Bragg ptychography. *Nature Mater.* **16**, 244–251.

S.-h. Huang, W.-j. Wang, C.-s. Chang, Y.-k. Hwu, F.-g. Tseng, J.-j. Kai and F.-r. Chen (2006). The fabrication and application of Zernike electrostatic phase plate. *J. Electron Microsc.* **55**, 273–280.

G. Hubert, B. Krisch and D. Willasch (1978). Results on obtaining structural and analytical information with high lateral resolution in a STEM. ICEM-9, Toronto, **1**, 22–23.

C. J. Humphreys (1981). Fundamental concepts of STEM imaging. *Ultramicroscopy* **7**, 7–12.

M. J. Humphry, B. Kraus, A. C. Hurst, A. M. Maiden and J. M. Rodenburg (2012). Ptychographic electron microscopy using high-angle dark-field scattering for sub-nanometre resolution imaging. *Nature Commun.* **3**, 730, 7 pp.

M. Hÿtch and C. Gatel (2021). Phase detection limits in off-axis electron holography from pixelated detectors: gain variations, geometric distortion and failure of reference-hologram correction. *Microscopy* **70**, 47–58.

H. Iijima, N. Hosogi, H. Shigematsu, H. Terashima, M. Homma and K. Nagayama (2010). Zernike phase contrast cryo-electron tomography of flagellar hook-based bodies from *Vibrio alginolyticus*. *Microsc. Microanal.* **16** (Suppl. 2), 558–559.

H. IIjima, S. Motoki, F. Hosokawa and Y. Ohkura (2012). Aberration corrected Zernike phase contrast TEM. *Microsc. Microanal.* **18** (Suppl. 2), 492–493.

H. Iijima, H. Minoda, T. Tamai, Y. Kondo, T. Fukuda and F. Hosokawa (2015). Development of phase contrast scanning transmission electron microscopy. *Microsc. Microanal.* **21** (Suppl. 3), 1943–1944.

H. Iijima, Y. Konyuba, N. Hosogi, Y. Ohkura, H. Jinnai and T. Higuchi (2016). Contrast enhancement of long-range periodic structures using hole-free plate. *Microsc. Microanal.* **22** (Suppl. 3), 60–61.

Y. Inayoshi, H. Minoda, Y. Arai and K. Nagayama (2012). Direct observation of biological molecules in liquid by environmental phase-plate transmission electron microscopy. *Micron* **43**, 1091–1098.

V. Intaraprasonk, H. L. Xin and D. A. Muller (2008). Analytic derivation of optimal imaging conditions for incoherent imaging in aberration-corrected electron microscopes. *Ultramicroscopy* **108**, 1454–1466.

M. Isaacson, M. Utlaut and D. Kopf (1980). Analog computer processing of scanning transmission electron microscope imagers. In *Computer Processing of Electron Microscope Images* (P. W. Hawkes, Ed.), 257–283 (Springer, Berlin and New York).

T. Ishida, A. Shinozaki, M. Kuwahara, T. Miyoshi, K. Saitoh and Y. Arai (2021a). Performance of a silicon-on-insulator direct electron detector in a low-voltage transmission electron microscope. *Microscopy* **70**, 321–325.

T. Ishida, A. Shinozaki, M. Kuwahara, T. Miyoshi, K. Saitoh and Y. Arai (2021b). Performance evaluation of a silicon-on-insulator direct electron detector by a low-voltage TEM. *Kenbikyo* **56** (Suppl. 1), 121.

R. Ishikawa, E. Okunishi, H. Sawada, Y. Kondo, F. Hosokawa and E. Abe (2011). Direct imaging of hydrogen-atom columns in a crystal by annular bright-field electron microscopy. *Nature Mater.* **10**, 278–281.

K. Ishizuka (1980). Contrast transfer of crystal images in TEM. *Ultramicroscopy* **5**, 55–65.

K. Ishizuka (1989). Coherence and simulation of a high-resolution image. In *Computer Simulation of Electron Microscope Diffraction and Images* (W. Krakow and M. O'Keefe, Eds), 43–55 (Minerals, Metals and Materials Society, Warrendale PA).

K. Ishizuka (1990). New form of transmission cross coefficient for high-resolution imaging. ICEM-12, Seattle, WA, **1**, 60–61.

Y. Ito, J. H. Paterson, A. L. Bleloch and L. M. Brown (1992). Electron diffraction from nanostructures fabricated by a finely focussed electron beam. EUREM-10, Granada, **2**, 195–196.

Y. Ito, A. L. Bleloch, J. R. Granleese and L. M. Brown (1993a). Electron phase gratings by electron beam nanolithography. EMAG, Liverpool, 507–510.

Y. Ito, A. L. Bleloch, J. H. Paterson and L. M. Brown (1993b). Electron diffraction from gratings fabricated by electron beam nanolithography. *Ultramicroscopy* **52**, 347–352.

Y. Ito, A. L. Bleloch and L. M. Brown (1998). Nanofabrication of solid-state Fresnel lenses for electron optics. *Nature* **394**, 49–52.

K. Izui, S. Furuno and H. Otsu (1977). Observations of crystal structure images of silicon. *J. Electron Microsc.* **26**, 129–132.

K. Izui, S. Furuno, T. Nishida, H. Otsu and S. Kuwabara (1978). High resolution electron microscopy of images of atoms in silicon crystal oriented in (110). *J. Electron Microsc.* **27**, 171–179.

D. Jannis, C. Hofer, C. Gao, X. Xie, A. Béché, T. J. Pennycook and J. Verbeeck (2022). Event driven 4D STEM acquisition with a Timepix3 detector: microsecond dwell time and faster scans for high precision and low dose applications. *Ultramicroscopy* **233**, 113423, 9 pp.

D. Janoschka, P. Dreher, A. Rödl, T. Franz, O. Schaff, M. Horn-von Hagen and F.-J. Meyer zu Heringdorf (2021). Implementation and operation of a fiber-coupled CMOS detector in a low energy electron microscope. *Ultramicroscopy* **221**, 113180, 6 pp.

R. Janzen, J. Schundelmeier, S. Hettler, M. Dries and D. Gerthsen (2016). Towards understanding of charging effects of conductive thin-film based phase plates. *Microsc. Microanal.* **22** (Suppl. 3), 78–79.

W. K. Jenkins (1979). Contrast transfer in bright field electron microscopy of amorphous specimens. Dissertation, Cambridge.

W. K. Jenkins and R. H. Wade (1977). Contrast transfer in the electron microscope for tilted and conical bright field illumination. EMAG, Glasgow, 115–118.

D. E. Jesson and S. J. Pennycook (1990). Atomic imaging of crystals using large-angle electron scattering in STEM. ICEM-12, Seattle, WA, **1**, 74–75.

D. E. Jesson and S. J. Pennycook (1993). Incoherent imaging of thin specimens using coherently scattered electrons. *Proc. R. Soc. Lond. A* **441**, 261–281.

C.-l. Jia, M. Lentzen and K. Urban (2004). High-resolution transmission electron microscopy using negative spherical aberration. *Microsc. Microanal.* **10**, 174–184.

C.-l. Jia, L. Houben, A. Thust and J. Barthel (2010). On the benefit of the negative-spherical-aberration imaging technique for quantitative HRTEM. *Ultramicroscopy* **110**, 500–505.

C. L. Jia, S. B. Mi, J. Bartel, D. W. Wang, R. E. Dunin-Borkowski, K. W. Urban and A. Thust (2014). Determination of the 3D shape of a nanoscale crystal with atomic resolution from a single image. *Nature Mater.* **13**, 1044–1049.

Y. Jiang, Z. Chen, Y. Han, P. Deb, H. Gao, S. Xie, P. Purohit, M. W. Tate, J. Park, S. M. Gruner, V. Elser and D. A. Muller (2018). Electron ptychography of 2D materials to deep sub-ångström resolution. *Nature* **559**, 343–349.

I. J. Johnson, K. C. Bustillo, J. Ciston, B. R. Draney, P. Ercius, E. Fong, A. Goldschmidt, J. M. Joseph, J. R. Lee, A. M. Minor, C. Ophus, A. Selvarajan, D. E. Skinner, T. Stezelberger, C. S. Tindall and P. Denes (2018). A next generation electron microscopy detector aimed at enabling new scanning diffraction techniques and online data reconstruction. *Microsc. Microanal.* **24** (Suppl. 1), 166–167.

A. V. Jones (1988). New imaging modes in STEM. EMSA, Milwaukee, **46**, 648–649.

L. Jones (2021). Practical aspects of quantitative and high-fidelity STEM data recording. In *Scanning Transmission Electron Microscopy* (A. Bruma, Ed.), 1–39 (CRC Press, Boca Raton, FL).

A. V. Jones and M. Haider (1989). Modular detector system for scanning transmission electron microscope. *Scanning Microsc.* **3** (1), 33–42.

A. V. Jones and K. R. Leonard (1978). Scanning transmission electron microscopy of unstained biological sections. *Nature, 659–660.*

A. V. Jones, J.-C. Homo, B. M. Unitt and N. Webster (1985). The CryoSTEM: a STEM with superconducting objective lens. *J. Microsc. Spectrosc. Electron* **10**, 361–370.

L. Jones, H. Yang, T. J. Pennycook, M. S. J. Marshall, S. Van Aert, N. D. Browning, M. R. Castell and P. D. Nellist (2015). Smart Align—a new tool for robust non-rigid registration of scanning microscope data. *Adv. Struct. Chem. Imaging* **1**, 8, 16 pp.

L. Jones, A. Varambhia, H. Sawada and P. D. Nellist (2018). An optical configuration for fastidious STEM detector calibration and the effect of the objective-lens pre-field. *J. Microsc. (Oxford)* **270**, 176–187.

A. S. Jurling and J. R. Fienup (2014). Applications of algorithmic differentiation to phase retrieval algorithms. *J. Opt. Soc. Amer. A* **31**, 1348–1359.

R. Juškaitis and T. Wilson (1994). Scanning interference and confocal microscopy. *J. Microsc. (Oxford), 188–194.*

F. Kahl and H. Rose (1995). Theoretical concepts of electron holography. *Adv. Imaging Electron Phys.* **94**, 197–257.

F. Kahl, V. Gerheim, L. M., H. Müller, R. Schillinger and S. Uhlemann (2019). Test and characterization of a new post-column imaging energy filter. *Adv. Imaging. Electron Phys.* **212**, 35–70.

K. Kanaya and H. Kawakatsu (1958). Electron phase microscope. ICEM-4, Berlin, **1**, 308–316.

K. Kanaya, Y. Inoue and A. Ishikawa (1954). Image formation of electron microscopes from the view-point of wave-optics. ICEM-3, London, 46–60.

K. Kanaya, H. Kawakatsu and A. Ishikawa (1957). Preliminary experiment on the electron phase microscope. *Bull. Electrotech. Lab.* **21**, 825–833.

K. Kanaya, H. Kawakatsu and H. Yotsumoto (1958a). On the electron phase microscope. *J. Electronmicrosc.* **6**, 1–4.

K. Kanaya, H. Kawakatsu, K. Ito and H. Yotsumoto (1958b). Experiment on the electron phase microscope. *J. Appl. Phys.* **29**, 1046–1049.

K. Kanaya, E. Oho, M. Naka, T. Koyanagi and T. Sasaki (1985). An image processing method for scanning electron microscopy based on the information transmission theory. *J. Electron Microsc. Tech.* **2**, 73–87.

K. Kanaya, N. Baba, E. Oho and T. Sasaki (1986). Image formation of STEM based on the information transmission theory. ICEM-11, Kyoto, **1**, 435–436. Also published in *Proceedings of the International Symposium on Electron Optics, ISEOB* (J.-y. Ximen, Ed.), 435–436 (Institute of Electronics, Academia Sinica, Beijing).

S. Kandel, S. Maddali, M. Allain, S. O. Hruszkewycz, C. Jacobsen and Y. S. G. Nashed (2019). Using automatic differentiation as a general framework for ptychographic reconstruction. *Opt. Express* **27**, 18653–18672.

Y. Kaneko, R. Danev and K. Nitta (2005). In vivo subcellular ultrastructures recognized with Hilbert differential contrast transmission electron microscopy. *J. Electron Microsc.* **54**, 79–84.

Y. Kaneko, R. Danev, K. Nagayama and H. Nakamoto (2006). Intact carboxysomes in a cyanobacterial cell visualized by Hilbert differential contrast transmission electron microscopy. *J. Bacteriol.* **188**, 805–808.

Y. Kaneko, K. Nitta and K. Nagayama (2007). Observation of in vivo DNA in ice embedded whole cyanobacterial cells by Hilbert differential contrast transmission electron microscopy (HDC–TEM). *Plasma & Fusion. Res.* **2**, S1007, 4 pp.

P. L. Kapitza and P. A. M. Dirac (1933). The reflection of electrons from standing light waves. *Proc. Cambridge Philos. Soc.* **29**, 297–300.

M. Kawasaki, M. Malac, P. Li, H. Qian and R. Egerton (2009). Convenient electron optics set up for Zernike phase microscopy in TEM. *Microsc. Microanal.* **15** (Suppl. 2), 1234–1235.

T. Kawasaki, T. Matsutani, T. Ikuta, M. Ichihashi and T. Tanji (2010). Simulation of a hollow cone-shaped probe in aberration-corrected STEM for high-resolution tomography. *Ultramicroscopy* **110**, 1332–1337.

M. Kawasaki, T. Ishida, T. Kodama, T. Tanji and T. Ikuta (2016). STEM phase imaging by annular pixel array detector (A-PAD) combined with quasi-Bessel beam. *Microsc. Microanal.* **22** (Suppl. 3), 480–481.

S. M. Kay (1998). *Fundamentals of Statistical Signal Processing. Vol. II. Detection Theory* (Prentice-Hall, Upper Saddle River, NJ).

S. Keramati, E. Jones, J. Armstrong and H. Batelaan (2020). Partially coherent quantum degenerate electron matter waves. *Adv. Imaging Electron Phys.* **213**, 3–26.

D. Kermisch (1977). Principle of equivalence between scanning and conventional imaging systems. *J. Opt. Soc. Amer.* **67**, 1357–1360.

H. A. Khant, C. Fu, S. Motoki, M. H. Sullivan, G. DeRose and W. Chiu (2015). Optimization of JEM2200FS for Zernike phase contrast Cryo-EM. *Microsc. Microanal.* **21** (Suppl. 3), 1577–1578.

M. Khoshouei, R. Danev, G. Gerisch, M. Ecke, J. Plitzko and W. Baumeister (2014). Phase contrast cryo-electron tomography with a new phase plate. IMC-18, Prague, 3212.

M. Khoshouei, M. Radjainia, A. J. Phillips, J. A. Gerrard, A. K. Mitra, J. M. Plitzko, W. Baumeister and R. Danev (2016). Volta phase plate cryo-EM of the small protein complex Prx3. *Nature Commun.* **7**, 10534.

M. Khoshouei, M. Radjainia, W. Baumeister and R. Danev (2017). Cryo-EM structure of haemoglobin at 3.2Å determined with the Volta phase plate. *Nature Commun.* **8**, 16099, 6 pp.

Y.-m. Kim, S. J. Pennycook and A. Y. Borisevich (2017). Quantitative comparison of bright field and annular bright field imaging modes for characterization of oxygen octahedral tilts. *Ultramicroscopy* **181**, 1–7.

K. Kimoto (2014). Practical aspects of monochromators developed for transmission electron microscopy. *Microscopy* **63**, 337–344.

K. Kimoto, H. Sawada, T. Sasaki, Y. Sato, T. Nagai, M. Ohwada, K. Suenaga and K. Ishizuka (2013). Quantitative evaluation of temporal partial coherence using 3D Fourier transforms of through-focus TEM images. *Ultramicroscopy* **134**, 86–93.

E. J. Kirkland and M. G. Thomas (1996). A high efficiency annular dark field detector for STEM. *Ultramicroscopy* **62**, 79–88.

E. J. Kirkland, R. F. Loane and J. Silcox (1987). Simulation of annular dark field STEM images using a modified multislice method. *Ultramicroscopy* **23**, 77–96.

A. I. Kirkland, W. O. Saxton, K.-l. Chau, K. Tsuno and M. Kawasaki (1995). Super-resolution by aperture synthesis: tilt series reconstruction in CTEM. *Ultramicroscopy* **57**, 355–374.

A. I. Kirkland, W. O. Saxton and G. Chand (1997). Multiple beam tilt microscopy for super resolved imaging. *J. Electron Microsc.* **46**, 11–22.

A. G. Kiselev and M. B. Sherman (1976). Space spectra of electron microscope images obtained by primary beam tilt. *Optik* **46**, 55–60.

M. Kläui, C. A. F. Vaz, L. J. Heyderman, U. Rüdiger and J. A. C. Bland (2005). Spin switching phase diagram of mesoscopic ring magnets. *J. Magn. Magn. Mater.* **290–291**, 61–67.

C. T. Koch and W. Van den Broek (2014). Measuring three-dimensional positions of atoms to the highest accuracy with electrons. *C.R. Phys.* **15**, 119–125.

P. J. B. Koeck (2015). Improved Hilbert phase contrast for transmission electron microscopy. *Ultramicroscopy* **154**, 37–41.

P. J. B. Koeck (2017). An aperture design for single side band imaging in the transmission electron microscope. *Ultramicroscopy* **182**, 81–84.

P. J. B. Koeck (2018). Design of an electrostatic phase shifting device for biological transmission electron microscopy. *Ultramicroscopy* **187**, 107–112.

P. J. B. Koeck (2019). Design of a charged particle beam phase plate for transmission electron microscopy. *Ultramicroscopy* **205**, 62–69.

H. Kohl (1986). A transfer function approach to image formation by inelastically scattered electrons. ICEM-11, Kyoto, **1**, 777–778.

H. Kohl and H. Rose (1985). Theory of image formation by inelastically scattered electrons in the electron microscope. *Adv. Electron Electron Phys.* **65**, 173–227.

A. Köhler (1893). Ein neues Beleuchtungsverfahren für mikrophotographische Zwecke. *Z. Wiss. Mikrosk.* **10**, 433–440, Translated into English as *Proc. R. Microsc. Soc.* **28** (1993) 181–185.

A. Köhler (1899). Beleuchtungsapparat für gleichmäßige Beleuchtung mikroskopischer Objekte mit beliebigem einfarbigen Licht. *Z. Wiss. Mikrosk.* **16**, 1–28.

A. Köhler and W. Loos (1941). Das Phasenkontrastverfahren und seine Anwendungen in der Mikroskopie. *Naturwissenschaften* **29**, 49–61.

H. Koike, Y. Harada, T. Goto, Y. Kokubo, K. Yamada, T. Someya and M. Watanabe (1974). Development of 100 kV field emission scanning electron microscope. ICEM-8, Canberra, **1**, 42–43.

P. C. Konda, L. Loetgering, K. C. Zhou, S. Xu, A. R. Harvey and R. Horstmeyer (2020). Fourier ptychography: current applications and future promises. *Opt. Express* **28**, 9603–9630.

A. P. Konijnenberg, W. Coene, S. F. Pereira and H. P. Urbach (2016). Combining ptychographical algorithms with the hybrid input-output (HIO) algorithm. *Ultramicroscopy* **171**, 41–54.

Y. Konyuba, H. Iijima, Y. Abe, M. Suga and Y. Ohkura (2014a). A thin film Zernike phase plate micro fabricated using MEMS technology. IMC-18, Prague, 1545.

Y. Konyuba, H. Iijima, Y. Abe, M. Suga and Y. Ohkura (2014b). High throughput fabrication process of a Zernike phase plate. *Microsc. Microanal.* **20** (Suppl. 3), 222–223.

Y. Konyuba, H. Iijima, N. Hosogi, Y. Abe, I. Ishikawa and Y. Ohkura (2015). Development of amorphous carbon thin film phase plate. *Microsc. Microanal.* **21** (Suppl. 3), 1573–1574.

M. V. Korzhik, O. V. Misevich and A. A. Fyodorov (1992). $YAlO_3$:Ce scintillators: application for X- and soft γ-ray detection. *Nucl. Instrum. Meth, Phys. Res. B* **72**, 499–501.

A. Kotani, K. Harada, M. Malac, M. Salomons, M. Hayashida and S. Mori (2018a). Observation of FeGe skyrmions by electron phase microscopy with hole-free phase plate. *AIP Adv.* **6**, 055216.

A. Kotani, M. Malac, K. Harada and S. Mori (2018b). Observation of magnetic nanostructures by phase plate microscopy with hole-free phase plate. IMC-19, Sydney.

A. Kotani, M. Malac, K. Harada and S. Mori (2018c). Observation of the magnetic skyrmion by phase microscopy with hole-free phase plate. *Kenbikyo* **53** (Suppl. 1), 118.

M. Krajnak, D. McGrouther, D. Maneuski, V. O'Shea and S. McVitie (2016). Pixelated detectors and improved efficiency for magnetic imaging in STEM differential phase contrast. *Ultramicroscopy* **165**, 42–50.

W. Krakow (1976a). Computer experiments for tilted beam dark-field imaging. *Ultramicroscopy* **1**, 203–221.

W. Krakow (1976b). The use of tilted beam bright-field illumination in high resolution electron microscopy. EMSA, Miami Beach, FL, **34**, 566–567.

W. Krakow (1977). Electronic cone illumination with the conventional transmission electron microscope. EMSA, Boston, MA, **35**, 72–73.
W. Krakow (1978). Applications of electronically controlled illumination in the conventional transmission electron microscope. *Ultramicroscopy* **3**, 291–301.
W. Krakow (1982). The investigation of electron microscope images with partial coherence and virtual apertures. ICEM-10, Hamburg, **1**, 203–204.
W. Krakow (1984). Computer simulation and analysis of high-resolution electron microscope images and diffraction patterns with partial coherence, hollow cone illumination, and virtual apertures. *J. Electron Microsc. Tech.* **1**, 107–130.
W. Krakow and L. A. Howland (1976). A method for producing hollow cone illumination electronically in the conventional transmission microscope. *Ultramicroscopy* **2**, 53–67.
W. Krakow and B. M. Siegel (1975). Phase contrast in electron microscope images with the electrostatic phase plate. *Optik* **42**, 245–268.
W. Krakow, D. C. Ast, W. Goldfarb and B. M. Siegel (1976). Origin of the fringe structure observed in high resolution bright-field electron micrographs of amorphous materials. *Philos. Mag.* **33**, 985–1014.
F. F. Krause and A. Rosenauer (2017). Reciprocity relations in transmission electron microscopy: a rigorous derivation. *Micron* **92**, 1–5.
F. F. Krause, M. Schowalter, T. Mehrtens, K. Müller-Caspary, A. Béché, K. W. H.[sic] van den Bos, S. Van Aert, J. Verbeeck and A. Rosenauer (2016). ISTEM: a realisation of incoherent imaging for ultra-high resolution TEM beyond the classical limit. EMC-16, Lyon, **1**, 501–502.
F. F. Krause, A. Rosenauer and D. Van Dyck (2017). Imaging theory for the ISTEM imaging mode. *Ultramicroscopy* **181**, 107–116.
M. Krielaart and P. Kruit (2019). Potentially programmable virtual phase plate for electron beams. *Microsc. Microanal.* **25** (Suppl. 1), 92–93.
B. Krisch, K.-H. Müller, V. Rindfleisch and R. Schliepe (1975). Signalelektronik für hochauflösende Durchstrahlungs-Rastermikroskopie. *Beiträge zur Elektronenmikroskopischen Direktabbildung von Oberflächen* **8**, 325–330.
B. Krisch, K.-H. Müller, R. Schliepe and D. Willasch (1976a). The ELMISKOP ST 100F, a high performance transmission scanning electron microscope. *Siemens Rev.* **43**, 390–393.
B. Krisch, K.-H. Müller, R. Schliepe and D. Willasch (1976b). ELMISKOP ST100F, ein Durchstrahlungs-Rasterelektronenmikroskop höchster Leistung. *Siemens Z.* **50**, 47–50.
B. Krisch, K.-H. Müller, M. V. Rauch, R. Schliepe and H. M. Thieringer (1977). Analytical microscopy using a high-resolution STEM. *Scanning Electron Microsc., 423–430.*
O. L. Krivanek (1975). The influence of beam intensity on the electron microscope contrast transfer function. *Optik* **43**, 361–372.
O. L. Krivanek (1976). Studies of the envelope of the EM contrast transfer function. EUREM-6, Jerusalem, **1**, 263–264.
O. L. Krivanek (1978). EM contrast transfer functions for tilted illumination imaging. ICEM-9, Toronto, **1**, 168–169.
O. L. Krivanek (2021). Aberration correction in electron microscopy and spectroscopy. *Microsc. Microanal.* **27** (Suppl. 1), 3474–3478.
O. L. Krivanek and A. Howie (1975). Kinematical theory of images from polycrystalline and random-network structures. *J. Appl. Cryst.* **8**, 213–219.
O. L. Krivanek, P. H. Gaskell and A. Howie (1976). Seeing order in 'amorphous' materials. *Nature* **262**, 454–457.
O. Krivanek, N. Dellby, R. J. Keyse, M. Murfitt, C. Own and Z. Szilagyi (2008a). Advances in aberration-corrected scanning transmission electron microscopy and electron spectroscopy. *Adv. Imaging Electron Phys.* **153**, 121–160.
O. L. Krivanek, G. J. Corbin, N. Dellby, B. F. Elston, R. J. Keyse, M. F. Murfitt, C. S. Own, Z. S. Szilagyi and J. W. Woodruff (2008b). An electron microscope for the aberration-corrected era. *Ultramicroscopy* **108**, 179–195.
W. Kühlbrandt (2014). The resolution revolution. *Science* **343**, 1443–1444.
W. Kunath (1976). Probe forming by hollow cone beams. EUREM-6, Jerusalem, **1**, 340–341.

W. Kunath (1979). Signal-to-noise enhancement by superposition of bright-field images obtained under different illumination tilts. *Ultramicroscopy* **4**, 3–7.

W. Kunath and H. Gross (1985). New high resolution imaging methods as tools for the observation of small clusters. *Ultramicroscopy* **16**, 349–356.

W. Kunath and K. Weiss (1980). Hollow cone illumination in bright field imaging of ferritin. EUREM-7, The Hague, **1**, 114–115.

W. Kunath, K. Weiss and E. Zeitler (1981). Experiments in bright-field imaging with hollow-cone illumination. EMSA, Atlanta, GA, **39**, 226–227.

W. Kunath, F. Zemlin and K. Weiss (1985). Apodization in phase-contrast electron microscopy realized with hollow-cone illumination. *Ultramicroscopy* **16**, 123–138.

W. Kunath, P. Gao, P. Schiske and K. Weiss (1986). Suppression of phase-contrast artefacts in high-resolution imaging. ICEM-11, Kyoto, **1**, 775–776.

P. Kurth, S. Pattai, D. Rudolph, J. Wamser and S. Irsen (2012). Silicon-based thin film phase plates for 300 kV: first installation in an FEI Titan Krios. EMC-15, Manchester, **2**, 507–508.

P. Kurth, S. Pattai, D. Rudolph, J. Overbuschmann, J. Wamser and S. Irsen (2014). Artifact-free, long-lasting phase plate. *Microsc. Microanal.* **29** (Suppl. 3), 220–221.

W. Kuypers, M. N. Thompson and W. H. J. Andersen (1973). A scanning transmission electron microscope. *Scanning Electron Microsc., 9–16.*

A. Laberrigue, G. Balossier, A. Beorchia, P. Bonhomme, N. Bonnet and M. Troyon (1980). Traitement direct en m. e. t., deconvolution holographique à l'aide de l'Electrotitus. Utilisation d'un diaphragme de phase de type électrostatique. *J. Microsc. Spectrosc. Electron* **5**, 655–663.

S. Lackovic, M. T. Browne and R. E. Burge (1979). Data recording and replay in STEM. *Scanning Electron Microsc.* (I), 137–144.

M. N. Landauer and J. M. Rodenburg (1995). Experimental tests of double-resolution imaging with quadrant detectors in the STEM. EMAG, Birmingham, 281–284.

M. N. Landauer, B. C. McCallum and J. M. Rodenburg (1995). Double resolution imaging of weak phase specimens with quadrant detectors in the STEM. *Optik* **100**, 37–46.

R. Langer and W. Hoppe (1967a). Die Erhöhung von Auflösung und Kontrast im Elektronenmikroskop mit Zonenkorrekturplatten. II. Der Einfluß von Strahlspannung und von sekundären Abbildungsfehlern. *Optik* **25**, 413–428.

R. Langer and W. Hoppe (1967b). Die Erhöhung von Auflösung und Kontrast im Elektronenmikroskop mit Zonenkorrekturplatten. III. Abbildung von Atomen bei nahezu inkohärenter beleuchtung. *Optik* **25**, 507–522.

A. Lannes, M. Tanaka and P. Temple (1981). Point-spread and transfer functions for low-contrast objects in bright-field microscopy. *Optik* **60**, 1–28.

I. Lazić and E. G. T. Bosch (2017). Analytical review of direct STEM imaging techniques for thin samples. *Adv. Imaging Electron Phys.* **199**, 75–184.

I. Lazić, E. G. T. Bosch and S. Lazar (2016). Phase contrast STEM for thin samples: integrated differential phase contrast. *Ultramicroscopy* **160**, 265–280.

L. [sic] Léauté (1946). Les applications du microscope électronique å la métallurgie. In *l'Optique Electronique* (L. de Broglie, Ed.), 209–220 (Editions de la Revue d'Optique Théorique et Instrumentale, Paris).

J. M. LeBeau and S. Stemmer (2008). Experimental quantification of annular dark-field images in scanning transmission electron microscopy. *Ultramicroscopy* **108**, 1635–1658.

S. Lee, Y. Oshima, E. Hosono, H. Zhou and K. Takayanagi (2013). Reversible contrast in focus series of annular bright field images of a crystalline $LiMn_2O_4$ nanowire. *Ultramicroscopy* **125**, 43–48.

Z. Lee, U. Kaiser and H. Rose (2017). Advantages of nonstandard imaging methods in TEM and STEM. MC-2017, Lausanne, 509–510.

Z. Lee, U. Kaiser and H. Rose (2019a). Prospects of annular differential phase contrast applied for optical sectioning in STEM. *Ultramicroscopy* **196**, 58–66.

Z. Lee, Y. Li, J. Biskupek, H. Rose and U. Kaiser (2019b). TEM imaging with hollow-cone bright-field illumination at low voltages. MC-2019, Berlin, 438.

M. Lentzen (2004). The tuning of a Zernike phase plate with defocus and variable spherical aberration and its use in HRTEM imaging. *Ultramicroscopy* **99**, 211–220.

M. Lentzen (2006). Progress in aberration-corrected high-resolution transmission electron microscopy using hardware aberration correction. *Microsc. Microanal.* **12**, 181–205.

M. Lentzen (2008). Contrast transfer and resolution limits for sub-Angstrom high-resolution transmission electron microscopy. *Microsc. Microanal.* **14**, 16–26.

M. Lentzen and A. Thust (2005). Contrast transfer theory for transmission electron microscopes equipped with a Wien-filter monochromator. *Microsc. Microanal.* **11** (Suppl. 2), 2154–2155.

M. Lentzen and K. Urban (2006). Contrast transfer and resolution limits for sub-Ångström high-resolution transmission electron microscopy. *Microsc. Microanal.* **12** (Suppl. 2), 1456–1457.

M. Lentzen, B. Jahnen, C. L. Jia, A. Thust, K. Tillmann and K. Urban (2002). High-resolution imaging with an aberration-corrected transmission electron microscope. *Ultramicroscopy* **92**, 233–242.

M. Lentzen, C.-l. Jia and K. Urban (2003a). Atomic structure imaging using an aberration-corrected transmission electron microscope. *Microsc. Microanal.* **9** (Suppl. 3), 48–49.

M. Lentzen, C.-l. Jia and K. Urban (2003b). HRTEM imaging of weakly scattering atom columns using a negative spherical aberration combined with an overfocus. *Microsc. Microanal.* **9** (Suppl. 2), 932–933.

M. Lentzen, A. Thust and K. Urban (2004). The error of aberration measurement in HRTEM using Zemlin tableaus. *Microsc. Microanal.* **10** (Suppl. 2), 980–981.

F. Lenz (1963). Zonenplatten zur Öffnungsfehlerkorrektur und zur Kontrasterhöhung. *Z. Phys.* **172**, 498–502.

F. Lenz (1964). Dimensionierung und Anordnung von Hoppe-Platten und Kontrastplatten in starken Magnetlinsen. *Optik* **21**, 489–493.

F. Lenz (1965). The influence of lens imperfections on image formation. *Lab. Invest.* **14**, 808–818.

F. A. Lenz (1971a). Transfer of image information in the electron microscope. In *Electron Microscopy in Material Science* (U. Valdrè, Ed.), 540–569 (Academic Press, New York and London).

F. Lenz (1971b). Fourier electron optics. EMAG, Cambridge, 224–229.

F. Lenz and W. Scheffels (1958). Das Zusammenwirken von Phasen- und Amplitudenkontrast in der elektronenmikroskopischen Abbildung. *Z. Naturforsch.* **13a**, 226–230.

Z. G. Li and D. Dorignac (1986). High resolution STEM: CTF optimization by detector modulation. ICEM-11, Kyoto, **1**, 427–428.

Z. Li, J. Biskupek, U. Kaiser and H. Rose (2021). A novel method for improving resolution and image contrast in integrated differential phase-contrast (IDPC)-STEM. MC-2021, Vienna (virtual), 554–555.

M. Liao, E. Cao, D. Julius and Y. Cheng (2013). Structure of the TRPV1 ion channel determined by electron cryo-microscopy. *Nature* **504**, 107–112.

E. H. Linfoot (1956). Transmission factors and optical design. *J. Opt. Soc. Amer.* **46**, 740–752.

E. H. Linfoot (1957). Image quality and optical resolution. *Opt. Acta* **4**, 12–16.

E. H. Linfoot (1960). *Qualitätsbewertung optischer Bilder* (Vieweg, Braunschweig).

E. H. Linfoot (1964). *Fourier Methods in Optical Image Evaluation* (Focal Press, London and New York).

J. Liu (2021). Advances and applications of atomic-resolution scanning transmission electron microscopy. *Microsc. Microanal.* **27**, 943–995.

J. Liu and J. M. Cowley (1993). High-resolution scanning transmission electron microscopy. *Ultramicroscopy* **52**, 335–346.

R. F. Loane, E. J. Kirkland and J. Silcox (1988). Visibility of single heavy atoms on thin crystalline silicon in simulated annular dark-field STEM images. *Acta Cryst. A* **44**, 912–927.

R. F. Loane, P. Xu and J. Silcox (1992). Incoherent imaging of zone axis crystals with ADF STEM. *Ultramicroscopy* **40**, 121–138.

J.-P. Locquet, J. Perret, J. Fompeyrine, E. Mächler, J. W. Seo and G. Van Tendeloo (1998). Doubling the critical temperature of $La_{1.9}Sr_{0.1}CuO_4$ using epitaxial strain. *Nature* **394**, 453–456.

M. Locquin (1954). L'influence des modifications pupillaires de l'objectif sur les contrastes en microscopie électronique. ICEM-3, London, 285–289.

M. Locquin (1955). Premiers essais de contraste de phase en microscopie électronique. *Z. Wiss. Mikrosk.* **62**, 220–223.

M. Locquin (1956). Contraste de phase et contraste interchromatique. Etude comparée des méthodes. EUREM-1, Stockholm, 78–79.

S. Löffler, D. Boya and M. Stöger-Pollach (2021). Higher moments in DPC. MC-2021, Vienna (virtual), 567.

S. Lopatin, Yu. P. Ivanov, J. Kosel and A. Chuvilin (2016). Multiscale differential phase contrast analysis with a unitary detector. *Ultramicroscopy* **162**, 74–81.

A. Lubk (2018). Holography and tomography with electrons. *Adv. Imaging Electron Phys.* **206**, 1–318.

X. Ma, Z. Zhang, M. Yao, J. Peng and J. Zhong (2018). Spatially-incoherent annular illumination microscopy for bright-field optical sectioning. *Ultramicroscopy* **195**, 74–84.

K. E. MacArthur, A. J. D'Alfonso, D. Ozkaya, L. J. Allen and P. D. Nellist (2015). Optimal ADF STEM imaging parameters for tilt-robust image quantification. *Ultramicroscopy* **156**, 1–8.

I. Maclaren, E. Frutos-Myro, D. McGrouther, S. McFadzean, J. K. Weiss, D. Cosart, J. Portillo, A. Robins, S. Nicolopoulos, E. N. del Busto and R. Skogeby (2020). A comparison of a direct electron detector and a high-speed video camera for a scanning precession electron diffraction phase and orientation mapping. *Microsc. Microanal.* **26**, 1110–1116.

J. Mahamid, S. Pfeffer, M. Schaffer, E. Villa, R. Danev, L. K. Cuellar, F. Förster, A. A. Hyman, J. M. Plitzko and W. Baumeister (2016). Visualizing the molecular sociology at the HeLa cell nuclear periphery. *Science* **351**, 969–972.

C. Mahr, F. F. Krause, B. Gerken, M. Schowalter, D. Van Dyck and A. Rosenauer (2018). ISTEM – strongly incoherent imaging for ultra-high resolution TEM. IMC-19, Sydney.

A. M. Maiden, M. J. Humphry and J. M. Rodenburg (2012). Ptychographic transmission microscopy in three dimensions using a multi-slice approach. *J. Opt. Soc. Amer. A* **29**, 1606–1614.

A. M. Maiden, M. C. Sarahan, M. D. Stagg, S. M. Schramm and M. J. Humphry (2015). Quantitative electron phase imaging with high sensitivity and an unlimited field of view. *Sci. Rep.* **5**, 14690, 8 pp.

A. Maiden, D. Johnson and F. Li (2017). Further improvements to the ptychographical iterative engine. *Optica* **4**, 736–745.

E. Majorovits (2002). Assessing phase contrast electron microscopy with electrostatic and Zernike-type phase plates. Dissertation, Heidelberg.

E. Majorovits and R. R. Schröder (2002). Improved information recovery in phase contrast EM for non-twofold symmetric Boersch phase plate geometry. ICEM-15, Durban, **3**, 305–306.

E. Majorovits, B. Barton, K. Schultheiß, F. Pérez-Willard, D. Gerthsen and R. R. Schröder (2007). Optimizing phase contrast in transmission electron microscopy with an electrostatic (Boersch) phase plate. *Ultramicroscopy* **107**, 213–226.

M. Malac (2018). Progress toward quantitative hole-free phase plate imaging in a TEM. IMC-19, Sydney.

M. Malac, M. Beleggia, R. Egerton and Y. Zhu (2008). Imaging of radiation sensitive samples in transmission electron microscopes equipped with Zernike phase plates. *Ultramicroscopy* **108**, 126–140.

M. Malac, M. Kawasaki, M. Beleggia, P. Li and R. F. Egerton (2010). Convenient contrast enhancement by hole-free phase plate in a TEM. *Microsc. Microanal.* **16** (Suppl. 2), 526–527.

M. Malac, M. Beleggia, M. Kawasaki, P. Li and R. Egerton (2012a). Convenient contrast enhancement by a hole-free phase plate. *Ultramicroscopy* **118**, 77–89.

M. Malac, M. Bergen, R. Egerton, M. Kawasaki, M. Beleggia, H. Furukawa and M. Shimizu (2012b). Practical hole-free phase plate imaging: principles, advantages and pitfalls. *Microsc. Microanal.* **18** (Suppl. 2), 484–485.

M. Malac, M. Beleggia, R. Egerton, M. Kawasaki, M. Berge, Y. Okura, I. Ishikawa and S. Motoki (2014a). Charging of thin film phase plates under electron beam irradiation. *Microsc. Microanal.* **20** (Suppl. 3), 230–231.

M. Malac, M. Kawasaki, M. Beleggia, S. Pollard, Y. Zhu, R. Egerton and Y. Okura (2014b). Operating principles and practical applications of hole-free phase plate. IMC-18, Prague, 1720.

M. Malac, M. Beleggia, T. Rowan, R. Egerton, M. Kawasaki, Y. Okura and R. A. McLeod (2015). Electron beam-induced charging and modifications of thin films. *Microsc. Microanal.* **21** (Suppl. 3), 1385–1388.

M. Malac, S. Hettler, M. Hayashida, M. Kawasaki, Y. Konyuba, Y. Okura, H. Iijima, I. Ishikawa and M. Beleggia (2017a). Computer simulations analysis for determining the polarity of charge generated by high energy electron irradiation of a thin film. *Micron* **100**, 10–22.
M. Malac, E. Kano, M. Hayashida, M. Kawasaki, S. Motoki, R. Egerton, I. Ishikawa, Y. Okura and M. Beleggia (2017b). Hole-free phase plate energy filtering imaging of graphene: toward quantitative hole-free phase plate imaging in a TEM. *Microsc. Microanal.* **23** (Suppl. 1), 842–843.
M. Malac, K. Harada, M. Hayashida, K. Shimada, K. Niitsu, T. Rowan and M. Beleggia (2018a). Observation of interaction between direct and diffracted electron beams by phase grating with hole free phase plate. *Kenbikyo* **53** (Suppl. 1), 118.
M. Malac, M. Hayashida, K. Harada, K. Shimada, H. Niimi, T. Rowan and M. Beleggia (2018b). Hole-free phase plate imaging of a phase grating. *Microsc. Microanal.* **24** (Suppl. 1), 894–895.
M. Malac, S. Hettler, M. Hayashida, E. Kano, R. F. Egerton and M. Beleggia (2021). Phase plates in the transmission electron microscope: operating principles and applications. *Microscopy* **70**, 75–115.
M. Marko and C. Hsieh (2015). Initial experience with the Volta phase plate. *Microsc. Microanal.* **21** (Suppl. 3), 1579–1580.
M. Marko, A. Leith, C. Hsieh and R. Danev (2011). Retrofit implementation of Zernike phase plate imaging for cryo-TEM. *J. Struct. Biol.* **174**, 400–412.
M. Marko, X. Meng, C. Hsieh, J. Roussie and C. Striemer (2013). Methods for testing Zernike phase plates and a report on silicon-based phase plates with reduced charging and improved ageing characteristics. *J. Struct. Biol.* **184**, 237–244.
M. Marko, C. Hsieh, E. Leith, D. Mastronarde and S. Motoki (2016). Practical experience with hole-free phase plates for cryo electron microscopy. *Microsc. Microanal.* **22**, 1316–1328.
D. Marquardt, M. Schowalter, F. F. Krause, T. Grieb, C. Mahr, T. Mehrtens and A. Rosenauer (2021). Accuracy and precision of position determination in ISTEM imaging of $BaTiO_3$. *Ultramicroscopy* **227**, 113325, 8 pp.
G. T. Martinez, A. Rosenauer, A. De Backer, J. Verbeeck and S. Van Aert (2014a). Quantitative composition determination at the atomic level using model-based high-angle annular dark field scanning transmission electron microscopy. *Ultramicroscopy* **137**, 12–19.
G. T. Martinez, A. De Backer, A. Rosenauer, J. Verbeeck and S. Van Aert (2014b). The effect of probe inaccuracies on the quantitative model-based analysis of high angle annular dark field scanning transmission electron microscopy images. *Micron* **63**, 57–63.
B. März, A. Bangun, O. Melnyk, A. Clausen, D. Weber, R. E. Dunin-Borkowski, F. Frank and K. Müller-Caspary (2021). Comparison of phase retrieval algorithms applied to electron ptychography. MC-2021, Vienna (virtual), 605–606.
T. Matsumoto and A. Tonomura (1996). The phase constancy of electron waves traveling through Boersch's electrostatic phase plate. *Ultramicroscopy* **63**, 5–10.
G. Matteucci, G. F. Missiroli and G. Pozzi (1982). A new electrostatic phase-shifting effect. *Ultramicroscopy* **10**, 247–252.
G. Matteucci, F. F. Medina and G. Pozzi (1992). Electron-optical analysis of the electrostatic Aharonov–Bohm effect. *Ultramicroscopy* **41**, 255–268.
B. C. McCallum and J. M. Rodenburg (1992). Two-dimensional demonstration of Wigner phase-retrieval microscopy in the STEM configuration. *Ultramicroscopy* **45**, 371–380.
B. McCallum and J. Rodenburg (1993a). Simultaneous reconstruction of object and aperture functions from multiple far-field intensity measurements. *J. Opt. Soc. Amer. A* **10**, 231–239.
B. McCallum and J. M. Rodenburg (1993b). Error analysis of crystalline ptychography in the STEM mode. *Ultramicroscopy* **52**, 85–99.
B. C. McCallum, M. N. Landauer and J. M. Rodenburg (1995). Complex image reconstruction of weak specimens from a three-sector detector in the STEM. *Optik* **101**, 53–62.
S. C. McFarlane (1975). The imaging of amorphous specimens in a tilted-beam electron microscope. *J. Phys. C: Solid State Phys* **8**, 2819–2836.
S. C. McFarlane and W. Cochrane (1975). Lattice fringes from amorphous Ge: fact or artefact? *J. Phys. C: Solid State Phys* **8**, 1311–1321.

S. McGregor, W. C.-w. Huang, B. A. Shadwick and H. Batelaan (2015). Spin-dependent two-color Kapitza–Dirac effects. *Phys. Rev. A* **92**, 023834, 11 pp.

R. A. McLeod and M. Malac (2013). Characterization of detector modulation-transfer function with noise, edge, and holographic methods. *Ultramicroscopy* **129**, 42–52.

G. McMullan and A. R. Faruqi (2008). Electron microscope imaging of single particles using the Medipix2 detector. *Nucl. Instrum. Meth, Phys. Res. A* **591**, 129–133.

G. McMullan, D. M. Cattermole, S. Chen, R. Henderson, X. Llopart, C. Summerfield, L. Tlustos and A. R. Faruqi (2007). Electron imaging with Medipix2 hybrid pixel detector. *Ultramicroscopy* **107**, 401–413.

G. McMullan, S. Chen, R. Henderson and A. R. Faruqi (2009a). Detective quantum efficiency of electron area detectors in electron microscopy. *Ultramicroscopy* **109**, 1126–1143.

G. McMullan, A. T. Clark, R. Turchetta and A. R. Faruqi (2009b). Enhanced imaging in low dose electron microscopy using electron counting. *Ultramicroscopy* **109**, 1411–1416.

G. McMullan, A. R. Faruqi, D. Clare and R. Henderson (2014). Comparison of optimal performance at 300 keV of three direct electron detectors for use in low dose electron microscopy. *Ultramicroscopy* **147**, 156–163.

G. McMullan, A. R. Faruqi and R. Henderson (2016). Direct electron detectors. In *The Resolution Revolution: Recent Advances in CryoEM* (R. A. Crowther, Ed.), 1–17; *Methods in Enzymology* vol. 579.

S. McVitie and J. N. Chapman (1990). Measurement of domain wall widths in permalloy using differential phase contrast imaging in STEM. *J. Magn. Magn. Mater.* **83**, 97–98.

S. McVitie, D. McGrouther, S. McFadzean, D. A. Maclaren, K. J. O'Shea and M. J. Benitez (2015). Aberration corrected Lorentz scanning transmission electron microscopy. *Ultramicroscopy* **152**, 57–62.

S. McVitie, S. Hughes, K. Fallon, S. McFadzean, D. McGrouther, M. Krajnak, W. Legrand, D. Maccariello, S. Collin, K. Garcia, N. Reyren, V. Cros, A. Fert, K. Zeissler and C. H. Marrows (2018). A transmission electron microscope study of Néel skyrmion magnetic textures in multilayer thin film systems with large interfacial chiral interaction. *Sci. Rep.* **8**, 5703, 10 pp.

R. R. Meyer, A. I. Kirkland, R. E. Dunin-Borkowski and J. L. Hutchison (2000). Experimental characterisation of CCD cameras for HREM at 300 kV. *Ultramicroscopy* **85**, 9–13.

A.-C. Milazzo, P. Leblanc, F. Duttweiler, L. Jin, J. C. Bouwer, S. Peltier, M. Ellisman, F. Bieser, H. S. Matis, H. Wieman, P. Denes, S. Kleinfelder and N.-h. Xuong (2005). Active pixel sensor array as a detector for electron microscopy. *Ultramicroscopy* **104**, 152–159.

A.-C. Milazzo, G. Moldovan, J. Lanman, L. Jin, J. C. Bouwer, S. Klienfelder, S. T. Peltier, M. H. Ellisman, A. I. Kirkland and N.-h. Xuong (2010). Characterization of a direct detection device imaging camera for transmission electron microscopy. *Ultramicroscopy* **110**, 741–744.

A.-C. Milazzo, A. Cheng, A. Moeller, D. Lyumkis, E. Jacovetty, J. Polukas, M. H. Ellisman, N.-h. Xuong, B. Carragher and C. S. Potter, (2011). Initial evaluation of a direct detection device detector for single particle cryo-electron microscopy. *J. Struct. Biol.* **176**, 404–408.

H. Minoda, T. Okabe and H. Iijima (2011). Contrast enhancement in the phase plate transmission electron microscopy using an objective lens with a long focal length. *J. Electron Microsc.* **60**, 337–343.

H. Minoda, A. Yada, Y. Kawana, H. Iijima and Y. Konyuba (2013). Development of a new type of thin film phase plate and its application for in-situ observation. *Microsc. Microanal.* **19** (Suppl. 2), 478–479.

H. Minoda, T. Tamai, H. Iijima, F. Hosokawa and Y. Kondo (2014). First demonstration of phase contrast scanning transmission electron microscopy. *Microsc. Microanal.* **20** (Suppl. 3), 224–225.

H. Minoda, T. Tamai, H. Iijima, F. Hosokawa and Y. Kondo (2015a). Phase-contrast scanning transmission electron microscopy. *Microscopy* **64**, 181–187.

H. Minoda, T. Tamai, H. Iijima and Y. Kondo (2015b). Development of phase contrast scanning transmission electron microscopy and its application. *Microsc. Microanal.* **21** (Suppl. 3), 2301–2302.

H. Minoda, T. Tamai, H. Iijima and Y. Kondo (2016). Contrast enhancement of nano-materials using phase plate STEM. *Microsc. Microanal.* **22** (Suppl. 3), 62–63.

H. Minoda, T. Tamai, Y. Ohmori and H. Iijima (2017). Contrast enhancement of nanomaterials using phase plate STEM. *Ultramicroscopy* **182**, 163–168.

H. Minoda, Y. Ohmori and K. Yunoki (2018a). Design of phase plate for high contrast imaging in phase plate scanning transmission electron microscopy. IMC-19, Sydney.

H. Minoda, Y. Ohmori and K. Yunoki (2018b). Design of phase plate for high contrast imaging in phase plate electron microscopy. *Kenbikyo* **53** (Suppl. 1), 15.

J. Mir, J. Mir [sic], R. Clough, R. MacInnes, C. Gough, R. Plackett, H. Sawada, I. MacLaren, D. Maneuski, V. O'Shea, D. McGrouther and A. Kirkland (2016). Characterisation of the Medipix3 detector for electron imaging. EMC-16, Lyon, **1**, 350–351.

J. A. Mir, R. Clough, R. MacInnes, C. Gough, R. Plackett, I. Shipsey, H. Sawada, I. MacLaren, R. Ballabriga, D. Maneuski, V. O'Shea, D. McGrouther and A. I. Kirkland (2017). Characterisation of the Medipix3 detector for 60 and 80 keV electrons. *Ultramicroscopy* **182**, 44–53.

D. L. Misell (1971). Image formation in the electron microscope. I. The application of transfer theory to a consideration of elastic electron scattering. II. The application of transfer theory to a consideration of inelastic electron scattering. *J. Phys. A: Gen. Phys.* **4**, 782–797 and 798–812.

D. L. Misell (1973). Image resolution and image contrast in the electron microscope. I. Elastic scattering and coherent illumination. II. Elastic scattering and incoherent illumination. *J. Phys. A: Math. Nucl. Gen.* **6**, 62–78 and 205–217.

D. L. Misell (1977). Conventional and scanning transmission electron microscopy: image contrast and radiation damage. *J. Phys. D: Appl. Phys* **10**, 1085–1107.

D. L. Misell and A. J. Atkins (1973). Image resolution and image contrast in the electron microscope. III. Inelastic scattering and coherent illumination. *J. Phys. A: Math. Nucl. Gen.* **6**, 218–235.

D. L. Misell, G. W. Stroke and M. Halioua (1974). Coherent and incoherent imaging in the scanning transmission electron microscope. *J. Phys. D: Appl. Phys* **7**, L113–L117.

K. Mitsuishi and T. Sannomiya (2021). Effect and correction of intermediate lens defocus on phase by ptychographical iterative engine. *Kenbikyo* **56** (Suppl. 1), 112.

K. Mitsuishi and M. Takeguchi (2015). Scanning confocal electron microscopy. In *Scanning Transmission Electron Microscopy of Materials* (N. Tanaka, Ed.), 345–382 (Imperial College Press, London).

K. Mitsuishi, K. Iakoubovskii, M. Takeguchi, M. Shimojo, A. Hashimoto and K. Furuya (2008). Bloch wave-based calculation of imaging properties of high-resolution scanning confocal electron microscopy. *Ultramicroscopy* **108**, 981–988.

K. Mitsuishi, A. Hashimoto, M. Takeguchi, M. Shimojo and K. Ishizuka (2010). Imaging properties of bright-field and annular-dark-field scanning confocal electron microscopy. *Ultramicroscopy* **111**, 20–26.

K. Mitsuishi, A. Hashimoto, M. Takeguchi, M. Shimojo and K. Ishizuka (2012). Imaging properties of bright-field and annular-dark-field scanning confocal electron microscopy: II. Point spread function analysis. *Ultramicroscopy* **112**, 53–60.

G. Möllenstedt, R. Speidel, W. Hoppe, R. Langer, K.-H. Katerbau and F. Thon (1968). Electron microscopical imaging using zonal correction plates. EUREM-4, Rome, **1**, 125–126.

G. R. Morrison and J. N. Chapman (1981). STEM imaging with a quadrant detector. EMAG, Cambridge, 329–332.

G. R. Morrison and J. N. Chapman (1982). A comparison of split and quadrant detector systems for STEM imaging. ICEM-10, Hamburg, **1**, 211–212.

G. R. Morrison and J. N. Chapman (1983). A comparison of three differential phase contrast systems suitable for use in STEM. *Optik* **64**, 1–12.

G. R. Morrison, J. N. Chapman and A. J. Craven (1979). Applications of a STEM equipped with a quadrant detector. EMAG, Brighton, 257–260.

C. Mory and C. Colliex (1985). Experimental study of the resolution, contrast and signal/noise ratio in different STEM imaging modes. *J. Microsc. Spectrosc. Electron* **10**, 389–394.

C. Mory, C. Colliex and J. M. Cowley (1987a). Optimum defocus for STEM imaging and microanalysis. *Ultramicroscopy* **21**, 171–177.

C. Mory, N. Bonnet, C. Colliex, H. Kohl and M. Tencé (1987b). Evaluation and optimization of the performance of elastic and inelastic scanning transmission electron microscope imaging by correlation analysis. *Scanning Microsc.* (Suppl. 2), 329–342.

A. Mostaed, A. I. Kirkland and P. D. Nellist (2021). The correlation between ptychographic phase and ADF intensity – a new approach for quantitative STEM. MC-2021, Vienna (virtual), 596–597.

S. Motoki, F. Hosokawa, Y. Arai, R. Danev and Y. Nagatani (2005). 200 kV TEM with a Zernike phase plate. *Microsc. Microanal.* **11** (Suppl. 2), 708–709.

S. Motoki, Y. Fukuda, H. Suga, Y. Okura, R. Danev, J. Brink and B. L. Armbruster (2010). Design evolution of the Zernike phase contrast transmission electron microscope. *Microsc. Microanal.* **16** (Suppl. 2), 530–531.

D. A. Muller (2021). Imaging atoms and fields by electron ptychography. MC-2021, Vienna (virtual), 8–9.

K.-H. Müller (1971). Elektronen-Mikroschreiber mit gechwindigkeitsgesteuerter Strahlführung. I, II. *Optik* **33**, 296–311 and 331–343.

K.-H. Müller (1976). Phasenplatten für Elektronenmikroskope. *Optik* **45**, 73–85.

K.-H. Müller and V. Rindfleisch (1971). Applications of a micro-recording device. EMSA, Boston, MA, **29**, 48–49.

H. Müller, J. Jin, R. Danev, J. Spence, H. Padmore and R. M. Glaeser (2010). Design of an electron microscope phase plate using a focused continuous-wave laser. *New J. Phys.* **12**, 073011, 10 pp.

H. Müller, V. Gerheim, J. Zach and M. Haider (2012). A quadrupole optics with large aspect ratio for an anamorphotic electrostatic phase plate without beam blocking. *Microsc. Microanal.* **18** (Suppl. 2), 494–495.

K. Murata, X. Liu, R. Danev, J. Jakana, M. F. Schmid, J. King, K. Nagayama and W. Chiu (2010). Zernike phase contrast cryo-electron microscopy and tomography for structure determination at nanometer and subnanometer resolutions. *Structure* **18**, 903–912.

K. Murata, N. Miyazaki and K. Nagayama (2015). Zernike phase contrast electron microscopy: observation of the image formation and improvement of the image quality using direct detector. *Microsc. Microanal.* **21** (Suppl. 3), 2141–2142.

Y. Nagatani and K. Nagayama (2011). Complex observation in electron microscopy. VII. Iterative phase retrieval for strong-phase objects by plural Hilbert differential contrast experiments. *J. Phys. Soc. Jpn.* **80**, 094402, 8 pp.

K. Nagayama (1999a). Complex observation in electron microscopy. I. Basic scheme to surpass the Scherzer limit. *J. Phys. Soc. Jpn.* **68**, 811–822.

K. Nagayama (1999b). Complex observation in electron microscopy. V. Phase retrieval for strong objects with Foucault knife-edge scanning. *J. Phys. Soc. Jpn.* **73**, 2725–2731.

K. Nagayama (2005). Phase contrast enhancement with phase plates in electron microscopy. *Adv. Imaging Electron Phys.* **138**, 69–146.

K. Nagayama (2008). Development of phase plates for electron microscopes and their biological application. *Eur. Biophys. J.* **37**, 345–358.

K. Nagayama (2011). Another 60 years in electron microscopy: development of phase-plate electron microscopy and biological applications. *J. Electron Microsc.* **60**, S43–S62.

K. Nagayama and R. Danev (2008). Phase contrast electron microscopy: development of thin-film phase plates and biological applications. *Philos. Trans. R. Soc. Lond. B* **363**, 2153–2162.

A. Nakamura, K. Ooe, T. Seki, Y. Kohno, H. Sawada and N. Shibata (2021). Higher dose-efficient atomic resolution STEM imaging using segmented detector. *Kenbikyo* **56** (Suppl. 1), 37.

P. D. Nellist (2011). The principles of STEM imaging. In *Scanning Transmission Electron Microscopy* (S. J. Pennycook and P. D. Nellist, Eds), 91–115 (Springer, New York).

P. D. Nellist (2019). Scanning transmission electron microscopy. In *Handbook of Microscopy* (P. W. Hawkes and J. C. H. Spence, Eds), 49–99 (Springer, Cham).

P. D. Nellist and S. J. Pennycook (1999). Incoherent imaging using dynamically scattered coherent electrons. *Ultramicroscopy* **78**, 111–124.

P. D. Nellist and S. J. Pennycook (2000). The principles and interpretation of annular dark-field Z-contrast imaging. *Adv. Imaging Electron Phys.* **113**, 147–203.

P. D. Nellist and J. M. Rodenburg (1998). Electron ptychography. I. Experimental demonstration beyond the conventional resolution limits. *Acta Cryst. A* **54**, 49–60. Part II is Plamann and Rodenburg (1998).

P. D. Nellist and P. Wang (2012a). Energy-filtered scanning confocal electron microscopy. *Microsc. Microanal.* **18** (Suppl. 2), 1962–1963.

P. D. Nellist and P. Wang (2012b). Optical sectioning and confocal imaging and analysis in the transmission electron microscope. *Annu. Rev. Mater. Res.* **42**, 125–143.

P. D. Nellist, B. McCallum and J. M. Rodenburg (1995). Resolution beyond the 'information limit' in transmission electron microscopy. *Nature* **374**, 630.

P. D. Nellist, G. Behan, A. I. Kirkland and C. J. D. Hetherington (2006). Confocal operation of a transmission electron microscope with two aberration correctors. *Appl. Phys. Lett.* **89**, 124105.

P. D. Nellist, E. C. Cosgriff, G. Behan and A. I. Kirkland (2008a). Imaging modes for scanning confocal electron microscopy in a double aberration-corrected transmission electron microscope. *Microsc. Microanal.* **14**, 82–88.

P. D. Nellist, E. C. Cosgriff, G. Behan, A. I. Kirkland, A. J. d'Alfonso, S. D. Findlay and L. J. Allen (2008b). Three-dimensinal imaging and analysis by optical sectioning in the aberration-corrected scanning transmission and scanning confocal electron microscopes. *Microsc. Microanal.* **14** (Suppl. 2), 104–105.

P. D. Nellist, P. Wang, G. Behan, A. I. Kirkland, A. Hashimoto, M. Shimojo, K. Mitsuishi, M. Takeguchi, E. C. Cosgriff, A. J. d'Alfonso, L. J. Allen and S. D. Findlay (2010). Three-dimensional resolution limits and image contrast mechanisms in scanning confocal electron microscopy. *Microsc. Microanal.* **16** (Suppl. 2), 1834–1835.

T. Nguyen, Y. Xue, Y. Li, L. Tian and G. Nehmetallah (2018). Deep learning approach for Fourier ptychography microscopy. *Opt. Express* **26**, 26470–26484.

H. Niehrs (1973). Zur Formulierung der Bildintensität bei ringförmiger Objektbestrahlung in der Elektronen-Mikroskopie. *Optik* **38**, 44–63.

H. Nishihara and T. Suhara (1987). Micro Fresnel lenses. *Prog. Opt.* **24**, 1–37.

A. Nishikawa, S. Toyama, T. Seki, A. Kumamoto, Y. Ikuhara and N. Shibata (2021). Development of method to separate electric field and magnetic field using DPC STEM. *Kenbikyo* **56** (Suppl. 1), 110.

K. Nitta, K. Nagayama, R. Danev and Y. Kaneko (2009). Visualization of BrdU-labelled DNA in cyanobacterial cells by Hilbert differential contrast transmission electron microscopy. *J. Microsc. (Oxford)* **234**, 118–123.

M. Nord, R. W. H. Webster, K. A. Paton, S. McVitie, D. McGrouther, I. MacLaren and G. W. Paterson (2020). Fast pixelated detectors in scanning transmission electron microscopy. Part I: Data acquisition, live processing, and storage. *Microsc. Microanal.* **26**, 653–666, for Part II, see Paterson et al. (2020).

H. J. Nussbaumer (1982). *Fast Fourier Transform and Convolution Algorithms* (Springer, Berlin and New York).

M. A. O'Keefe (1979). Resolution-damping functions in non-linear images. EMSA, San Antonio, TX, **37**, 556–557.

M. A. O'Keefe (1992). "Resolution" in high-resolution electron microscopy. *Ultramicroscopy* **47**, 282–297.

M. A. O'Keefe and J. C. H. Spence (1991). "Resolution" in atomic-resolution electron microscopy. EMSA, San Jose, CA, **49**, 498–499.

C. M. O'Leary, G. T. Martinez, E. Liberti, M. J. Humphry, A. I. Kirkland and P. D. Nellist (2019a). Contrast transfer and noise minimization in electron ptychography. *Microsc. Microanal.* **25** (Suppl. 2), 64–65.

C. M. O'Leary, E. Liberti, S. M. Collins, D. N. Johnstone, M. Rothmann, J. Hou, C. S. Allen, J. S. Kim, T. D. Bennett, P. A. Midgley, A. I. Kirkland and P. D. Nellist (2019b). Electron ptychography using fast binary 4D STEM data. *Microsc. Microanal.* **25** (Suppl. 2), 1662–1663.

C. M. O'Leary, C. S. Allen, C. Huang, J. S. Kim, E. Liberti, P. D. Nellist and A. I. Kirkland (2020). Phase reconstruction using fast binary 4D STEM data. *Appl. Phys. Lett.* **116**, 124101, 4 pp.

C. M. O'Leary, G. T. Martinez, E. Liberti, J. Humphry, A. I. Kirkland and P. D. Nellist (2021). Contrast transfer and noise considerations in focused-probe electron ptychography. *Ultramicroscopy* **221**, 113189, 11 pp.

E. L. O'Neill (1963). *Introduction to Statistical Optics* (Addison-Wesley, Reading MA and London).

M. Obermair, S. Hettler, M. Dries and D. Gerthsen (2017). Electrostatic Zach phase plates for transmission electron microscopy: status and future investigations. *Microsc. Microanal.* **23** (Suppl. 1), 828–829.

M. Obermair, S. Hettler, M. Dries, C. Hsieh, M. Marko and D. Gerthsen (2018a). Comparison of hole-free phase plates and electrostatic Zach phase plates for cryo-electron microscopy of biological specimens. IMC-19, Sydney.

M. Obermair, M. Marko, S. Hettler, C. Hsieh and D. Gerthsen (2018b). Physical phase plates for cryo-electron microscopy of biological specimens: comparison of hole-free phase plates and Zach electrostatic phase plates. *Microsc. Microanal.* **24** (Suppl. 1), 892–893.

M. Obermair, S. Hettler and D. Gerthsen (2019). Gradual Zernike phase plates for phase contrast imaging with reduced fringing artifacts in transmission electron microscopy. MC-2019, Berlin, 434–435.

M. Obermair, S. Hettler, C. Hsieh, M. Dries, M. Marko and D. Gerthsen (2020). Analyzing contrast in cryo-transmission electron microscopy: comparison of electrostatic Zach phase plates and hole-free phase plates. *Ultramicroscopy* **218**, 113086, 14 pp.

M. Ohtsuki and A.V. Crewe (1980). Application of the STEM optimum imaging method on a biological macromolecule. EMSA, San Francisco, CA, **38**, 62–63.

M. Ohtsuki, M. S. Isaacson and A. V. Crewe (1979). Dark field imaging of biological macromolecules with the scanning transmission electron microscope. *Proc. Natl. Acad. Sci. U.S.A.* **76**, 1228–1232.

H. Okamoto (2008). Noise suppression by active optics in low-dose electron microscopy. *Appl. Phys. Lett.* **92**, 063901.

H. Okamoto (2010). Adaptive quantum measurement for low-dose electron microscopy. *Phys. Rev. A* **81**, 043807, 16 pp.

H. Okamoto (2012). Possible use of a Cooper-pair box for low-dose electron microscopy. *Phys. Rev. A* **85**, 043810, 13 pp. Errata, **88** (2013) 049907.

H. Okamoto (2014). Measurement errors in entanglement-assisted electron microscopy. *Phys. Rev. A* **89**, 063828, 17 pp.

H. Okamoto (2015). Quantum interface to charged particles in a vacuum. *Phys. Rev. A* **92**, 053805, 10 pp.

H. Okamoto and Y. Nagatani (2014). Entanglement-assisted electron microscopy based on a flux qubit. *Appl. Phys. Lett.* **104**, 062604, 4 pp.

H. Okamoto, T. Latychevskaia and H.-W. Fink (2006). A quantum mechanical scheme to reduce radiation damage in electron microscopy. *Appl. Phys. Lett.* **88**, 164103.

E. Okunishi, I. Ishikawa, H. Sawada, F. Hosokawa, Hori. and Y. Kondo (2009). Visualization of light elements at ultrahigh resolution by STEM annular bright field microscopy. *Microsc. Microanal.* **15** (Suppl. 2), 164–165.

K. Ooe, T. Seki, Y. Ikuhara and N. Shibata (2019). High contrast STEM imaging for light elements by an annular segmented detector. *Ultramicroscopy* **202**, 148–155.

K. Ooe, T. Seki, Y. Ikuhara and N. Shibata (2021a). Ultra-high contrast STEM imaging for segmented/pixelated detectors by maximizing the signal-to-noise ratio. *Ultramicroscopy* **220**, 113133, 11 pp.

K. Ooe, T. Seki, Y. Kohno, Y. Ikuhara and N. Shibata (2021b). Direct observation of atomic structures in beam sensitive materials using optimum bright-field STEM technique. *Kenbikyo* **56** (Suppl. 1), 38.

C. Ophus, J. Ciston, J. Pierce, T. R. Harvey, J. Chess, B. J. McMorran, C. Czarnik, H. H. Rose and P. Ercius (2016). Efficient linear phase contrast in scanning transmission electron microscopy with matched illumination and detector interferometry. *Nature Commun.* **7**, 10719, 7 pp.

C. Ophus, T. R. Harvey, F. S. Yasin, H. G. Brown, P. M. Pelz, B. H. Savitzky, J. Ciston and B. J. McMorran (2019). Advanced phase reconstruction methods enabled by four-dimensional scanning transmission electron microscopy. *Microsc. Microanal.* **25** (Suppl. 2), 10–11.

X. Ou, R. Horstmeyer, G. Zheng and C. Yang (2015). High numerical aperture Fourier ptychography: principle, implementation and characterization. *Opt. Express* **23**, 3472–3491.

M. P. Oxley and O. E. Dyck (2020). The importance of temporal and spatial incoherence in quantitative interpretation of 4D-STEM. *Ultramicroscopy* **215**, 113015, 5 pp.

M. Oxley, D. Mukherjee and J. Hachtel (2020). Asymmetry and 4D-STEM: When the phase object approximation is qualitatively incorrect. *Microsc. Microanal.* **26** (Suppl. 2), 1910–1911.

A. Pakzad and R. Dos Reis (2021). Application of Stela hybrid pixel electron detector for 4D STEM characterisation of 2D materials at low energies. MC-2021, Vienna (virtual), 608.

J. R. Parsons and C. W. Hoelke (1974). Electron microscopy of plasmons. *Philos. Mag.* **30**, 135–143.

D. F. Parsons and H. M. Johnson (1972). Possibility of a phase contrast electron microscope. *Appl. Opt.* **11**, 2840–2843.

G. W. Paterson, R. W. H. Webster, A. Ross, K. A. Paton, T. A. Macgregor, D. McGrouther, I. MacLaren and M. Nord (2020). Fast pixelated detectors in scanning transmission electron microscopy. Part II: Post-acquisition data processing, visualization, and structural characterization. *Microsc. Microanal.* **26**, 944–963. Part I is Nord et al. (2020).

G. W. Paterson, G. M. Macauley, S. McVitie and Y. Togawa (2021). Parallel mode differential phase contrast in transmission electron microscopy, I: Theory and analysis. II. K_2CuF_4 phase transition. *Microsc. Microanal.* **27**, 1113–1122 and 1123–1132.

K. A. Paton, M. C. Veale, X. Mu, C. S. Allen, D. Maneuski, C. Kübel, V. O'Shea, A. I. Kirkland and D. McGrouther (2021). Quantifying the performance of a hybrid pixel detector with GaAs:Cr sensor for transmission electron microscopy. *Ultramicroscopy* **227**, 113298, 13 pp.

J. Pawley, Ed. (2006). *Handbook of Biological Confocal Microscopy* (Springer, New York).

P. M. Pelz, W. X. Qiu, R. Bücker, G. Kassier and R. J. D. Miller (2017). Low-dose cryo electron ptychography via non-convex Bayesian optimization. *Sci. Rep.* **7**, 9883, 13 pp.

P. Pelz, R. Buecker, G. Ramm, G. Kassier and R. D. Miller (2018). Low-dose cryo electron ptychography. IMC-19, Sydney.

P. Pelz, P. Ercius, C. Ophus, I. Johnson and M. Scott (2021). Real-time interactive ptychography from electron event representation data. *Microsc. Microanal.* **27** (Suppl. 1), 188–189.

S. J. Pennycook (1981). Study of supported ruthenium catalysts by STEM. *J. Microsc. (Oxford)* **124**, 15–22.

S. J. Pennycook (1989). Z-contrast STEM for materials science. *Ultramicroscopy* **30**, 58–69.

S. J. Pennycook (1992). Z-contrast transmission electron microscopy: direct atomic imagery of materials. *Annu. Rev. Mat. Sci.* **22**, 171–195.

S. J. Pennycook (2011). A scan through the history of STEM. In *Scanning Transmission Electron Microscopy, Imaging and Analysis* (S. J. Pennycook and P. D. Nellist, Eds), 1–90 (Springer, New York).

S. J. Pennycook and L. A. Boatner (1988). Chemically sensitive structure-imaging with a scanning transmission electron microscope. *Nature* **336**, 565–567.

S. J. Pennycook and D. E. Jesson (1990). High-resolution incoherent imaging of crystals. *Phys. Rev. Lett.* **64**, 938–941.

S. J. Pennycook and D. E. Jesson (1992). High-resolution imaging in the scanning transmission electron microscope. In *Electron Microscopy in Materials Science* (P. G. Merli and M. V. Antisari, Eds), 333–362 (World Scientific, Singapore, River Edge NJ and London).

S. J. Pennycook and P. D. Nellist, Eds (2011). *Scanning Transmission Electron Microscopy, Imaging and Analysis* (Springer, New York).

S. J. Pennycook, A. J. Craven and L. M. Brown (1977). Cathodoluminescence on a scanning transmission electron microscope. EMAG, Glasgow, 69–72.

S. J. Pennycook, D. E. Jesson, M. F. Chisholm, A. G. Ferridge and M. J. Seddon (1991). Sub-angstrom microscopy through incoherent imaging and image reconstruction. *Scanning Microsc.* (Suppl. 6, published 1994), 233–243; Proceedings of the 10th Pfefferkorn Conference, Cambridge, 1991.

C. R. Perrey, M. Lentzen and C. B. Carter (2003). Distinct as snowflakes: the shapes of silicon nanoscale particles. *Microsc. Microanal.* **9** (Suppl. 2), 958–959.

J. Perry-Houts, B. Barton, A. K. Schmid, N. Andresen and C. Kisielowski (2012). Novel long-lived electrostatic work function phase plates for TEM. *Microsc. Microanal.* **18** (Suppl. 2), 476–477.

H.-C. Pfeiffer (1967). Wechselwirkung freier Elektronen mit Licht. Dissertation, Technische Universität, Berlin.

H.-C. Pfeiffer (1968). Experimentelle Prüfung der Streuwahrscheinlichkeit für Elektronen beim Kapitza–Dirac-Effekt. *Phys. Lett. A* **26**, 362–363.

T. Plamann and J. M. Rodenburg (1992). Three-dimensional scattering effects in phase-retrieval microscopy. EUREM-10, Granada, **1**, 659–660.

T. Plamann and J. Rodenburg (1994). Double resolution imaging with infinite depth of focus in single lens scanning microscopy. *Optik* **96**, 31–36.

T. Plamann and J. Rodenburg (1998). Electron ptychography. II. Theory of three-dimensional propagation effects. *Acta Cryst. A* **54**, 61–73. Part I is Nellist and Rodenburg (1998).

J. Plitzko and W. Baumeister (2019). Cryo-electron tomography. In *Handbook of Microscopy* (P. W. Hawkes and J. C. H. Spence, Eds), 189–228 (Springer Nature, Cham).

B. Plotkin-Swing, G. J. Corbin, S. de Carlo, N. Dellby, C. Hoermann, M. V. Hoffman, T. C. Lovejoy, C. E. Meyer, A. Mittelberger, R. Pantelic, L. Piazza and O. L. Krivanek (2020). Hybrid pixel direct detector for electron energy loss spectroscopy. *Ultramicroscopy* **217**, 113067, 10 pp.

S. Pollard, M. Malac, M. Beleggia, M. Kawasaki and Y. Zhu (2013). Magnetic imaging with a Zernike-type phase plate in a transmission electron microscope. *Appl. Phys. Lett.* **102**, 192401.

S. Pollard, M. Malac, M. Beleggia, M. Kawasaki and Y. Zhu (2014). Magnetic imaging with a novel hole-free phase plate. *Microsc. Microanal.* **20** (Suppl. 3), 250–251.

S. Pöllath, F. Schwarzhuber and J. Zweck (2021). The differential phase contrast uncertainty relation: connection between electron dose and field resolution. *Ultramicroscopy* **228**, 113342, 9 pp.

K. C. Prabhat, K. A. Mohan, C. Phatak, C. Bouman and M. de Graef (2017). 3D reconstruction of the magnetic vector potential using model based iterative reconstruction. *Ultramicroscopy* **182**, 131–144.

R. Pretzsch, M. Dries, S. Hettler, M. Obermair and D. Gerthsen (2018). Investigation of parameters that influence the performance of a hole-free phase plate and its application on a carbon nanotube sample. IMC-19, Sydney.

R. Pretzsch, M. Dries, S. Hettler, M. Spiecker, M. Obermair and D. Gerthsen (2019). Investigation of hole-free phase plate performance in transmission electron microscopy under different operation conditions by experiments and simulations. *Adv. Struct, Chem. Imaging* **5**, 5, 11 pp.

W. Probst, G. Benner, J. Bihr, R. Bauer and E. Weimer (1991a). Koehler illumination– advantages for imaging and diffraction in TEM. *Microscopia Elettronica* **12** (Pt 2 (Suppl.)), 369–370.

W. Probst, R. Bauer, G. Benner and J.L. Lehman (1991b). Koehler illumination, advantages for imaging in TEM, EMSA 49, San Jose, CA, 1010–1011.

C. T. Putkunz, A. J. d'Alfonso, A. J. Morgan, M. Weyland, C. Dwyer, L. Bourgeois, J. Etheridge, A. Roberts, R. E. Scholten, K. A. Nugent and L. J. Allen (2012). Atom-scale ptychographic electron diffractive imaging of boron nitride cones. *Phys. Rev. Lett.* **108**, 073901, 4 pp.

H. Qian, H. Furukawa, M. Shimizu, M. Kawasaki and M. Shiojiri (2012). Study for automated imaging with phase plate electron microscopy and suggestions for the future instrumentation. *Microsc. Microanal.* **18** (Suppl. 2), 500–501.

I. L. F. Ray, I. W. Drummond and J.R. Banbury (1975). A high resolution field emission scanning transmission electron microscope with energy analysis. EMAG, Bristol, 11–14.

I. S. Reed, T. K. Truong, Y. S. Kwoh and E. L. Hall (1977). In *Image Science Mathematics* (C. O. Wilde and E. Barrett, Eds), 229–233 (Western Periodicals, North Hollywood).

R. Reichelt and A. Engel (1985). Quantitative scanning transmission electron microscopy in biology. *J. Microsc. Spectrosc. Electron* **10**, 491–498.

R. Reichelt and A. Engel (1986). Contrast and resolution of scanning transmission electron microscope imaging modes. *Ultramicroscopy* **19**, 43–56.

L. Reimer and H. G. Badde (1970). Einfluss der Elektronenstreuung in einer Phasenplatte auf die Abbildung von Einzelatomen. ICEM-7, Grenoble, **1**, 15–16.

L. Reimer and P. Hagemann (1977). Anwendung eines Rasterzusatzes zu einem Transmissionselektronenmikroskop. II. Abbildung kristalliner Objekte. *Optik* **47**, 325–336.

L. Reimer and H. Kohl (2008). *Transmission Electron Microscopy* (Springer, New York).

L. Reimer, P. Gentsch and P. Hagemann (1975). Anwendung eines Rasterzusatzes zu einem Transmissionselektronenmikroskop. I. Grundlagen und Abbildung amorpher Objekte. *Optik* **43**, 431–452.

D. Rhinow (2016). Towards an optimum design for thin film phase plates. *Ultramicroscopy* **160**, 1–6.

H. L. Robert, F. Winkler, I. Lobato, S. Van Aert, F. J. Lyu, Q. Chen and K. Müller-Caspary (2021). Impact of defocus in high-resolution momentum-resolved STEM. MC-2021, Vienna (virtual), 558–559.

D.W. Robinson (1979). A 2-D computer simulation of partially coherent image formation, in a multiple, centro-symmetric detector STEM, of heavy atom model compounds. EMAG, Brighton, 253–256.

J. M. Rodenburg (1989a). The phase problem, microdiffraction and wavelength-limited resolution – a discussion. *Ultramicroscopy* **27**, 413–422.

J. M. Rodenburg (1989b). Higher spatial resolution via signal processing of the microdiffraction plane. EMAG, London, **1**, 103–106.

J. Rodenburg (2008). Ptychography and related diffractive imaging methods. *Adv. Imaging Electron Phys.* **150**, 87–184.

J. M. Rodenburg and R. H. T. Bates (1992). The theory of super-resolution electron microscopy via Wigner-distribution deconvolution. *Philos. Trans. R. Soc. Lond. A* **339**, 521–553.

J. Rodenburg and A. Maiden (2019). Ptychography. In *Handbook of Microscopy* (P. W. Hawkes and J. C. H. Spence, Eds), 819–904 (Springer, Cham).

J. M. Rodenburg and B. C. McCallum (1991). A robust solution to the super-resolution phase problem in scanning transmission electron microscopy. In Hawkes et al. (1992), 223–232.

J. M. Rodenburg and B. C. McCallum (1992). Super-resolution structure determination in STEM. EUREM-10, Granada, **1**, 125–129.

J. M. Rodenburg, B. C. McCallum and P. D. Nellist (1993). Experimental tests on double-resolution coherent imaging via STEM. *Ultramicroscopy* **48**, 304–314.

J. Rodenburg, A. Hurst and A. Cullis (2007). Transmission microscopy without lenses for objects of unlimited size. *Ultramicroscopy* **107**, 227–231.

R. Röhler (1967). *Informationstheorie in der Optik* (Wissenschaftliche Verlagsgesellschaft, Stuttgart).

A. Rohou and N. Grigorieff (2015). CTFFIND4: fast and accurate defocus estimation from electron micrographs. *J. Struct. Biol.* **192**, 216–221.

H. Rose (1974a). Phase contrast in scanning transmission electron microscopy. *Optik* **39**, 416–436.

H. Rose (1974b). Resolution and contrast transfer in fixed beam and scanning transmission electron microscopy. ICEM-8, Canberra, **1**, 212–213.

H. Rose (1975). Zur Theorie der Bildentstehung im Elektronen-Mikroskop. I. *Optik* **42**, 217–244. Part II is Fertig and Rose (1979).

H. Rose (1977). Nonstandard imaging methods in electron microscopy. *Ultramicroscopy* **2**, 251–267.

H. Rose (1978). Effects of inelastically scattered electrons in various CTEM and STEM imaging modes. *Ann. N. Y. Acad. Sci.* **306**, 47–61.

H. Rose (1984). Information transfer in transmission electron microscopy. *Ultramicroscopy* **15**, 173–191.

H. Rose (2010). Theoretical aspects of image formation in the aberration-corrected electron microscope. *Ultramicroscopy* **110**, 488–499.

H. Rose (2022). Minimum-dose phase-contrast tomography by successive numerical optical sectioning employing the aberration-corrected STEM and a pixelated detector. *Ultramicroscopy* (in press).

H. Rose and J. Fertig (1976). Influence of detector geometry on image properties of the STEM for thick objects. *Ultramicroscopy* **2**, 77–87.

H. Rose and J. Fertig (1977). Improvement of TEM bright-field images by hollow-cone illumination. EMSA, Boston, MA, **35**, 200–201.

A. Rosenauer (2021). Understanding quantitative STEM – how far have we come?. MC-2021, Vienna (virtual), 6–7.

A. Rosenauer and M. Schowalter (2007). STEMSIM – a new software for simulation of STEM HAADF *Z*-contrast imaging. *Springer Proc. Phys.* **120**, 169–172.

A. Rosenauer, B. Kraus, K. Müller, M. Schowalter and T. Mehrtens (2014). Conventional transmission electron microscopy imaging beyond the diffraction and information limits. *Phys. Rev. Lett.* **113**, 096101, 5 pp.

R. S. Ruskin, Z. Yu and N. Grigorieff (2013). Quantitative characterization of electron detectors for transmission electron microscopy. *J. Struct. Biol.* **184**, 385–393.

H. Ryll, M. Simson, R. Hartmann, P. Holl, M. Huth, S. Ihle, Y. Kondo, P. Kotula, A. Liebel, K. Müller-Caspary, A. Rosenauer, R. Sagawa, J. Schmidt, H. Soltau and R. Strüder (2016). A pnCCD-based, fast direct single electron imaging camera for TEM and STEM. *J. Instrument.* **11**, P04006, 18 pp.

K. Sader, B. Buijsse, G. van Duinen and R. Danev (2014). Challenges in phase plate development and applications. IMC-18, Prague, 1982.

K. Sader, B. Buijsse, I. Peschiera and I. Ferlenghi (2015). Applications and new investigations of the Volta phase plate. *Microsc. Microanal.* **21** (Suppl. 3), 1837–1838.

R. Sagawa, H. Hashiguchi and Y. Kondo (2018). Low dose STEM imaging by ptychography using pixelated STEM detector. *Kenbikyo* **53** (Suppl. 1), 6.

R. Sagawa, A. Yasuhara, H. Hashiguchi, T. Naganuma, S. Tanba, T. Ishikawa, T. Riedel, P. Hartel, M. Linck, S. Uhlemann, H. Müller and H. Sawada (2022). Exploiting the full potential of the advanced two-hexapole corrector for STEM exemplified at 60kV. *Ultramicroscopy* **233**, 113440, 8 pp.

Z. Saghi, X.-j. Xu and G. Möbus (2009). Model based atomic resolution tomography. *J. Appl. Phys.* **106**, 024304.

K. Saitoh and N. Tanaka (2005). Detection of interstitial atoms by hollow-cone illumination HAADF-STEM. *Microsc. Microanal.* **11** (Suppl. 2), 710–711.

C. W. Sandweg, N. Wiese, D. McGrouther, S. J. Hermsdoerfer, H. Schultheiss, B. Leven, S. McVitie, B. Hillebrands and J. N. Chapman (2008). Direct observation of domain wall structures in curved permalloy wires containing an antinotch. *J. Appl. Phys.* **103**, 093906.

T. Sannomiya, Y. Haga, Y. Nakamura and O. Nittono (2004). Observation of magnetic structures in Fe granular films by differential phase contrast scanning transmission electron microscopy. *J. Appl. Phys.* **95**, 214–218.

T. Sannomiya, J. Junesch, F. Hosokawa, K. Nagayama, Y. Arai and Y. Kayama (2014). Multi-pore carbon phase plate for phase-contrast transmission electron microscopy. *Ultramicroscopy* **146**, 91–96.

M. C. Sarahan, B. Kraus, M. J. Humphry and A. M. Maiden (2012). Electron ptychography: applications of the electron wave phase. *Microsc. Microanal.* **18** (Suppl. 2), 502–503.

M. Sarikaya and J. M. Howe (1992). Resolution in conventional transmission electron microscopy. *Ultramicroscopy* **47**, 145–161.

W. O. Saxton (1977). Coherence in bright field microscopy of weak objects. EMAG, Glasgow, 111–114.

W. O. Saxton (1986). Focal series restoration in HREM. ICEM-11, Kyoto, post-deadline paper 1.

W. O. Saxton (1987). The ultimate reliability of high-resolution electron imaging. EMSA, **45**, 10–13.

W. O. Saxton and D. J. Smith (1979). Bright-field hollow cone illumination – theory and experiment. EMAG, Brighton, 265–268.

W.O. Saxton, A. Howie, A. Mistry and A. Pitt (1977). Fact and artefact in high resolution microscopy. EMAG, Glasgow, 119–122.

W. O. Saxton, W. K. Jenkins, L. A. Freeman and D. J. Smith (1978). TEM observations using bright field hollow cone illumination. *Optik* **49**, 505–510.

O. Scherzer (1949). The theoretical resolution limit of the electron microscope. *J. Appl. Phys.* **20**, 20–29.

O. Scherzer (1970). Die Strahlenschädigung der Objekte als Grenze für die hochauflösende Elektronenmikroskopie. *Ber. Bunsen-Ges. Phys. Chem.* **74**, 1154–1167.

P. Schiske (1973). Fourier methods for the treatment of images with coma and distortion. In *Image Processing and Computer-Aided Design in Electron Optics* (P. W. Hawkes, Ed.), 54–65 (Academic Press, London and New York).

R. Schliepe (1978). *Das ELMISKOP ST 100F. Aufbau, Funktionsweise und Anwendungen eines hochauflösenden Durchstrahlungs-Rasterelektronenmikroskops (STEM).* Typescript, Siemens Archives.

M. Schloz, T. C. Pekin, Z. Chen, W. Van den Broek, D. A. Muller and C. T. Koch (2020). Overcoming information reduced data and experimentally uncertain parameters in ptychography with regularized optimization. *Opt. Express* **28**, 28306–28323.

R. R. Schröder, B. Barton, H. H. Rose and G. Benner (2007). Contrast enhancement by anamorphotic phase plates in an aberration-corrected TEM. *Microsc. Microanal.* **13** (Suppl. 2), 136–137, and (Suppl. 3), 8–9.

K. Schultheiß (2010). Entwicklung und Anwendung elektrostatischer Phasenplatten in der Transmissionselektronenmikroskopie. Dissertation, Karlsruhe Institute of Technology.

K. Schultheiß, F. Pérez-Willard, B. Barton, D. Gerthsen and R. R. Schröder (2006). Fabrication of a Boersch phase plate for phase contrast in a transmission electron microscope. *Rev. Sci. Instrum.* **77**, 033701, 4 pp.

K. Schultheiß, J. Zach, B. Gamm, M. Dries, N. Frindt, R. Schröder and D. Gerthsen (2010a). New electrostatic phase plate for transmission electron microscopy and its application for wave-function reconstruction. *Microsc. Microanal.* **16** (Suppl. 2), 536–537.

K. Schultheiß, J. Zach, B. Gamm, M. Dries, N. Frindt, R. R. Schröder and D. Gerthsen (2010b). New electrostatic phase plate for phase-contrast transmission electron microscopy and its application to wave-function reconstruction. *Microsc. Microanal.* **16**, 785–794.

O. Schwartz, J. J. Axelrod, D. R. Tuthill, P. Haslinger, C. Ophus, R. M. Glaeser and H. Müller (2017). Near-concentric Fabry–Pérot cavity for continuous-wave laser control of electron waves. *Opt. Express* **25**, 14453.

O. Schwartz, J. J. Axelrod, S. L. Campbell, C. Turnbaugh, R. M. Glaeser and H. Müller (2019). Laser phase plate for transmission electron microscopy. *Nature Methods* **16**, 1016–1020.

F. Schwarzhuber, P. Melzl and J. Zweck (2017). On the achievable field sensitivity of a segmented annular detector for differential phase contrast measurements. *Ultramicroscopy* **177**, 97–105.

F. Schwarzhuber, P. Melzl, S. Pöllath and J. Zweck (2018). Introducing a non-pixelated and fast centre of mass detector for differential phase contrast microscopy. *Ultramicroscopy, 21–28.*

S. Seifer, L. Houben and M. Elbaum (2021). Flexible STEM with simultaneous phase and depth contrast. *Microsc. Microanal.* **27**, 1476–1487.

T. Seki, G. Sánchez-Santolino, R. Ishikawa, S. D. Findlay, Y. Ikuhara and N. Shibata (2017). Quantitative electric field mapping in thin specimens using a segmented detector: revisiting the transfer function for differential phase contrast. *Ultramicroscopy* **182**, 258–263.

T. Seki, N. Takanashi and E. Abe (2018a). Integrated contrast-transfer-function for aberration-corrected phase-contrast STEM. *Ultramicroscopy* **194**, 193–198.

T. Seki, Y. Ikuhara and M. Shibano (2018b). Theoretical framework of statistical noise in scanning transmission electron microscopy. *Ultramicroscopy* **193**, 118–125.

T. Seki, Y. Ikuhara and N. Shibata (2021). Toward quantitative electromagnetic field imaging by differential-phase-contrast scanning transmission electron microscopy. *Microscopy* **70**, 148–160.

D. G. Sentürk, A. De Backer and S. Van Aert (2021). Optimal experiment design for counting atoms of different chemical nature using 4D scanning transmission electron microscopy. MC-2021, Vienna (virtual), 581–582.

M. Setou, R. Danev, K. Atsuzawa, I. Yao, T. Fukuda and N. Usuda (2006). Mammalian cell nano structures visualized by cryo Hilbert differential contrast transmission electron microscopy. *Med. Mol. Morphology* **39**, 176–180.

T. H. Sharp, F. G. A. Faas, A. J. Koster and P. Gros (2017). Imaging complement by phase-plate cryo-electron tomography from initiation to pore formation. *J. Struct. Biol.* **197**, 155–162.

C. J. R. Sheppard (1977). The use of lenses with annular aperture in scanning optical microscopy. *Optik* **48**, 329–334.

C. J. R. Sheppard (1987). Scanning optical microscopy. *Adv. Opt. Electron Microsc.* **10**, 1–98.

C. J. R. Sheppard (1989). Axial resolution of confocal fluorescence microscopy. *J. Microsc. (Oxford)* **154**, 237–241.

C. J. R. Sheppard (2003). Scanning confocal microscopy. In *Encyclopedia of Optical Engineering* (R. G. Driggers, Ed.), **3**, 2525–2544 (Marcel Dekker, New York).

C. J. R. Sheppard and A. Choudhury (1977). Image formation in the scanning microscope. *Opt. Acta* **24**, 1051–1073.

C. J. R. Sheppard and C. J. Cogswell (1990a). Three-dimensional image formation in confocal microscopy. *J. Microsc. (Oxford)* **159**, 179–194.

C. J. R. Sheppard and C. J. Cogswell (1990b). Confocal microscopy with detector arrays. *J. Mod. Opt.* **37**, 267–279.

C. J. R. Sheppard and M. Gu (1991a). Improvement of axial resolution in confocal microscopy using an annular pupil. *Opt. Commun.* **84**, 8–12.

C. J. R. Sheppard and M. Gu (1991b). Optical sectioning in confocal microscopes with annular pupil. *Optik* **86**, 169–172.

C. J. R. Sheppard and M. Gu (1992). The significance of 3-D transfer functions in confocal scanning microscopy. *J. Microsc. (Oxford)* **165**, 377–390.

C. J. R. Sheppard and D. R. Hamilton (1984). Edge enhancement by defocusing of confocal images. *Opt. Acta* **31**, 723–727.

C. J. R. Sheppard and T. Wilson (1979a). Effect of spherical aberration on the imaging properties of scanning optical microscopes. *Appl. Opt.* **18**, 1058–1063.

C. J. R. Sheppard and T. Wilson (1979b). Imaging properties of annular lenses. *Appl. Opt.* **18**, 3764–3769.

C. J. R. Sheppard and T. Wilson (1980a). Fourier imaging of phase information in scanning and conventional optical microscopes. *Philos. Trans. R. Soc. Lond. A* **295**, 513–536.

C. J. R. Sheppard and T. Wilson (1980b). Image formation in confocal scanning microscopes. *Optik* **55**, 331–342.

C. J. R. Sheppard and T. Wilson (1981). The theory of the direct-view confocal microscope. *J. Microsc. (Oxford)* **124**, 107–117.

C. J. R. Sheppard and T. Wilson (1986). On the equivalence of scanning and conventional microscopes. *Optik* **73**, 39–43.

C. J. R. Sheppard, D. K. Hamilton and I. J. Cox (1983). Optical microscopy with extended depth of field. *Proc. R. Soc. Lond. A* **387**, 171–186.

N. Shibata, Y. Kohno, S. D. Findlay, H. Sawada, Y. Kondo and Y. Ikuhara (2010). New area detector for atomic-resolution scanning transmission electron microscopy. *J. Electron Microsc.* **59**, 473–479.

N. Shibata, S. D. Findlay, Y. Kohno, H. Sawada, Y. Kondo and Y. Ikuhara (2012). Differential phase-contrast microscopy at atomic resolution. *Nature Phys.* **8**, 611–615.

H. Shigematsu, T. Sokabe, R. Danev, M. Tominaga and K. Nagayama (2010). A 30.5-nm structure of rat TRPV4 cation channel revealed by Zernike phase-contrast cryoelectron microscopy. *J. Biol. Chem.* **285**, 11210–11218.

R. Shiloh, Y. Lilach and A. Arie (2014). Sculpturing [sic] the electron wave function using nanoscale phase masks. *Ultramicroscopy* **144**, 26–31.

R. Shiloh, R. Remez and A. Arie (2016). Prospects for electron beam aberration correction using sculpted phase masks. *Ultramicroscopy* **163**, 69–74.

R. Shiloh, R. Remez, P.-h. Lu, Y. Lereah, A. H. Tavabi, R. E. Dunin-Borkowki and A. Arie (2018). Spherical aberration correction in a scanning transmission electron microscope using a sculpted thin film. *Ultramicroscopy* **189**, 46–53, Corrigendum *ibid.* **216** (2020) 112965, 1 p.

J. Shiue and S.-k. Hung (2010). A TEM phase plate loading system with loading monitoring and nano-positioning functions. *Ultramicroscopy* **110**, 1238–1242.

J. Shiue, C.-s. Chang, S.-h. Huang, C.-h. Hsu, J.-s. Tsai and F.-w. Chen (2009). Phase TEM for biological imaging utilizing a Boersch electrostatic phase plate: theory and practice. *J. Electron Microsc.* **58**, 137–145.

P. Sieber and K. Tonar (1975). Test of electron microscopes by lattice imaging in the 0.1 nm domain. *Optik* **42**, 375–380.

P. Sieber and K. Tonar (1976). A possibility of misinterpreting lattice images taken with axial illumination. *Optik* **44**, 361–364.

B. M. Siegel, C. B. Eisenhandler and M. G. Coan (1966). Ultimate resolution by phase contrast imaging of molecular objects. ICEM-6, Kyoto **1**, 41–42.

I. Sikharulidze, R. van Gastel, S. Schramm, J. P. Abrahams, B. Poelsema, R. M. Tromp and S. J. van der Molen (2011). Low energy electron microscopy imaging using Medipix2 detector. *Nucl. Instrum. Meth, Phys. Res. A* **633**, S239–S242.

K. C. A. Smith (2009). STEM at Cambridge University: reminiscences and reflections from the 1950s and 1960s. *Adv. Imaging Electron Phys.* **159**, 387–406.

K. C. A. Smith and S. J. Erasmus (1982). A configurable STEM detector. *J. Microsc. (Oxford)* **122**, RP1–RP2.

T. Someya, T. Goto, Y. Harada, K. Yamada, H. Koike, Y. Kokubo and M. Watanabe (1974). On the development of a 100 kV field emission electron microscope. *Optik* **41**, 225–244.

J. Song, Z. Ding, C. S. Allen, H. Sawada, F. Zhang, X. Pan, J. Warner, A. I. Kirkland and P. Wang (2018a). Hollow electron ptychographic diffractive imaging. *Phys. Rev. Lett.* **121**, 146101, 6 pp.

J. Song, B. Song, L. Zou, C. Allen, H. Sawada, F. Zhang, X. Pan, A. I. Kirkland and P. Wang (2018b). Fast and low-dose electron ptychography. *Microsc. Microanal.* **24** (Suppl. 1), 224–225.

J. Song, C. S. Allen, S. Gao, C. Huang, H. Sawada, X. Pan, J. Warner, P. Wang and A. I. Kirkland (2019). Atomic resolution defocused electron ptychography at low dose with a fast, direct electron detector. *Sci. Rep.* **9**, 3919, 8 pp.

J. C. H. Spence (1978). Practical phase determination of inner dynamical reflections in STEM. *Scanning Electron Microsc.* (I), 61–68.

J. C. H. Spence (1992). Convergent-beam nano-diffraction, in-line holography and coherent shadow imaging. *Optik* **92**, 57–68.

J. C. H. Spence (2013). *High-Resolution Electron Microscopy* (Oxford University Press, New York and Oxford).

J. C. H. Spence (2019). Diffractive imaging of single particles. In *Handbook of Microscopy* (P. W. Hawkes and J. C. H. Spence, Eds), 1009–1036 (Springer, Cham).

J. C. H. Spence and J. M. Cowley (1978). Lattice imaging in STEM. *Optik* **50**, 129–142.

J. C. H. Spence and J. M. Zuo (1992). *Electron Microdiffraction* (Plenum, New York and London).

J. W. Steeds and E. Carlino (1992). Electron crystallography. In *Electron Microscopy in Materials Science* (P. G. Merli and M. V. Antisari, Eds), 279–313 (World Scientific, Singapore, River Edge NJ and London).

A. Strauch, D. Weber, A. Clausen, A. Lesnichaia, A. Bangun, B. März, F. J. Lyu, Q. Chen, A. Rosenauer, R. Dunin-Borkowski and K. Müller-Caspary (2021). Live processing of momentum-resolved STEM data for first moment imaging and ptychography. *Microsc. Microanal.* **27**, 1078–1092.

K. Strehl (1902). Über Luftschlieren und Zonenfehler. *Z. Instrumentenkde* **22**, 213–217.

H. Struve (1882). Beitrag zur Theorie der Diffraction an Fernröhren. *Ann. Phys. Chem.* **253** (17), 1008–1016.

S. Sugitani and K. Nagayama (2002). Complex observation in electron microscopy. III. Inverse theory of observation-scheme dependent information transfer. *J. Phys. Soc. Jpn.* **71**, 744–756.

S. Sugitani and K. Nagayama (2006). Complex observation in electron microscopy. VI. Comparison of information transfer reliability between Zernike complex observation and transport of intensity equation method. *J. Phys. Soc. Jpn.* **75**, 084401.

J. Sun, C. Zuo and Q. Chen (2019). Improved multi-slice Fourier ptychographic microscopy technique for high-accuracy three-dimensional tomography under oblique illuminations. *Proc. SPIE* **11205**, 11205-0E, 5 pp.

J. Sun, C. Zuo and Q. Chen (2020). Improved multi-slice Fourier ptychographic diffraction tomography based on high-numerical-aperture illuminations. *Proc. SPIE* **11571**, 11571-0R.

T. Susi, J. C. Meyer and J. Kotakoski (2019). Quantifying transmission electron microscopy irradiation effects using two-dimensional materials. *Nature Rev. Phys.* **1**, 397–405. Correction *ibid.* 635.

T. Taguchi, L. Piazza and S. de Cario (2021). A novel hybrid-pixel detector revolutionizing STEM–EELS in materials science. *Kenbikyo* **56** (Suppl. 1), 120.

M. Takeguchi, A. Hashimoto, M. Shimojo, K. Mitsuishi and K. Furuya (2008). Development of a stage-scanning system for high-resolution confocal STEM. *J. Electron Microsc.* **57**, 123–127.

M. Takeguchi, A. Hashimoto, K. Mitsuishi and M. Shimojo (2009). Development of annular dark field confocal scanning transmission electron microscopy. *Microsc. Microanal.* **15** (Suppl. 2), 612–613.

M. Takeguchi, M. Okuda, A. Hashimoto, K. Mitsuishi, M. Shimojo, X. Zhang, P. Wang, P. D. Nellist and A. I. Kirkland (2010). Three dimensional characterization of a silica hollow sphere with an iron oxide core by annular dark field scanning confocal electron microscopy. *Microsc. Microanal.* **16** (Suppl. 2), 1836–1837.

M. Takeguchi, X. Zhang, A. Hashimoto, K. Mitsuishi, M. Shimojo, P. Wang, P. D. Nellist and A. I. Kirkland (2012). Scanning confocal electron microscopy (SCEM) combined with deconvolution technique. *Microsc. Microanal.* **18** (Suppl. 2), 332–333.

H. Tamaki, H. Kasai, K. Harada, Y. Takahashi and R. Nishi (2013). Development of a contact-potential-type phase plate. *Microsc. Microanal.* **19** (Suppl. 2), 1148–1149.

N. Tanaka (2015a). Historical survey of the development of STEM instruments. In *Scanning Transmission Electron Microscopy of Materials* (N. Tanaka, Ed.), 9–38 (Imperial College Press, London).

N. Tanaka (2015b). Electron tomography in STEM. In *Scanning Transmission Electron Microscopy of Materials* (N. Tanaka, Ed.), 383–401 (Imperial College Press, London).

N. Tanaka (2015c). Electron holography and Lorentz electron microscopy in STEM. In *Scanning Transmission Electron Microscopy of Materials* (N. Tanaka, Ed.), 403–424 (Imperial College Press, London).

N. Tanaka and N. Ikarashi (2021). Theoretical basis of observation of charge distribution in DPC electron microscopy. *Kenbikyo* **56** (Suppl. 1), 111.

T. Tanji, U. Ikeda, H. Niimi and J. Usukura (2014). Electron differential phase microscopy with an A–B effect phase plate. IMC-18, Prague, 2469.

M. Taya, T. Ikuta and Y. Takai (2008). Wave field restoration using focal-depth extension techniques under dynamic hollow-cone illumination. *Optik* **119**, 153–160.

P. Thibault and A. Menzel (2013). Reconstructing state mixtures from diffraction measurements. *Nature* **494**, 68–71.

M. N. Thompson (1973a). A scanning transmission electron microscope: some techniques and applications. EMAG, Newcastle-upon-Tyne, 176–181.

M. N. Thompson (1973b). The application of SEM techniques to the study of crystal defects. EMSA, New Orleans, **31**, 154–155.

M. N. Thompson (1975). Calibration and selection of STEM detector collection angles. EMAG, Bristol, 31–34.

M. G. R. Thomson (1973). Resolution and contrast in the conventional and the scanning high resolution transmission electron microscopes. *Optik* **39**, 15–38.

F. Thon (1965). Elektronenmikroskopische Untersuchungen an dünnen Kohlefolien. *Z. Naturforsch.* **20a**, 154–155.

F. Thon (1966a). On the defocusing dependence of phase contrast in electron microscopical images. *Z. Naturforsch.* **21a**, 476–478.

F. Thon (1966b). Imaging properties of the electron microscope near the theoretical limit of resolution. ICEM-6, Kyoto, **1**, 23–24.

F. Thon (1967). Probleme der Hochauflösungs-Elektronenmikroskopie. *Phys. Blätt* **23**, 450–458.

F. Thon (1968a). Zur Deutung der Bildstrukturen in hochaufgelösten elektronenmikroskopischen Anfnahmen dünner amorpher Objekte. Dissertation, Tübingen.

F. Thon (1968b). Hochauflösende elektronenmikroskopische Abbildung amorpher Objekte mittels Zweistrahlinterferenzen. EUREM-4, Rome, **1**, 127–128.

F. Thon (1971). Phase contrast electron microscopy. In *Electron Microscopy in Material Science* (U. Valdrè, Ed.), 570–625 (Academic Press, New York and London).

Thon, F. (1974). In-line and a posteriori improvement of high resolution electron microscopical images. ICEM-8, Canberra, **1**, 238–239.

F. Thon and D. Willasch (1970). Hochauflösungs-Elektronenmikroskopie mit Spezialaperturblenden und Phasenplatten. ICEM-7, Grenoble, **1**, 3–4.

F. Thon and D. Willasch (1971a). High resolution electron microscopy using phase plates. EMSA, Boston, **29**, 38–39.

F. Thon and D. Willasch (1971b). A calibration method usable in the production of phase plates for high resolution electron microscopy. EMSA, Boston, **29**, 46–47.

F. Thon and D. Willasch (1972a). Imaging of heavy atoms in darkfield electron microscopy using hollow cone illumination. *Optik* **36**, 55–58.

F. Thon and D. Willasch (1972b). Recent results in high resolution microscopy with special apertures. EUREM-5, Manchester, 650–651.

A. Thust, J. Barthel, L. Houben, C.-l. Jia, M. Lentzen, K. Tillmann and K. Urban (2005). Strategies for aberration control in sub-Angstrom HRTEM. *Microsc. Microanal.* **11** (Suppl. 2), 58–59.

W. Tichelaar, W. J. H. Hagen, T. E. Gorelik, L. Xue and J. Mahamid (2020). TEM bright field imaging of thick specimens: nodes in Thon ring patterns. *Ultramicroscopy* **216**, 113023, 9 pp.

H. Tochigi, H. Nakatsuka, A. Fukami and K. Kanaya (1970). The improvement of the image contrast by using the phase plate in the transmission electron microscope. ICEM-7, Grenoble, **1**, 73–74.

H. Tochigi, A. Ishikawa and Y. Satake (1974). Newly designed scanning transmission electron microscope. ICEM-8, Canberra, **1**, 50–51.

M. Tomita, Y. Nagatani, K. Murata and A. Momose (2020). Enhancement of low-spatial-frequency components by a new phase-contrast STEM using a probe formed with an amplitude Fresnel zone plate. *Ultramicroscopy* **218**, 113089, 10 pp.

A. Töpler (1864). *Beobachtungen nach einer neuen optischen Methode: Ein Beitrag zur Experimentalphysik* (Cohen, Bonn).

A. Töpler (1866). Ueber die Methode der Schlierenbeobachtung als mikroskopisches Hülfsmittel, nebst Bemerkungen zur Theorie schiefe Beleuchtung. *Ann. Phys. Chem.* **203** (127), 557–580.

M. Tosaka, R. Danev and K. Nagayama (2005). Application of phase contrast transmission microscopic methods to polymer materials. *Macromolecules* **36**, 7884–7886.

M. M. J. Treacy (1981). Imaging with Rutherford scattered electrons in the STEM. *Scanning Electron Microsc.* (I), 185–197.

M. M. J. Treacy and J. M. Gibson (1981). Detection of interstitial impurities using a high angle annular detector in the STEM. *Ultramicroscopy* **7**, 109.

M. M. J. Treacy, A. Howie and C. J. Wilson (1978). *Z* contrast of platinum and palladium catalysts. *Philos. Mag.* **A38**, 569–585.

M. M. J. Treacy, A. Howie and S.J. Pennycook (1979). Z contrast of supported catalyst particles on the STEM. EMAG, Brighton, 261–264.

M. Tsubouchi and H. Minoda (2019). Phase plate STEM imaging using two dimensional electron detector. *Microsc. Microanal.* **25** (Suppl. 2), 84–85.

M. Tsubouchi and H. Minoda (2020a). Phase plate STEM imaging using two dimensional electron detector. *Microsc. Microanal.* **26** (Suppl. 2), 84–85.

M. Tsubouchi and H. Minoda (2020b). Enhancement of phase contrast at low spatial frequency using phase plate STEM. *Microsc. Microanal.* **26** (Suppl. 2), 228–229.

M. Tsubouchi and H. Minoda (2021a). Increased efficiency of phase plate STEM using 2D detector. *Microsc. Microanal.* **27** (Suppl. 1), 2192–2193.

M. Tsubouchi and H. Minoda (2021b). Phase plate STEM imaging using 2D electron detector. *Kenbikyo* **56** (Suppl. 1), 162.

K. Tsuno (1993). Resolution limit of a transmission electron microscope with an uncorrected conventional magnetic objective lens. *Ultramicroscopy* **50**, 245–253.

D. Typke and D. Köstler (1976). Reply to notes on transfer functions in electron microscopy with tilted illumination. *Optik* **45**, 495–498.

D. Typke and D. Köstler (1977). Determination of the wave aberration of electron lenses from superposition diffractograms of images with differently tilted illumination. *Ultramicroscopy* **2**, 285–295.

T. Uhlig and J. Zweck (2004). Direct observation of switching processes in permalloy rings with Lorentz microscopy. *Phys. Rev. Lett.* **93**, 047203.

T. Uhlig, M. Rahm, C. Dietrich, R. Hollinger, M. Heumann, D. Weiss and J. Zweck (2005). Shifting and pinning of a magnetic vortex core in a permalloy dot by a magnetic field. *Phys. Rev. Lett.* **95**, 237205.

M. Unser, B. L. Trus and A. C. Steven (1987a). A new resolution criterion based on spectral signal-to-noise ratios. *Ultramicroscopy* **23**, 39–51.

M. Unser, B. L. Trus and A. C. Steven (1987b). Resolution assessment from spectral signal-to-noise ratios. EMSA, Baltimore, **45**, 746–747.

P. N. T. Unwin (1970a). An electrostatic phase plate for the electron microscope. *Ber. Bunsen-Ges. Phys. Chem.* **74**, 1137–1141.

P. N. T. Unwin (1970b). Interference microscopy with a six lens electron microscope. ICEM-7, Grenoble, **1**, 65–66.

P. N. T. Unwin (1971). Phase contrast and interference microscopy with the electron microscope. *Philos. Trans. R. Soc. Lond. B* **261**, 95–104.

P. N. T. Unwin (1972). Electron microscopy of biological specimens by means of an electrostatic phase plate. *Proc. R. Soc. Lond. A* **329**, 327–359.

P. N. T. Unwin (1973). Phase contrast electron microscopy of biological materials. *J. Microsc. (Oxford)* **98**, 299–312.

P. N. T. Unwin (1974). A new electron microscope imaging method for enhancing detail in thin biological specimens. *Z. Naturforsch.* **29a**, 158–163.

K. Urban and M. Lentzen (2002). Application of aberration-corrected transmission electron microscopy to materials science. *Microsc. Microanal.* **8** (Suppl. 2), 8–9.

K. W. Urban, C.-l. Jia, L. Houben, M. Lentzen, S.-b. Mi and K. Tillmann (2009). Negative spherical aberration ultrahigh-resolution imaging in corrected transmission electron microscopy. *Philos. Trans. R. Soc. Lond. A, 3735–3753.*

U. Valdrè (1974). Recenti sviluppi della microscopia elettronica a contrasto di fase controllato, su immagini in fuoco. In *Atti del Convegno del Gruppo di Strumentazione e Tecniche Nonbiologiche della Società Italiana di Microscopia Elettronica* (P. G. Merli, A. Armigliato and L. Pedulli, Eds), 37–56 (CLUE, Bologna).

U. Valdrè (1979). Electron microscope stage design and applications. *J. Microsc. (Oxford)* **117**, 55–75.

L. Valzania, J. Dong and S. Gigan (2021). Accelerating ptychographic reconstructions using spectral initializations. *Opt. Lett.* **46**, 1357–1360.

S. Van Aert (2012). Statistical parameter estimation theory – a tool for quantitative electron microscopy. In *Handbook of Nanoscopy* (G. Van Tendeloo, D. Van Dyck and S. J. Pennycook, Eds), 281–308 (Wiley-VCH, Weinheim).

S. Van Aert (2019). Model-based electron microscopy. In *Handbook of Microscopy* (P. W. Hawkes and J. C. H. Spence, Eds), 605–624 (Springer, Cham).

S. Van Aert, A. J. den Dekker, D. Van Dyck and A. van den Bos (2002a). High-resolution electron microscopy and electron tomography: resolution vs precision. *J. Struct. Biol.* **138**, 21–33.

S. Van Aert, A. J. den Dekker, A. van den Bos and D. Van Dyck (2002b). High-resolution electron microscopy: from imaging toward measuring. *IEEE Trans. Instrum. Meas.* **51**, 611–615.

S. Van Aert, A. J. den Dekker, D. Van Dyck and A. van den Bos (2002c). Optimal experimental design of STEM measurement of atom column positions. *Ultramicroscopy* **90**, 273–289.

S. Van Aert, A. J. den Dekker and D. Van Dyck (2004). Statistical experimental design for quantitative atomic resolution transmission electron microscopy. *Adv. Imaging Electron Phys.* **130**, 1–164.

S. Van Aert, A. J. den Dekker, A. van den Bos, D. Van Dyck and J. H. Chen (2005). Maximum likelihood estimation of structure parameters from high resolution electron microscopy images. Part II: A practical example. *Ultramicroscopy* **104**, 107–125, Part I is den Dekker et al. (2005).

S. Van Aert, D. Van Dyck and A. J. den Dekker (2006). Resolution of coherent and incoherent imaging systems reconsidered – classical criteria and a statistical alternative. *Opt. Express* **14**, 3830–3939.

S. Van Aert, J. Verbeeck, R. Erni, S. Bals, M. Luysberg, D. Van Dyck and G. Van Tendeloo (2009). Quantitative atomic resolution mapping using high-angle annular dark field scanning transmission electron microscopy. *Ultramicroscopy* **109**, 1236–1244.

S. Van Aert, K. J. Batenburg, M. D. Rossell, R. Erni and G. Van Tendeloo (2011). Three-dimensional atomic imaging of crystalline nanoparticles. *Nature* **470**, 374–377.

S. Van Aert, W. Van den Broek, P. Goos and D. Van Dyck (2012a). Model-based electron microscopy: from images toward precise numbers for unknown structure parameters. *Micron* **43**, 509–515.

S. Van Aert, S. Turner, R. Delville, D. Schryvers, G. Van Tendeloo and E. K. H. Salje (2012b). Direct observation of ferrielectricity at ferroelastic domain boundaries in $CaTiO_3$ by electron microscopy. *Adv. Mater.* **24**, 523–527.

S. Van Aert, A. De Backer, G. T. Martinez, B. Goris, S. Bals, G. Van Tendeloo and A. Rosenauer (2013). Procedure to count atoms with trustworthy single-atom sensitivity. *Phys. Rev. B* **87**, 064107, 6 pp.

S. Van Aert, A. De Backer, G. T. Martinez, A. J. den Dekker, D. Van Dyck, S. Bals and G. Van Tendeloo (2016). Advanced electron crystallography through model-based imaging. *IUCrJ* **3**, 71–83.

S. Van Aert, J. Fatermans, A. De Backer, K.H.W. van den Bos, C.M. O'Leary, K. Müller-Caspary, L. Jones, I. Lobato, A. Béché, A. J. den Dekker, S. Bals and P. D. Nellist (2018a). Maximising dose efficiency in quantitative STEM to reveal the 3D atomic structure of nanomaterials. IMC-19, Sydney.

S. Van Aert, J. Fatermans, A.J. den Dekker, K. Müller-Caspary and I. Lobato (2018b). The maximum a posteriori probability rule to detect single atoms from low signal-to-noise ratio scanning transmission electron microscopy images. IMC-19, Sydney.

S. Van Aert, A. De Backer, L. Jones, G. T. Martinez, A. Béché and P. D. Nellist (2019). Control of knock-on damage for 3D atomic scale quantification of nanostructures: making every electron count in scanning transmission electron microscopy. *Phys. Rev. Lett.* **122**, 066101, 6 pp.

S. Van Aert, A. De Backer, A. De wael, J. Fatermans, T. Friedrich, I. Lobato, C. O'Leary, A. Varambhia, T. Altantzis, I. Jones, A. den Dekker, P. Nellist and S. Bals (2020). 3D atomic scale quantification of nanostructures and their dynamics using model-based STEM. *Microsc. Microanal.* **26** (Suppl. 2), 2606–2608.

S. Van Aert, A. De wael, E. Arslan Irmak, A. De Backer, Z. Zhang, T. Friedrich, D.G. Sentürk, I. Lobato, P. Liu, T. Altantzis, L. Jones, P.D. Nellist and S. Bals (2021). Novel approaches to investigate the 3D atomic structure of nanomaterials and their dynamics from STEM images using statistical parameter estimation and deep convolutional neural networks. MC-2021, Vienna (virtual), 4–5.

K. van Benthem, N. de Jonge, A. Y. Borisevich, M. P. Oxley and S. J. Pennycook (2006). 3D imaging with single atom sensitivity using confocal STEM. *Microsc. Microanal.* **12** (Suppl. 2), 1562–1563.

A. van den Bos (1991). Ultimate resolution: a mathematical framework. EMSA, San Jose, CA, **49**, 492–493.

A. van den Bos (1992). Ultimate resolution: a mathematical framework. *Ultramicroscopy* **47**, 298–306.

A. van den Bos (2007). *Parameter Estimation for Scientists and Engineers* (Wiley, Chichester).

A. van den Bos and A. J. den Dekker (2001). Resolution reconsidered—Conventional approaches and an alternative. *Adv. Imaging Electron Phys.* **117**, 241–360.

K. H. W. van den Bos, A. De Backer, G. T. Martinez, N. Winckelmans, S. Bals, P. D. Nellist and S. Van Aert (2016a). The atomic lensing model: extending HAADF STEM atom counting from homogeneous to heterogeneous nanostructures. EMC-16, Lyon, **1**, 499–500.

K. H. W. van den Bos, A. De Backer, G. T. Martinez, N. Winckelmans, S. Bals, P. D. Nellist and S. Van Aert (2016b). Unscrambling mixed elements using high angle annular dark field scanning transmission electron microscopy. *Phys. Rev. Lett.* **116**, 246101, 6 pp.

K. H. W. van den Bos, F. F. Krause, A. Béché, J. Verbeeck, A. Rosenauer and S. Van Aert (2016c). Precise atomic column position measurement using ISTEM. EMC-16, Lyon, **1**, 559–560.

K. H. W. van den Bos, F. F. Krause, A. Béché, J. Verbeeck, A. Rosenauer and S. Van Aert (2017). Locating light and heavy atomic column positions with picometer precision using ISTEM. *Ultramicroscopy* **172**, 75–81.

W. Van den Broek, S. Van Aert, P. Goos and D. Van Dyck (2011). Throughput maximization of particle radius measurements through balancing size vs current of the electron probe. *Ultramicroscopy* **111**, 940–947.

W. Van den Broek, S. Van Aert and D. van Dyck (2012). Fully automated measurement of the modulation transfer function of charge-coupled devices above the Nyquist frequency. *Microsc. Microanal.* **18**, 336–342.

A. C. van Dorsten, J. E. Mellema and H. F. Premsela (1968). Possible procedures for dealing with the problem of simultaneous occurrence of amplitude and phase contrast in images of macromolecules. EUREM-4, Rome, **2**, 103–104.

D. Van Dyck (1989). Future prospects in high resolution electron microscopy of amorphous materials. *Microscopia Elettronica* **10** (Pt 2 (Suppl.)), 245–246.

D. Van Dyck (1991a). Research thrusts and the BRITE–EURAM project. EMSA, San Jose, CA, **49**, 448–449.

D. Van Dyck (1991b). Ultimate resolution and information in electron microscopy. EMSA, San Jose, CA, **49**, 496–497.

D. Van Dyck (1992). High resolution electron microscopy. In *Electron Microscopy in Materials Science* (P. G. Merli and M. V. Antisari, Eds), 193–268 (World Scientific, Singapore, River Edge NJ and London).

D. Van Dyck (2002). High-resolution electron microscopy. *Adv. Imaging Electron Phys.* **123**, 105–171.

D. Van Dyck (2010). Wave reconstruction in TEM using a variable phase plate. *Ultramicroscopy* **110**, 571–572.

D. Van Dyck and W. Coene (1987). A new procedure for wave function restoration in high resolution electron microscopy. *Optik* **77**, 125–128.

D. Van Dyck and A. F. de Jong (1992). Ultimate resolution and information in electron microscopy: general principles. *Ultramicroscopy* **47**, 266–281.

D. Van Dyck, M. Op de Beeck and W. Coene (1993). A new approach to object wavefunction reconstruction in electron microscopy. *Optik* **93**, 103–107.

D. Van Dyck, E. Bettens, J. Sijbers, A. J. den Dekker, A. van den Bos, M. Op de Beeck, J. Jansen and H. Zandbergen (1997). Resolving atoms: what do we have?, what do we want?. EMAG, Cambridge, 95–100.

D. Van Dyck, S. Van Aert, A. J. den Dekker and A. van den Bos (2003). Is atomic resolution transmission electron microscopy able to resolve and refine amorphous structures? *Ultramicroscopy* **98**, 27–42.

R. van Gastel, I. Sikharulidze, S. Schramm, J. P. Abrahams, B. Poelsema, R. M. Tromp and S. J. van der Molen (2009). Medipix 2 detector applied to low energy electron microscopy. *Ultramicroscopy* **110**, 33–35.

M. G. van Heel (1978). On the imaging of relatively strong objects in partially coherent illumination in optics and electron optics. *Optik* **49**, 389–408.

J. P. van Schayck, E. van Genderen, E. Maddox, L. Roussel, H. Boulanger, E. Fröjdh, J.-P. Abrahams, P. J. Peters and R. B. G. Ravelli (2020). Sub-pixel electron detection using a convolutional neural network. *Ultramicroscopy* **218**, 113091, 10 pp.

P. van Toorn and D. W. Robinson (1980). Computer simulation of partially coherent image formation in two dimensions. *Optik* **56**, 323–331.

A. M. Varambhia, L. Jones, A. De Backer, V. I. Fauske, S. Van Aert, D. Ozkaya and P. D. Nellist (2016). Quantifying a heterogeneous Ru catalyst on carbon black using ADF STEM. *Part. Part. Syst. Charact.* **33**, 438–444.

L. H. Veneklasen (1975). On line holographic imaging in the scanning transmission electron microscope. *Optik* **44**, 447–468.

J. Verbeeck, A. Béché, G. Guzzinati, D. Jannis and K. Müller-Caspary (2018a). Design and realisation of pixelated programmable phase plate for electrons. IMC-19, Sydney.

J. Verbeeck, A. Béché, K. Müller-Caspary, G. Guzzinati, M. A. Luong and M. den Hertog (2018b). Demonstration of a 2 x 2 programmable phase plate for electrons. *Ultramicroscopy* **190**, 58–65.

M. von Ardenne (1938a). Das Elektronen-Rastermikroskop. Theoretische Grundlagen. *Z. Phys.* **109**, 553–572.

M. von Ardenne (1938b). Das Elektronen-Rastermikroskop. Praktische Ausführung. *Z. Tech. Phys.* **19**, 407–416.

B. von Borries and F. Lenz (1956). Über die Entstehung des Kontrastes im elektronenmikroskopischen Bild. EUREM-1, Stockholm, 60–64.

H. S. von Harrach (1994). Medium-voltage field-emission STEM – the ultimate AEM. *Microsc. Microanal. Microstruct.* **5**, 153–164.

H. S. von Harrach (1995). Instrumental factors in high-resolution FEG STEM. *Ultramicroscopy* **58**, 1–5.

H. S. von Harrach (2009). Development of the 300-kV Vacuum Generators STEM (1985–1996). *Adv. Imaging Electron Phys.* **159**, 287–323.

H.S. von Harrach and D. Krause (1994). Maximum Entropy Reconstruction of Atomic Resolution STEM Images. ICEM-13, Paris, **1**, 479–480.

H. S. von Harrach, D. C. Joy, G. E. Verney and A. R. Walker (1974). Ein Hochauflösungs-Raster-Elektronenmikroskop. *Beiträge zur Elektronenmikroskopischen Direktabbildung von Oberflächen* **7**, 139–147.

H. S. von Harrach, A. W. Nicholls, D. E. Jesson and S. J. Pennycook (1993). First results of a 300 kV high resolution field-emission STEM. EMAG, Liverpool, 499–502.

H. von Helmholtz (1887). Ueber die physikalische Bedeutung des Princips der kleinsten Wirking. *J. Reine. Angew. Math. (Crelle's J.)* **100**, 137–166 and 213–222.

O. von Loeffelholz, G. Papai, R. Danev, A. G. Myasnikov, S. K. Natchiar, I. Hazemann, J.-F. Ménétret and B. P. Klaholz (2018). Volta phase plate data collection facilitates image processing and cryo-EM structure determination. *J. Struct. Biol.* **202**, 191–199.

E. M. Waddell and J. N. Chapman (1979). Linear imaging of strong phase objects using asymmetrical detectors in STEM. *Optik* **54**, 83–96.

E. M. Waddell, J. N. Chapman and R. P. Ferrier (1977). Linear imaging of strong phase objects by differential phase contrast. EMAG, Glasgow, 267–270.

E. M. Waddell, J. N. Chapman and R. P. Ferrier (1978). The role of differential phase contrast in electron microscopy. ICEM-9, Toronto, **1**, 176–177.

R. H. Wade (1976a). Concerning tilted beam electron microscope transfer functions. *Optik* **45**, 87–91.

R. H. Wade (1976b). Tilted-beam electron microscopy of amorphous films. *Phys. Status Solidi A* **37**, 247–256.

R. H. Wade (1978). The phase contrast characteristics in bright field electron microscopy. *Ultramicroscopy* **3**, 329–334.

R. H. Wade and J. Frank (1977). Electron microscope transfer functions for partially coherent axial illumination and chromatic defocus spread. *Optik* **49**, 81–92.

R. H. Wade and W. K. Jenkins (1978). Tilted beam electron microscopy: the effective coherent aperture. *Optik* **50**, 1–17.

A. Walter, H. Muzik, H. Vieker, A. Turchanin, A. Beyer, A. Gölzhäuser, M. Lacher, S. Steltenkamp, S. Schmitz, P. Holik, W. Kühlbrandt and D. Rhinow (2012). Practical aspects of Boersch phase contrast electron microscopy of biological specimens. *Ultramicroscopy* **116**, 62–72.

A. Walter, S. Steltenkamp, S. Schmitz, P. Holik, E. Pakanavicius, R. Sachser, M. Huth, D. Rhinow and W. Kühlbrandt (2015). Towards an optimum design for electrostatic phase plates. *Ultramicroscopy* **153**, 22–31.

Z. L. Wang (1994). Dislocation contrast in high-angle hollow-cone dark-field TEM. *Ultramicroscopy* **53**, 73–90.

S.-y. Wang, M. Mankos and J.M. Cowley (1992). Configured detectors in STEM holography. EMSA 50, Boston, MA, 982–983.

Y. C. Wang, T. M. Chou, M. Libera and T. F. Kelly (1997). Transmission electron holography of silicon nanospheres with surface oxide layers. *Appl. Phys. Lett.* **70**, 1296–1298.

P. Wang, G. Behan, A. I. Kirkland and P. D. Nellist (2009). Energy filtered scanning confocal electron microscopy in a double aberration-corrected transmission electron microscope. *Microsc. Microanal.* **15** (Suppl. 2), 42–43.

P. Wang, G. Behan, M. Takeguchi, A. Hashimoto, K. Mitsuishi, M. Shimojo, A. I. Kirkland and P. D. Nellist (2010). Nanoscale energy-filtered scanning confocal electron microscopy using a double-aberration-corrected transmission electron microscope. *Phys. Rev. Lett.* **104**, 218101.

P. Wang, G. Behan, A. I. Kirkland, P. D. Nellist, E. C. Cosgriff, A. J. d'Alfonso, A. J. Morgan, L. J. Allen, A. Hashimoto, M. Takeguchi, K. Mitsuishi and M. Shimojo (2011). Bright-field scanning confocal electron microscopy using a double aberration-corrected transmission electron microscope. *Ultramicroscopy* **111**, 877–886.

P. Wang, A. Hashimoto, M. Takeguchi, K. Mitsuishi, M. Shimojo, Y. Zhu, M. Okuda, A. I. Kirkland and P. D. Nellist (2012a). Three-dimensional elemental mapping of hollow Fe_2O_3@SiO_2 mesoporous spheres using scanning confocal electron microscopy. *Appl. Phys. Lett.* **100**, 213117.

P. Wang, A. I. Kirkland, P. D. Nellist, A. J. d'Alfonso, A. J. Morgan, L. J. Allen, A. Hashimoto, M. Takeguchi, K. Mitsuishi and M. Shimojo (2012b). Current developments of scanning confocal electron microscopy in a double aberration-corrected transmission electron microscope. *Microsc. Microanal.* **18** (Suppl. 2), 532–533.

A. Wang, S. Van Aert, P. Goos and D. Van Dyck (2012c). Precision of three-dimensional atomic scale measurements from HRTEM images: what are the limits? *Ultramicroscopy* **114**, 20–30.

P. Wang, A. J. d'Alfonso, A. Hashimoto, A. J. Morgan, M. Takeguchi, K. Mitsuishi, M. Shimojo, A. I. Kirkland, L. J. Allen and P. D. Nellist (2013). Contrast in atomically resolved EF-SCEM imaging. *Ultramicroscopy* **134**, 185–192.

P. Wang, C. B. Boothroyd, R. E. Dunin-Borkowski, A. I. Kirkland and P. D. Nellist (2014a). Towards 4-D EEL spectroscopic scanning confocal electron microscopy (SCEM-EELS) optical sectioning on a C_c and C_s double-corrected transmission electron microscope. IMC-18, Prague, 2435.
P. Wang, A. I. Kirkland, P. D. Nellist, A. J. d'Alfonso, A. J. Morgan, L. J. Allen, A. Hashimoto, M. Takeguchi, K. Mitsuishi and M. Shimojo (2014b). Atomically resolved scanning confocal electron microscopy using a double aberration-corrected transmission electron microscope. *Microsc. Microanal.* **20** (Suppl. 3), 376–377.
P. Wang, F. Zhang, S. Gao, M. Zhang and A. I. Kirkland (2017). Electron ptychographic diffractive imaging of boron atoms in LaB6 crystals. *Sci. Rep.* **7**, 2857, 8 pp.
G. Wang, G. B. Giannakis and Y. C. Eldar (2018). Solving systems of random quadratic equations via truncated amplitude flow. *IEEE Trans. Information Theory* **64**, 773–794.
J.F.L. Ward, M.T. Browne and R.E. Burge (1979). A multi-channel detector system for STEM. EMAG, Brighton, 85–88.
I. R. M. Wardell (1981). Some design considerations concerning a STEM probe forming system. *Ultramicroscopy* **7**, 39–44.
I. R. M. Wardell and P. E. Bovey (2009). A history of Vacuum Generators' 100-kV scanning transmission electron microscope. *Adv. Imaging Electron Phys.* **159**, 221–285.
I. R. M. Wardell, J. Morphew and P. E. Bovey (1973). Results and performance of a high resolution STEM. EMAG, Newcastle-upon-Tyne, 182–185.
J. H. Warner, Z. Liu, K. He, A. W. Robertson and K. Suenaga (2013). Sensitivity of graphene edge states to surface adatom interactions. *Nano Lett.* **13**, 4820–4826.
K. Watanabe (2015). Theory for HAADF-STEM and its image simulation. In *Scanning Transmission Electron Microscopy of Materials* (N. Tanaka, Ed.), 179–216 (Imperial College Press, London).
G. N. Watson (1922). *A Treatise on the Theory of Bessel Functions* (Cambridge University Press, Cambridge).
D. Weber, A. Clausen and R. E. Dunin-Borkowski (2020). Next-generation information technology systems for fast detectors in electron microscopy. In *Handbook on Big Data and Machine Learning in the Physical Sciences* (S. Kalinin, Ed.), 83–120 (World Scientific, Singapore).
D. Weber, A. Strauch, A. Clausen, A. Bangun, A. Lesnichaia and K. Müller-Caspary (2021). Live scanning ptychography with the LiberTEM software framework. MC-2021, Vienna (virtual), 575–576.
X. Wei, H. P. Urbach, P. van der Walle and W. M. J. Coene (2021). Parameter retrieval of small particles in dark-field Fourier ptychography and a rectangle in real-space ptychography. *Ultramicroscopy* **229**, 113335, 11 pp.
W. T. Welford (1972). On the relationship between the modes of image formation in scanning microscopy and conventional microscopy. *J. Microsc. (Oxford)* **96**, 105–107.
Y. Wen, C. Ophus, C. S. Allen, S. Fang, J. Chen, E. Kaxiras, A. I. Kirkland and J. H. Warner (2019). Simultaneous identification of low and high atomic number atoms in monolayer 2D materials using 4D scanning transmission electron microscopy. *Nano Lett.* **19**, 6482–6491.
T. Weßels, S. Däster, Y. Murooka, B. Zingsem, V. Migunov, M. Kruth, S. Finizio, P-H. Lu, A. Kovács, A. Oelsner, K. Müller-Caspary, Y. Acremann and Rafal E. Dunin-Borkowski (2022). Continuous illumination picosecond imaging using a delay line detector in a transmission electron microscope. *Ultramicroscopy* 113392 (in press).
D. Willasch (1973). Versuche zur Kontrastverbesserung in der Elektronenmikroskopie durch Hellfeldabbildung mittels Phasenplatten und Dunkelfeldabbildung bei hohlkegelförmiger Beleuchtung. Dissertation, Tübingen.
D. Willasch (1975a). High resolution electron microscopy with profiled phase plates. *Optik* **44**, 17–36.
D. Willasch (1975b). Recent developments in high resolution electron microscopy. EMAG, Bristol, 185–190.
D. Willasch (1975c). Das neue Transmissions-Rasterelektronenmikroskop ELMISKOP ST100F. *Beiträge zur Elektronenmikroskopischen Direktabbildung von Oberflächen* **8**, 757–770.
D. Willasch (1976a). High resolution scanning transmission electron microscopy: aspects and trends of a new method. *J. Microsc. Spectrosc. Electron* **1**, 505–506.
D. Willasch (1976b). Features and applications of a high resolution scanning transmission electron microscope. In *Proceedings of the 15th Annual Conference on Electron Microscopy*, 1–2 (Society of Southern Africa, Johannesburg).

T. Wilson (1990). The role of detector geometry in confocal imaging. *J. Microsc. (Oxford)* **158**, 133–144.

T. Wilson and A. R. Carlini (1989). The effect of aberrations on the axial response of confocal imaging systems. *J. Microsc. (Oxford)* **154**, 243–256.

T. Wilson and D. R. Hamilton (1984). Difference confocal scanning microscopy. *Opt. Acta* **31**, 453–465.

F. Winkler, A. H. Tavabi, J. Barthel, M. Duchamp, E. Yucelen, S. Borghardt, B. E. Kardynal and R. E. Dunin-Borkowki (2017). Quantitative measurement of mean inner potential and specimen thickness from high-resolution off-axis electron holograms of ultrathin layered WSe_2. *Ultramicroscopy* **178**, 38–47.

D. Wolf, A. Lubk, A. Lenk, S. Sturm and H. Lichte (2013). Tomographic investigation of Fermi level pinning at focused ion beam milled semiconductor surfaces. *Appl. Phys. Lett.* **103**, 264104.

Y.-m. Wu, C.-h Wang, J.-w. Chang, Y.-y. Chen, N. Miyazaki, K. Murata, K. Nagayama and W.-h. Chang (2013). Zernike phase contrast cryo-electron microscopy reveals 100 kDa component in a protein complex. *J. Phys. D.: Appl. Phys* **46**, 494008.

H. L. Xin and D. A. Muller (2009). Aberration-corrected ADF-STEM depth sectioning and prospects for reliable 3D imaging in S/TEM. *J. Electron Microsc.* **58**, 157–165.

H. L. Xin, V. Intaraprasonk and D. A. Muller (2008). Depth sectioning of individual dopant atoms with aberration-corrected scanning transmission electron microscopy. *Appl. Phys. Lett.* **92**, 013125.

H. L. Xin, C. Dwyer, D. A. Muller, H. Zheng and P. Ercius (2012). Scanning confocal electron energy loss microscopy in TEAM 1.0 with post specimen C_c correction. *Microsc. Microanal.* **18** (Suppl. 2), 400–401.

H. L. Xin, C. Dwyer, D. A. Muller, H. Zheng and P. Ercius (2013a). Scanning confocal electron energy-loss microscopy using valence-loss signals. *Microsc. Microanal.* **19**, 1036–1049.

H. L. Xin, H. Zheng and P. Ercius (2013b). Post-specimen C_c correction enabled scanning confocal electron energy loss microscopy for high-throughput 3-D spectroscopic imaging of nanomaterials. *Microsc. Microanal.* **19** (Suppl. 2), 536–537.

R. Xu, C.-c Chen, L. Wu, M. C. Scott, W. Theis, C. Ophus, M. Bartels, Y. Yang, H. Ramezani-Dakhel, M. R. Sawaya, H. Heinz, L. D. Marks, P. Ercius and J. Miao (2015). Three-dimensional coordinates of individual atoms in materials revealed by electron tomography. *Nature Mater.* **14**, 1099–1103.

M. Yamaguchi, R. Danev, K. Nishiyama, K. Sugawara and K. Nagayama (2008). Zernike phase contrast electron microscopy of ice-embedded influenza A virus. *J. Struct. Biol.* **162**, 271–276.

H. Yang, P. Ercius, P. D. Nellist and C. Ophus (2016). Enhanced phase contrast transfer using ptychography combined with a pre-specimen phase plate in a scanning transmission electron microscope. *Ultramicroscopy* **171**, 117–125.

H. Yang, I. MacLaren, L. Jones, G. T. Martinez, M. Simson, M. Huth, H. Ryll, H. Soltau, R. Sagawa, Y. Kondo, C. Ophus, P. Ercius, L. Jin, A. Kovács and P. D. Nellist (2017). Electron ptychographic phase imaging of light elements in crystalline materials using Wigner distribution deconvolution. *Ultramicroscopy* **180**, 173–179.

Y. Yang, C.-s Kim, R. G. Hobbs, P. D. Keathley and K. K. Berggren (2020). Nanostructured-membrane electron phase plates. *Ultramicroscopy* **217**, 113053, 7 pp.

A. B. Yankovich, B. Berkels, W. Dahmen, P. Binev, S. I. Sanchez, S. A. Bradley, A. Li, I. Szlufarska and P. M. Voyles (2014). Picometre-precision analysis of scanning transmission electron microscopy images of platinum nanocatalysts. *Nature Commun.* **5**, 4155, 7 pp.

A. B. Yankovich, B. Berkels, W. Dahmen, P. Binev and P. M. Voyles (2015). High-precision scanning transmission electron microscopy at coarse pixel sampling for reduced electron dose. *Adv. Struct, Chem. Imaging* **1**, 2, 5 pp.

R. Yu, M. Lentzen and J. Zhu (2012). Effective object planes for aberration-corrected transmission electron microscopy. *Ultramicroscopy* **112**, 15–21.

M. Yu, A. B. Yankovich, A. Kaczmarowski, D. Morgan and P. M. Voyles (2016). Integrated computational and experimental structure refinement for nanoparticles. *ACS Nano* **10**, 4031–4038.

E. Yücelen, I. Lazić and E. G. T. Bosch (2018). Phase contrast scanning transmission electron microscopy imaging of light and heavy atoms at the limit of contrast and resolution. *Sci. Rep.* **8**, 2676, 10 pp.

H. Yui, H. Minamikawa, R. Danev, K. Nagayama, S. Kamiya and T. Shimizu (2008). Growth process and molecular packing of a self-assembled lipid nanotube: phase-contrast transmission electron microscopy and XRD analyses. *Langmuir* **24**, 709–713.

N. J. Zaluzec (2003). The scanning confocal electron microscope. *Microsc. Today* **11**, 8–13.

N. J. Zaluzec (2007). Scanning confocal electron microscopy. *Microsc. Microanal.* **13** (Suppl. 2), 1560–1561.

N. J. Zaluzec, M. Weyland and J. Etheridge (2009). Scanning confocal electron microscopy in a FEI double corrected Titan TEM/STEM. *Microsc. Microanal.* **15** (Suppl. 2), 614–615.

E. Zeitler (1975). Coherence in scanning transmission microscopy. *Scanning Electron Microsc.* 671–678.

E. Zeitler and M.G.R. Thomson (1969). The theoretical contrast in scanning and conventional high resolution microscopes. EMSA, St Paul, MN, **27**, 170–171.

E. Zeitler and M. G. R. Thomson (1970). Scanning transmission electron microscopy. I, II. *Optik* **31**, 258–280 and 359–366.

S. E. Zeltmann, A. Müller, K. C. Bustillo, B. Savitzky, L. Hughes, A. M. Minor and C. Ophus (2020). Patterned probes for high precision 4D-STEM Bragg measurement. *Ultramicroscopy* **209**, 112890, 9 pp.

F. Zemlin and P. Schiske (1980). Measurement of the phase contrast transfer function and the cross-correlation peak using Young interference fringes. *Ultramicroscopy* **5**, 139–145.

F. Zemlin and K. Weiss (1993). Young's interference fringes in electron microscopy revisited. *Ultramicroscopy* **50**, 123–126.

F. Zemlin, W. Kunath and K. Weiss (1982). Optimization of phase contrast imaging by hollow cone illumination and simultaneous image translation. ICEM-10, Hamburg, **1**, 199–200.

F. Zernike (1933). Een nieuwe methode van microscopische waarneming. In *Handeling van het Natuur- en Heelkundig Congres*, **24**, 100. Also in Zernike's collected papers, No. 35, p. 484.

F. Zernike (1934). Beugungstheorie des Schneidenverfahrens und seiner verbesserten Form, der Phasekontrastmethode. *Physica* **1**, 689–704.

F. Zernike (1935). Das Phasenkontrastverfahren bei der mikroskopischen Beobachtung. *Z. Tech. Phys.* **16**, 454–457, Also published in *Phys Z.* **36** (1935) 848–851.

F. Zernike (1942a). De theorie van de microscopische beeldforming. *Nederlandsch Tijdschrift voor Natuurkunde* **9**, 357–377.

F. Zernike (1942b). Phase contrast, a new method for the microscopic observation of transparent objects. *Physica* **9**, 686–698 and 974–986. Shorter version in A. Bouwers, *Achievements in Optics* (Elsevier, New York & Amsterdam, 1946).

F. Zernike (1953). How I discovered phase contrast. In *Nobel Lectures, Physics, 1942–1962*, 235–249. (Elsevier, Amsterdam), Republished by World Scientific, 1998.

F. Zernike and F. J. M. Stratton (1934). Diffraction theory of the knife-edge test and its improved form, the phase contrast method. *Monthly Not. R. Astron. Soc.* **94**, 377–384.

K. Zhang (2016). Gctf: real-time CTF determination and correction. *J. Struct. Biol.* **193**, 1–12.

X. Zhang, M. Takeguchi, A. Hashimoto, K. Mitsuishi and M. Shimojo (2010). Deconvolution method used in improving the depth resolution of three-dimensional images taken by scanning confocal electron microscopy. *Microsc. Microanal.* **16** (Suppl. 2), 290–291.

X. Zhang, M. Takeguchi, A. Hashimoto, K. Mitsuishi, M. Tezuka and M. Shimojo (2012). Improvement of depth resolution of ADF-STEM by deconvolution: effect of electron energy loss and chromatic aberration on depth resolution. *Microsc. Microanal.* **18**, 603–611.

G. Zheng (2016). *Fourier Ptychographic Imaging. A MATLAB tutorial* (Morgan & Claypool, San Rafael, CA), eBook).

C. L. Zheng and J. Etheridge (2013). Measurement of chromatic aberration in STEM and SCEM by coherent convergent beam electron diffraction. *Ultramicroscopy* **125**, 49–58.

G. Zheng, R. Horstmeyer and C. Yang (2013). Wide-field, high-resolution Fourier ptychographic microscopy. *Nature Photonics* **7**, 739–745.

C. Zheng, Y. Zhu, S. Lazar and J. Etheridge (2014a). Fast imaging with inelastically scattered electrons by off-axis chromatic confocal electron microscopy. *Phys. Rev. Lett.* **112**, 166101, 5 pp.

C. Zheng, Y. Zhu, S. Lazar and J. Etheridge (2014b). Fast imaging with inelastically scattered electrons by off-axis chromatic confocal electron microscopy. IMC-18, Prague, 2674.
C. Zheng, Y. Zhu, S. Lazar and J. Etheridge (2015). Off-axis chromatic scanning confocal electron microscopy for inelastic imaging with atomic resolution. *Microsc. Microanal.* **21** (Suppl. 3), 2175–2176.
C.-l. Zheng, Y. Zhu, S. Lazar and J. Etheridge (2016). Three dimensional confocal imaging using coherent elastically scattered electrons. EMC-16, Lyon, **1**, 19–20.
C. Zheng, L. Sorin, Y. Zhu and J. Etheridge (2017). Three-dimensional confocal imaging using coherent elastically scattered electrons. *Microsc. Microanal.* **23** (Suppl. 1), 450–451.
C. Zheng, T. Petersen, H. Kirmse, W. Neumann, M. Morgan and J. Etheridge (2018). Shaping electron beam using magnetic vortex. IMC-19, Sydney.
J. M. Zuo and J. C. H. Spence (2017). *Advanced Transmission Electron Microscopy* (Springer, New York).
J. Zweck, F. Schwarzhuber, J. Wild and V. Galioit (2016). On detector linearity and precision of beam shift detection for quantitative differential phase contrast applications. *Ultramicroscopy* **168**, 53–64.

Conference Proceedings

1. International Congresses on Electron Microscopy, later International Microscopy Congresses.
2. European Regional Congresses on Electron Microscopy, later European Microscopy Congresses.
3. Asia-Pacific Congresses on Electron Microscopy, later Asia–Pacific Microscopy Congresses.
4. Charged Particle Optics Conferences.
5. High-voltage Electron Microscopy Conferences.
6. EMAG [Electron Microscopy and Analysis Group of the Institute of Physics] meetings.
7. Multinational Congresses on (Electron) Microscopy (MCEM, MCM).
8. The Dreiländertagungen (Germany, Austria, Switzerland) and related meetings.
9. Recent Trends in Charged Particle Optics and Surface Physics Instrumentation (Skalský Dvůr).
10. SPIE Proceedings.
11. Soviet All-Union Conferences on Electron Microscopy, later Russian Conferences on Electron Microscopy.
12. Problems of Theoretical and Applied Electron Optics [Problemyi Teoreticheskoi i Prikladnoi Elektronnoi Optiki].
13. Mathematical Morphology and Its Applications to Signal and Image Processing.
14. Related Meetings.

* * *

The following list gives full publishing details of the series of International and Regional conferences on Electron Microscopy. The South American (CIASEM) conferences are not listed as they contain little optics. For the reader's convenience, a few other meetings are included, in particular those on charged particle optics, the Multinational Conferences on Electron Optics (now Multinational Conferences on Microscopy), the Dreiländertagungen (now Microscopy Conferences) and the conferences organized by the Electron Microscopy and Analysis Group (EMAG) of the British Institute of Physics. In the lists of references, these are referred to by their acronyms and venue. The irregular, short-lived series of meetings on high-voltage electron microscopy is identified by the acronym HVEM.

The list does not include the proceedings of the annual meetings of the Electron Microscopy Society of America, which are identified in the reference lists by EMSA or MSA, venue and the meeting number. Proceedings were first issued for the 25th meeting (1967) and have been published ever since, at first in print and more recently on-line, today as Supplements to *Microscopy and Microanalysis*. For full details, see the lists published by Hawkes in *Advances in Imaging and Electron Physics*, Vol. 117 (2003) 203–379 and Vol. 190 (2015) 143–175.

Many other national electron microscopy societies publish proceedings of their major meetings but few contain much optics. A notable exception is the series of All-Union meetings held in Russia, the proceedings of which have mostly been published in *Izv. Akad. Nauk (Ser. Fiz.)*, translated as *Bull. Acad. Sci. USSR (Phys. Ser.)*, now *Izv. Ross. Akad. Nauk (Fiz.)*, translated as *Bull. Russ. Acad. Sci (Phys.)*; a few papers appeared in *Radiotekhnika i Elektronika (Radio Engineering and Electronic Physics* and later *Soviet Journal of Communications Technology and Electronics)*. Brief details of these are given at the end of the main list. The other noteworthly exception is Japan; abstracts of Japanese national meetings were published regularly and rapidly in the *Journal of Electron Microscopy* and now appear in a supplement to *Kenbikyo*.

Details of other related meetings are to be found in the articles by Hawkes mentioned above, notably the International Congresses on X-ray Optics and Microscopy (ICXOM), the Low-energy Electron Microscopy and Photoemission Electron Microscopy (LEEM, PEEM) meetings and Frontiers of Aberration-corrected Electron Microscopy (PICO).

1. International Congresses on Electron Microscopy, later International Microscopy Congresses

ICEM-1, Delft, 1949: *Proceedings of the Conference on Electron Microscopy*, Delft, 4–8 July 1949 (A.L. Houwink, J.B. Le Poole and W.A. Le Rütte, Eds) Hoogland, Delft, 1950. Available at Delft-1949_Proceedings EMConference.pdf.

ICEM-2, Paris, 1950: *Comptes Rendus du Premier Congrès International de Microscopie Electronique*, Paris, 14–22 September 1950. Editions de la Revue d'Optique Théorique et Instrumentale, Paris, 1953.

ICEM-3, London, 1954: *The Proceedings of the Third International Conference on Electron Microscopy*, London, 1954 (R. Ross, Ed.) Royal Microscopical Society, London, 1956.

ICEM-4, Berlin, 1958: *Vierter Internationaler Kongress für Elektronenmikroskopie*, Berlin, 10–17 September, 1958, *Verhandlungen* (W. Bargmann, G. Möllenstedt, H. Niehrs, D. Peters, E. Ruska and C. Wolpers, Eds) Springer, Berlin, 1960. 2 Vols; on-line via SpringerLink.

ICEM-5, Philadelphia, PA, 1962: *Electron Microscopy. Fifth International Congress for Electron Microscopy*, Philadelphia, PA, Pennsylvania, 29 August to 5 September, 1962 (S.S. Breese, Ed.) Academic Press, New York, 1962. 2 Vols.

ICEM-6, Kyoto, 1966: *Electron Microscopy 1966. Sixth International Congress for Electron Microscopy*, Kyoto (R. Uyeda, Ed.) Maruzen, Tokyo, 1966. 2 Vols.

ICEM-7, Grenoble, 1970: *Microscopie Electronique 1970. Résumés des Communications Présentées au Septième Congrès International*, Grenoble (P. Favard, Ed.) Société Française de Microscopie Electronique, Paris, 1970. 3 Vols.

ICEM-8, Canberra, 1974: *Electron Microscopy 1974. Abstracts of Papers Presented to the Eighth International Congress on Electron Microscopy*, Canberra (J. V. Sanders and D. J. Goodchild, Eds) Australian Academy of Science, Canberra, 1974. 2 Vols.

ICEM-9, Toronto, 1978: *Electron Microscopy 1978. Papers Presented at the Ninth International Congress on Electron Microscopy*, Toronto (J. M. Sturgess, Ed.) Microscopical Society of Canada, Toronto, 1978. 3 Vols.

ICEM-10, Hamburg, 1982: *Electron Microscopy, 1982. Papers Presented at the Tenth International Congress on Electron Microscopy*, Hamburg, 17–24 August 1982. Deutsche Gesellschaft für Elektronenmikroskopie, Frankfurt, 1982. 3 Vols.

ICEM-11, Kyoto, 1986: *Electron Microscopy 1986. Proceedings of the XIth International Congress on Electron Microscopy*, Kyoto, 31 August–7 September 1986 (T. Imura, S. Maruse and T. Suzuki, Eds). Japanese Society of Electron Microscopy, Tokyo. 4 Vols; published as a supplement to *J. Electron Microsc.* **35** (1986).

ICEM-12, Seattle, WA, 1990: *Electron Microscopy 1990. Proceedings of the XIIth International Congress for Electron Microscopy*, Seattle WA, 12–18 August 1990 (L. D. Peachey and D. B. Williams, Eds). San Francisco Press, San Francisco, CA. 4 Vols. See also *Ultramicroscopy* **36** (1991) Nos 1–3, 1–274.

ICEM-13, Paris 1994: *Electron Microscopy 1994. Proceedings of the 13th International Congress on Electron Microscopy*, Paris, 17–22 July 1994 [B. Jouffrey, C. Colliex, J. P. Chevalier, F. Glas, P. W. Hawkes, D. Hernandez–Verdun, J. Schrevel and D. Thomas, (Vol 1), B. Jouffrey, C. Colliex, J. P. Chevalier, F. Glas and P. W. Hawkes (Vols 2A and 2B) and B. Jouffrey, C. Colliex, D. Hernandez–Verdun, J. Schrevel and D. Thomas (Vols 3A and 3B), Eds]. Editions de Physique, Les Ulis, 1994. The complete list of contents is available on the website of the TIB, Hannover.

ICEM-14, Cancún 1998: *Electron Microscopy 1998. Proceedings of the 14th International Congress on Electron Microscopy*, Cancún, 31 August–4 September 1998 [Memorias del 14to Congreso Internacional de Microscopía Electrónica celebrado en Cancún (México) del 31 de Agosto al 4 de Septiembre de 1998] (H. A. Calderón Benavides and M. J. Yacamán, Eds). Institute of Physics Publishing, Bristol and Philadelphia, PA 1998. 4 Vols. See also *Micron* **31** (2000), No. 5.

ICEM-15, Durban 2002: *Electron Microscopy 2002. Proceedings of the 15th International Congress on Electron Microscopy*, International Convention Centre, Durban, 1–6 September 2002 [R. Cross, P. Richards, M. Witcomb and J. Engelbrecht (vol. 1, Physical, Materials and Earth Sciences), R. Cross, P. Richards, M. Witcomb and T. Sewell (vol. 2, Life Sciences) and R. Cross, P. Richards, M. Witcomb, J. Engelbrecht and T. Sewell (vol. 3, Interdisciplinary), Eds]. Microscopy Society of Southern Africa, Onderstepoort, 2002.

IMC-16, Sapporo 2006: *Proceedings 16th International Microscopy Conference, "Microscopy for the 21st Century"*, Sapporo, 3–8 September 2006 (H. Ichinose and T. Sasaki, Eds). Vol. 1, Biological and Medical Science; Volume 2, Instrumentation; Volume 3, Materials Science. Publication Committee of IMC16, Sapporo, 2006.

IMC-17, Rio de Janeiro 2010: *Proceedings IMC17, The 17th IFSM International Microscopy Congress*, Rio de Janeiro, 19–24 September 2010 (G. Solórzano and W. de Souza, Eds). Sociedade Brasileira de Microscopia e Microanálise, Rio de Janeiro, 2010.

IMC-18, Prague 2014: Prague Convention Centre, 7–12 September 2014. Proceedings open-access at http://www.microscopy.cz/proceedings/all.html, edited by P. Hozak.

IMC-19, Sydney, International Convention Centre, 9–14 September 2018.

IMC-20, Busan (South Korea), 10–15 September 2023.

2. European Regional Congresses on Electron Microscopy, later European Microscopy Congresses

EUREM-1, Stockholm, 1956: *Electron Microscopy. Proceedings of the Stockholm Conference*, September, 1956 (F. J. Sjöstrand and J. Rhodin, Eds) Almqvist and Wiksells, Stockholm, 1957).

EUREM-2, Delft, 1960: *The Proceedings of the European Regional Conference on Electron Microscopy*, Delft, 1960 (A. L. Houwink and B. J. Spit, Eds) Nederlandse Vereniging voor Elektronenmicroscopie, Delft, n.d. 2 Vols.

EUREM-3, Prague, 1964: *Electron Microscopy 1964. Proceedings of the Third European Regional Conference*, Prague (M. Titlbach, Ed.) Publishing House of the Czechoslovak Academy of Sciences, Prague, 1964. 2 Vols.

EUREM-4, Rome, 1968: *Electron Microscopy 1968. Pre-Congress Abstracts of Papers Presented at the Fourth Regional Conference*, Rome (D. S. Bocciarelli, Ed.) Tipografia Poliglotta Vaticana, Rome, 1968. 2 Vols.

EUREM-5, Manchester, 1972: *Electron Microscopy 1972. Proceedings of the Fifth European Congress on Electron Microscopy*, Manchester (Institute of Physics, London, 1972).

EUREM-6, Jerusalem, 1976: *Electron Microscopy 1976. Proceedings of the Sixth European Congress on Electron Microscopy*, Jerusalem [D.G. Brandon (Vol. I) and Y. Ben-Shaul (Vol. II), Eds] Tal International, Jerusalem, 1976. 2 Vols.

EUREM-7, The Hague, 1980: *Electron Microscopy 1980. Proceedings of the Seventh European Congress on Electron Microscopy*, The Hague [P. Brederoo and G. Boom (Vol. I), P. Brederoo and W. de Priester (Vol. II), P. Brederoo and V.E. Cosslett (Vol. III) and P. Brederoo and J. van Landuyt (Vol. IV), Eds]. Vols. I and II contain the proceedings of the Seventh European Congress on Electron Microscopy, vol. III those of the Ninth International Conference on X-Ray Optics and Microanalysis, and Vol. IV those of the Sixth International Conference on High Voltage Electron Microscopy. Seventh European Congress on Electron Microscopy Foundation, Leiden, 1980.

EUREM-8, Budapest, 1984: *Electron Microscopy 1984. Proceedings of the Eighth European Congress on Electron Microscopy*, Budapest 13–18 August 1984 (A. Csanády, P. Röhlich and D. Szabó, Eds) Programme Committee of the Eighth European Congress on Electron Microscopy, Budapest, 1984. 3 Vols.

EUREM-9, York, 1988: *Proceedings of the Ninth European Congress on Electron Microscopy*, York, 4–9 September, 1988 (P.J. Goodhew and H.G. Dickinson, Eds) Institute of Physics, Bristol and Philadelphia, PA, 1988. Conference Series **93**, 3 Vols.

EUREM-10, Granada, 1992: *Electron Microscopy 92. Proceedings of the 10th European Congress on Electron Microscopy*, Granada, 7–11 September 1992 [A. Ríos, J. M. Arias, L. Megías-Megías and A. López-Galindo (Vol. I), A. López-Galindo and M.I. Rodríguez-García (Vol. II) and L. Megías-Megías, M. I. Rodríguez-García, A. Ríos and J. M. Arias, (Vol. III), Eds]. Secretariado de Publicaciones de la Universidad de Granada, Granada. 3 Vols.

EUREM-11, Dublin, 1996: *Electron Microscopy 1996. Proceedings of the 11th European Conference on Electron Microscopy*, Dublin, 26–30 August 1996, distributed on CD-ROM. Subsequently published in book form by CESM, the Committee of European Societies of Microscopy, Brussels, 1998. 3 vols.

EUREM-12, Brno, 2000: *Electron Microscopy 2000. Proceedings of the 12th European Conference on Electron Microscopy*, Brno, 9–14 July 2000. (L. Frank and F. Čiampor, General Eds; vol. I, Biological Sciences, S. Čech and R. Janisch, Eds; vol. II, Physical Sciences, (J. Gemperlová and I. Vávra, Eds; vol. III, Instrumentation and Methodology, (P. Tománek and R. Kolařík, Eds); vol. IV, Supplement, (L. Frank and F. Čiampor, Eds); vols I–III also distributed on CD-ROM). Czechoslovak Society of Electron Microscopy, Brno, 2000.

EMC-13, Antwerp, 2004: *Proceedings European Microscopy Congress*, Antwerp, 23–27 August 2004. (D. Schryvers, J.-P. Timmermans and E. Pirard, General Eds; Biological Sciences, (J.-P. Verbelen and E. Wisse, Eds); Materials Sciences, (G. van Tendeloo and C. van Haesendonck, Eds); Instrumentation and Methodology, (D. van Dyck and P. van Oostveldt, Eds). Belgian Society for Microscopy, Liège, 2004.

EMC-14, Aachen 2008: *Proceedings EMC 2008, 14th European Microscopy Congress*, Aachen, 1–5 September 2008. Volume 1, Instrumentation and Methods (M. Luysberg and K. Tillmann, Eds); Volume 2, Materials Science (S. Richter and A. Schwedt, Eds); Volume 3, Life Science (A. Aretz, B. Hermanns-Sachweh and J. Mayer, Eds). Springer, Berlin, 2008.

EMC-15, Manchester, 2012: *Proceedings EMC2012, 15th European Microscopy Congress*, Manchester, 16–21 September 2012. Volume 1, Physical Sciences: Applications (D. J. Stokes and W. M. Rainforth, Eds); Volume 2, Physical Sciences: Tools and Techniques (D. J. Stokes and J. L. Hutchison, Eds); Volume 3, Life Sciences (D. J. Stokes, P. J. O'Toole and T. Wilson, Eds). Royal Microscopical Society, Oxford, 2012.

EMC-16, Lyon, 2016: 28 August–2 September 2016: *European Microscopy Congress 2016*. Vol. 1: Instrumentation and Methods (O. Stéphan, M. Hÿtch, B. Satiat–Jeunemaître, C. Venien-Bryan, P. Bayle-Guillemaud and T. Epicier, Eds); Vols 2.1 and 2.2: Materials Science (O. Stéphan, M. Hÿtch and T. Epicier, Eds); Vol. 3: Life Sciences (B. Satiat-Jeunemaître, C, Venien-Bryan and T. Epicier, Eds). Wiley-VCH, Weinheim, 2016.

EMC-17, Copenhagen, 2020, postponed to 2024; partly replaced by a Virtual Early Career European Microscopy Conference, 24–26 November 2020 (online).

3. Asia-Pacific Congresses on Electron Microscopy, later Asia–Pacific Microscopy Congresses

APEM-1, Tokyo, 1956: *Electron Microscopy. Proceedings of the First Regional Conference in Asia and Oceania*, Tokyo, 1956. Electrotechnical Laboratory, Tokyo, 1957.

APEM-2, Calcutta, 1965: *Proceedings of the Second Regional Conference on Electron Microscopy in Far East and Oceania*, Calcutta 2–6 February 1965. Electron Microscopy Society of India, Calcutta.

APEM-3, Singapore, 1984: *Conference Proceedings 3rd Asia Pacific Conference on Electron Microscopy*, Singapore, 29 August–3 September 1984 (Chung Mui Fatt, Ed.) Applied Research Corporation, Singapore.

APEM-4, Bangkok, 1988: *Electron Microscopy 1988. Proceedings of the IVth Asia-Pacific Conference and Workshop on Electron Microscopy*, Bangkok, 26 July–4 August 1988 (V. Mangclaviraj, W. Banchorndhevakul and P. Ingkaninun, Eds) Electron Microscopy Society of Thailand, Bangkok, 1988.

APEM-5, Beijing, 1992: *Electron Microscopy I and II. 5th Asia-Pacific Electron Microscopy Conference*, Beijing, 2–6 August 1992 (K. H. Kuo and Z. H. Zhai, Eds). World Scientific, Singapore, River Edge NJ, London and Hong Kong, 1992. 2 Vols. See also *Ultramicroscopy* **48** (1993) No. 4, 367–490.

APEM-6, Hong Kong, 1996: *Proceedings of the 6th Asia–Pacific Conference on Electron Microscopy*, Hong Kong, 1–5 July 1996 (D. Barber, P. Y. Chan, E. C. Chew, J. S. Dixon, and J. K. L. Lai, Eds). Chinetek Promotion, Kowloon, Hong Kong, 1996.

APEM-7, Singapore, 2000: *Proceedings of the 7th Asia–Pacific Conference on Electron Microscopy*, Singapore International Convention & Exhibition Centre, Suntec City, Singapore, 26–30 June 2000 (two volumes and CD-ROM, Y. T. Yong, C. Tang, M. Leong, C. Ng and P. Netto, Eds). 7th APEM Committee, Singapore, 2000.

APEM-8, Kanazawa, 2004: *Proceedings 8th Asia–Pacific Conference on Electron Microscop*y (8APEM), Kanazawa, Ishikawa Prefecture, 7–11 June 2004. Full proceedings on CD-ROM, Japanese Society of Microscopy, Tokyo, 2004.

APEM-9, Jeju, 2008: *Proceedings of the Ninth Asia–Pacific Microscopy Conference* (APMC9) Jeju, Korea, 2–7 November 2008. (H.-c. Lee, D. H. Kim, Y.-w. Kim, I. J. Rhyu, and H.-t. Jeong, Eds). *Korean J. Microsc.* **38** (2008), No. 4, Supplement, on CD-ROM only.

APEM-10, Perth, 2012: *Proceedings of the Tenth Asia–Pacific Microscopy Conference* (APMC-10) Perth, Australia, 5–9 February 2012 (B. Griffin, L. Faraone and M. Martyniuk, Eds). Held in conjunction with the 2012 International Conference on Nanoscience and Nanotechnology (ICONN2012) and the 22nd Australian Conference on Microscopy and Microanalysis (ACMM22).

APEM-11, Phuket 2016: *11th Asia-Pacific Microscopy Conference (APMC-11)*, Phuket, Thailand, 23–27 May 2016. Held in conjunction with the 33rd Annual Conference of the Microscopy Society of Thailand (MST-33) and the 39th Annual Conference of the Anatomy Association of Thailand (AAT-39). Selected articles published in *Siriraj Medical Journal* **8** (3), Suppl. 1 (2016) and *J. Microsc. Soc. Thailand* **29** (2016) No. 1.

APEM-12, Hyderabad, Hyderabad International Convention Centre, 3–7 February 2020.

APEM-13, Brisbane, 4–8 February 2024.

4. Charged Particle Optics Conferences

CPO-1, Giessen, 1980: *Proceedings of the First Conference on Charged Particle Optics*, Giessen, 8–11 September 1980 (H. Wollnik, Ed.) *Nucl. Instrum. Meth.* **187** (1981) 1–314.

CPO-2, Albuquerque, 1986: *Proceedings of the Second International Conference on Charged Particle Optics*, Albuquerque, 19–23 May 1986 (S. O. Schriber and L. S. Taylor, Eds) *Nucl. Instrum. Meth. Phys. Res. A* **258** (1987) 289–598.

CPO-3, Toulouse, 1990: *Proceedings of the Third International Conference on Charged Particle Optics*, Toulouse, 24–27 April 1990 (P. W. Hawkes, Ed.) *Nucl. Instrum. Meth. Phys. Res. A* **298** (1990) 1–508.

CPO-4, Tsukuba 1994: *Proceedings of the Fourth International Conference on Charged Particle Optics*, Tsukuba, 3–6 October 1994 (K. Ura, M. Hibino, M. Komuro, M. Kurashige, S. Kurokawa, T. Matsuo, S. Okayama, H. Shimoyama and K. Tsuno, Eds) *Nucl. Instrum. Meth. Phys. Res. A* **363** (1995) 1–496.

CPO-5, Delft 1998: *Proceedings of the Fifth International Conference on Charged Particle Optics*, Delft, 14–17 April 1998 (P. Kruit and P. W. van Amersfoort, Eds). *Nucl. Instrum. Meth. Phys. Res. A* **427** (1999) 1–422.

CPO-6, College Park 2002: *Proceedings of the Sixth International Conference on Charged Particle Optics*, Marriott Hotel, Greenbelt MD, 21–25 October 2002 (A. Dragt and J. Orloff, Eds). *Nucl. Instrum. Meth. Phys. Res. A* **519** (2004) 1–487.

CPO-7. Cambridge 2006: *Charged Particle Optics. Proceedings of the Seventh International Conference on Charged Particle Optics*, Trinity College, Cambridge, 24–28 July 2006 (E. Munro and J. Rouse, Eds). *Physics Procedia* **1** (2008) 1–572.

CPO-8, Singapore 2010: *Proceedings of the Eighth International Conference on Charged Particle Optics*, Suntec Convention Centre, Singapore, 12–16 July 2010 (A. Khursheed, P. W. Hawkes and M. B. Osterberg, Eds). *Nucl. Instrum. Meth. Phys. Res. A* **645** (2011) 1–354.

CPO-9, Brno 2014: *Proceedings of the Ninth International Conference on Charged Particle Optics*, Brno, 31 August–5 September 2014 (L. Frank, P. W. Hawkes and T. Radlička, Eds). *Microsc. Microanal.* **21** (2015) Suppl. 4.

CPO-10, Key West 2018: Proceedings published in *Advances in Imaging and Electron Physics*, volume 212, 2019 and *Int. J. Mod. Phys. A* **34** (2019) No. 35 (M. Berz and K. Makino, Eds).

CPO-11, Japan, postponed *sine die*.

5. High-voltage Electron Microscopy Conferences

HVEM Monroeville, 1969: *Current Developments in High Voltage Electron Microscopy (First National Conference)*, Monroeville, 17–19 June 1969. Proceedings not published but *Micron* **1** (1969) 220–307 contains official reports of the meeting based on the session chairmen's notes.

HVEM Stockholm, 1971: *The Proceedings of the Second International Conference on High-Voltage Electron Microscopy*, Stockholm, 14–16 April 1971; published as *Jernkontorets Annaler* **155** (1971) No. 8.

HVEM Oxford, 1973: *High Voltage Electron Microscopy. Proceedings of the Third International Conference*, Oxford, August 1973 (P. R. Swann, C. J. Humphreys and M. J. Goringe, Eds) Academic Press, London and New York, 1974.

HVEM Toulouse, 1975: *Microscopie Electronique à Haute Tension. Textes des Communications Présentées au Congrès International*, Toulouse, 1–4 Septembre 1975 (B. Jouffrey and P. Favard, Eds) SFME Paris, 1976.

HVEM Kyoto, 1977: *Proceedings of the Fifth International Conference on High Voltage Electron Microscopy* (T. Imura and H. Hashimoto, Eds) (Japanese Society of Electron Microscopy, Tokyo). Also published as a Supplement to *J. Electron Microsc.* **26** (1977).

HVEM The Hague, 1980 see EUREM-7, The Hague, 1980.

HVEM Berkeley, 1983: *Proceedings of the Seventh International Conference on High Voltage Electron Microscopy*, Berkeley, 16–19 August 1983 (R. M. Fisher, R. Gronsky and K.H. Westmacott, Eds). Published as a Lawrence Berkeley Laboratory Report, LBL-16031, UC-25, CONF-830819.

6. EMAG [Electron Microscopy and Analysis Group of the Institute of Physics] Meetings

EMAG, 1971: *Electron Microscopy and Analysis. Proceedings of the 25th Anniversary Meeting of the Electron Microscopy and Analysis Group of the Institute of Physics*, Cambridge, 29 June–1 July 1971 (W. C. Nixon, Ed.) Institute of Physics, London, 1971. Conference Series **10.**

EMAG, 1973: *Scanning Electron Microscopy: Systems and Applications*, Newcastle-upon-Tyne, 3–5 July 1973 (W. C. Nixon, Ed.) Institute of Physics, London, 1973. Conference Series **18**.

EMAG, 1975: *Developments in Electron Microscopy and Analysis. Proceedings of EMAG 75*, Bristol, 8–11 September 1975 (J. A. Venables, Ed.) Academic Press, London and New York, 1976.

EMAG, 1977: *Developments in Electron Microscopy and Analysis. Proceedings of EMAG* 77, Glasgow, 12–14 September 1977 (D.L. Misell, Ed.) Institute of Physics, Bristol, 1977. Conference Series **36**.

EMAG, 1979: *Electron Microscopy and Analysis, 1979. Proceedings of EMAG 79*, Brighton, 3–6 September 1979 (T. Mulvey, Ed.) Institute of Physics, Bristol, 1980. Conference Series **52**.

EMAG, 1981: *Electron Microscopy and Analysis, 1981. Proceedings of EMAG 81*, Cambridge, 7–10 September 1981 (M. J. Goringe, Ed.) Institute of Physics, Bristol, 1982. Conference Series **61**.

EMAG, 1983: *Electron Microscopy and Analysis, 1983. Proceedings of EMAG 83*, Guildford, 30 August–2 September 1983 (P. Doig, Ed.) Institute of Physics, Bristol, 1984. Conference Series **68**.

EMAG, 1985: *Electron Microscopy and Analysis, 1985. Proceedings of EMAG 85*. Newcastle-upon-Tyne, 2–5 September 1985 (G. J. Tatlock, Ed.) Institute of Physics, Bristol, 1986. Conference Series **78**.

EMAG, 1987: *Electron Microscopy and Analysis, 1987. Proceedings of EMAG 87*, Manchester, 8–9 September 1987 (L. M. Brown, Ed.) Institute of Physics, Bristol and Philadelphia, PA, 1987. Conference Series **90**.

EMAG, 1989: *EMAG-MICRO 89. Proceedings of the Institute of Physics Electron Microscopy and Analysis Group and Royal Microscopical Society Conference*, London, 13–15 September 1989 (P. J. Goodhew and H. Y. Elder, Eds) Institute of Physics, Bristol and New York, 1990. Conference Series **98**, 2 Vols.

EMAG, 1991: *Electron Microscopy and Analysis 1991. Proceedings of EMAG 91*, Bristol, 10–13 September 1991 (F. J. Humphreys, Ed.) Institute of Physics, Bristol, Philadelphia, PA and New York, 1991. Conference Series **119**.

EMAG, 1993: *Electron Microscopy and Analysis 1993. Proceedings of EMAG 93*, Liverpool, 15–17 September 1993 (A. J. Craven, Ed.) Institute of Physics, Bristol, Philadelphia, PA and New York, 1994. Conference Series **138**.

EMAG 1995: *Electron Microscopy and Analysis 1995. Proceedings of EMAG 95*. Birmingham, 12–15 September 1995 (D. Cherns, Ed.) Institute of Physics, Bristol, Philadelphia, PA and New York, 1995. Conference Series **147**.

EMAG 1997: *Electron Microscopy and Analysis 1997. Proceedings of EMAG 97*, Cavendish Laboratory, Cambridge, 2–5 September 1997 (J. M. Rodenburg, Ed) Institute of Physics, Bristol and Philadelphia, PA, 1997. Conference Series **153**.

EMAG 1999: *Electron Microscopy and Analysis 1999. Proceedings of EMAG 99*, University of Sheffield, 25–27 August 1999 (C. J. Kiely, Ed.); Institute of Physics, Bristol and Philadelphia, PA, 1999. Conference Series **161**.

EMAG 2001: *Electron Microscopy and Analysis 2001. Proceedings of the Institute of Physics Electron Microscopy and Analysis Group Conference*, University of Dundee, 5–7 September 2001 (M. Aindow and C. J. Kiely, Eds) Institute of Physics Publishing, Bristol and Philadelphia, PA 2002) Conference Series **168**.

EMAG 2003: *Electron Microscopy and Analysis 2003. Proceedings of the Institute of Physics Electron Microscopy and Analysis Group Conference*, Examination Schools, University of Oxford, 3–5 September 2003 (S. McVitie and D. McComb, Eds) Institute of Physics Publishing, Bristol and Philadelphia, PA 2004. Conference Series **179**.

EMAG–NANO 2005: University of Leeds, 31 August–2 September 2005 (P.D. Brown, R. Baker and B. Hamilton, Eds). *J. Phys.: Conf.* **26** (2006).

EMAG 2007: Caledonian University and University of Glasgow, 3–7 September 2007 (R. T. Baker, G. Möbus and P. D. Brown, Eds). *J. Phys.: Conf.* **126** (2008).

EMAG 2009: University of Sheffield, 8–11 September 2009 (R. T. Baker, Ed.). *J. Phys.: Conf.* **241** (2010).

EMAG 2011: University of Birmingham (R. T. Baker, P. D. Brown and Z. Li, Eds). *J. Phys.: Conf.* **371** (2012).

EMAG 2013: University of York, 3–6 September 2013 (P. Nellist, Ed.). *J. Phys.: Conf.* **522** (2014).

EMAG 2015, Manchester, 29 June–2 July 2015, joint with the Microscience Microscopy Conference (Royal Microscopical Society) (I. MacLaren, Ed.). *J. Phys.: Conf.* **644** (2015).

EMAG 2016, Durham, 7–8 April 2016 (no publication).

EMAG 2017, Manchester, 3–6 July 2017, joint with the Microscience Microscopy Conference (Royal Microsopical Society). *J. Phys.: Conf.* **902** (2017).

EMAG 2018, Electron Microscopy of Beam-sensitive Materials. *Micron* (2019). The articles appear in several different volumes, a list with references can be found on the *Micron* website.

EMAG 2019, Manchester, 1–4 July 2019. *J. Microsc.* (Oxford) **279** (3), 139–281 (2020)

EMAG 2020, Microscopy enabled by Direct Electron Detection. Virtual meeting, 6–8 July 2020. http://emag2020.iopconfs.org/agenda.

EMAG 2021, Manchester 5–8 July 2021 (virtual). Link to abstracts database available at https://www.mmc-series.org.uk/.

EMAG 2023, Manchester, 3–6 July 2023.

7. Multinational Congresses on (Electron) Microscopy (MCEM, MCM)

The first of these meetings brought together the Italian, Hungarian, Czechoslovak and Slovenian Societies. For subsequent congresses, these were joined by the Austrian and Croatian societies.

MCEM-93. *Multinational Congress on Electron Microscopy*, Parma, 13–17 September 1993; *Proceedings* issued as Supplement to **14** (2) of *Microscopia Elettronica.*

MCEM -95: *Proceedings Multinational Conference on Electron Microscopy*, Stará Lesná (High Tatra Mountains), 16–20 October 1995. Slovak Academic Press, Bratislava, 1995.

MCEM-97. *Proceedings Multinational Congress on Electron Microscopy*, Portorož (Slovenia), 5–8 October 1997. Part I, Microscopy Applications in the Life Sciences; Part II, Microscopy Applications in the Material Sciences; Part III, Microscopy Methods and Instrumentation. *J. Computer-assisted Microsc.* **8** (1996) No. 4 and **9** (1997) Nos 1 and 2.

MCEM-99. *Proceedings 4th Multinational Congress on Electron Microscopy*, Veszprém (Hungary), 5–8 September 1999 (K. Kovács, Ed.). University of Veszprém, 1999.

MCEM-5. Proceedings of the 5th Multinational Congress on Electron Microscopy, Department of Biology, University of Lecce (Italy), 20–25 September 2001 (L. Dini and M. Catalano, Eds). Rinton Press, Princeton, NJ, 2001.

MCM-6. Proceedings of the Sixth Multinational Congress on Electron Microscopy, Pula (Croatia), 1–5 June 2003 (O. Milat and D. Ježek, Eds). Croatian Society for Electron Microscopy, Zagreb, 2003.

MCM-7. *Proceedings of the 7th Multinational Congress on Microscopy*, Portorož, (Slovenia) 26–30 June 2005 (M. Čeh, G. Dražič, and S. Fidler, Eds). Slovene Society for Microscopy and Department for Nanostructured Materials, Jožef Stefan Institute, Ljubljana, 2005.

MCM-8. *Proceedings 8th Multinational Congress on Microscopy*, Prague (Czech Republic), 17–21 June 2007 (J. Nebesářova and P. Hozák, Eds). Czechoslovak Microscopy Society, Prague 2007.

MC 2009 incorporating MCM-9. Microscopy Conference, Graz, Austria 30 August–4 September 2009. *Proceedings First Joint Meeting of Dreiländertagung & Multinational Congress on Microscopy*. Volume 1, Instrumentation and Methodology (G. Kothleitner and M. Leisch, Eds); Volume 2, Life Sciences (M.A. Pabst and G. Zellnig, Eds); Volume 3, Materials Science (W. Grogger, F. Hofer and P. Pölt, Eds). Verlag der Technischen Universität, Graz, 2009.

MCM-10. *Proceedings 10th Multinational Conference on Microscopy*, Urbino, 4–9 September 2011 (E. Falcieri, Ed.). Società Italiana di Scienze Microscopiche (SISM), 2011.

MC-2013, Regensburg, 25–30 August 2013. *Joint Meeting of Dreiländertagung & Multinational Congress on Microscopy*, together with the Serbian and Turkish Microscopy Societies. Proceedings can be downloaded from http://www.mc2013.de. urn:nbn:de:bvb:355-epub-287343 (R. Rachel, J. Schröder, R. Witzgall and J. Zweck, Eds).

MCM-12, *Multinational Conference on Microscopy*, Eger (Hungary), 23–29 August 2015. Webarchive.

MCM-13, *Multinational Conference on Microscopy*, Rovinj (Croatia), 24–29 September 2017. Abstracts available at mcm2017.irb.hr.

MCM-14, *Multinational Conference on Microscopy*, Belgrade, 15–20 September 2019. Abstracts available at http://www.sdm.rs.

MCM-15, Vienna, 22–26 August 2021, *Joint Meeting of Dreiländertagung & Multinational Congress on Microscopy.*

MCM-16, Brno, 15–20 September 2022.

8. The Dreiländertagungen (Germany, Austria, Switzerland) and related meetings

These conferences are organized in turn by the Austrian, German and Swiss Microscopy societies; originally designed for German-speaking microscopists, they now tend to use English and attract a wider participation.

Dreiländertagung für Elektronenmikroskopie: Konstanz, 15–21 September 1985. *Optik* (1985) Suppl. 1 or *Eur. J. Cell Biol.* (1985) Suppl. 10. See also *Beiträge zur Elektronenmikroskopischen Direktabbildung von Oberflächen* **18** (1985).

Dreiländertagung für Elektronenmikroskopie: Salzburg, 10–16 September 1989. *Optik* **83** (1989) Suppl. 4 or *Eur. J. Cell Biol.* **49** (1989) Suppl. 27.

Dreiländertagung für Elektronenmikroskopie: Zürich, 5–11 September 1993. *Optik* **94** (1993) Suppl. 5 or *Eur. J. Cell Biol.* **61** (1993) Suppl. 39.

Dreiländertagung für Elektronenmikroskopie: Regensburg, 7–12 September 1997. *Optik* **106** (1997) Suppl. 7 or *Eur. J. Cell Biol.* **74** (1997) Suppl. 45.

Dreiländertagung für Elektronenmikroskopie: Innsbruck, 9–14 September 2001. Abstracts book (168 pp) not published as a Supplement to *Optik* or *Eur. J. Cell Biol.*

MC-2003, Dresden 7–12 September 2003. *Microsc. Microanal.* **9** (2003) Suppl. 3 (T. Gemming, M. Lehmann, H. Lichte and K. Wetzig, Eds).

Dreiländertagung für Elektronenmikroskopie: Microscopy Conference 2005, Paul Scherrer Institute, Davos, 25 August–2 September 2005. *Paul-Scherrer-Institute Proceedings* **PSI 05–01**, 2005.

MC-2007, Saarbrücken, 2–7 September 2007. *Microsc. Microanal.* **13** (2007) Suppl. 3 (T. Gemming, U. Hartmann, P. Mestres and P. Walther, Eds).

Microscopy Conference (MC 2009), Graz, 30 August–4 September 2009. *First Joint Meeting of Dreiländertagung & Multinational Congress on Microscopy.* Volume 1, Instrumentation and Methodology (G. Kothleitner and M. Leisch, Eds); Volume 2, Life Sciences (M.A. Pabst and G. Zellnig, Eds); Volume 3, Materials Science (W. Grogger, F. Hofer and P. Pölt, Eds). Verlag der Technischen Universität, Graz, 2009.

MC-2011, Kiel, 28 August–2 September 2011. *Joint meeting of the German Society (DGE, the Nordic Microscopy Society (SCANDEM), and the Polish Microscopy Society (PTMi) with participation of microscopists from Estonia, Latvia, Lithuania and St Petersburg, Russia.* Proceedings published in 3 volumes by the German Society for Electron Microscopy and also distributed as a USB key (W. Jäger, W. Kaysser, W. Benecke, W. Depmeier, S. Gorb, L. Kienle, M. Mulisch, D. Häußler and A. Lotnyk, Eds).

MC-2013, Regensburg, 25–30 August 2013. *Joint Meeting of Dreiländertagung & Multinational Congress on Microscopy, together with the Serbian and Turkish Microscopy Societies.* Proceedings can be downloaded from http://www.mc2013.de. urn:nbn:de:bvb:355-epub-287343 (R. Rachel, J. Schröder, R. Witzgall and J. Zweck, Eds).

MC-2015, Georg-August-Universität, Göttingen, 6–11 September 2015. Proceedings at http://www.mc2015.de.

MC-2017, Lausanne, 21–25 August 2017. Proceedings available at epub.uni-regensburg.de/36099/.

MC-2019, Berlin, 1–5 September 2019. Proceedings available at epub.uni-regensburg.de/40685/.

MC-2021, Vienna, 22–26 August 2021, *Joint Meeting of Dreiländertagung & Multinational Congress on Microscopy.*

MC-2023, Darmstadt, 26 February–2 March 2023.

9. Recent Trends in Charged Particle Optics and Surface Physics Instrumentation (Skalský Dvůr)

1989: First Seminar, Brno, 4–6 September 1989 (no proceedings).

1990: Second Seminar, Brno, 27–29 September 1990 (no proceedings).

1992: Third Seminar, Skalský Dvůr (near Brno), 15–19 June 1992 (no proceedings).

1994: Fourth Seminar, Skalský Dvůr, 5–9 September 1994 (no proceedings).

1996: Fifth Seminar, Skalský Dvůr, 24–28 June 1996. (I. Müllerová and L. Frank, Eds). 92 pp.

1998: Sixth Seminar, Skalský Dvůr, 29 June–3 July 1998. (I. Müllerová and L. Frank, Eds). 84 pp Published by the CSEM (Brno 1998).

2000: Seventh Seminar, Skalský Dvůr, 15–19 July 2000. No proceedings book.

2002: Eighth Seminar, Skalský Dvůr, 8–12 July 2002. (L. Frank, Ed.). 96 pp + Supplement, 6 pp Published by the CSMS (Brno 2002).

2004: Ninth Seminar, Skalský Dvůr, 12–16 July 2004. (I. Müllerová, Ed.). Published by the CSMS (Brno 2004).

2006: 10th Seminar, Skalský Dvůr, 22–26 May 2006. (I. Müllerová, Ed.). Published by the CSMS (Brno 2006).

2008: 11th Seminar, Skalský Dvůr, 14–18 July 2008. (F. Mika, Ed.). Published by the CSMS (Brno 2008).

2010: 12th Seminar, Skalský Dvůr, 31 May–4 June 2010. (F. Mika, Ed.). Published by the CSMS (Brno 2010).

2012: 13th Seminar, Skalský Dvůr, 25–29 June 2012. (F. Mika, Ed.). Published by the CSMS (Brno 2012).

2014: 14th Seminar, incorporated in CPO-9, Brno, see Section 5.1.

2016: 15th Seminar, Skalský Dvůr, 25–29 June 2012. (F. Mika, Ed.). Published by the CSMS (Brno 2016).

2018: 16th Seminar, Skalský Dvůr, 4–8 June 2018. (F. Mika and Z. Pokorná, Eds) Published by the Institute of Scientific Instruments, Czech Academy of Sciences (Brno 2018).

10. SPIE Proceedings

1. *Charged Particle Optics*, San Diego, CA, 15 July 1993 (W.B. Thompson, M. Sato and A.V. Crewe, Eds). *Proc. SPIE* **2014** (1993).
2. *Electron-beam Sources, and Charged Particle Optics*, San Diego, CA, 19–14 July 1995 (E. Munro and H. P. Freund, Eds). *Proc. SPIE* **2522** (1995).
3. *Charged Particle Optics II*, Denver CO, 5 August 1996 (E. Munro, Ed.). *Proc. SPIE* **2858** (1996).
4. *Charged Particle Optics III*, San Diego, CA, 27–28 July 1997 (E. Munro, Ed.). *Proc. SPIE* **3155** (1997).
5. *Charged Particle Optics IV*, Denver CO, 22–23 July 1999 (E. Munro, Ed.). *Proc. SPIE* **3777** (1999).
6. *Charged Particle Beam Optics Imaging*, San Diego, CA, 30 July 2001. In Charged Particle Detection, Diagnostics and Imaging (O. Delage, E. Munro and J.A. Rouse, Eds). *Proc. SPIE* **4510** (2001) 71–236.

11. Soviet All-Union Conferences on Electron Microscopy, later Russian Conferences on Electron Microscopy

The Proceedings of the Soviet All-Union conferences on electron microscopy are to be found in the volumes of *Izv. Akad. Nauk (Ser. Fiz.)* or *Bull. Acad. Sci. (Phys. Ser.)* indicated:

1. Moscow 15–19 December1950; **15** (1951) Nos 3 and 4 (no English translation).
2. Moscow 9–13 May 1958; **23** (1959) Nos 4 and 6.
3. Leningrad 24–29 October 1960; **25** (1961) No. 6.
4. Sumy 12–14 March 1963; **27** (1963) No. 9.
5. Sumy 6–8 July 1965; **30** (1966) No. 5.
6. Novosibirsk 11–16 July 1967; **32** (1968) Nos 6 and 7.
7. Kiev 14–21 July 1969; **34** (1970) No. 7.
8. Moscow 15–20 November 1971; **36** (1972) Nos 6 and 9.
9. Tbilisi 28 October–2 November 1973; **38** (1974) No. 7.
10. Tashkent 5–8 October 1976; **41** (1977) Nos 5 and 11.
11. Tallin October 1979; **44** (1980) Nos 6 and 10.
12. Sumy 1982; **48** (1984), No. 2.
13. Sumy October 1987; **52** (1988) No. 7 and **53** (1989) No. 2.
14. Suzdal, October and November 1990; **55** (119) No. 8.

From now on, the names of the journal and its English translation are *Izv. Ross. Akad. Nauk (Ser. Fiz.)* and *Bull. Russ. Acad. Sci. (Phys.)*.

15. Chernogolovka, May 1994; **59** (1995) No. 2.
16. Chernogolovka, December 1996; **61** (1997) No. 10.
17. Chernogolovka, June 1998; 63 (1999) No. 7.
18. Chernogolovka, 5–8 June 2000; **65** (2001) No. 9.
19. Chernogolovka, 27–31 May 2002; **67** (2003) No. 4.
20. Chernogolovka, 1– June 2004; **69** (2005) No. 4.
21. Chernogolovka, 5–10 June 2006; *Izv. Ross. Akad. Nauk (Ser. Fiz.)* or *Bull. Russ. Acad. Sci. (Phys.)* **71** (2007) No. 10.
22. Chernogolovka, 2008; **73** (2009) No. 4; also *Poverkhnost'* (2009), No. 10, *J. Surface Invest. X-Ray Synchrotron Neutron Techs* **3** (2009) No. 5.

23. Chernogolovka, 2010; **75** (2011) No. 9; also *Poverkhnost'* (2011), No. 10.
24. Chernogolovka, 2012; **77** (2013) No. 8.
25. Chernogolovka, 2014; **79** (2015) No. 5.
26. Chernogolovka, 2016; **81** (2017).
27. Chernogolovka, 26–30 August, 2018.

12. Problems of Theoretical and Applied Electron Optics [Problemyi Teoreticheskoi i Prikladnoi Elektronnoi Optiki][1]

1. *Proceedings of the First All-Russia Seminar*, Scientific Research Institute for Electron and Ion Optics, Moscow, 1996. *Prikladnaya Fizika* (1996) No. 3.
2. *Proceedings of the Second All-Russia Seminar*, Scientific Research Institute for Electron and Ion Optics, Moscow, 25 April 1997. *Prikladnaya Fizika* (1997) No. 2–3.
3. *Proceedings of the Third All-Russia Seminar*, Scientific Research Institute for Electron and Ion Optics, Moscow, 31 March–2 April 1998. *Prikladnaya Fizika* (1998) Nos 2 and 3/4.
4. *Proceedings of the Fourth All-Russia Seminar*, Scientific Research Institute for Electron and Ion Optics, Moscow, 21–22 October 1999. *Prikladnaya Fizika* (2000) Nos 2 and 3; *Proc. SPIE* **4187** (2000), edited by A.M. Filachev and I.S. Gaidoukova.
5. *Proceedings of the Fifth All-Russia Seminar*, Scientific Research Institute for Electron and Ion Optics, Moscow, 14–15 November 2001. *Prikladnaya Fizika* (2002) No. 3; *Proc. SPIE* **5025** (2003), edited by A.M. Filachev.
6. *Proceedings of the Sixth All-Russia Seminar*, Scientific Research Institute for Electron and Ion Optics, Moscow, 28–30 May 2003. *Prikladnaya Fizika* (2004) No. 1 and *Proc. SPIE* **5398** (2004), edited by A.M. Filachev and I.S. Gaidoukova.
7. *Proceedings of the Seventh All-Russia Seminar*, Scientific Research Institute for Electron and Ion Optics, Moscow, 25–27 May 2005. *Prikladnaya Fizika* (2006) No. 3 and *Proc. SPIE* **6278** (2004), edited by A.M. Filachev and I.S. Gaidoukova.
8. *Proceedings of the Eighth All-Russia Seminar*, Scientific Research Institute for Electron and Ion Optics, Moscow, 29–31 May 2007. *Prikladnaya Fizika* (2008) No. 2 and *Proc. SPIE* **7121** (2008), edited by A.M. Filachev and I.S. Gaidoukova.
9. *Proceedings of the Ninth All-Russia Seminar*, Scientific Research Institute for Electron and Ion Optics, Moscow, 28–31 May 2009. *Prikladnaya Fizika* (2010) No. 3, pp 31–115, edited by A.M. Filachev and I. S. Gaidoukova.
10. *Proceedings of the Tenth All-Russia Seminar*, Scientific Research Institute for Electron and Ion Optics, Moscow, 24–26 May 2011. *Prikladnaya Fizika* (2012) No. 2, edited by A.L. Dirochka and A.M. Filachev.
11. *Proceedings of the Eleventh All-Russia Seminar*, Scientific Research Institute for Electron and Ion Optics, Moscow, 28–30 May 2013. *Uspekhi Prikladnoi Fiziki* **1** (2013) No. 5, pp 571–600.
12. *Proceedings of the Twelfth All-Russia Seminar*, Scientific Research Institute for Electron and Ion Optics, Moscow, 10 December 2015. *Uspekhi Prikladnoi Fiziki*. Papers are not collected in a single issue, see **4** (2016) No. 1.

13. Mathematical Morphology and Its Applications to Signal and Image Processing

ISMM-1, J. Serra, P. Salembier, Eds, *Mathematical Morphology and Its Applications to Signal Processing*, Barcelona, May, 1993 (Universitat Politècnic de Catalunya, Barcelona, 1993).

ISMM-2, J. Serra, P. Soille, Eds, *Mathematical Morphology and Its Applications to Image Processing*, Fontainebleau, September, 1994 (Kluwer, Dordrecht, Boston, MA, London, 1994).

[1] For *Prikladnaya Fizika*, see applphys.vimi.ru

ISMM-3, P. Maragos, R. W. Schafer, M. A. Butt, Eds, *Mathematical Morphology and Its Applications to Image and Signal Processing*, Atlanta, GA, May 1996 (Kluwer, London, Dordrecht and Boston, MA, 1996).

ISMM-4, H. J. A. M. Heijmans, J. B. T. M. Roerdink, Eds, *Mathematical Morphology and Its Applications to Image and Signal Processing*, Amsterdam, June 1998 (Kluwer, Dordrecht, Boston, MA, London, 1998).

ISMM-5, J. Goutsias, L. Vincent, D. S. Bloomberg, Eds, *Mathematical Morphology and Its Applications to Image and Signal Processing*, Palo Alto, 26–29 June 2000 (Kluwer, Boston, MA, 1999).

ISMM-6, H. Talbot, R. Beare, Eds, *Mathematical Morphology, Proceedings of the VIth International Symposium*, Sydney, 3–5 April 2002 (CSIRO Publishing, Collingwood, 2002).

ISMM-7, C. Ronse, L. Najman, E. Decencière, Eds, *Mathematical Morphology 40 Years On. Proceedings of the Seventh International Symposium on Mathematical Morphology*, Paris, 18–20 April 2005 (Springer, Berlin, 2005).

ISMM-8, G. J. F. Banon, J. Barrera, U. de Mendonça Braga-Neto, Eds, *Mathematical Morphology and Its Applications to Signal and Image Processing. Proceedings of the 8th Symposium on Mathematical Morphology*, Rio de Janeiro, 10–13 October 2007 (MCT/INPE 2007).

ISMM-9, M. H. F. Wilkinson, J. B. T. M. Roerdink, Eds, *Mathematical Morphology and Its Application to Signal and Image Processing*, Groningen, 24–27 August 2009 (Springer, Heidelberg, 2009). Lecture Notes in Computer Science No. 5720.

ISMM-10, P. Soille, M. Pesaresi, G. K. Ouzounis, Eds, *Mathematical Morphology and Its Applications to Image and Signal Processing*, Verbania-Intra, 6–8 July 2011, 2009 (Springer, Berlin, 2011). Lecture Notes in Computer Science No. 6671.

ISMM-11, C. L. L. Hendriks, G. Borgefors and R. Strand, Eds, *Mathematical Morphology and Its Applications to Signal and Image Processing*, Uppsala, 27–29 May 2013 (Springer, Heidelberg, 2013. Lecture Notes in Computer Science No. 7883.

ISMM-12, J. A. Benediktsson, J. Chanussot, L. Najman, H. Talbot, Eds, *Mathematical Morphology and Its Applications to Signal and Image Processing*, Reykjavik, 27–29 May 2015, 2009 (Springer, Cham, 2015). Lecture Notes in Computer Science No. 9082.

ISMM-13, J. Angulo, S. Velasco-Forero, F. Meyer, Eds, *Mathematical Morphology and Its Applications to Signal and Image Processing*, Fontainebleau, 15–17 May 2017 (Springer, Cham, 2017). Lecture Notes in Computer Science No. 10225.

ISMM-14, B. Burgeth, A. Klefeld, B. Naegel, N. Passat and B. Perret, Eds, *Mathematical Morphology and Its Applications to Signal and Image Processing*, Saarbrücken, 8–10 July 2019 (Springer, Cham, 2019). Lecture Notes in Computer Science No. 11564.

DGMM 2021, J. Lindblad, F. Malmberg and N. Sladoje, Eds, *Discrete Geometry and Mathematical Morphology*, Uppsala, 24–27 May 2021 (Springer Nature, Cham, 2021). Lecture Notes in Computer Science No. 12708.

DGMM 2022, Second IAPR International Conference on Discrete Geometry and Mathematical Morphology, Strasbourg, 24–27 October 2022.

14. Related Meetings

Washington, 1951: *Electron Physics. Proceedings of the NBS Semicentennial Symposium on Electron Physics*, Washington, DC, 5–7 November 1951. Issued as National Bureau of Standards Circular 527 (1954).

Gent, 1954: *Rapport Europees Congrès Toegepaste Electronenmicroscopie*, Gent, 7–10 April 1954, edited and published by G. Vandermeersche (Uccle-Bruxelles, 1954).

Toulouse, 1955: *Les Techniques Récentes en Microscopie Electronique et Corpusculaire*, Toulouse, 4–8 April 1955 (C.N.R.S., Paris, 1956).

Ocean City, 1984: *Electron Optical Systems for Microscopy, Microanalysis and Microlithography. Proceedings of the 3rd Pfefferkorn Conference*, Ocean City, MD, 9–14 April 1984 (J. J. Hren, F. A. Lenz, E. Munro and P. B. Sewell, Eds) Scanning Electron Microscopy, AMF O'Hare, IL.

Beijing, 1986: *Proceedings of the International Symposium on Electron Optics*, Beijing, 9–13 September 1986 (J.-y. Ximen, Ed.) Institute of Electronics, Academia Sinica, 1987.

Index

Note: Page numbers followed by "*f*" and "*t*" refer to figures and tables, respectively.

B

C

E

F

G

H

Q

R

S

T

Printed in the United States
by Baker & Taylor Publisher Services